Digital Design with CPLD Applications and VHDL, 2nd edition

Digital Design with CPLD Applications and VHDL, 2nd edition

Robert K. Dueck
Red River College, Winnipeg, Manitoba

THOMSON
™
DELMAR LEARNING

Australia Canada Mexico Singapore Spain United Kingdom United States

THOMSON

DELMAR LEARNING

Digital Design with CPLD Applications and VHDL, 2e
Robert K. Dueck

Vice President, Technology and Trades SBU:
Alar Elken

Editorial Director:
Sandy Clark

Senior Acquisitions Editor:
Steve Helba

Senior Development Editor:
Michelle Ruelos Cannistraci

Marketing Director:
Dave Garza

Channel Manager:
Fair Huntoon

Marketing Coordinator:
Casey Bruno

Production Director:
Mary Ellen Black

Production Manager:
Larry Main

Senior Project Editor:
Christopher Chien

Art/Design Coordinator:
Francis Hogan

Technology Project Manager:
Kevin Smith

Technology Project Specialist:
Linda Verde

Senior Editorial Assistant:
Dawn Daugherty

Library of Congress Cataloging-in-Publication Data:

Dueck, Robert K.
   Digital design with CPLD applications and VHDL / Robert K. Dueck.—2nd ed.
      p. cm.
   Includes index.
   ISBN 1-4018-4030-2 (alk. paper)
   1. Programmable logic devices—Design and construction,
   2. Programmable array logic. 3. Logic design. 4. VHDL (Computer hardware description language) I. Title.

TK7872.L64D84 2004
621.39'5—dc22

                              2004045514

NOTICE TO THE READER

Publisher does not warrant or guarantee any of the products described herein or perform any independent analysis in connection with any of the product information contained herein. Publisher does not assume, and expressly disclaims, any obligation to obtain and include information other than that provided to it by the manufacturer.

The reader is expressly warned to consider and adopt all safety precautions that might be indicated by the activities herein and to avoid all potential hazards. By following the instructions contained herein, the reader willingly assumes all risks in connection with such instructions.

The publisher makes no representation or warranties of any kind, including but not limited to, the warranties of fitness for particular purpose or merchantability, nor are any such representations implied with respect to the material set forth herein, and the publisher takes no responsibility with respect to such material. The publisher shall not be liable for any special, consequential, or exemplary damages resulting, in whole or part, from the readers' use of, or reliance upon, this material.

Contents

Preface   ix

1

Basic Principles of Digital Systems   2

1.1   Digital Versus Analog Electronics   4
1.2   Digital Logic Levels   5
1.3   The Binary Number System   6
1.4   Hexadecimal Numbers   14
1.5   Digital Waveforms   18

2

Logic Functions and Gates   30

2.1   Basic Logic Functions   32
2.2   Derived Logic Functions   38
2.3   DeMorgan's Theorems and Gate Equivalence   42
2.4   Logic Switches and LED Indicators   45
2.5   Enable and Inhibit Properties of Logic Gates   48
2.6   Integrated Circuit Logic Gates   55

3

Boolean Algebra and Combinational Logic   70

3.1   Boolean Expressions, Logic Diagrams, and Truth Tables   72
3.2   Sum-of-Products and Product-of-Sums Forms   80
3.3   Theorems of Boolean Algebra   86
3.4   Simplifying SOP and POS Expressions   100
3.5   Simplification by the Karnaugh Map Method   104
3.6   Simplification by DeMorgan Equivalent Gates   120
3.7   Universal Property of NAND/NOR Gates   122
3.8   Practical Circuit Implementation in SSI Logic   128
3.9   Pulsed Operation of Logic Circuits   131
3.10   A General Approach to Logic Circuit Design   133

4

Introduction to PLDs and Quartus II   164

4.1   What is a PLD?   166
4.2   Programmable Sum-of-Products Arrays   169
4.3   PAL Fuse Matrix and Combinational Outputs   171
4.4   PAL Outputs with Programmable Polarity   175
4.5   Programming CPLDs Using Quartus II   178
4.6   Quartus II Design Flow and Graphical User Interface   182
4.7   Creating a Quartus II Project and Block Diagram File   185
4.8   Compiling and Simulating a Design in Quartus II   195
4.9   Transferring a Design to a Target CPLD   205
4.10   Using the Quartus II Block Editor to Create a Hierarchical Design   216

5

Introduction to VHDL   232

5.1   VHDL Basics   234
5.2   Making a VHDL File in Quartus II   237
5.3   VHDL Syntax for Port, Mode, and Type   244
5.4   Signals in VHDL   254

6

Combinational Logic Functions   266

6.1   Decoders   268
6.2   Encoders   291
6.3   Multiplexers   297
6.4   Demultiplexers   313
6.5   Magnitude Comparators   318
6.6   Parity Generators and Checkers   325

7

Digital Arithmetic and Arithmetic Circuits   346

7.1   Digital Arithmetic   348
7.2   Representing Signed Binary Numbers   351
7.3   Signed Binary Arithmetic   353
7.4   Hexadecimal Arithmetic   360
7.5   Numeric and Alphanumeric Codes   362
7.6   Binary Adders and Subtractors   367
7.7   BCD Adders   388
7.8   Carry Generation in Quartus II   393

8

Introduction to Sequential Logic   408

8.1   Latches   411
8.2   NAND/NOR Latches   414
8.3   Gated Latches   429
8.4   Edge-Triggered D Flip-Flops   447

8.5 Edge-Triggered JK Flip-Flops 452
8.6 Edge-Triggered T Flip-Flops 460
8.7 Flip-Flops in PLDs (Registered Outputs) 462
8.8 Generic Array Logic 467
8.9 MAX 7000S CPLD 471
8.10 FLEX 10K CPLD 473

9
Counters and Shift Registers 500

9.1 Basic Concepts of Digital Counters 503
9.2 Synchronous Counters 508
9.3 Design of Synchronous Counters 514
9.4 Programming Binary Counters for CPLDs 523
9.5 Control Options for Synchronous Counters 534
9.6 Programming Presettable and Bidirectional Counters for CPLDs 552
9.7 Shift Registers 575
9.8 Programming Shift Registers in VHDL 589
9.9 Shift Register Counters 601

10
State Machine Design 626

10.1 State Machines 628
10.2 State Machines with No Control Inputs 629
10.3 State Machines with Control Inputs 636
10.4 Switch Debouncer for a Normally Open Pushbutton Switch 645
10.5 Unused States in State Machines 653
10.6 Traffic Light Controller 659

11
Logic Gate Circuitry 668

11.1 Electrical Characteristics of Logic Gates 670
11.2 Propagation Delay 674
11.3 Flip-Flop Timing Parameters 676
11.4 Fanout 679
11.5 Power Dissipation 684
11.6 Noise Margin 688
11.7 Interfacing TTL and CMOS Gates 690
11.8 Internal Circuitry of TTL Gates 693
11.9 Internal Circuitry of CMOS Gates 714
11.10 TTL and CMOS Variations 726

12
Interfacing Analog and Digital Circuits 744

12.1 Analog and Digital Signals 746
12.2 Digital-to-Analog Conversion 761
12.3 Analog-to-Digital Conversion 780
12.4 Data Acquisition 805

13
Memory Devices and Systems 828

13.1 Basic Memory Concepts 830
13.2 Random Access Read/Write Memory (RAM) 838
13.3 Read Only Memory (ROM) 844
13.4 Sequential Memory: FIFO and LIFO 851
13.5 Dynamic RAM Modules 853
13.6 Memory Systems 855

14
Introduction to Microprocessors 864

14.1 Basic Structure of a Microcomputer 866
14.2 Register Level Structure of a Microcomputer System 870
14.3 Tristate Busses in Altera CPLDs 885
14.4 Quartus II Implementation of the RISC8v1 MCU 896
14.5 Creating New Instructions (RISC8v2) 914
14.6 Branch Instructions (RISC8v3) 917
14.7 RISC8 Test Circuit 922
14.8 Possible Enhancements to RISC8v3 MCU 924

Appendices

Appendix A Converting MAX+PLUS II Projects to Quartus II 937

Answers to Selected Odd-Numbered Problems 955

Index 997

Lab Manual Contents

All labs are available electronically on the accompanying Lab.Source CD.

Lab 1   DIP Integrated Circuits

Lab 2   Expanding Logic Gates

Lab 3   Pulsed Operation of Logic Gates

Lab 4   Boolean Algebra: SOP Forms

Lab 5   Boolean Algebra: DeMorgan Equivalent Forms

Lab 6   Installing and Licensing Quartus II on Your PC (Tutorial)

Lab 7   Introduction to Quartus II (Tutorial)

Lab 8   Introduction to VHDL

Lab 9   Standard Wiring Configuration for Altera UP-1/UP-2 and RSR PLDT-2 Boards

Lab 10   Binary Decoders (Block Diagram File)

Lab 11   Binary and Seven-Segment Decoders in VHDL

Lab 12   Priority Encoders

Lab 13   Multiplexers

Lab 14   Multiplexer Applications

Lab 15   Demultiplexers

Lab 16   Magnitude Comparators

Lab 17   Parity Generators and Checkers

Lab 18   Full Adder and Parallel Binary Adder

Lab 19   Two's Complement Adder/Subtractor

Lab 20   NAND Latches, D Latches, and D Flip-Flops

Lab 21   JK and T Flip-Flops

Lab 22   Binary Counters (Block Diagram File)

Lab 23   Binary and BCD Counters with Control Inputs (VHDL)

Lab 24   Parameterized Counters

Lab 25   Digital Stopwatch (Lab Project 1)

Lab 26   Shift Registers

Lab 27   Parameterized Shift Registers

Lab 28   Time Division Multiplexing (Lab Project 2)

Lab 29   State Machines

Lab 30   Traffic Light Controller (Lab Project 3)

Lab 31   Digital-to-Analog Conversion

Lab 32   DAC-Based Function Generator

Lab 33   Analog-to-Digital Conversion

Lab 34   RAM and ROM in Altera CPLDs

Lab 35   Tristate Bussing in Altera CPLDs

Lab 36   CPLD-Based Microprocessor Core

Preface

INTENDED AUDIENCE

This book is intended as a textbook for an introductory course in digital electronics in an electronics engineering technologist (EET) or computer engineering technologist (CET) program. There is sufficient material for a second course that expands upon the principles of an introductory course.

No prior knowledge of digital systems is assumed. Prerequisite or corequisite courses in basic DC circuits and introductory college algebra, while not strictly necessary, allow the student to derive maximum benefit from the course of study laid out by this book. A working knowledge of transistors and operational amplifiers is very helpful for the chapters on Logic Gate Circuitry (Chapter 11) and Interfacing Analog and Digital Circuits (Chapter 12). These topics would usually be covered in a second course in digital electronics.

ABOUT THIS BOOK: PROGRAMMABLE LOGIC AS A VEHICLE FOR TEACHING DIGITAL DESIGN

Historically, digital logic or digital design courses at the EET level have primarily focussed on using fixed function TTL and CMOS Small and Medium Scale Integration (SSI and MSI) devices as the vehicle for teaching principles of logic design. However, the digital design field has changed; more and more, digital designs are being implemented in Programmable Logic Devices (PLDs), rendering many of the popular fixed function devices obsolete. The new devices require a different teaching paradigm for which there are few EET-level resources at present.

In the past, many digital components were so-called "fixed function" devices; a manufacturer would have a range of functional offerings (counters, decoders, shift registers, and so on), each of which was manufactured into a single chip. Digital systems were constructed of multiple devices, which sometimes resulted in inefficient designs with large package counts. It was important for technicians and technologists to understand how these devices worked and how to use them in digital circuits. This is the assumption behind many of the classic digital electronics textbooks.

Programmable Logic Devices (PLDs) allow the user to define a function on single or multiple chips, rather than work around standard offerings from device manufacturers. Designs for these devices are developed in concert with sophisticated computer programs that take advantage of the modern PC or workstation environment. The programmable nature and high capacity of these devices result in increased design efficiency and flexibility, as well as decreased time to market.

PLDs have been entering the digital curriculum slowly for a number of years, but have always been treated as one, often optional, topic among many, rather than as an underlying foundation for digital design. Two changes in recent years have made a new approach possible.

First, many of the new Complex PLDs (CPLDs) can be programmed, erased, and reprogrammed via a connection to a PC serial or parallel port, without removing them from the circuit in which they are installed. This feature, known as In-System Programmability (ISP), eliminates the need for separate, expensive programming hardware and reduces the mechanical wear-and-tear and risk of damage due to electrostatic discharge (ESD) that always attends multiple insertion and removal cycles. Second, the average PC user has access to much more powerful computers than in the past. These computers have sufficient resources to run the design and programming software required by the new CPLDs. Every student with a PC and a CPLD board can now have a complete design and prototyping station at school *and at home,* which has never been possible until now.

This book focuses on the new digital paradigm by using Altera's University Program Laboratory Design Package (UP-2), which includes the Web Edition of Quartus II, Altera's Programmable Logic Development Software. This Windows-based software allows the student to design, test and program CPLD designs in text-based (VHDL) and graphical (schematic entry) formats. (VHDL, an industry-standard language for PLD design, stands for VHSIC Hardware Description Language. VHSIC is an acronym for Very High Speed Integrated Circuit.) The elementary functions of VHDL and Quartus II are simple enough to be used successfully in a first course in digital electronics, yet there is enough scope for the software to be useful to students at the senior design project level. A Quartus II Web Edition CD, which can be legally copied for students' home use, is included with the book.

The Altera UP-2 circuit board is available on a donation basis to educational institutions that belong to the Altera University Program and can be purchased by students for $149. It contains two Altera PLDs and a number of standard input and output devices (DIP switches, pushbuttons, LEDs, and seven-segment displays). For more information, see Altera's University Program web page at: http://www.altera.com/education/univ/unv-index.html.

Two other boards with the EPM7128SLC84 device are widely used: the eSOC board, used by programs at DeVry Institutes and the RSR PLDT-2 board, manufactured and sold by Electronix Express in New Jersey (http://www.elexp.com/). These boards offer less-expensive alternatives to the Altera UP-2 board, with some advantages in terms of wiring requirements and active levels of the LED displays. A CPLD-to-breadboard adapter is also available from HVW Technologies (http://www.hvwtech.com/) in Calgary, Canada. Photographs of all these boards appear in Chapter 4 of this text.

Any of these boards can be used for the CPLD designs described in Chapters 4–13. The material in Chapter 14 requires the use of the Altera UP-1 or UP-2 board, since it uses the FLEX 10K device. (If students are purchasing one of the less-expensive boards, one solution might be for faculty to request a donation from Altera of several UP-2 boards for use with the Chapter 14 material.)

WHY QUARTUS II?

The largest change from the first edition of this book is the switch from Altera's MAX+PLUS II CPLD design software to Quartus II, also by Altera. Users of the previous edition may wonder why the switch, especially so soon after they had gone to all the trouble to make extensive changes to their course work and lab equipment. Fair questions! The main reason for the change in software is that Altera will probably not continue to support MAX+PLUS II through the lifetime of the present edition of this book. I had been told at one point that Altera had hoped that all new designs would be done using Quartus II by the end of 2003, half a year before the intended publication of this text. It seemed prudent, to say the least, to rework the book examples to use the Quartus II software.

If you are a present user of MAX+PLUS II, should you worry? No, for three reasons: availability, ease of use, and compatibility.

- Any school using Altera tools for teaching or research can become a member of the Altera University Program. As such, the school has access to the full version of Quartus II on a donation basis, so software replacement won't cost anything.

- The user interface and design flow of Quartus II are different from MAX+PLUS II, but the major functions can be accomplished with a few simple menu options and toolbar commands. Some learning is required, but no more than for MAX+PLUS II. A tutorial for basic use of the Quartus II software is included in Chapter 4 and additional tips are included as needed with later material.
- MAX+PLUS II projects can be converted to Quartus II format by following the steps outlined in Appendix A. All Quartus II examples, except some in Chapters 13 and 14, will work on any CPLD test board with an EPM7128SLC84 device, such as the Altera UP-1 or UP-2 boards, the RSR PLDT-2 board, or the DeVry eSOC board. Thus, the same lab equipment can be used as for the first edition.

For those who do not upgrade to the Quartus II software at this time, the following resources are provided on the e.resource CD (ISBN: 1401840329) available to instructors.

- Chapter 4 from the first edition of this book, containing tutorials on MAX+PLUS II and VHDL
- Chapter 5.1, Decoders, from Dueck 1/e, containing a tutorial on using the MAX+PLUS II simulator
- Lab 4, Intro to MAX+PLUS II, from the first edition of the lab manual for this book
- Lab 5, Intro to VHDL, from Dueck 1/e lab manual
- Lab 6, Binary and Seven-Segment Decoders, from Dueck 1/e lab manual

Also, the problems in the second edition can be worked out using MAX+PLUS II. The answers in the solutions manual might not look absolutely identical for the graphic/block diagram files or simulations. The VHDL code *would* be absolutely identical in both cases. In short, the book can be used with either version of the software.

ORGANIZATION

Although this text is based on CPLD implementation of digital design it does not neglect underlying digital fundamentals. The coverage of binary and hexadecimal numbers, basic logic functions, Boolean algebra, logic minimization, and simple combinational circuits is still present (Chapters 1–3), but application emphasis in later chapters is shifted away from fixed-function SSI and MSI devices and toward CPLDs.

The text introduces CPLDs early in the teaching sequence (Chapter 4) and continues to provide VHDL and Quartus II applications throughout the book. Students learn the new digital paradigm as an integral part of their training, not just as an optional add-on. Chapters on the Quartus II design environment (Chapter 4), VHDL syntax (Chapter 5), combinational logic functions (Chapter 6), arithmetic circuits (Chapter 7), latches and flip-flops (Chapter 8), counters and shift registers (Chapter 9), state machines (Chapter 10), and microprocessors (Chapter 14) are all based primarily on CPLDs and VHDL programming. Sections in Chapter 12 describe CPLD interfacing to digital-to-analog and analog-to-digital converters. Additional topics include electrical characteristics of TTL and CMOS logic, including low-voltage CMOS (Chapter 11), digital-to-analog and analog-to-digital conversion (Chapter 12), and memory devices (Chapter 13). Appendix A describes techniques for converting MAX+PLUS II projects to Quartus II. Additional appendices, including a VHDL language reference with examples, and selected data sheets, are included on the CD that accompanies this book.

Emphasis on CPLDs is not merely descriptive, but experiential. An accompanying lab manual, written by the same author as the text is included on the CD at the back of the book. The labs are tightly integrated with the main text and contain practical exercises that allow students to design, program, and construct CPLD-based circuits, rather than just read about them. The labs have been broken into smaller units than those in the first edition of the lab manual, making them more suitable for single lab sessions. Also, within the main text, end-of-chapter problems require students to use the Quartus II design software to create schematic and VHDL design and simulations.

NEW IN THE SECOND EDITION

Simulation Criteria

Students have always found simulation to be the most difficult part of the CPLD design flow, not because they cannot master the mechanics of creating a simulation, but because they don't know what to look for when creating it. Students are encouraged throughout the text to think about what a design is supposed to do, how it might fail, and to write simulation criteria that could be used to test for those functional and failure modes. Repeated practice in writing these criteria helps students develop critical thinking skills and increases the effectiveness of their simulations. This process is formalized under the heading of Simulation Criteria throughout the text.

Index of VHDL Constructs

A summary of VHDL constructs is included in the front of the book, indicating locations of first use or important use of VHDL constructs. Every chapter includes a list of VHDL constructs for that chapter.

Lab.Source

Introducing new free lab exercises with Lab.Source. The entire lab manual is included on the *Lab.Source* CD that accompanies this book. There are 36 labs, including three tutorials and four lab projects that should take several lab periods or would require out-of-lab time. The remaining labs are shorter than those in the lab manual for the first edition of this book and are generally suitable for a single lab period of 2–3 hours. Most of the labs are suitable for use with the Altera UP-1 or UP-2 board, the RSR PLDT-2 board, or the DeVry eSOC board. Labs 34 and 36 require the use of the Altera UP-1 or UP-2 board, as they use the FLEX 10K chip, which is not found on the other boards. A special CD icon identifies which labs best reinforce the theory within the chapter.

➡ *LAB SOURCE*
Lab 11, Binary and Seven-Segment Decoders in VHDL

Additional Chapter Material

A number of chapters contain expanded or substantially modified material.

Chapter 3

Expanded coverage of combinational logic, including:

- a simplification technique using DeMorgan equivalent gates
- the universal property of NAND and NOR gates
- practical implementation of combinational logic using SSI logic
- pulsed behavior of combinational logic circuits, and
- a general approach to logic design that teaches students to translate word problems into Boolean equations for combinational circuits.

Chapter 4

- Quartus II tutorial, including simulation.
- Additional material on programmable logic, transferred from Chapter 8 in the first edition.

Chapter 5

- New chapter on VHDL that gives a more detailed introduction to the subject than the material in the first edition.

Chapter 6

- Expanded treatment of VHDL as applied to combinational logic functions.
- Introduction of VHDL signal attributes in the section on parity.

Chapter 8

- Introduction of megafunctions, as used to instantiate latch components.
- Material on registered PLD outputs and macrocells, as well as CPLD architecture.

Chapter 9

- Use of megafunctions to instantiate counters.
- Use of std_logic_unsigned and std_logic_arith packages in counter design.
- Use of signal attributes in counter and shift register design.
- Use of parameters to make counters and shift registers with generic characteristics.

Chapter 12

- Expanded coverage of unipolar and bipolar data coding for analog-to-digital and digital-to-analog conversion.
- New section on sigma-delta ADC.

Chapter 14

- New chapter on microprocessors. Develops a register-level processor that can be programmed into a FLEX 10K CPLD on the Altera UP-1 or UP-2 board.

HOW TO USE THIS BOOK

Special features of this book include:

Chapter Opener

Each chapter begins with an **Outline** and list of **Objectives** to prepare students for major concepts to be studied and learned.

Key Terms and Notes

Definitions of **Key Terms** are placed at the beginning of each section. All key terms for a chapter are listed in an end-of-chapter **Glossary.** First use in context of each term is also indicated in boldface. **Notes** are text boxes that contain suggestions, hints, and tips.

VHDL Examples

Numerous solved-problem **Examples** are placed throughout each chapter. Many examples of VHDL programming teach an industry-standard digital hardware design method. Examples are stepped in difficulty from basic to more advanced making learning of new material easier.

Screen Captures and Schematics

Frequent use of schematics, illustrations, tables and screen captures help understanding of the digital principles. Screen captures show exactly what to expect in the Altera™ Quartus™ II design environment.

CHAPTER OPENER

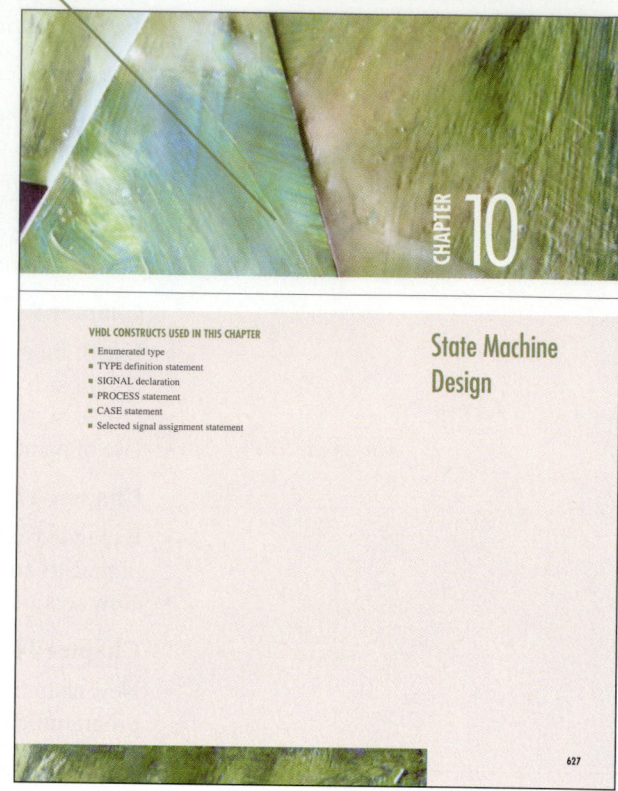

KEY TERMS
Bubble
Conditional Transition
Control Input
Enumerated Type
Form A Contact
Form B Contact
Form C Contact
Mealy Machine
Moore Machine
State Machine
State Variables
Unconditional Transition

OUTLINE
10.1 State Machines
10.2 State Machines with No Control Inputs
10.3 State Machines with Control Inputs
10.4 Switch Debouncer for a Normally Open Pushbutton Switch
10.5 Unused States in State Machines
10.6 Traffic Light Controller

CHAPTER OBJECTIVES

Upon successful completion of this chapter you will be able to:

■ Describe the components of a state machine.
■ Distinguish between Moore and Mealy implementations of state machines.
■ Draw the state diagram of a state machine from a verbal description.
■ Use the "classical" (state table) method of state machine design to determine the Boolean equations of the state machine.
■ Translate the Boolean equations of a state machine into a Block Diagram File in Altera's Quartus II software.
■ Write VHDL code to implement state machines.
■ Create simulations in Quartus II to verify the function of a state machine design.
■ Determine whether the output of a state machine is vulnerable to asynchronous changes of input.
■ Design state machine applications, such as a switch debouncer, a single-pulse generator, and a traffic light controller.

626

VHDL CONSTRUCTS USED IN THIS CHAPTER

■ Enumerated type
■ TYPE definition statement
■ SIGNAL declaration
■ PROCESS statement
■ CASE statement
■ Selected signal assignment statement

CHAPTER 10

State Machine Design

627

VHDL EXAMPLES

SECTION REVIEW PROBLEMS

The statements for **p(2)** and **p(3)** can be written more generally as:

p(i) <= p(i - 1) xor d(i);

where **i** is an **index variable** that shows the relationship between a specific instance of **p** and other hardware constructs in the design. For repetitive hardware such as this circuit, we can use a VHDL **GENERATE statement** that has the general form:

```
__generate_label:
FOR __index variable IN __range GENERATE
    __statement;
    __statement;
END GENERATE;
```

For example, the equations for **p(2)** and **p(3)** can be created with the following GENERATE statement:

```
parity_generate:
FOR i IN 2 to 3 GENERATE
    P(i) <= p(i-1) xor d(i);
END GENERATE;
```

The complete VHDL file for this design is shown below.

```
-- 4-bit parity generator using GENERATE statement.
LIBRARY ieee;
USE ieee.std_logic_1164.ALL;

ENTITY parity4_gen IS
    PORT(
        d  : IN  STD_LOGIC_VECTOR(0 TO 3);
        pe : OUT STD_LOGIC);
END parity4_gen;

ARCHITECTURE parity OF parity4_gen IS
    SIGNAL p : STD_LOGIC_VECTOR(1 to 3);
BEGIN
    p(1) <= d(0) xor d(1);

    parity_generate:
    FOR i IN 2 to 3 GENERATE
        p(i) <= p(i-1) xor d(i);
    END GENERATE;

    pe <= p(3);
END parity;
```

ALTERA MAX+PLUS II AND QUARTUS II FILES

parity_gen.vhd
parity_gen.vwf

EXAMPLE 6.25

Expand the circuit of the 4-bit parity generator in Figure 6.65 to make an 8-bit parity generator. Write a VHDL file that uses a GENERATE statement to implement the design.

■ **Solution**

Figure 6.66 shows the circuit for the 8-bit parity generator. The VHDL file for this design follows. The file is the same as for the 4-bit parity generator, except that the vector sizes and index variable are changed.

(continued)

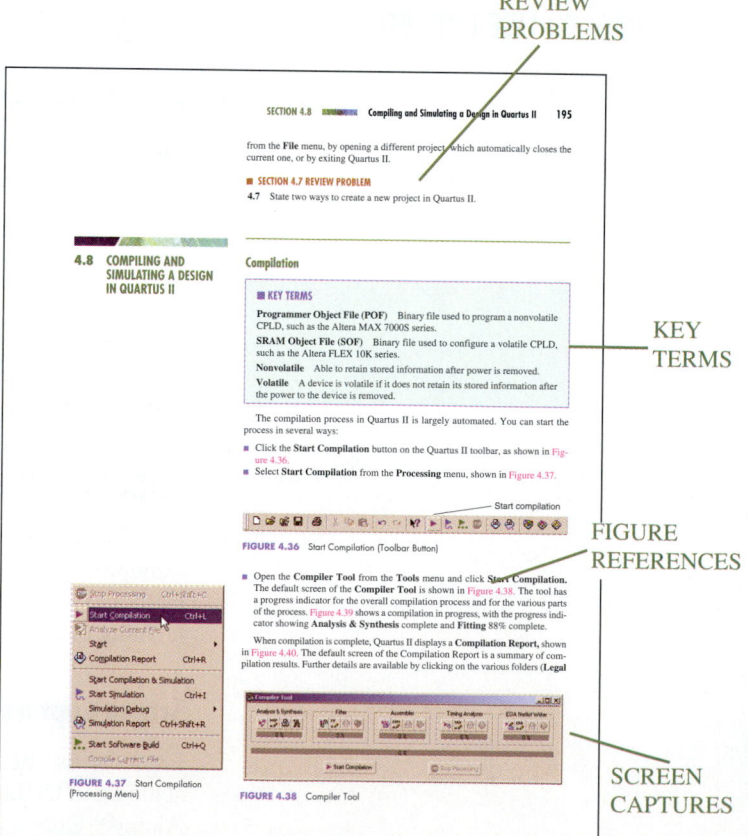

from the **File** menu, by opening a different project which automatically closes the current one, or by exiting Quartus II.

SECTION 4.7 REVIEW PROBLEM

4.7 State two ways to create a new project in Quartus II.

4.8 COMPILING AND SIMULATING A DESIGN IN QUARTUS II

Compilation

KEY TERMS

Programmer Object File (POF) Binary file used to program a nonvolatile CPLD, such as the Altera MAX 7000S series.

SRAM Object File (SOF) Binary file used to configure a volatile CPLD, such as the Altera FLEX 10K series.

Nonvolatile Able to retain stored information after power is removed.

Volatile A device is volatile if it does not retain its stored information after the power to the device is removed.

The compilation process in Quartus II is largely automated. You can start the process in several ways:

■ Click the **Start Compilation** button on the Quartus II toolbar, as shown in Figure 4.36.
■ Select **Start Compilation** from the **Processing** menu, shown in Figure 4.37.

Start compilation

FIGURE 4.36 Start Compilation (Toolbar Button)

KEY TERMS

FIGURE REFERENCES

■ Open the **Compiler Tool** from the **Tools** menu and click **Start Compilation.** The default screen of the **Compiler Tool** is shown in Figure 4.38. The tool has a progress indicator for the overall compilation process and for the various parts of the process. Figure 4.39 shows a compilation in progress, with the progress indicator showing **Analysis & Synthesis** complete and **Fitting** 88% complete.

When compilation is complete, Quartus II displays a **Compilation Report,** shown in Figure 4.40. The default screen of the Compilation Report is a summary of compilation results. Further details are available by clicking on the various folders (**Legal**

FIGURE 4.37 Start Compilation (Processing Menu)

FIGURE 4.38 Compiler Tool

SCREEN CAPTURES

SCHEMATICS

EXAMPLES

CHAPTER
PROBLEM
SETS

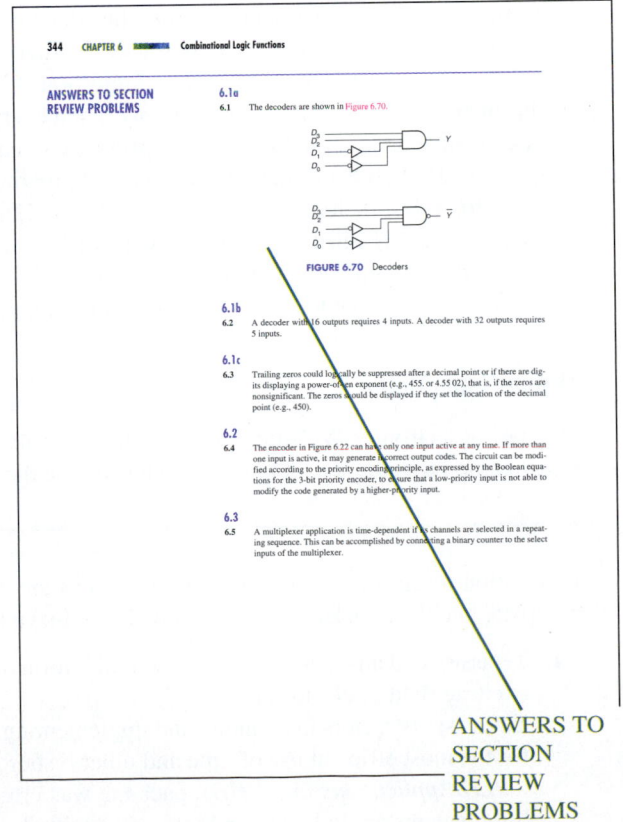

ANSWERS TO
SECTION
REVIEW
PROBLEMS

Section Review Problems

Numerous section review problems assist in retention of recently learned material. Answers to section review problems are placed at the end of each chapter for quick assessment.

Summary

A chapter **Summary** provides a recapitulation of the key topics covered in the chapter.

Problem Set

Questions and problems at the end of each chapter are separated into sections. Basic and advanced problems are designated clearly. Answers to selected odd numbered problems are placed at the end of the book.

Altera™ Quartus™ II Files

The accompanying CD includes **Altera™ Quartus™ II Web Edition** CPLD design and programming software along with block diagram files, VHDL files and simulation files from examples. Students can run simulations or program CPLDs with existing error-free design files at home or at school. Students can use existing files as templates for their own modifications. Filenames are identified by a CD icon beside specific examples throughout the text. For this edition, equivalent MAX+PLUS II files are included, where appropriate.

What Is on the CD?

The accompanying CD features:

Lab.Source. Save your students from the added expense of purchasing a separate printed lab manual. With Lab.Source, learners will be provided with valuable hands-on experience they need to reinforce all key digital topics covered throughout the text. In today's economy when every dollar counts, Lab.Source is a sensible alternative to expensive printed lab manuals.
Quartus II Web Edition. CPLD design and programming software from Altera.
Quartus Files. Includes block diagram files, VHDL files, and simulation files.
MAX+PLUS II Files. Includes equivalent files for this edition.
Appendices. Includes a VHDL language reference, data sheets, handling precautions for CMOS, and EPROM data for a digital function generator.

Online Companion™

Delmar's Electronics Website is located at www.electronictech.com. The Online Companion provides access to text updates, online quizzes, and more.

THE LEARNING PACKAGE

In addition to this book, the following materials are available to instructors. The complete ancillary package was developed to achieve two goals:

1. To assist students in learning the essential information needed to prepare for the exciting field of electronics.
2. To assist instructors in planning and implementing their instructional programs for the most efficient use of time and other resources. The *Digital Design with CPLD Applications and VHDL* package was created as an integrated whole. Supplements are linked to and integrated with the text to create a comprehen-

sive supplement package that supports students and instructors. All supplements are available on the e.resource.

e.resource™

This electronic Instructor's Management System is an educational resource that creates a truly electronic classroom. It is a CD-ROM containing tools and instructional resources that enrich your classroom and make your preparation time shorter. The elements of *e.resource* link directly to the text and tie together to provide a unified instructional system.
ISBN: 1401840329
Features contained in *e.resource* include:

Instructor's Guide

This comprehensive ancillary contains solutions to all end of chapter problems from the textbook.

PowerPoint Presentation Slides

These slides provide the basis for a lecture outline that helps you to present concepts and material. Key points and concepts can be graphically highlighted for student retention. There are over 700 slides, covering every chapter in the text.

ExamView

This computerized testbank includes over 1200 questions of varying levels of difficulty provided in multiple formats to assess student comprehension.

Image Library

Images from the textbook allow you to customize power point presentations, or use them as transparency masters. Image Library comes with the ability to browse and search images with key words and allows quick and easy use.

Altera™ Quartus™ II Files

Quartus II schematics, VHDL, and simulation files for examples, end-of-chapter problems, and labs. Error-free files can be used as a basis for comparison with student or instructor-developed files.

Lab.Builder™

Lab.Builder™ gives the instructor the advantage of being able to create their own labs electronically.

MAX+PLUS II Resources

Includes MAX+PLUS II tutorials and labs.

Electronics Technology Website

Visit www.electronictech.com and to the textbook's Online Companion for additional resources.

ACKNOWLEDGMENTS

My thanks go out to all those who helped in the production of this book. Greg Clayton of Thomson Delmar Learning has been a constant inspiration in the writing of this book. I wish him well in his new endeavors at Delmar. Thanks also to his successors Dave Garza and Steve Helba. Michelle Ruelos Cannistraci, my developmental editor at Delmar, was, as always, a pleasure to work with, a true collaborator in the making of this project. Thanks also to Francis Hogan, Christopher Chien, and Larry Main for their help in the production process.

Thanks to Mary Grivetti and all at Shepherd Inc. for their tireless work to copyedit and typeset this book under a very tight schedule.

Mike Phipps of Altera Corporation has been generous in his support by providing the hardware and software platforms that were so essential to the writing of this book. Thanks to Hugh McLaren of St. Lawrence College for his work on the PowerPoint presentation on the e.resource, along with William Wagner of DeVry University for creating ExamView questions, and to Ken Reid of Indiana University Purdue University for his work on the Image Library.

Thanks to the following reviewers whose comments and suggestions helped to improve the final form of this book:

Rasoul Esfahani, DeVry University, Columbus, OH
Richard Henderson, DeVry University, Kansas City, MO
Hugh McLaren, St. Lawrence College, Ontario
Mike Phipps, Altera
Ken Reid, Indiana University Purdue University
Genevieve Sapijaszko, DeVry University, Orlando, FL
Gilbert Seah, DeVry University, Ontario
Lloyd Stallkamp, Montana State University, Havre, MT
Leslie Taylor, DeVry University, North Brunswick, NJ
William Wagner, DeVry University, Philadelphia, PA
Jamie Zipay, Oregon Institute of Technology, Klamath Falls, OR

Best wishes to my former colleague Mike Gale in his retirement. Thanks to my colleagues Alex McIlraith for bringing fresh ideas to my teaching and to Glenn Duncan for asking a lot of clarifying questions and for pointing out yet another undetected error in the first edition of this book.

Thanks always to my students. I don't know how anyone could write this kind of a book without also teaching the material. The questions you ask always sharpen my own understanding of the material.

My wife, Joan Duerksen, deserves thanks beyond my mere ability to convey them. For supporting and challenging me, Joan, my deepest gratitude.

The author can be reached by e-mail at bdueck@rrc.mb.ca.

Bob Dueck
Winnipeg, Manitoba, Canada

ABOUT THE AUTHOR

Robert Dueck received his degree in Electrical Engineering from the University of Manitoba, in Winnipeg, Canada and worked for several years as a design engineer at Motorola Canada in Toronto. He began his teaching career in 1986, specializing in digital and microcomputer subjects in the Electronics and Computer Engineering Technology programs at Seneca College in Toronto. His first book, *Fundamentals of Digital Electronics,* was published in 1994. He now teaches digital design and related courses at Red River College in Winnipeg. Mr. Dueck is a member of the Association of Professional Engineers and Geoscientists of Manitoba (APEGM) and the Institute of Electrical and Electronics Engineers (IEEE). He is active in the Winnipeg Section of IEEE, serving as Section Chair in 2002 and Branch Counselor of the Red River College Student Branch since 1997.

Summary of VHDL Constructs

Construct	Page Reference
Architecture body	235
Array	244, 246
BIT and BIT_VECTOR types	246, 247
CASE statement	289, 290, 633, 880, 909–11
Clearing of vector using (others => '0')	526, 559–60, 597
Comments	245
Component declaration statement	376, 378–80, 561, 590, 594–6
Component instantiation statement, including GENERIC MAP and PORT MAP	376–7, 378–80, 433–5, 527, 561, 565–6, 590, 594–6
Concatenation operator (&)	256–7, 280, 554, 594–6, 895, 906
Concurrent signal assignment statements	237
Conditional signal assignment statement	281, 295
Entity declaration	235, 559
Enumerated type	633–5, 640, 909–11
Explicit port mapping	377–8, 565–6, 597, 598
GENERATE statement	331, 378, 383, 435, 590
GENERIC clause (entity declaration)	379, 559–60, 595–6
IF statement	287–8, 290, 295, 431, 557–8
Implicit port mapping	377–8, 561, 590
INTEGER, NATURAL, and POSITIVE types	253, 254
LIBRARY declaration	251, 525, 526
Library of Parameterized Modules (LPM)	433–5, 436–44, 564–6, 597–601
Mode	235, 245–6
Naming conventions	244–5
others clause	250, 252
Port definitions	Section 5.3
Port mode BUFFER	525–6, 559, 590
Primitives	431–2, 435, 590
PROCESS statement	289, 431, 552–3, 557–8, 633
Range (to and downto)	244, 246, 253–4, 524, 590
Reserved keywords	245
Selected signal assignment statement	249–50, 280, 301, 547
Setting vector to high-impedance state using (others => 'Z';)	552–3, 895–6, 906–7

Signal assignment operator <=	235–6, 280, 524–5, 911
Signal attribute ('EVENT)	524, 557–8
Signal attributes ('HIGH, 'LOW, 'LEFT, 'RIGHT)	333–4
SIGNAL declaration	255–9, 591, 634–5
State machine encoding	633–5, 640, 909–11
STD_LOGIC and STD_LOGIC_VECTOR types	251–2
Type	Section 5.3, 633–5
Type conversion function CONV_STD_LOGIC	559–60, 596
TYPE definition statement	633–5, 640, 909–11
USE clause	251
Variable assignment operator :=	524
VARIABLE declaration	524–5
VHDL package: altera.maxplus2	431, 434, 590
VHDL package: ieee.std_logic_1164	251, 559, 590, 880
VHDL package: ieee.std_logic_arith	559, 596, 880
VHDL package: ieee.std_logic_signed	880
VHDL package: ieee.std_logic_unsigned	525–6, 559
VHDL package: lpm.lpm_components	435, 526–7, 564–6, 597, 599, 600

Digital Design with CPLD Applications and VHDL, 2nd edition

KEY TERMS

Amplitude
Analog
Aperiodic waveform
Binary number system
Binary point
Bit
Continuous
Digital
Digital waveform
Discrete
Duty cycle (*DC*)
Edge
Fall time (t_f)
Falling edge
Frequency (*f*)
Hexadecimal number system
Leading edge
Least significant bit (LSB)
Logic HIGH
Logic level
Logic LOW
Most significant bit (MSB)
Negative logic
Period (*T*)
Periodic waveform
Positional notation
Positive logic
Pulse
Pulse width (t_w)
Radix point
Rise time (t_r)
Rising edge
Time HIGH (t_h)
Time LOW (t_l)
Trailing edge

OUTLINE

1.1 Digital Versus Analog Electronics
1.2 Digital Logic Levels
1.3 The Binary Number System
1.4 Hexadecimal Numbers
1.5 Digital Waveforms

CHAPTER OBJECTIVES

Upon successful completion of this chapter, you will be able to:

■ Describe some differences between analog and digital electronics.

■ Understand the concept of HIGH and LOW logic levels.

■ Explain the basic principles of a positional notation number system.

■ Translate logic HIGHs and LOWs into binary numbers.

■ Distinguish between the most significant bit and least significant bit of a binary number.

■ Count in binary, decimal, or hexadecimal.

■ Convert a number in binary, decimal, or hexadecimal to any of the other number bases.

■ Calculate the fractional binary equivalent of any decimal number.

■ Describe the difference between periodic, aperiodic, and pulse waveforms.

■ Calculate the frequency, period, and duty cycle of a periodic digital waveform.

■ Calculate the pulse width, rise time, and fall time of a digital pulse.

CHAPTER 1

Basic Principles of Digital Systems

Digital electronics is the branch of electronics based on the combination and switching of voltages called logic levels. Any quantity in the physical world, such as temperature, pressure, or voltage, can be symbolized in a digital circuit by a group of logic voltages that, taken together, represent a binary number.

Each logic level corresponds to a digit in the binary (base 2) number system. The binary dig*its,* or bits, 0 and 1, are sufficient to write any number, given enough places. The hexadecimal (base 16) number system is also important in digital systems. Since every combination of four binary digits can be uniquely represented as a hexadecimal digit, this system is often used as a compact way of writing binary information.

Inputs and outputs in digital circuits are not always static. Often they vary with time. Time-varying digital waveforms can have three forms:

1. Periodic waveforms, which repeat a pattern of logic 1s and 0s
2. Aperiodic waveforms, which do not repeat
3. Pulse waveforms, which produce a momentary variation from a constant logic level ■

1.1   DIGITAL VERSUS ANALOG ELECTRONICS

The study of electronics often is divided into two basic areas: **analog** and **digital** electronics. Analog electronics has a longer history and can be regarded as the "classical" branch of electronics. Digital electronics, although newer, has achieved greater prominence through the advent of the computer age. The modern revolution in microcomputer chips, as part of everything from personal computers to cars and coffee makers, is founded almost entirely on digital electronics.

The main difference between analog and digital electronics can be stated simply. Analog voltages or currents are **continuously** variable between defined values, and digital voltages or currents can vary only by distinct, or **discrete,** steps.

Some keywords highlight the differences between digital and analog electronics:

Analog	Digital
Continuously variable	Discrete steps
Amplification	Switching
Voltages	Numbers

An example often used to illustrate the difference between analog and digital devices is the comparison between a light dimmer and a light switch. A light dimmer is an analog device, since it can make the light it controls vary in brightness anywhere within a defined range of values. The light can be fully on, fully off, or at some brightness level in between. A light switch is a digital device, since it can turn the light on or off, but there is no value in between those two states.

The light switch/light dimmer analogy, although easy to understand, does not show any particular advantage to the digital device. If anything, it makes the digital device seem limited.

One modern application in which a digital device is clearly superior to an analog device is digital audio reproduction. Compact disc players have achieved their high level of popularity because of the accurate and noise-free way in which they reproduce recorded music. This high quality of sound is possible because the music is stored not as a physical copy of the sound waves (as in a vinyl record) or as a magnetic copy (as in an analog tape), but as a coded series of numbers that represent the amplitude steps in the sound waves.

Figure 1.1a shows a sound waveform represented as a voltage. Such a waveform could be measured at the output of an amplified microphone. Figure 1.1b and Figure 1.1c show the voltage as an analog copy and a digital copy of the sound waveform.

The analog copy, shown in Figure 1.1b, shows distortion with respect to the original. The distortion is introduced by both the analog storage and playback processes.

A digital audio system doesn't make a copy of the waveform, but rather stores a code (a series of amplitude numbers) that tells the compact disc player how to recreate the original sound every time a disc is played. During the recording process, the sound waveform is "sampled" at precise intervals. That is, the voltage of the

a. Original audio source

b. Analog reproduction (shows distortion)

c. Digital reproduction (simplified)

FIGURE 1.1 Digital and Analog Sound Reproduction

waveform is measured at certain intervals and each measurement is converted to a representative binary number. A typical encoding scheme might assign the voltage to a value between 0 and 65,535. Such a large number of possible values means the voltage difference between any two consecutive digital numbers is very small. The numbers can thus correspond extremely closely to the actual amplitude of the sound waveform.

Digital representations of physical quantities are also superior to analog in that they can easily be stored, transferred, and copied without the distortion accompanying analog processes. The digital values can be stored in a variety of media, such as optical (CD or DVD), magnetic (hard drive on a PC) or semiconductor (flash memory). They can be transmitted over electronic communications systems such as fiber optic, radio, or telephone. As long as the integrity of the digital data is maintained, any copy of the data is as good as any other. Copies can be made from other copies without deterioration between copy generations.

■ **SECTION 1.1 REVIEW PROBLEM**

1.1 What is the basic difference between analog and digital audio reproduction?

1.2 DIGITAL LOGIC LEVELS

■ **KEY TERMS**

Logic Level   A voltage level that represents a defined digital state in an electronic circuit.

Logic LOW (or **logic 0**)   The lower of two voltages in a digital system with two logic levels.

Logic HIGH (or **logic 1**)   The higher of two voltages in a digital system with two logic levels.

Positive Logic   A system in which logic LOW represents binary digit 0 and logic HIGH represents binary digit 1.

Negative Logic   A system in which logic LOW represents binary digit 1 and logic HIGH represents binary digit 0.

Digitally represented quantities, such as the amplitude of an audio waveform, are usually represented by binary, or base 2, numbers. When we want to describe a digital

FIGURE 1.2   Logic Levels Based on +5 V and 0 V

quantity electronically, we need to have a system that uses voltages or currents to symbolize binary numbers.

The binary number system has only two digits, 0 and 1. Each of these digits can be denoted by a different voltage called a **logic level.** For a system having two logic levels, the lower voltage (usually 0 volts) is called a **logic LOW** or **logic 0** and represents the digit 0. The higher voltage (traditionally 5 V, but in some systems a specific value such as 1.8 V, 2.5 V or 3.3 V) is called a **logic HIGH** or **logic 1,** which symbolizes the digit 1. Except for some allowable tolerance, as shown in Figure 1.2, the range of voltages between HIGH and LOW logic levels is undefined.

> **NOTE . . .**
>
> For the voltages in Figure 1.2:
>
> $$+5 \text{ V} = \text{Logic HIGH} = 1$$
> $$0 \text{ V} = \text{Logic LOW} = 0$$

The system assigning the digit 1 to a logic HIGH and digit 0 to logic LOW is called **positive logic.** Throughout the remainder of this text, logic levels will be referred to as HIGH/LOW or 1/0 interchangeably. (A complementary system, called **negative logic,** also exists that makes the assignment the other way around.)

1.3   THE BINARY NUMBER SYSTEM

> **■ KEY TERMS**
>
> **Binary Number System**   A number system used extensively in digital systems, based on the number 2. It uses two digits, 0 and 1, to write any number.
>
> **Bit**   *B*inary dig*it*. A 0 or a 1.
>
> **Positional Notation**   A system of writing numbers where the value of a digit depends not only on the digit, but also on its placement within a number.

Positional Notation

The **binary number system** is based on the number 2. This means that we can write any number using only two binary digits (or **bits**), 0 and 1. Compare this to the decimal system, which is based on the number 10, where we can write any number with only ten decimal digits, 0 to 9.

The binary and decimal systems are both **positional notation** systems; the value of a digit in either system depends on its placement within a number. In the decimal number 845, the digit 4 really means 40, whereas in the number 9426, the digit 4 really means 400 ($845 = 800 + 40 + 5$; $9426 = 9000 + 400 + 20 + 6$). The value of the digit is determined by *what* the digit is as well as *where* it is.

In the decimal system, a digit in the position immediately to the left of the decimal point is multiplied by 1 (10^0). A digit two positions to the left of the decimal point is multiplied by 10 (10^1). A digit in the next position left is multiplied by 100 (10^2). The positional multipliers, as you move left from the decimal point, are ascending powers of 10.

The same idea applies in the binary system, except that the positional multipliers are powers of 2 ($2^0 = 1$, $2^1 = 2$, $2^2 = 4$, $2^3 = 8$, $2^4 = 16$, $2^5 = 32$, . . .). For example, the binary number 101 has the decimal equivalent:

$$
\begin{aligned}
&(1 \times 2^2) + (0 \times 2^1) + (1 \times 2^0) \\
=\ &(1 \times 4)\ \ + (0 \times 2)\ \ + (1 \times 1) \\
=\ &\ \ \ 4\ \ \ \ +\ \ \ 0\ \ \ +\ \ \ 1 \\
=\ &\ \ \ 5
\end{aligned}
$$

EXAMPLE 1.1

Calculate the decimal equivalents of the binary numbers 1010, 111, and 10010.

■ **Solutions**

$$
\begin{aligned}
1010 &= (1 \times 2^3) + (0 \times 2^2) + (1 \times 2^1) + (0 \times 2^0) \\
&= (1 \times 8) + (0 \times 4) + (1 \times 2) + (0 \times 1) \\
&= 8 + 2 = 10 \\
111 &= (1 \times 2^2) + (1 \times 2^1) + (1 \times 2^0) \\
&= (1 \times 4) + (1 \times 2) + (1 \times 1) \\
&= 4 + 2 + 1 = 7 \\
10010 &= (1 \times 2^4) + (0 \times 2^3) + (0 \times 2^2) + (1 \times 2^1) + (0 \times 2^0) \\
&= (1 \times 16) + (0 \times 8) + (0 \times 4) + (1 \times 2) + (0 \times 1) \\
&= 16 + 2 = 18
\end{aligned}
$$

Binary Inputs

■ **KEY TERMS**

Most Significant Bit   The leftmost bit in a binary number. This bit has the number's largest positional multiplier.

Least Significant Bit   The rightmost bit of a binary number. This bit has the number's smallest positional multiplier.

A major class of digital circuits, called combinational logic, operates by accepting logic levels at one or more input terminals and producing a logic level at an output. In the analysis and design of such circuits, it is frequently necessary to find the output logic level of a circuit for all possible combinations of input logic levels.

The digital circuit in the black box in Figure 1.3 has three inputs. Each input can have two possible states, LOW or HIGH, which can be represented by positive logic as 0 or 1. The number of possible input combinations is $2^3 = 8$. (In general, a circuit with n binary inputs has 2^n input combinations, ranging from 0 to 2^n-1.) Table 1.1 shows a list of these combinations, both as logic levels and binary numbers, and their decimal equivalents.

Inputs   Output

Digital circuit

A (MSB)
B   Y
C (LSB)

FIGURE 1.3   3-input Digital Circuit

TABLE 1.1   Possible Input Combinations for a 3-Input Digital Circuit

Logic Level			Binary Value			Decimal Equivalent
A	**B**	**C**	**A**	**B**	**C**	
L	L	L	0	0	0	0
L	L	H	0	0	1	1
L	H	L	0	1	0	2
L	H	H	0	1	1	3
H	L	L	1	0	0	4
H	L	H	1	0	1	5
H	H	L	1	1	0	6
H	H	H	1	1	1	7

A list of output logic levels corresponding to all possible input combinations, applied in ascending binary order, is called a truth table. This is a standard form for showing the function of a digital circuit.

The input bits on each line of Table 1.1 can be read from left to right as a series of 3-bit binary numbers. The numerical values of these eight input combinations range from 0 to 7 (2^n possible input combinations, having decimal equivalents ranging from 0 to 2^n-1) in decimal.

Bit A is called the **most significant bit (MSB),** and bit C is called the **least significant bit (LSB).** As these terms imply, a change in bit A is more significant, since it has the greatest effect on the number of which it is part.

Table 1.2 shows the effect of changing each of these bits in a 3-bit binary number and compares the changed number to the original by showing the difference in magnitude. A change in the MSB of any 3-bit number results in a difference of 4. A change in the LSB of any binary number results in a difference of 1. (Try it with a few different numbers.)

TABLE 1.2   Effect of Changing the LSB and MSB of a Binary Number

	A	**B**	**C**	**Decimal**	
Original	0	1	1	3	
Change MSB	1	1	1	7	Difference = 4
Change LSB	0	1	0	2	Difference = 1

EXAMPLE 1.2

Figure 1.4 shows a 4-input digital circuit. List all the possible binary input combinations to this circuit and their decimal equivalents. What is the value of the MSB?

FIGURE 1.4   Example 1.2: 4-Input Digital Circuit

■ Solution

Since there are four inputs, there will be $2^4 = 16$ possible input combinations, ranging from 0000 to 1111 (0 to 15 in decimal). Table 1.3 shows the list of all possible input combinations.

The MSB has a value of 8 (decimal).

TABLE 1.3   Possible Input Combinations for a 4-Input Digital Circuit

A	B	C	D	Decimal
0	0	0	0	0
0	0	0	1	1
0	0	1	0	2
0	0	1	1	3
0	1	0	0	4
0	1	0	1	5
0	1	1	0	6
0	1	1	1	7
1	0	0	0	8
1	0	0	1	9
1	0	1	0	10
1	0	1	1	11
1	1	0	0	12
1	1	0	1	13
1	1	1	0	14
1	1	1	1	15

Knowing how to construct a binary sequence is a very important skill when working with digital logic systems. Two ways to do this are:

1. *Learn to count in binary.* You should know all the binary numbers from 0000 to 1111 and their decimal equivalents (0 to 15). *Make this your first goal in learning the basics of digital systems.*

Each binary number is a unique representation of its decimal equivalent. You can work out the decimal value of a binary number by adding the weighted values of all the bits.

For instance, the binary equivalent of the decimal sequence 0, 1, 2, 3 can be written using two bits: the 1's bit and the 2's bit. The binary count sequence is:

$$00 \ (= 0 + 0)$$
$$01 \ (= 0 + 1)$$
$$10 \ (= 2 + 0)$$
$$11 \ (= 2 + 1)$$

To count beyond this, you need another bit: the 4's bit. The decimal sequence 4, 5, 6, 7 has the binary equivalents:

$$100 \ (= 4 + 0 + 0)$$
$$101 \ (= 4 + 0 + 1)$$
$$110 \ (= 4 + 2 + 0)$$
$$111 \ (= 4 + 2 + 1)$$

The two least significant bits of this sequence are the same as the bits in the 0 to 3 sequence; a repeating pattern has been generated.

The sequence from 8 to 15 requires yet another bit: the 8's bit. The three LSBs of this sequence repeat the 0 to 7 sequence. The binary equivalents of 8 to 15 are:

$$1000 \ (= 8 + 0 + 0 + 0)$$
$$1001 \ (= 8 + 0 + 0 + 1)$$
$$1010 \ (= 8 + 0 + 2 + 0)$$
$$1011 \ (= 8 + 0 + 2 + 1)$$
$$1100 \ (= 8 + 4 + 0 + 0)$$
$$1101 \ (= 8 + 4 + 0 + 1)$$
$$1110 \ (= 8 + 4 + 2 + 0)$$
$$1111 \ (= 8 + 4 + 2 + 1)$$

Practice writing out the binary sequence, as listed in Table 1.3, until it becomes familiar. In the 0 to 15 sequence, it is standard practice to write each number as a 4-bit value, as in Example 1.2, so that all numbers have the same number of bits. Numbers up to 7 have leading zeros to pad them out to 4 bits.

This convention has developed because each bit has a physical location in a digital circuit; we know a particular bit is logic 0 because we can measure 0 V at a particular point in a circuit. A bit with a value of 0 doesn't go away just because there is not a 1 at a more significant location.

While you are still learning to count in binary, you can use a second method.

2. *Follow a simple repetitive pattern.* Look at Table 1.1 and Table 1.3 again. Notice that the least significant bit follows a pattern. The bits alternate with every line, producing the pattern 0, 1, 0, 1, ... The 2's bit alternates every two lines: 0, 0, 1, 1, 0, 0, 1, 1, ... The 4's bit alternates every four lines: 0, 0, 0, 0, 1, 1, 1, 1, ... This pattern can be expanded to cover any number of bits, with the number of lines between alternations doubling with each bit to the left.

Decimal-to-Binary Conversion

There are two methods commonly used to convert decimal numbers to binary: sum of powers of 2 and repeated division by 2.

Sum of Powers of 2

You can convert a decimal number to binary by adding up powers of 2 by inspection, adding bits as you need them to fill up the total value of the number. For example, convert 57_{10} to binary.

$$64_{10} > 57_{10} > 32_{10}$$

■ We see that 32 ($= 2^5$) is the largest power of two that is smaller than 57. Set the 32's bit to 1 and subtract 32 from the original number, as shown.

$$57 - 32 = 25$$

■ The largest power of two that is less than 25 is 16. Set the 16's bit to 1 and subtract 16 from the accumulated total.

$$25 - 16 = 9$$

■ 8 is the largest power of two that is less than 9. Set the 8's bit to 1 and subtract 8 from the total.

$$9 - 8 = 1$$

■ 4 is greater than the remaining total. Set the 4's bit to 0.
■ 2 is greater than the remaining total. Set the 2's bit to 0.
■ 1 is left over. Set the 1's bit to 1 and subtract 1.

$$1 - 1 = 0$$

■ Conversion is complete when there is nothing left to subtract. Any remaining bits should be set to 0.

32	16	8	4	2	1	
1						$57 - 32 = 25$

32	16	8	4	2	1	
1	1					$25 - 16 = 9$

32	16	8	4	2	1	
1	1	1				$9 - 8 = 1$

32	16	8	4	2	1	
1	1	1	0	0	1	$1 - 1 = 0$

$$57_{10} = 111001_2$$

EXAMPLE 1.3

Convert 92_{10} to binary using the sum-of-powers-of-2 method.

■ **Solution**

$$128 > 92 > 64$$

64	32	16	8	4	2	1	
1							$92 - 64 = 28$

64	32	16	8	4	2	1	
1	0	1					$28 - 16 = 12$

64	32	16	8	4	2	1	
1	0	1	1				$12 - 8 = 4$

64	32	16	8	4	2	1	
1	0	1	1	1	0	0	$4 - 4 = 0$

$$92_{10} = 1011100_2$$

Repeated Division by 2

Any decimal number divided by 2 will leave a remainder of 0 or 1. Repeated division by 2 will leave a string of 0 and 1 remainders that become the binary equivalent of the decimal number. Let us use this method to convert 46_{10} to binary.

1. Divide the decimal number by 2 and note the remainder.

$$46/2 = 23 + \text{remainder } 0 \text{ (LSB)}$$

The remainder is the least significant bit of the binary equivalent of 46.

2. Divide the quotient from the previous division and note the remainder. The remainder is the second LSB.

$$23/2 = 11 + \text{remainder } 1$$

3. Continue this process until the *quotient* is 0. The last remainder is the most significant bit of the binary number.

$$11/2 = 5 + \text{remainder } 1$$
$$5/2 = 2 + \text{remainder } 1$$
$$2/2 = 1 + \text{remainder } 0$$
$$1/2 = 0 + \text{remainder } 1 \text{ (MSB)}$$

To write the binary equivalent of the decimal number, read the remainders from the bottom up.

$$46_{10} = 101110_2$$

EXAMPLE 1.4

Use repeated division by 2 to convert 115_{10} to a binary number.

■ **Solution**

$$115/2 = 57 + \text{remainder } 1 \text{ (LSB)}$$
$$57/2 = 28 + \text{remainder } 1$$
$$28/2 = 14 + \text{remainder } 0$$
$$14/2 = 7 + \text{remainder } 0$$
$$7/2 = 3 + \text{remainder } 1$$
$$3/2 = 1 + \text{remainder } 1$$
$$1/2 = 0 + \text{remainder } 1 \text{ (MSB)}$$

Read the remainders from bottom to top: 1110011.

$$115_{10} = 1110011_2$$

In any decimal-to-binary conversion, the number of bits in the binary number is the exponent of the smallest power of 2 that is larger than the decimal number. For example, for the numbers 92_{10} and 46_{10}

$$2^7 = 128 > 92 \qquad \text{7 bits: } 1011100$$
$$2^6 = 64 > 46 \qquad \text{6 bits: } 101110$$

Fractional Binary Numbers

■ **KEY TERMS**

Binary Point   A period (".") that marks the dividing line between positional multipliers that are positive and negative powers of 2 (e.g., first multiplier right of binary point = 2^{-1}; first multiplier left of binary point = 2^0).

> **Radix Point** The generalized form of a decimal point. In any positional number system, the radix point marks the dividing line between positional multipliers that are positive and negative powers of the system's number base.

In the decimal system, fractional numbers use the same digits as whole numbers, but the digits are written to the right of the decimal point. The multipliers for these digits are negative powers of 10—10^{-1} (1/10), 10^{-2} (1/100), 10^{-3} (1/1000), and so on.

So it is in the binary system. Digits 0 and 1 are used to write fractional binary numbers, but the digits are to the right of the **binary point**—the binary equivalent of the decimal point. (The decimal point and binary point are special cases of the **radix point,** the general name for any such point in any number system.)

Each digit is multiplied by a positional factor that is a negative power of 2. The first four multipliers on either side of the binary point are:

$$
\begin{array}{cccccccc}
 & & & & \text{binary} & & & \\
 & & & & \text{point} & & & \\
2^3 & 2^2 & 2^1 & 2^0 & . & 2^{-1} & 2^{-2} & 2^{-3} & 2^{-4} \\
= 8 & = 4 & = 2 & = 1 & & = 1/2 & = 1/4 & = 1/8 & = 1/16
\end{array}
$$

EXAMPLE 1.5

Write the binary fraction 0.101101 as a decimal fraction.

■ **Solution**

$$
\begin{aligned}
1 \times 1/2 &= 1/2 \\
0 \times 1/4 &= 0 \\
1 \times 1/8 &= 1/8 \\
1 \times 1/16 &= 1/16 \\
0 \times 1/32 &= 0 \\
1 \times 1/64 &= 1/64 \\
1/2 + 1/8 + 1/16 + 1/64 &= 32/64 + 8/64 + 4/64 + 1/64 \\
&= 45/64 \\
&= 0.703125_{10}
\end{aligned}
$$

Fractional-Decimal-to-Fractional-Binary Conversion

Simple decimal fractions such as 0.5, 0.25, and 0.375 can be converted to binary fractions by a sum-of-powers method. These decimal numbers can also be written 0.5 = 1/2, 0.25 = 1/4, and 0.375 = 3/8 = 1/4 + 1/8. These numbers can all be represented by negative powers of 2. Thus, in binary,

$$
\begin{aligned}
0.5_{10} &= 0.1_2 \\
0.25_{10} &= 0.01_2 \\
0.375_{10} &= 0.011_2
\end{aligned}
$$

The conversion process becomes more complicated if we try to convert decimal fractions that cannot be broken into powers of 2. For example, the number 1/5 = 0.2_{10} cannot be exactly represented by a sum of negative powers of 2. (Try it.) For this type of number, we must use the method of repeated multiplication by 2.

Method:

1. Multiply the decimal fraction by 2 and note the integer part. The integer part is either 0 or 1 for any number between 0 and 0.999. . . . The integer part of the product is the first digit to the left of the binary point.

$$0.2 \times 2 = 0.4 \quad \text{Integer part: 0}$$

2. Discard the integer part of the previous product. Multiply the fractional part of the previous product by 2. Repeat step 1 until the fraction repeats or terminates.

$$0.4 \times 2 = 0.8 \qquad \text{Integer part: 0}$$
$$0.8 \times 2 = 1.6 \qquad \text{Integer part: 1}$$
$$0.6 \times 2 = 1.2 \qquad \text{Integer part: 1}$$
$$0.2 \times 2 = 0.4 \qquad \text{Integer part: 0}$$

(Fraction repeats; product is same as in step 1)

Read the integer parts from top to bottom to obtain the fractional binary number. Thus, $0.2_{10} = 0.00110011 \ldots ._2 = 0.\overline{0011}_2$. The bar shows the portion of the digits that repeats.

EXAMPLE 1.6

Convert 0.95_{10} to its binary equivalent.

■ **Solution**

$$0.95 \times 2 = 1.90 \qquad \text{Integer part: 1}$$
$$0.90 \times 2 = 1.80 \qquad \text{Integer part: 1}$$
$$0.80 \times 2 = 1.60 \qquad \text{Integer part: 1}$$
$$0.60 \times 2 = 1.20 \qquad \text{Integer part: 1}$$
$$0.20 \times 2 = 0.40 \qquad \text{Integer part: 0}$$
$$0.40 \times 2 = 0.80 \qquad \text{Integer part: 0}$$
$$0.80 \times 2 = 1.60 \qquad \text{Fraction repeats last four digits}$$

$$0.95_{10} = 0.11\overline{1100}_2$$

■ **SECTION 1.3 REVIEW PROBLEMS**

1.2   How many different binary numbers can be written with 6 bits?

1.3   How many can be written with 7 bits?

1.4   Write the sequence of 7-bit numbers from 1010000 to 1010111.

1.5   Write the decimal equivalents of the numbers written for Problem 1.4.

1.4   HEXADECIMAL NUMBERS

■ **KEY TERM**

Hexadecimal Number System   Base-16 number system. Hexadecimal numbers are written with sixteen digits, 0–9, and A–F, with power-of-16 positional multipliers.

After binary numbers, hexadecimal (base 16) numbers are the most important in digital applications. Hexadecimal, or hex, numbers are primarily used as a shorthand form of binary notation. Since 16 is a power of 2 ($2^4 = 16$), each hexadecimal digit can be converted directly to four binary digits. Hex numbers can pack more digital information into fewer digits.

Hex numbers have become particularly popular with the advent of small computers, which use binary data having 8, 16, or 32 bits. Such data can be represented by 2, 4, or 8 hexadecimal digits, respectively.

TABLE 1.4 Hex Digits and Their Binary and Decimal Equivalents

Hex	Decimal	Binary
0	0	0000
1	1	0001
2	2	0010
3	3	0011
4	4	0100
5	5	0101
6	6	0110
7	7	0111
8	8	1000
9	9	1001
A	10	1010
B	11	1011
C	12	1100
D	13	1101
E	14	1110
F	15	1111

Counting in Hexadecimal

The positional multipliers in the hex system are powers of sixteen: $16^0 = 1$, $16^1 = 16$, $16^2 = 256$, $16^3 = 4096$, and so on.

We need 16 digits to write hex numbers; the decimal digits 0 through 9 are not sufficient. The usual convention is to use the capital letters A through F, each letter representing a number from 10_{10} through 15_{10}. Table 1.4 shows how hexadecimal digits relate to their decimal and binary equivalents.

> **NOTE...**
>
> Counting Rules for Hexadecimal Numbers:
>
> 1. Count in sequence from 0 to F in the least significant digit.
> 2. Add 1 to the next digit to the left and set the least significant digit to 0.
> 3. Repeat in all other columns.

For instance, the hex numbers between 19 and 22 are 19, 1A, 1B, 1C, 1D, 1E, 1F, 20, 21, 22. (The decimal equivalents of these numbers are 25_{10} through 34_{10}.)

EXAMPLE 1.7

What is the next hexadecimal number after 999? After 99F? After 9FF? After FFF?

■ **Solution**

The hexadecimal number after 999 is 99A. The number after 99F is 9A0. The number after 9FF is A00. The number after FFF is 1000.

EXAMPLE 1.8

List the hexadecimal digits from 190_{16} to 200_{16}, inclusive.

■ **Solution**

The numbers follow the counting rules: Use all the digits in one position, add 1 to the digit one position left, and start over.
For brevity, we will list only a few of the numbers in the sequence:

190, 191, 192, . . . , 199, 19A, 19B, 19C, 19D, 19E, 19F,
1A0, 1A1, 1A2, . . . , 1A9, 1AA, 1AB, 1AC, 1AD, 1AE, 1AF,
1B0, 1B1, 1B2, . . . , 1B9, 1BA, 1BB, 1BC, 1BD, 1BE, 1BF,
1C0, . . . , 1CF, 1D0, . . . , 1DF, 1E0, . . . , 1EF, 1F0, . . . , 1FF, 200

■ **SECTION 1.4A REVIEW PROBLEMS**

1.6 List the hexadecimal numbers from FA9 to FB0, inclusive.

1.7 List the hexadecimal numbers from 1F9 to 200, inclusive.

Hexadecimal-to-Decimal Conversion

To convert a number from hex to decimal, multiply each digit by its power-of-16 positional multiplier and add the products. In the following examples, hexadecimal numbers are indicated by a final "H" (e.g., 1F7H), rather than a "16" subscript.

EXAMPLE 1.9

Convert 7C6H to decimal.

■ **Solution**

$$7 \times 16^2 = 7_{10} \times 256_{10} = 1792_{10}$$
$$C \times 16^1 = 12_{10} \times 16_{10} = 192_{10}$$
$$6 \times 16^0 = 6_{10} \times 1_{10} = \underline{6_{10}}$$
$$1990_{10}$$

EXAMPLE 1.10

Convert 1FD5H to decimal.

■ **Solution**

$$1 \times 16^3 = 1_{10} \times 4096_{10} = 4096_{10}$$
$$F \times 16^2 = 15_{10} \times 256_{10} = 3840_{10}$$
$$D \times 16^1 = 13_{10} \times 16_{10} = 208_{10}$$
$$5 \times 16^0 = 5_{10} \times 1_{10} = \underline{5_{10}}$$
$$8149_{10}$$

■ **SECTION 1.4B REVIEW PROBLEM**

1.8   Convert the hexadecimal number A30FH to its decimal equivalent.

Decimal-to-Hexadecimal Conversion

Decimal numbers can be converted to hex by the sum-of-weighted-hex-digits method or by repeated division by 16. The main difficulty we encounter in either method is remembering to convert decimal numbers 10 through 15 into the equivalent hex digits, A through F.

Sum of Weighted Hexadecimal Digits

This method is useful for simple conversions (about three digits). For example, the decimal number 35 is easily converted to the hex value 23.

$$35_{10} = 32_{10} + 3_{10} = (2 \times 16) + (3 \times 1) = 23H$$

EXAMPLE 1.11

Convert 175_{10} to hexadecimal.

■ **Solution**

$$256_{10} > 175_{10} > 16_{10}$$

Since $256 = 16^2$, the hexadecimal number will have two digits.

$$(11 \times 16) > 175 > (10 \times 16)$$

16	1
A	

$175 - (A \times 16) = 175 - 160 = 15$

16	1
A	F

$175 - ((A \times 16) + (F \times 1))$
$= 175 - (160 + 15) = 0$

Repeated Division by 16

Repeated division by 16 is a systematic decimal-to-hexadecimal conversion method that is not limited by the size of the number to be converted.

It is similar to the repeated-division-by-2 method used to convert decimal numbers to binary. Divide the decimal number by 16 and note the remainder, making sure to express it as a hex digit. Repeat the process until the quotient is zero. The last remainder is the most significant digit of the hex number.

EXAMPLE 1.12

Convert 31581_{10} to hexadecimal.

■ Solution

$$31581/16 = 1973 + \text{remainder } 13 \text{ (D) (LSD)}$$
$$1973/16 = 123 + \text{remainder } 5$$
$$123/16 = 7 + \text{remainder } 11 \text{ (B)}$$
$$7/16 = 0 + \text{remainder } 7 \text{ (MSD)}$$
$$31581_{10} = 7\text{B5DH}$$

■ SECTION 1.4C REVIEW PROBLEM

1.9 Convert the decimal number 8137 to its hexadecimal equivalent.

Conversions Between Hexadecimal and Binary

Table 1.4 shows all 16 hexadecimal digits and their decimal and binary equivalents. Note that for every possible 4-bit binary number, there is a hexadecimal equivalent.

Binary-to-hex and hex-to-binary conversions simply consist of making a conversion between each hex digit and its binary equivalent.

EXAMPLE 1.13

Convert 7EF8H to its binary equivalent.

■ Solution

Convert each digit individually to its equivalent value:

$$7\text{H} = 0111_2$$
$$\text{EH} = 1110_2$$
$$\text{FH} = 1111_2$$
$$8\text{H} = 1000_2$$

The binary number is all of these binary numbers in sequence:

$$7\text{EF8H} = 111|1110|1111|1000_2$$

The leading zero (the MSB of 0111) has been left out.

■ SECTION 1.4D REVIEW PROBLEMS

1.10 Convert the hexadecimal number 934BH to binary.

1.11 Convert the binary number 11001000001101001001 to hexadecimal.

1.5 DIGITAL WAVEFORMS

■ **KEY TERM**

Digital Waveform   A series of logic 1s and 0s plotted as a function of time.

The inputs and outputs of digital circuits often are not fixed logic levels but **digital waveforms,** where the input and output logic levels vary with time. There are three possible types of digital waveforms. *Periodic* waveforms repeat the same pattern of logic levels over a specified period of time. *Aperiodic* waveforms do not repeat. *Pulse* waveforms follow a HIGH-LOW-HIGH or LOW-HIGH-LOW pattern and may be periodic or aperiodic.

Periodic Waveforms

■ **KEY TERMS**

Periodic Waveform   A time-varying sequence of logic HIGHs and LOWs that repeats over a specified period of time.

Time LOW (t_l) Time during one period that a waveform is in the LOW state. Unit: seconds *(s)*.

Time HIGH (t_h) Time during one period that a waveform is in the HIGH state. Unit: seconds *(s)*.

Period *(T)*   Time required for a periodic waveform to repeat. Unit: seconds *(s)*.

Frequency *(f)*   Number of times per second that a periodic waveform repeats. $f = 1/T$ Unit: Hertz (Hz).

Duty cycle *(DC)*   Fraction of the total period that a digital waveform is in the HIGH state. $DC = t_h/T$ (often expressed as a percentage: $\%DC = t_h/T \times 100\%$).

Periodic waveforms repeat the same pattern of HIGHs and LOWs over a specified period of time. The waveform may or may not be symmetrical; that is, it may or may not be HIGH and LOW for equal amounts of time.

 EXAMPLE 1.14

Calculate the **time LOW, time HIGH, period, frequency,** and percent **duty cycle** for each of the periodic waveforms in Figure 1.5.

FIGURE 1.5   Example 1.14: Periodic Digital Waveforms

How are the waveforms similar? How do they differ?

■ **Solution**

a. Time LOW: $t_l = 3$ ms
Time HIGH: $t_h = 1$ ms
Period: $T = t_l + t_h = 3$ ms $+ 1$ ms $= 4$ ms
Frequency: $f = 1/T = 1/(4$ ms$) = 0.25$ kHz $= 250$ Hz
Duty cycle: $\%DC = (t_h/T) \times 100\% = (1$ ms$/4$ ms$) \times 100\%$
$\qquad\qquad\qquad = 25\%$
(1 ms = 1/1000 second; 1 kHz = 1000 Hz.)

b. Time LOW: $t_l = 2$ ms
Time HIGH: $t_h = 2$ ms
Period: $T = t_l + t_h = 2$ ms $+ 2$ ms $= 4$ ms
Frequency: $f = 1/T = 1/(4$ ms$) = 0.25$ kHz $= 250$ Hz
Duty cycle: $\%DC = (t_h/T) \times 100\% = (2$ ms$/4$ ms$) \times 100\%$
$\qquad\qquad\qquad = 50\%$

c. Time LOW: $t_l = 1$ ms
Time HIGH: $t_h = 3$ ms
Period: $T = t_l + t_h = 1$ ms $+ 3$ ms $= 4$ ms
Frequency: $f = 1/T = 1/(4$ ms$) = 0.25$ kHz $= 250$ Hz
Duty cycle: $\%DC = (t_h/T) \times 100\% = (3$ ms$/4$ ms$) \times 100\%$
$\qquad\qquad\qquad = 75\%$

The waveforms all have the same period but different duty cycles. A square waveform, shown in Figure 1.5b, has a duty cycle of 50%.

Aperiodic Waveforms

■ **KEY TERM**

Aperiodic Waveform A time-varying sequence of logic HIGHs and LOWs that does not repeat.

An **aperiodic waveform** does not repeat a pattern of 0s and 1s. Thus, the parameters of time HIGH, time LOW, frequency, period, and duty cycle have no meaning for an aperiodic waveform. Most waveforms of this type are one-of-a-kind specimens. (It is also worth noting that most digital waveforms are aperiodic.)

Figure 1.6 shows some examples of aperiodic waveforms.

FIGURE 1.6 Aperiodic Digital Waveforms

EXAMPLE 1.15

A digital circuit generates the following strings of 0s and 1s:

a. 00111111011010110100000110000
b. 00110011001100110011001100110011
c. 00000000111111110000000001111
d. 101110111011101110111011101110 11

The time between two bits is always the same. Sketch the resulting digital waveform for each string of bits. Which waveforms are periodic and which are aperiodic?

■ **Solution**

Figure 1.7 shows the waveforms corresponding to the strings of bits just mentioned. The waveforms are easier to draw if you break up the bit strings into smaller groups of, say, 4 bits each. For instance:

a. 0011 1111 0110 1011 0100 0011 0000

All of the waveforms except waveform (a) are periodic.

FIGURE 1.7   Example 1.15: Waveforms

Pulse Waveforms

■ **KEY TERMS**

Pulse   A momentary variation of voltage from one logic level to the opposite level and back again.

Rising Edge   The part of a pulse where the logic level is in transition from a LOW to a HIGH.

Falling Edge   The part of a pulse where the logic level is in transition from a HIGH to a LOW.

Amplitude   The instantaneous voltage of a waveform. Often used to mean maximum amplitude, or peak voltage, of a pulse.

Pulse Width (t_w)   Elapsed time from the 50% point of the leading edge of a pulse to the 50% point of the trailing edge.

Rise Time (t_r)   Elapsed time from the 10% point to the 90% point of the rising edge of a pulse.

Fall Time (t_f)   Elapsed time from the 90% point to the 10% point of the falling edge of a pulse.

Leading Edge   The edge of a pulse that occurs earliest in time.

Trailing Edge   The edge of a pulse that occurs latest in time.

Edge   The part of the pulse that represents the transition from one logic level to the other.

Figure 1.8 shows the forms of both an ideal and a nonideal **pulse**. The **rising and falling edges** of an ideal pulse are vertical. That is, the transitions between logic HIGH and LOW levels are instantaneous. There is no such thing as an ideal pulse in a real digital circuit. That is, an **edge** in a real pulse is never absolutely vertical.

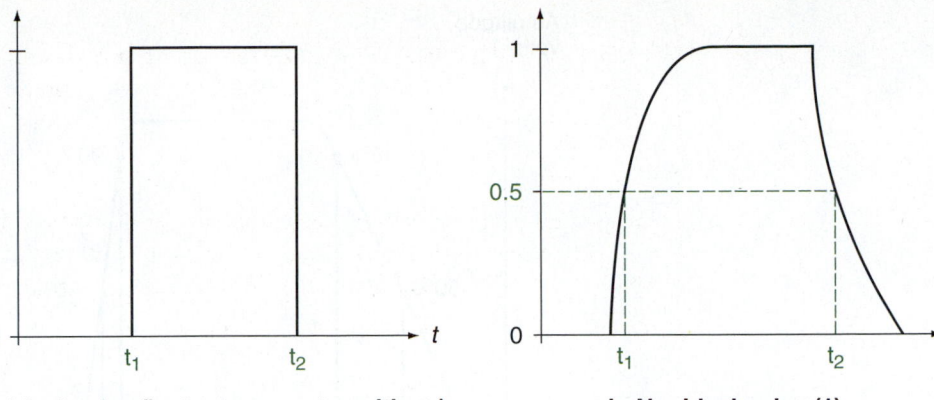

a. Ideal pulse (instantaneous transitions)   **b. Nonideal pulse (1)**

c. Nonideal pulse (2)

FIGURE 1.8   Ideal and Nonideal Pulses

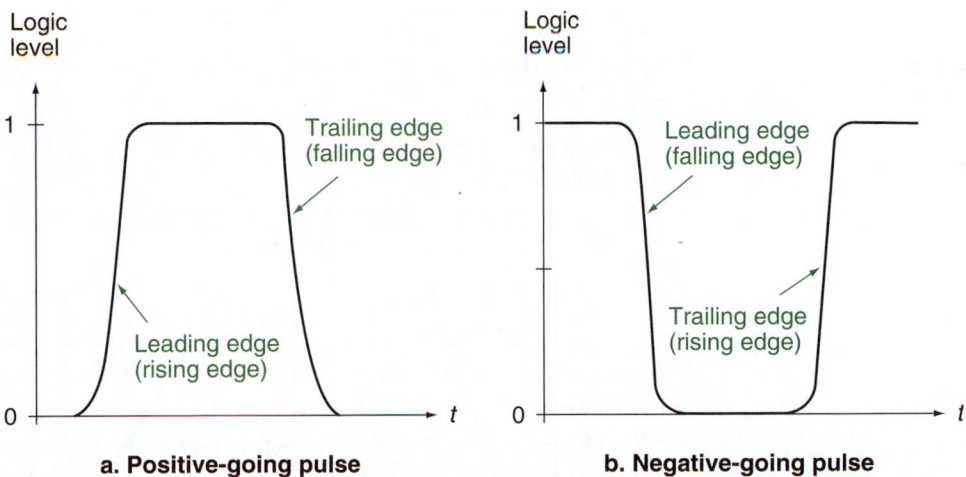

a. Positive-going pulse   **b. Negative-going pulse**

FIGURE 1.9   Pulse Edges

Circuit capacitance, inductance, and other factors make the pulse more like the non-ideal pulses in Figure 1.8b and Figure 1.8c.

Pulses can be either positive-going or negative-going, as shown in Figure 1.9. In a positive-going pulse, the measured logic level is normally LOW, goes HIGH for the duration of the pulse, and returns to the LOW state. A negative-going pulse acts in the opposite direction.

FIGURE 1.10 Pulse Width, Rise Time, Fall Time

Nonideal pulses are measured in terms of several timing parameters. Figure 1.10 shows the 10%, 50%, and 90% points on the rising and falling edges of a nonideal pulse. (100% is the maximum **amplitude** of the pulse.)

The 50% points are used to measure **pulse width** because the edges of the pulse are not vertical. Without an agreed reference point, the pulse width is indeterminate. The 10% and 90% points are used as references for the **rise and fall times** since the edges of a nonideal pulse are nonlinear. Most of the nonlinearity is below the 10% or above the 90% point.

 EXAMPLE 1.16

Calculate the pulse width, rise time, and fall time of the pulse shown in Figure 1.11.

FIGURE 1.11 Example 1.16: Pulse

■ **Solution**

From the graph in Figure 1.11, read the times corresponding to the 10%, 50%, and 90% values of the pulse on both the **leading and trailing edges.**

Leading edge: 10%:   2 μs   Trailing edge: 90%:   20 μs
                 50%:   5 μs   50%:   25 μs
                 90%:   8 μs   10%:   30 μs

Pulse width: 50% of leading edge to 50% of trailing edge.
$$t_w = 25 \text{ μs} - 5 \text{ μs} = 20 \text{ μs}$$
Rise time: 10% of rising edge to 90% of rising edge.
$$t_r = 8 \text{ μs} - 2 \text{ μs} = 6 \text{ μs}$$
Fall time: 90% of falling edge to 10% of falling edge.
$$t_f = 30 \text{ μs} - 20 \text{ μs} = 10 \text{ μs}$$

■ SECTION 1.5 REVIEW PROBLEMS

A digital circuit produces a waveform that can be described by the following periodic bit pattern: 0011001100110011.

1.12 What is the duty cycle of the waveform?

1.13 Write the bit pattern of a waveform with the same duty cycle and twice the frequency of the original.

1.14 Write the bit pattern of a waveform having the same frequency as the original and a duty cycle of 75%.

SUMMARY

1. The two basic areas of electronics are analog and digital electronics. Analog electronics deals with continuously variable quantities; digital electronics represents the world in discrete steps.

2. Digital logic uses defined voltage levels, called logic levels, to represent binary numbers within an electronic system.

3. The higher voltage in a digital system represents the binary digit 1 and is called a logic HIGH or logic 1. The lower voltage in a system represents the binary digit 0 and is called a logic LOW or logic 0.

4. The logic levels of multiple locations in a digital circuit can be combined to represent a multibit binary number.

5. Binary is a positional number system (base 2) with two digits, 0 and 1, and positional multipliers that are powers of 2.

6. The bit with the largest positional weight in a binary number is called the most significant bit (MSB); the bit with the smallest positional weight is called the least significant bit (LSB). The MSB is also the leftmost bit in the number; the LSB is the rightmost bit.

7. A decimal number can be converted to binary by sum of powers of 2 (add place values to get a total) or repeated division by 2 (divide by 2 until quotient is 0; remainders are the binary value).

8. The positional multipliers in a fractional binary number are negative powers of 2.

9. The hexadecimal number system is based on 16. It uses 16 digits, from 0–9 and A–F, with power-of-16 multipliers.

10. Each hexadecimal digit uniquely corresponds to a 4-bit binary value. Hex digits can thus be used as shorthand for binary.

11. A digital waveform is a sequence of bits over time. A waveform can be periodic (repetitive), aperiodic (nonrepetitive), or pulsed (a single variation and return between logic levels).

12. Periodic waveforms are measured by period (T: time for one cycle), time HIGH (t_h), time LOW (t_l), frequency (f: number of cycles per second), and duty cycle (DC or $\%DC$: fraction of cycle in HIGH state).

13. Pulse waveforms are measured by pulse width (t_w: time from 50% of leading edge of 50% of trailing edge), rise time (t_r: time from 10% to 90% of rising edge) and fall time (t_f: time from 90% to 10% of falling edge).

GLOSSARY

Amplitude   The instantaneous voltage of a waveform. Often used to mean maximum amplitude, or peak voltage, of a pulse.

Analog   A way of representing some physical quantity, such as temperature or velocity, by a proportional continuous voltage or current. An analog voltage or current can have any value within a defined range.

Aperiodic Waveform   A time-varying sequence of logic HIGHs and LOWs that does not repeat.

Binary Number System   A number system used extensively in digital systems, based on the number 2. It uses two digits to write any number.

Binary Point   A period (".") that marks the dividing line between positional multipliers that are positive and negative powers of 2 (e.g., first multiplier right of binary point = 2^{-1}; first multiplier left of binary point = 2^0).

Bit   *Bi*nary dig*it*. A 0 or a 1.

Continuous   Smoothly connected. An unbroken series of consecutive values with no instantaneous changes.

Digital   A way of representing a physical quantity by a series of binary numbers. A digital representation can have only specific discrete values.

Digital Waveform   A series of logic 1s and 0s plotted as a function of time.

Discrete   Separated into distinct segments or pieces. A series of discontinous values.

Duty Cycle (*DC*)   Fraction of the total period that a digital waveform is in the HIGH state. $DC = t_h/T$ (often expressed as a percentage: $\%DC = t_h/T \times 100\%$).

Edge   The part of the pulse that represents the transition from one logic level to the other.

Fall Time (*t_f*)   Elapsed time from the 90% point to the 10% point of the falling edge of a pulse.

Falling Edge   The part of a pulse where the logic level is in transition from a HIGH to a LOW.

Frequency (*f*)   Number of times per second that a periodic waveform repeats. $f = 1/T$ Unit: Hertz (Hz).

Hexadecimal Number System   Base-16 number system. Hexadecimal numbers are written with sixteen digits, 0–9 and A–F, with power-of-16 positional multipliers.

Leading Edge   The edge of a pulse that occurs earliest in time.

Least Significant Bit (LSB)   The rightmost bit of a binary number. This bit has the number's smallest positional multiplier.

Logic HIGH   The higher of two voltages in a digital system with two logic levels.

Logic Level   A voltage level that represents a defined digital state in an electronic circuit.

Logic LOW   The lower of two voltages in a digital system with two logic levels.

Most Significant Bit (MSB)   The leftmost bit in a binary number. This bit has the number's largest positional multiplier.

Negative Logic   A system in which logic LOW represents binary digit 1 and logic HIGH represents binary digit 0.

Period (*T*)   Time required for a periodic waveform to repeat. Unit: seconds (s).

Periodic Waveform   A time-varying sequence of logic HIGHs and LOWs that repeats over a specified period of time.

Positional Notation   A system of writing numbers in which the value of a digit depends not only on the digit, but also on its placement within a number.

Positive Logic   A system in which logic LOW represents binary digit 0 and logic HIGH represents binary digit 1.

Pulse   A momentary variation of voltage from one logic level to the opposite level and back again.

Pulse Width (t_w)   Elapsed time from the 50% point of the leading edge of a pulse to the 50% point of the trailing edge.

Radix Point   The generalized form of a decimal point. In any positional number system, the radix point marks the dividing line between positional multipliers that are positive and negative powers of the system's number base.

Rise Time (t_r)   Elapsed time from the 10% point to the 90% point of the rising edge of a pulse.

Rising Edge   The part of a pulse where the logic level is in transition from a LOW to a HIGH.

Time HIGH (t_h)   Time during one period that a waveform is in the HIGH state. Unit: seconds (s).

Time LOW (t_l)   Time during one period that a waveform is in the LOW state. Unit: seconds (s).

Trailing Edge   The edge of a pulse that occurs latest in time.

PROBLEMS

Problem numbers set in color indicate more difficult problems.

1.1   Digital Versus Analog Electronics

1.1   Which of the following quantities is analog in nature and which digital? Explain your answers.

    a. Water temperature at the beach

    b. Weight of a bucket of sand

    c. Grains of sand in a bucket

    d. Waves hitting the beach in one hour

    e. Height of a wave

    f. People in a square mile

1.2   Digital Logic Levels

1.2   A digital logic system is defined by the voltages 3.3 volts and 0 volts. For a positive logic system, state which voltage corresponds to a logic 0 and which to a logic 1.

1.3   The Binary Number System

1.3   Calculate the decimal values of each of the following binary numbers:

a. 100	**f.** 11101
b. 1000	**g.** 111011
c. 11001	**h.** 1011101
d. 110	**i.** 100001
e. 10101	**j.** 10111001

1.4   Translate each of the following combinations of HIGH (H) and LOW (L) logic levels to binary numbers using positive logic:

a. H H L H	**d.** L L L H
b. L H L H	**e.** H L L L
c. H L H L	

1.5   List the sequence of binary numbers from 101 to 1000.

1.6   List the sequence of binary numbers from 10000 to 11111.

1.7   What is the decimal value of the most significant bit for the numbers in Problem 1.6?

1.8   Convert the following decimal numbers to binary. Use the sum-of-powers-of-2 method for parts a, c, e, and g. Use the repeated-division-by-2 method for parts b, d, f, and h.

    a. 75_{10} **b.** 83_{10}

     c. 237_{10} f. 64_{10}

     d. 198_{10} g. 4087_{10}

     e. 63_{10} h. 8193_{10}

1.9 Convert the following fractional binary numbers to their decimal equivalents.

     a. 0.101   c. 0.1101

     b. 0.011

1.10 Convert the following fractional binary numbers to their decimal equivalents.

     a. 0.01   c. 0.010101

     b. 0.0101   d. 0.01010101

1.11 The numbers in Problem 1.10 are converging to a closer and closer binary approximation of a simple fraction that can be expressed by decimal integers a/b. What is the fraction?

1.12 What is the simple decimal fraction (a/b) represented by the repeating binary number 0.101010 . . .?

1.13 Convert the following decimal numbers to their binary equivalents. If a number has an integer part larger than 0, calculate the integer and fractional parts separately.

     a. 0.75_{10} e. 1.75_{10}

     b. 0.625_{10} f. 3.95_{10}

     c. 0.1875_{10} g. 67.84_{10}

     d. 0.65_{10}

1.4   Hexadecimal Numbers

1.14 Write all the hexadecimal numbers in sequence from 308H to 321H inclusive.

1.15 Write all the hexadecimal numbers in sequence from 9F7H to A03H inclusive.

1.16 Convert the following hexadecimal numbers to their decimal equivalents.

     a. 1A0H   e. F3C8H

     b. 10AH   f. D3B4H

     c. FFFH   g. C000H

     d. 1000H   h. 30BAFH

1.17 Convert the following decimal numbers to their hexadecimal equivalents.

     a. 709_{10} e. 10128_{10}

     b. 1889_{10} f. 32000_{10}

     c. 4095_{10} g. 32768_{10}

     d. 4096_{10}

1.18 Convert the following hexadecimal numbers to their binary equivalents.

     a. F3C8H   e. 30ACH

     b. D3B4H   f. 3E7B6H

     c. 8037H   g. 743DCFH

     d. FABDH

1.19 Convert the following binary numbers to their hexadecimal equivalents.

     a. 101111010000110_2 e. 10101011110000101_2

     b. 10110110101010_2 f. 11001100010110111_2

     c. 110001011011_2 g. 101000000000000000_2

     d. 110101111000100_2

1.5   Digital Waveforms

1.20 Calculate the time LOW, time HIGH, period, frequency, and percent duty cycle for the waveforms shown in Figure 1.12. How are the waveforms similar? How do they differ?

FIGURE 1.12 Problem 1.20: Periodic Waveforms

1.21 Which of the waveforms in Figure 1.13 are periodic and which are aperiodic? Explain your answers.

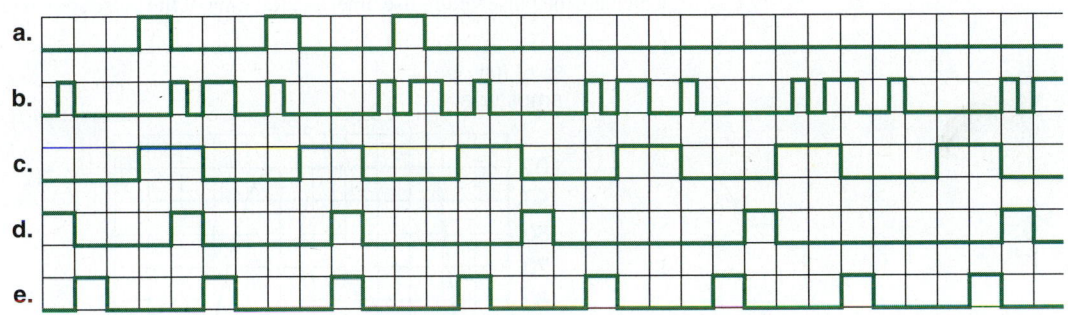

FIGURE 1.13 Problem 1.21: Aperiodic and Periodic Waveforms

1.22 Sketch the pulse waveforms represented by the following strings of 0s and 1s. State which waveforms are periodic and which are aperiodic.

  a. 110011110011101100000000110110101

  b. 111000111000111000111000111000111

  c. 111111110000000001111111111111111

  d. 011001100110011001100110011001100110

  e. 011101101001101001011010011101110

1.23 Classify each of the waveforms in Figure 1.14 as aperiodic or periodic. For the periodic waveforms, calculate time HIGH, time LOW, period, frequency, and duty cycle.

FIGURE 1.14 Problem 1.23: Aperiodic and Periodic Waveforms

1.24 For each of the periodic waveforms shown in Figure 1.15, calculate the period, frequency, time HIGH, time LOW and percent duty cycle. (The time scale is shown in nanoseconds; 1 ns = 10^{-9} seconds.)

FIGURE 1.15   Problem 1.24: Periodic Waveforms

1.25   Calculate the pulse width, rise time, and fall time of the pulse shown in Figure 1.16.

FIGURE 1.16   Problem 1.25: Pulse

1.26   Repeat Problem 1.25 for the pulse shown in Figure 1.17.

FIGURE 1.17   Problem 1.26: Pulse

1.27 Calculate the pulse width, rise time, and fall time of the pulse shown in Figure 1.18.

FIGURE 1.18 Problem 1.27: Periodic Waveform

ANSWERS TO SECTION REVIEW PROBLEMS

1.1

1.1 An analog audio system makes a direct copy of the recorded sound waves. A digital system stores the sound as a series of binary numbers.

1.3

1.2 64
1.3 128
1.4 1010000, 1010001, 1010010, 1010011, 1010100, 1010101, 1010110, 1010111
1.5 80, 81, 82, 83, 84, 85, 86, 87

1.4a

1.6 FA9, FAA, FAB, FAC, FAD, FAE, FAF, FB0
1.7 1F9, 1FA, 1FB, 1FC, 1FD, 1FE, 1FF, 200

1.4b

1.8 41743_{10}

1.4c

1.9 1FC9H

1.4d

1.10 1001001101001011
1.11 C8349H

1.5

1.12 50%
1.13 0101010101010101
1.14 0111011101110111

KEY TERMS

Active HIGH
Active level
Active LOW
AND gate
Ball grid array (BGA)
Boolean algebra
Boolean expression
Boolean variable
Breadboard
Bubble
Buffer
Bus
Chip
Coincidence gate
Complement form
Complementary metal-oxide-
    semiconductor (CMOS)
Data book
Datasheet
DeMorgan equivalent forms
DeMorgan's theorems
Digital signal (or pulse waveform)
Distinctive-shape symbols
Dual in-line package (DIP)
Enable
Exclusive NOR (XNOR) gate
Exclusive OR (XOR) gate
High-impedance state
IEEE/ANSI Standard 91-1984
In phase
Inhibit (or disable)
Integrated circuit (IC)
Inverter
Large scale integration (LSI)
LED (light-emitting diode)
Logic function
Logic gate

Logical product
Logical sum
Medium scale integration (MSI)
NAND gate
NOR gate
OR gate
Out of phase
Plastic leaded chip carrier (PLCC)
Portable document format (PDF)
Printed circuit board (PCB)
Pull-up resistor
Quad flat pack (QFP)
Qualifying symbol
Rectangular-outline symbols
Small outline IC (SOIC)
Small scale integration (SSI)
Surface-mount technology (SMT)
Thin shrink small outline package (TSSOP)
Through-hole
Transistor-transistor logic (TTL)
Tristate buffer
True form
Truth table
V_{CC}
Very large scale integration (VLSI)
Wire-wrap

OUTLINE

2.1	Basic Logic Functions
2.2	Derived Logic Functions
2.3	DeMorgan's Theorems and Gate Equivalence
2.4	Logic Switches and LED Indicators
2.5	Enable and Inhibit Properties of Logic Gates
2.6	Integrated Circuit Logic Gates

CHAPTER 2

Logic Functions and Gates

CHAPTER OBJECTIVES

Upon successful completion of this chapter, you will be able to:

- Describe the basic logic functions: AND, OR, and NOT
- Draw simple switch circuits to represent AND, OR, and Exclusive OR functions.
- Draw simple logic switch circuits for single-pole single-throw (SPST) and normally open and normally closed pushbutton switches.
- Describe the use of light-emitting diodes (LEDs) as indicators of logic HIGH and LOW states.
- Describe those logic functions derived from the basic ones: NAND, NOR, Exclusive OR, and Exclusive NOR.
- Explain the concept of active levels and identify active LOW and HIGH terminals of logic gates.
- Choose appropriate logic functions to solve simple design problems.
- Draw the truth table of any logic gate.
- Draw any logic gate, given its truth table.
- Draw the DeMorgan equivalent form of any logic gate.
- Determine when a logic gate will pass a digital waveform and when it will block the signal.
- Describe the behavior of tristate buffers.
- Describe several types of integrated circuit packaging for digital logic gates.

All digital logic functions can be synthesized by various combinations of the three basic logic functions: AND, OR, and NOT. These so-called Boolean functions are the basis for all further study of combinational logic circuitry. (Combinational logic circuits are digital circuits whose outputs are functions of their inputs, regardless of the order the inputs are applied.) Standard circuits, called logic gates, have been developed for these and for more complex digital logic functions.

Logic gates can be represented in various forms. A standard set of distinctive-shape symbols has evolved as a universally understandable means of representing the various functions in a circuit. A useful pair of mathematical theorems, called DeMorgan's theorems, enables us to draw these gate symbols in different ways to represent different aspects of the same function. A newer way of representing standard logic gates is outlined in IEEE/ANSI Standard 91-1984, a standard copublished by the Institute of Electrical and Electronic Engineers and the American National Standards Institute. It uses a set of symbols called rectangular-outline symbols.

Simple switches can be configured to apply digital logic levels to a circuit. A Single-pole single-throw (SPST) switch and a resistor can be connected to the power supply and ground of a logic circuit to produce logic HIGHs and LOWs in opposite switch positions. Normally open (NO) and normally closed (NC) pushbuttons can also be used for this purpose.

A light-emitting diode (LED) and a series resistor can be used to indicate the logic level at a particular point in a logic circuit. Depending on the configuration of the LED, it can be used to indicate a logic HIGH or a logic LOW when illuminated.

Logic gates can be used as electronic switches to block or allow passage of digital waveforms. Each logic gate has a different set of properties for enabling (passing) or inhibiting (blocking) digital waveforms.

Data flow can also be controlled by tristate buffers. These devices have three output states: logic HIGH, logic LOW, and high-impedance. When enabled by a control input, the tristate output is either HIGH or LOW. When disabled, the output is in the high-impedance state, which is like an open circuit. In this latter state, the gate output is electrically isolated from the rest of the circuit and does not act like a HIGH or a LOW.

Logic gates are available in a variety of packages for use in electronic circuits. For many years, the standard packaging option was the dual in-line package (DIP), with two rows of pins that would be inserted into circuit board holes and soldered to allow connection to the gate inputs and outputs. More recently, other packaging options have become available which allow the devices to be mounted directly on the surface of a circuit board. These surface-mount devices are typically smaller in profile than the older DIP varieties and thus can be more densely packed onto a circuit board. ■

2.1   BASIC LOGIC FUNCTIONS

■ KEY TERMS

Boolean Algebra   A system of algebra that operates on Boolean variables. The binary (two-state) nature of Boolean algebra makes it useful for analysis, simplification, and design of combinational logic circuits.

Boolean Expression   An algebraic expression made up of Boolean variables and operators, such as AND, OR, or NOT. Also referred to as a **Boolean function** or a **logic function.**

Boolean Variable   A variable having only two possible values, such as HIGH/LOW, 1/0, On/Off, or True/False.

Logic Gate   An electronic circuit that performs a Boolean algebraic function.

At its simplest level, a digital circuit works by accepting logic 1s and 0s at one or more inputs and producing 1s or 0s at one or more outputs. A branch of mathematics known as **Boolean algebra** (named after nineteenth-century mathematician George Boole) describes the relation between inputs and outputs of a digital circuit. We call these input and output values **Boolean variables** and the functions **Boolean** expressions, logic functions, or **Boolean functions.** The distinguishing characteristic of these functions is that they are made up of variables and constants that can have only two possible values: 0 or 1.

All possible operations in Boolean algebra can be created from three basic logic functions: AND, OR, and NOT.[1] Electronic circuits that perform these logic functions are called **logic gates.** When we are analyzing or designing a digital circuit, we usually don't concern ourselves with the actual circuitry of the logic gates, but treat them as black boxes that perform specified logic functions. We can think of each variable in a logic function as a circuit input and the whole function as a circuit output.

In addition to gates for the three basic functions, there are also gates for compound functions that are derived from the basic ones. NAND gates combine the NOT and AND functions in a single circuit. Similarly, NOR gates combine the NOT and OR functions. Gates for more complex functions, such as Exclusive OR and Exclusive NOR, also exist. We will examine all of these devices later in the chapter.

NOT, AND, and OR Functions

■ KEY TERMS

Truth Table   A list of all possible input values to a digital circuit, listed in ascending binary order, and the output response for each input combination.

Inverter   Also called a NOT gate or an inverting buffer. A logic gate that changes its input logic level to the opposite state.

Distinctive-Shape Symbols   Graphic symbols for logic circuits that show the function of each type of gate by a special shape.

Bubble   A small circle indicating logical inversion on a circuit symbol.

Rectangular-Outline Symbols   Rectangular logic gate symbols that conform to IEEE/ANSI Standard 91-1984.

IEEE/ANSI Standard 91-1984   A standard format for drawing logic circuit symbols as rectangles with logic functions shown by a standard notation inside the rectangle for each device.

Qualifying Symbol   A symbol in IEEE/ANSI logic circuit notation, placed in the top center of a rectangular symbol, that shows the function of a logic gate. Some of the qualifying symbols include: 1 = "buffer"; & = "AND"; ≥1 = "OR"

Buffer   An amplifier that acts as a logic circuit. Its output can be inverting or noninverting.

NOT Function

TABLE 2.1   NOT Function Truth Table

A	Y
0	1
1	0

The NOT function, the simplest logic function, has one input and one output. The input can be either HIGH or LOW (1 or 0), and the output is always the opposite logic level. We can show these values in a **truth table,** a list of all possible input values and the output resulting from each one. Table 2.1 shows a truth table for a NOT function, where A is the input variable and Y is the output.

The NOT function is represented algebraically by the Boolean expression:

$$Y = \overline{A}$$

This is pronounced "Y equals NOT A" or "Y equals A bar." We can also say "Y is the complement of A."

---

[1]Words in uppercase letters represent either logic functions (AND, OR, NOT) or logic levels (HIGH, LOW). The same words in lowercase letters represent their conventional nontechnical meanings.

$A \longrightarrow\!\!\!\!\triangleright\!\!\circ \longrightarrow Y = \overline{A}$

$A \longrightarrow\!\!\circ\!\!\triangleright \longrightarrow Y = \overline{A}$

a. Distinctive-shape

$A \longrightarrow \boxed{1}\!\!\triangleright \longrightarrow Y = \overline{A}$

$A \triangleright\!\!\boxed{1} \longrightarrow Y = \overline{A}$

b. Rectangular-outline
(IEEE Std. 91-1984)

FIGURE 2.1   Inverter Symbols

TABLE 2.2a   Partial Truth Table for a 2-Input AND Gate

A	B	Y
0	0	
0	1	
1	0	
1	1	1

TABLE 2.2b   Complete Truth Table for a 2-Input AND Gate

A	B	Y
0	0	0
0	1	0
1	0	0
1	1	1

$A \atop B$ $\longrightarrow\!\!\!\!\!\!\!\!\Large)\!\!\!\longrightarrow Y = AB$

a. Distinctive-shape

$A \atop B$ $\longrightarrow \boxed{\&} \longrightarrow Y = AB$

b. Rectangular-outline

FIGURE 2.2   2-Input AND Gate Symbols

The circuit that produces the NOT function is called the NOT gate or, more usually, the **inverter.** Several possible symbols for the inverter, all performing the same logic function, are shown in Figure 2.1.

The symbols shown in Figure 2.1a are the standard **distinctive-shape symbols** for the inverter. The triangle represents an amplifier circuit, and the **bubble** (the small circle on the input or output) represents inversion. There are two symbols because sometimes it is convenient to show the inversion at the input and sometimes it is convenient to show it at the output.

Figure 2.1b shows the **rectangular-outline** inverter symbol specified by **IEEE/ANSI Standard 91-1984.** This standard is most useful for specifying the symbols for more complex digital devices. We will show the basic gates in both distinctive-shape and rectangular-outline symbols, although most examples will use the distinctive-shape symbols.

The "1" in the top center of the IEEE rectangle symbol is a **qualifying symbol,** indicating the logic gate function. In this case, it shows that the circuit is a **buffer,** an amplifying circuit used as a digital logic element. The arrows at the input and output of the two IEEE symbols show inversion, like the bubbles in the distinctive-shape symbols.

AND Function

> **■ KEY TERMS**
>
> **Logical Product**   AND function.
>
> **AND Gate**   A logic circuit whose output is HIGH when all inputs (e.g., A AND B AND C) are HIGH.

The AND function combines two or more input variables so that the output is HIGH only if all inputs (e.g. A and B) are HIGH. A sentence that describes the behavior of the AND gate is, "**All inputs HIGH make output HIGH.**" The partially filled truth table in Table 2.2a shows the part of the table for which this condition is true ($A = 1$, $B = 1$, $Y = 1$.) Since the gate output can only be 1 or 0, all remaining conditions must have a 0 output, as shown in the complete truth table of Table 2.2b.

We can replace the boldface words in the descriptive sentence to describe almost any type of logic gate. We will repeatedly use this systematic "fill-in-the-blanks" method to give us a reliable analytical tool for determining the behavior of logic gates.

Algebraically, the AND function is written:

$$Y = A \cdot B$$

Pronounce this expression "Y equals A AND B." The AND function is similar to multiplication in linear algebra and thus is sometimes called the **logical product.** The dot between variables may or may not be written, so it is equally correct to write $Y = AB$. The logic circuit symbol for an **AND gate** is shown in Figure 2.2 in both distinctive-shape and IEEE/ANSI rectangular-outline form. The qualifying symbol in IEEE/ANSI notation is the ampersand (&).

We can also represent the AND function as a set of switches in series, as shown in Figure 2.3. The circuit consists of a voltage source, a lamp, and two series switches.

FIGURE 2.3   AND Function Represented by Switches

TABLE 2.3 3-Input AND Function Truth Table

A	B	C	Y
0	0	0	0
0	0	1	0
0	1	0	0
0	1	1	0
1	0	0	0
1	0	1	0
1	1	0	0
1	1	1	1

The lamp turns on when switches *A* AND *B* are both closed. For any other condition of the switches, the lamp is off.

Table 2.3 shows the truth table for a 3-input AND function. Each of the three inputs can have two different values, which means the inputs can be combined in $2^3 = 8$ different ways. In general, *n* binary (i.e., two-valued) variables can be combined in 2^n ways. The condition "**all** inputs **HIGH** make output **HIGH**" is satisfied only by the last line in the truth table, where $Y = 1$. In all other lines, $Y = 0$.

Figure 2.4 shows the logic symbols for the device. The output is HIGH only when all inputs are HIGH.

A 3-input AND gate can be created by using two 2-input AND gates, as shown in Figure 2.5. The output of the first gate ($A \cdot B$) is combined with the third variable (*C*) in the second gate to give the output expression $(A \cdot B) \cdot C = A \cdot B \cdot C$. The circuit in Figure 2.5 is logically equivalent to the gates shown in Figure 2.4.

a. Distinctive-shape

b. Rectangular-outline

FIGURE 2.4 3-Input AND Gate Symbols

FIGURE 2.5 3-Input AND Function from 2-Input AND Gates

OR Function

> **■ KEY TERMS**
>
> **Logical Sum**   OR function.
>
> **OR Gate**   A logic circuit whose output is HIGH when at least one input (e.g., *A* OR *B* OR *C*) is HIGH.

TABLE 2.4a Partial Truth Table for a 2-Input OR Gate

A	*B*	*Y*
0	0	
0	0	1
1	0	1
1	1	1

TABLE 2.4b Complete Truth Table for a 2-Input OR Gate

A	*B*	*Y*
0	0	0
0	1	1
1	0	1
1	1	1

The 2-input OR function has an output *Y* that is HIGH when either or both of inputs *A* OR *B* are HIGH. Thus we can say, "**At least one** input **HIGH** makes output **HIGH**." This condition is shown in the partial truth table of Table 2.4a. The condition is satisfied for all but one line of the table. Since 0 and 1 are the only possible outputs, the remaining line must have an output value of 0, as shown in the complete truth table of Table 2.4b.

The algebraic expression for the OR function is:

$$Y = A + B$$

which is pronounced "*Y* equals *A* OR *B*." This is similar to the arithmetic addition function, but it is not the same. The last line of the truth table tells us that $1 + 1 = 1$ (pronounced "1 OR 1 equals 1"), which is not what we would expect in standard arithmetic. The similarity to the addition function leads to the name **logical sum.** (This is different from the "arithmetic sum," where, of course, $1 + 1$ *does not* equal 1.)

Figure 2.6 shows the logic circuit symbols for an **OR gate.** The qualifying symbol for the OR function in IEEE/ANSI notation is "≥1," which tells us that *one or more* inputs must be HIGH to make the output HIGH.

The OR function can be represented by a set of switches connected in parallel, as in Figure 2.7. The lamp is on when either switch *A* OR switch *B* is closed. (Note

a. Distinctive-shape

b. Rectangular-outline

FIGURE 2.6 2-Input OR Gate Symbols

TABLE 2.5 3-Input OR Function Truth Table

A	B	C	Y
0	0	0	0
0	0	1	1
0	1	0	1
0	1	1	1
1	0	0	1
1	0	1	1
1	1	0	1
1	1	1	1

FIGURE 2.7 OR Function Represented by Switches

that the lamp is also on if *both* A and B are closed. This property distinguishes the OR function from the Exclusive OR function, which we will study later in this chapter.)

Like AND gates, OR gates can have several inputs, such as the 3-input OR gates shown in Figure 2.8. Table 2.5 shows the truth table for this gate. Again, three inputs can be combined in eight different ways. The output is HIGH when at least one input is HIGH. This condition is satisfied by all but the first line in the table.

We can create a 3-input OR function from two 2-input OR gates, as shown in Figure 2.9. The first gate combines the inputs A and B to get $(A + B)$. This result is combined in the second gate with input C to get $(A + B) + C = A + B + C$. This is the equivalent function to the gates in Figure 2.8.

a. Distinctive-shape

b. Rectangular-outline

FIGURE 2.8 3-Input OR Gate Symbols

FIGURE 2.9 3-Input OR Function from 2-Input OR Gates

EXAMPLE 2.1

State which logic function is most suitable for the following operations. Draw a set of switches to represent each function.

1. A manager and one other employee both need a key to open a safe.
2. A light comes on in a storeroom when either (or both) of two doors is open. (Assume the switch closes when the door opens.)
3. For safety, a punch press requires two-handed operation.

■ Solution

1. Both keys are required, so this is an AND function. Figure 2.10a shows a switch representation of the function.
2. One or more switches closed will turn on the lamp. This OR function is shown in Figure 2.10b.
3. Two switches are required to activate a punch press, as shown in Figure 2.10c. This is an AND function.

a. Two keys to open a safe (AND)

b. One or more switches turn on a lamp (OR)

c. Two switches are required to activate a punch press (AND)

FIGURE 2.10   Example 2.1: Switches Used for AND and OR Functions

Active Levels

■ KEY TERMS

Active Level   A logic level defined as the "ON" state for a particular circuit input or output. The active level can be either HIGH or LOW.

Active LOW   An active-LOW terminal is considered "ON" when it is in the logic LOW state, indicated by a bubble at the terminal in distinctive-shape symbols.

Active HIGH   An active-HIGH terminal is considered "ON" when it is in the logic HIGH state, indicated by the absence of a bubble at the terminal in distinctive-shape symbols.

An **active level** of a gate input or output is the logic level, either HIGH or LOW, of the terminal when it is performing its designated function. An **active LOW** is shown by a bubble or an arrow symbol on the affected terminal. If there is no bubble or arrow, we assume the terminal is **active HIGH.**

The AND function has active-HIGH inputs and an active-HIGH output. To make the output HIGH, inputs A AND B must *both* be HIGH. The gate performs its designated function only when *all* inputs are HIGH.

The OR gate requires input *A* OR input *B* to be HIGH for its output to be HIGH. The HIGH active levels are shown by the absence of bubbles or arrows on the terminals.

■ **SECTION 2.1 REVIEW PROBLEM**

A 4-input gate has input variables *A*, *B*, *C*, and *D* and output *Y*. Write a descriptive sentence for the active output state(s) if the gate is

2.1   AND.

2.2   OR.

2.2   DERIVED LOGIC FUNCTIONS

■ **KEY TERMS**

NAND Gate   A logic circuit whose output is LOW when all inputs are HIGH. (A combination of NOT and AND.)

NOR Gate   A logic circuit whose output is LOW when at least one input is HIGH. (A combination of NOT and OR.)

Exclusive OR (XOR) Gate   A 2-input logic circuit whose output is HIGH when one input (but not both) is HIGH.

Exclusive NOR (XNOR) Gate   A 2-input logic circuit whose output is the complement of an Exclusive OR gate.

Coincidence Gate   An Exclusive NOR gate.

The basic logic functions, AND, OR, and NOT can be combined to make any other logic function. Special logic gates exist for several of the most common of these derived functions. In fact, for reasons we will discover later, two of these derived-function gates, NAND and NOR, are the most common of all gates, and *each* can be used to create any logic function.

NAND and NOR Functions

The names NAND and NOR are contractions of NOT AND and NOT OR, respectively. The NAND is generated by inverting the output of an AND function. The symbols for the **NAND gate** and its equivalent circuit are shown in Figure 2.11.

The algebraic expression for the NAND function is:

$$Y = \overline{A \cdot B}$$

The NAND gate has active-HIGH inputs and an active-LOW output, shown by the bubble. Since the gate has an AND shape, these conditions lead to the descriptive sentence, "**All** inputs **HIGH** make output **LOW**." This condition is satisfied only by the last line of the gate's truth table. The partial truth table in Table 2.6a shows this condition: when *A* = 1 AND *B* = 1, output *Y* = 0. Since the remaining lines do not satisfy this condition, the output is opposite (*Y* = 1) for all these lines, as shown in the complete truth table of Table 2.6b.

TABLE 2.6a   Partial Truth Table for a 2-Input NAND Gate

A	*B*	*Y*
0	0	
0	1	
1	0	
1	1	**0**

TABLE 2.6b   Complete Truth Table for a 2-Input NAND Gate

A	*B*	*Y*
0	0	1
0	1	1
1	0	1
1	1	**0**

a. Distinctive-shape

b. Rectangular-outline

c. Equivalent circuit

FIGURE 2.11   2-Input NAND Gate Symbols

FIGURE 2.12 2-Input NOR Gate Symbols

TABLE 2.7a Partial Truth Table of a NOR Gate

A	B	Y
0	0	
0	1	0
1	0	0
1	1	0

TABLE 2.7b Complete Truth Table of a NOR Gate

A	B	Y
0	0	1
0	1	0
1	0	0
1	1	0

Figure 2.12 shows the logic symbols for the **NOR gate.** Because the gate is OR-shaped, with active-HIGH inputs (no bubbles) and an active-LOW output (bubble), it can be described by the sentence, "**At least one** input **HIGH** makes output **LOW.**" Table 2.7a shows the lines on the truth table for which this condition is satisfied: at least one input is HIGH in all lines but the first. For each of these lines, $Y = 0$, since the output is active-LOW. The remaining line ($A = 0$, $B = 0$) does not satisfy the condition. Therefore, for this line $Y = 1$, the opposite level from the other lines.

The algebraic expression for the NOR function is:

$$Y = \overline{A + B}$$

Multiple-Input NAND and NOR Gates

Table 2.8 shows the truth tables of the 3-input NAND and NOR functions. The logic circuit symbols for these gates are shown in Figure 2.13.

TABLE 2.8 3-Input NAND and NOR Function Truth Tables

A	B	C	$\overline{A \cdot B \cdot C}$	$\overline{A + B + C}$
0	0	0	1	1
0	0	1	1	0
0	1	0	1	0
0	1	1	1	0
1	0	0	1	0
1	0	1	1	0
1	1	0	1	0
1	1	1	0	0

FIGURE 2.13 3-Input NAND and NOR Gates

The truth tables of these gates can be generated from the active levels of their inputs and outputs, as well as their shape (AND = "all," OR = "at least one"). For the NAND gate, we can say, "**All inputs HIGH** make output **LOW.**" This is shown in the last line of the NAND truth table. All other lines have an output with the opposite logic level. For the NOR gate, we can say, "**At least one** input **HIGH** makes output **LOW.**" This condition is met in all lines but the first.

FIGURE 2.14 Expanding a NAND Gate from Two Inputs to Three

Expanding NAND and NOR Gates

Recall that we could use two 2-input AND gates to make a 3-input AND, and two 2-input OR gates to make a 3-input OR. We can also use 2-input gates to make 3-input NAND and NOR gates, but not quite so simply. We must keep in mind that a NAND gate combines all of its inputs in an AND function, then inverts the total result. Similarly, a NOR combines all of its inputs in an OR function, then inverts the result. Therefore, inversion must not be done until the very last step before the output.

Figure 2.14a shows how a 3-input NAND can be created using a 2-input AND and a 2-input NAND. The AND gate combines A and B. The NAND combines the compound AB with C, then inverts the total result. This is equivalent to the 3-input NAND gate in Figure 2.13. Trying to make the 3-input NAND with two 2-input NANDs, as shown in Figure 2.14b, does not work. In this case, we end up inverting a partial result $(\overline{AB})$ before all inputs can be combined in the AND function. The result $(\overline{\overline{AB}C})$ is not equivalent to the 3-input NAND function.

Figure 2.15 shows a similar configuration for the 3-input NOR function. Figure 2.15a shows the correct way to get a 3-input NOR from 2-input gates. The OR gate combines A and B. This intermediate result is ORed with C and then the total result is inverted. This is equivalent to the 3-input NOR gate shown in Figure 2.13. Figure 2.15b shows an incorrect connection for a 3-input NOR function. The first NOR combines A OR B, then inverts this partial result $(\overline{A + B})$. When this is combined with C in the second NOR gates, we get $\overline{\overline{A + B} + C}$, which is not equivalent to the 3-input OR function.

FIGURE 2.15 Expanding a NOR Gate from Two Inputs to Three

NAND and NOR Gates as Inverters

NAND and NOR gates can be used as inverters if we short-circuit their inputs, as shown in Figure 2.16. Reexamine the NAND truth table of Table 2.6b. If the NAND inputs are shorted, as in Figure 2.16a, then the only lines on the truth table that can be used are the lines where A and B are the same logic level, that is, the first and last lines. In the first line, both inputs are LOW and the output is HIGH. In the last line, both inputs are HIGH and the output is LOW. This has the effect of inverting the

FIGURE 2.16 Three Equivalent Ways of Inverting an Input

single input that is applied to both inputs of the gate. In a similar way, if we short the NOR inputs, both inputs are the same, yielding the result that if the inputs are LOW, the output is HIGH, and vice versa.

Exclusive OR and Exclusive NOR Functions

The Exclusive OR function (abbreviated XOR) is a special case of the OR function. The output of a 2-*input* XOR gate is HIGH when *one and only one* of the inputs is HIGH. (Multiple-input XOR circuits do not expand as simply as other functions. As we will see in a later chapter, an XOR output is HIGH when an *odd number* of inputs is HIGH.)

A HIGH at both inputs makes the output LOW. (We could say that the case in which both inputs are HIGH is excluded.)

The gate symbol for the **Exclusive OR (XOR) gate** is shown in Figure 2.17. Table 2.9 shows the truth table for the XOR function.

Another way of looking at the Exclusive OR gate is that its output is HIGH when the inputs are different and LOW when they are the same. This is a useful property in some applications, such as error detection in digital communication systems. (Transmitted data can be compared with received data. If they are the same, no error has been detected.)

The XOR function is expressed algebraically as:

$$Y = A \oplus B$$

The Exclusive NOR (XNOR) function is the complement of the Exclusive OR function and shares some of the same properties. The symbol, shown in Figure 2.18, is an XOR gate with a bubble on the output, implying that the entire function is inverted. Table 2.10 shows the Exclusive NOR truth table.

The algebraic expression for the Exclusive NOR function is:

$$Y = \overline{A \oplus B}$$

The output of the **Exclusive NOR (XNOR) gate** is HIGH when the inputs are the same and LOW when they are different. For this reason, the XNOR gate is also called a **coincidence gate.** This same/different property is similar to that of the Exclusive OR gate, only opposite in sense. Many of the applications that make use of this property can use either the XOR or the XNOR gate.

TABLE 2.9 Exclusive OR Function Truth Table

A	B	Y
0	0	0
0	1	1
1	0	1
1	1	0

TABLE 2.10 Exclusive NOR Function Truth Table

A	B	Y
0	0	1
0	1	0
1	0	0
1	1	1

a. Distinctive-shape

b. Rectangular-outline

FIGURE 2.17 Exclusive OR Gate

a. Distinctive-shape

b. Rectangular-outline

FIGURE 2.18 Exclusive NOR Gate

■ SECTION 2.2 REVIEW PROBLEMS

The output of a logic gate turns on an active-HIGH light when it is HIGH. The gate has two inputs, each of which is connected to a logic switch, as shown in Figure 2.19.

2.3 What type of gate will turn on the light when the switches are in opposite positions?

2.4 Which gate will turn off the light only when both switches are HIGH?

➡ **LAB SOURCE**
Lab 1, DIP
Integrated Circuits

FIGURE 2.19 Section Review Problems: Logic Gate Properties

➡ **LAB SOURCE**
Lab 2, Expanding Logic Gates

2.5   What type of gate turns off the light when at least one switch is HIGH?

2.6   Which gate turns on the light when the switches are in the same position?

2.3 DEMORGAN'S THEOREMS AND GATE EQUIVALENCE

> ■ **KEY TERMS**
>
> **DeMorgan Equivalent Forms**   Two gate symbols, one AND-shaped and one OR-shaped, that are equivalent according to DeMorgan's theorems.
>
> **DeMorgan's Theorems**   Two theorems in Boolean algebra that allow us to transform any gate from an AND-shaped to an OR-shaped gate and vice versa.

TABLE 2.11 NAND Truth Table

A	B	Y
0	0	1
0	1	1
1	0	1
1	1	0

Recall the description of a 2-input NAND gate: "**All** inputs **HIGH** make output **LOW**." This condition is satisfied in the last line of the 2-input NAND truth table, repeated in Table 2.11. We could also describe the gate function by saying, "**At least one** input **LOW** makes output **HIGH**." This condition is satisfied by the first three lines of Table 2.11.

The gates in Figure 2.20 represent positive and negative forms of a NAND gate. Figure 2.21 shows the logic equivalents of these gates. In the first case, we combine the inputs in an AND function, then invert the result. In the second case, we invert the input variables, then combine the inverted inputs in an OR function.

The Boolean function for the AND-shaped gate is given by:

$$Y = \overline{A \cdot B}$$

The Boolean expression for the OR-shaped gate is:

$$Y = \overline{A} + \overline{B}$$

FIGURE 2.20 NAND Gate and DeMorgan Equivalent (positive and negative NAND)

a. AND then invert   **b. Invert then OR**

FIGURE 2.21 Logic Equivalents of Positive and Negative NAND Gates

The gates shown in Figure 2.20 are called **DeMorgan equivalent forms.** Both gates have the same truth table, but represent different aspects or ways of looking at the NAND function. We can extend this observation to state that *any* gate (except XOR and XNOR) has two equivalent forms, one AND, one OR.

A gate can be categorized by examining three attributes: *shape, input,* and *output.* A question arises from each attribute:

1. **What is its shape (AND/OR)?**
   AND: *all*
   OR: *at least one*
2. **What active level is at the gate inputs (HIGH/LOW)?**
3. **What active level is at the gate output (HIGH/LOW)?**

The answers to these questions characterize any gate and allow us to write a descriptive sentence and a truth table for that gate. The DeMorgan equivalent forms of the gate will yield opposite answers to each of these questions.

Thus the gates in Figure 2.20 have the following complementary attributes:

	Basic Gate	**DeMorgan Equivalent**
Boolean Expression	$\overline{A \cdot B}$	$\overline{A} + \overline{B}$
Shape	AND	OR
Input Active Level	HIGH	LOW
Output Active Level	LOW	HIGH

EXAMPLE 2.2

Analyze the shape, input, and output of the gates shown in Figure 2.22 and write a Boolean expression, a descriptive sentence, and a truth table of each one. Write an asterisk beside the active output level on each truth table. Describe how these gates relate to each other.

$$
A \quad \text{——} \quad Y \qquad\qquad A \quad \text{——} \quad Y
$$
$$
B \quad \text{——} \qquad\qquad\qquad B \quad \text{——}
$$
      a.   **b.**

FIGURE 2.22 Example 2.2: Logic Gates

■ **Solution**

a. **Boolean expression:** $Y = \overline{A + B}$
   Shape: OR *(at least one)*
   Input: HIGH
   Output: LOW
   Descriptive sentence: At least one input **HIGH** makes output **LOW.**
   Truth table:

TABLE 2.12 Truth Table of Gate in Figure 2.22a

A	B	Y
0	0	1
0	1	0*
1	0	0*
1	1	0*

b. **Boolean expression:** $Y = \overline{A} \cdot \overline{B}$
   Shape: AND *(all)*
   Input: LOW
   Output: HIGH
   Descriptive sentence: All inputs LOW make output HIGH.
   Truth table:

TABLE 2.13   Truth Table of Gate in Figure 2.22b

A	B	Y
0	0	1*
0	1	0
1	0	0
1	1	0

Both gates in this example yield the same truth table. Therefore they are DeMorgan equivalents of one another (positive- and negative-NOR gates).

The gates in Figures 2.20 and 2.22 yield the following algebraic equivalencies:

$$\overline{A \cdot B} = \overline{A} + \overline{B}$$
$$\overline{A + B} = \overline{A} \cdot \overline{B}$$

These equivalencies are known as **DeMorgan's theorems.** (You can remember how to use DeMorgan's theorems by a simple rhyme: "Break the line and change the sign.")

It is tempting to compare the first gate in Figure 2.20 and the second in Figure 2.22 and declare them equivalent. Both gates are AND-shaped, both have inversions. However, the comparison is false. The gates have different truth tables, as we have found in Tables 2.11 and 2.13. Therefore they have different logic functions and are not equivalent. The same is true of the OR-shaped gates in Figures 2.20 and 2.22. The gates may look similar, but since they have different truth tables, they have different logic functions and are therefore not equivalent.

The confusion arises when, after changing the logic input and output levels, you forget to change the shape of the gate. This is a common, but serious, error. These inequalities can be expressed as follows:

$$\overline{A \cdot B} \neq \overline{A} \cdot \overline{B}$$
$$\overline{A + B} \neq \overline{A} + \overline{B}$$

As previously stated, any AND- or OR-shaped gate can be represented in its DeMorgan equivalent form. All we need to do is analyze a gate for its shape, input, and output, then *change everything*.

EXAMPLE 2.3

FIGURE 2.23   Example 2.3: Logic Gates

Analyze the gate in Figure 2.23 and write a Boolean expression, descriptive sentence, and truth table for the gate. Mark active output levels on the truth table with asterisks. Find the DeMorgan equivalent form of the gate and write its Boolean expression and description.

■ **Solution**

Boolean expression: $Y = \overline{\overline{A} + \overline{B} + \overline{C}}$
Shape: OR *(at least one)*
Input: *LOW*
Output: *LOW*
Descriptive sentence: At least one input **LOW** makes output **LOW**.
Truth table:

TABLE 2.14   Truth Table of Gate
in Figure 2.23

A	B	C	Y
0	0	0	0*
0	0	1	0*
0	1	0	0*
0	1	1	0*
1	0	0	0*
1	0	1	0*
1	1	0	0*
1	1	1	1

Figure 2.24 shows the DeMorgan equivalent form of the gate in Figure 2.23. To create this symbol, we change the shape from OR to AND and invert the logic levels at both input and output. The result is an AND gate.

Boolean expression: $Y = ABC$
Descriptive sentence: All inputs **HIGH** make output **HIGH.**

FIGURE 2.24   Example 2.3: DeMorgan Equivalent of Gate in Figure 2.23

■ **SECTION 2.3 REVIEW PROBLEM**

2.7   The output of a gate is described by the following Boolean expression:

$$Y = \overline{A} + \overline{B} + \overline{C} + \overline{D}$$

Write the Boolean expression for the DeMorgan equivalent form of this gate.

2.4   LOGIC SWITCHES AND LED INDICATORS

Before continuing on, we should examine a few simple circuits that can be used for input or output in a digital circuit. Single-pole single-throw (SPST) and pushbutton switches can be used, in combination with resistors, to generate logic voltages for circuit inputs. Light-emitting diodes (LEDs) can be used to monitor outputs of circuits.

Logic Switches

■ **KEY TERMS**

Pull-up Resistor   A resistor connected from a point in an electronic circuit to the power supply of that circuit.

V_{CC} The power supply voltage in a transistor-based electronic circuit. The term often refers to the power supply of digital circuits.

Figure 2.25a shows a single-pole single-throw (SPST) switch connected as a logic switch. An important premise of this circuit is that the input of the digital circuit to which it is connected has a very high resistance to current. When the switch is open, the current flowing through the **pull-up resistor** from V_{CC} to the digital circuit is very small. Since the current is small, Ohm's law states that very little voltage drops across the pull-up resistor; the voltage is about the same at one end as at the other. Therefore, an open switch generates a logic HIGH at point X.

When the switch is closed, the majority of current flows to ground, limited only by the value of the pull-up resistor. (A pull-up resistor is typically between 1 kΩ and 10 kΩ. Therefore, the LOW-state current in the resistor is about 0.5 mA to 5 mA for a circuit where $V_{CC} = 5.0$ V.) Point X is approximately at ground potential, or logic LOW. Thus the switch generates a HIGH when open and a LOW when closed. The pull-up resistor provides a connection to V_{CC} in the HIGH state and limits power supply current in the LOW state. Figure 2.25b shows the voltage levels when the switch is closed and when it is open.

Figure 2.26 shows how pushbuttons can be used as logic inputs. Figure 2.26a shows a normally open pushbutton and a pull-up resistor. The pushbutton has a

a. Circuit   **b. Logic levels**

FIGURE 2.25   SPST Logic Switch

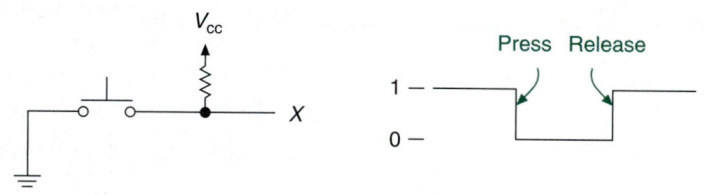

a. Normally open pushbutton

b. Normally closed pushbutton

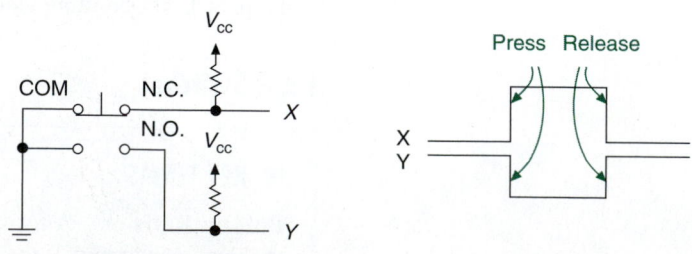

c. Two-pole pushbutton

FIGURE 2.26   Pushbuttons as Logic Switches

spring-loaded plunger that makes a connection between two internal contacts when pressed. When released, the spring returns the plunger to the "normal" (open) state. The logic voltage at X is normally HIGH, but LOW when the button is pressed.

Figure 2.26b shows a normally closed pushbutton. The internal spring holds the plunger so that the connection is normally made between the two contacts. When the button is pressed, the connection is broken and the resistor pulls up the voltage at X to a logic HIGH. At rest, X is grounded and the voltage at X is LOW.

It is sometimes desirable to have normally HIGH and normally LOW levels available from the same switch. The two-pole pushbutton in Figure 2.26c provides such a function. The switch has a normally open and a normally closed contact. One contact of each switch is connected to the other, in an internal COMMON connection, allowing the switch to have three terminals rather than four. The circuit has two pull-up resistors, one for X and one for Y. X is normally HIGH and goes LOW when the switch is pressed. Y is opposite.

LED Indicators

> ### ■ KEY TERM
>
> **LED (Light-Emitting Diode)**   An electronic device that conducts current in one direction only and illuminates when it is conducting.

A device used to indicate the status of a digital output is the **light-emitting diode** or **LED.** This is sometimes pronounced as a word ("led") and sometimes said as separate initials ("ell ee dee"). This device comes in a variety of shapes, sizes, and colors, some of which are shown in the photo in Figure 2.27. The circuit symbol, shown in Figure 2.28, has two terminals, called the anode (positive) and cathode (negative). The arrow coming from the symbol indicates emitted light.

The electrical requirements for the LED are simple: current flows through the LED if the anode is more positive than the cathode by more than a specified value (about 1.5 volts). If enough current flows, the LED illuminates. If more current flows, the illumination is brighter. (If too much flows, the LED burns out, so a series resistor is used to keep the current in the required range.) Figure 2.29 shows a circuit in which an LED illuminates when a switch is closed.

Figure 2.30a shows an AND gate driving an LED. The LED is on when Y is HIGH (5 volts), since the anode of the LED is more positive than the cathode.

FIGURE 2.27   LEDs

FIGURE 2.28 Light-Emitting Diode (LED)

FIGURE 2.29 Condition for LED Illumination

a. LED on when *Y* is HIGH

b. LED on when *Y* is LOW

FIGURE 2.30 Logic Gate Driving an LED

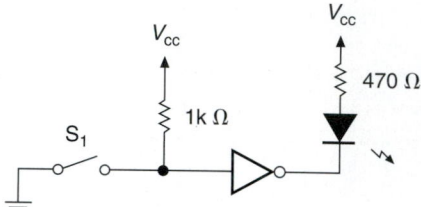

FIGURE 2.31 LED Indicates Status of Switch

In Figure 2.30b, the LED is driven by a NAND gate, which has an active-LOW output. The direction of the LED is such that it turns on when *Y* is LOW, again since the anode is more positive than the cathode. Note that for either case, the LED is on when *A* AND *B* are both HIGH.

Figure 2.31 shows a circuit in which an LED indicates the status of a logic switch. When the switch is open, the 1 kΩ pull-up applies a HIGH to the inverter input. The inverter output is LOW, turning on the LED (anode is more positive than cathode). When the switch is closed, the inverter input is LOW. The inverter output is HIGH (same value as V_{CC}), making anode and cathode voltages equal. No current flows through the LED, and it is therefore off. Thus, the LED is on for a HIGH state at the switch and off for a LOW. Note, however, that the LED is *on* when the inverter output is *LOW*.

■ **SECTION 2.4 REVIEW PROBLEM**

2.8   A single-pole single-throw switch is connected such that one end is grounded and the other end is connected to a 1 kΩ pull-up resistor. The other end of the resistor connects to the circuit power supply, V_{CC}. What logic level does the switch provide when it is open? When it is closed?

2.5 ENABLE AND INHIBIT PROPERTIES OF LOGIC GATES

■ **KEY TERMS**

Digital Signal (or **Pulse Waveform**)   A series of 0s and 1s plotted over time.

Enable   A logic gate is enabled if it allows a digital signal to pass from an input to the output in either true or complement form.

> **Inhibit** (or **Disable**)   A logic gate is inhibited if it prevents a digital signal from passing from an input to the output.
>
> **In Phase**   Two digital waveforms are in phase if they are always at the same logic level at the same time.
>
> **True Form**   Not inverted.
>
> **Complement Form**   Inverted.
>
> **Out of Phase**   Two digital waveforms are out of phase if they are always at opposite logic levels at any given time.

In Chapter 1, we saw that a **digital signal** is just a string of bits (0s and 1s) generated over time. A major task of digital circuitry is the direction and control of such signals. Logic gates can be used to **enable** (pass) or **inhibit** (block) these signals. (The word "gate" gives a clue to this function; the gate can "open" to allow a signal through or "close" to block its passage.)

AND and OR Gates

The simplest case of the enable and inhibit properties is that of an AND gate used to pass or block a logic signal. Figure 2.32 shows the output of an AND gate under different conditions of input A when a digital signal (an alternating string of 0s and 1s) is applied to input B.

Recall the properties of an AND gate: both inputs must be HIGH to make the output HIGH. Thus, if input A is LOW, the output must always be LOW, regardless of the state of input B. The digital signal applied to B has no effect on the output, and we say that the gate is inhibited or disabled. This is shown in the first half of the timing diagram in Figure 2.32.

If A AND B are HIGH, the output is HIGH. When A is HIGH and B is LOW, the output is LOW. Thus, output Y is the same as input B if input A is HIGH; that is, Y and B are **in phase** with each other. The input waveform is passed to the output in **true form,** and we say the gate is enabled. The last half of the timing diagram in Figure 2.32 shows this waveform.

It is convenient to define terms for the A and B inputs. Since we apply a digital signal to B, we will call it the Signal input. Since input A controls whether or not the signal passes to the output, we will call it the Control input. These definitions are illustrated in Figure 2.33.

Each type of logic gate has a particular set of enable/inhibit properties that can be predicted by examining the truth table of the gate. Let us examine the truth table of the AND gate to see how the method works.

FIGURE 2.32   Enable/Inhibit Properties of an AND Gate

FIGURE 2.33   Control and Signal Inputs of an AND Gate

TABLE 2.15   AND Truth Table Showing Enable/Inhibit Properties

A	B	Y	
0	0	0	$(Y = 0)$
0	1	0	Inhibit
1	0	0	$(Y = B)$
1	1	1	Enable

Divide the truth table in half, as shown in Table 2.15. Since we have designated A as the Control input, the top half of the truth table shows the inhibit function ($A = 0$), and the bottom half shows the enable function ($A = 1$). To determine the gate properties, we compare input B (the Signal input) to the output in each half of the table.

Inhibit mode: If $A = 0$ and B is pulsing (B is continuously going back and forth between the first and second lines of the truth table), output Y is always 0. Since the Signal input has no effect on the output, we say that the gate is disabled or inhibited ($Y = 0$).

Enable mode: If $A = 1$ and B is pulsing (B is going continuously between the third and fourth lines of the truth table), the output is the same as the Signal input. Since the Signal input affects the output, we say that the gate is enabled ($Y = B$).

EXAMPLE 2.4

Use the method just described to draw the output waveform of an OR gate if the input waveforms of A and B are the same as in Figure 2.32. Indicate the enable and inhibit portions of the timing diagram.

■ Solution

Divide the OR gate truth table in half. Designate input A the Control input and input B the Signal input.

As shown in Table 2.16, when $A = 0$ and B is pulsing, the output is the same as B and the gate is enabled. When $A = 1$, the output is always HIGH. (At least one input HIGH makes the output HIGH.) Since B has no effect on the output, the gate is inhibited. This is shown in Figure 2.34 in graphical form.

TABLE 2.16   OR Truth Table Showing Enable/Inhibit Properties

A	B	Y	
0	0	0	$(Y = B)$
0	1	1	Enable
1	0	1	$(Y = 1)$
1	1	1	Inhibit

FIGURE 2.34   Example 2.4: OR Gate Enable/Inhibit Waveform

Example 2.4 shows that a gate can be in the inhibit state even if its output is HIGH. It is natural to think of the HIGH state as "ON," but this is not always the case. Enable or inhibit states are determined by the effect the Signal input has on the gate's output. If an input signal does not affect the gate output, the gate is inhibited. If the Signal input does affect the output, the gate is enabled.

NAND and NOR Gates

When inverting gates, such as NAND and NOR, are enabled, they will invert an input signal before passing it to the gate output. In other words, they transmit the signal in **complement form.** Figure 2.35 and Figure 2.36 show the output waveforms of a NAND and a NOR gate when a square waveform is applied to input B and input A acts as a Control input.

The truth table for the NAND gate is shown in Table 2.17, divided in half to show the enable and inhibit properties of the gate.

Table 2.18 shows the NOR gate truth table, divided in half to show its enable and inhibit properties.

Figure 2.35 and Figure 2.36 show that when the NAND and NOR gates are enabled, the Signal and output waveforms are opposite to one another; we say that they are **out of phase.**

Compare the enable/inhibit waveforms of the AND, OR, NAND, and NOR gates. Gates of the same shape are enabled by the same Control level. AND and NAND gates are enabled by a HIGH on the Control input and inhibited by a LOW. OR and NOR are the opposite. A HIGH Control input inhibits the OR/NOR; a LOW Control input enables the gate.

TABLE 2.17 NAND Truth Table Showing Enable/Inhibit Properties

A	B	Y	
0	0	1	$(Y = 1)$
0	1	1	Inhibit
1	0	1	$(Y = \bar{B})$
1	1	0	Enable

TABLE 2.18 NOR Truth Table Showing Enable/Inhibit Properties

A	B	Y	
0	0	1	$(Y = \bar{B})$
0	1	0	Enable
1	0	0	$(Y = 0)$
1	1	0	Inhibit

FIGURE 2.35 Enable/Inhibit Properties of a NAND Gate

FIGURE 2.36 Enable/Inhibit Properties of a NOR Gate

$A = 0$ Y = ⎍⎍ $A = 1$ Y = ⎍⎍

$B =$ ⎍⎍ $B =$ ⎍⎍

Gate enabled (true)   Gate enabled (complement)

FIGURE 2.37   Dynamic Properties of an Exclusive OR Gate

Exclusive OR and Exclusive NOR Gates

Neither the XOR nor the XNOR gate has an inhibit state. The Control input on both of these gates acts only to determine whether the output waveform will be in or out of phase with the input signal. Figure 2.37 shows the dynamic properties of an XOR gate.

The truth table for the XOR gate, showing the gate's dynamic properties, is given in Table 2.19.

Notice that when $A = 0$, the output is in phase with B and when $A = 1$, the output is out of phase with B. A useful application of this property is to use an XOR gate as a programmable inverter. When $A = 1$, the gate is an inverter; when $A = 0$, it is a noninverting buffer.

The XNOR gate has properties similar to the XOR gate. That is, an XNOR has no inhibit state, and the Control input switches the output in and out of phase with the Signal waveform, although not the same way as an XOR gate does. You will derive these properties in one of the end-of-chapter problems.

Table 2.20 summarizes the enable/inhibit properties of the six gates previously examined.

TABLE 2.19   XOR Truth Table Showing Dynamic Properties

A	B	Y	
0	0	0	$(Y = B)$
0	1	1	Enable
1	0	1	$(Y = \bar{B})$
1	1	0	Enable

TABLE 2.20   Summary of Enable/Inhibit Properties

Control	AND	OR	NAND	NOR	XOR	XNOR
$A = 0$	$Y = 0$	$Y = B$	$Y = 1$	$Y = \bar{B}$	$Y = B$	$Y = \bar{B}$
$A = 1$	$Y = B$	$Y = 1$	$Y = \bar{B}$	$Y = 0$	$Y = \bar{B}$	$Y = B$

➡ **LAB SOURCE**
Lab #3, Pulsed Operation
of Logic Gates

■ **SECTION 2.5 REVIEW PROBLEM**

2.9   Briefly explain why an AND gate is inhibited by a LOW Control input and an OR gate is inhibited by a HIGH Control input.

Tristate Buffers

■ **KEY TERMS**

High-Impedance State   The output state of a tristate buffer that is neither logic HIGH nor logic LOW, but is electrically equivalent to an open circuit. (Abbreviation: Hi-Z.)

Tristate Buffer   A gate having three possible output states: logic HIGH, logic LOW, and high-impedance.

Bus   A common wire or parallel group of wires connecting multiple circuits.

IN ▷ OUT

$\overline{OE}$

a. Noninverting

IN ▷ OUT

$\overline{OE}$

b. Inverting

FIGURE 2.38   Tristate Buffers

In the previous section, logic gates were used to enable or inhibit signals in digital circuits. For the AND, NAND, NOR, and OR gates, however, in the inhibit state the output was always logic HIGH or LOW. In some cases, it is desirable to have an output state that is neither HIGH nor LOW, but acts to electrically disconnect the gate output from the circuit. This third state is called the **high-impedance state** and is one of three available states in a class of devices known as **tristate buffers.**

Figure 2.38 shows the logic symbols for two tristate buffers, one with a noninverting output and one with an inverting output. The third input, $\overline{OE}$ (Output enable), is an active-LOW signal that enables or disables the buffer output.

When $\overline{OE} = 0$, as shown in Figure 2.39a, the noninverting buffer transfers the input value directly to the output as a logic HIGH or LOW. When $\overline{OE} = 1$, as in Figure 2.39b, the output is electrically disconnected from any circuit to which it is connected. (The open switch in Figure 2.39b does not literally exist. It is shown as a symbolic representation of the electrical disconnection of the output in the high-impedance state.)

This type of enable/disable function is particularly useful when digital data are transferred from more than one source to one or more destinations along a common wire (or **bus**), as shown in Figure 2.40. (This is the underlying principle in modern computer systems, where multiple components use the same bus to pass data back and forth.) The destination circuit in Figure 2.40 can receive data from source 1 or source 2. If the source circuits were directly connected to the bus, they could produce contradictory logic levels at the destination. To prevent this, only one source is enabled at a time, with control of this switching left to the two tristate buffers. For example, to transfer data from source 1 to the destination, we make $\overline{OE}_1 = 0$ and $\overline{OE}_2 = 1$. Data is transferred from source 1 to the bus and thus to the destination, whereas source 2 is electrically disconnected from the bus. In this way the data from source 1 and source 2 do not interfere with one another.

IN ▷ OUT = IN   IN ▷ ⚬ ⚬ OUT = Hi-Z

$\overline{OE} = 0$ $\overline{OE} = 1$

a. Output enabled   **b. Output disabled**

FIGURE 2.39   Electrical Equivalent of Tristate Operation

FIGURE 2.40   Using Tristate Buffers to Switch Two Sources to a Single Destination

Octal Tristate Buffers

Sometimes tristate buffers are packaged in multiples that make it convenient to enable or disable an entire multibit group of signals. The 74LS244 octal tristate buffer, shown in Figure 2.41, is such a device. It contains two groups of four non-inverting tristate buffers, with each group controlled by a separate $\overline{G}$ input. ("Octal" means "eight." 74LS244 is an industry standard part number. We shall see more about such numbers in the next section.)

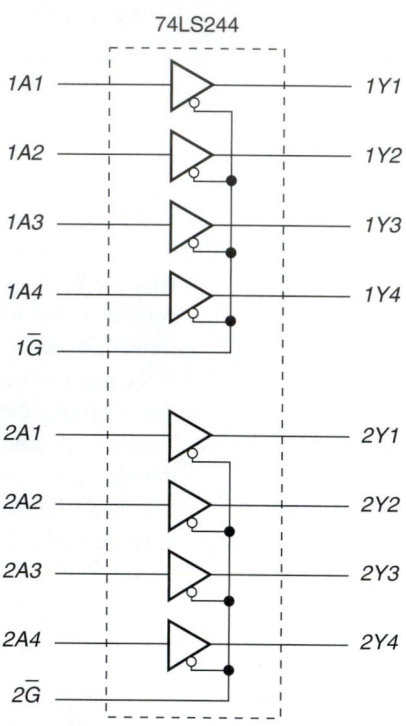

FIGURE 2.41   Octal Tristate Buffer

When $1\overline{G} = 0$, then $1Y = 1A$. Otherwise, $1Y$ = Hi-Z, where Hi-Z is an abbreviation for the high-impedance state. ($1A$ and $1Y$ are the 4-bit values consisting of $1A1$ through $1A4$ and $1Y1$ through $1Y4$. Thus, a single $\overline{G}$ input controls four Y outputs simultaneously.)

Similarly, when $2\overline{G} = 0$, $2Y = 2A$. When $2\overline{G} = 1$, $2Y$ = Hi-Z.

EXAMPLE 2.5

Draw a logic circuit showing how a 74LS244 octal tristate buffer can be connected to make a data bus where one of two 4-bit numbers can be transferred to a 4-bit output.

◼ Solution

Refer to Figure 2.42. The tristate outputs $1Y1$ through $1Y4$ are connected to outputs $2Y1$ through $2Y4$. The inverter connects to the $2\overline{G}$ input to keep it opposite from the $1\overline{G}$ input. This ensures that only one group of four buffers is enabled at any time. When $SELECT = 0$, the A inputs connect to Y, and B is in the Hi-Z state. When $SELECT = 1$, $Y = B$, and A is in the Hi-Z state.

FIGURE 2.42 Octal Tristate Buffer Connected as a 4-Bit Data Bus Driver

2.6 INTEGRATED CIRCUIT LOGIC GATES

■ **KEY TERMS**

Integrated Circuit (IC)   An electronic circuit having many components, such as transistors, diodes, resistors, and capacitors, in a single package.

Small Scale Integration (SSI)   An integrated circuit having 12 or fewer gates in one package.

Chip   An integrated circuit. Specifically, a chip of silicon on which an integrated circuit is constructed.

Medium Scale Integration (MSI)   An integrated circuit having the equivalent of 12 to 100 gates in one package.

Large Scale Integration (LSI)   An integrated circuit having from 100 to 10,000 equivalent gates.

Transistor-Transistor Logic (TTL)   A family of digital logic devices whose basic element is the bipolar junction transistor.

Complementary Metal-Oxide-Semiconductor (CMOS)   A family of digital logic devices whose basic element is the metal-oxide-semiconductor field effect transistor (MOSFET).

Dual In-Line Package (DIP)   A type of IC with two parallel rows of pins for the various circuit inputs and outputs.

Printed Circuit Board (PCB)   A circuit board in which connections between components are made with lines of copper on the surfaces of the circuit board.

Breadboard   A circuit board for wiring temporary circuits, usually used for prototypes or laboratory work.

Wire-Wrap   A circuit construction technique in which the connecting wires are wrapped around the posts of a special chip socket, usually used for prototyping or laboratory work.

Small Outline IC (SOIC)   An IC package similar to a DIP, but smaller, which is designed for automatic placement and soldering on the surface of a circuit board. Also called gull-wing, for the shape of the package leads.

Thin Shrink Small Outline Package (TSSOP)   A thinner version of an SOIC package.

Plastic Leaded Chip Carrier (PLCC)   A square IC package with leads on all four sides designed for surface mounting on a circuit board. Also called J-lead, for the profile shape of the package leads.

Quad Flat Pack (QFP)   A square surface-mount IC package with gull-wing leads.

Ball Grid Array (BGA)   A square surface-mount IC package with rows and columns of spherical leads underneath the package.

Surface-Mount Technology (SMT)   A system of mounting and soldering integrated circuits on the surface of a circuit board, as opposed to inserting their leads through holes on the board.

Datasheet   A printed specification giving details of the pin configuration, electrical properties, and mechanical profile of an electronic device.

Data Book   A bound collection of datasheets. A digital logic data book usually contains datasheets for a specific logic family or families.

Portable Document Format (PDF)   A format for storing published documents in compressed form.

Very Large Scale Integration (VLSI)   An integrated circuit having more than 10,000 equivalent gates.

Through-Hole   A means of mounting DIP ICs on a circuit board by inserting the IC leads through holes in the board and soldering them in place.

All the logic gates we have looked at so far are available in **integrated circuit** form. Most of these **small scale integration (SSI)** functions are available either in **transistor-transistor logic (TTL)** or **complementary metal-oxide-semiconductor (CMOS)** technologies. TTL and CMOS devices differ not in their logic functions, but in their construction and electrical characteristics.

TTL and CMOS **chips** are designated by an industry-standard numbering system, as shown in the following illustration. This system is often referred to as 74-series or 7400-series logic. In the past it was exclusively applied to TTL, but more recently has been used to designate high-speed CMOS devices, as well. Other, more complex, TTL and CMOS devices such as **medium scale integration (MSI)** and some **large scale integration (LSI)** devices also adopt this numbering system. (An MSI device has between 12 and 100 equivalent gates. An LSI device has between 100 and 10,000 equivalent gates.)

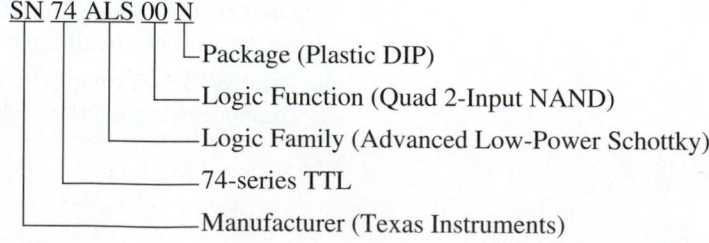

The portions of interest in a part number are those that designate the logic family, which specifies the component's electrical characteristics, and the logic function. For example, in the part number shown, the designation *ALS* indicates that the component belongs to the advanced low-power Schottky TTL family. The digits *00* indicate that the component is a quadruple 2-input NAND gate; that is, a package that contains four NAND gates (indicated by "quadruple"), each with two inputs.

Earlier versions of CMOS had a different set of unrelated numbers of the form 4*NNN*B or 4*NNN*UB where *NNN* was the logic function designator. The suffixes B and UB stand for buffered and unbuffered, respectively. Other, more-specialized **very large scale integration (VLSI)** chips have different standard numbering systems (e.g., 27C64 for a 64-kilobit EPROM [a type of memory chip]) or part numbers that are not industry-standard, but relate solely to the products of a particular manufacturer (e.g., EPM7128SLC84-7 for a complex programmable logic device made by Altera).

Table 2.21 lists the quadruple 2-input NAND function as implemented in different logic families. These devices all have the same logic function, but different electrical characteristics.

TABLE 2.21   Part Numbers for a Quad 2-Input NAND Gate in Different Logic Families

Part Number	Logic Family
74LS00	Low-power Schottky TTL
74ALS00	Advanced low-power Schottky TTL
74F00	FAST TTL
74HC00	High-speed CMOS
74HCT00	High-speed CMOS (TTL-compatible inputs)
74LVX00	Low-voltage CMOS
74ABT00	Advanced BiCMOS (TTL/CMOS hybrid)

Table 2.22 lists several logic functions available in the high-speed CMOS family. These devices all have the same electrical characteristics, but different logic functions.

TABLE 2.22   Part Numbers for Different Functions Within a Logic Family (High-Speed CMOS)

Part Number	Function
74HC00	Quadruple 2-input NAND
74HC02	Quadruple 2-input NOR
74HC04	Hex inverter
74HC08	Quadruple 2-input AND
74HC32	Quadruple 2-input OR
74HC86	Quadruple 2-input XOR

Until recently, the most common way to package logic gates has been in a plastic or ceramic **dual in-line package,** or **DIP,** which has two parallel rows of pins. The standard spacing between pins in one row is 0.1″ (or 100 mil). For packages having fewer than 28 pins, the spacing between rows is 0.3″ (or 300 mil). For larger packages, the rows are spaced by 0.6″ (600 mil).

This type of package is designed to be inserted in a **printed circuit board** in one of two ways: (a) the pins are inserted through holes in the circuit board and soldered in place; or (b) a socket is soldered to the circuit board and the IC is placed in the socket. Method (a) is referred to a **through-hole** placement. The latter method is more expensive, but makes chip replacement much easier. A socket can occasionally cause its own problems by making a poor connection to the pins of the IC.

FIGURE 2.43 14-Pin DIP (top view)

FIGURE 2.44 84-Pin PLCC Package (top view)

The DIP is also convenient for laboratory and prototype work, as it can also be inserted easily into a **breadboard,** a special type of temporary circuit board with internal connections between holes of a standard spacing. It is also convenient for **wire-wrapping,** a technique in which a special tool is used to wrap wires around posts on the underside of special sockets.

The outline of a 14-pin DIP is shown in Figure 2.43. There is a notch on one end to show the orientation of the pins. When the IC is oriented as shown and viewed from above, pin 1 is at the top left corner and the pins number counterclockwise from that point.

Figure 2.44 shows the outline of another common IC package, the 84-pin plastic leaded chip carrier (PLCC). A PLCC component can be mounted on the surface of a circuit board or inserted in a socket. The package has pins equally distributed on four sides, with pin 1 placed in the center of one of the rows, as indicated by a dot on the package. Pins number counterclockwise from this point. The orientation of the chip is also shown by a cutoff corner, which is at the top left when looking down at the chip from above.

Besides DIP and PLCC packages, there are numerous other types of packages for digital ICs, including, among others, **small outline IC (SOIC), thin shrink small outline package (TSSOP), plastic leaded chip carrier (PLCC), quad flat pack (QFP),** and **ball grid array (BGA)** packages. They are used in applications where circuit board space is at a premium and in manufacturing processes relying on **surface-mount technology (SMT).** In fact, these devices represent the majority of IC packages found in new designs. Some of these IC packaging options are shown in Figure 2.45.

SMT is a sophisticated technology which relies on automatic placement of chips and soldering of pins onto the surface of a circuit board, not through holes in the circuit board. This technique allows a manufacturer to mount components on both sides of a circuit board.

Primarily due to the great reduction in board space requirements, many new ICs are available only in the newer surface-mount packages and are not being offered at all in the DIP package. However, we will look at DIP offerings in logic gates because they are inexpensive and easy to use with laboratory breadboards and therefore useful as a learning tool.

Logic gates come in packages containing several gates. Common groupings available in DIP packages are six 1-input gates, four 2-input gates, three 3-input gates, or two 4-input gates, although other arrangements are available. The usual way of stating the number of logic gates in a package is to use the numerical prefixes hex (6), quad or quadruple (4), triple (3), or dual (2).

a. SOIC   b. TSSOP   c. PLCC

d.QFP   e. DIP   f. BGA

FIGURE 2.45 Some IC Packaging Options. *Sources:* (SOIC, TSSOP, PLCC and DIP) From National Semiconductor. (QFP) From Xilinx. (BGA) From Motorola. Used with permission.

Some common gate packages are listed in Table 2.23.

TABLE 2.23 Some Common Logic Gate ICs

Gate	Family	Function
74HC00A	High-speed CMOS	Quad 2-input NAND
74HC02	High-speed CMOS	Quad 2-input NOR
74ALS04	Advanced low-power Schottky TTL	Hex inverter
74LS11	Low-power Schottky TTL	Triple 3-input AND
74F20	FAST TTL	Dual 4-input NAND
74HC27	High-speed CMOS	Triple 3-input NOR

Information about pin configurations, electrical characteristics, and mechanical specifications of a part is available in a **datasheet** provided by the chip manufacturer. A collection of datasheets for a particular logic family is often bound together in a **data book.** More recently, device manufacturers have been making datasheets available on their corporate World Wide Web sites in **portable document format (PDF),** readable by a special program such as Adobe Acrobat Reader. Links to some of these manufacturers can be found on the Online Companion Web Site for this book. (http://www.electronictech.com)

Figure 2.46 shows the internal diagrams of the gates listed in Table 2.23. Notice that the gates can be oriented inside a chip in a number of ways. That is why it is important to confirm pin connections with a datasheet.

In addition to the gate inputs and outputs there are two more connections to be made on every chip: the power (V_{CC}) and ground connections. In TTL, connect V_{CC} to +5 Volts and GND to ground. In CMOS, connect the V_{CC} pin to the supply voltage (+3 V to +6 V) and GND to ground. The gates won't work without these connections.

Every chip requires power and ground. This might seem obvious, but it's surprising how often it is forgotten, especially by students who are new to digital electronics. Probably this is because most digital circuit diagrams don't show the power connections, but assume that you know enough to make them.

FIGURE 2.46   Pinouts of ICs Listed in Table 2.23

The only place a chip gets its required power is through the V_{CC} pin. Even if the power supply is connected to a logic input as a logic HIGH, you still need to connect it to the power supply pin.

Even more important is a good ground connection. A circuit with no power connection will not work at all. A circuit without a ground may appear to work, but it will often produce bizarre errors that are very difficult to detect and repair.

In later chapters, we will work primarily with complex ICs in PLCC packages. The power and ground connections are so important to these chips that they will not be left to chance; they are provided on a specially designed circuit board. Only input and output pins are accessible for connection by the user.

As digital designs become more complex, it is increasingly necessary to follow good practices in board layout and prototyping procedure to ensure even minimal functionality. Thus, hardware platforms for prototype and laboratory work will need to be at least partially constructed by the board manufacturer in order to supply the requirements of a stable circuit configuration.

■ SECTION 2.6 REVIEW PROBLEM

2.10   How are the pins numbered in a DIP? How are the pins numbered in a PLCC package?

SUMMARY

1. Digital systems can be analyzed and designed using Boolean algebra, a system of mathematics that operates on variables that have one of two possible values.

2. Any Boolean expression can be constructed from the three simplest logic functions: NOT, AND, and OR.

3. A NOT gate, or inverter, has an output state that is in the opposite logic state of the input.

4. The main logic functions are described by the following sentences:

   AND: All inputs **HIGH** make output **HIGH.**
   OR: At least one input **HIGH** makes output **HIGH.**
   NAND: All inputs **HIGH** make output **LOW.**
   NOR: At least one input **HIGH** makes output **LOW.**
   XOR: Output is **HIGH** if **one** input is **HIGH,** but **not both.**
   XNOR: Output is **LOW** if **one** input is **HIGH,** but **not both.**

5. The function of a logic gate can be represented by a truth table, a list of all possible inputs in binary order and the output corresponding to each input state.

6. A 3-input AND function can be made using two 2-input AND gates, where the output of one gate connects to one input of the next gate. The same configuration is possible with OR gates to make a 3-input OR function.

7. A 3-input NAND function can be made using a 2-input AND gate whose output connects to one input of a 2-input NAND gate. A similar connection with an OR and a NOR gate can be used to make a 3-input NOR function. In both cases, the inversion must be the last step in the process. In other words, the AND and NAND are not interchangeable and the OR and NOR are not interchangeable.

8. An inverter can be made from a NAND gate by shorting its inputs together. A NOR gate can also be used this way.

9. DeMorgan's theorems $(\overline{A \cdot B} = \overline{A} + \overline{B}$ and $\overline{A + B} = \overline{A} \cdot \overline{B})$ allow us to represent any gate in an AND form and an OR form.

10. To change a gate into its DeMorgan equivalent form, change its shape from AND to OR or vice versa and change the active levels of inputs and output.

11. A logic switch can be created from a single-pole single-throw switch by grounding one end and tying the other end to V_{CC} through a pull-up resistor. The logic level is available on the same side of the switch as the resistor. An open switch is HIGH and a closed switch is LOW. A similar circuit can be made with a pushbutton switch.

12. A light-emitting diode (LED) can be used to indicate logic HIGH or LOW levels. To indicate a HIGH, ground the cathode through a series resistor (about 470 Ω for a 5-volt power supply) and apply the logic level to the anode. To indicate a LOW, tie the anode to V_{CC} through a series resistor and apply the logic level to the cathode.

13. Logic gates can be used to pass or block digital signals. For example, an AND gate will pass a digital signal applied to input B if the input A is HIGH ($Y = B$). If input A is LOW, the signal is blocked and the gate output is always LOW ($Y = 0$). Similar properties apply to other gates, as summarized in Table 2.20.

14. Tristate buffers have outputs that generate logic HIGH and LOW when enabled and a high-impedance state when disabled. The high-impedance state is electrically equivalent to an open circuit.

15. Logic gates are available as integrated circuits in a variety of packages. Packages that have fewer than 12 gates are called small scale integration (SSI) devices.

16. Many logic functions have an industry-standard part number of the form 74*XXNN,* where *XX* is an alphabetic family designator and *NN* is a numeric function designator (e.g. 74HC02 = quadruple 2-input NOR gate in the high-speed CMOS family).

17. Some common IC packages include dual in-line package (DIP), small outline IC (SOIC), thin shrink small outline package (TSSOP), plastic leaded chip carrier (PLCC), quad flat pack (QFP), and ball grid array (BGA) packages.

18. Most new IC packages are for surface mounting on a printed circuit board. These have largely replaced DIPs in through-hole circuit boards, due to better use of board space.

19. IC pin connections and functional data can be determined from manufacturers' datasheets, available in paper format or electronically via the Internet.

20. All ICs require power and ground, which must be applied to special power supply pins on the chip.

GLOSSARY

Active HIGH   An active-HIGH terminal is considered "ON" when it is in the logic HIGH state, indicated by the absence of a bubble at the terminal in distinctive-shape symbols.

Active Level   A logic level defined as the "ON" state for a particular circuit input or output. The active level can be either HIGH or LOW.

Active LOW   An active-LOW terminal is considered "ON" when it is in the logic LOW state, indicated by a bubble at the terminal in distinctive-shape symbols.

AND Gate   A logic circuit whose output is HIGH when *all* inputs (e.g., *A* AND *B* AND *C*) are HIGH.

Ball Grid Array (BGA)   A square surface-mount IC package with rows and columns of spherical leads underneath the package.

Boolean Algebra   A system of algebra that operates on Boolean variables. The binary (two-state) nature of Boolean algebra makes it useful for analysis, simplification, and design of combinational logic circuits.

Boolean Expression   An algebraic expression made up of Boolean variables and operators, such as AND (·), OR (+), or NOT (⁻). Also referred to as a **Boolean function** or a **logic function.**

Boolean Variable   A variable having only two possible values, such as HIGH/LOW, 1/0, On/Off, or True/False.

Breadboard   A circuit board for wiring temporary circuits, usually used for prototypes or laboratory work.

Bubble   A small circle indicating logical inversion on a circuit symbol.

Buffer   An amplifier that acts as a logic circuit. Its output can be inverting or noninverting.

Bus   A common wire or parallel group of wires connecting multiple circuits.

Chip   An integrated circuit. Specifically, a chip of silicon on which an integrated circuit is constructed.

Clock Generator   A circuit that generates a periodic digital waveform.

Coincidence Gate   An Exclusive NOR gate.

Complement Form   Inverted.

Complementary Metal-Oxide-Semiconductor (CMOS)   A family of digital logic devices whose basic element is the metal-oxide-semiconductor field effect transistor (MOSFET).

Data Book   A bound collection of datasheets. A digital logic data book usually contains datasheets for a specific logic family or families.

Datasheet   A printed specification giving details of the pin configuration, electrical properties, and mechanical profile of an electronic device.

DeMorgan Equivalent Forms   Two gate symbols, one AND-shaped and one OR-shaped, that are equivalent according to DeMorgan's theorems.

DeMorgan's Theorems   Two theorems in Boolean algebra that allow us to transform any gate from an AND-shaped to an OR-shaped gate and vice versa.

Digital Signal (or **Pulse Waveform**)   A series of 0s and 1s plotted over time.

Distinctive-Shape Symbols   Graphic symbols for logic circuits that show the function of each type of gate by a special shape.

Dual In-Line Package (DIP)   A type of IC with two parallel rows of pins for the various circuit inputs and outputs.

Enable   A logic gate is enabled if it allows a digital signal to pass from an input to the output in either true or complement form.

Exclusive NOR Gate   A two-input logic circuit whose output is the complement of an Exclusive OR gate.

Exclusive OR Gate   A two-input logic circuit whose output is HIGH when one input (but not both) is HIGH.

High-Impedance State   The output state of a tristate buffer that is neither logic HIGH nor logic LOW, but is electrically equivalent to an open circuit. (Abbreviation: Hi-Z.)

IEEE/ANSI Standard 91-1984   A standard format for drawing logic circuit symbols as rectangles with logic functions shown by a standard notation inside the rectangle for each device.

In Phase   Two digital waveforms are in phase if they are always at the same logic level at the same time.

Inhibit (or **Disable**)   A logic gate is inhibited if it prevents a digital signal from passing from an input to the output.

Integrated Circuit (IC)   An electronic circuit having many components, such as transistors, diodes, resistors, and capacitors, in a single package.

Inverter Also called a NOT gate or an inverting buffer. A logic gate that changes its input logic level to the opposite state.

Large Scale Integration (LSI)   An integrated circuit having from 100 to 10,000 equivalent gates.

LED   Light-emitting diode. An electronic device that conducts current in one direction only and illuminates when it is conducting.

Logic Function   See **Boolean expression.**

Logic Gate   An electronic circuit that performs a Boolean algebraic function.

Logical Product   AND function.

Logical Sum   OR function.

Medium Scale Integration (MSI)   An integrated circuit having the equivalent of 12 to 100 gates in one package.

NAND Gate   A logic circuit whose output is LOW when *all* inputs are HIGH.

NOR Gate   A logic circuit whose output is LOW when *at least one* input is HIGH.

OR Gate   A logic circuit whose output is HIGH when *at least one* input (e.g., *A* OR *B* OR *C*) is HIGH.

Out of Phase   Two digital waveforms are out of phase if they are always at opposite logic levels at any given time.

Plastic Leaded Chip Carrier (PLCC)   A square IC package with leads on all four sides designed for surface mounting on a circuit board. Also called J-lead, for the profile shape of the package leads.

Portable Document Format (PDF)   A format for storing published documents in compressed form.

Printed Circuit Board (PCB)   A circuit board in which connections between components are made with lines of copper on the surfaces of the circuit board.

Pull-Up Resistor   A resistor connected from a point in an electronic circuit to the power supply of that circuit. In a digital circuit it supplies the required logic level in a HIGH state and limits current from the power supply in the LOW state.

Quad Flat Pack (QFP)   A square surface-mount IC package with gull-wing leads.

Qualifying Symbol   A symbol in IEEE/ANSI logic circuit notation, placed in the top center of a rectangular symbol, that shows the function of a logic gate. Some qualifying symbols include: 1 = "buffer"; & = "AND"; ≥1 = "OR"

Rectangular-Outline Symbols   Rectangular logic gate symbols that conform to IEEE/ANSI Standard 91-1984.

Small Outline IC (SOIC)   An IC package similar to a DIP, but smaller, which is designed for automatic placement and soldering on the surface of a circuit board. Also called gull-wing, for the shape of the package leads.

Small Scale Integration (SSI)   An integrated circuit having 12 or fewer gates in one package.

Surface-Mount Technology (SMT)   A system of mounting and soldering integrated circuits on the surface of a circuit board, as opposed to inserting their leads through holes on the board.

Thin Shrink Small Outline Package (TSSOP)   A thinner version of an SOIC package.

Through-Hole   A means of mounting DIP ICs on a circuit board by inserting the IC leads through holes in the board and soldering them in place.

Transistor-Transistor Logic (TTL)   A family of digital logic devices whose basic element is the bipolar junction transistor.

Tristate Buffer   A gate having three possible output states: logic HIGH, logic LOW, and high-impedance.

True Form   Not inverted.

Truth Table   A list of all possible input values to a digital circuit, listed in ascending binary order, and the output response for each input combination.

V_{CC} The power supply voltage in a transistor-based electronic circuit. The term often refers to the power supply of digital circuits.

Very Large Scale Integration (VLSI)   An integrated circuit having more than 10,000 equivalent gates.

Wire-Wrap   A circuit construction technique in which the connecting wires are wrapped around the posts of a special chip socket, usually used for prototyping or laboratory work.

PROBLEMS

Problem numbers set in color indicate more difficult problems.

2.1   Basic Logic Functions

2.1   Draw the symbol for the NOT gate (inverter) in both rectangular-outline and distinctive-shape forms.

2.2   Draw the distinctive-shape and rectangular-outline symbols for a 3-input AND gate.

2.3   Draw the distinctive-shape and rectangular-outline symbols for a 3-input OR gate.

2.4   Write a sentence that describes the operation of a 4-input AND gate that has inputs *P, Q, R,* and *S* and output *T.* Make the truth table of this gate and draw an asterisk beside the line(s) of the truth table indicating when the gate output is in its active state.

2.5   Write a sentence that describes the operation of a 4-input OR gate with inputs *J, K, L,* and *M* and output *N.* Make the truth table of this gate and draw an asterisk beside the line(s) of the truth table indicating when the gate output is in its active state.

2.6   State how three switches must be connected to represent a 3-input AND function. Draw a circuit diagram showing how this function can control a lamp.

2.7   State how four switches must be connected to represent a 4-input OR function. Draw a circuit diagram showing how this function can control a lamp.

2.8   Draw the circuit of a 3-input AND function, made using only 2-input logic gates.

2.9   Draw the circuit of a 3-input OR function, made using only 2-input logic gates.

2.2   Derived Logic Functions

2.10   For a 4-input NAND gate with inputs *A, B, C,* and *D* and output *Y:*

   a. Write the truth table and a descriptive sentence.

   b. Write the Boolean expression.

   c. Draw the logic circuit symbol in both distinctive-shape and rectangular-outline symbols.

2.11   Repeat Problem 2.10 for a 4-input NOR gate.

2.12   State the active levels of the inputs and outputs of a NAND gate and a NOR gate.

2.13 Write a descriptive sentence of the operation of a 5-input NAND gate with inputs A, B, C, D, and E and output Y. How many lines would the truth table of this gate have?

2.14 Repeat Problem 2.13 for a 5-input NOR gate.

2.15 A pump motor in an industrial plant will start only if the temperature and pressure of liquid in a tank exceed a certain level. The temperature sensor and pressure sensor, shown in Figure 2.47 each produce a logic HIGH if the measured quantities exceed this value. The logic circuit interface produces a HIGH output to turn on the motor. Draw the symbol and truth table of the gate that corresponds to the action of the logic circuit.

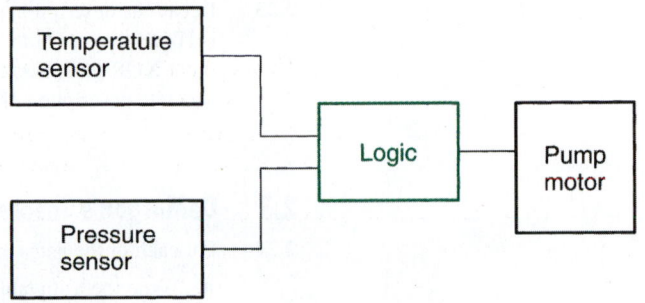

FIGURE 2.47   Problem 2.15: Temperature and Pressure Sensors

2.16 Repeat Problem 2.15 for the case in which the motor is activated by a logic LOW.

2.17 Draw the circuit of a 3-input NAND function, made using only 2-input logic gates.

2.18 Draw the circuit of a 4-input NAND function, made using only 2-input logic gates.

2.19 Draw the circuit of a 3-input NOR function, made using only 2-input logic gates.

2.20 Draw the circuit of a 4-input NOR function, made using only 2-input logic gates.

2.21 Figure 2.48 shows a circuit for a two-way switch for a stairwell. This is a common circuit that allows you to turn on a light from either the top or the bottom of the stairwell and off at the other end. The circuit also allows anyone coming along after you to do the same thing, no matter which direction they are coming from.

FIGURE 2.48   Problem 2.21: Circuit for Two-Way Switch

The lamp is ON when the switches are in the same positions and OFF when they are in opposite positions. What logic function does this represent? Draw the truth table of the function and use it to explain your reasoning.

2.22   Find the truth table for the logic circuit shown in Figure 2.49.

FIGURE 2.49   Problem 2.22: Logic Circuit

2.23   Recall the description of a 2-input Exclusive OR gate: "Output is HIGH if one input is HIGH, but not both." This is not the best statement of the operation of a multiple-input XOR gate. Look at the truth table derived in Problem 2.22 and write a more accurate description of *n*-input XOR operation.

2.3   DeMorgan's Theorems and Gate Equivalence

2.24   For each of the gates in Figure 2.50:

   a. Write the truth table.

   b. Indicate with an * which lines on the truth table show the gate output in its active state.

   c. Convert the gate to its DeMorgan equivalent form.

   d. Rewrite the truth table and indicate which lines on the truth table show output active states for the DeMorgan equivalent form of the gate.

FIGURE 2.50   Problem 2.24: Logic Gates

2.25   Refer to Figure 2.51. State which two gates of the three shown are DeMorgan equivalents of each other. Explain your choice.

FIGURE 2.51   Problem 2.25: Logic Gates

2.4 Logic Switches and LED Indicators

2.26 Sketch the circuit of a single-pole single-throw (SPST) switch used as a logic switch. Briefly explain how it works.

2.27 Refer to Figure 2.26 (logic pushbuttons). Should the normally open pushbutton be considered an active HIGH or active LOW device? Briefly explain your choice.

2.28 Should the normally closed pushbutton be considered an active HIGH or active LOW device? Why?

2.29 Briefly state what is required for an LED to illuminate.

2.30 Briefly state the relationship between the brightness of an LED and the current flowing through it. Why is a series resistor required?

2.31 Draw a circuit showing how an OR-gate output will illuminate an LED when the gate output is LOW. Assume the required series resistor is 470 Ω.

2.5 Enable and Inhibit Properties of Logic Gates

2.32 Draw the output waveform of the Exclusive NOR gate when a square waveform is applied to one input and

a. The other input is held LOW

b. The other input is held HIGH

How does this compare to the waveform that would appear at the output of an Exclusive OR gate under the same conditions?

2.33 Sketch the input waveforms represented by the following 32-bit sequences (use 1/4-inch graph paper, 1 square per bit. Spaces are provided for readability only):

a. 0000 0000 0000 1111 1111 1111 1111 0000

b. 1010 0111 0010 1011 0101 0011 1001 1011

Assume that these waveforms represent inputs to a logic gate. Sketch the waveform for gate output Y if the gate function is:

a. AND

b. OR

c. NAND

d. NOR

e. XOR

f. XNOR

2.34 Repeat Problem 2.33 for the waveforms shown in Figure 2.52.

FIGURE 2.52 Problem 2.34: Input Waveforms

2.35   The *A* and *B* waveforms shown in Figure 2.53 are inputs to an OR gate. Complete the sketch by drawing the waveform for output *Y*.

FIGURE 2.53   Problem 2.35: Waveforms

2.36   Repeat Problem 2.35 for a NOR gate.

2.37   Figure 2.54 shows a circuit that will make a lamp flash at 3 Hz when the gasoline level in a car's gas tank drops below a certain point. A float switch in the tank monitors the level of gasoline. What logic level must the float switch produce to make the light flash when the tank is approaching empty? Why?

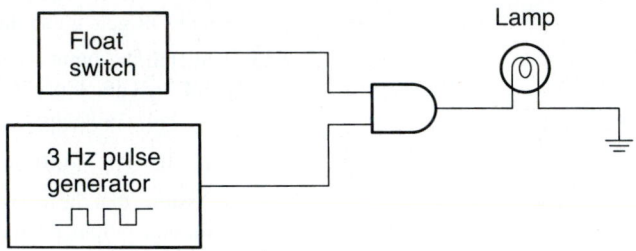

FIGURE 2.54   Problem 2.37: Gasoline Level Circuit

2.38   Repeat Problem 2.37 for the case where the AND gate is replaced by a NOR gate.

2.39   Will the circuit in Figure 2.54 work properly if the AND gate is replaced by an Exclusive OR gate? Why or why not?

2.40   Make a truth table for the tristate buffers shown in Figure 2.38. Indicate the high-impedance state by the notation "Hi-Z." How do the enable properties of these gates differ from gates such as AND and NAND?

2.6   Integrated Circuit Logic Gates

2.41   Name two logic families used to implement digital logic functions. How do they differ?

2.42   List the industry-standard numbers for a quadruple 2-input NAND gate in low-power Schottky TTL and high-speed CMOS technologies.

2.43   Repeat Problem 2.42 for a quadruple 2-input NOR gate. How does each numbering system differentiate between the NAND and NOR functions?

2.44   List six types of packaging that a logic gate could come in.

ANSWERS TO SECTION REVIEW PROBLEMS

2.1

2.1 AND: "*A* AND *B* AND *C* AND *D* must be HIGH to make *Y* HIGH."

2.2 OR: "*A* OR *B* OR *C* OR *D* must be HIGH to make *Y* HIGH."

2.2

2.3 XOR

2.4 NAND

2.5 NOR

2.6 XNOR

2.3

2.7 $Y = \overline{A\,B\,C\,D}$

2.4

2.8 When the switch is open, it provides a logic HIGH because of the pull-up resistor. A closed switch is LOW, due to the connection to ground.

2.5

2.9 An AND needs two HIGH inputs to make a HIGH output. If the Control input is LOW, the output can never be HIGH; the output remains LOW. An OR output is HIGH if one input is HIGH. If the Control input is HIGH, the output is always HIGH, regardless of the level at the Signal input. In both cases, the output is "stuck" at one level, signifying that the gate is inhibited.

2.6

2.10 DIP: Viewed from above, with the notch in the package away from you, pin 1 is on the left side at the far end. The pins are numbered counterclockwise from that point. PLCC: When viewed from above, with the cutoff corner at the top left, pin 1 is the center pin in the top row. The pins number counterclockwise from there.

KEY TERMS

Adjacent cell
Associative property
Bubble-to-bubble convention
Bus form
Cell
Combinational logic
Combinatorial logic
Commutative property
Distributive property
Don't care state
Double-rail inputs
Karnaugh map
Levels of gating
Logic diagram
Logic gate network
Maximum POS simplification
Maximum SOP simplification
Maxterm
Minterm
Octet
Order of precedence
Pair
Product-of-sums (POS)
Product term
Quad
Sequential logic
Sum-of-products (SOP)
Sum term
Synthesis

OUTLINE

3.1 Boolean Expressions, Logic Diagrams, and Truth Tables
3.2 Sum-of-Products and Product-of-Sums Forms
3.3 Theorems of Boolean Algebra
3.4 Simplifying SOP and POS Expressions
3.5 Simplification by the Karnaugh Map Method
3.6 Simplification by DeMorgan Equivalent Gates
3.7 Universal Property of NAND/NOR Gates
3.8 Practical Circuit Implementation in SSI Logic
3.9 Pulsed Operation of Logic Gates
3.10 A General Approach to Logic Circuit Design

CHAPTER OBJECTIVES

Upon successful completion of this chapter you will be able to:

- Explain the relationship between the Boolean expression, logic diagram, and truth table of a logic gate network and be able to derive any one from either of the other two.

- Draw logic gate networks in such a way as to cancel out internal inversions automatically (bubble-to-bubble convention).

- Write the sum of products (SOP) or product of sums (POS) forms of a Boolean equation.

- Use rules of Boolean algebra to simplify the Boolean expressions derived from logic diagrams and truth tables.

- Apply the Karnaugh map method to reduce Boolean expressions and logic circuits to their simplest forms.

- Use a graphical technique based on DeMorgan equivalent gates to simplify logic diagrams.

- Redraw a logic diagram using all-NAND or all-NOR implementations.

- Draw logic circuits that account for the practical limitations of commercially available logic gates.
- Predict the response of a logic circuit to a time-varying input signal.
- Apply analysis tools to convert word problems to Boolean equations for the purposes of logic circuit design.

Boolean Algebra and Combinational Logic

In Chapter 3, we will examine the rudiments of combinational logic. A combinational logic circuit is one in which two or more gates are connected together to combine several Boolean inputs. These circuits can be represented several ways: as a logic diagram, truth table, or Boolean expression.

A Boolean expression for a network of logic gates is often not in its simplest form. In such a case, we may be using more components than would be required for the job, so it is of benefit to us if we can simplify the Boolean expression. Several tools are available to us, such as Boolean algebra and a graphical technique known as Karnaugh mapping. We can also simplify the Boolean expression by taking care to draw the logic diagrams in such a way as to automatically eliminate inverting functions within the circuit.

Any combinational logic circuit can be implemented using only NAND or only NOR gates. This property can be useful for circuit simplification or for best use of available gate resources. When using commercially available logic gates, we must account for their limitations, such as availability of gates with a particular number of inputs and what to do with unused gates in a package. Sometimes, selection of a particular circuit configuration, such as NAND-NAND, as opposed to AND-OR, can make the practical implementation simpler.

Logic gate networks are often used to control the transmission of time-varying waveforms. We saw in Chapter 2 how this could be done with single gates. In this chapter, we will examine the behavior of logic gate networks with pulsed inputs.

One goal in the design of combinational logic circuits is to translate word problems into gate networks by deriving Boolean expressions for the networks. Once we have a network's Boolean expression, we can apply our tools of simplification and circuit synthesis to complete our design. We will examine a general method of analysis that allows us to derive that initial descriptive equation.

3.1 BOOLEAN EXPRESSIONS, LOGIC DIAGRAMS, AND TRUTH TABLES

■ KEY TERMS

Logic Gate Network   Two or more logic gates connected together.

Logic Diagram   A diagram, similar to a schematic, showing the connection of logic gates.

Combinational Logic   Digital circuitry in which an output is derived from the combination of inputs, independent of the order in which they are applied.

Combinatorial Logic   Another name for combinational logic.

Sequential Logic   Digital circuitry in which the output state of the circuit depends not only on the states of the inputs, but also on the sequence in which they reached their present states.

In Chapter 2, we examined the functions of single logic gates. However, most digital circuits require multiple gates. When two or more gates are connected together, they form a **logic gate network.** These networks can be described by a truth table, a **logic diagram** (i.e., a circuit diagram), or a Boolean expression. Any one of these can be derived from any other.

A digital circuit built from gates is called a **combinational** (or **combinatorial**) **logic** circuit. The output of a combinational circuit depends on the *combination* of inputs. The inputs can be applied in any sequence and still produce the same result. For example, an AND gate output will always be HIGH if all inputs are HIGH, regardless of the order in which they became HIGH. This is in contrast to **sequential logic,** in which sequence matters; a sequential logic output may have a different value with two identical sets of inputs if those inputs were applied in a different order. We will study sequential logic in a later chapter.

Boolean Expressions from Logic Diagrams

■ KEY TERMS

Bubble-to-Bubble Convention   The practice of drawing gates in a logic diagram so that inverting outputs connect to inverting inputs and noninverting outputs connect to noninverting inputs.

Order of Precedence   The sequence in which Boolean functions are performed, unless otherwise specified by parentheses.

Writing the Boolean expression of a logic gate network is similar to finding the expression for a single gate. The difference is that in a multiple gate network, the inputs will usually not consist of single variables, but compound expressions that represent outputs of previous gates.

These compound expressions are combined according to the same rules as single variables. In an OR gate, with inputs x and y, the output will always be $x + y$ regardless of whether x and y are single variables (e.g., $x = A$, $y = B$, output = $A + B$) or compound expressions (e.g., $x = AB$, $y = AC$, output = $AB + AC$).

Figure 3.1 shows a simple logic gate network, consisting of a single AND and a single OR gate. The AND gate combines inputs A and B to give the output expression AB. The OR combines the AND function and input C to yield the compound expression $AB + C$.

FIGURE 3.1   Boolean Expression from a Gate Network

EXAMPLE 3.1

Derive the Boolean expression of the logic gate network shown in Figure 3.2a.

a. Logic gate network

$Y = AB + \overline{CD}$

b. Boolean expression from logic gate network

FIGURE 3.2 Example 3.1: Boolean Expression from a Logic Gate Network

■ **Solution**

Figure 3.2b shows the gate network with the output terms indicated for each gate. The AND and NAND functions are combined in an OR function to yield the output expression:

$$Y = AB + \overline{CD}$$

The Boolean expression in Example 3.1 includes a NAND function. It is possible to draw the NAND in its DeMorgan equivalent form. If we choose the gate symbols so that outputs with bubbles connect to inputs with bubbles, we will not have bars over groups of variables, except possibly one bar over the entire function. In a circuit with many inverting functions (NANDs and NORs), this results in a cleaner notation and often a clearer idea of the function of the circuit. We will follow this notation, which we will refer to as the **bubble-to-bubble convention,** as much as possible. A more detailed treatment of the bubble-to-bubble convention follows in Section 3.6 of this chapter.

EXAMPLE 3.2

Redraw the circuit in Figure 3.2 to conform to the bubble-to-bubble convention. Write the Boolean expression of the new logic diagram.

$Y = AB + \overline{C} + \overline{D}$

FIGURE 3.3 Example 3.2: Using DeMorgan Equivalents to Simplify a Circuit

■ **Solution**

Figure 3.3 shows the new circuit. The NAND has been converted to its DeMorgan equivalent so that its active-HIGH output drives an active-HIGH input on the OR gate. The new Boolean expression is $Y = AB + \overline{C} + \overline{D}$.

a. No parentheses required (AND, then OR)

b. Parentheses required (OR, then AND)

FIGURE 3.4   Order of Precedence

Boolean functions are governed by an **order of precedence.** Unless otherwise specified, AND functions are performed first, followed by ORs. This order results in a form similar to that of linear algebra, where multiplication is performed before addition, unless otherwise specified.

Figure 3.4 shows two logic diagrams, one whose Boolean expression requires parentheses and one that does not.

The AND functions in Figure 3.4a are evaluated first, eliminating the need for parentheses in the output expression. The expression for Figure 3.4b requires parentheses since the ORs are evaluated first.

Write the Boolean expression for the logic diagrams in Figure 3.5.

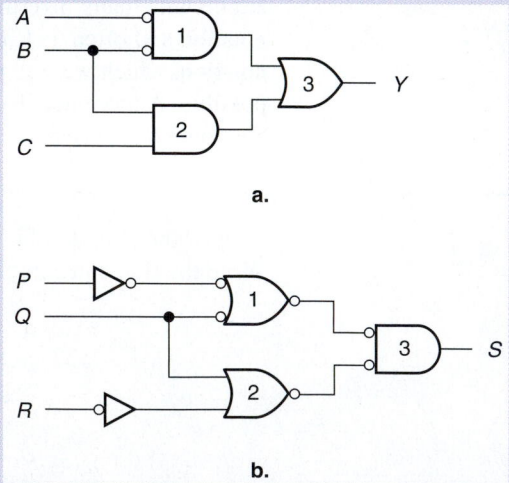

FIGURE 3.5   Example 3.3: Order of Precedence

■ **Solution**

Examine the output of each gate and combine the resultant terms as required.

Figure 3.5a: Gate 1: $\overline{A} \cdot \overline{B}$

Gate 2: $B \cdot C$

Gate 3: $Y =$ Gate 1 $+$ Gate 2 $= \overline{A} \cdot \overline{B} + B \cdot C$

Figure 3.5b: Gate 1: $\overline{\overline{\overline{P}}+Q} = \overline{P+\overline{Q}}$

Gate 2: $\overline{Q+\overline{R}}$

Gate 3: $S = \overline{\text{Gate 1} \cdot \text{Gate 2}} = \overline{(\overline{\overline{P}+\overline{Q}})(\overline{Q+\overline{R}})} = (P+\overline{Q})(Q+\overline{R})$

Note that when two bubbles touch, they cancel out, as in the doubly inverted P input or the connection between the outputs of gates 1 and 2 and the inputs of gate 3. *In the resultant Boolean expression, bars of the same length cancel; bars of unequal length do not.*

■ SECTION 3.1A REVIEW PROBLEM

3.1 Write the Boolean expression for the logic diagrams in Figure 3.6, paying attention to the rules of order of precedence.

a.

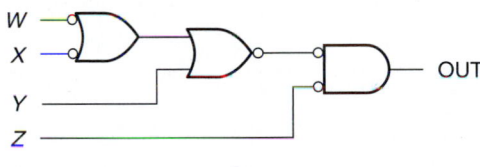

b.

FIGURE 3.6 Section Review Problem 3.1

Logic Diagrams from Boolean Expressions

■ KEY TERMS

Levels of Gating   The number of gates through which a signal must pass from input to output of a logic gate network.

Double-Rail Inputs   Boolean input variables that are available to a circuit in both true and complement form.

Synthesis   The process of creating a logic circuit from a description such as a Boolean equation or truth table.

We can derive a logic diagram from a Boolean expression by applying the order of precedence rules. We examine an expression to create the first **level of gating** from the circuit inputs, then combine the output functions of the first level in the second level gates, and so forth. Input inverters are often not counted as a gating level, as we usually assume that each variable is available in both true (noninverted) and complement (inverted) form. When input variables are available to a circuit in true and complement form, we refer to them as **double-rail inputs.**

The first level usually will be AND gates if no parentheses are present, and OR gates if parentheses are used. (Not always, however; parentheses merely tell us which functions to **synthesize** first.) Although we will try to eliminate bars over groups of variables by use of DeMorgan's theorems and the bubble-to-bubble convention, we should recognize that a bar over a group of variables is the same as having those variables in parentheses.

Let us examine the Boolean expression $Y = AC + BD + AD$. Order of precedence tells us that we synthesize the AND functions first. This yields three 2-input AND gates, with outputs AC, BD, and AD, as shown in Figure 3.7a. In the next step, we combine these AND functions in a 3-input OR gate, as shown in Figure 3.7b.

When the expression has OR functions in parentheses, we synthesize the ORs first, as for the expression $Y = (A + B)(A + C + D)(B + C)$. Figure 3.8 shows this process. In the first step, we synthesize three OR gates for the terms $(A + B)$, $(A + C + D)$, and $(B + C)$. We then combine these terms in a 3-input AND gate.

a. ANDs first

a. ORs first

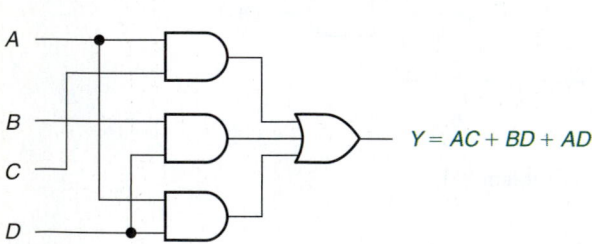

b. Combine ANDs in an OR gate

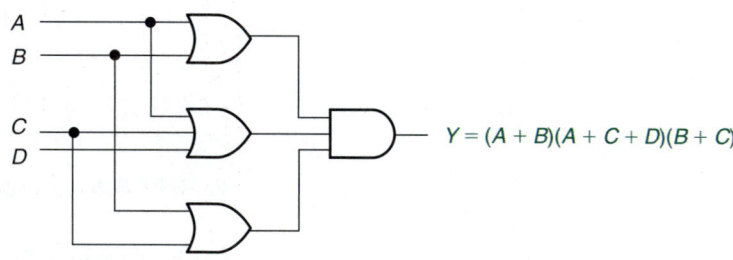

b. Combine ORs in an AND gate

FIGURE 3.7   Logic Diagram for $Y = AC + BD + AD$

FIGURE 3.8   Logic Diagram for $Y = (A + B) (A + C + D) (B + C)$

 EXAMPLE 3.4

Synthesize the logic diagrams for the following Boolean expressions:

1. $P = P = Q\overline{RS} + \overline{S}T$
2. $X = (W + Z + Y)\overline{V} + (\overline{W} + V)\overline{Y}$

■ **Solution**

1. Recall that a bar over two variables acts like parentheses. Thus the $Q\overline{RS}$ term is synthesized from a NAND, then an AND, as shown in Figure 3.9a. Also shown is the second AND term, $\overline{S}T$.

   Figure 3.9b shows the terms combined in an OR gate.

2. Figure 3.10 shows the synthesis of the second logic diagram in three stages. Figure 3.10a shows how the circuit inputs are first combined in two OR

gates. We do this first because the ORs are in parentheses. In Figure 3.10b, each of these functions is combined in an AND gate, according to the normal order of precedence. The AND outputs are combined in a final OR function, as shown in Figure 3.10c.

FIGURE 3.9 Example 3.4: Logic Diagram of $P = Q\overline{RS} + \overline{S}T$

a. Combine inputs (NAND, then AND)

b. First and second level gates combined in an OR

a. ORs first (parentheses)

b. Combine with ANDs (order of precedence)

c. Find output (OR)

FIGURE 3.10 Example 3.4: Logic Diagram for $X = (W + Z + Y)\overline{V} + (\overline{W} + V)\overline{Y}$

EXAMPLE 3.5

Use DeMorgan's theorem to modify the Boolean equation in part 1 of Example 3.4 so that there is no bar over any group of variables. Redraw Figure 3.9b to reflect the change.

■ **Solution**

$$P = Q\overline{RS} + \overline{S}T = Q(\overline{R} + \overline{S}) + \overline{S}T$$

Figure 3.11a shows the modified logic diagram. The levels of gating could be further reduced from three to two (not counting input inverters) by "multiplying through" the parentheses to yield the expression:

$$P = Q\overline{R} + Q\overline{S} + \overline{S}T$$

Figure 3.11b shows the logic diagram for this form. We will examine this simplification procedure more formally in a later section of this chapter.

a. Logic diagram of $P = Q(\overline{R} + \overline{S}) + \overline{S}T$

b. Logic diagram of $P = Q\overline{R} + Q\overline{S} + \overline{S}T$

FIGURE 3.11   Example 3.5: Reworking Figure 3.9b

Truth Tables from Logic Diagrams or Boolean Expressions

There are two basic ways to find a truth table from a logic diagram. We can examine the output of each gate in the circuit and develop its truth table. We then use our knowledge of gate properties to combine these intermediate truth tables into the final output truth table. Alternatively, we can develop a Boolean expression for the logic diagram and by examining the expression fill in the truth table in a single step. The former method is more thorough and probably easier to understand when you are learning the technique. The latter method is more efficient, but requires some practice and experience. We will look at both.

Examine the logic diagram in Figure 3.12. Since there are three binary inputs, there will be eight ways those inputs can be combined. Thus, we start by making an 8-line truth table, as in Table 3.1.

The OR gate output will describe the function of the whole circuit. In order to assess the OR function, we must first evaluate the AND output. We add a column to the truth table for the AND gate and look for the lines in the table where both A AND B equal logic 1 (in this case, the last two rows). For these lines, we write a 1 in the AB column. Next, we look at the values in column C and the AB column. If there is a 1 in either column, we write a 1 in the column for the final output.

TABLE 3.1   Truth Table for Figure 3.12

A	B	C	AB	$AB + C$
0	0	0	0	0
0	0	1	0	1
0	1	0	0	0
0	1	1	0	1
1	0	0	0	0
1	0	1	0	1
1	1	0	1	1
1	1	1	1	1

FIGURE 3.12 Logic Diagram for $AB + C$

EXAMPLE 3.6

Derive the truth table for the logic diagram shown in Figure 3.13.

FIGURE 3.13 Example 3.6: Logic Diagram

■ **Solution**

The Boolean equation for Figure 3.13 is $(\overline{A} + \overline{B})(A + C)$. We will create a column for each input variable and for each term in parentheses, as well as a column for the final output. Table 3.2 shows the result. For the lines where A OR B is 0, we write a 1 in the $(\overline{A} + \overline{B})$ column. Where A OR C is 1, we write a 1 in the $(A + C)$ column. For the lines where there is a 1 in both the $(\overline{A} + \overline{B})$ AND $(A + C)$ columns, we write a 1 in the final output column.

TABLE 3.2 Truth Table for Figure 3.13

A	B	C	$(\overline{A} + \overline{B})$	$(A + C)$	$(\overline{A} + \overline{B})\ (A + C)$
0	0	0	1	0	0
0	0	1	1	1	1
0	1	0	1	0	0
0	1	1	1	1	1
1	0	0	1	1	1
1	0	1	1	1	1
1	1	0	0	1	0
1	1	1	0	1	0

Another approach to finding a truth table involves analysis of the Boolean expression of a logic diagram. The logic diagram in Figure 3.14 can be described by the Boolean expression $Y = \overline{A}BC + \overline{A}\ \overline{C} + \overline{B}\ \overline{D}$.

We can examine the Boolean expression to determine that the final output of the circuit will be HIGH under one of the following conditions:

$A = 0$ AND $B = 1$ AND $C = 1$
OR $A = 0$ AND $C = 0$
OR $B = 0$ AND $D = 0$.

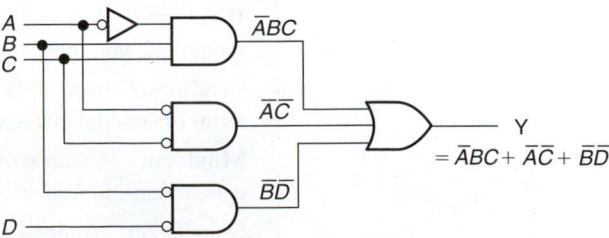

FIGURE 3.14 Logic Diagram

All we have to do is look for these conditions in the truth table and write a 1 in the output column whenever a condition is satisfied. Table 3.3 shows the result of this analysis with each line indicating which term, or terms, contribute to the HIGH output.

TABLE 3.3   Truth Table for Figure 3.14

A	B	C	D	Y	terms
0	0	0	0	1	$\overline{A}\,\overline{C},\ \overline{B}\,\overline{D}$
0	0	0	1	1	$\overline{A}\,\overline{C}$
0	0	1	0	1	$\overline{B}\,\overline{D}$
0	0	1	1	0	
0	1	0	0	1	$\overline{A}\,\overline{C}$
0	1	0	1	1	$\overline{A}\,\overline{C}$
0	1	1	0	1	$\overline{A}\,BC$
0	1	1	1	1	$\overline{A}\,BC$
1	0	0	0	1	$\overline{B}\,\overline{D}$
1	0	0	1	0	
1	0	1	0	1	$\overline{B}\,\overline{D}$
1	0	1	1	0	
1	1	0	0	0	
1	1	0	1	0	
1	1	1	0	0	
1	1	1	1	0	

■ **SECTION 3.1B REVIEW PROBLEM**

3.2   Find the truth table for the logic diagram shown in Figure 3.15.

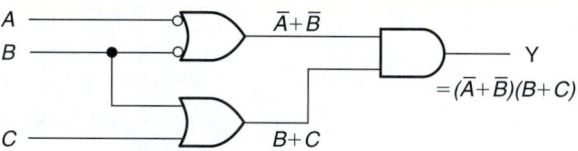

FIGURE 3.15   Section Review Problem 3.2

3.2 SUM-OF-PRODUCTS AND PRODUCT-OF-SUMS FORMS

■ **KEY TERMS**

Minterm   A product term in a Boolean expression where all possible variables appear once in true or complement form (e.g., $\overline{A}\,\overline{B}\,\overline{C}$; $A\,\overline{B}\,\overline{C}$).

Product Term   A term in a Boolean expression where one or more true or complement variables are ANDed (e.g., $\overline{A}\,\overline{C}$).

Sum-of-Products (SOP)   A type of Boolean expression where several product terms are summed (ORed) together (e.g., $\overline{A}\,B\,\overline{C} + \overline{A}\,\overline{B}\,C + A\,B\,C$).

Bus Form   A way of drawing a logic diagram so that each true and complement input variable is available along a continuous conductor called a bus.

Product-of-Sums (POS)   A type of Boolean expression where several sum terms are multiplied (ANDed) together (e.g., $(\overline{A}+\overline{B}+C)(A+\overline{B}+\overline{C})(\overline{A}+\overline{B}+\overline{C})$).

Maxterm   A sum term in a Boolean expression where all possible variables appear once, in true or complement form (e.g., $(\overline{A}+\overline{B}+C)$; $(A+\overline{B}+C)$).

Sum Term   A term in a Boolean expression where one or more true or complement variables are ORed (e.g., $\overline{A}+B+\overline{D}$).

Suppose we have an unknown digital circuit, represented by the block in Figure 3.16. All we know is which terminals are inputs, which are outputs, and how to connect the power supply. Given only that information, we can find the Boolean expression of the output.

The first thing to do is find the truth table by applying all possible input combinations in binary order and reading the output for each one. Suppose the unknown circuit in Figure 3.16 yields the truth table shown in Table 3.4.

FIGURE 3.16 Digital Circuit with Unknown Function

TABLE 3.4 Truth Table for Figure 3.16

A	B	C	Y
0	0	0	1
0	0	1	0
0	1	0	0
0	1	1	1
1	0	0	1
1	0	1	0
1	1	0	0
1	1	1	0

The truth table output is HIGH for three conditions:

1. When A AND B AND C are all LOW, OR
2. When A is LOW AND B AND C are HIGH, OR
3. When A is HIGH AND B AND C are LOW.

Each of those conditions represents a **minterm** in the output Boolean expression. A minterm is a **product term** (AND term) that includes all variables *(A, B, C)* in true or complement form. The variables in a minterm are written in complement form (with an inversion bar) if the variable is a 0 in the corresponding line of the truth table, and in true form (no bar) if the variable is a 1. The minterms are:

1. $\overline{A}\,\overline{B}\,\overline{C}$
2. $\overline{A}\,B\,C$
3. $A\,\overline{B}\,\overline{C}$

Since condition 1 OR condition 2 OR condition 3 produces a HIGH output from the circuit, the Boolean function Y consists of all three minterms summed (ORed) together, as follows:

$$Y = \overline{A}\,\overline{B}\,\overline{C} + \overline{A}\,B\,C + A\,\overline{B}\,\overline{C}$$

This expression is in a standard form called **sum-of-products (SOP)** form. Figure 3.17 shows the equivalent logic circuit.

FIGURE 3.17 Logic Circuit for $Y = \overline{A}\,\overline{B}\,\overline{C} + \overline{A}\,B\,C + A\,\overline{B}\,\overline{C}$

The inputs A, B, and C and their complements are shown in **bus form.** Each variable is available, in true or complement form, at any point along a conductor. This is a useful, uncluttered notation for circuits that require several of the input variables more than once. The inverters are shown with the bubbles on their inputs to indicate that the complement line is looking for a LOW input to activate it. The true line (no inverter) is looking for a HIGH.

> **NOTE . . .**
>
> We can derive an SOP expression from a truth table as follows:
>
> 1. Every line on the truth table that has a HIGH output corresponds to a minterm in the truth table's Boolean expression.
> 2. Write all truth table variables for every minterm in true or complement form. If a variable is 0, write it in complement form (with a bar over it); if it is 1, write it in true form (no bar).
> 3. Combine all minterms in an OR function.

EXAMPLE 3.7

Tables 3.5 and 3.6 show the truth tables for the Exclusive OR and the Exclusive NOR functions. Derive the sum-of-products expression for each of these functions and draw the logic diagram for each one.

TABLE 3.5 XOR Truth Table

A	B	$A \oplus B$
0	0	0
0	1	1
1	0	1
1	1	0

TABLE 3.6 XNOR Truth Table

A	B	$\overline{A \oplus B}$
0	0	1
0	1	0
1	0	0
1	1	1

■ **Solution**

XOR: The truth table yields two product terms: $\overline{A}B$ and $A\overline{B}$. Thus, the SOP form of the XOR function is $A \oplus = B = \overline{A}B + A\overline{B}$. Figure 3.18 shows the logic diagram for this equation.

FIGURE 3.18 Example 3.7: SOP Form of XOR Function

XNOR: The product terms for this function are: $\overline{A}\,\overline{B}$ and AB. The SOP form of the XNOR function is $A \oplus B = \overline{A}\,\overline{B} + AB$. The logic diagram in Figure 3.19 represents the XNOR function.

$$\overline{A \oplus B} = \overline{A}\,\overline{B} + AB$$

FIGURE 3.19 Example 3.7: SOP Form of XNOR Function

We can also find the Boolean function of a truth table in **product-of-sums (POS)** form. The product-of-sums form of a Boolean expression consists of a number of **maxterms** (i.e., **sum terms,** OR terms, containing all variables in true or complement form) that are ANDed together. To find the POS form of Y, we will find the SOP expression for $\overline{Y}$ and apply DeMorgan's theorems.

Recall DeMorgan's theorems:

$$\overline{x + y + z} = \overline{x}\ \overline{y}\ \overline{z}$$

$$\overline{x\, y\, z} = \overline{x} + \overline{y} + \overline{z}$$

When the theorems were introduced, they were presented as two-variable theorems, but in fact they are valid for any number of variables.

Let's reexamine Table 3.4. To find the sum-of-products expression for Y, we wrote a minterm for each line where $Y = 1$. To find the SOP expression for $\overline{Y}$, *we must write a minterm for each line where $Y = 0$.* Variables A, B, and C must appear in each minterm, in true or complement form. A variable is in complement form (with a bar over the top) if its value is 0 in that minterm, and it is in true form (no bar) if its value is 1.

We get the following minterms for $\overline{Y}$:

$$\overline{A}\ \overline{B}\ C$$
$$\overline{A}\ B\ \overline{C}$$
$$A\ \overline{B}\ C$$
$$A\ B\ \overline{C}$$
$$A\ B\ C$$

Thus, the SOP form of $\overline{Y}$ is

$$\overline{Y} = \overline{A}\ \overline{B}\ C + \overline{A}\ B\ \overline{C} + A\ \overline{B}\ C + A\ B\ \overline{C} + A\ B\ C$$

To get Y in POS form, we must invert both sides of the above expression and apply DeMorgan's theorems to the right-hand side.

$$Y = \overline{\overline{Y}} = \overline{\overline{A}\ \overline{B}\ C + \overline{A}\ B\ \overline{C} + A\ \overline{B}\ C + A\ B\ \overline{C} + A\ B\ C}$$
$$= (\overline{\overline{A}\ \overline{B}\ C})(\overline{\overline{A}\ B\ \overline{C}})(\overline{A\ \overline{B}\ C})(\overline{A\ B\ \overline{C}})(\overline{A\ B\ C})$$
$$= (A + B + \overline{C})(A + \overline{B} + C)(\overline{A} + B + \overline{C})(\overline{A} + \overline{B} + C)(\overline{A} + \overline{B} + \overline{C})$$

FIGURE 3.20 Logic Circuit for $Y = (A + B + \bar{C})(A + \bar{B} + C)(\bar{A} + B + \bar{C})(\bar{A} + \bar{B} + C)(\bar{A} + \bar{B} + \bar{C})$

This Boolean expression can be implemented by the logic circuit in Figure 3.20.

We don't have to go through the whole process just outlined every time we want to find the POS form of a function. We can find it directly from the truth table, using the procedure summarized in the following Note. Use this procedure to find the POS form of the expression given by Table 3.4. The terms in this expression are the same as those derived by DeMorgan's theorem.

NOTE . . .

Deriving a POS expression from a truth table:

1. Every line on the truth table that has a LOW output corresponds to a maxterm in the truth table's Boolean expression.
2. Write all truth table variables for every maxterm in true or complement form. If a variable is 1, write it in complement form (with a bar over it); if it is 0, write it in true form (no bar).
3. Combine all maxterms in an AND function.

Note that these steps are all opposite to those used to find the SOP form of the Boolean expression.

EXAMPLE 3.8

Find the Boolean expression, in both SOP and POS forms, for the logic function represented by Table 3.7. Draw the logic circuit for each form.

■ **Solution**

All minterms (for SOP form) and maxterms (for POS form) are shown in the last two columns of Table 3.5.

Boolean Expressions:

SOP form:

$$Y = \bar{A}\,\bar{B}\,\bar{C}\,\bar{D} + \bar{A}\,\bar{B}\,\bar{C}\,D + \bar{A}\,\bar{B}\,C\,D + A\,\bar{B}\,\bar{C}\,\bar{D} + A\,\bar{B}\,C\,\bar{D} + A\,B\,\bar{C}\,\bar{D}$$
$$+ A\,B\,\bar{C}\,D + A\,B\,C\,\bar{D}$$

POS form:

$$Y = (A + B + \bar{C} + D)(A + \bar{B} + C + D)(A + \bar{B} + C + \bar{D})(A + \bar{B} + \bar{C} + D)$$
$$(A + \bar{B} + \bar{C} + \bar{D})(\bar{A} + B + C + \bar{D})(\bar{A} + B + \bar{C} + \bar{D})$$
$$(\bar{A} + \bar{B} + \bar{C} + \bar{D})$$

The logic circuits are shown in Figures 3.21 and 3.22.

TABLE 3.7 Truth Table for Example 3.8 (with Minterms and Maxterms)

A	B	C	D	Y	Minterms	Maxterms
0	0	0	0	1	$\bar{A}\,\bar{B}\,\bar{C}\,\bar{D}$	
0	0	0	1	1	$\bar{A}\,\bar{B}\,\bar{C}\,D$	
0	0	1	0	0		$A + B + \bar{C} + D$
0	0	1	1	1	$\bar{A}\,\bar{B}\,C\,D$	
0	1	0	0	0		$A + \bar{B} + C + D$
0	1	0	1	0		$A + \bar{B} + C + \bar{D}$
0	1	1	0	0		$A + \bar{B} + \bar{C} + D$
0	1	1	1	0		$A + \bar{B} + \bar{C} + \bar{D}$
1	0	0	0	1	$A\,\bar{B}\,\bar{C}\,\bar{D}$	
1	0	0	1	0		$\bar{A} + B + C + \bar{D}$
1	0	1	0	1	$A\,\bar{B}\,C\,\bar{D}$	
1	0	1	1	0		$\bar{A} + B + \bar{C} + \bar{D}$
1	1	0	0	1	$A\,B\,\bar{C}\,\bar{D}$	
1	1	0	1	1	$A\,B\,\bar{C}\,D$	
1	1	1	0	1	$A\,B\,C\,\bar{D}$	
1	1	1	1	0		$\bar{A} + \bar{B} + \bar{C} + \bar{D}$

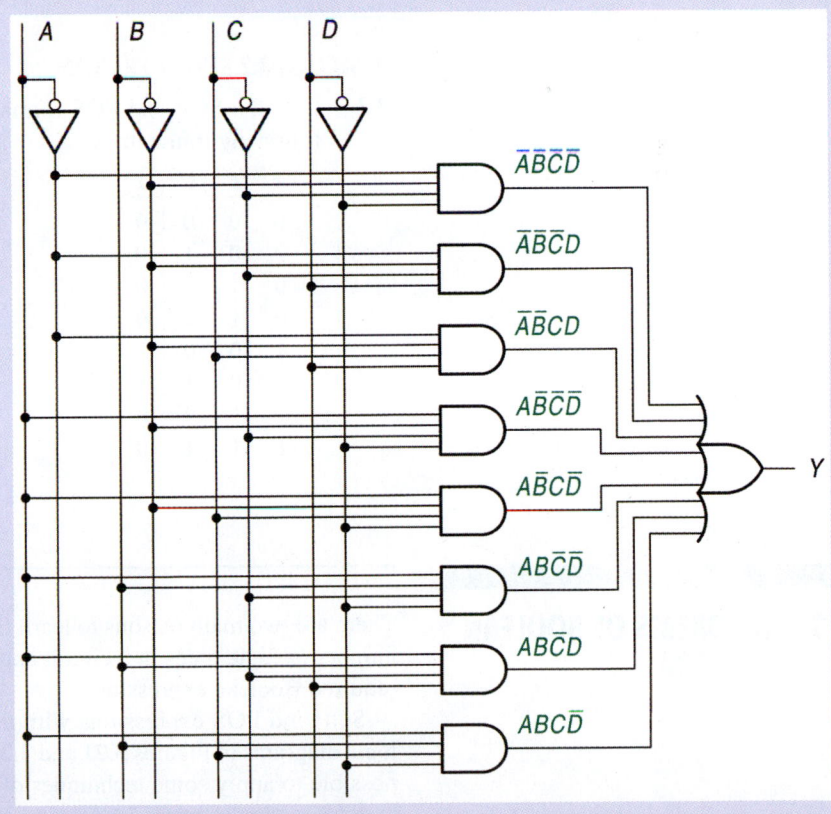

FIGURE 3.21 Example 3.8: SOP Form

FIGURE 3.22   Example 3.8: POS Form

■ **SECTION 3.2 REVIEW PROBLEM**

3.3   Find the SOP and POS forms of the Boolean functions represented by the following truth tables.

a.	A	B	C	Y
	0	0	0	0
	0	0	1	0
	0	1	0	0
	0	1	1	0
	1	0	0	1
	1	0	1	1
	1	1	0	0
	1	1	1	0

b.	A	B	C	Y
	0	0	0	1
	0	0	1	0
	0	1	0	0
	0	1	1	0
	1	0	0	1
	1	0	1	1
	1	1	0	1
	1	1	1	0

3.3   THEOREMS OF BOOLEAN ALGEBRA

There are two main reasons to learn Boolean algebra: to learn how to minimize the number of logic gates in a circuit and to best fit the available logic to implement a required Boolean expression.

SOP and POS expressions with many terms, such as those represented by the logic diagrams in Figures 3.21 and 3.22, are seldom in their simplest form. It is often possible to apply some techniques of Boolean algebra to derive a simpler form of expression that requires fewer gates to implement.

For example, the logic circuit in Figure 3.21 requires eight 4-input AND gates and an 8-input OR gate. Using Boolean algebra, we can reduce its Boolean expression to $Y = A\bar{D} + \bar{A}\,\bar{B}\,\bar{C} + \bar{A}\,\bar{B}\,D + A\,B\,\bar{C}$. This form can be implemented with four AND gates and a 4-input OR. You will use a simplification technique for this example in an end-of-chapter problem.

Sometimes, due to the devices we have available, we must implement a logic circuit in SOP form, even if the POS form of the equation would yield a circuit with fewer gates. A common example is that of a programmable logic device (PLD). One type of PLD is a device whose internal circuits are based on an SOP array that can be programmed by the user to make any desired logic function. In such a device, within certain limits, the number of gates in a circuit is less important than their configuration.

For example, one theorem of Boolean algebra states that $(x + y)(w + z) = xw + xz + yw + yz$. This theorem converts a POS network to an SOP. The POS circuit requires two 2-input OR gates and one 2-input AND gate. The SOP requires four 2-input AND gates and one 4-input OR. Even so, it may still be necessary to use the SOP form because of the constraints of the device we use to implement it. A PLD with an SOP array would require the SOP form. The internal configuration of the PLD may be such that the number of gates in this circuit is not a limiting factor.

We will study PLDs and their applications in some detail in later chapters of this book. In the meantime, let us examine some basic rules of Boolean algebra.

Commutative, Associative, and Distributive Properties

> ### ■ KEY TERMS
>
> **Commutative Property** A mathematical operation is commutative if it can be applied to its operands in any order without affecting the result. For example, addition is commutative ($a + b = b + a$), but subtraction is not ($a - b \neq b - a$).
>
> **Associative Property** A mathematical function is associative if its operands can be grouped in any order without affecting the result. For example, addition is associative ($(a + b) + c = a + (b + c)$), but subtraction is not ($(a - b) - c \neq a - (b - c)$).
>
> **Distributive Property** Full name: distributive property of multiplication over addition. The property that allows us to distribute ("multiply through") an AND across several OR functions. For example, $a(b + c) = ab + ac$.

AND and OR functions are both **commutative** and **associative**. The commutative property states that AND and OR operations are independent of input order. For inputs x and y,

Theorem 1 $xy = yx$
and

Theorem 2 $x + y = y + x$
The associative property allows us to perform several two-input AND or OR functions in any order. In other words,

Theorem 3 $(xy)z = x(yz) = (xz)y$
and

Theorem 4 $(x + y) + z = x + (y + z) = (x + z) + y$
The **distributive** property allows us to "multiply through" an AND function across several OR functions. For example,

Theorem 5 $x(y + z) = xy + xz$
and

Theorem 6 $(x + y)(w + z) = xw + xz + yw + yz$

Figure 3.23 shows the logic gate equivalents of Theorems 5 and 6.

a. $x(y + z) = xy + xz$ (Theorem 5)

b. $(x + y)(w + z) = xw + xz + yw + yz$ (Theorem 6)

FIGURE 3.23   Distributive Properties

EXAMPLE 3.9

Find the Boolean expression of the POS circuit in Figure 3.24a. Apply the distributive property to transform the circuit to an SOP form.

a. POS form

FIGURE 3.24   Example 3.9: Distributive Property

b. SOP form

■ **Solution**

The Boolean expression for Figure 3.24a is $Y = (\overline{A} + B)(\overline{C} + D)$. Using the distributive property, we get the expression $Y = \overline{A}\,\overline{C} + B\overline{C} + \overline{A}\,D + BD$. The logic diagram for this expression is shown in Figure 3.24b.

In Example 3.9, we see that the distributive property can be used to convert a POS circuit to SOP or vice versa. In this case, the circuit was not simplified, just transformed. This type of transformation is useful for fitting a Boolean expression to an SOP-based programmable logic device, as is the transformation in Example 3.10.

EXAMPLE 3.10

Write the Boolean expression for the circuit in Figure 3.25a. Use the distributive property to convert this to an SOP circuit.

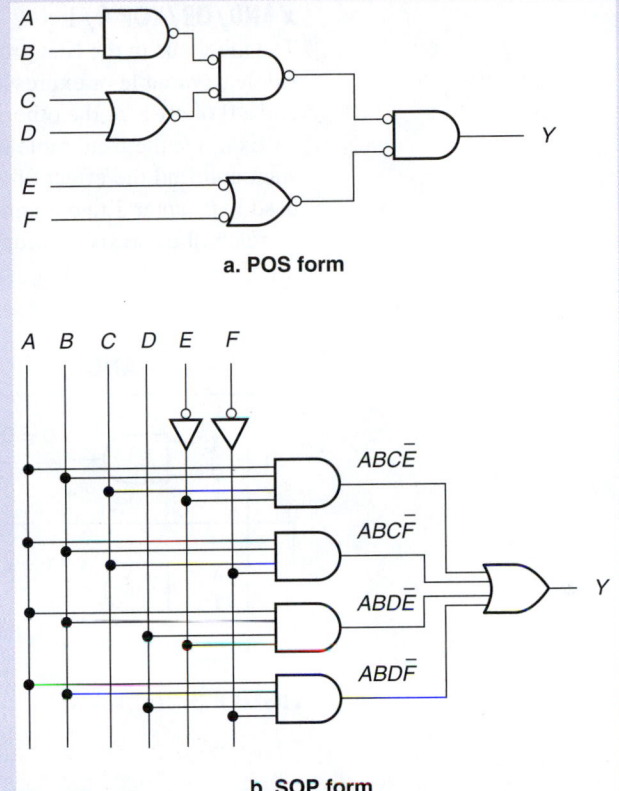

a. POS form

b. SOP form

FIGURE 3.25 Example 3.10: Distributive Property

■ Solution

The Boolean expression for Figure 3.25a is $AB(C + D)(\bar{E} + \bar{F})$. The distributive property can be applied in two stages:

$$Y = (ABC + ABD)(\bar{E} + \bar{F})$$

$$= ABC\bar{E} + ABC\bar{F} + ABD\bar{E} + ABD\bar{F}$$

The logic diagram for this equation is shown in Figure 3.25b. This results in a network that is "wider" (more gates on one level), but also "flatter" (fewer levels). The advantage of the second circuit is that signals would pass through the network faster, since it has fewer levels of gating. Also, since the second circuit has an equal-length path for most signals, they will tend to arrive at the output at the same time. A circuit with different-length signal paths, such as in Figure 3.25a, is vulnerable to "skew," a phenomenon that results when signals arrive at an output at different times, possibly causing momentary unwanted output states ("glitches").

Single-Variable Theorems

There are thirteen theorems that can be used to manipulate a single variable in a Boolean expression. An easy way to remember these theorems is to divide them into three groups:

1. Six theorems: x AND/OR/XOR 0/1
2. Six theorems: x AND/OR/XOR $x/\bar{x}$
3. One theorem: Double Inversion

x AND/OR/XOR 0/1

The theorems in the first group can be generated by asking what happens when x, a Boolean variable or expression, is at one input of an AND, an OR, or an XOR gate and a 0 or a 1 is at the other.

Examine the truth table of the gate in question. Hold one input of the gate constant and find the effect of the other on the output. This is the same procedure we used in Chapter 2 to examine the enable/inhibit properties of logic gates.

Each of these six theorems can be represented by a logic gate, as shown in Figure 3.26.

	AND	**OR**	**XOR**
0	$x \cdot 0 = 0$	$x + 0 = x$	$x \oplus 0 = x$
1	$x \cdot 1 = x$	$x + 1 = 1$	$x \oplus 1 = \bar{x}$

FIGURE 3.26 x AND/OR/XOR 0/1

Theorem 7 $x \cdot 0 = 0$

$x \cdot 0$:

A	x	Y
0	0	0
0	1	0
1	0	0
1	1	1

If $x = 0$, $Y = 0$

If $x = 1$, $Y = 0$

(Can never have both inputs HIGH, therefore output is always LOW.)

Theorem 8 $x + 0 = x$

$x + 0$:

A	x	Y
0	0	0
0	1	1
1	0	1
1	1	1

If $x = 0$, $Y = 0$

If $x = 1$, $Y = 1$

(LOW input enables OR gate.)

Theorem 9 $x \oplus 0 = x$

$x \oplus 0$:

A	x	Y
0	0	0
0	1	1
1	0	1
1	1	0

If $x = 0$, $Y = 0$

If $x = 1$, $Y = 1$

(XOR acts as a noninverting buffer.)

Theorem 10 $x \cdot 1 = x$

$x \cdot 1$:

A	x	Y
0	0	0
0	1	0
1	0	0
1	1	1

If $x = 0$, $Y = 0$

If $x = 1$, $Y = 1$

(HIGH input enables AND gate.)

Theorem 11 $x + 1 = 1$

$x + 1$:

A	x	Y
0	0	0
0	1	1
1	0	1
1	1	1

If $x = 0$, $Y = 1$

If $x = 1$, $Y = 1$

(One input always HIGH, therefore output is always HIGH.)

Theorem 12 $x \oplus 1 = \bar{x}$

$x \oplus 1$:

A	x	Y
0	0	0
0	1	1
1	0	1
1	1	0

If $x = 0$, $Y = 1$

If $x = 1$, $Y = 0$

(XOR acts as an inverting buffer.)

x AND/OR/XOR *x*/*x̄*

Six theorems are generated by combining a Boolean variable or expression, x, with itself or its complement in an AND, an OR, or an XOR function.

FIGURE 3.27   x AND/OR/XOR x/x̄

Again, we can use the AND, OR, and XOR truth tables. For the first three theorems, we look only at the lines where both inputs are the same. For the other three, we use the lines where the inputs are different.

Figure 3.27 shows the logic gates that represent these theorems.

Theorem 13 $x \cdot x = x$

$x \cdot x$:

A	x	Y
0	0	0
0	1	0
1	0	0
1	1	1

If $x = 0$, $Y = 0$

If $x = 1$, $Y = 1$

Theorem 14 $x + x = x$

$x + x$:

A	x	Y
0	0	0
0	1	1
1	0	1
1	1	1

If $x = 0$, $Y = 0$

If $x = 1$, $Y = 1$

Theorem 15 $x \oplus x = 0$

$x \oplus x$:

A	x	Y
0	0	0
0	1	1
1	0	1
1	1	0

If $x = 0$, $Y = 0$

If $x = 1$, $Y = 0$

(Output is LOW if neither input is HIGH or if both are.)

Theorem 16 $x \cdot \bar{x} = 0$

$x \cdot \bar{x}$:

A	x	Y
0	0	0
0	1	0
1	0	0
1	1	1

If $x = 0$, $Y = 0$

If $x = 1$, $Y = 0$

(Since inputs are opposite, can never have both HIGH, output is always LOW.)

Theorem 17 $x + \bar{x} = 1$

$x + \bar{x}$:

A	x	Y
0	0	0
0	1	1
1	0	1
1	1	1

If $x = 0$, $Y = 1$

If $x = 1$, $Y = 1$

(Since inputs are opposite, one input always HIGH. Therefore, output is always HIGH.)

Theorem 18 $x \oplus \bar{x} = 1$

$x \oplus \bar{x}$:

A	x	Y
0	0	0
0	1	1
1	0	1
1	1	0

If $x = 0$, $Y = 1$

If $x = 1$, $Y = 1$

(One input HIGH, but not both.)

Double Inversion

The final single-variable theorem is just common sense. It states that a variable or expression inverted twice is the same as the original variable or expression. It is given by:

Theorem 19 $\bar{\bar{x}} = x$

This theorem is illustrated by the two inverters in Figure 3.28.

FIGURE 3.28   Double Inversion

Multivariable Theorems

There are numerous multivariable theorems we could learn, but we will look only at five of the most useful.

DeMorgan's Theorems

We have already seen DeMorgan's theorems. We will list them again, but will not comment further on them at this time.

Theorem 20 $\overline{xy} = \bar{x} + \bar{y}$

Theorem 21 $\overline{x + y} = \bar{x}\,\bar{y}$

Other Multivariable Theorems

Theorem 22 $x + xy = x$

Proof

$$x + xy = x\,(1 + y) \text{ (Distributive property)}$$
$$= x \cdot 1 \quad (1 + y = 1; \text{ Theorem 11})$$
$$= x \quad (x \cdot 1 = x; \text{ Theorem 10})$$

Figure 3.29 illustrates the circuit in this theorem. Note that the equivalent is not a circuit at all, but a single, unmodified variable. Thus, the circuit shown need never be built.

FIGURE 3.29   Theorem 22

   EXAMPLE 3.11

Simplify the following Boolean expressions, using Theorem 22 and other rules of Boolean algebra. Draw the logic circuits of the unsimplified and simplified expressions.

   a. $H = K\bar{L} + K$

   b. $Y = \overline{(A + B)}CD + \overline{(A + B)}$

   c. $W = (PQR + \bar{P}\,\bar{Q})(S + T) + (\bar{P} + \bar{Q})(S + T) + (S + T)$

■ Solution

Figure 3.30 shows the logic circuits for the unsimplified and simplified versions of expressions a through c.

   a.   Let $x = K$, let $y = \bar{L}$:

$$H = x + xy = K + K\bar{L}$$

   Theorem 22 states $x + xy = x$. Therefore $K + K\bar{L} = K$.

   b.   Let $x = \overline{(A + B)}$, let $y = CD$:

$$Y = x + xy = x = \overline{A + B}$$

   c.   Let $x = (S + T)$, let $y = (\bar{P} + \bar{Q})$:
   Since $x + xy = x$, $(\bar{P} + \bar{Q})(S + T) + (S + T) = (S + T)$.

$$W = (PQR + \bar{P}\,\bar{Q})(S + T) + (S + T)$$

Let $x = (S + T)$, let $y = (PQR + \bar{P}\,\bar{Q})$

$$W = x + xy = x = S + T$$

Alternate Method

$$W = (PQR + \bar{P}\bar{Q})(S + T) + (\bar{P} + \bar{Q})(S + T) + (S + T)$$

By the distributive property:

$$W = ((PQR + \bar{P}\,\bar{Q} + (\bar{P} + \bar{Q}))(S + T) + (S + T)$$

Let $x = (S + T)$, let $y = ((PQR + \bar{P}\bar{Q}) + (\bar{P} + \bar{Q}))$:

$$W = x + xy = x = S + T$$

FIGURE 3.30 Example 3.11: Logic Circuits for Unsimplified and Simplified Expressions

Theorem 23 $(x + y)(x + z) = x + yz$

Proof $(x + y)(x + z) = xx + xz + xy + yz$ (distributive property)
$= (x + xy) + xz + yz$ ($xx = x$; associative property)
$= x + xz + yz$ ($x + xy = x$ (Theorem 22))
$= (x + xz) + yz$ (associative property)
$= x + yz$ (Theorem 22)

Figure 3.31 shows the logic circuits for the left and right sides of the equation for Theorem 23. This theorem is a special case of one of the distributive properties, Theorem 6, where $w = x$.

$$(x + y)(x + z) = x + yx$$

FIGURE 3.31   Theorem 23

 EXAMPLE 3.12

Simplify the following Boolean expressions, using Theorem 23 and other rules of Boolean algebra. Draw the logic circuits of the unsimplified and simplified expressions.

a. $L = (M + \bar{N})(M + \bar{P})$

b. $Y = \overline{A + B} + AB)\overline{A + B} + C)$

■ **Solution**

Figure 3.32 shows the logic circuits for the unsimplified and simplified versions of expressions a and b.

FIGURE 3.32   Example 3.12: Logic Circuits for Unsimplified and Simplified Expressions

Theorem 23: $(x + y)(x + z) = x + yz$

a. Let $x = M$, let $y = N$, let $z = \bar{P}$:

$$L = (x + y)(x + z) = x + yz = M + \bar{N}\,\bar{P}$$

b. Let $x = \overline{A + B}$, let $y = AB$, let $z = C$:

$$Y = (x + y)(x + z) = x + yz = \overline{A + B} + ABC = \bar{A}\,\bar{B} + ABC$$

Theorem 24 $x + \bar{x}y = x + y$

Proof   Since $(x + y)(x + z) = x + yz$, then for $y = \bar{x}$:

$$x + \bar{x}y = (x + \bar{x})(x + y)$$
$$= 1 \cdot (x + y) \qquad (x + \bar{x} = 1)$$
$$= x + y$$

Figure 3.33 illustrates Theorem 24 with a logic circuit.

FIGURE 3.33   Theorem 24

NOTE . . .

Here is another way to remember Theorem 24:
If a variable (x) is ORed with a term consisting of a different variable (y)
AND the first variable's complement ($\bar{x}$), the complement disappears.

$$x + \bar{x}y = x + y$$

EXAMPLE 3.13

Simplify the following Boolean expressions, using Theorem 24 and other rules
of Boolean algebra. Draw the logic circuits of the unsimplified and simplified
forms of the expressions.

a. $W = \bar{U} + U\bar{V}$

b. $P = Q\bar{R}S + (\bar{Q} + R + \bar{S})\,T$

c. $J = \overline{KM}\,(\bar{K} + L + M) + KM$

■ Solution

Figure 3.34 shows the circuits for the unsimplified and simplified expressions.

Theorem 24: $x + \bar{x}y = x + y$

a. Let $x = \bar{U}$, let $y = \bar{V}$:

$$W = x + \bar{x}y = x + y = \bar{U} + \bar{V}$$

b. $P = Q\bar{R}S + (\bar{Q} + R + \bar{S})\,T$

$$= Q\bar{R}S + \overline{Q\,\bar{R}S}\,T \quad \text{(DeMorgan's theorem)}$$

Let $x = Q\bar{R}S$, let $y = T$:

$$P = x + \bar{x}y = x + y = Q\bar{R}S + T$$

FIGURE 3.34   Example 3.13: Logic Circuits for Unsimplified and Simplified Expressions

c. Let $x = KM$, let $y = (\bar{K} + L + M)$:

$$J = x + \bar{x}y = x + y = KM + \bar{K} + L + M$$
$$= \bar{K} + L + (M + KM) \quad \text{(associative property)}$$
$$= \bar{K} + L + M \quad \text{(Theorem 22)}$$

The rules of Boolean algebra are summarized in Table 3.8. Don't try to memorize all these rules. The commutative, associative, and distributive properties are the same as their counterparts in ordinary algebra. The single-variable theorems can be reasoned out by your knowledge of logic gate operation. That leaves only five multivariable theorems, including DeMorgan's theorems and three others. Memorize these three if you wish, but it is better to understand how they are derived.

TABLE 3.8 Theorems of Boolean Algebra

Commutative Properties

1. $x + y = y + x$
2. $x \cdot y = y \cdot x$

Associative Properties

3. $x + (y + z) = (x + y) + z$
4. $x(yz) = (xy)z$

Distributive Properties

5. $x(y + z) = xy + xz$
6. $(x + y)(w + z) = xw + xz + yw + yz$

x AND/OR/XOR 0/1

7. $x \cdot 0 = 0$
8. $x + 0 = x$
9. $x \oplus 0 = x$
10. $x \cdot 1 = x$
11. $x + 1 = 1$
12. $x \oplus 1 = \bar{x}$

x AND/OR/XOR $x/\bar{x}$

13. $x \cdot x = x$
14. $x + x = x$
15. $x \oplus x = 0$
16. $x \cdot \bar{x} = 0$
17. $x + \bar{x} = 1$
18. $x \oplus \bar{x} = 1$

Double Inversion

19. $\bar{\bar{x}} = x$

DeMorgan's Theorems

20. $\overline{xy} = \bar{x} + \bar{y}$
21. $\overline{x + y} = \bar{x}\,\bar{y}$

Other Multivariable Theorems

22. $x + xy = x$
23. $(x + y)(x + z) = x + yz$
24. $x + \bar{x}y = x + y$

■ **SECTION 3.3 REVIEW PROBLEM**

3.4 Use theorems of Boolean algebra to simplify the following Boolean expressions.

    a. $Y = \overline{AC} + (\bar{A} + \bar{C})D$
    b. $Y = \bar{A} + \bar{C} + ACD$
    c. $Y = (A\bar{B} + \bar{B}C)(A\bar{B} + \bar{C})$

3.4 SIMPLIFYING SOP AND POS EXPRESSIONS

TABLE 3.9   Truth Table for the SOP and POS Networks in Figure 3.35

A	B	C	Y
0	0	0	1
0	0	1	0
0	1	0	0
0	1	1	1
1	0	0	1
1	0	1	0
1	1	0	0
1	1	1	0

Earlier in this chapter, we discovered that we can generate a Boolean equation from a truth table and express it in sum-of-products (SOP) or product-of-sums (POS) form. From this equation, we can develop a logic circuit diagram. The next step in the design or analysis of a circuit is to simplify its Boolean expression as much as possible, with the ultimate aim of producing a circuit that has fewer physical components than the unsimplified circuit.

In Section 3.2, we found the SOP and POS forms of the Boolean expression represented by Table 3.9. These forms yield the logic diagrams shown in Figures 3.17 and 3.20. For convenience, the circuits are illustrated again in Figure 3.35. The corresponding algebraic expressions can be simplified by the rules of Boolean algebra to give us a simpler circuit in each case.

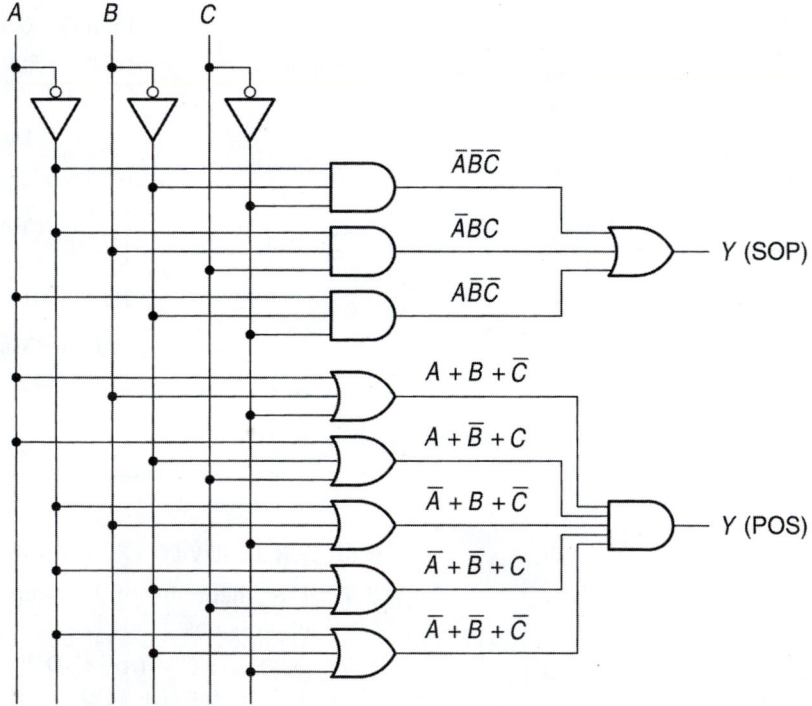

FIGURE 3.35   Unsimplified SOP and POS Networks

The sum-of-products and product-of-sums expressions represented by Table 3.9 are:

$$Y = \bar{A}\,\bar{B}\,\bar{C} + \bar{A}\,B\,C + A\,\bar{B}\,\bar{C} \text{ (SOP)}$$

and

$$Y = (A + B + \bar{C})(A + \bar{B} + C)(\bar{A} + B + \bar{C})(\bar{A} + \bar{B} + C)(\bar{A} + \bar{B} + \bar{C}) \text{ (POS)}$$

The SOP form is fairly easy to simplify:

$$Y = \overline{A}\,\overline{B}\,\overline{C} + \overline{A}\,B\,C + A\,\overline{B}\,\overline{C}$$
$$= (\overline{A} + A)\,\overline{B}\,\overline{C} + \overline{A}\,B\,C \quad \text{(distributive property)}$$
$$= 1 \cdot \overline{B}\,\overline{C} + \overline{A}\,B\,C \quad (x + \bar{x} = 1)$$
$$= \overline{B}\,\overline{C} + \overline{A}\,B\,C \quad (x \cdot 1 = x)$$

➡ **LAB SOURCE**
Lab #4, Boolean Algebra:
SOP forms

Since we cannot cancel any more SOP terms, we can call this final form the **maximum SOP simplification.** The logic diagram for the simplified expression is shown in Figure 3.36.

FIGURE 3.36 Simplified SOP Circuit

NOTE . . .

Two terms in an SOP expression can be reduced to one if they are identical except for one variable that is in true form in one term and complement form in the other. Such a grouping of a variable and its complement always cancels.

$$xy\bar{z} + xyz = xy(\bar{z} + z) = xy$$

There is a similar procedure for the POS form. Examine the following expression:

$$Y = (A + B + \overline{C})(A + B + C)$$

Recall Theorem 23: $(x + y)(x + z) = x + yz$.

Let $x = A + B$, let $y = \overline{C}$, let $z = C$.

$$Y = (A + B) + \overline{C}C \quad \text{(Theorem 23)}$$
$$= (A + B) + 0 \quad (x\bar{x} = 0)$$
$$= (A + B) \quad (x + 0 = x)$$

NOTE . . .

A POS expression can be simplified by grouping two terms that are identical except for one variable that is in true form in one term and complement form in the other.

$$(x + y + \bar{z})(x + y + z) = (x + y) + \bar{z}z = x + y$$

Let us use this procedure to simplify the POS form of the previous Boolean expression, shown again here with the terms numbered for our reference. The numbered value of each term corresponds to the binary value of the line in the truth table from which it is derived.

$$\overset{(1)}{} \quad \overset{(2)}{} \quad \overset{(5)}{} \quad \overset{(6)}{} \quad \overset{(7)}{}$$
$$Y = (A + B + \overline{C})\,(A + \overline{B} + C)\,(\overline{A} + B + \overline{C})\,(\overline{A} + \overline{B} + C)\,(\overline{A} + \overline{B} + \overline{C})$$

FIGURE 3.37 Simplified POS Circuit

There can be more than one way to simplify an expression. The following grouping of the numbered POS terms is one possibility.

$$(1)(5): (A + B + \bar{C})\,(\bar{A} + B + \bar{C}) = B + \bar{C}$$
$$(2)(6): (A + \bar{B} + C)\,(\bar{A} + \bar{B} + C) = \bar{B} + C$$
$$(6)(7): (\bar{A} + \bar{B} + C)\,(\bar{A} + \bar{B} + \bar{C}) = \bar{A} + \bar{B}$$

Combining these terms, we get the expression:

$$Y = (B + \bar{C})(\bar{B} + C)(\bar{A} + \bar{B})$$

Figure 3.37 shows the logic diagram for this expression. Compare this logic diagram and that of Figure 3.36 with the unsimplified circuits of Figure 3.35. Since there are no more cancellations of POS terms possible, we can call this the **maximum POS simplification.** We can, however, apply other rules of Boolean algebra and simplify further.

$$
\begin{aligned}
Y &= (B + \bar{C})(\bar{B} + C)(\bar{A} + \bar{B}) \\
  &= (\bar{B} + \bar{A}C)(B + \bar{C}) && \text{(Theorem 23)} \\
  &= \bar{B}\,B + \bar{B}\,\bar{C} + \bar{A}\,B\,C + \bar{A}\,C\,\bar{C} && \text{(distributive property)} \\
  &= \bar{B}\,\bar{C} + \bar{A}\,B\,C && (x \cdot \bar{x} = 0)
\end{aligned}
$$

This is the same result we got when we simplified the SOP form of the expression.

To be sure you are getting the maximum SOP or POS simplification, you should be aware of the following guidelines:

1. Each term must be grouped with another, if possible.
2. When attempting to group all terms, it is permissible to group a term more than once, such as term (6) above. The theorems $x \cdot x = x$ (POS forms) and $x + x = x$ (SOP forms) imply that using a term more than once does not change the Boolean expression.
3. Each pair of terms should have at least one term that appears only in that pair. Otherwise, you will have redundant terms that will need to be canceled later. For example, another possible group in the POS simplification above is terms (5) and (7). But since both these terms are in other groups, this pair is unnecessary and would yield a term you would have to cancel.

EXAMPLE 3.14

Find the maximum SOP simplification for the Boolean function represented by Table 3.10. Draw the logic diagram for the simplified expression.

■ **Solution**

SOP form:

$$
Y = \overset{(8)}{A\,\bar{B}\,\bar{C}\,\bar{D}} + \overset{(9)}{A\,\bar{B}\,\bar{C}\,D} + \overset{(10)}{A\,\bar{B}\,C\,\bar{D}} + \overset{(11)}{A\,\bar{B}\,C\,D} + \overset{(12)}{A\,B\,\bar{C}\,\bar{D}} + \overset{(14)}{A\,B\,C\,\bar{D}}
$$

TABLE 3.10 Truth Table for Example 3.14

A	B	C	D	Y
0	0	0	0	0
0	0	0	1	0
0	0	1	0	0
0	0	1	1	0
0	1	0	0	0
0	1	0	1	0
0	1	1	0	0
0	1	1	1	0
1	0	0	0	1
1	0	0	1	1
1	0	1	0	1
1	0	1	1	1
1	1	0	0	1
1	1	0	1	0
1	1	1	0	1
1	1	1	1	0

Group the terms as follows:

$$(8)+(9):\ A\bar{B}\bar{C}\bar{D}+A\bar{B}\bar{C}D=A\bar{B}\bar{C}$$
$$(10)+(11):\ A\bar{B}C\bar{D}+A\bar{B}CD=A\bar{B}C$$
$$(12)+(14):\ AB\bar{C}\bar{D}+ABC\bar{D}=AB\bar{D}$$

Combine the simplified groups and apply techniques of Boolean algebra to simplify further:

$$Y=A\bar{B}\bar{C}+A\bar{B}C+AB\bar{D}$$
$$=A\bar{B}(\bar{C}+C)+AB\bar{D}$$
$$=A\bar{B}+AB\bar{D}$$
$$=A(\bar{B}+B\bar{D})\ \text{(distributive property)}$$
$$=A(\bar{B}+\bar{D})\quad\text{(Theorem 24: } x+\bar{x}y=x+y)$$
$$=A\bar{B}+A\bar{D}$$

Figure 3.38 shows the logic diagram of the simplified expression.

FIGURE 3.38 Example 3.14: Simplified SOP Circuit

EXAMPLE 3.15

TABLE 3.11 Truth Table for Example 3.15

A	B	C	Y
0	0	0	1
0	0	1	1
0	1	0	0
0	1	1	1
1	0	0	1
1	0	1	0
1	1	0	0
1	1	1	0

Find the maximum SOP simplification for the Boolean expression represented by the truth table in Table 3.11.

■ **Solution**

The best solution is not the one that is most immediately apparent. Let us select terms as would seem most obvious, then reevaluate our selection to get the best grouping (i.e., the grouping that yields the fewest terms).

First, group each term with another, if possible, recalling that it is OK to use a term more than once.

$$\text{Terms: } (0)+(1)=\bar{A}\bar{B}\bar{C}+\bar{A}\bar{B}C=\bar{A}\bar{B}$$
$$(1)+(3)=\bar{A}\bar{B}C+\bar{A}BC=\bar{A}C$$
$$(0)+(4)=\bar{A}\bar{B}\bar{C}+A\bar{B}\bar{C}=\bar{B}\bar{C}$$

Notice that the terms in the first pair, (0) and (1), are each part of another group. This tells us that the simplified term for this pair ($\bar{A}\bar{B}$) is redundant. In order to use the simplest form of the expression, we should only use the last two pairs of terms, which gives us the simplified expression $Y=\bar{A}C+\bar{B}\bar{C}$.

■ **SECTION 3.4 REVIEW PROBLEM**

3.5 Find the maximum SOP and POS simplifications for the function represented by Table 3.12.

TABLE 3.12 Truth Table for Section 3.4 Review Problem

A	B	C	Y
0	0	0	0
0	0	1	1
0	1	0	1
0	1	1	1
1	0	0	0
1	0	1	0
1	1	0	1
1	1	1	0

3.5 SIMPLIFICATION BY THE KARNAUGH MAP METHOD

■ **KEY TERMS**

Karnaugh Map   A graphical tool for finding the maximum SOP or POS simplification of a Boolean expression. A Karnaugh map (or K-map) works by arranging the terms of an expression in such a way that variables can be canceled by grouping minterms or maxterms.

Cell   The smallest unit of a Karnaugh map, corresponding to one line of a truth table. The input variables are the cell's coordinates, and the output variable is the cell's contents.

Adjacent Cell   Two cells are adjacent if there is only one variable that is different between the coordinates of the two cells. For example, the cells for minterms ABC and $\overline{A}BC$ are adjacent.

Pair   A group of two adjacent cells in a Karnaugh map. A pair cancels one variable in a K-map simplification.

Quad   A group of four adjacent cells in a Karnaugh map. A quad cancels two variables in a K-map simplification.

Octet   A group of eight adjacent cells in a Karnaugh map. An octet cancels three variables in a K-map simplification.

In Example 3.14, we derived a sum-of-products Boolean expression from a truth table and simplified the expression by grouping minterms that differed by one variable. We made this task easier by breaking up the truth table into groups of four lines. (It is difficult for the eye to grasp an overall pattern in a group of 16 lines.) We chose groups of four because variables A and B are the same in any one group and variables C and D repeat the same binary sequence in each group. This allows us to see more easily when we have terms differing by only one variable.

The **Karnaugh map,** or K-map, is a graphical tool for simplifying Boolean expressions that uses a similar idea. A K-map is a square or rectangle divided into smaller squares called **cells,** each of which represents a line in the truth table of the Boolean expression to be mapped. Thus, the number of cells in a K-map is always a power of 2, usually 4, 8, or 16. The coordinates of each cell are the input variables of the truth table. The cell content is the value of the output variable on that line of the truth table. Figure 3.39 shows the formats of Karnaugh maps for Boolean expressions having two, three, and four variables, respectively.

a. Two-variable forms

b. Three-variable forms

c. Four-variable forms

FIGURE 3.39 Karnaugh Map Formats

There are two equivalent ways of labeling the cell coordinates: numerically or by true and complement variables. We will use the numerical labeling since it is always the same, regardless of the chosen variables.

The cells in the Karnaugh maps are set up so that the coordinates of any two **adjacent cells** differ by only one variable. By grouping adjacent cells according to specified rules, we can simplify a Boolean expression by canceling variables in their true and complement forms, much as we did algebraically in the previous section.

Two-Variable Map

Table 3.13 shows the truth table of a two-variable Boolean expression.

The Karnaugh map shown in Figure 3.40 is another way of showing the same information as the truth table. Every line in the truth table corresponds to a cell, or square, in the Karnaugh map.

The coordinates of each cell correspond to a unique combination of input variables (*A, B*). The content of the cell is the output value for that input combination. If the truth table output is 1 for a particular line, the content of the corresponding cell is also 1. If the output is 0, the cell content is 0.

TABLE 3.13 Truth Table for a Two-Variable Boolean Expression

A	B	Y
0	0	1
0	1	1
1	0	0
1	1	0

FIGURE 3.40 Karnaugh Map for Table 3.13

The SOP expression of the truth table is

$$Y = \overline{A}\,\overline{B} + \overline{A}\,B$$

which can be simplified as follows:

$$Y = \overline{A}\,(\overline{B} + B)$$
$$= \overline{A}$$

We can perform the same simplification by grouping the adjacent **pair** of 1s in the Karnaugh map, as shown in Figure 3.41.

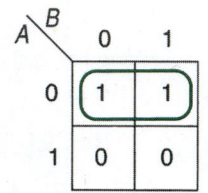

FIGURE 3.41 Grouping a Pair of Adjacent Cells

> **NOTE...**
>
> When we circle a pair of 1s in a K-map, we are grouping the common variable in two minterms, then factoring out and canceling the complements.

To find the simplified form of the Boolean expression represented in the K-map, we examine the coordinates of all the cells in the circled group. We retain coordinate variables that are the same in all cells and eliminate coordinate variables that are different in different cells.

In this case:

$\overline{A}$ is a coordinate of both cells of the circled pair. (Keep $\overline{A}$.)

$\overline{B}$ is a coordinate of one cell of the circled pair, and B is a coordinate of the other. (Discard $B/\overline{B}$.)

$$Y = \overline{A}$$

Three- and Four-Variable Maps

Refer to the forms of three- and four-variable Karnaugh maps shown in Figure 3.39. Each cell is specified by a unique combination of binary variables. This implies that the three-variable map has 8 cells (since $2^3 = 8$) and the four-variable map has 16 cells (since $2^4 = 16$).

The variables specifying the row (both maps) or the column (the four-variable map) do not progress in binary order; they advance such that there is only *one change of variable per row or column*. For example, the numbering of the rows is 00, 01, 11, 10, rather than the binary order 00, 01, 10, 11. If we were to use binary order, adjacent cells in rows 2 and 3 or 3 and 4 would differ by two variables, meaning we could not factor out and cancel a pair of complements by grouping these cells. For instance, we cannot cancel complementary variables from the pair $\overline{A}\,B\,C + A\,\overline{B}\,C$, which differs by two variables.

> **NOTE...**
>
> The number of cells in a group must be a power of 2, such as 1, 2, 4, 8, or 16.

FIGURE 3.42 Quad

FIGURE 3.43 Octet

A group of four adjacent cells is called a **quad.** Figure 3.42 shows a Karnaugh map for a Boolean function whose terms can be grouped in a quad. The Boolean expression displayed in the K-map is:

$$Y = \bar{A}\,\bar{B}\,C + \bar{A}\,B\,C + A\,B\,C + A\,\bar{B}\,C$$

A and B are both part of the quad coordinates in true and complement form. (Discard A and B.)

C is a coordinate of *each cell* in the quad. (Keep C.)

$$Y = C$$

Grouping cells in a quad is equivalent to factoring two complementary pairs of variables and canceling them.

$$Y = (A + \bar{A})(B + \bar{B})C = C$$

You can verify that this is the same as the original expression by multiplying out the terms.

An **octet** is a group of eight adjacent cells. Figure 3.43 shows the Karnaugh map for the following Boolean expression:

$$Y = \bar{A}\,B\,\bar{C}\,\bar{D} + \bar{A}\,B\,\bar{C}\,D + \bar{A}\,B\,C\,D + \bar{A}\,B\,C\,\bar{D}$$
$$+ A\,B\,\bar{C}\,\bar{D} + A\,B\,\bar{C}\,D + A\,B\,C\,D + A\,B\,C\,\bar{D}$$

Variables A, C, and D are all coordinates of the octet cells in true and complement form. (Discard A, C, and D.)

B is a coordinate of *each* cell. (Keep B.)

$$Y = B$$

The algebraic equivalent of this octet is an expression where three complementary variables are factored out and canceled.

$$Y = (A + \bar{A})B(C + \bar{C})(D + \bar{D}) = B$$

> **NOTE . . .**
>
> A Karnaugh map completely filled with 1s implies that all input conditions yield an output of 1. For a Boolean expression Y, $Y = 1$.

Grouping Cells Along Outside Edges

The cells along an outside edge of a three- or four-variable map are adjacent to cells along the opposite edge (only one change of variable). Thus we can group cells "around the outside" of the map to cancel variables. In the case of the four-variable map, we can also group the four corner cells as a quad, since they are all adjacent to one another.

 EXAMPLE 3.16

Use Karnaugh maps to simplify the following Boolean expressions:

a. $Y = \bar{A}\,\bar{B}\,\bar{C} + \bar{A}\,\bar{B}\,C + A\,\bar{B}\,\bar{C} + A\,\bar{B}\,C$

b. $Y = \bar{A}\,\bar{B}\,\bar{C}\,\bar{D} + \bar{A}\,\bar{B}\,C\,\bar{D} + A\,\bar{B}\,\bar{C}\,\bar{D} + A\,\bar{B}\,C\,\bar{D}$

■ **Solutions**

Figure 3.44 shows the Karnaugh maps for the Boolean expressions labeled **a** and **b.** Cells in each map are grouped in a quad.

FIGURE 3.44   Example 3.16: K-maps

a. *A* and *C* are both coordinates of two cells in true form and two cells in complement form. (Discard *A* and *C*.)
$\bar{B}$ is a coordinate of each cell. (Keep $\bar{B}$.)

$$Y = \bar{B}$$

b. *A* and *C* are both coordinates of two cells in true form and two cells in complement form. (Discard *A* and *C*.)
$\bar{B}$ and $\bar{D}$ are coordinates of each cell. (Keep $\bar{B}$ and $\bar{D}$.)

$$Y = \bar{B}\,\bar{D}$$

Loading a K-Map from a Truth Table

NOTE...

We don't need a Boolean expression to fill a Karnaugh map if we have the function's truth table.

Figure 3.45 and Figure 3.46 show truth table and Karnaugh map forms for three- and four-variable Boolean expressions. The numbers in parentheses show the order of terms in binary sequence for both forms.

The Karnaugh map is not laid out in the same order as the truth table. That is, it is not laid out in a binary sequence. This is due to the criterion for cell adjacency: no more than one variable change between rows or columns is permitted.

Filling in a Karnaugh map from a truth table is easy when you understand a system for doing it quickly. For the three-variable map, fill row 1, then row 2, skip to

A	B	C	Y
0	0	0	(0)
0	0	1	(1)
0	1	0	(2)
0	1	1	(3)
1	0	0	(4)
1	0	1	(5)
1	1	0	(6)
1	1	1	(7)

AB\C	0	1
00	(0)	(1)
01	(2)	(3)
11	(6)	(7)
10	(4)	(5)

a. Truth table   **b. K-map**

FIGURE 3.45   Order of Terms (Three-Variable Function)

A	B	C	D	Y
0	0	0	0	(0)
0	0	0	1	(1)
0	0	1	0	(2)
0	0	1	1	(3)
0	1	0	0	(4)
0	1	0	1	(5)
0	1	1	0	(6)
0	1	1	1	(7)
1	0	0	0	(8)
1	0	0	1	(9)
1	0	1	0	(10)
1	0	1	1	(11)
1	1	0	0	(12)
1	1	0	1	(13)
1	1	1	0	(14)
1	1	1	1	(15)

a. Truth table   b. K-map

FIGURE 3.46 Order of Terms (Four-Variable Function)

row 4, then go back to row 3. By doing this, you trace through the cells in binary order. Use the mnemonic phrase "1, 2, skip, back" to help you remember this.

The system for the four-variable map is similar but must account for the columns as well. The rows get filled in the same order as the three-variable map, but within each row, fill column 1, then column 2, skip to column 4, then go back to column 3. Again, "1, 2, skip, back."

The four-variable map is easier to fill from the truth table if we break up the truth table into groups of four lines, as we have done in Figure 3.46. Each group is one row in the Karnaugh map. Following this system will quickly fill the cells in binary order.

Go back and follow the order of terms on the four-variable map in Figure 3.46, using this system. (Remember, for both rows and columns, "1, 2, skip, back.")

Multiple Groups

> **NOTE . . .**
>
> If there is more than one group of 1s in a K-map simplification, each group is a term in the maximum SOP simplification of the mapped Boolean expression. The resulting terms are ORed together.

EXAMPLE 3.17

Use the Karnaugh map method to simplify the Boolean function represented by Table 3.14.

■ Solution

Figure 3.47 shows the Karnaugh map for the truth table in Table 3.14. There are two groups of 1s—a pair and a quad.

(continued)

TABLE 3.14 Truth Table for Example 3.17

A	B	C	D	Y
0	0	0	0	1
0	0	0	1	0
0	0	1	0	1
0	0	1	1	0
0	1	0	0	0
0	1	0	1	1
0	1	1	0	0
0	1	1	1	1
1	0	0	0	0
1	0	0	1	0
1	0	1	0	0
1	0	1	1	0
1	1	0	0	0
1	1	0	1	1
1	1	1	0	0
1	1	1	1	1

FIGURE 3.47 Example 3.17: K-Map

Pair:

Variables $\overline{A}$, $\overline{B}$, and $\overline{D}$ are coordinates of both cells. (Keep $\overline{A}\,\overline{B}\,\overline{D}$.) C is a coordinate of one cell and $\overline{C}$ is a coordinate of the other. (Discard C.)
Term: $\overline{A}\,\overline{B}\,\overline{D}$

Quad:

Both A and C are coordinates of two cells in true form and two cells in complement form. (Discard A and C.)
B and D are coordinates of all four cells. (Keep $B\,D$.)
Term: $B\,D$

Combine the terms in an OR function:

$$Y = \overline{A}\,\overline{B}\,\overline{D} + B\,D$$

Overlapping Groups

> **NOTE . . .**
>
> A cell may be grouped more than once. The only condition is that every group must have at least one cell that does not belong to any other group. Otherwise, redundant terms will result.

EXAMPLE 3.18

TABLE 3.15 Truth Table for Example 3.18

A	B	C	Y
0	0	0	1
0	0	1	1
0	1	0	0
0	1	1	1
1	0	0	0
1	0	1	0
1	1	0	1
1	1	1	1

Simplify the function represented by Table 3.15.

■ **Solution**

The Karnaugh map for the function in Table 3.15 is shown in Figure 3.48, with two different groupings of terms.

a. The simplified Boolean expression drawn from the first map has three terms.

$$Y = \overline{A}\,\overline{B} + A\,B + B\,C$$

b. The second map yields an expression with four terms.

$$Y = \overline{A}\,\overline{B} + A\,B + B\,C + \overline{A}\,C$$

One of the last two terms is redundant, as neither of the pairs corresponding to these terms has a cell belonging only to that pair. We could retain either pair of cells and its corresponding term, but not both.

a. $Y = \bar{A}\bar{B} + AB + BC$

b. $Y = \bar{A}\bar{B} + AB + BC + \bar{A}C$

One of these is redundant

Only one of these terms is necessary

FIGURE 3.48 Example 3.18: K-Maps

We can show algebraically that the last term is redundant and thus make the expression the same as that in part a.

$$Y = \bar{A}\,\bar{B} + A\,B + B\,C + \bar{A}\,C$$
$$= \bar{A}\,\bar{B} + A\,B + B\,C + \bar{A}\,(B + \bar{B})C$$
$$= \bar{A}\,\bar{B} + A\,B + B\,C + \bar{A}\,B\,C + \bar{A}\,\bar{B}\,C$$
$$= \bar{A}\,\bar{B}\,(1 + C) + A\,B + B\,C\,(1 + \bar{A})$$
$$= \bar{A}\,\bar{B} + A\,B + B\,C$$

Conditions for Maximum Simplification

> **NOTE . . .**
>
> The maximum simplification of a Boolean expression is achieved only if the circled groups of cells in its K-map are as large as possible and there are as few groups as possible.

EXAMPLE 3.19

Find the maximum SOP simplification of the Boolean function represented by Table 3.16.

■ Solution

The values of Table 3.16 are loaded into the three K-maps shown in Figure 3.49. Three different ways of grouping adjacent cells are shown. One results in maximum simplification; the other two do not.

We get the maximum SOP simplification by grouping the two octets shown in Figure 3.49a. The resulting expression is

  a. $Y = \bar{B} + D$

Figure 3.49b and Figure 3.49c show two simplifications that are less than the maximum because the chosen cell groups are smaller than they could be. The resulting expressions are:

(continued)

TABLE 3.16 Truth Table for Example 3.19

A	B	C	D	Y
0	0	0	0	1
0	0	0	1	1
0	0	1	0	1
0	0	1	1	1
0	1	0	0	0
0	1	0	1	1
0	1	1	0	0
0	1	1	1	1
1	0	0	0	1
1	0	0	1	1
1	0	1	0	1
1	0	1	1	1
1	1	0	0	0
1	1	0	1	1
1	1	1	0	0
1	1	1	1	1

a. **Maximum simplification**
$Y = \bar{B} + D$

b. **Less than maximum simplification**
$Y = \bar{A}\bar{B} + A\bar{B} + D$

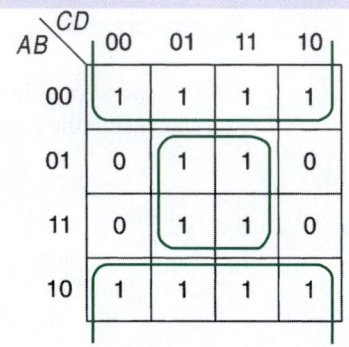

c. **Less than maximum simplification**
$Y = \bar{B} + BD$

FIGURE 3.49   Example 3.19: K-Maps

b. $Y = \bar{A}\,\bar{B} + A\,\bar{B} + D$
c. $Y = \bar{B} + B\,D$

Neither of these expressions is the simplest possible, since both can be reduced by Boolean algebra to the form in Figure 3.49a.

Using K-Maps for Partially Simplified Circuits

Figure 3.50 shows a logic diagram that can be further simplified. If we want to use a Karnaugh map for this process, we must do one of two things:

1. Fill in the K-map from the existing product terms. Each product term that is not a minterm will represent more than one cell in the Karnaugh map. When the map is filled, regroup the cells for maximum simplification; or
2. Expand the sum-of-products expression of the circuit to get a sum-of-minterms form. Each minterm represents one cell in the K-map. Group the cells for maximum simplification.

Method 1: Figure 3.51 shows the K-map derived from the existing circuit and the regrouped cells that yield the maximum simplification.

Method 2: The algebraic method requires us to expand the existing Boolean expression to get a sum of minterms. The original expression is:

$$Y = \bar{A}\,B\,\bar{C}\,\bar{D} + A\,B\,\bar{D} + \bar{A}\,C$$

FIGURE 3.50 Logic Diagram That Can Be Further Simplified

a. K-map from logic diagram (Figure 3.50)

b. Maximum simplification

FIGURE 3.51 Further Simplification of Logic Diagram (Figure 3.50)

FIGURE 3.52 Simplified Circuit

The theorem $(x + \bar{x}) = 1$ implies that we can AND a variable with a term in true and complement form without changing the term. The expanded expression is:

$$Y = \bar{A}\,B\,\bar{C}\,\bar{D} + A\,B\,(C + \bar{C})\,\bar{D} + \bar{A}\,(B + \bar{B})\,C\,(D + \bar{D})$$

$$= \bar{A}\,B\,\bar{C}\,\bar{D} + A\,B\,C\,\bar{D} + A\,B\,\bar{C}\,\bar{D}$$

$$+ \bar{A}\,B\,C\,D + \bar{A}\,B\,C\,\bar{D} + \bar{A}\,\bar{B}\,C\,D + \bar{A}\,\bar{B}\,C\,\bar{D}$$

The terms of this expression can be loaded into a K-map and simplified, as shown in Figure 3.51b. Figure 3.52 shows the logic diagram for the simplified expression.

EXAMPLE 3.20

Use a Karnaugh map to find the maximum SOP simplification of the circuit shown in Figure 3.53.

FIGURE 3.53   Example 3.20: Circuit to Be Simplified

■ **Solution**

Figure 3.54a shows the Karnaugh map of Figure 3.53 with terms grouped as shown in the original circuit. Figure 3.54b shows the terms regrouped for the maximum simplification, which is given by:

$$Y = \bar{A}\,D + \bar{B}D + \bar{A}\,B\,\bar{C}$$

Alternate Method   The Boolean expression for the circuit in Figure 3.53 is:

$$Y = \bar{A}\,B\,\bar{C} + \bar{A}\,C\,D + \bar{B}\,\bar{C}\,D + A\,\bar{B}\,C\,D$$

a. K-map from Figure 3.53   **b. Maximum simplification**

FIGURE 3.54   Example 3.20: Maximum Simplification of Figure 3.53

This expands to the following expression:

$$Y = \bar{A}\,B\,\bar{C}\,(D + \bar{D}) + \bar{A}\,(B + \bar{B})\,C\,D + (A + \bar{A})\,\bar{B}\,\bar{C}\,D + A\,\bar{B}\,C\,D$$
$$= \bar{A}\,B\,\bar{C}\,D + \bar{A}\,B\,\bar{C}\,\bar{D} + \bar{A}\,B\,C\,D + \bar{A}\,\bar{B}\,C\,D + A\,\bar{B}\,\bar{C}\,D$$
$$+ \bar{A}\,\bar{B}\,\bar{C}\,D + A\,\bar{B}\,C\,D$$

This expression can be loaded directly into the K-map and simplified, as shown in Figure 3.54b. The logic diagram for the simplified expression is shown in Figure 3.55.

FIGURE 3.55   Example 3.20: Simplified Circuit

Don't Care States

> **■ KEY TERM**
>
> **Don't Care State**   An output state that can be regarded as either HIGH or LOW, as is most convenient. A don't care state is the output state of a circuit for a combination of inputs that will never occur under stated design conditions.

Sometimes a digital circuit will be intended to work only for certain combinations of inputs; any other input values will never be applied to the circuit.

In such a case, it may be to our advantage to use so-called **don't care states** to simplify the circuit. A don't care state is shown in a K-map cell as an "X" and can be either a 0 or a 1, depending on which case will yield the maximum simplification.

A common application of the don't care state is a digital circuit designed for binary-coded decimal (BCD) inputs. In BCD, a decimal digit (0–9) is encoded as a 4-bit binary number (0000–1001). This leaves six binary states that are never used (1010, 1011, 1100, 1101, 1110, 1111). In any circuit designed for BCD inputs, these states are don't care states.

All cells containing 1s must be grouped if we are looking for a maximum SOP simplification. (If necessary, a group can contain one cell.) The don't care states can be used to maximize the size of these groups. We need not group all don't care states, only those that actually contribute to a maximum simplification.

Notice that each X must be treated *either* as a 0 or a 1. When we group the Xs in a K-map, each circled X becomes a 1; all others must be treated as 0s.

EXAMPLE 3.21

The circuit in Figure 3.56 is designed to accept binary-coded decimal inputs. The output is HIGH when the input is the BCD equivalent of 5, 7, or 9. If the BCD equivalent of the input is not 5, 7 or 9, the output is LOW. The output is not defined for input values greater than 9.

Find the maximum SOP simplification of the circuit.

■ Solution

The Karnaugh map for the circuit is shown in Figure 3.57a.

(continued)

FIGURE 3.56 Example 3.21: Circuit to Be Simplified

a. **Karnaugh map**   b. **Logic diagram**

FIGURE 3.57   Example 3.21: Karnaugh Map and Logic Diagram

We can designate three of the don't care cells as 1s—those corresponding to input states 1011, 1101, and 1111. This allows us to group the 1s into two overlapping quads, which yield the following simplification.

$$Y = D_4 D_1 + D_3 D_1$$

The ungrouped don't care states are treated as 0s. The corresponding circuit is shown in Figure 3.57b.

 EXAMPLE 3.22

One type of decimal code is called 2421 code, so called because of the positional weights of its bits. Compare this to BCD code, described in Example 3.21, which is sometimes called 8421 code. For example, 1101 in BCD has the decimal value $8 + 4 + 0 + 1 = 13$, whereas the same bits in 2421 code have the decimal value $2 + 4 + 0 + 1 = 7$. Table 3.17 shows how this code compares to its equivalent decimal digits and to the BCD code used in Example 3.21.

2421 code is sometimes used because it is "self-complementing," a property that BCD code does not have, but that is useful in digital decimal arithmetic circuits.

The bits of the BCD code are designated $D_4 D_3 D_2 D_1$. The bits of the 2421 code are designated $Y_4 Y_3 Y_2 Y_1$.

Use the Karnaugh map method to design a logic circuit that accepts any BCD input and generates an output in 2421 code, as specified by Table 3.17.

TABLE 3.17   BCD and 2421 Code

Decimal Equivalent	BCD Code				2421 Code			
	D_4	D_3	D_2	D_1	Y_4	Y_3	Y_2	Y_1
0	0	0	0	0	0	0	0	0
1	0	0	0	1	0	0	0	1
2	0	0	1	0	0	0	1	0
3	0	0	1	1	0	0	1	1
4	0	1	0	0	0	1	0	0
5	0	1	0	1	1	0	1	1
6	0	1	1	0	1	1	0	0
7	0	1	1	1	1	1	0	1
8	1	0	0	0	1	1	1	0
9	1	0	0	1	1	1	1	1

◼ Solution

The required circuit is called a code converter. A block symbol representing the code converter is shown in Figure 3.58. Each 4-bit BCD input corresponds to a 4-bit 2421 output. Thus, we must find four Boolean expressions, one for each bit of the 2421 code. We can derive each Boolean expression from a truth table represented by the corresponding output column in Table 3.17.

We can load the 2421 values into four different Karnaugh maps, as shown in Figure 3.59. The cells corresponding to the unused input BCD codes 1010, 1011, 1100, 1101, 1110, and 1111 are don't care states in each map.

The K-maps yield the following simplifications:

$$Y_4 = D_4 + D_3\,D_2 + D_3\,D_1$$
$$Y_3 = D_4 + D_3\,D_2 + D_3\,\bar{D}_1$$
$$Y_2 = D_4 + \bar{D}_3\,D_2 + D_3\,\bar{D}_2\,D_1$$
$$Y_1 = D_1$$

Figure 3.60 shows the logic diagram for these equations.

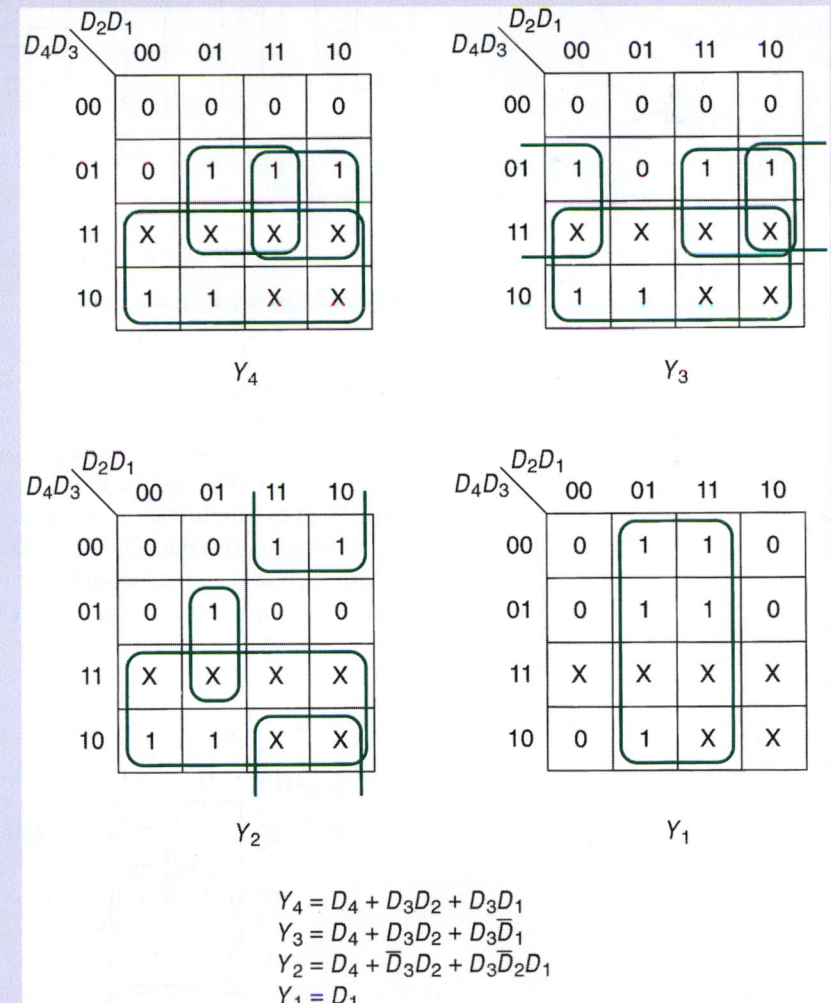

$$Y_4 = D_4 + D_3 D_2 + D_3 D_1$$
$$Y_3 = D_4 + D_3 D_2 + D_3 \bar{D}_1$$
$$Y_2 = D_4 + \bar{D}_3 D_2 + D_3 \bar{D}_2 D_1$$
$$Y_1 = D_1$$

FIGURE 3.58 BCD-to-2421 Code Converter

FIGURE 3.59 Example 3.22: K-maps: BCD to 2421

(continued)

FIGURE 3.60   Example 3.22: BCD-to-2421 Code Converter

POS Simplification

Until now, we have looked only at obtaining the maximum SOP simplification from a Karnaugh map. It is also possible to find the maximum POS simplification from the same map.

Figure 3.61 shows a Karnaugh map with the cells grouped for an SOP simplification and a POS simplification. The SOP simplification is shown in Figure 3.61a and the POS simplification in Figure 3.61b.

When we derive the POS form of an expression from a truth table, we use the lines where the output is 0 and we use the complements of the input variables on these lines as the elements of the selected maxterms. The same principle applies here.

a. SOP simplification
$Y = \overline{A}C + B$

b. POS simplification
$Y = (\overline{A} + B)(B + C)$

FIGURE 3.61   SOP and POS Forms on K-Map

The maxterms are:

$(A + B + C)$	Top left cell
$(\bar{A} + B + C)$	Bottom left cell
$(\bar{A} + B + \bar{C})$	Bottom right cell

The variables are canceled in much the same way as in the SOP form. Remember, however, that the POS variables are the complements of the variables written beside the Karnaugh map.

If there is more than one simplified term, the terms are ANDed together, as in a full POS form.

Cancellations:

Left-hand pair: A is present in both true and complement form in the pair. (Discard A.)

       B and C are present in both cells of the pair. (Keep B and C.) Term: $B + C$

Bottom pair: $\bar{A}$ and B are present in both cells of the pair. (Keep $\bar{A}$ and B.) C is present in both true and complement form in the pair. (Discard C.) Term: $\bar{A} + B$

Maximum POS simplification:

$$Y = (\bar{A} + B)(B + C)$$

Compare this with the maximum SOP simplification:

$$Y = \bar{A}\,C + B$$

By the Boolean theorem $(x + y)(x + z) = x + yz$, we see that the SOP and POS forms are equivalent.

EXAMPLE 3.23

Find the maximum POS simplification of the logic function represented by Table 3.18.

TABLE 3.18   Truth Table for Example 3.23

A	B	C	D	Y
0	0	0	0	0
0	0	0	1	0
0	0	1	0	0
0	0	1	1	0
0	1	0	0	1
0	1	0	1	1
0	1	1	0	1
0	1	1	1	1
1	0	0	0	0
1	0	0	1	1
1	0	1	0	0
1	0	1	1	1
1	1	0	0	1
1	1	0	1	0
1	1	1	0	1
1	1	1	1	1

■ **Solution**

Figure 3.62 shows the Karnaugh map from the truth table in Table 3.18. The cells containing 0s are grouped in two quads and there is a single 0 cell left over.

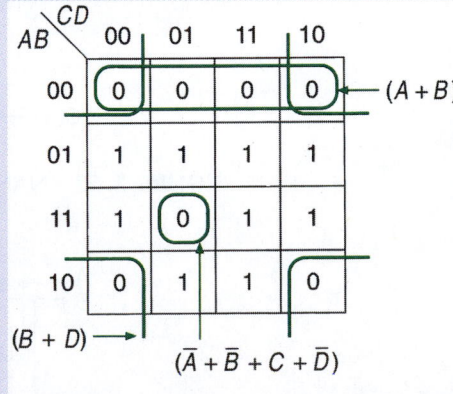

FIGURE 3.62   Example 3.23: POS Simplification of Table 3.18

(continued)

Simplification:

Corner quad: $(B + D)$

Horizontal quad: $(A + B)$

Single cell: $(\bar{A} + \bar{B} + C + \bar{D})$

$$Y = (A + B)(B + D)(\bar{A} + \bar{B} + C + \bar{D})$$

3.6   SIMPLIFICATION BY DEMORGAN EQUIVALENT GATES

In Example 3.2, we showed how a logic gate network could be redrawn to make it conform to the bubble-to-bubble convention. The aim was to cancel the internal inversions within the logic circuit and therefore make the Boolean expression easier to write and to understand.

For example, suppose we have a circuit like the one shown in Figure 3.63. We can write the Boolean expression of the circuit as follows:

$$Y = \overline{\overline{A B \bar{C}} \; \overline{\bar{A} \bar{B} C}}$$

This expression is very difficult to interpret in terms of what the Boolean function is actually supposed to do. The levels of inversion bars make it difficult to simplify and to make a truth table. We can use theorems of Boolean algebra, such as DeMorgan's theorems, to simplify the expression as follows:

$$Y = \overline{\overline{A B \bar{C}} \; \overline{\bar{A} \bar{B} C}}$$
$$= \overline{\overline{A B \bar{C}}} + \overline{\overline{\bar{A} \bar{B} C}}$$
$$= A B \bar{C} + \bar{A} \bar{B} C$$

Alternatively, you could redraw the last NAND gate (the one driving output Y), in its DeMorgan equivalent form, as shown in Figure 3.64. The inputs of this gate are now active-LOW, as are the outputs of the two gates driving them. These two inversions automatically cancel, giving the simplified output expression without having to do the algebra.

FIGURE 3.63   NAND Circuit

FIGURE 3.64   NAND Circuit Simplified Using a DeMorgan Equivalent Gate

Apply the following guidelines when making a circuit conform to the bubble-to-bubble convention:

1. Start at the output and work back. There are always two possible choices for the output gate: a positive- or negative-logic form. One of these will be AND-shaped and one will be OR-shaped. Choose the OR-shaped gate for an SOP result and the AND-shaped gate for a POS result. If the active level of the output is important, choose the desired level, converting the output gate if necessary.
2. Go back toward the circuit inputs to the next level of gating. Match the output levels of these gates to the input levels of the gate described in Step 1, converting the next-level gates to their DeMorgan equivalent if necessary.
3. Repeat Step 2 until you reach the circuit inputs.

There are always two different correct results for this procedure, one with an active-HIGH output and one with an active-LOW.

EXAMPLE 3.24

a. Write the Boolean expression for the logic diagram shown in Figure 3.65.
b. Redraw the circuit in Figure 3.65 to make it conform to the bubble-to-bubble convention. Show two different solutions.
c. Write the Boolean expressions of the redrawn circuits.

FIGURE 3.65 Example 3.24: Original Circuit

■ **Solution**

a. The Boolean expression for the logic diagram in Figure 3.65 is:

$$\overline{\overline{(\overline{A}\,B)(\overline{\overline{A}}\,C)} + \overline{(C+D)}}$$

b. Figure 3.66 shows the logic diagram with the output gate converted to its DeMorgan equivalent form and the remaining gates changed to conform to the bubble-to-bubble convention. The changed gates are shown in color.

FIGURE 3.66 Example 3.24: DeMorgan Equivalent Circuit

(continued)

FIGURE 3.67   Example 3.24: DeMorgan Equivalent Circuit

Figure 3.67 shows the second solution. The output gate is left in its original form and the two gates at the next level, shown in color, are changed to suit. The two inverters are also redrawn. Even though their function does not change, it indicates that the inputs driving the inverters are active-LOW.

  c. The Boolean expressions for the redrawn circuits are:

Figure 3.66: $Y = (A + \overline{B})(A + C)(C + D)$

Figure 3.67: $Y = \overline{\overline{\overline{A}\,B} + \overline{\overline{A}\,\overline{C}} + \overline{\overline{C}\,\overline{D}}}$ or $\overline{Y} = \overline{A}\,B + \overline{A}\,\overline{C} + \overline{C}\,\overline{D}$

⇒ LAB SOURCE
Lab # 5, Boolean Algebra:
DeMorgan Equivalent
Forms

■ SECTION 3.6 REVIEW PROBLEM

3.6   Redraw the logic diagram in Figure 3.68 to make it conform to the bubble-to-bubble convention. Write the Boolean expression of the redrawn network.

FIGURE 3.68   Section Review Problem 3.6

3.7   UNIVERSAL PROPERTY OF NAND/NOR GATES

At the beginning of Chapter 2, we stated that any Boolean function can be realized by a combination of AND, OR, or NOT functions. We can go one step further and state that any function can be implemented using only NAND functions (NOT and AND) or only NOR functions (NOT and OR).

   We can create the basic logic functions from NAND or NOR gates by applying DeMorgan's and other Boolean theorems.

$$\overline{x\,y} = \overline{x} + \overline{y}$$
$$\overline{x + y} = \overline{x}\,\overline{y}$$

All-NAND Forms

NOT: An inverter can be constructed from a single NAND gate by connecting both inputs together, as shown in Figure 3.69.
        Algebraic expression: $\overline{x \cdot x} = \overline{x}$

FIGURE 3.69 NOT from NAND

AND: The AND function is created by inverting a NAND function, as in Figure 3.70.

Algebraic expression: $\overline{\overline{x \cdot y}} = x \cdot y$

FIGURE 3.70 AND from NANDs

OR: The OR function requires three NANDs, as shown in Figure 3.71a. The circuit can be redrawn to get the same algebraic result a little more directly. This equivalent circuit is shown in Figure 3.71b.

Algrebraic expressions: **a.** $\overline{\overline{x} \cdot \overline{y}} = \overline{\overline{x + y}} = x + y$ or

b. $\overline{\overline{x}} + \overline{\overline{y}} = x + y$

a. Standard form

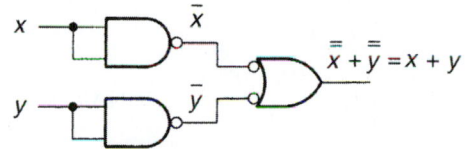

b. DeMorgan equivalent form

FIGURE 3.71 OR from NANDs

NOR: Figure 3.72a and Figure 3.72b show two forms of the NOR circuit, each using four NAND gates. The NOR is implemented by inverting the output of the OR circuit with a NAND inverter. Note that the OR and NOR circuits in Figure 3.71b and Figure 3.72b are drawn with matching input and output active levels in all gates.

Algebraic expressions: **a.** $\overline{\overline{\overline{x} \cdot \overline{y}}} = \overline{\overline{\overline{x + y}}} = \overline{x + y}$ or

b. $\overline{\overline{\overline{x}} + \overline{\overline{y}}} = \overline{x + y}$

a. Standard form

b. DeMorgan equivalent form

FIGURE 3.72 NOR from NANDs

All-NOR Forms

NOT: The NOT function is formed by connecting the inputs of a NOR gate together, as shown in Figure 3.73.

Algebraic expressions: $\overline{x + x} = \overline{x}$

FIGURE 3.73 NOT from NOR

OR: The OR function is created by inverting the NOR function with a NOR gate inverter. This is shown in Figure 3.74.

Algebraic expression: $\overline{\overline{x+y}} = x + y$

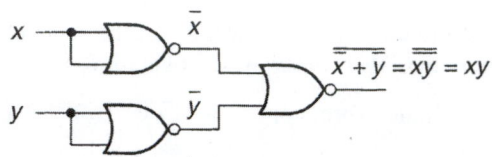

FIGURE 3.74 OR from NORs

AND: The AND function is synthesized from three NOR gates. Two forms of the circuit are shown in Figure 3.75.

Algebraic expressions: **a.** $\overline{\overline{x}+\overline{y}} = \overline{\overline{x}\cdot\overline{y}} = x \cdot y$ or

b. $\overline{\overline{x}} \cdot \overline{\overline{y}} = x \cdot y$

a. Standard form   b. DeMorgan equivalent form

FIGURE 3.75 AND from NORs

NAND: The NAND function is created by inverting the previously derived AND function. This requires four NOR gates, as shown in Figure 3.76a and Figure 3.76b. Note that the AND and NAND circuits in Figure 3.75b and Figure 3.76b have matching input and output active levels in all gates.

Algebraic expressions: **a.** $\overline{\overline{\overline{x}+\overline{y}}} = \overline{\overline{\overline{x}\cdot\overline{y}}} = \overline{x \cdot y}$ or

b. $\overline{\overline{\overline{x}} \cdot \overline{\overline{y}}} = \overline{x \cdot y}$

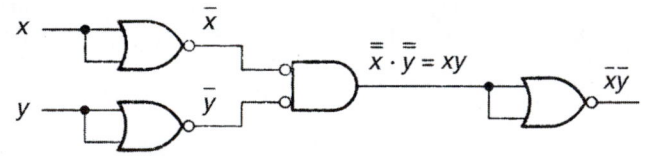

a. Standard form   b. DeMorgan equivalent form

FIGURE 3.76 NAND from NORs

There are several ways we can make a circuit with only NANDs or only NORs.

1. If the circuit is in SOP form, we can change the ANDs and ORs to NANDs directly by drawing a pair of bubbles between each AND output and each OR input. The bubbles will cancel, giving us the original Boolean expression.

   If the SOP circuit must be built with only NORs, we convert the AND-OR (SOP) gates to NOR gates in standard or DeMorgan equivalent form and insert NOR inverters to adjust the input and output logic levels.

2. If the circuit is in POS form, we can replace the ORs and ANDs directly with NOR gates in standard or DeMorgan equivalent form by inserting a pair of bubbles between each OR output and each AND input. We can also replace the OR-AND gates (i.e., the POS function) with NAND gates and use NAND inverters to correct the logic levels of the inputs and output.

3. If the circuit is in neither SOP nor POS form, we can apply rules of Boolean algebra or graphical techniques to get it into such a form, then apply one of the previously mentioned techniques to implement and all-NAND or all-NOR circuit.

As implied by points 1 and 2, the SOP form is best-suited to all-NAND implementation and the POS to the all-NOR form. If we use these forms, we will often be able to make a circuit more efficiently, that is, with fewer logic gate packages.

EXAMPLE 3.25

FIGURE 3.77 Example 3.25: SOP Circuit

Convert the logic diagram of Figure 3.77:
a. To an all-NAND circuit
b. To an all-NOR circuit

If the only gates available are the TTL logic gates shown in Figure 3.78, state how many logic gate packages are required to make the original circuit in Figure 3.77 and the two converted forms. State which of the forms is most efficient to build.

FIGURE 3.78 Example 3.25: TTL Logic Gates

(continued)

■ **Solution**

a. Figure 3.79a shows the all-NAND equivalent of the SOP circuit of Figure 3.77. All gates are NAND gates in either standard or DeMorgan equivalent form. The inverter is in the same place as in the original circuit.

b. Figure 3.79b shows the same circuit in all-NOR form. The circuit is constructed around two levels of AND-shaped and OR-shaped NORs, which preserves the SOP function. We must add three inverters to the circuit to get the Boolean expression back to its original form. The inverters are placed on inputs B and C and the circuit output. There is no inverter on A. Notice that this is the exact opposite of where we placed inverters for the all-NAND form.

In both circuits the inverters are drawn so that active levels of gate inputs and outputs are matched.

The circuit in Figure 3.77 requires the following logic gate packages:
    1—74LS08 AND (2 of 4 gates used)
    1—74LS32 OR (1 of 4 gates used)
    1—74LS04 inverter (1 of 6 gates used)

The all-NAND circuit in Figure 3.79a requires:
    1—74LS00 NAND gate (4 of 4 gates used)

The all-NOR circuit in Figure 3.79b requires:
    2—74LS02 NOR gates (6 of 8 gates used)
    The all-NAND circuit is most efficient, as it uses only one gate package, whereas the original and all-NOR use three or two packages for the same function.

Note that although the four gates in Figure 3.79a do not all look identical, they all have the same function according to DeMorgan's theorem. Thus, the four gates in the 74LS00 NAND package can have either form and can be used as shown in Figure 3.79a. In the same way, although the gates in Figure 3.79b do not look the same, they all have the same function. Therefore, any gate from a 74LS02 NOR package can be used for any of the gates in Figure 3.79b.

FIGURE 3.79 Example 3.25: All-NAND and ALL-NOR Forms of an SOP Circuit

Convert the circuit of Figure 3.80:

a. To an all-NOR circuit
b. To an all-NAND circuit

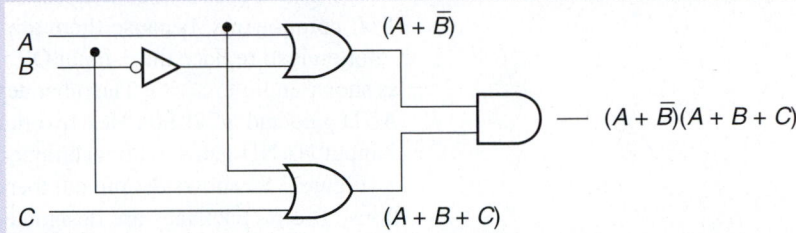

FIGURE 3.80 Example 3.26: POS Circuit

■ **Solution**

a. Figure 3.81a shows the all-NOR implementation of the POS circuit in Figure 3.80. We have converted the AND and OR gates to NORs by adding bubbles to the AND inputs and the OR outputs. In the configuration shown, the bubbles cancel, leaving us with the original Boolean expression. The inverter stays at input B.

b. Figure 3.81b shows the equivalent all-NAND circuit. We retain the POS function by drawing the NANDs in their OR shape, followed by an AND-shaped gate. The active levels of the circuit must be adjusted by adding NAND inverters as shown. The inverters are at locations complementary to those of the all-NOR circuit.

FIGURE 3.81 Example 3.26: All-NAND and All-NOR Forms of a POS Circuit

3.8 PRACTICAL CIRCUIT IMPLEMENTATION IN SSI LOGIC

Until now, we have been drawing logic circuits without regard to whether we can actually build them or not. Even a fairly elementary circuit may be impossible to construct without some modification because the required parts are not available.

For instance, examine the circuit in Figure 3.82: You can't build this circuit from TTL components, because there are no 3-input OR gates available in TTL. One solution is to replace the 3-input OR with two gates of a 74LS32 quad 2-input OR, as shown in Figure 3.83. The other devices in the circuit are a 74LS11 triple 3-input AND gate and a 74LS04 Hex Inverter. (Alternatively, we could use 74LS10 triple 3-input NAND gates, with techniques from Section 3.7.)

Figure 3.83 shows the pin numbers for each gate, as well as the gate logic functions. The IC packages are designated U1, U2, and U3, with the individual gates shown as U1A, U1B, and so on.

In a practical design, we have to account for the gates left over in each package, since all gates in a package have an effect on various electrical parameters, such as power consumption, whether they are used in a circuit or not. A commercial design would connect the inputs of the unused gates to either a logic HIGH or a logic LOW to protect the other gates in the package from excessive noise. This is particularly important for CMOS components, where unconnected inputs can cause damage by static and overheating.

FIGURE 3.82   Circuit That Can't Be Built from TTL Parts

FIGURE 3.83   Practical Implementation of Figure 3.82

It doesn't matter at which logic level the input is held for CMOS gates, but generally a TTL gate consumes less power if the output is in the HIGH state. For example, we can connect the inputs of AND and NAND gates as shown in Figure 3.84. The AND inputs are tied HIGH and the NAND inputs are tied LOW. Both connections generate logic HIGH outputs, resulting in lower power consumption than LOW outputs.

Table 3.19 shows the availability of different gates for the basic logic functions in two popular technologies: low-power Schottky TTL and High-Speed CMOS.

The choice of which technology to use will probably depend on factors other than gate availability, as there are a number of ways to produce any logic function with the gates you do have. In fact, we know we can implement any Boolean function with only NAND or NOR gates. Given that, it is interesting to note that the largest selections of gate configurations are for the NAND and NOR functions.

Pin assignments for TTL gates are shown in Figure 3.85. V_{CC} (power) is connected to pin 14 and ground is connected to pin 7 for each of these gates.

Connections for most High-Speed CMOS chips are the same as those for TTL, as this family was designed to supply a CMOS alternative to TTL and thus be completely pin-compatible. Exceptions to this general case are those High-Speed CMOS chips whose functions are not available in TTL, such as 74HC4002 and 74HC4075.

a. AND gate

b. NAND gate

FIGURE 3.84 TTL NAND and AND Gates Connected for Minimum Power Consumption

TABLE 3.19 Available SSI Gates

| Logic Function | Configuration | | Logic Family | |
			LSTTL	High-Speed CMOS
NAND	Quadruple	2-input	74LS00	74HC00
	Triple	3-input	74LS10	74HC10
	Dual	4-input	74LS20	74HC20
	Single	8-input	74LS30	74HC30
	Single	13-input	74LS133	74HC133
NOR	Quadruple	2-input	74LS02	74HC02
	Triple	3-input	74LS27	74HC27
	Dual	4-input		74HC4002
	Dual (with strobe)	4-input	74LS25	
AND	Quadruple	2-input	74LS08	74HC08
	Triple	3-input	74LS11	74HC11
	Dual	4-input	74LS21	
OR	Quadruple	2-input	74LS32	74HC32
	Triple	3-input		74HC4075

FIGURE 3.85   Pinouts for TTL and High-Speed CMOS Gates

EXAMPLE 3.27

The circuit in Figure 3.86 represents the maximum SOP simplification of a Boolean function. Redraw the circuit, including pin numbers and unused gates, for an actual implementation in High-Speed CMOS.

FIGURE 3.86   Example 3.27: Original Circuit

■ **Solution**

There are no 4-input AND or OR gates available in High-Speed CMOS. One simple solution is to use an all- (well, almost all-) NAND implementation, as shown in Figure 3.87. This requires only three IC packages—a 74HC04 inverter, a 74HC10 3-input NAND, and a 74HC20 4-input NAND. One of the 4-input NANDs (U3B) is drawn in DeMorgan equivalent form.

FIGURE 3.87   Example 3.27: Practical Implementation of Figure 3.86 in High-Speed CMOS

3.9 PULSED OPERATION OF LOGIC CIRCUITS

In Chapter 2, we looked at the enable and inhibit properties of various individual logic gates. We used these gates to pass or block pulsed digital signals by applying the signal to one input and controlling the logic levels of the remaining inputs.

We can devise more-complex schemes for controlling pulsed signals, using multi-gate logic results. Figure 3.88 shows one such example. The circuit in Figure 3.88 will tell us the period of time, in milliseconds, that one or both of the switches on A and B are open (i.e., in the HIGH state). The counter is a digital circuit whose internal binary value advances by 1 with every cycle of waveform Y (i.e., it counts the number of cycles at output Y).

The output Y will be pulsing at 1 kHz, provided the AND gate is in the enable condition, that is, if its Control input is HIGH. This occurs if either A or B is HIGH.

Figure 3.88b shows one possible timing diagram of the circuit. Assume that the counter is reset to zero before either A or B goes HIGH. The third line in the timing

a. Enable circuit

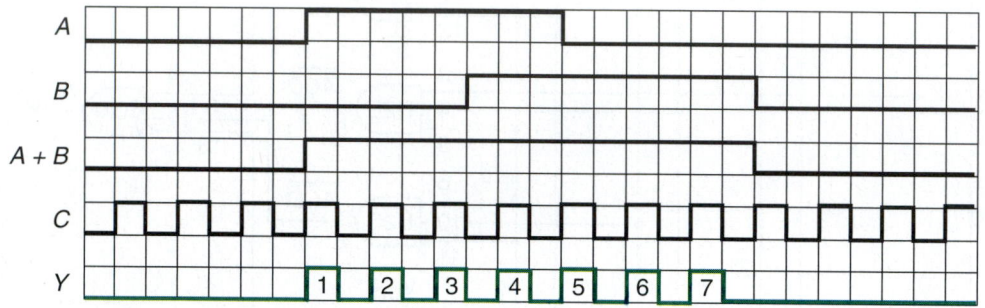

"One or both switches open for 7ms."

b. Timing diagram

FIGURE 3.88   Multigate Logic Circuit

diagram, the one showing $(A + B)$, is the enable/inhibit Control input of the AND gate. The expression $A + B$ goes HIGH as a result of either switch A or switch B opening and enables the pulse waveform for seven cycles, or 7 ms.

 EXAMPLE 3.28

A lamp mounted in a car's instrument panel warns the driver if the engine temperature is too high, the oil pressure is too low, or both. The engine temperature and oil pressure are each monitored by a corresponding active-HIGH sensor, as shown in Figure 3.89.

When the output NAND gate is enabled, it will drive the warning lamp at 3 Hz, alerting the driver to a failure condition. The circuit works only if the ignition switch is closed (logic LOW).

Draw the waveforms, showing a pulsing waveform at D and all possible conditions of inputs A, B, and C. Show the resulting waveform of Y.

■ Solution

The circuit output is enabled when the Control input of the NAND gate is HIGH. The conditions required are either low oil pressure ($A = 1$) OR high engine temperature ($B = 1$), AND ignition ON ($C = 0$).

The timing diagram in Figure 3.90 shows all possible combinations of A, B, and C and shows the output for all combinations. When Y is enabled, it is pulsing out of phase with D.

FIGURE 3.89 Example 3.28: Monitoring Sensors

FIGURE 3.90 Example 3.28: Timing Diagram

3.10 A GENERAL APPROACH TO LOGIC CIRCUIT DESIGN

In previous sections of this chapter, we have seen a number of example design problems, in which we start with a verbal description of a problem and translate it into a combinational logic circuit. Each of these examples has been used to explain a problem under study at the time, but in doing so, we have not followed any consistent method of design. In this section, we will try to systematize our design methodology.

When designing a combinational logic circuit, ultimately we are looking for a Boolean equation. Any Boolean combinational function can be written using AND, OR, and NOT functions, which can be arranged in sum-of-products or product-of-sums form, as derived from a truth table and simplified using Boolean algebra or a Karnaugh map. This is all just mechanics; the tricky bit is knowing how to translate the words of a design idea into that requisite Boolean equation.

The first step in the design process is to have *an accurate description of the problem*. What this generally consists of is understanding the effect of all inputs, singly and in combination, on the circuit output or outputs. These should be listed systematically and examined to make sure that all possible input combinations have been accounted for. Active levels of inputs and outputs should be properly specified, as should any special constraints on inputs (e.g., binary-coded decimal values only).

Once the problem has been stated, *each output of the circuit should be described, either verbally or with a truth table.* If using a verbal description, look for keywords, such as AND, OR, NOT that can be translated into a Boolean expression. If using a truth table, enter the requirements for each circuit output, as it relates to the circuit inputs.

Using Boolean algebra or Karnaugh maps, simplify the Boolean equations of the circuit.

EXAMPLE 3.29

A sealed tank in a chemical factory, shown in Figure 3.91, is used in an industrial process with specific constraints on the volume of liquid in the tank, as well as its pressure and its temperature. A level sensor determines if the volume of liquid in the tank is too large. Pressure is monitored by a pressure sensor and temperature by a temperature sensor. If any of these parameters is exceeded, the corresponding sensor generates a logic HIGH.

The three sensors are monitored by a safety system, shown in Figure 3.92, that controls the behavior of three active-HIGH indicator lights. A green light, designated "OK," indicates that the system is functioning normally. This is defined as the condition where no more than one sensor indicates an out-of-range condition. A yellow light, called the "ALERT" light, illuminates when two or more

FIGURE 3.91 Example 3.29: Level, Temperature, and Pressure Sensors in a Tank

FIGURE 3.92 Example 3.29: Active-HIGH Controller for Status Lamps

sensors indicate an out-of-range condition. The red light, designated as the "DANGER" condition, comes on when all sensors are out of range. (Note that in the final condition, both the yellow and red lights are on.)

Design a circuit for the system in Figure 3.92 that will fulfill the requirements for the three indicator lights.

■ **Solution**

Let's summarize the input and output requirements of the circuit:

1. There are three active-HIGH sensors for out-or-range parameters: level (L), temperature (T), and pressure (P).
2. There are three active-HIGH status lamps: green (OK), yellow (ALERT), and red (DANGER).
3. Table 3.20 shows the relationship between lamps and sensors.

TABLE 3.20 Lamps and Sensors for Figure. 3.92

State	Number of Sensors ON	Green Lamp	Yellow Lamp	Red Lamp
OK	None	ON	OFF	OFF
OK	One	ON	OFF	OFF
ALERT	Two	OFF	ON	OFF
DANGER	Three	OFF	ON	ON

Since both sensors and lamps are active-HIGH, an ON means a logic 1 in a truth table. Table 3.21 shows a truth table for this circuit. At this point, we are ready to use the mechanics of Boolean simplification to draw the logic diagram of the indicator lamp control circuit.

TABLE 3.21 Truth Table for the Circuit in Figure 3.92

L	T	P	OK	ALERT	DANGER
0	0	0	1	0	0
0	0	1	1	0	0
0	1	0	1	0	0
0	1	1	0	1	0
1	0	0	1	0	0
1	0	1	0	1	0
1	1	0	0	1	0
1	1	1	0	1	1

Figure 3.93 shows Karnaugh maps for the OK and ALERT expressions. The DANGER function can be derived directly from the truth table.

$$OK = \bar{L}\,\bar{T} + \bar{L}\,\bar{P} + \bar{T}\,\bar{P}$$

$$ALERT = LT + LP + TP$$

$$DANGER = LTP$$

These expressions can be read as follows:

OK: The active-HIGH green lamp is ON when any two sensors are OFF.

ALERT: The active-HIGH yellow lamp is ON when any two sensors are ON.

DANGER: The active-HIGH red lamp is ON when all three sensors are ON.

(continued)

FIGURE 3.93 Example 3.29: K-Maps for Active-HIGH Lamp Controller

Note that the outputs for the OK and ALERT lights are exactly opposite from one another. Therefore, an alternative implementation for the OK light would be:

$$OK = \overline{LT + LP + TP}$$

However, we will stick to the original equation, as its output is active-HIGH and thus more accurately represents the behavior of the green status lamp. The completed circuit is shown in Figure 3.94.

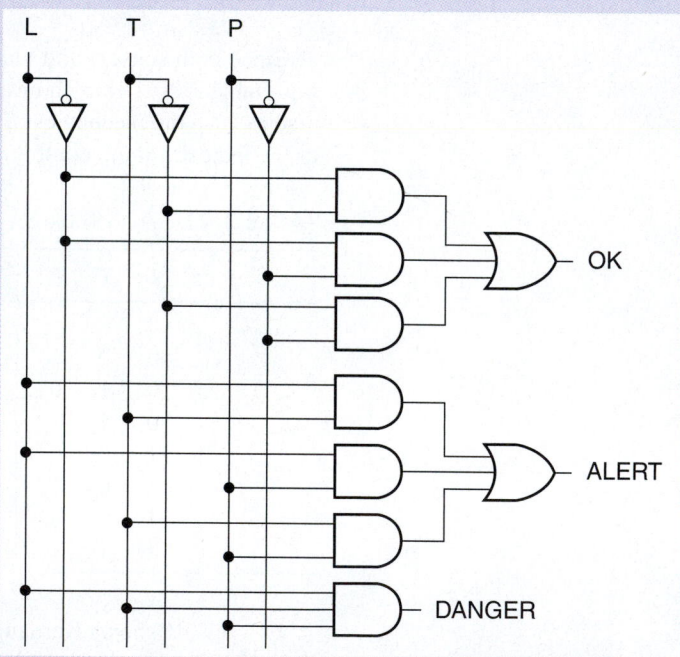

FIGURE 3.94 Example 3.29: Logic Diagram for Active-HIGH Lamp Controller

The controller in Example 3.29 uses active-HIGH inputs, which were chosen because they are easy to understand. However, for a variety of practical reasons, many electronic circuits have active-LOW outputs. In these cases, we can use inverting gates, such as NANDs and NORs, to drive our output loads. The design procedure for these circuits favors the 0s in the circuit truth table, rather than the 1s.

Truth tables also allow us to incorporate time-varying (pulsed) output into our designs. We can use methods similar to those in Chapter 2 for using gate truth tables to derive the enable and inhibit properties of gate networks.

EXAMPLE 3.30

The same industrial process as in Example 3.29 is being monitored, but with a different indicator lamp controller, shown in Figure 3.95.

FIGURE 3.95 Example 3.30: Active-LOW Controller for Status Lamps

The level *(L)*, temperature *(T)*, and pressure *(P)* sensors are active-HIGH, as in Example 3.29. However, the lamp behavior is different. The lamps are active-LOW and instead of three levels of information, we now have four: OK (no sensors active), ALERT (one sensor active), DANGER (two sensors), and CRITICAL (all three sensors). The DANGER and CRITICAL states both use the red lamp: DANGER is a steady ON-level on the red lamp; CRITICAL is red flashing at 3 Hz. Design a controller circuit for the new system.

■ **Solution**

The lamp operation is summarized in Table 3.22.

TABLE 3.22 Lamps and Sensors for Figure 3.95

State	Number of Sensors ON	Green Lamp	Yellow Lamp	Red Lamp
OK	None	ON	OFF	OFF
ALERT	One	OFF	ON	OFF
DANGER	Two	OFF	OFF	ON
CRITICAL	Three	OFF	OFF	Flashing

Since the lamps in the indicator panel are active-LOW, the truth table for the controller shows a 0 for each lamp's ON state. The active-HIGH sensor ON-states are represented by 1s. Since the OK lamp and the ALERT lamp depend only on the sensors, and not the pulse source, we can create a 3-input truth table for those lamps, as shown in Table 3.23.

The DANGER/CRITICAL lamp poses a more complicated problem. The lamp has three possible states: OFF, ON, and flashing. In addition to the sensor inputs, we must also account for the pulse input, which we can call *S* (for Signal). The truth table for this lamp is shown in Table 3.24.

When two sensors are active we want the lamp ON, regardless of the status of the pulse input. Thus the lamp is ON for *LTPS* = 0110 or 0111; for *LTPS* = 1010 or 1011; and for *LTPS* = 1100 or 1101. The lamp should flash when all sensors

TABLE 3.23 Truth Table for OK and ALERT Lamps

L	T	P	OK	ALERT
0	0	0	0	1
0	0	1	1	0
0	1	0	1	0
0	1	1	1	1
1	0	0	1	0
1	0	1	1	1
1	1	0	1	1
1	1	1	1	1

(continued)

TABLE 3.24   Truth Table for DANGER/CRITICAL Lamp

L	T	P	S	DANGER/CRITICAL
0	0	0	0	1
0	0	0	1	1
0	0	1	0	1
0	0	1	1	1
0	1	0	0	1
0	1	0	1	1
0	1	1	0	0
0	1	1	1	0
1	0	0	0	1
1	0	0	1	1
1	0	1	0	0
1	0	1	1	0
1	1	0	0	0
1	1	0	1	0
1	1	1	0	0
1	1	1	1	1

are active. Thus, when $LTPS = 1110$, the lamp output is 0 and when $LTPS = 1111$, the lamp output is 1. This allows the lamp to flash in phase wth S.

The Boolean equation for the OK lamp can be derived by inspection.

$$OK = L + T + P$$

or alternatively,

$$OK = \overline{\overline{L}\,\overline{T}\,\overline{P}}$$

This can be read as, "the active-LOW OK lamp is ON when all sensors are LOW."

a. K-map for ALERT lamp

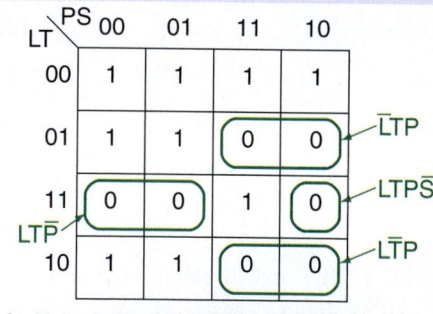

b. K-map for DANGER/CRITICAL lamp

FIGURE 3.96   Example 3.30: K-Maps for Active-LOW Lamp Controller

Figure 3.96 shows the K-maps for the ALERT and DANGER/CRITICAL lamps. Since our outputs are active LOW, we can derive the outputs from the groups of 0s in the K-maps. We can use the 0s to derive an active-HIGH product-of-sums expression, or we can derive product terms for an active-LOW sum-of-products expression. If we choose the latter technique, we must remember to invert the entire function. Since our lamps are active-LOW, we will use the K-maps to derive the active-LOW SOP expression.

Examination of the terms in the K-map for the ALERT lamp yields:

$$ALERT = \overline{\overline{L}\,\overline{T}\,P + \overline{L}\,T\,\overline{P} + L\,\overline{T}\,\overline{P}}$$

This can be read as, "the active-LOW ALERT lamp is ON if any one sensor is HIGH."

The K-map for the DANGER/CRITICAL lamp show three pairs of 0s, each of which corresponds to a unique 2-sensor active combination. The remaining 0, along with the horizontally adjacent 1, is the "flashing" term, activated by all three sensors. The 0 in this term could be combine with any of the other three, but for the sake of keeping the meaning clear, we will leave it separate, and hence unsimplified. The resulting Boolean equation is:

$$DANGER/CRITICAL = \overline{\overline{L}TP + L\overline{T}P + LT\overline{P} + LTP\overline{S}}$$

This can be read as, "the active-LOW DANGER/CRITICAL lamp is ON with any two sensors HIGH. If all three sensors are HIGH, the lamp follows the pulse at S." Note that the two-sensor terms are all disabled when all sensors are HIGH. This prevents the lamp from being stuck in the ON state when all sensors are out-of-range.

The circuit for the controller is shown in Figure 3.97.

FIGURE 3.97 Example 3.30: Logic Circuit for Active-LOW Lamp Controller

SUMMARY

1. Two or more gates connected together form a logic gate network or combinational logic circuit, which can be described by a truth table, a logic diagram, or a Boolean expression.

2. The output of a combinational logic circuit is always the same with the same combination of inputs, regardless of the order in which they are applied.

3. The order of precedence in a logic gate network is AND, then OR, unless otherwise indicated by parentheses.

4. DeMorgan's theorems: $\overline{x \cdot y} = \overline{x} + \overline{y}$

$$\overline{x + y} = \overline{x} \cdot \overline{y}$$

5. Inequalities: $\overline{x \cdot y} \neq \overline{x} \cdot \overline{y}$

$$\overline{x + y} \neq \overline{x} + \overline{y}$$

6. A logic gate network can be drawn to simplify its Boolean expression by ensuring that bubbled (active-LOW) outputs drive bubbled inputs and outputs with no bubble (active-HIGH) drive inputs with no bubble. Some gates might need to be drawn in their DeMorgan equivalent form to achieve this.

7. In Boolean expressions, logic inversion bars of equal lengths cancel; bars of unequal lengths do not. Bars of equal length represent bubble-to-bubble connections (e.g., $\overline{\overline{x + y + z}} = x + y + z$, but $\overline{\overline{x} + \overline{y} + \overline{z}} \neq x + y + z$).

8. A logic diagram can be derived from a Boolean expression by order of precedence rules: synthesize ANDs before ORs, unless parentheses indicate otherwise. Inversion bars act as parentheses for a group of variables.

9. A truth table can be derived from a logic gate network either by finding truth tables for intermediate points in the network and combining them by the laws of Boolean algebra, or by simplifying the Boolean expression into a form that can be directly written into a truth table.

10. A sum-of-products (SOP) network combines inputs in AND gates to yield a group of product terms that are combined in an OR gate (logical sum) output.

11. A product-of-sums (POS) network combines inputs in OR gates to yield a group of sum terms that are combined in an AND gate (logical product) output.

12. An SOP Boolean expression can be derived from the lines in a truth table where the output is at logic 1. Each product term contains all inputs in true or complement form, where inputs at logic 0 have a bar and inputs at logic 1 do not.

13. A POS expression is derived from the lines where the output is at logic 0. Each sum term contains all inputs in true or complement form, where inputs at logic 1 have a bar and inputs at logic 0 do not.

14. Theorems of Boolean algebra, summarized in Table 3.8, allow us to simplify logic gate networks.

15. SOP networks can be simplified by grouping pairs of product terms and applying the Boolean identity $xyz + xy\overline{z} = xy$.

16. POS networks can be simplified by grouping pairs of sum terms and applying the Boolean identity $(x + y + z)(x + y + \overline{z}) = (x + y)$.

17. To achieve maximum simplification of an SOP or POS network, each product or sum term should be grouped with another if possible. A product or sum term can be grouped more than once, as long as each group has a term that is only in that group.

18. A Karnaugh map can be used to graphically reduce a Boolean expression to its simplest form by grouping adjacent cells containing 1s. One cell is equivalent to one line of a truth table. A group of adjacent cells that contain 1s represents a simplified product term.

19. Adjacent cells in a K-map differ by only one variable. Cells around the outside of the map are considered adjacent. For example, a cell in the top row of the map is adjacent to a cell in the bottom row, provided they are in the same column. A cell in the far-left column of the map is adjacent to a cell in the far-right column, provided they are in the same row. The four corners of the map are adjacent to one another, as there is a single difference of input variable in any vertical or horizontal direction.

20. A group in a K-map must be a power of two in size: 1, 2, 4, 8, or 16. A group of two is called a pair, a group of four is a quad, and a group of eight is an octet.

21. A pair cancels one variable. A quad cancels two variables. An octet cancels three variables.

22. A K-map can have multiple groups. Each group represents one simplified product term in a sum-of-products expression.

23. Groups in K-maps can overlap as long as each group has one or more cells that appear only in that group.

24. Groups in a K-map should be as large as possible for maximum SOP simplification.

25. Don't care states represent output states of input combinations that will never occur in a circuit. They are represented by Xs in a truth table or K-map and can be used as 0s or 1s, whichever is most advantageous for the simplification of the circuit.

26. A Karnaugh map can be used to find the maximum POS simplification of a circuit by grouping the 0s in the map. The circled groups, which represent sum terms of the POS expression, are simplified by keeping common variables and discarding those that differ among the cells of the group. Those variables that are kept have an inversion bar where a variable is at the logic 1 level and no bar where the variable is at logic 0. This is the opposite of the SOP simplification.

27. A K-map can be used to obtain the maximum SOP simplification of a circuit with an active-LOW output by grouping the 0s in a K-map. Each group yields a product term in the same way as for an active-HIGH SOP simplification. However, the entire function must be inverted (e.g., $Y = \overline{A \, \overline{B} \, C + \overline{A} \, B \, \overline{D}}$.

28. When using DeMorgan equivalent gates to simplify a logic circuit, redraw the logic diagram by starting at the output and working back to the inputs. Use DeMorgan's theorems to modify gates as required to ensure that connections conform to the bubble-to-bubble convention.

29. Any combinational logic circuit can be made using only NAND or only NOR gates. NAND-NAND circuits are best for implementing SOP expressions. NOR-NOR forms are best for POS circuits.

30. We cannot always build an arbitrary combinational logic circuit using commercial logic gates. We are restricted by factors such as the number of inputs to an available logic gate (e.g., there are no 3-input OR gates available in TTL packages).

31. Practical methods of synthesizing unavailable gates include synthesis from simpler gates (e.g., 3-input OR from two 2-input ORs) and using DeMorgan equivalent forms (e.g., NAND-NAND circuit instead of AND-OR).

32. Unused gates in a logic gate package should have their inputs disabled by tying them HIGH or LOW to minimize power consumption and to protect the rest of the gates in the package from electrical noise. Unused TTL gates dissipate less power when their inputs are configured to make their outputs permanently HIGH (e.g., tie AND or OR gate inputs HIGH to make output HIGH; tie NAND or NOR inputs LOW to make output HIGH). *IMPORTANT: Unused outputs should be left unconnected.*

33. Logic circuits can be analyzed for pulsed operation by determining the logic levels at points in the circuit equivalent to the Control inputs of a gate that enables or inhibits the pulsed waveform. For example, a logic circuit with output $Y = AB + C$, where C is the pulsed waveform, can be analyzed to see how A and B affect output Y by passing or blocking waveform C. For this example, the waveform is inhibited when both A and B are HIGH, since under this condition $Y = 1 + C = 1$.

34. Combinational logic circuit design largely consists of translating word problems to Boolean equations, then applying tools of simplification and synthesis to make the best circuit.

35. Boolean equations required by a combinational logic design process can be derived from verbal descriptions by creating an accurate description of the problem that analyzes the effects of each of the inputs on each circuit output. These descriptions can then be translated into Boolean expressions either directly or by creating a truth table.

GLOSSARY

Adjacent Cell   Two cells in a K-map are adjacent if there is only one variable that is different between the coordinates of the two cells.

Associative Property   A mathematical function is associative if its operands can be grouped in any order without affecting the result. For example, addition is associative $((a + b) + c = a + (b + c))$, but subtraction is not $((a - b) - c \neq a - (b - c))$.

Bubble-to-Bubble Convention   The practice of drawing gates in a logic diagram so that inverting outputs connect to inverting inputs and noninverting outputs connect to noninverting inputs.

Bus Form   A way of drawing a logic diagram so that each true and complement input variable is available along a conductor called a bus.

Cell   The smallest unit of a Karnaugh map, corresponding to one line of a truth table. The input variables are the cell's coordinates and the output variable is the cell's contents.

Combinational Logic   Digital circuitry in which an output is derived from the combination of inputs, independent of the order in which they are applied.

Combinatorial Logic   Another name for combinational logic.

Commutative Property   A mathematical operation is commutative if it can be applied to its operands in any order without affecting the result. For example, addition is commutative $(a + b = b + a)$, but subtraction is not $(a - b \neq b - a)$.

Distributive Property   Full name: distributive property of multiplication over addition. The property that allows us to distribute ("multiply through") an AND across several OR functions. For example, $a(b + c) = ab + ac$.

Don't Care State   An output state that can be regarded either as HIGH or LOW, as is most convenient. A don't care state is the output state of a circuit for a combination of inputs that will never occur under stated design conditions.

Double-Rail Inputs   Boolean input variables that are available to a circuit in both true and complement form.

Karnaugh Map   A graphical tool for finding the maximum SOP or POS simplification of a Boolean expression. A Karnaugh map works by arranging the terms of an expression in such a way that variables can be cancelled by grouping minterms or maxterms.

Levels of Gating   The number of gates through which a signal must pass from input to output of a logic gate network.

Logic Diagram   A diagram, similar to a schematic, showing the connection of logic gates.

Logic Gate Network   Two or more logic gates connected together.

Maximum POS Simplification   The form of a POS Boolean expression that cannot be further simplified by cancelling variables in the sum terms. It may be possible to get an SOP form with fewer terms or variables.

Maximum SOP Simplification   The form of an SOP Boolean expression that cannot be further simplified by cancelling variables in the product terms. It may be possible to get a POS form with fewer terms or variables.

Maxterm   A sum term in a Boolean expression where all possible variables appear once in true or complement form.

Minterm   A product term in a Boolean expression where all possible variables appear once in true or complement form.

Octet   A group of eight cells in a Karnaugh map. An octet cancels three variables in a K-map simplification.

Order of Precedence   The sequence in which Boolean functions are performed, unless otherwise specified by parentheses.

Pair   A group of two cells in a Karnaugh map. A pair cancels one variable in a K-map simplification.

Product Term   A term in a Boolean expression where one or more true or complement variables are ANDed.

Product-of-Sums (POS)   A type of Boolean expression where several sum terms are multiplied (ANDed) together.

Quad   A group of four cells in a Karnaugh map. A quad cancels two variables in a K-map simplification.

Sequential Logic   Digital circuitry in which the output state of the circuit depends not only on the states of the inputs, but also on the sequence in which they reached their present states.

Sum Term   A term in a Boolean expression where one or more true or complement variables are ORed.

Sum-of-Products (SOP)   A type of Boolean expression where several product terms are summed (ORed) together.

Synthesis   The process of creating a logic circuit from a description such as a Boolean equation or truth table.

PROBLEMS

*Problem numbers set in color
indicate more difficult problems.*

3.1   Boolean Expressions, Logic Diagrams, and Truth Tables

3.1   Write the unsimplified Boolean expression for each of the logic gate networks shown in Figure 3.98.

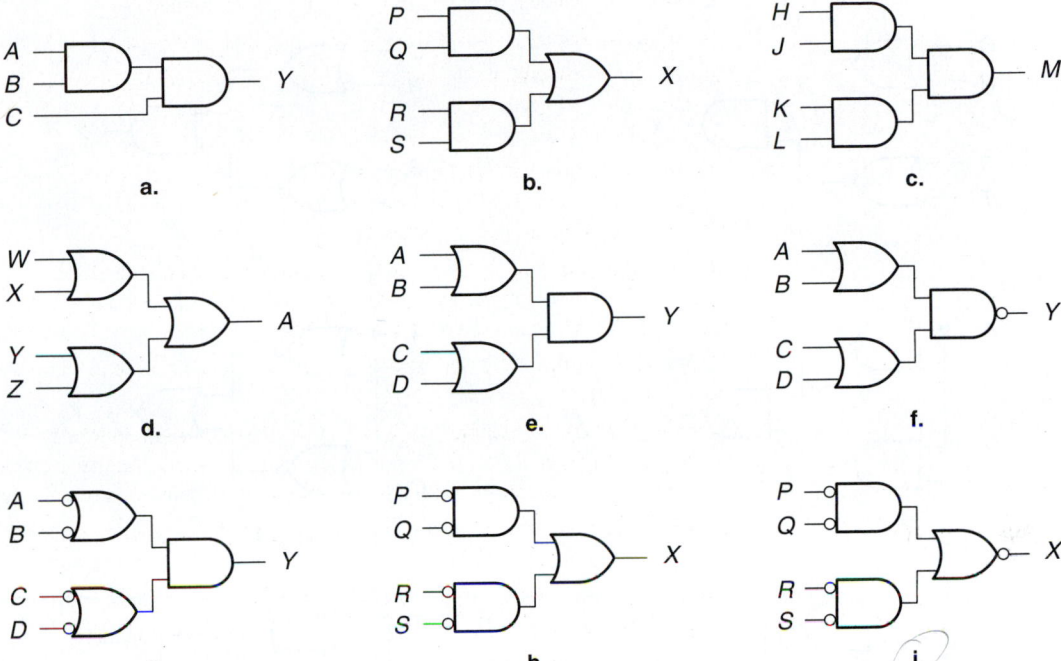

FIGURE 3.98   Problem 3.1: Logic Circuits

3.2   Write the unsimplified Boolean expression for each of the logic gate networks shown in Figure 3.99.

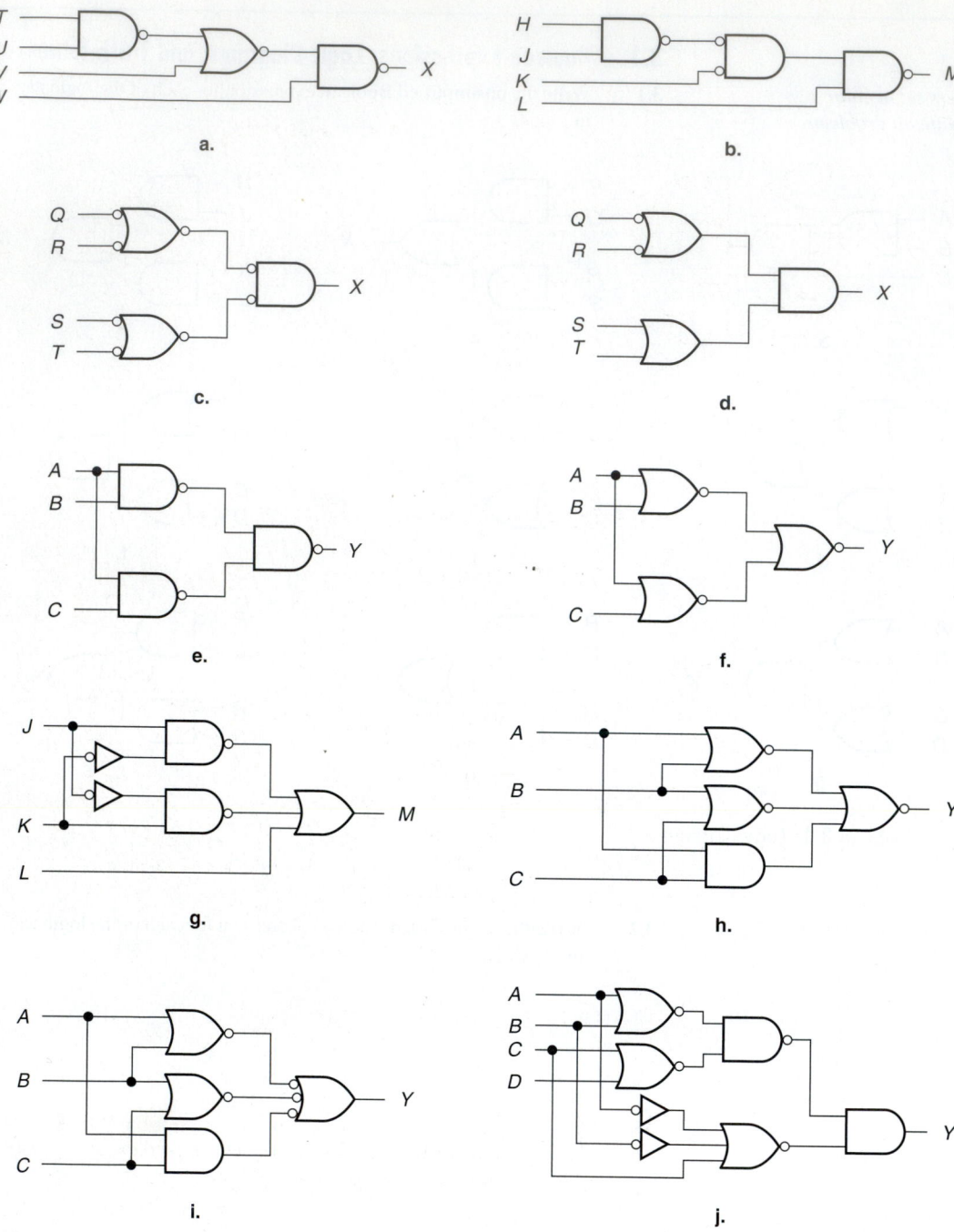

FIGURE 3.99   Problem 3.2: Logic Circuits

3.3   Redraw the logic diagrams of the gate networks shown in Figure 3.99a, e, f, h, i, and j so that they conform to the bubble-to-bubble convention. Rewrite the Boolean expression of each of the redrawn circuits.

3.4   **a.** Write the unsimplified Boolean equation for each of the logic diagrams in Figure 3.100.

b. Redraw each of the logic diagrams in Figure 3.100 so that they conform to the bubble-to-bubble convention. Rewrite the Boolean expression for each of the redrawn circuits.

FIGURE 3.100 Problem 3.4: Logic Circuits

3.5 **a.** Write the unsimplified Boolean equation for the logic diagram in Figure 3.101.

b. Redraw the logic diagram in Figure 3.101 so that it conforms to the bubble-to-bubble convention. Rewrite the Boolean expression of the redrawn circuit.

FIGURE 3.101 Problem 3.5: Logic Circuits

3.6 The circuit in Figure 3.102 is called a majority vote circuit. It will turn on an active-HIGH indicator lamp only if a majority of inputs (at least two out of three) are HIGH. Write the Boolean expression for the circuit.

FIGURE 3.102 Problem 3.6: Majority Vote Circuit

3.7   Suppose you wish to design a circuit that indicates when at least three out of four inputs are HIGH. The circuit has four inputs, D_3, D_2, D_1, and D_0 and an active-HIGH output, Y. Write the Boolean expression for the circuit and draw the logic circuit.

3.8   Draw the logic circuit for each of the following Boolean expressions:

    a. $Y = AB + BC$

    b. $Y = ACD + BCD$

    c. $Y = (A + B)(C + D)$

    d. $Y = A + BC + D$

    e. $Y = \overline{AC} + \overline{B + C}$

    f. $Y = \overline{\overline{AC} + \overline{B} + C}$

    g. $Y = \overline{\overline{\overline{ABD}} + \overline{BC} + \overline{A} + C}$

    h. $Y = \overline{\overline{\overline{AB}} + \overline{\overline{AC}} + \overline{BC}}$

    i. $Y = \overline{\overline{AB}} + \overline{\overline{\overline{AC}}} + \overline{BC}$

3.9   Use DeMorgan's theorems to modify the Boolean equations in Problem 3.8, parts e, f, g, h, and i, so that there is no bar over any group of variables. Redraw the logic diagrams of the circuits to reflect the changes. (The final circuit versions should conform to the bubble-to-bubble convention.)

3.10   Write the truth tables for the logic diagrams in Figure 3.98, parts b, e, f, and g.

3.11   Write the truth tables for the logic diagrams in Figure 3.99, parts a, h, i, and j.

3.12   Write the truth tables for the Boolean expression in Problem 3.8, parts c, d, e, f, h, and i.

3.2   Sum-of-Products (SOP) and Product-of-Sums (POS) Forms

3.13   Find the Boolean expression, in both sum-of-products (SOP) and product-of-sums (POS) forms, for the logic function represented by the following truth table. Draw the logic diagram for each form.

A	B	C	Y
0	0	0	1
0	0	1	1
0	1	0	1
0	1	1	1
1	0	0	0
1	0	1	0
1	1	0	0
1	1	1	0

3.14   Find the Boolean expression, in both sum-of-products (SOP) and product-of-sums (POS) forms, for the logic function represented by the following truth table. Draw the logic diagram for the SOP form only.

A	B	C	Y
0	0	0	0
0	0	1	1
0	1	0	0
0	1	1	0
1	0	0	1
1	0	1	0
1	1	0	1
1	1	1	0

3.15   Find the Boolean expression, in both sum-of-products (SOP) and product-of-sums (POS) forms, for the logic function represented by the following truth table. Draw the logic diagram for the POS form only.

A	B	C	Y
0	0	0	0
0	0	1	1
0	1	0	1
0	1	1	0
1	0	0	0
1	0	1	1
1	1	0	1
1	1	1	1

3.16   Find the Boolean expression, in both sum-of-products (SOP) and product-of-sums (POS) forms, for the logic function represented by the following truth table. Draw the logic diagram for the SOP form only.

A	B	C	D	Y
0	0	0	0	0
0	0	0	1	0
0	0	1	0	0
0	0	1	1	0
0	1	0	0	1
0	1	0	1	0
0	1	1	0	1
0	1	1	1	0
1	0	0	0	1
1	0	0	1	0
1	0	1	0	0
1	0	1	1	0
1	1	0	0	0
1	1	0	1	1
1	1	1	0	0
1	1	1	1	0

3.17   Use the truth table of a 2-input XOR gate to write its Boolean expression in POS form. Draw the logic diagram of the POS form of the XOR function.

3.18   Use the truth table of a 2-input XNOR gate to write its Boolean expression in POS form. Draw the logic diagram of the POS form of the XNOR function.

3.3   Theorems of Boolean Algebra

3.19   Write the Boolean expression for the circuit shown in Figure 3.103. Use the distributive property to transform the circuit into a sum-of-products (SOP) circuit.

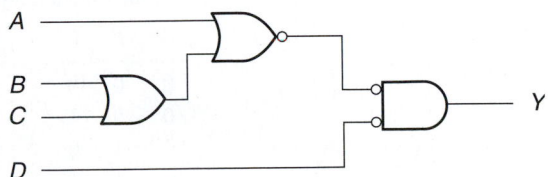

FIGURE 3.103   Problem 3.19: Logic Circuit

3.20   Write the Boolean expression for the circuit shown in Figure 3.104. Use the distributive property to transform the circuit into a sum-of-products (SOP) circuit.

FIGURE 3.104   Problem 3.20: Logic Circuit

3.21   Use the rules of Boolean algebra to simplify the following expressions as much as possible.

    a. $Y = A A B + C$

    b. $Y = A \bar{A} B + C$

    c. $J = K + L \bar{L}$

    d. $S = (T + U) V \bar{V}$

    e. $S = T + V \bar{V}$

    f. $Y = (A \bar{B} + \bar{C})(B \bar{D} + F)$

3.22   Use the rules of Boolean algebra to simplify the following expressions as much as possible.

    a. $M = P Q + \overline{P Q} R$

    b. $M = P Q + P Q R$

    c. $S = (\overline{T + U}) V + (T + U)$

    d. $Y = (\bar{A} + B + \bar{D}) A C + A \bar{B} D$

    e. $Y = (\bar{A} + B + \bar{D}) A C + \overline{A \bar{B} D}$

    f. $P = (\overline{Q R} + S T)(\overline{Q R} + Q)$

    g. $U = (X + \bar{Y} + \overline{W} Z)(W Y + Y + \overline{W} Z)$

3.23   Use the rules of Boolean algebra to simplify the following expressions as much as possible.

    a. $Y = \overline{A B} \, C D + (\bar{A} + \bar{B}) \, \overline{C + D} + \bar{A} + \bar{B}$

    b. $Y = \overline{A B} \, C D + (\bar{A} + \bar{B}) \, \overline{C + D} + \overline{\bar{A} + \bar{B}}$

    c. $K = (\bar{L} M + L \bar{M})(M \bar{N} + L M N) + M(\bar{N} + L)$

3.4   Simplifying SOP and POS Expressions

3.24   Use the rules of Boolean algebra to find the maximum SOP and POS simplifications of the function represented by the following truth table.

A	B	C	Y
0	0	0	0
0	0	1	1
0	1	0	0
0	1	1	1
1	0	0	0
1	0	1	1
1	1	0	0
1	1	1	1

3.25 Use the rules of Boolean algebra to find the maximum SOP and POS simplifications of the function represented by the following truth table.

A	B	C	Y
0	0	0	1
0	0	1	0
0	1	0	1
0	1	1	0
1	0	0	0
1	0	1	0
1	1	0	1
1	1	1	0

3.26 Use the rules of Boolean algebra to find the maximum SOP and POS simplifications of the function represented by the following truth table.

A	B	C	Y
0	0	0	0
0	0	1	1
0	1	0	0
0	1	1	1
1	0	0	0
1	0	1	1
1	1	0	0
1	1	1	0

3.27 Use the rules of Boolean algebra to find the maximum SOP simplification of the function represented by the following truth table.

A	B	C	D	Y
0	0	0	0	0
0	0	0	1	0
0	0	1	0	0
0	0	1	1	0
0	1	0	0	1
0	1	0	1	1
0	1	1	0	0
0	1	1	1	0
1	0	0	0	0
1	0	0	1	1
1	0	1	0	0
1	0	1	1	1
1	1	0	0	1
1	1	0	1	1
1	1	1	0	0
1	1	1	1	1

3.28　Use the rules of Boolean algebra to find the maximum SOP simplification of the function represented by the following truth table.

A	B	C	D	Y
0	0	0	0	1
0	0	0	1	0
0	0	1	0	1
0	0	1	1	0
0	1	0	0	1
0	1	0	1	0
0	1	1	0	0
0	1	1	1	0
1	0	0	0	0
1	0	0	1	0
1	0	1	0	0
1	0	1	1	0
1	1	0	0	1
1	1	0	1	1
1	1	1	0	0
1	1	1	1	0

3.29　Use the rules of Boolean algebra to find the maximum SOP simplification of the function represented by the following truth table.

A	B	C	D	Y
0	0	0	0	0
0	0	0	1	1
0	0	1	0	0
0	0	1	1	1
0	1	0	0	0
0	1	0	1	1
0	1	1	0	1
0	1	1	1	1
1	0	0	0	0
1	0	0	1	1
1	0	1	0	0
1	0	1	1	0
1	1	0	0	0
1	1	0	1	1
1	1	1	0	1
1	1	1	1	0

3.30　Use the rules of Boolean algebra to find the maximum SOP simplification of the function represented by the following truth table.

A	B	C	D	Y
0	0	0	0	1
0	0	0	1	0
0	0	1	0	0
0	0	1	1	1
0	1	0	0	1
0	1	0	1	0
0	1	1	0	0
0	1	1	1	0
1	0	0	0	1
1	0	0	1	0
1	0	1	0	1
1	0	1	1	1
1	1	0	0	1
1	1	0	1	0
1	1	1	0	1
1	1	1	1	1

3.31 Use the rules of Boolean algebra to find the maximum SOP simplification of the function represented by the following truth table.

A	B	C	D	Y
0	0	0	0	0
0	0	0	1	1
0	0	1	0	0
0	0	1	1	0
0	1	0	0	0
0	1	0	1	0
0	1	1	0	1
0	1	1	1	1
1	0	0	0	1
1	0	0	1	0
1	0	1	0	0
1	0	1	1	0
1	1	0	0	0
1	1	0	1	0
1	1	1	0	1
1	1	1	1	1

3.32 Use the rules of Boolean algebra to find the maximum SOP simplification of the function represented by the following truth table.

A	B	C	D	Y
0	0	0	0	0
0	0	0	1	0
0	0	1	0	0
0	0	1	1	1
0	1	0	0	0
0	1	0	1	1
0	1	1	0	0
0	1	1	1	1
1	0	0	0	0
1	0	0	1	0
1	0	1	0	0
1	0	1	1	1
1	1	0	0	0
1	1	0	1	1
1	1	1	0	0
1	1	1	1	1

3.5 Simplification by the Karnaugh Map Method

3.33 Use the Karnaugh map method to find the maximum SOP simplification of the logic diagram in Figure 3.21.

3.34 Use the Karnaugh map method to reduce the following Boolean expressions to their maximum SOP simplifications:

a. $Y = \bar{A}\,\bar{B}\,C + \bar{A}\,B\,C + A\,B\,C$

b. $Y = \bar{A}\,\bar{B}\,C + \bar{A}\,B\,C + A\,B\,\bar{C} + A\,B\,C + A\,\bar{B}\,C$

c. $Y = \bar{A}\,\bar{B}\,\bar{C} + \bar{A}\,B\,C + A\,B\,C + A\,\bar{B}\,C$

d. $Y = \bar{A}\,\bar{B}\,\bar{C}\,\bar{D} + \bar{A}\,\bar{B}\,\bar{C}\,D + \bar{A}\,\bar{B}\,C\,D + \bar{A}\,B\,C\,\bar{D} + \bar{A}\,B\,C\,\bar{D} + A\,B\,\bar{C}\,D + A\,B\,\bar{C}\,\bar{D}$
$+ A\,\bar{B}\,\bar{C}\,\bar{D} + A\,\bar{B}\,C\,\bar{D}$

3.35   Use the Karnaugh map method to reduce the Boolean expression represented by the following truth table to simplest SOP form.

A	B	C	D	Y
0	0	0	0	0
0	0	0	1	0
0	0	1	0	0
0	0	1	1	1
0	1	0	0	1
0	1	0	1	1
0	1	1	0	1
0	1	1	1	1
1	0	0	0	0
1	0	0	1	0
1	0	1	0	0
1	0	1	1	1
1	1	0	0	0
1	1	0	1	0
1	1	1	0	0
1	1	1	1	1

3.36   Use the Karnaugh map method to reduce the Boolean expression represented by the following truth table to simplest SOP form.

A	B	C	D	Y
0	0	0	0	0
0	0	0	1	1
0	0	1	0	0
0	0	1	1	1
0	1	0	0	0
0	1	0	1	1
0	1	1	0	0
0	1	1	1	1
1	0	0	0	1
1	0	0	1	0
1	0	1	0	0
1	0	1	1	0
1	1	0	0	0
1	1	0	1	1
1	1	1	0	0
1	1	1	1	1

3.37   Use the Karnaugh map method to reduce the Boolean expression represented by the following truth table to simplest SOP form.

A	B	C	D	Y
0	0	0	0	0
0	0	0	1	0
0	0	1	0	1
0	0	1	1	1
0	0	1	0	0
0	1	0	1	0
0	1	1	0	0
0	1	1	1	0
1	0	0	0	0
1	0	0	1	1
1	0	1	0	X
1	0	1	1	X
1	1	0	0	X
1	1	0	1	X
1	1	1	0	X
1	1	1	1	X

3.38   Use the Karnaugh map method to reduce the Boolean expression represented by the following truth table to simplest SOP form.

A	B	C	D	Y
0	0	0	0	1
0	0	0	1	1
0	0	1	0	1
0	0	1	1	1
0	1	0	0	0
0	1	0	1	0
0	1	1	0	1
0	1	1	1	0
1	0	0	0	1
1	0	0	1	0
1	0	1	0	1
1	0	1	1	0
1	1	0	0	0
1	1	0	1	1
1	1	1	0	1
1	1	1	1	0

3.39   Use the Karnaugh map method to reduce the Boolean expression represented by the following truth table to simplest SOP form.

A	B	C	D	Y
0	0	0	0	0
0	0	0	1	1
0	0	1	0	0
0	0	1	1	1
0	1	0	0	0
0	1	0	1	1
0	1	1	0	0
0	1	1	1	1
1	0	0	0	1
1	0	0	1	0
1	0	1	0	0
1	0	1	1	1
1	1	0	0	0
1	1	0	1	0
1	1	1	0	0
1	1	1	1	1

3.40   Use the Karnaugh map method to reduce the Boolean expression represented by the following truth table to simplest SOP form.

A	B	C	D	Y
0	0	0	0	1
0	0	0	1	1
0	0	1	0	0
0	0	1	1	0
0	1	0	0	1
0	1	0	1	1
0	1	1	0	0
0	1	1	1	1
1	0	0	0	0
1	0	0	1	0
1	0	1	0	1
1	0	1	1	1
1	1	0	0	0
1	1	0	1	1
1	1	1	0	1
1	1	1	1	1

3.41   Use the Karnaugh map method to reduce the Boolean expression represented by the following truth table to simplest SOP form.

A	B	C	D	Y
0	0	0	0	0
0	0	0	1	0
0	0	1	0	0
0	0	1	1	0
0	1	0	0	1
0	1	0	1	1
0	1	1	0	1
0	1	1	1	1
1	0	0	0	0
1	0	0	1	1
1	0	1	0	0
1	0	1	1	0
1	1	0	0	1
1	1	0	1	0
1	1	1	0	1
1	1	1	1	0

3.42   Use the Karnaugh map method to reduce the Boolean expression represented by the following truth table to simplest SOP form.

A	B	C	D	Y
0	0	0	0	1
0	0	0	1	0
0	0	1	0	1
0	0	1	1	1
0	1	0	0	0
0	1	0	1	0
0	1	1	0	0
0	1	1	1	1
1	0	0	0	0
1	0	0	1	0
1	0	1	0	1
1	0	1	1	1
1	1	0	0	0
1	1	0	1	0
1	1	1	0	0
1	1	1	1	1

3.43   Use the Karnaugh map method to reduce the Boolean expression represented by the following truth table to simplest SOP form.

A	B	C	D	Y
0	0	0	0	0
0	0	0	1	1
0	0	1	0	0
0	0	1	1	1
0	1	0	0	0
0	1	0	1	1
0	1	1	0	0
0	1	1	1	1
1	0	0	0	0
1	0	0	1	1
1	0	1	0	0
1	0	1	1	1
1	1	0	0	0
1	1	0	1	1
1	1	1	0	0
1	1	1	1	1

3.44   Use the Karnaugh map method to reduce the Boolean expression represented by the following truth table to simplest SOP form.

A	B	C	D	Y
0	0	0	0	1
0	0	0	1	1
0	0	1	0	1
0	0	1	1	1
0	1	0	0	0
0	1	0	1	0
0	1	1	0	1
0	1	1	1	1
1	0	0	0	1
1	0	0	1	1
1	0	1	0	1
1	0	1	1	1
1	1	0	0	0
1	1	0	1	0
1	1	1	0	0
1	1	1	1	0

3.45   Use the Karnaugh map method to reduce the Boolean expression represented by the following truth table to simplest SOP form.

A	B	C	D	Y
0	0	0	0	0
0	0	0	1	0
0	0	1	0	0
0	0	1	1	0
0	1	0	0	0
0	1	0	1	0
0	1	1	0	0
0	1	1	1	1
1	0	0	0	1
1	0	0	1	1
1	0	1	0	1
1	0	1	1	1
1	1	0	0	1
1	1	0	1	1
1	1	1	0	0
1	1	1	1	1

3.46   The circuit in Figure 3.105 represents the maximum SOP simplification of a Boolean function.

FIGURE 3.105   Problem 3.46: Logic Circuit

Use a Karnaugh map to derive the circuit for the maximum POS simplification.

3.47   Repeat Problem 3.46 for the circuit in Figure 3.106.

FIGURE 3.106   Problem 3.47: Logic Circuit

3.48   Refer to the BCD-to-2421 code converter developed in Example 3.22. Use a similar design procedure to develop the circuit of a 2421-to-BCD code converter.

3.49   Excess-3 code is a decimal code that is generated by adding 0011 ($= 3_{10}$) to a BCD code. Table 3.25 shows the relationship between a decimal digital code, natural BCD code, and Excess-3 code. Draw the circuit of a BCD-to-Excess-3 code converter, using the Karnaugh map method to simplify all Boolean expressions.

TABLE 3.25   BCD and Excess-3 Code

Decimal Equivalent	BCD Code				Excess-3			
	D_4	D_3	D_2	D_1	E_4	E_3	E_2	E_1
0	0	0	0	0	0	0	1	1
1	0	0	0	1	0	1	0	0
2	0	0	1	0	0	1	0	1
3	0	0	1	1	0	1	1	0
4	0	1	0	0	0	1	1	1
5	0	1	0	1	1	0	0	0
6	0	1	1	0	1	0	0	1
7	0	1	1	1	1	0	1	0
8	1	0	0	0	1	0	1	1
9	1	0	0	1	1	1	0	0

3.50   Repeat Problem 3.49 for an Excess-3-to-BCD code converter.

3.51   **a.** Write the unsimplified Boolean expression for the logic circuit shown in Figure 3.107.

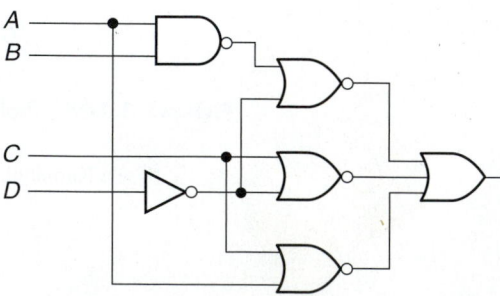

FIGURE 3.107   Problem 3.51: Logic Circuit

   b. Redraw the circuit in Figure 3.107 to conform to the bubble-to-bubble convention. Write the Boolean expression of the redrawn circuit.

3.52   **a.** Write the unsimplified Boolean expression for the logic circuit shown in Figure 3.108.

FIGURE 3.108   Problem 3.52: Logic Circuit

   b. Redraw the circuit in Figure 3.108 to conform to the bubble-to-bubble convention. Write the Boolean expression of the redrawn circuit.

3.53   **a.** Write the unsimplified Boolean expression for the logic circuit shown in Figure 3.109.

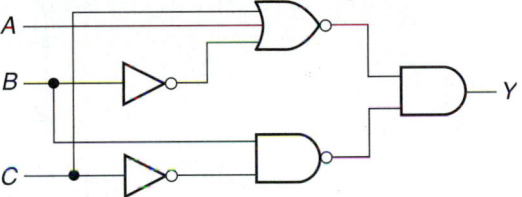

FIGURE 3.109   Problem 3.53: Logic Circuit

   b. Redraw the circuit in Figure 3.109 to conform to the bubble-to-bubble convention. Write the Boolean expression of the redrawn circuit.

3.7   Universal Property of NAND/NOR Gates

3.54   Redraw the circuit in Figure 3.110 as an all-NAND circuit and as an all-NOR circuit. In each case, simplify the circuit as much as possible.

FIGURE 3.110   Problem 3.54: Logic Circuit

3.55   Repeat Problem 3.54 for the circuit in Figure 3.111.

FIGURE 3.111   Problem 3.55: Logic Circuit

3.56   Repeat Problem 3.54 for the circuit in Figure 3.112.

FIGURE 3.112   Problem 3.56: Logic Circuit

3.57 Repeat Problem 3.54 for the circuit in Figure 3.113.

FIGURE 3.113 Problem 3.57: Logic Circuit

3.8 Practical Circuit Implementation

3.58 Redraw the circuit shown in Figure 3.114, including pin numbers and unused gates, for an implementation in High-Speed CMOS. You may use any available High-Speed CMOS gates. (Refer to Table 3.19 and Figure 3.85 for pinouts).

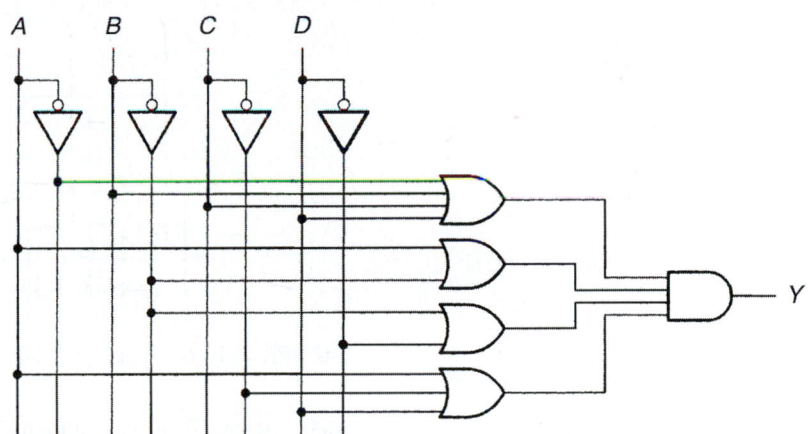

FIGURE 3.114 Problem 3.58: Logic Circuit

3.59 Repeat Problem 3.58 for an implementation in low-power Schottky (LS) TTL.

3.60 Redraw the circuit in Figure 3.112, including pin numbers and unused gates, for an implementation in High-Speed CMOS.

3.61 Redraw the circuit in Figure 3.113, including pin numbers and unused gates, for an implementation in High-Speed CMOS.

3.62 Repeat Problem 3.61 for an implementation in low-power Schottky (LS) TTL.

3.9   Pulsed Operation of Logic Circuits

3.63   Given inputs *A*, *B*, and *C* to the logic circuit shown in Figure 3.115, draw the wave-form for the circuit output, *Y*.

FIGURE 3.115   Problem 3.63: Logic Circuit

3.64   Repeat Problem 3.63 for the circuit in Figure 3.116.

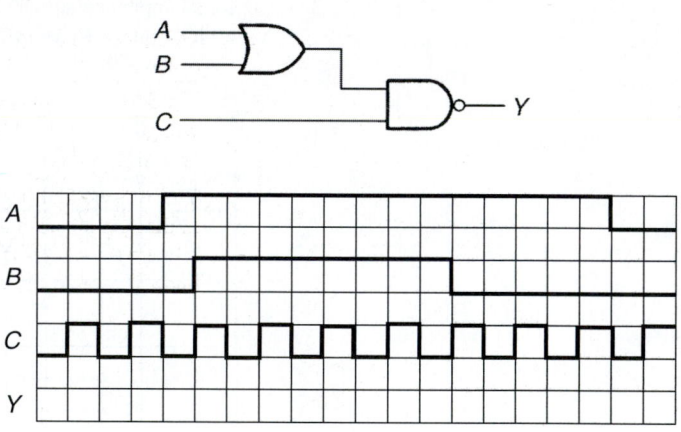

FIGURE 3.116   Problem 3.64: Logic Circuit

3.65   Repeat Problem 3.63 for the circuit in Figure 3.117.

FIGURE 3.117   Problem 3.65: Logic Circuit

3.66 Repeat Problem 3.63 for the circuit in Figure 3.118.

FIGURE 3.118 Problem 3.66: Logic Circuit

3.67 Part of a security system is shown in Figure 3.119. The system sounds a 200-Hz buzzer if certain security conditions are breached. The system is controlled by a conductive window strip, a door switch, and two related timers.

a. When the door opens, the door switch opens, producing a logic **HIGH**.

b. There are two timers, one for 30 seconds, one for 60 seconds. Each timer starts as soon as its START input goes HIGH (i.e., when the door opens).

c. The output, *A,* of the 60-second timer is normally LOW. When this timer starts, *A* goes HIGH for 60 seconds.

d. The 30-second timer is slightly different. Its output, *B,* is also normally LOW, but will go HIGH *after* the 30 seconds has elapsed. After this time, it will stay HIGH until the system is reset by pressing the RESET switch.

e. The 30-second timer can be reset any time after the door opens. If it is not reset in the first 30 seconds, B goes HIGH and a 200-Hz buzzer will sound. (We can prevent the buzzer from sounding by pressing the RESET switch within 30 seconds of opening the door.)

FIGURE 3.119 Problem 3.67: Security System

f. The buzzer sounds immediately if a conductive strip around a window (represented by a switch on input C) opens at any time, making input C HIGH.

Draw the timing diagram for all possible input conditions, and indicate the action taking place for each condition.

3.10   A General Approach to Logic Circuit Design

3.68   A digital circuit is required to display a digital signal on a bar-graph LED, similar to a sound meter on a stereo. You know that the input to your circuit has values from 0 to 3 coded in binary format. You need to indicate the value in a bar graph where progressively more LEDs turn on as the input number gets higher, as shown in Figure 3.120. When the highest number is reached, all LEDs are on.

a. Write the truth table for the circuit required to drive the LEDs. Assume that the inputs are called D_1 and D_0 and the outputs are Y_1, Y_2, and Y_3, where Y_1 is the lowest LED in the diagram in Figure 3.120 and Y_3 is the highest. Assume that the outputs are active-HIGH.

b. Write and simplify the Boolean equations for the three outputs. Draw a logic circuit to show how the simplified functions can be implemented.

FIGURE 3.120   Problem 3.68: Bar-Graph LED

3.69   Repeat Problem 3.68 for a 7-position bar-graph LED, as shown in Figure 3.121.

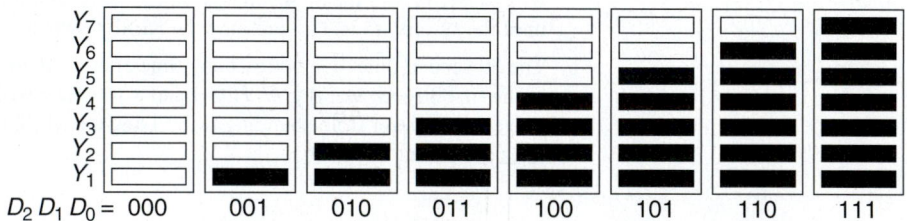

FIGURE 3.121   Problem 3.69: Bar-Graph LED

ANSWERS TO SECTION REVIEW PROBLEMS

3.1A

3.1a $Y = \overline{ABC + D}$

3.1b $\text{OUT} = (\overline{W} + \overline{X} + Y)\overline{Z}$

3.1B

3.2

A	B	C	Y
0	0	0	0
0	0	1	1
0	1	0	1
0	1	1	1
1	0	0	0
1	0	1	1
1	1	0	0
1	1	1	0

3.2

3.3a   SOP: $Y = A\bar{B}\bar{C} + A\bar{B}C$

POS: $Y = (A + B + C)(A + B + \bar{C})(A + \bar{B} + C)(A + \bar{B} + \bar{C})(\bar{A} + \bar{B} + C)(\bar{A} + \bar{B} + \bar{C})$

3.3b   SOP: $Y = \bar{A}\bar{B}\bar{C} + A\bar{B}\bar{C} + A\bar{B}C + AB\bar{C}$

POS: $Y = (A + B + \bar{C})\,(A + \bar{B} + C)\,(A + \bar{B} + \bar{C})\,(\bar{A} + \bar{B} + \bar{C})$

3.3

3.4a $Y = \overline{AC}$ or $Y = \bar{A} + \bar{C}$

3.4b $Y = \overline{AC} + D$ or $Y = \bar{A} + \bar{C} + D$

3.4c $Y = A\bar{B}$

3.4

3.5   SOP: $Y = \bar{A}C + B\bar{C}$ POS: $Y = (\bar{A} + \bar{C})(B + C)$

3.6

3.6   See Figure 3.122 and Figure 3.123 for two separate solutions.

FIGURE 3.122   Section Review Problem 3.6 (solution 1)

FIGURE 3.123   Section Review Problem 3.6 (solution 2)

KEY TERMS

Altera UP-2 board
Block Diagram File
ByteBlaster
ByteBlaster MV
Cell
Checksum
Complex PLD (CPLD)
Compile
Design Entry
Download
Fitting
Fixed-function Logic
Hardware Description Language (HDL)
Hierarchical Design
Hierarchy
Hierarchy Tree
Input Line
Input Line Number
Instance
ICR
ISP
JEDEC
JEDEC File
JTAG
JTAG Chain
Multiplexer
Nonvolatile
PAL
Primitives
Product Line
Product Line First Cell Number
Programmable Logic Device (PLD)

Programmer Object File (pof)
Programming
Project
Project Navigator Window
Schematic Entry
Simulation
Software Tools
SRAM Object File (sof)
Suite (of software tools)
Symbol File
Target device
Text File
TCK
TDI
TDO
TMS
Top Level (of a Hierarchy)
User Library
Vector Waveform File (vwf)
Volatile

OUTLINE

4.1   What Is a PLD?

4.2   Programmable Sum-of-Products Arrays

4.3   PAL Fuse Matrix and Combinational Outputs

4.4   PAL Outputs with Programmable Polarity

4.5   Programming CPLDs Using Quartus II

4.6   Quartus II Design Flow and Graphical User Interface

4.7   Creating a Quartus II Project and Block Diagram File

4.8   Compiling and Simulating a Design in Quartus II

4.9   Transferring a Design to a Target CPLD

4.10   Using the Quartus II Block Editor to Create a Hierarchical Design

CHAPTER 4

Introduction to PLDs and Quartus II

CHAPTER OBJECTIVES

Upon successful completion of this chapter you will be able to:

- Draw a diagram showing the basic hardware conventions for a sum-of-products-type programmable logic device.
- Describe the structure of a programmable array logic (PAL) AND matrix.
- Draw fuses on the logic diagram of a PAL to implement simple logic functions.
- Describe the structures of combinational PAL outputs, including those with programmable polarity.
- Determine the number and type of outputs from a PAL/GAL part number.
- Describe some advantages of programmable logic over fixed-function logic.
- Name some types of programmable logic devices (PLDs).
- Use Altera's Quartus II PLD design software to enter simple combinational circuits using schematic capture.
- Create circuit symbols from schematic designs and use them in hierarchical designs for PLDs.
- Assign device and pin numbers to schematic designs and compile them for programming Altera MAX 7000S or FLEX 10K devices.
- Program Altera PLDs via a JTAG interface and a ByteBlaster Parallel Port Download Cable.

In the first three chapters of this book, we examined logic gates and Boolean algebra. These basic foundations of combinational circuitry, as well as the sequential logic circuits we will study in a later chapter, form the fundamental building blocks of many digital integrated circuits (ICs).

In the past, such digital ICs were fixed in their logic functions; it was not possible to change designs without changing the chips in a circuit. Programmable logic offers the digital circuit designer the possibility of changing design function even after it has been built. A programmable logic device (PLD) can be programmed, erased, and reprogrammed many times, allowing easier prototyping and design modification. (The industry marketing buzz often refers to "rapid prototyping" and "reduced time to market.") The number of IC packages required to implement a design with one or more PLDs is often reduced, compared to a design fabricated using standard fixed-function ICs.

PLDs can be programmed from a personal computer (PC) or workstation running special software. This software is a set of programs that allows us to design circuits for various PLDs. Quartus II, owned by Altera Corporation, is such a software package. Quartus II allows us to enter PLD designs, either as schematics or in several hardware description languages (specialized computer languages for modeling and synthesizing digital hardware). A design can contain components that are in themselves complete digital circuits. Quartus II converts the design information into a binary form that can be transferred into a PLD via a special interface connected to the parallel port of a PC.

Before we examine the workings of the Quartus II software, it is useful to have an overview of the structure of some simple PLDs. Old-style PLDs, such as the PAL16L8, were structured as a sum-of-products (SOP) array that could be programmed by blowing metal fuse links. This structure was such that, once programmed, a device could not be erased and reprogrammed, thus giving rise to the designation of one-time programmable, or OTP.

Newer PLDs, such as the GAL22V10 and MAX 7000S series, are not programmed with fuses, but by the ON or OFF states of programmable transistors in an SOP array. These PLDs are based on electrically erasable programmable read-only memory (EEPROM) technology, a type of memory that retains its programming information even after power has been removed from the device.

The Altera FLEX series of CPLDs is based on another technology altogether. It stores logic functions in look-up tables (LUTs) that act as truth tables with four input bits. The main logic element of the FLEX series is the SRAM (static random-access memory) cell. SRAM-based CPLDs must have their programming data loaded every time they are powered up. They have the advantage of being faster than EEPROM devices, with a higher bit capacity.

We will examine GAL22V10, MAX 7000S, and FLEX 10K devices in a later chapter. ∎

4.1   WHAT IS A PLD?

> ∎ **KEY TERMS**
>
> **Programmable Logic Device (PLD)**   A digital integrated circuit that can be programmed by the user to implement any digital logic function.
>
> **Fixed-Function Logic**   A digital logic device or circuit whose function is determined when it is manufactured and, hence, cannot be changed.
>
> **Complex PLD (CPLD)**   A digital device consisting of several programmable sections with internal interconnections between the sections.
>
> **Schematic Entry**   A technique of entering CPLD design information by using a CAD (computer aided design) tool to draw a logic circuit as a schematic. The schematic can then be interpreted by design software to generate programming information for the CPLD.
>
> **Quartus II**   CPLD design and programming software owned by Altera Corporation.
>
> **Compile**   The process used by CPLD design software to interpret design information (such as a drawing or text file) and create required programming information for a CPLD.

One of the most far-reaching developments in digital electronics has been the introduction of **programmable logic devices (PLDs).** Prior to the development of PLDs, digital circuits were constructed in various scales of integrated circuit logic, such as small scale integration (SSI) and medium scale integration (MSI) devices. These devices contained logic gates and other digital circuits. The functions of these devices, which we can refer to as **fixed-function logic,** were determined at the time of manufacture and could not be changed. This necessitated the manufacture of a large number of device types, requiring shelves full of data books just to describe them. Also, if a designer wanted a device with a particular function that was not in a manufacturer's list of offerings, he or she was forced to make a circuit that used multiple devices, some of which might contain functions neither wanted nor needed, thus wasting circuit board space and design time.

Programmable logic provides a solution to these problems. A PLD is supplied to the user with no logic function programmed in at all. It is up to the designer to make the PLD perform in whatever way a design requires; only those functions required by the design need be programmed. Several functions can usually be combined in the design and programmed onto a single chip, so the package count and required board space can be reduced as well. Also, if a design needs to be changed, a PLD can be reprogrammed with the new design information, often without removing it from the circuit.

PLD is a generic term. There is a wide variety of PLD types, including PAL (programmable array logic), GAL (generic array logic), EPLD (erasable PLD), **CPLD (complex PLD),** FPGA (field-programmable gate array), as well as several others. We will focus on CPLDs as a representative type of PLD. Although terminology varies somewhat throughout the industry, we will use the term "CPLD" to mean a device based on a programmable sum-of-products (SOP) array, with several programmable sections that are connected internally. In effect, a CPLD is several interconnected PLDs on a single chip. This structure is not apparent to the user and doesn't really concern us at this time, except as background information. We will look at the structure of PALs, GALs, and CPLDs in Chapter 8. We will use the term "PLD" when we are referring to a generic device and "CPLD" as a more specific type of PLD.

A complication in the use of programmable logic is that we must use specialized computer software to design and program our circuit. Initially, this might seem as though we are adding another level of work to the design, but when these computer techniques are mastered, it shortens the design process greatly and yields a level of flexibility not otherwise available.

Let's look at two examples, comparing the use of SSI logic versus programmable logic.

EXAMPLE 4.1

FIGURE 4.1 Example 4.1: Majority Vote Circuit

Figure 4.1 shows a majority vote circuit, as described in Problem 3.6 of Chapter 3. This circuit will produce a HIGH output when at least two out of three inputs are HIGH. Write the Boolean equation for the circuit and state the minimum number and type of 74HC devices required to build the circuit. How many packages would be required to build two such circuits?

■ Solution

Boolean equation: $Y = AB + BC + AC$

Figure 4.2 shows the 74HC devices required to build the majority vote circuit: one 74HC08A quad 2-input AND gate and one 74HC4075 triple 3-input OR gate. (We could also use a 74HC00 quad 2-input NAND and a 74HC10 triple 3-input NAND for a NAND-NAND SOP implementation.) Figure 4.2 also shows connections between the AND and OR devices. Note that unused gate inputs are grounded and unused outputs are left open.

FIGURE 4.2   Example 4.1: 74HC Devices Required to Build a Majority Vote Circuit

Two majority vote circuits would require six ANDs and two ORs. This requires one more 74HC08A package.

 EXAMPLE 4.2

Show how a CPLD can be programmed with a majority vote function, using a **schematic entry** tool. State how many CPLDs would be required to build two majority vote circuits.

■ Solution

A CPLD can be programmed by entering the schematic directly, using PLD programming software such as Altera Corporation's **Quartus II.** Figure 4.3 shows the circuit as entered in a Quartus II Block Diagram File.

FIGURE 4.3   Example 4.2: Quartus II Block Diagram File of Majority Vote Circuit

The design can be **compiled** by Quartus II to create the information required to program the CPLD with the majority vote circuit. If a second copy of the circuit is required, the first circuit can easily be duplicated by a Copy and Paste procedure. The two circuits can than be compiled together and used to program a single CPLD.

■ **SECTION 4.1 REVIEW PROBLEM**

4.1 How does the digital logic function of a programmable logic device compare to fixed-function logic?

4.2 PROGRAMMABLE SUM-OF-PRODUCTS ARRAYS

■ KEY TERMS

Product Line A single line on a logic diagram used to represent all inputs to an AND gate (i.e., one product term) in a PLD sum-of-products array.

Input Line A line that applies the true or complement form of an input variable to the AND matrix of a PLD.

PAL Programmable array logic. Programmable logic with a fixed OR matrix and a programmable AND matrix.

The original programmable logic devices (PLDs) consisted of a number of AND and OR gates organized in sum-of-products (SOP) arrays in which connections were made or broken by a matrix of fuse links. An intact fuse allowed a connection to be made; a blown fuse would break a connection.

Figure 4.4a shows a simple fuse matrix connected to a 4-input AND gate. True and complement forms of two variables, A and B, can be connected to the AND gate in any combination by blowing selected fuses. In Figure 4.4a, fuses for $\overline{A}$ and B are blown. The output of the AND gate represents the product term $A\overline{B}$, the logical product of the intact fuse lines.

Figure 4.4b shows a more compact notation for the AND-gate fuse matrix. Rather than showing each AND input individually, a single line, called the **product line,** goes into the AND gate, crossing the true and complement **input lines.** An intact connection to an input line is shown by an "X" on the junction between the input line and the product line.

A symbol convention similar to Figure 4.4b has been developed for programmable logic. Figure 4.5 shows an example.

The circuit shown in Figure 4.5 is a sum-of-products network whose Boolean expression is given by:

$$F = \overline{A}\,\overline{B}C + A\,\overline{B}\,\overline{C}$$

The product terms are accumulated by the AND gates as in Figure 4.4b. A buffer having true and complement outputs applies each input variable to the AND matrix,

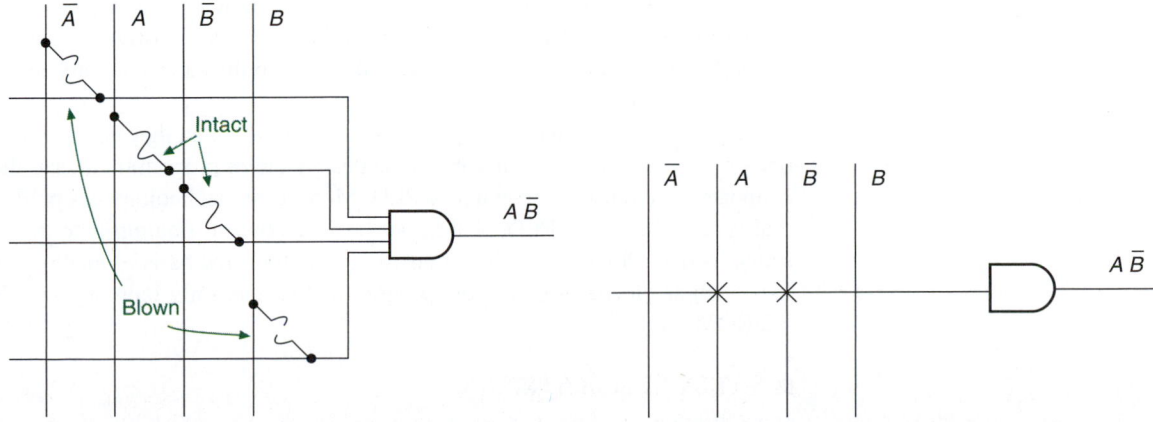

a. Crosspoint fuse matrix (A and $\overline{B}$ intact)   b. PLD notation for fuse matrix

FIGURE 4.4 Crosspoint Fuse Matrix

FIGURE 4.5   PLD Symbology

thus producing two input lines. Each product line can be joined to any input line by leaving the corresponding fuse intact at the junction between the input and product lines.

If a product line, such as for the third AND gate, has all its fuses intact, we do not show the fuses on that product line. Instead, this condition is indicated by an "X" through the gate. The output of the third AND gate is a logic 0 since $(\overline{A}\, A\, \overline{B}\, B\, \overline{C}\, C) = 0$. This is necessary to enable the OR gate output:

$$\overline{A}\,\overline{B}\,C + A\,\overline{B}\,\overline{C} + 0 = \overline{A}\,\overline{B}\,C + A\,\overline{B}\,\overline{C}$$

Unconnected inputs are HIGH (e.g., $A \cdot 1 \cdot B \cdot 1 \cdot 1 \cdot C = \overline{A}\,\overline{B}\,C$ for the first product line).

If the unused AND output was HIGH, the function F would be:

$$\overline{A}\,\overline{B}\,C + A\,\overline{B}\,\overline{C} + 1 = 1$$

The configuration in Figure 4.5, with a programmable AND matrix and a hardwired OR connection, is called **PAL (programmable array logic)** architecture.[1]

Since any combinational logic function can be written in SOP form, any Boolean function can be programmed into these PLDs by blowing selected fuses. The programming is done by special equipment and its associated software. The hardware and software selects each fuse individually and applies a momentary high-current pulse if the fuse is to be blown.

The main problem with fuse-programmable PLDs is that they can be programmed one time only; if there is a mistake in the design and/or programming or if the design is updated, we must program a new PLD. More recent technology has produced several types of erasable PLDs, based not on fuses but on floating-gate metal-oxide-semiconductor transistors. These transistors also form the basis of memory technologies such as electrically erasable programmable read-only memory (EEPROM or E[2]PROM).

■ SECTION 4.2 REVIEW PROBLEM

4.2   What type of Boolean expression represents the underlying structure of many

---

[1]PAL is a registered trademark of Lattice Semiconductor.

programmable logic devices?

4.3 PAL FUSE MATRIX AND COMBINATIONAL OUTPUTS

■ **KEY TERMS**

JEDEC   Joint Electron Device Engineering Council

JEDEC File   An industry-standard form of text file indicating which fuses are blown and which are intact in a programmable logic device.

Text File   An ASCII-coded document stored electronically.

Cell   A programmable location in a PLD, specified by the intersection of an input line and a product line.

Product Line First Cell Number   The lowest cell number on a particular product line in a PAL AND matrix where all cells are consecutively numbered.

Input Line Number   A number assigned to a true or complement input line in a PAL AND matrix.

Checksum   An error-checking code derived from the accumulated sum of the data being checked.

Multiplexer   A circuit that selects one of several signals to be directed to a single output.

In this section and the next, we will examine the structure and programming of two simple PLDs: the PAL16L8 and PAL20P8. Although these devices are essentially obsolete, they are worth examining since their basic structure is similar to more modern PLDs, yet easier to understand because of their relatively small scale. Improvements such as flexibility of output configuration, in-circuit programmability, and reprogrammability have rendered these devices less useful than they once were. However, they still form a useful foundation for understanding the devices that have superceded them.

Figure 4.6 shows the logic diagram of a PAL16L8 PAL circuit. This device can produce up to eight different sum-of-products expressions, one for each group of AND and OR gates. The device has active-LOW tristate outputs, as indicated by the "L" in the part number. The output enable of each tristate buffer is controlled by a product line from the related AND matrix.

The pins that can be used only as inputs or outputs are marked "I" or "O," respectively. Six of the pins can be used as inputs or outputs and are marked "I/O." The I/O pins can also feed back a derived Boolean expression into the matrix, where it can be employed as part of another function. A detail of an I/O section is shown in Figure 4.7.

The part number of a PAL device gives the designer information about the number of inputs and outputs and their configurations, as follows:

```
                    ┌──────── Number of inputs
                   ┌──────── Output type:
                   │           H = Active HIGH
                   │           L = Active LOW
                   │           P = Programmable polarity
                   │           R = Registered (D flip-flop)
                   │           X = XOR registered
                   │           C = Complementary (both HIGH and LOW)
                   │        ┌─ Number of (registered) outputs
PAL 16 R 8
```

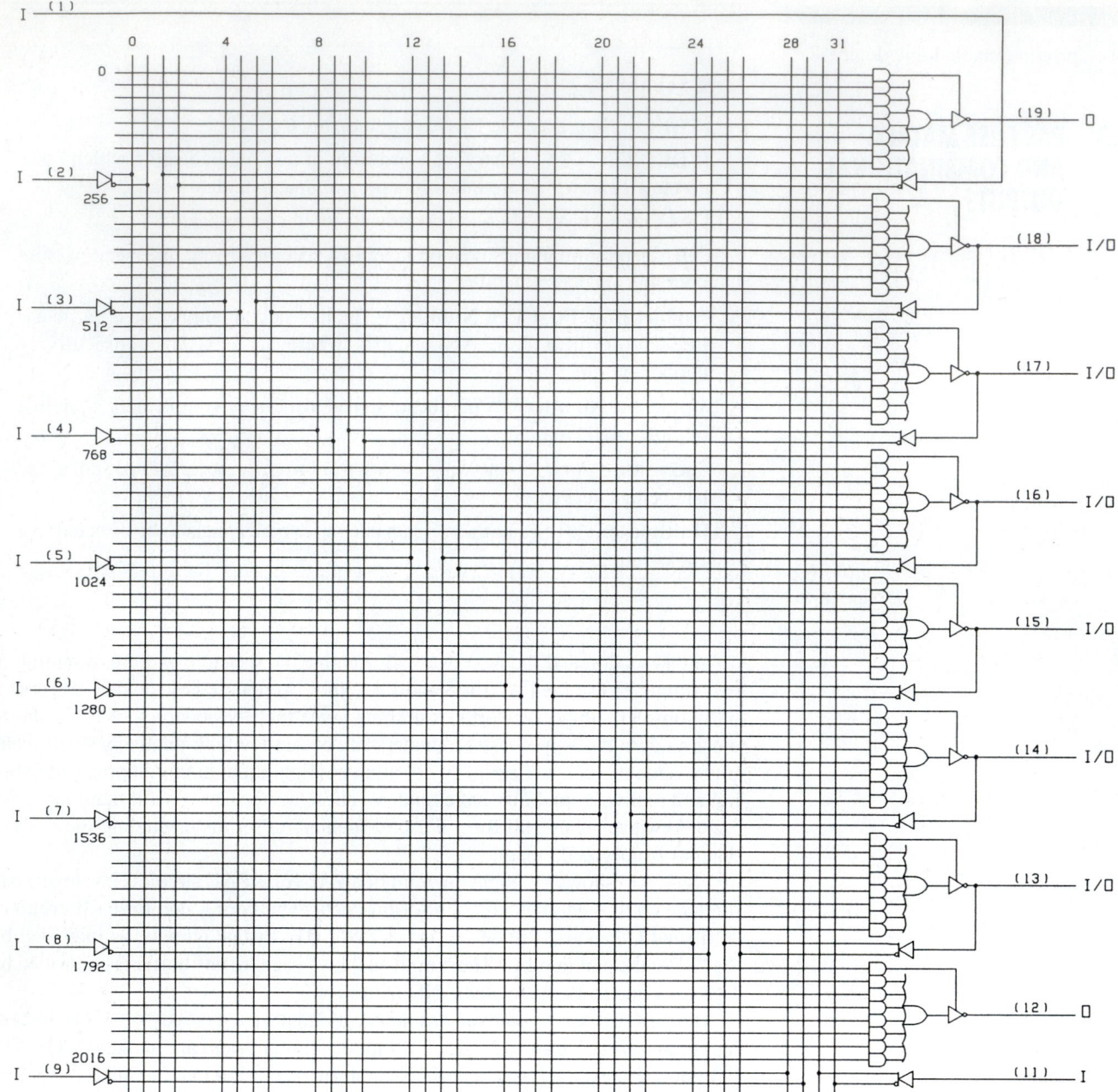

FIGURE 4.6   Unprogrammed PAL16L8

The numbering system has some potential ambiguities. For example, it is not possible to use 16 inputs and 8 outputs in a PAL16L8 device at the same time; 6 of the inputs are actually input/output pins. Some possible configurations are as follows:

16 inputs (10 dedicated + 6 I/O) and 2 dedicated outputs
10 dedicated inputs and 8 outputs (2 dedicated + 6 I/O)
12 inputs (10 dedicated + 2 I/O) and 6 outputs (2 dedicated + 4 I/O)

Each of the outputs of the PAL16L8 is buffered by a tristate inverter, whose *ENABLE* input is controlled by its own product line. When the *ENABLE* line of the tristate inverter is HIGH, the inverter output is the same as it would normally be—a logic HIGH or LOW, determined by the state of the corresponding OR gate output.

FIGURE 4.7 PAL16L8 I/O Section

When the *ENABLE* line is LOW, the inverter output is in the high-impedance state. The output acts as an open circuit, neither HIGH nor LOW; it is as though the output was completely disconnected from the circuit. The inverter is always enabled if all fuses on the *ENABLE* product line are blown, and always disabled if these fuses are all intact.

Published logic diagrams of PAL devices generally do not have fuses drawn on them. This allows us to draw fuses for any application. In practice, PLDs have become too complex to manually draw fuse maps for most applications.

Historically, PLD programming would begin with fuses drawn on a logic diagram, and each fuse would be selected and blown individually by someone operating a hardware device constructed for such a purpose.

Fuse assignment is now done with special software such as ABEL, CUPL, or PALASM. These programs will take inputs such as Boolean equations, truth tables, or other forms and produce the simplest SOP solution to the particular problem. (Quartus II is not configured to generate programming data for low-density PALs.)

The end result of such software is a **JEDEC file,** an industry-standard way of listing which fuses in the PLD should remain intact and which should be blown. This file format has been standardized by **JEDEC** (the Joint Electron Device Engineering Council), an industry group that regulates technical matters in the semiconductor industry. The JEDEC file is stored electronically as an ASCII **text file.** Most PLD programmers will accept the JEDEC file and use it as a template for blowing fuses in the target device.

Fuse locations, called **cells,** are specified by two numbers: the **product line first cell number,** shown along the left side of the diagram, and the **input line number,** shown along the top. The address of any particular fuse is the sum of its product line first cell number and its input line number. The fuses on the PAL16L8 device are numbered from 0000 to 2047 (= 2016 + 31).

Figure 4.8 shows an example of a JEDEC file for a PAL16L8 application. The file starts with an ASCII "Start Text" character (^B). Next is some information required by the PAL programmer about the type of device (PAL16L8), number of fuses (2048), and so forth. The fuse information starts with the line L0000, which is the first product line. The 1s and 0s which follow show the programmed state of each cell in each product line; a 1 is a blown fuse and a 0 is an intact fuse. In other words, each 0 in the JEDEC file represents an X in the same position on the PAL logic diagram.

```
^B
PAL16L8
*
QF2048*QP20*F0*
L0000
1111 1111 1111 1111 1111 1111 1111 1111
1111 1111 0111 1111 1111 1111 0111 1111
1111 1111 0111 1111 1111 1111 1101 1111
1111 1111 1111 1111 1111 1111 0101 1111*
L0256
1111 1111 1111 1111 1111 1111 1111 1111
1110 1111 1111 1011 1111 1111 1111 0111
1110 1111 1111 0111 1111 1111 1111 1011
1101 1111 1111 1011 1111 1111 1111 1011
1101 1111 1111 0111 1111 1111 1111 0111*
L0512
1111 1111 1111 1111 1111 1111 1111 1111
1101 1111 1111 0111 1111 1111 1111 1111
1101 1111 1111 1111 1111 1111 1111 0111
1111 1111 1111 0111 1111 1111 1111 0111*
L0768
1111 1111 1111 1111 1111 1111 1111 1111
1011 1111 1101 1111 1011 1111 1111 1111
1011 1111 1110 1111 0111 1111 1111 1111
0111 1111 1110 1111 1011 1111 1111 1111
0111 1111 1101 1111 0111 1111 1111 1111*
L1024
1111 1111 1111 1111 1111 1111 1111 1111
0111 1111 1111 1111 0111 1111 1111 1111
0111 1111 1101 1111 1111 1111 1111 1111
1111 1111 1101 1111 0111 1111 1111 1111*
L1280
1111 1111 1111 1111 1111 1111 1111 1111
1111 1011 1111 1111 1101 1011 1111 1111
1111 1011 1111 1111 1110 0111 1111 1111
1111 0111 1111 1111 1110 1011 1111 1111
1111 0111 1111 1111 1101 0111 1111 1111*
L1536
1111 1111 1111 1111 1111 1111 1111 1111
1111 0111 1111 1111 1111 0111 1111 1111
1111 0111 1111 1111 1101 1111 1111 1111
1111 1111 1111 1111 1101 0111 1111 1111*
L1792
1111 1111 1111 1111 1111 1111 1111 1111
1111 1111 1011 1111 1111 1111 1001 1111
1111 1111 1011 1111 1111 1111 0110 1111
1111 1111 0111 1111 1111 1111 1010 1111
1111 1111 0111 1111 1111 1111 0101 1111*
L2048
1111 1111*
C8DCF*
^C0000
```

FIGURE 4.8   Sample JEDEC File

The product terms for first sum-of-products output are set by the states of fuses 0000 to 0255 (eight product lines). In the file shown, all fuses are blown in the first product line, the second product line shows three intact fuses, and so forth. Since all fuses are intact in the last three lines, they need not be shown in the JEDEC file.

Whenever some unprogrammed product lines are omitted from the fuse map, the last fuse line shown ends with an asterisk (*). The next line with programmed fuses is indicated by a new fuse number. For example, the second group of fuses (0256 to 0511) in Figure 4.8 begins after the line marked L0256 in the JEDEC file. The remaining fuse lines are similarly indicated.

a. Active-HIGH output   b. Active-LOW output

FIGURE 4.9 4-to-1 Multiplexer Circuits

The JEDEC file in Figure 4.8 ends with a hexadecimal **checksum** (C8DCF), an error-checking code derived from the programming data, and an ASCII "End Text" code (^C).

In order to examine the general principle of fuse programming, let us develop the programmed logic diagram for a common combinational circuit: a 4-to-1 **multiplexer.** (After developing the fuse maps for several examples, we will not refer to this technique again.)

This circuit, shown in Figure 4.9, directs one of four input logic signals, D_0 to D_3, to output Y, depending on the state of two select inputs S_0 and S_1. The circuit works on the enable/inhibit principle; each AND gate is enabled by a different combination of $S_1 S_0$. The binary state of the select inputs is the same as the decimal subscript of the selected data input. For instance, $S_1 S_0 = 10$ selects data input D_2; the AND gate corresponding to D_2 is enabled and the other three ANDs are inhibited.

The logic equation for output Y is given by:

$$Y = D_0 \overline{S}_1 \overline{S}_0 + D_1 \overline{S}_1 S_0 + D_2 S_1 \overline{S}_0 + D_3 S_1 S_0$$

Since the outputs of the PAL16L8 are active LOW, as illustrated in Figure 4.9b, we should rewrite the equation as follows:

$$\overline{Y} = \overline{D}_0 \overline{S}_1 \overline{S}_0 + \overline{D}_1 \overline{S}_1 S_0 + \overline{D}_2 S_1 \overline{S}_0 + \overline{D}_3 S_1 S_0$$

The D inputs must be complemented to reverse the effect of the active-LOW output. The output is enabled when the *EN* input is HIGH. Figure 4.10 shows the PAL16L8A logic diagram with fuses for the multiplexer application.

■ **SECTION 4.3 REVIEW PROBLEM**

4.3 How is fuse assignment generally done in a modern PLD design?

4.4 PAL OUTPUTS WITH PROGRAMMABLE POLARITY

The multiplexer application developed in the previous section uses a PAL device whose output is always fixed at the active-LOW polarity. This fixed polarity is suitable for most applications, but Boolean functions that would normally have active-HIGH outputs must be implemented in DeMorgan equivalent form, which is not always very straightforward.

Some applications require both active-HIGH and active-LOW outputs. In such cases it is useful to have a device whose output polarity is fuse programmable.

FIGURE 4.10 Programmed Logic Diagram for a 4-to-1 Multiplexer

Figure 4.11 shows the logic diagram of a PAL20P8 PAL device. This device is the same as a PAL16L8, except that there are four more dedicated inputs, and the polarity of each output is programmable. The Exclusive OR gate on each output is programmed to act as either an inverter or a buffer. When its associated fuse is intact, the XOR input is grounded and passes the output of its related SOP network in true form. When combined with the output inverter, this produces an active-LOW output. When the polarity fuse is blown, the fused XOR input floats to the HIGH state, inverting the SOP output; the output pin becomes active HIGH.

The polarity fuses are given numbers higher than those of the main fuse array. In this case, the product line fuses are numbered 0000 to 2559 and the output polarity fuses are numbered 2560 to 2567.

FIGURE 4.11 PAL20P8 Logic Diagram

Figure 4.12 illustrates the selection of output polarity. Two Boolean functions, *F1* and *F2,* are programmed into the fuse array, with output at pins (17) and (15), respectively. The equations are:

$$F1 = A\,B + \bar{A}\,\bar{B}$$
$$F2 = \overline{A\,B + \bar{A}\,\bar{B}}$$

We could, if we chose, rewrite *F2* to show the output as active LOW:

$$\overline{F2} = AB + \bar{A}\,\bar{B}$$

The portion of the PAL20P8 logic diagram shown in Figure 4.12 represents the fuses required to program *F1* and *F2*. Pins (14) and (16) supply inputs *A* and *B* to the matrix. The *ENABLE* lines of the tristate output buffers float HIGH, since all fuses are blown on the corresponding product lines, thus permanently enabling the output buffers.

The fuses numbered 2565 and 2567 select the polarity at pins (15) and (17). Fuse 2565 is blown. The fused input to the corresponding XOR gate floats HIGH, thus

FIGURE 4.12   PAL Outputs with Programmable Polarity

making the gate into an inverter. Combined with the tristate buffer, this makes pin (17) active HIGH.

Fuse 2567 is intact. This grounds the input to the corresponding XOR gate, making the gate into a noninverting buffer. Combined with the tristate output buffer, this makes pin (15) active LOW.

■ SECTION 4.4 REVIEW PROBLEM

4.4   State the name of the logic function used to select the polarity of a PLD output.

4.5 PROGRAMMING CPLDs USING QUARTUS II

■ KEY TERMS

Design Entry   The process of using software tools to describe the design requirements of a PLD. Design entry can be done by entering a schematic or a text file that describes the required digital function.

Simulation   Verifying design function by specifying a set of inputs and observing the resultant outputs. Simulation is generally shown as a series of input and output waveforms.

Fitting   Assigning internal PLD circuitry, as well as input and output pins, for a PLD design. Also called "place and route".

Programming   Transferring design information from the computer running PLD design software to the actual PLD chip.

Suite (of Software Tools)   A related collection of tools for performing specific tasks. **Quartus II** is a suite of tools for designing and programming digital functions in a PLD.

Software Tools   Specialized computer programs used to perform specific functions such as design entry, compiling, fitting, and so on. (Sometimes just called "tools.")

Target Device   The specific PLD for which a digital design is intended.

Altera UP-2 Board   A circuit board, part of Altera's University Program Design Laboratory Package, containing two CPLDs and a number of input and output devices.

Download   Program a PLD from a computer running PLD design and programming software.

In order to take a digital design from the idea stage to the programmed silicon chip, we must go through a series of steps known as the CPLD Design Cycle or Design Flow. These include **design entry, simulation,** compiling, **fitting,** and **programming.** All steps require the use of PLD software, such as Altera's Quartus II, a **suite** of **software tools,** to perform the various tasks of the design cycle. Some tasks, such as design entry, require a great deal of attention; others, such as fitting a design to a specified CPLD, are done automatically during the compiling process.

We will be using Quartus II as a vehicle for learning the concepts that relate to CPLD design and programming. The **target device** for most of the designs in this book will be the Altera EPM7128SLC84 CPLD, a member of Altera's MAX 7000S family of devices. This device has the capacity for 128 different programmable functions, 60 pins for user-defined input and output, and can be erased and reprogrammed without removing it from the circuit.

Several available laboratory circuit boards use the EPM7128SLC84 device. The first of these boards, the UP-1 University Program Design Laboratory Package, was introduced in 1997 by Altera. It has a number of DIP switches and pushbuttons for input and LEDs and numerical displays for output. The board has a second device, the Altera EPF10K20RC240 (a member of the FLEX 10K family), which is a more advanced device with a greater programming capacity. Recently, the UP-1 board has been superceded by the **Altera UP-2 board,** which has the same layout, but an upgraded FLEX 10K device, the EPF10K70RC240. A photograph of the Altera UP-2 board is shown in Figure 4.13a.

Two other boards with the EPM7128SLC84 device are widely used: the eSOC board, used by programs at DeVry Institutes, and the RSR PLDT-2 board, manufactured and sold by Electronix Express in New Jersey (*http://www.elexp.com/*). These boards are shown in Figures 4.13b and c. The eSOC and RSR boards offer less-expensive alternatives to the Altera UP-2 board, with some advantages in terms of wiring requirements and active levels of the LED displays. The Altera UP-2 board is available for donation to schools belonging to the Altera University Program and for sale to students.

A CPLD-to-breadboard adapter is also available from HVW Technologies (*http://www.hvwtech.com/*) in Calgary, Canada. This is shown in Figure 4.13d.

FIGURE 4.13A Altera UP-2 Board

FIGURE 4.13B DeVry eSOC Board

Any of these boards can be used for the CPLD designs described in Chapters 4–13. The material in Chapter 14 requires the use of the Altera UP-1 or UP-2 board because it uses the FLEX 10K device. (If students are purchasing one of the less-expensive boards, one solution might be for faculty to request a donation of several UP-2 boards for use with the Chapter 14 material.)

Figure 4.14 shows photos of the two CPLDs used in the Altera UP-2 Board. Figure 4.14a shows the CPLD from the MAX 7000S family, part number EPM7128SLC84-7. Figure 4.14b shows the CPLD from Altera's FLEX 10K series, part number EPF10K70RC240-4. These part numbers will be discussed in detail in Chapter 8.

FIGURE 4.13C RSR Electronics PLDT-2 Board

FIGURE 4.13D HVW Technologies Intro-FPGA Board

In the remaining part of this chapter, we will learn how to enter a design in the Quartus II Block Editor, how to compile the design, and how to **download** it into either one of the CPLDs on the Altera UP-2 circuit board.

Treat this design example as a tutorial in Quartus II. Follow along with all the steps on your own computer to get the maximum benefit from the chapter. (These are the identical steps as for Lab #7 on the accompanying CD.) If you do not have access to the Altera UP-2 board or an equivalent, you can still follow through most of the steps.

■ SECTION 4.5 REVIEW PROBLEM

4.5 What is a target device in a PLD design?

FIGURE 4.14A   Altera MAX 7000S Device

FIGURE 4.14B   Altera FLEX 10K Device

4.6   QUARTUS II DESIGN FLOW AND GRAPHICAL USER INTERFACE

■ KEY TERMS

Project   A collection of files associated with a PLD design in Quartus II.

Block Diagram File   A design file in which PLD design information is entered as a schematic or a block diagram.

Hardware Description Language (HDL)   A computer language used to design digital hardware.

Design Flow

Figure 4.15 shows a simplified flow chart indicating the design flow of a **project** in Quartus II. A project is a collection of files associated with a particular design entity. In the design flow, we enter information that describes our design, either in the form

FIGURE 4.15 Simplified Design Flow for Quartus II

of a schematic or block diagram in a **Block Diagram File,** or in text form, using a **Hardware Description Language (HDL),** such as AHDL or VHDL. (We will look at VHDL in detail in later chapters. For now, we will concentrate our efforts on creating a project with the Quartus II Block Editor.)

When we have entered the design (or any time after creating the design file), we create a project in Quartus II, using the design file as a starting point. Subsequent files are either added automatically or can be added to the project when they are saved or at any other time via the **Project** menu or the **Settings** dialog box. In the process of creating a project, we can select a target device, a PLD that will eventually hold our design.

After the design has been entered and a project created, we compile and simulate the design. The compilation process creates programming information for a target CPLD. Simulation tests the design concept before we program a chip, using a series of stimulus waveforms to create a timing diagram for the design.

If the simulation indicates that there are errors in our design, we make corrections and repeat the compile/simulate process until we are satisfied that our project is conceptually correct. We then assign pin numbers for the PLD inputs and outputs and recompile the project to update our programming information. Finally, we move to the programming step, where our design information is loaded into the target CPLD.

Graphical User Interface

If you have not done so, install Quartus II on your computer and obtain a license file from Altera. The installation and licensing procedure is found in the Altera document *Quartus® II Installation & Licensing for PCs,* as well as in Lab #6 on the accompanying CD.

Figure 4.16 shows the startup screen for Quartus II. Figure 4.17 shows the default toolbar configuration. The toolbars have a number of buttons that pertain to the design flow of a Quartus II project. The operations performed by these buttons can all be done through the regular menus of Quartus II, but the toolbar offers a quick way to access many available functions. Not all buttons on the toolbar in Figure 4.17 are labeled, just the ones that you will find particularly convenient at this time. You can find out the function of any button by placing the mouse cursor on the button without clicking, then reading a pop-up description of the function.

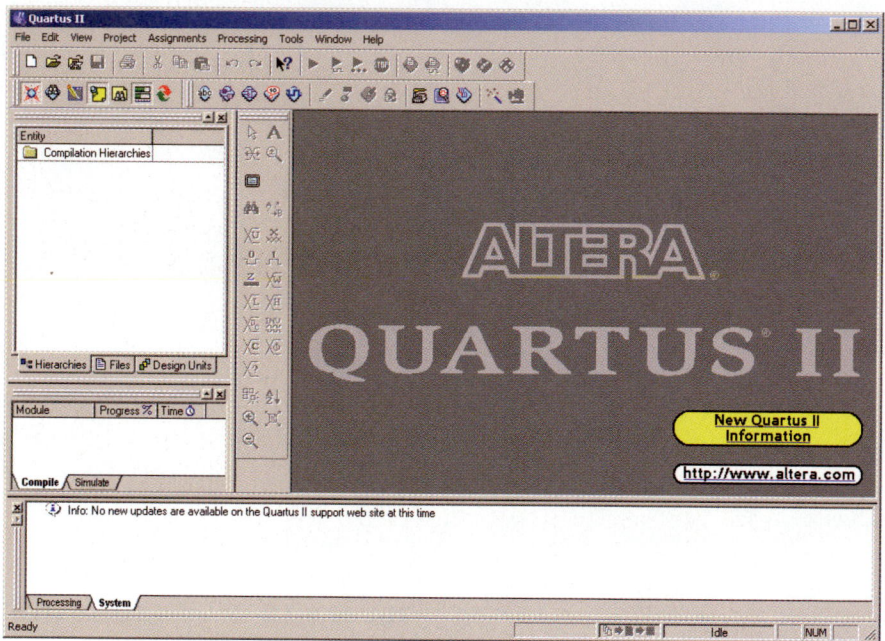

FIGURE 4.16 Quartus II Startup Screen

FIGURE 4.17 Quartus II Toolbar

In addition to the design flow functions, the toolbars also contain buttons that create, open, and save files (standard Windows icons). New files can be created by the standard new file icon or by the specialized icons that create new Text, Block Diagram, or Vector Waveform Files. Additional toolbars can be added or customized by selecting **Customize** from the **Tools** menu.

Each new project should be created in a new Windows folder. Since Quartus II creates many files in the design and compilation process, the folders would become unmanageable if designs were not kept in separate folders. Also, Quartus II can misinterpret design files that are left in an already-used folder, incorrectly adding them to a project other than the one intended. (A folder can contain multiple design files if they are all part of the same project.)

Quartus II installs a folder for working with design files called **qdesigns.** The examples in this text will be created in a subfolder of **qdesigns.** If you are working in a situation where many people share a computer (e.g., a computer lab) and you have access to a network drive of your own, you should keep your working files in a **qdesigns** folder on the network drive. Avoid storing your working files on a local hard drive unless you are the only one with regular access to the computer. Examples in this book will not specify a drive letter, but will indicate *drive:\qdesigns\folder\filename.*

Most of these examples are available on the CD that accompanies this book in a sub-folder of **qdesigns** called **textbook.** A special icon, shown in the margin, will indicate the example filename. These files can be copied into the **qdesigns** folder. The general pathname for the files will then be something like:

drive:\qdesigns\textbook\chxx\project_folder\filename.ext,

where *drive* is the drive where the files are stored, *xx* is the chapter number, and *ext* is the file extension, which depends on the file type.

For this edition of the book, MAX+PLUS II versions of the files are also included, where applicable.

> **NOTE . . .**
>
> When you copy the example files from the CD to your computer, remove the read-only attributes from all files before using them with Quartus II. To do so, open Windows Explorer and click on the folder containing the example files (e.g., *drive:\qdesigns\textbook*). Click CTRL+A to highlight all files in the folder. Right-click on the highlighted file names and select **Properties.** In the resulting dialog box, uncheck the box labeled **Read-only** under **Attributes.** Click **OK** to close the box.

■ SECTION 4.6 REVIEW PROBLEM

4.6 Name two methods of design entry for Quartus II.

4.7 CREATING A QUARTUS II PROJECT AND BLOCK DIAGRAM FILE

> **■ KEY TERMS**
>
> **Hierarchy**   The levels or layers of a hierarchical design.
>
> **Hierarchical Design**   A PLD design that is ordered in layers or levels. The highest level of design contains components that are themselves complete designs. These components may, in turn, have lower-level designs embedded within them.

To create a new Block Diagram File, select **New** from the **File** menu. From the dialog box, shown in Figure 4.18, choose the **Device Design Files** tab and select

FIGURE 4.18   New File Dialog Box (Block Diagram File)

FIGURE 4.19   Block Diagram File (Blank)

Block Diagram/Schematic File. The Quartus II Block Editor will open, as shown in Figure 4.19.

Before entering any design information, we will create a new project, with the new Block Diagram File as the top-level file in a design **hierarchy.** (This terminology indicates that anything contained within the Block Diagram File is a subsystem of the design we are going to enter. In other words, if we would create several Block Diagram Files or Text Files, we could use each file as a component in a larger **hierarchical design,** which would be the top level of the design hierarchy. (We will be working with designs having components in later chapters.)

FIGURE 4.20 Save As Dialog Box (New Project)

FIGURE 4.21 New Project Query

To create the new project, first save the blank Block Diagram File, using the **Save As** dialog, shown in Figure 4.20. Change the file name to **majority_vote.bdf** and save the file in the folder *drive:***qdesigns\textbook\ch04\majority\.** Make sure that the box labeled **Create new project based on this file** is checked. The dialog box in Figure 4.21 will appear, asking you to confirm your choice. Click **Yes.** You can also create a new project by choosing **New Project Wizard** from the **File** menu.

After saving your file, the New Project Wizard will automatically appear. The New Project Wizard is a series of setup screens that asks the user for information about a project. Initially, there is an introductory screen, shown in Figure 4.22. The next screen, in Figure 4.23, asks for information about the directory the project will use, project name, and name of the top-level design entity. A directory name should be unique for each project and can be created by typing it in or selecting it from an existing directory list. By default, the screen in Figure 4.23 contains the directory where the new Block Diagram File was stored and its file name. For most cases, simply accept the default names and click **Next.**

The next screen, shown in Figure 4.24, lists the files currently included in the project and asks if any more should be added. At a later point, we will add a file for simulation, but for now, just click **Next.** The screen in Figure 4.25 allows the user

FIGURE 4.22   New Project Wizard (intro)

FIGURE 4.23   New Project Wizard (Directory, Entity)

FIGURE 4.24 New Project Wizard (Files)

FIGURE 4.25 New Project Wizard (EDA Tools)

FIGURE 4.26 New Project Wizard (Device Family)

FIGURE 4.27 New Project Wizard (Target Device)

FIGURE 4.28 New Project Wizard (Summary)

to select various third-party software tools as plug-ins to the Quartus II software. We will not be using this feature, so click **Next** to see the next screen.

The screens in Figures 4.26 and 4.27 select the device family (MAX 7000S) and target device (EPM7128SLC84-7) for the project. The MAX 7000S is a family of sum-of-products CPLDs manufactured by Altera. The EPM7128SLC84-7 is a device in this family with 128 user-programmable locations. Since there are only 84 pins on the device, about half of these locations are for logic internal to the chip only. We select this device because it is the one provided on the Altera UP-1 and UP-2 boards. The EPM7128SLC84-6 device can also be selected, if necessary.

The final screen, shown in Figure 4.28 is a summary of project settings provided by the user. Click **Finish** to exit the wizard. The wizard can be exited at any point if all the relevant information has been entered. Project settings can be altered at any time via the **Settings** dialog box, accessible from the **Assignments** menu.

Entering Components

■ KEY TERMS

Primitives   Basic functional blocks, such as logic gates, used in PLD design files.

Instance   A single copy of a component in a PLD design file.

A Block Diagram File can be created from a Graphic Design File in MAX+PLUS II (an older Altera CPLD design tool) or created as new in the Quartus II Block Editor. Procedures to convert a MAX+PLUS II project to Quartus II are found in Appendix A.

The first step in entering the majority vote circuit in the Quartus II Block Editor is to lay out and align the required components. We require three 2-input AND gates, a 3-input

FIGURE 4.29 Edit Menu (Insert Symbol)

Check this box for multiple copies of component.

Component name

FIGURE 4.30 Symbol Dialog Box (and2)

OR gate, three input pins, and one output pin. These basic components are referred to as **primitives.** Let us start by entering three copies of the AND gate primitive, called **and2.**

Open the **Edit** menu, shown in Figure 4.29, and select **Insert Symbol,** or simply double-click on the Block Editor desktop. In the **Symbol** dialog box (Figure 4.30) type **and2** in the box labeled **Name.** The **and2** symbol appears in the desktop area on the right. Since we want to enter three **instances** of the symbol, check the box that is labeled **Repeat-insert mode.** Click **OK.**

Click on the desktop in the Block Editor window to place an instance of the **and2** component. Don't worry about its exact placement for the moment. Click two more times to place two more gates, then use the **ESC** key to exit the insert-repeat mode.

Enter the remaining components by following the **Insert Symbol** procedure outlined previously. The primitives are called **or3, input,** and **output.** The insert-repeat mode is not necessary for the **or3** and **output** components, as there is only one of each. When all components are entered we can align them, as in Figure 4.31, by highlighting then dragging each one to a desired location.

FIGURE 4.31 Aligned Components

FIGURE 4.32 Dragging a Line to Connect Components

FIGURE 4.33 Making a 90° Bend and a Connection

Connecting Components

To connect components, click over one end of one component and drag a line to one end of a second component, as shown in Figure 4.32. When you hover over a line end, the cursor changes from an arrow to a crosshair with a right-angle symbol. When you drag the line, a horizontal and a vertical grid helps you align connections properly.

A line will automatically make a connection to a perpendicular line, as shown in Figure 4.33.

A line can have one 90° bend, as at the inputs of the AND gates. If a line requires two bends, such as shown at the OR inputs in Figure 4.34, you must draw two separate lines.

Assigning Pin Names

Before a design can be compiled, its inputs and outputs must be assigned names. We could also specify pin numbers, if we wished to make the design conform to a

FIGURE 4.34 Making Two 90° Bends

FIGURE 4.35 Assigning Pin Names

particular CPLD, but it is not necessary to do so at this stage. It may not even be desirable to assign pin numbers, since the design we enter can be used as a component or subdesign of a larger circuit. We may also wish Quartus II to assign pins to make the best use of the CPLD's internal resources. At any rate, we will leave this step out for now.

Figure 4.35 shows the naming procedure. Pins **a** and **b** have already been assigned names. Double-click the pin name (not the pin symbol) to highlight the name. Type in the new name **(c).** Also change the output pin name to **y.**

If there are several pins that are spaced one above the other, such as **a, b,** and **c** in Figure 4.35, you can highlight the top pin name, as described previously, then highlight successive pin names by using the **Enter** key.

Working with Previously Defined Projects

It is important to note again that Quartus II performs all functions, such as compilation and simulation, on a project, not on a file. Therefore, it is not enough to open a file when we wish to work on a design. Rather, we must open its project. We can do so via the **File** menu, where we can use **Open Project** to select a project from a Windows folder or **Recent Projects** to see a list of the last several projects opened. When you are finished with a project, it can be closed by selecting **Close Project** from the **File** menu, by opening a different project, which automatically closes the current one, or by exiting Quartus II.

■ **SECTION 4.7 REVIEW PROBLEM**

4.7   State two ways to create a new project in Quartus II.

4.8 COMPILING AND SIMULATING A DESIGN IN QUARTUS II

Compilation

> ■ **KEY TERMS**
>
> **Programmer Object File (pof)**   Binary file used to program a nonvolatile CPLD, such as the Altera MAX 7000S series.
>
> **SRAM Object File (sof)**   Binary file used to configure a volatile CPLD, such as the Altera FLEX 10K series.
>
> **Nonvolatile**   Able to retain stored information after power is removed.
>
> **Volatile**   A device is volatile if it does not retain its stored information after the power to the device is removed.

The compilation process in Quartus II is largely automated. You can start the process in several ways:

■ Click the **Start Compilation** button on the Quartus II toolbar, as shown in Figure 4.36.
■ Select **Start Compilation** from the **Processing** menu, shown in Figure 4.37.

Start compilation

FIGURE 4.36   Start Compilation (Toolbar Button)

■ Open the **Compiler Tool** from the **Tools** menu and click **Start Compilation.** The default screen of the **Compiler Tool** is shown in Figure 4.38. The tool has a progress indicator for the overall compilation process and for the various parts of the process. Figure 4.39 shows a compilation in progress, with the progress indicator showing **Analysis & Synthesis** complete and **Fitting** 88% complete.

When compilation is complete, Quartus II displays a **Compilation Report,** shown in Figure 4.40. The default screen of the Compilation Report is a summary of compilation results. Further details are available by clicking on the various folders (**Legal**

FIGURE 4.37   Start Compilation (Processing Menu)

FIGURE 4.38   Compiler Tool

FIGURE 4.39 Compiler Tool (Compilation in Progress)

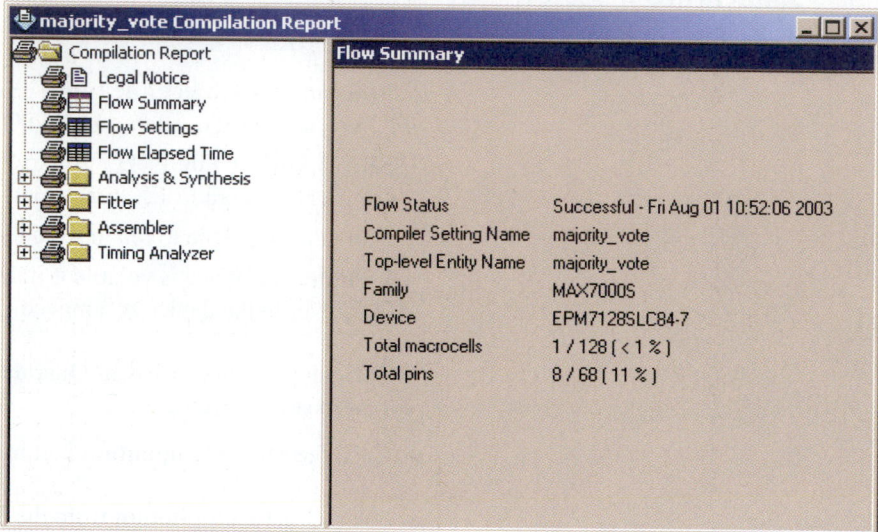

FIGURE 4.40 Compilation Report Window

Notice, **Flow Summary,** etc.) on the left side of the window. Some folders (**Analysis & Synthesis, Fitter,** etc.) must be expanded by clicking the + symbol to the left of the folder icons, much like expanding a Windows directory tree.

Depending on the device chosen, the compiler generates either a **Programmer Object File (pof)** or **SRAM Object File (sof).** The **pof** is used to *program* a sum-of-products device, such as the MAX 7000S family of CPLDs. The **sof** is used to *configure* a look-up table device, such as the FLEX 10K family of CPLDs. The difference is that the MAX 7000S device is **nonvolatile;** that is, it retains its programming information after the power has been removed. The FLEX 10K device is **volatile,** meaning that its programming information must be loaded each time the device powers up.

Simulation

> ■ **KEY TERMS**
>
> **Vector Waveform File (vwf)**   A file that forms the input to the Quartus II simulator, containing the inputs and outputs to be simulated as graphical waveforms.

Simulation is one of the most important parts of the CPLD design flow. It is also one of the most difficult, not because of the mechanics of the process, but because in order to create an effective simulation, we must carefully evaluate all possible operation and failure modes and test them intelligently. This involves understanding the required operation of our design, creating a set of criteria for testing correct operation and possible failure modes, and then, ideally, testing all possibilities.

Before we create a simulation for the majority vote circuit of our tutorial example, we should write a set of simulation criteria for the circuit. The circuit must generate a HIGH output when a majority of inputs is HIGH. Since we have three inputs, this means two or more inputs must be HIGH. The most thorough and systematic way to test the circuit is to apply all possible input combinations to the circuit in an ascending binary sequence, in other words, to take its truth table. We can summarize the criteria as follows:

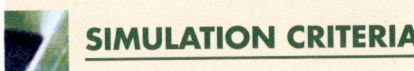 **SIMULATION CRITERIA**

- Apply all possible inputs in an ascending binary sequence.

- For any input combination having two or more HIGH inputs (011, 101, 110, 111), the output must be HIGH.

- For any other input combination (000, 001, 010, 100), the output must be LOW.

A simulation is based on a **Vector Waveform File (vwf),** which contains simulation input and output values in the form of graphical waveforms. A **vwf** can be created from a MAX+PLUS II file or created as new. We will look at the MAX+PLUS II conversion in Appendix A.

To create a new **vwf,** select **New** from the **File** menu or click the appropriate toolbar button. From the box shown in Figure 4.41, click the **Other Files** tab and select **Vector Waveform File.**

The default window of the Quartus II Waveform Editor will appear, as shown in Figure 4.42.

To add waveforms to the window, we can use the **Node Finder.** Start the Node Finder by clicking on the toolbar button shown in Figure 4.17 or selecting **Utility Windows, Node Finder** from the **View** menu. The default window opens, as shown in Figure 4.43. Select **Design Entry (all Names)** from the drop-down menu labeled **Filter.** Click **Start** to display a list of nodes, or signal points, for the project. The window pictured in Figure 4.44 shows a list of available nodes for the

FIGURE 4.41 New File Dialog Box (Vector Waveform File)

FIGURE 4.42   Waveform Editor (Default)

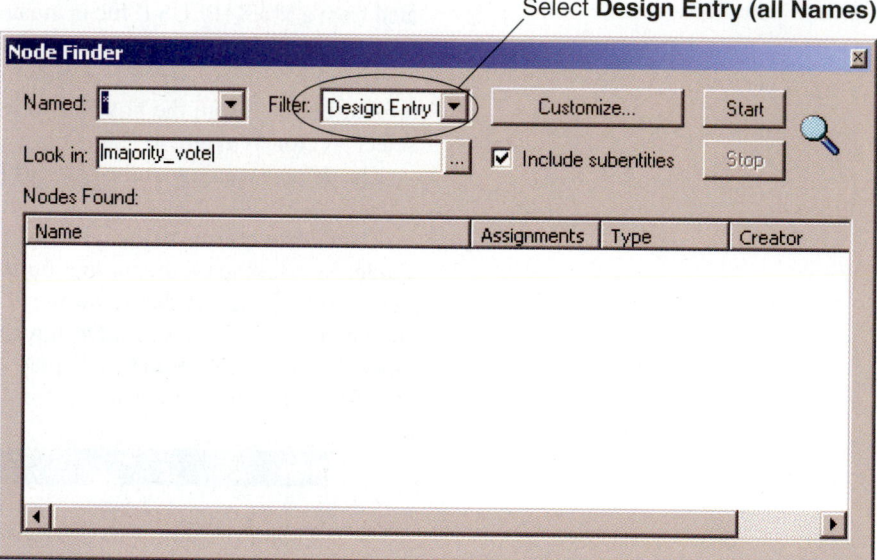

FIGURE 4.43   Node Finder (Default)

FIGURE 4.44   Node Finder (Nodes Selected)

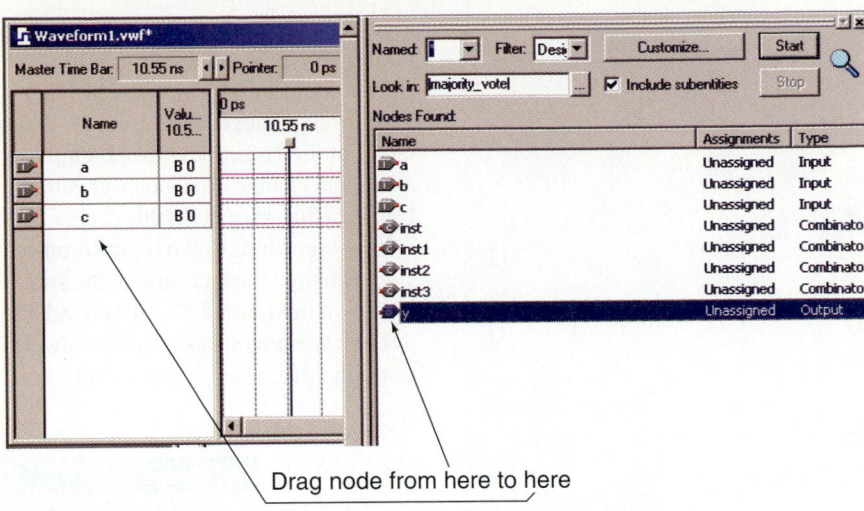

Drag node from here to here

FIGURE 4.45   Dragging from Node Finder to Vector Waveform File

This box should be checked.

FIGURE 4.46   Save as Dialog Box (Adding vwf to Project)

project. We are only interested in **a, b, c,** and **y;** the others represent logic levels internal to the circuit.

Add the required nodes to the Vector Waveform File by dragging them from the Node Finder to the Waveform Editor window. Do this by clicking on a waveform to highlight it, then drag and drop to place it in the Waveform Editor, as shown in Figure 4.45.

Once the waveforms have been added to the Waveform Editor, close the Node Finder window and save the Vector Waveform File, using the **Save As** dialog box as shown in Figure 4.46. Make sure that the box labeled **Add file to current project** is checked. (This step could be done earlier, such as immediately after creating a new Vector Waveform File.)

Once we have entered the input and output nodes, we must determine the length of time our simulation should run. The default time is 100 ns. The target CPLD has

FIGURE 4.47   Edit Menu (End Time)

a delay time from input to output of about 7 ns. (This is indicated by the –7 at the end of its part number.) If we changed our inputs every 10 ns, we would have sufficient room in the simulation to display ten input changes, which is enough to fit in the eight changes we need, plus a little more. However, due to the input-to-output delay in the target device, the output would not change until it is nearly time for a new input change. This delayed output would produce an offset waveform that would be confusing to read. In this case, we would be better off with a simulation time that is long compared to the input-to-output delay of the target device.

To change the end time of the simulation, select **End Time** from the **Edit** menu, as shown in Figure 4.47. In the **End Time** dialog box, shown in Figure 4.48, change the unit from **ns** to **us** (microseconds). This is an easy change that gives a simulation time that is long compared to device delay time.

FIGURE 4.48   End Time Dialog Box

To see the entire waveform file, select **Fit in Window** from the **View** menu. Alternatively, you can customize the Waveform Editor toolbar to have buttons for the zoom tools Zoom In, Zoom Out, and Fit in Window, as shown in Figure 4.49. (The buttons can be found under **Tools, Toolbars, Commands, Zoom** in Quartus II. Drag and drop the buttons onto the Waveform Editor toolbar.)

FIGURE 4.49   Waveform Editor Toolbar (Customized to Show Zoom Tools)

We could enter waveforms on **a, b,** and **c** individually to create an increasing 3-bit binary sequence, but it is easier and more accurate to group these waveforms together and apply the sequence to the whole group. Figure 4.50 shows a highlighted group of waveforms in the Quartus II Waveform Editor. To highlight a group, click on the top waveform **(a),** then drag the cursor to the last waveform of the group **(c).** Right-click on the highlighted group of waveforms and select **Group . . .** from the pop-up menu shown in Figure 4.51. Type the group name (**Inputs**) in the dialog box of Figure 4.52. Select the radix (i.e., the number system base) as **Binary.** Click **OK.** The grouped waveforms appear as shown in Figure 4.53.

Click and drag from a to c to highlight the group.

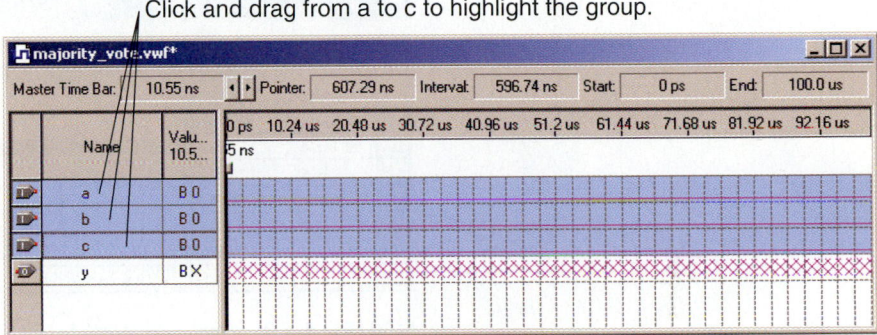

FIGURE 4.50 Highlighting a Group of Waveforms

FIGURE 4.51 Pop-up Menu (Group)

FIGURE 4.52 Group Name and Radix

FIGURE 4.53 Grouped Waveforms

Click on the waveform group (**Inputs**) to highlight it and apply an increasing binary count to the group by clicking the **Count Value** toolbar button, shown in Figure 4.54, or by selecting **Value, Count Value** from the **Edit** menu. In the **Counting** tab of the **Count Value** dialog box, shown in Figure 4.55, select **Radix: Binary; Start value: 000; Increment by: 1;** and **Count type: Binary. End value** is calculated from the other parameters of the simulation.

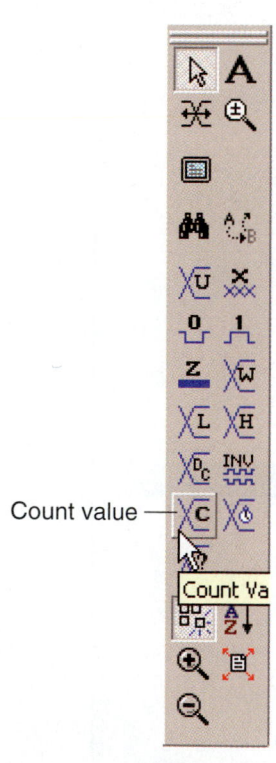

Count value ——

FIGURE 4.54 Waveform Editor Toolbar (Count Value)

FIGURE 4.55 Count Waveform Dialog (Counting)

In the **Timing** tab, shown in Figure 4.56, select **Start time: 0 ps; End time: 100 us; Count every: 10.24 us;** and **Multiplied by: 1.** The count interval of 10.24 μs is selected to match the spacing of the Waveform Editor timing grid. Click **OK** to accept the values. The count waveforms will appear as shown in Figure 4.57. The binary value on **Inputs** corresponds to the combined HIGH and LOW values on inputs **a, b,** and **c.**

FIGURE 4.56 Count Waveform Dialog (Timing)

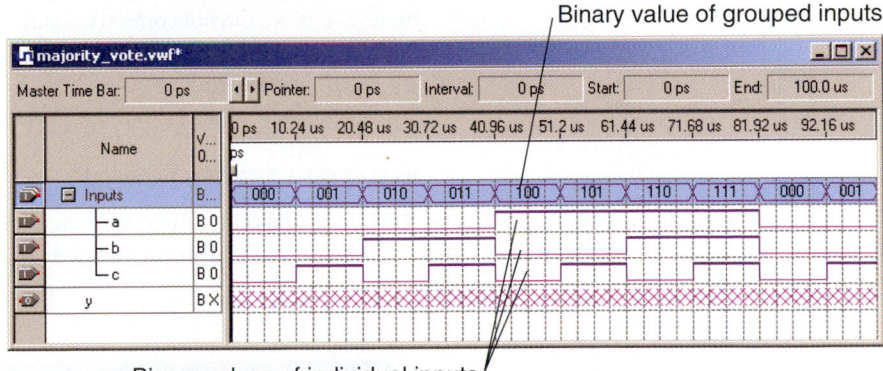

FIGURE 4.57 Count Waveform on Inputs

Start the simulation by clicking the toolbar button shown in Figure 4.58 or by selecting **Start Simulation** from the **Processing** menu, shown in Figure 4.59.

Start simulation

FIGURE 4.59 Start Simulation (Processing Menu)

FIGURE 4.58 Start Simulation (Toolbar Button)

Figure 4.60 shows the simulation result for the majority vote circuit example. The simulation result is not written back into the Vector Waveform File, but rather is part of the Simulation Report automatically generated when the simulator is run. Zoom back to see the whole file. You can see a larger image by selecting **Full Screen** from the **View** menu.

We can see from the simulation waveforms that the output **y** is HIGH whenever any two inputs are HIGH, and LOW otherwise, thus verifying the correctness of the design we entered in the Block Diagram File.

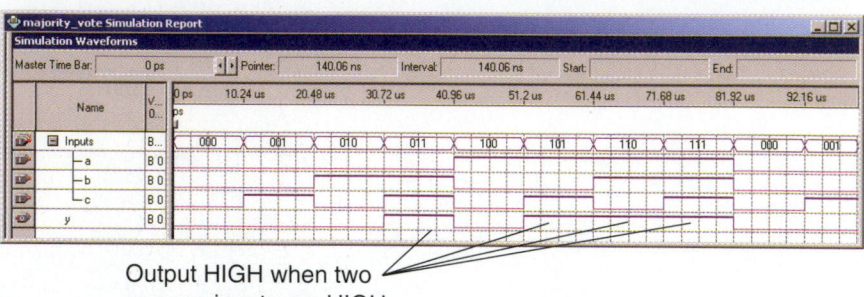

Output HIGH when two
or more inputs are HIGH.

FIGURE 4.60   Simulation Result

Simulation as a Troubleshooting Tool

Simulation can be very valuable as a means of detecting errors in the design entry of a project. For example, suppose, upon simulating the majority vote circuit, instead of the waveforms in Figure 4.60, we end up with those in Figure 4.61. The waveforms in Figure 4.61 do not show the output going HIGH when the inputs have the value 011. In other words, the circuit does not respond correctly to the case where inputs **b** and **c** are HIGH and input **a** is LOW. All other cases respond correctly. This indicates to us that either our design for a majority vote circuit is wrong, or more likely, we did not correctly connect the part of the circuit that gives the response to HIGH inputs on **b** and **c.** Upon investigation, we find that we accidentally entered the schematic shown in Figure 4.62. Compare this to the correct schematic of Figure 4.3 to see if you can find the mistake.

■ SECTION 4.8 REVIEW PROBLEM

4.8   State the requirements for creating an effective simulation of a design entered using a CPLD design tool, such as Quartus II.

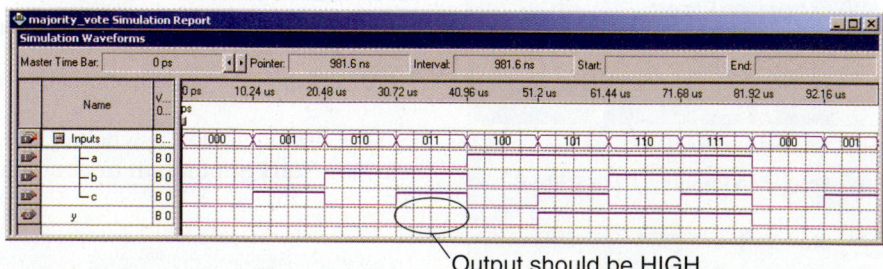

Output should be HIGH.

FIGURE 4.61   Simulation Waveforms Showing an Error at Input = 011

FIGURE 4.62 Block Diagram File with Incorrect Connection

4.9 TRANSFERRING A DESIGN TO A TARGET CPLD

After we have completed the compilation and simulation steps and are satisfied that our design is free of conceptual errors, we can convert the project to a physical design within our target CPLD. This involves three steps.

1. Since CPLD are shipped blank, we must decide which of the available pins are inputs and which are outputs. Thus, we must assign pin numbers to each of the pin names previously assigned.
2. We must recompile the file so as to make the programming information correspond to the new pin assignments.
3. We must use the Quartus II programming tool to transfer the design from our PC to the target device.

> **NOTE . . .**
>
> In order to assign pins and program the CPLD, your project must have the correct device assignment. Check this by opening the **Device** dialog box from the Quartus II **Assignments** menu. If using the Altera UP-1 or UP-2 board, the correct MAX 7000S device is EPM7128SLC84-7. The RSR PLDT-2 board and the DeVry eSOC board use the EPM7128SLC84-10 or -15. The FLEX 10K device is EPF10K20RC240-4 for the UP-1 board and EPF10K70RC240-4 for the UP-2. For other boards, check the part number on the CPLD, but it is probably EPM7128SLC84-7, -10, or -15. The last numbers (-7, -10, -15) don't matter for programming. They only affect the delay time in the Quartus II simulation.

Assigning Pins

We can assign pins either from a dialog box dedicated to this task or from the Quartus II Assignment Editor. Both procedures are outlined, but we only need to select one of them. Choose whichever one you prefer.

We will use the following pin assignments for the majority vote example:

Pin Name	UP-2	PLDT-2	eSOC
a	34	34	50
b	33	33	51
c	36	36	52
y	44	44	4

Menu Items (Figure 4.63)

- Device...
- Assign Pins...
- Timing Settings...
- EDA Tool Settings...
- Settings...   Ctrl+Shift+E
- Timing Wizard...
- Compiler Settings Wizard...
- Simulator Settings Wizard...
- Software Build Settings Wizard...
- Assignment Editor   Ctrl+Shift+A
- Remove Assignments...
- Demote Assignments...
- Back-Annotate Assignments...
- Import MAX+PLUS II Assignments...
- Last Compilation Floorplan
- Timing Closure Floorplan
- Chip Editor
- LogicLock Regions Window   Alt+L
- Import LogicLock Regions...
- Export LogicLock Regions...

FIGURE 4.63 Assignments Menu (Assign Pins)

Using the Assign Pins Dialog Box

Open the **Assign Pins** dialog box from the **Assignments** menu, as shown in Figure 4.63. Figure 4.64 shows this dialog with two assignments already made. The **Assign Pins** dialog box shows a list of pin numbers available on the selected device, any of which can be chosen for an input or output. To assign a pin number, click the line showing that number and type the name of the input or output in the **Pin name** box, then click **Add.** (If you click **OK** at this point instead of **Add,** the dialog box will close and you will have to open it again to assign another pin.)

Once you have assigned all pins, click **OK** to close the dialog box. The **Settings** dialog box, shown in Figure 4.65, also opens with the **Assign Pins** dialog. Close this box by clicking **OK.** Recompile the project.

Using the Quartus II Assignment Editor

Open the **Assignment Editor** from the **Assignments** menu, as shown in Figure 4.66. The **Assignment Editor** default screen, shown in Figure 4.67, will appear. Click the **Pin** button to see the pin assignment screen, shown in Figure 4.68.

Under the **Name** column, type a pin name. Select the name from the drop-down box, as shown in Figure 4.69. Figure 4.70 shows the **Assignment Editor** with pins **a** and **b** assigned. When you have made all assignments, close the assignment editor and confirm the assignments by clicking **Yes** in the dialog box shown in Figure 4.71. Recompile the project.

Once the pins have been assigned, they will appear in the Block Diagram File, as shown in Figure 4.72, provided this display option has been selected from the **View** menu.

1. Select pin number by clicking here

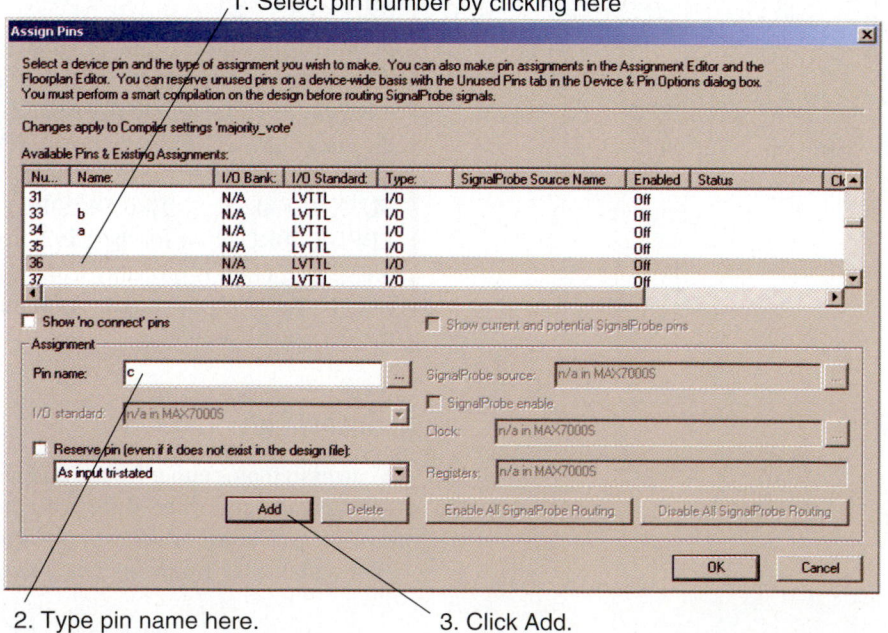

2. Type pin name here.   3. Click Add.

FIGURE 4.64 Assign Pins Dialog

FIGURE 4.65 Settings Dialog Box

FIGURE 4.66 Assignments Menu
(Assignment Editor)

FIGURE 4.67 Assignment Editor (Default Screen)

Click pin to edit pin assignments

FIGURE 4.68 Assignment Editor (Pin Assignments)

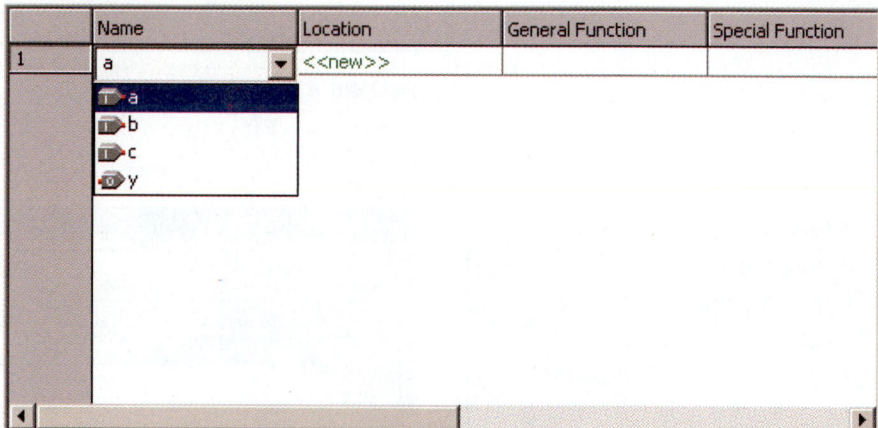

FIGURE 4.69 Adding Pin Names from a Drop-Down Menu

FIGURE 4.70 Two Pin Numbers Assigned

FIGURE 4.71 Pin Assignment Confirmation Dialog

FIGURE 4.72 Majority Vote Circuit, Showing Pin Numbers

Programming CPLDs on the Altera UP-1 or UP-2 Circuit Board

■ **KEY TERMS**

ByteBlaster An Altera ribbon cable and connector used to program or configure Altera CPLDs via the parallel port (LPT port) of a PC.

ByteBlaster MV A newer version of the ByteBlaster. MV stands for Multi-Volt, indicating that it can operate either at 5 volts or 3.3 volts.

JTAG Joint Test Action Group. A standards body that developed the format for testing and programming devices while they are installed in a system. This format has been adopted by the Institute of Electrical and Electronics Engineers as IEEE Standard 1149.1.

ISP In-System Programmability. The ability of a nonvolatile PLD (such as a MAX 7000S) to be programmed without removing it from a circuit board.

ICR In-Circuit Reconfigurability. The ability of a volatile PLD (such as a FLEX 10K) to be configured without removing it from a circuit board.

TDI Test Data In. In a JTAG port, the serial input data to a device.

TDO Test Data Out. In a JTAG port, the serial output data from a device.

TMS Test Mode Select. The JTAG signal that controls the downloading of test or programming data.

TCK Test Clock. The JTAG signal that drives the JTAG downloading process from one state to the next.

JTAG Chain Multiple JTAG-compliant devices whose TDI and TDO ports form a continuous chain connection. Such a chain allows multi-device programming.

FIGURE 4.73 ByteBlaster Parallel Port Download Cable (Courtesy of Altera Corporation)

The CPLDs on the Altera UP-1 and UP-2 circuit boards are programmed via the programming software in Quartus II and a ribbon cable called the **ByteBlaster** or **ByteBlaster MV.** The ByteBlaster, shown in Figure 4.73, connects to the parallel port of a PC running Quartus II to a 10-pin male socket that complies with the **JTAG** standard. This standard specifies a four-wire interface, originally developed for testing chips without removing them from a circuit board, but can also be used to program or configure PLDs.

PLDs that can be programmed or configured while installed on a circuit board are called **In-System Programmable (ISP)** or **In-Circuit Reconfigurable (ICR).** ISP is used to refer to nonvolatile devices, such as the MAX 7000S family; ICR refers to volatile devices, such as the FLEX 10K family.

The JTAG interface has four wires, as well as power and ground connections, as shown in Figure 4.74. Data are sent to a device from a JTAG controller (i.e. the PC) via the **TDI (Test Data In)** line and return from the device via **TDO (Test Data Out).** The data transfer is controlled by **TMS (Test Mode Select).** The process is

FIGURE 4.74 MAX 9000, MAX 7000S, and MAX 7000A Programming with the Byteblaster Cable (Courtesy of Altera Corporation)

FIGURE 4.75 JTAG Chain Device Programming and Configuration with the ByteBlaster Cable (Courtesy of Altera Corporation)

driven from one step to the next by a periodic digital signal called **TCK (Test Clock).** The process advances by one step with each period of TCK.

Multiple devices can be programmed in a **JTAG chain,** as shown in Figure 4.75. This connection allows both CPLDs on the Altera UP-1 or UP-2 board to be programmed at the same time. The UP-1 and UP-2 boards also have a female 10-pin socket labeled **JTAG OUT,** which allows two or more boards to be chained together. The choice of programming one or more CPLDs, or the CPLDs on one or more UP-1 or UP-2 boards, is determined by the placement of four on-board jumpers. These jumper positions are explained in the *Altera University Program Design Laboratory Package User Guide* that comes with the UP-1 or UP-2 board. A copy of the user guide is available on Altera's website at *http://www.altera.com/literature/univ/ upds.pdf.*

The operation of the JTAG port is controlled automatically by Quartus II, so further details are not necessary at this time. For more information on the JTAG interface, refer to *Altera Application Note 39, JTAG Boundary-Scan Testing in Altera Devices,* included in the Altera Documentation folder on the accompanying CD.

Quartus II Programmer

To program a device on the Altera UP-1 or UP-2 board, set the jumpers to program the EPM7128S or configure the EPF10K20 or EPF10K70, as shown in the *Altera University Program Design Laboratory Package User Guide.* Connect the ByteBlaster cable from the parallel port of the PC running Quartus II to the 10-pin JTAG header. (You may have to run a 25-wire straight-through parallel extension cable, with male D-connector to female D-connector, to make it reach.) Plug an AC adapter (7.5- to 9-volt dc output) into the power jack of the UP-1 or UP-2 board.

> **NOTE . . .**
>
> Connection procedures may differ for boards other than the Altera UP-1 or UP-2. For example, some boards incorporate the ByteBlaster circuitry directly on the CPLD board and plug directly into the PC parallel port without a ByteBlaster cable. Consult the manual for your CPLD board for details. Other than connection to the PC, programming procedures are identical for all boards.

Start the Quartus II Programmer, either with the toolbar button shown in Figure 4.76 or by selecting **Programmer** from the **Tools** menu (Figure 4.77). The programmer dialog box (Figure 4.78) will open, showing the programming file for the top-level file of the open project.

> **NOTE . . .**
>
> If you have never programmed a CPLD with your particular version of Quartus II, you will need to set up the programming hardware before proceeding. If you have at any time programmed a CPLD with your PC running this version of Quartus II, you do not need to perform the following hardware setup steps.

Programmer

FIGURE 4.76   Programmer Toolbar Button

FIGURE 4.77   Tools Menu (Programmer)

Click to set up Programmer hardware (first time only).

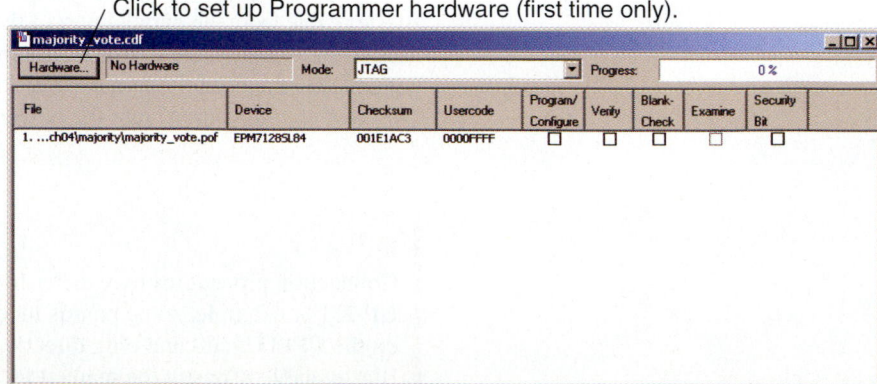

FIGURE 4.78   Programmer Dialog Box

Click the **Hardware** button at the top left corner of the Programmer dialog box. The **Hardware Setup** dialog box, shown in Figure 4.79, will open. Click the **Add Hardware** . . . button. In the **Add Hardware** dialog box shown in Figure 4.80, select **Hardware type: ByteBlasterMV** or **ByteBlaster II** and **Port: LPT1** (or another LPT port, if appropriate). Click **OK** to accept the choices and close the box. In the **Hardware Setup** dialog box, shown in Figure 4.81, highlight **ByteBlasterMV** in the **Available hardware items** box by clicking the item, then click **Select Hardware.** Click **Close** to return to the Programmer dialog.

FIGURE 4.79 Hardware Setup Dialog Box

FIGURE 4.80 Add Hardware Dialog Box

NOTE . . .

The selected LPT port must be configured as ECP mode to program a CPLD. This probably will not need to be set up, but if it does, make the changes in the BIOS settings of your PC.

FIGURE 4.81   Hardware Setup Dialog Box (New Hardware Showing)

Check this box before programming device.

FIGURE 4.82   Programmer Dialog (Hardware Selected)

— Start programming

FIGURE 4.83   Start Programming
(Toolbar Button)

Figure 4.82 shows the Programmer dialog box, now with the programming hardware selected. In order to program the CPLD, highlight the required programming file by clicking it, then select the checkbox for **Program/Configure.**

Start programming the CPLD by clicking the **Start Programming** toolbar button, shown in Figure 4.83, or by selecting **Start Programming** from the Quartus II **Processing** menu (Figure 4.84). Programming progress is shown by the **Progress** indicator in the Programmer dialog box in Figure 4.85.

The majority vote circuit can be tested on a CPLD board, such as the Altera UP-1 or UP-2 board or equivalent. Connect short lengths of #22 wire from three of the input switches on the board to pins 34, 33, and 36 of the CPLD prototyping headers (a series of four dual in-line female connectors surrounding the CPLD). Connect another length of wire from pin 44 of the CPLD header to an LED indicator. These connections are shown in Figure 4.86. (The pin numbers are printed on the board next to the headers.) When two of the three switches are HIGH, there should be a HIGH at the CPLD output.

(STOP) Stop Processing	Ctrl+Shift+C
▶ Start Compilation	Ctrl+L
☑ Analyze Current File	
Start	▶
◈ Compilation Report	Ctrl+R
Start Compilation & Simulation	
▶ Start Simulation	Ctrl+I
Simulation Debug	▶
▤ Simulation Report	Ctrl+Shift+R
▶ Start Software Build	Ctrl+Q
Compile Current File	
▶ Start Programming	
▶ Stop Programming	
▥ Auto Detect	

FIGURE 4.84 Start Programming (Processing Menu)

Programming progress bar

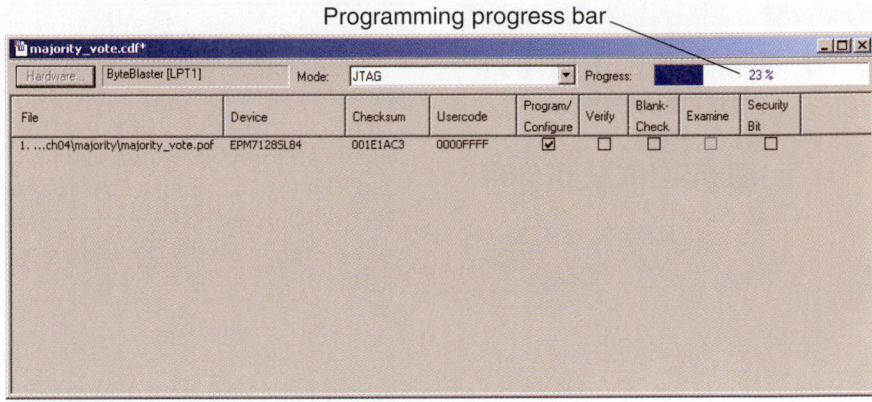

FIGURE 4.85 Programmer Dialog (Programming in Progress)

FIGURE 4.86 Wiring the Altera UP-2 Board

NOTE . . .

The LEDs on the Altera UP-1 and UP-2 boards are configured as active-LOW, so when an LED is on, it indicates a LOW at the CPLD output. When it is off, the LED indicates a HIGH. Some equivalent CPLD boards have LEDs that are active-HIGH. Check the manual for your board if in doubt.

■ **SECTION 4.9 REVIEW PROBLEM**

4.9 State the reason that we must assign pin numbers to a CPLD design before programming.

4.10 USING THE QUARTUS II BLOCK EDITOR TO CREATE A HIERARCHICAL DESIGN

A Quartus II Block Diagram File can be used as part of a hierarchical design. That is, it can be represented as a component in a higher-level design. Figure 4.87 shows a Block Diagram File that is constructed as a hierarchical design. It contains two majority vote circuits whose outputs are combined in an AND gate. Thus, output **z** is HIGH if two out of three inputs are HIGH on *both* symbols labeled **majority_vote.** These symbols are complete designs in their own right, and thus form a lower level of the design hierarchy. In the next section, we will create this design hierarchy in Quartus II.

Symbol Files and User Libraries

> ■ **KEY TERMS**
>
> **Symbol File**   A file that represents a PLD design as a graphic symbol, showing only the design's inputs and outputs. The symbol can be used as a component in any Block Diagram File.
>
> **User Library**   A folder containing design files that can be used in a hierarchical design.
>
> **Top Level (of a Hierarchy)**   The file in a hierarchy that contains components specified in other design files and is not itself a component of a higher-level file.

Before we use the majority vote circuit as a component in a higher-level file, we must create a **symbol file.** To do so, we open the project for the majority vote circuit of Figure 4.3, then create the symbol by selecting **Create/Update** from the Quartus II **File** menu, as shown in Figure 4.88. From the secondary menu that appears, select **Create Symbol Files for Current File,** as shown in Figure 4.89. This action will create a symbol file with the same name as the Block Diagram File and the extension **bsf** (for Block Symbol File). Before creating the symbol, make sure that the Block Diagram File for the majority vote circuit is saved and compiled. The newly created symbol can be embedded into a Block Diagram File as in Figure 4.87.

Before we can compile the Block Diagram File that uses the new symbol, we must make sure that Quartus II knows where to find the symbol. Quartus II looks for a component first in the present project's working directory, then in the **user library** folders in the order of priority listed in the **User Libraries** dialog in the **Settings** dialog box.

FIGURE 4.87   Two-Level Majority Vote Circuit

FIGURE 4.88 File Menu (Create/Update)

FIGURE 4.89 Create/Update Menu

NOTE . . .

If you are using Quartus II on a network, you may not be able to set the user library, as this feature requires you to have write privileges to certain folders that may not be available to general users. Check with your system administrator if this becomes a problem. If you are unable to set the user library path, simply store all required design files in the working folder for the project.

To create a path to a user library, first create a project containing the **top level** of the new design hierarchy. Do this by creating a new Block Diagram File called *drive:***\qdesigns\textbook\ch04\two_votes\two_votes.bdf.** When you save the file, make sure the box labeled **Create new project based on this file** is checked.

In the **Add Files** screen of the New Project Wizard, shown in Figure 4.90, click the button called **User Library Pathnames** . . . The box shown in Figure 4.91 will

FIGURE 4.90 Adding a User Library Pathname

FIGURE 4.91 User Library Pathname Dialog

appear. Click the button labeled (**...**) to browse the directory tree to find the folder containing the **majority_vote** design. When you have found the folder in the **Select Directory** dialog, click **Open.** The folder name will now appear in the **Libraries** box. Click **OK.**

NOTE . . .

When browsing for the user library pathname, you are looking for a *folder* name, not a *file* name.

NOTE . . .

If you are using Quartus II on a shared computer (e.g., in a computer lab), you should be aware that a library path that points to another user's directory can cause Quartus II to look there before (or instead of) looking in your directory, resulting in the apparent inability of Quartus II to find your file.

For example, suppose your project has a component in a file called **g:\qdesigns\my_file.bdf,** where g:\ is a network drive mapped exclusively to your user account. (i.e., everyone has a g:\ drive mapping, unique to their user account.) Further suppose that another user, against standard lab protocol, has created a file with the same name on the local hard drive: **c:\qdesigns\my_file.bdf.** (Don't think this doesn't happen. It does.)

At compile time, Quartus II will look for **my_file.bdf** first in the directory where the active project resides, then in the folders specified in the user library paths. If the user library path **c:\qdesigns** has a higher priority than **g:\qdesigns\,** it will compile the version of **my_file.bdf** found on the c:\ drive. When you make changes to the copy on the g:\ drive, they will not take effect because the file on g:\ is not being compiled.

To remedy this, delete the user libraries that point to local drives, such as a:\ or c:\. If you have no assigned network drive on your system, delete all user libraries except for your own. Since a user library is just the name of a folder where Quartus II should look for files, this won't do any great harm.

Creating a Design Hierarchy

To create the two-level majority vote circuit, we insert the symbols in the Block Diagram File, as before. We can find the symbol for the majority vote circuit in the folder where its Block Symbol File was saved. This is shown in Figure 4.92.

After the circuit of Figure 4.87 has been entered, the design must be compiled to update the changes to the circuit. If you have not properly specified the library path to the **majority_vote** component, you may get an error message like the one in Figure 4.93. (**Error: Node instance inst instantiates undefined entity majority_vote.**) This error means the compiler cannot locate the file **majority_vote.** To remedy the problem, add the new library under **User Libraries** in the **Settings** dialog box (Figure 4.94) or move the file to the project working folder.

Quartus II displays the hierarchy of a design in a window at the left side of the Quartus II workspace, called the **Project Navigator Window.** Figure 4.95 shows the hierarchy of the two-level majority vote circuit, **two_votes.** The compilation **hierarchy tree** indicates the relationship between the different levels of the hierarchy. The display can be collapsed or expanded by clicking the – or + symbols on the

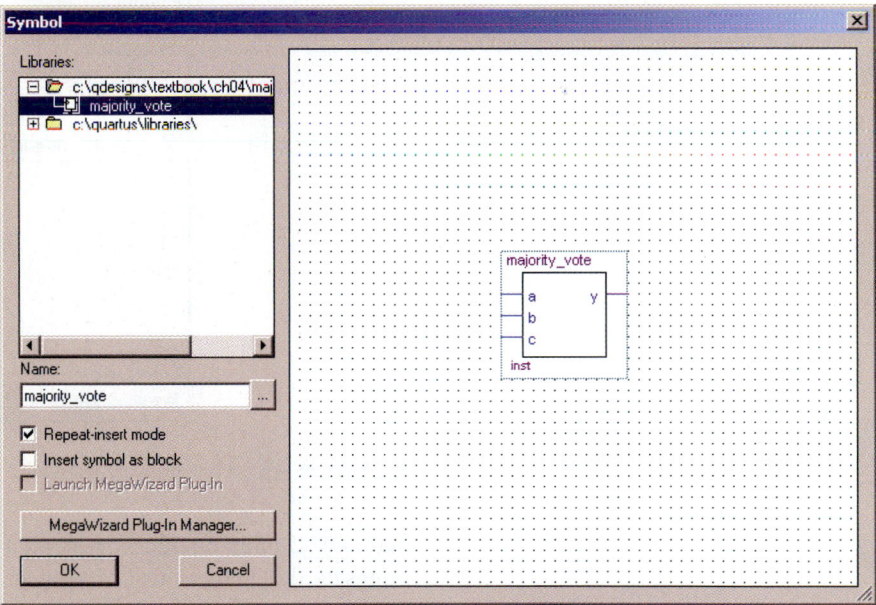

FIGURE 4.92   Inserting a Symbol from a User Library

Info: Command: quartus_map --import_settings_files=on --export_settings_files=off two_votes -c two_votes
Info: Found 1 design units and 1 entities in source file C:\qdesigns\textbook\ch04\two_votes\two_votes.bdf
Error: Node instance inst instantiates undefined entity majority_vote
Error: Quartus II Analysis & Synthesis was unsuccessful. 1 error, 0 warnings
Info: Writing report file two_votes.map.rpt

FIGURE 4.93   Error Message for Undefined User Library Path

FIGURE 4.94 Adding a User Library from the Settings Dialog

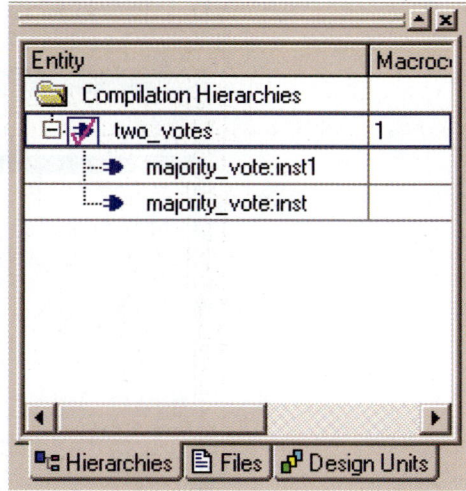

FIGURE 4.95 Hierarchy Display

diagram. Double-clicking on the name of a component in the hierarchy display brings the design file of that component to the front layer of the Quartus II workspace.

Figure 4.96 shows a simulation of the two-level majority vote circuit. Inputs **a1, b1,** and **c1** are grouped together, as are inputs **a2, b2,** and **c2.** The simulation works through all 64 possible input combinations, with the result that output **z** is HIGH only when there are majorities indicated by both of the **majority_vote** components. Figure 4.97 shows a detail of the simulation, where it is possible to read the binary values of the **majority_1** group. Note that **z** is LOW for the whole time that **majority_2** = 100, as there is no majority from the second group of inputs (**a2, b2,** and **c2**). When **majority_2** = 101, and there is a majority from **a2, b2,** and **c2,** then **z** is HIGH when a majority of **a1, b1,** and **c1** is HIGH (**majority_1** = 011, 101, 110, or 111). Similar comparisons can be made for the rest of the simulation.

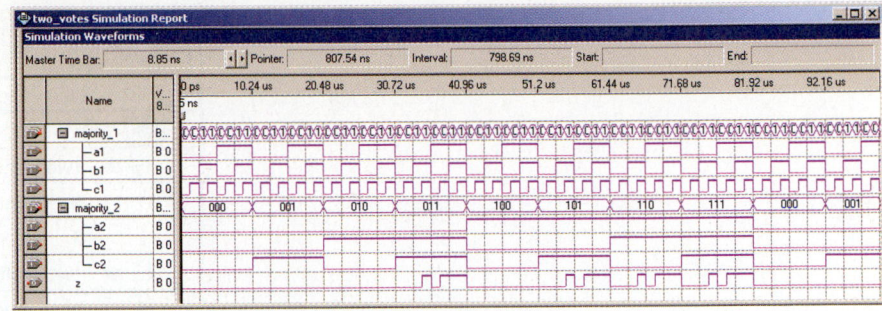

FIGURE 4.96 Simulation Waveforms for Two-Level Majority Vote Circuit

FIGURE 4.97 Simulation Detail for Two-Level Majority Vote Circuit

FIGURE 4.98 Hierarchical Design with Two Instances of Two_Votes

We can extend the hierarchy further by making a symbol for **two_votes.bdf** and embedding it in a higher-level file called **four_votes.bdf,** shown in Figure 4.98. This circuit generates a HIGH output if (at least two out of three of **a11, b11,** and **c11** are HIGH AND at least two out of three of **a21, b21,** and **c21** are HIGH) OR (the same is true for **a12, b12,** and **c12** AND **a22, b22,** and **c22**). The Boolean expression for this circuit is:

$$Y = (a11 \cdot b11 + b11 \cdot c11 + a11 \cdot c11)(a12 \cdot b12 + b12 \cdot c12 + a12 \cdot c12)$$
$$+ (a21 \cdot b21 + b21 \cdot c21 + a21 \cdot c21)(a22 \cdot b22 + b22 \cdot c22 + a22 \cdot c22)$$

FIGURE 4.99 Hierarchy Display for Four_Votes

Figure 4.99 shows the hierarchy for the project **four_votes.** Note that the highest level has two subdesigns, each of which breaks down further into two subdesigns. If we double-click on either instance of the name **two_votes** in the Project Navigator Window, the Block Editor will bring the file **two_votes.bdf** to the foreground. Double-clicking on any instance of the name **majority_vote** will open the file **majority_vote.bdf** and bring it to the front of the desktop.

Using hierarchical design and symbol files for Block Design Files or other design files allows us to create multiple instances of a basic design (**majority_vote.bdf**) and use it in many places.

■ SECTION 4.10 REVIEW PROBLEM

4.10 When would it *not* be necessary to specify a user library path when creating a hierarchical design in Quartus II?

SUMMARY

1. A programmable logic device (PLD) is a digital device that is shipped blank and whose function is determined by the end user.

2. PLDs offer design flexibility, reduce board space and package count, and can be used to develop digital designs more quickly than fixed-function logic.

3. Some types of PLDs include PAL (programmable array logic), GAL (generic array logic), EPLD (erasable PLD), CPLD (complex PLD), and FPGA (field-programmable gate array).

4. Complex PLDs (CPLDs) are sum-of-products devices with several programmable sections that are interconnected inside the chip.

5. Programmable logic devices (PLDs) are structured in two basic architectures: sum-of-products (SOP), which usually consist of a series of programmable AND/OR circuits, and look-up table (LUT), which stores the truth table of a Boolean function in a small memory.

6. Programmable array logic (PAL) is an SOP-type architecture in which there are a series of programmable AND gates that have a fixed connection to an OR-gate output.

7. Connections from PLD inputs to PAL AND arrays were historically made by leaving intact selected fuses in a crosspoint fuse array. In modern PLDs, these connections are made by programming EEPROM (electrically erasable programmable read-only memory) cells.

8. An AND-gate input in a PAL array is called a product line.

9. A PAL16L8 PLD is an SOP device with up to 16 inputs and up to 8 outputs. There are 10 dedicated inputs, 2 dedicated outputs, and 6 pins that can be configured as input or output. All outputs in the PAL16L8 are active-LOW.

10. A PAL is programmed by a computer and programmer hardware that uses a JEDEC file as a template for determining which fuses to blow and which to leave intact.

11. Some PAL devices have programmable-polarity outputs. This is achieved with an XOR gate that has a programmable cell or fuse on one input to switch the output between inverting and noninverting levels.

12. PLDs that can be programmed while installed in a circuit are called In-System Programmable (ISP) or In-Circuit Reconfigurable (ICR).

13. PLD design and programming requires special software, such as Altera's Quartus II.

14. A PLD design can be entered as a schematic or block diagram (Block Diagram Files) or using text-based languages, such as Altera Hardware Description Language (AHDL) and VHSIC Hardware Description Language (VHDL). (VHSIC = Very High Speed Integrated Circuit.)

15. Quartus II organizes PLD design files in a project, a collection of all files associated with the design.

16. A project can be created using the Quartus II New Project Wizard.

17. When working on a design file, open its project, not the file itself. If necessary, open the file separately.

18. A Quartus II Block Diagram File (**bdf**) consists of graphical symbols of components that are interconnected by lines drawn between the components.

19. A perpendicular line that terminates on another line is automatically connected in the Quartus II Block Editor. A line can be drawn with one 90° bend. A line with two 90° bends must be drawn as two lines.

20. Circuit input and output pins in a **bdf** have special symbols. The input and output pins must be named, but need not be numbered in the first stages of a design.

21. The Quartus II compiler translates the design information from a **bdf** or text file into binary data that can be downloaded into a PLD. For a MAX 7000S, the compiler generates a Programmer Object File (**pof**) to *program* the device. For a FLEX 10K, an SRAM Object File (**sof**) is generated to *configure* the device.

22. MAX 7000S devices are nonvolatile; they stay programmed when the power is removed from the chip. FLEX 10K devices are volatile; they lose their programming data when power is removed.

23. If a CPLD part number is not specified, the Quartus II compiler will automatically select one. It is good practice to assign the part number of the device before compiling, as this can affect the accuracy of the certain parts of the design process, such as simulation. The CPLDs on the Altera UP-1 board are: EPM7128SLC84-7 and EPF10K20RC240-4. On the Altera UP-2 board, the CPLDs are: EPM7128SLC84-7 and EPF10K70RC240-4.

24. Simulation is an important part of the CPLD design flow that verifies the correctness of the CPLD design by creating a timing diagram. The user enters a set of stimulus waveforms which the simulator uses to calculate a set of output waveforms, based on the design equations.

25. Ideally, simulation must test all possible operational and failure modes of a design. It is important to write a set of simulation criteria before creating the actual simulation.

26. In Quartus II, a Vector Waveform File holds the input and output nodes and the input waveforms of a simulation. The total simulation is shown in the **Simulation Waveforms** window.

27. Waveforms can be added to the Vector Waveform File by using the **Node Finder** utility.

28. Waveforms can be grouped so that values can be added to several waveforms at a time.

29. Various simulation parameters such as end time, count time, and so forth can be set before simulation.

30. Pin numbers must be assigned to a design before it can be downloaded to a CPLD. Pins can be assigned in the **Assign Pins** dialog box or from the Quartus II **Assignment Editor.** After pin numbers have been assigned, the project must be recompiled.

31. An Altera CPLD can be programmed directly from a PC parallel port via a ByteBlaster cable.

32. The ByteBlaster cable implements a programming interface specified by a standard (IEEE Std. 1149.1) of the Joint Test Action Group (JTAG).

33. A JTAG port is a 4-wire interface for loading test and programming information into one or more JTAG-compliant devices. It consists of an input (TDI), output (TDO), mode select (TMS), and clock (TCK).

34. Quartus II files can be arranged in a design hierarchy. That is, a Quartus II design file can contain components that are complete Quartus designs in and of themselves.

35. A file that contains other designs, but is not part of a higher-level design, is called the top level of a hierarchy.

36. If the top level of a hierarchy is a **bdf,** lower level designs are embedded in the **bdf** as symbols that are created from the original design files of the components.

37. Quartus II looks for symbol files in the present working directory, then in the directories specified as User Libraries.

38. The compilation hierarchy of a project is displayed in the Quartus II Project Navigator Window. The design file for any level of the hierarchy can be opened by double-clicking the design file name in the Project Navigator Window.

GLOSSARY

AHDL (Altera Hardware Description Language)   Altera's proprietary text-entry design tool for PLDs.

Altera UP-1 Board   The predecessor to the Altera UP-2 board, with the same layout and similar CPLDs.

Altera UP-2 Board   A circuit board, part of Altera's University Program Design Laboratory Package, containing two CPLDs and a number of input and output devices.

Block Diagram File   A design file in which PLD design information is entered as a schematic or a block diagram.

ByteBlaster   An Altera ribbon cable and connector used to program or configure Altera CPLDs via the parallel port (LPT port) of an IBM PC or compatible.

ByteBlaster MV   A newer version of the ByteBlaster. MV stands for Multi-Volt, indicating that it can operate either at 5 volts or 3.3 volts.

Cell   A fuse location in a programmable logic device, specified by the intersection of an input line and a product line.

Checksum   An error-checking code derived from the accumulating sum of the data being checked.

Compile   The process used by CPLD design software to interpret design information (such as a schematic or text file) and create required programming information for a CPLD.

Complex PLD (CPLD)   A digital device consisting of several programmable sections with internal interconnections between the sections.

Design Entry   The process of using software tools to describe the design requirements of a PLD. Design entry can be done by entering a schematic or a text file that describes the required digital function.

Download   Program a PLD from a computer running PLD design and programming software.

Fitting   Assigning internal PLD circuitry, as well as input and output pins, to a PLD design.

Fixed-Function Logic   A digital logic device or circuit whose function is determined when it is manufactured and, hence, cannot be changed.

Hardware Description Language (HDL)   A computer language used to design digital hardware.

Hierarchical Design   A PLD design that is ordered in layers or levels. The highest level of design contains components that are themselves complete designs. These components may, in turn, have lower level designs embedded within them.

Hierarchy   The levels or layers of a hierarchical design.

Hierarchy Tree   A structure describing the relationship of the different levels of a design hierarchy to one another. In graphical form, it looks like the directory tree structure of folders in a program such as Windows Explorer.

ICR   In-Circuit Reconfigurability. The ability of a volatile PLD (such as a FLEX 10K) to be configured without removing it from a circuit board.

Input Line   A line that applies the true or complement form of an input variable to the AND matrix of a PLD.

Input Line Number   A number assigned to a true or complement input line in a PAL AND matrix.

Instance   A single copy of a component in a PLD design file.

ISP   In-System Programmability. The ability of a nonvolatile PLD (such as a MAX 7000S) to be programmed without removing it from a circuit board.

JEDEC   Joint Electron Device Engineering Council.

JEDEC File   An industry standard form of text file indicating which fuses are blown and which are intact in a programmable logic device.

JTAG   Joint Test Action Group. A standards body that developed the format for testing and programming devices while they are installed in a system. This format has been adopted by the Institute of Electrical and Electronics Engineers as IEEE Standard 1149.1.

JTAG Chain   Multiple JTAG-compliant devices whose TDI and TDO ports form a continuous chain connection. Such a chain allows multi-device programming.

Multiplexer   A circuit that selects one of several signals to be directed to a single output.

Nonvolatile   Able to retain stored information after power is removed.

PAL   Programmable array logic. Programmable logic with a fixed OR matrix and a programmable AND matrix.

Primitives   Basic functional blocks, such as logic gates, used in PLD design files.

Product Line   A single line on a logic diagram used to represent all inputs to an AND gate (i.e., one product term) in a PLD sum-of-products array.

Product Line First Cell Number   The lowest cell number on a particular product line in a PAL AND matrix where all cells are consecutively numbered.

Programmable Logic Device (PLD)   A digital integrated circuit that can be programmed by the user to implement any digital logic function.

Programmer Object File (pof)   Binary file used to program a nonvolatile CPLD, such as the Altera MAX 7000S series.

Programming   Transferring design information from the computer running PLD design software to the actual PLD chip.

Project   A collection of files associated with a PLD design in Quartus II.

Project Navigator Window   A window in the Quartus II workspace that shows the components of a project hierarchy in relation to one another in the form of a graphical hierarchy tree.

Quartus II   CPLD design and programming software owned by Altera Corporation.

Schematic Entry   A technique of entering CPLD design information by using a CAD (computer aided design) tool to draw a logic circuit as a schematic. The schematic can then be interpreted by design software to generate programming information for the CPLD.

Simulation   Testing design function by specifying a set of inputs and observing the resultant outputs. Simulation is generally shown as a series of input and output waveforms.

Software Tools   Specialized computer programs used to perform specific functions such as design entry, compiling, fitting, and so on. (Sometimes just called "tools.")

SRAM Object File (sof)   Binary file used to configure a volatile CPLD, such as the Altera FLEX 10K series.

Suite (of Software Tools)   A related collection of tools for performing specific tasks. Quartus II is a suite of tools for designing and programming digital functions in a PLD.

Symbol File   A file that represents a PLD design as a graphic symbol, showing only the design's inputs and outputs. The symbol can be used as a component in any Block Diagram File.

Target Device   The specific PLD for which a digital design is intended.

TCK   Test Clock. The JTAG signal that drives the JTAG downloading process from one state to the next.

TDI   Test Data In. In a JTAG port, the serial input data to a device.

TDO   Test Data Out. In a JTAG port, the serial output data from a device.

Text File   An ASCII-coded document stored electronically.

TMS   Test Mode Select. The JTAG signal that controls the downloading of test or programming data.

Top Level (of a Hierarchy)   The file in a hierarchy that contains components specified in other design files and is not itself a component of a higher-level file.

User Library   A folder containing symbols that can be used in a **gdf** file.

Vector Waveform File (vwf)   A file that forms the input to the Quartus II simulator, containing the inputs and outputs to be simulated as graphical waveforms.

Volatile   A device is volatile if it does not retain its stored information after the power to the device is removed.

PROBLEMS

Problem numbers set in color indicate more difficult problems.

4.1   What Is a PLD?

4.1   List some of the advantages of programmable logic over fixed-function logic.

4.2   What does CPLD stand for? How is it different from the term PLD?

4.3   List some types of PLDs other than CPLDs.

4.4   Figure 4.100 shows a 4-to-1 multiplexer circuit. (The circuit switches one of four digital inputs to a single output, depending on the states of two "select inputs.") State the number of 74HC type devices required to make this circuit. You may use the following devices: 74HC04 hex inverter; 74HC11 triple 3-input AND gate; 74HC4002 dual 4-input NOR gate (there are no 4-input OR devices available in the 74HC family). State how many devices are required to make two multiplexers.

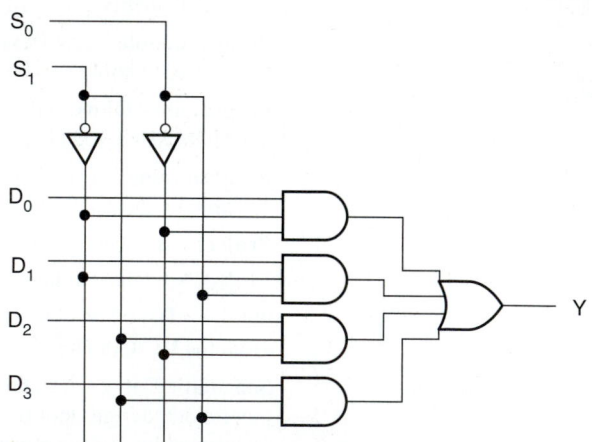

FIGURE 4.100   Problem 4.4: 4-to-1 Multiplexer

4.2   Programmable Sum-of-Products Arrays

4.5   Why are the inputs to AND gates in a programmable SOP array shown in a logic diagram as a single product line?

4.6   What does PAL stand for? Briefly describe the architecture of a PAL.

4.7   Briefly describe how the AND array inputs are programmed in a PLD having PAL architecture.

4.8   Why is the SOP array particularly well-suited for use in a PLD?

4.3 PAL Fuse Matrix and Combinational Outputs

4.9 How many of each of the following pin types are there in a PAL16L8 PLD?

a. dedicated input pins

b. dedicated output pins

c. pins that can be either an input or an output

4.10 How are the output enables of the tristate buffers controlled in a PAL16L8?

4.11 State how a tristate buffer on a PAL16L8 can be always enabled or always disabled.

4.12 How is programming information stored for a PAL device? How are intact and blown fuses indicated in this format?

4.4 PAL Outputs with Programmable Polarity

4.13 Draw a diagram showing the basic configuration and symbology for a PLD sum-of-products array.

4.14 Draw a basic PAL circuit having four inputs, eight product terms, and one active-LOW combinational output. Draw fuses on your diagram showing how to make the following Boolean expression:

$$\overline{F} = \overline{A}\,B\,\overline{C} + \overline{B}\,C\,D + \overline{A}\,C\,D + A\,\overline{C}\,D$$

4.15 Modify the PAL circuit drawn in Problem 4.14 to make two outputs having eight product terms and programmable polarity. Draw fuses on the diagram for each of the following functions:

$$F1 = \overline{A}\,B\,\overline{C} + \overline{B}\,C\,D + \overline{A}\,C\,D + A\,\overline{C}\,D$$
$$\overline{F2} = \overline{A}\,B\,\overline{C} + \overline{B}\,C\,D + \overline{A}\,C\,D + A\,\overline{C}\,D$$

4.16 Make a photocopy of Figure 4.11 (PAL20P8 logic diagram). Draw fuses on the PAL20P8 logic diagram showing how to make a BCD-to-2421 code converter, as developed in Example 3.22.

Table 4.1 shows how the two codes relate to each other.

TABLE 4.1 BCD and 2421 Code

Decimal Equivalent	BCD Code				2421 Code			
	D_4	D_3	D_2	D_1	Y_4	Y_3	Y_2	Y_1
0	0	0	0	0	0	0	0	0
1	0	0	0	1	0	0	0	1
2	0	0	1	0	0	0	1	0
3	0	0	1	1	0	0	1	1
4	0	1	0	0	0	1	0	0
5	0	1	0	1	1	0	1	1
6	0	1	1	0	1	1	0	0
7	0	1	1	1	1	1	0	1
8	1	0	0	0	1	1	1	0
9	1	0	0	1	1	1	1	1

The Boolean equations for the BCD-to-2421 decoder are:

$$Y_4 = D_4 + D_3 D_2 + D_3 D_1$$
$$Y_3 = D_4 + D_3 D_2 + D_3 \overline{D}_1$$
$$Y_2 = D_4 + \overline{D}_3 D_2 + D_3 \overline{D}_2 D_1$$
$$Y_1 = D_1$$

4.17 Repeat Problem 4.16 for a 2421-to-BCD code converter. (You will have to derive the equations for this design.)

4.5   Programming PLDs in Quartus II

4.18   What do we call the sequence of steps required to design and program a CPLD?

4.19   Why must we assign a target device to our CPLD design?

4.6   Quartus II Design Flow and Graphical User Interface

4.20   Briefly describe the difference between a design file and a project in Quartus II.

4.21   State two ways to create a project in Quartus II.

4.22   Why should each project in Quartus II have its own Windows folder?

4.7   Creating a Quartus II Project and Block Diagram File

4.23   State the definitions of the following terms:

    a. primitives

    b. instance

NOTE . . .

Problems 4.24 to 4.28 use techniques from Sections 4.7 to 4.10. They all require you to enter a design using the Quartus II Block Editor and compile it. In some problems, you will be asked to write a set of simulation criteria, create a simulation, add pin numbers to the project, or create a symbol file from the design file in that problem and use it in a hierarchical design in another problem. If you have not reached a particular design stage in your study, just do up to the part you have completed (e.g., design entry, simulation, or whatever) and return to the problem when you have completed the rest of the topics. Pin numbers, where required, are given for the Altera UP-1/UP-2 board (Figure 4.13a), the RSR PLDT-2 board (Figure 4.13c), and DeVry eSOC board (Figure 4.13b).

4.24   Use Quartus II to create a Block Diagram File for the multiplexer circuit shown in Figure 4.100.

    Save the file as ***drive:\qdesigns\textbook\ch04\problems\4to1mux\4to1mux.bdf.*** Assign pins as in Table 4.2. Save and compile the project. Write a set of simulation criteria and use them to create a simulation that verifies the operation of the circuit.

TABLE 4.2   Pin Assignments for Multiplexer Circuit

Function	Pin		
	UP-2	PLDT-2	eSOC
S1	34	34	50
S0	33	33	51
D0	37	37	55
D1	40	40	56
D2	39	39	57
D3	41	41	58
Y	44	44	4

4.25   Figure 4.101 shows the circuit for a 4-channel demultiplexer, which switches a digital input to one of four outputs, depending in the states of two "select inputs."

FIGURE 4.101 Problem 4.25: 4-Channel Demultiplexer

Use Quartus II to create a Block Diagram File for the demultiplexer circuit. Save the file as **drive:\qdesigns\textbook\ch04\problems\4ch_dmux\4ch_dmux.bdf.** Assign pins as in Table 4.3. Save and compile the project.

TABLE 4.3 Pin Assignments for Demultiplexer Circuit

	Pin		
Function	**UP-2**	**PLDT-2**	**eSOC**
S1	34	34	50
S0	33	33	51
D	41	41	58
Y0	44	44	4
Y1	45	45	5
Y2	46	46	8
Y3	48	48	9

4.26 Repeat Problem 4.25 for the 2-bit equality comparator in Figure 4.102. This circuit generates a HIGH output when the two 2-bit numbers A_2A_1 and B_2B_1 are equal. Save the file as **drive:\qdesigns\textbook\ch04\problems\eq_comp\eq_comp.bdf.** Use the pin assignments in Table 4.4.

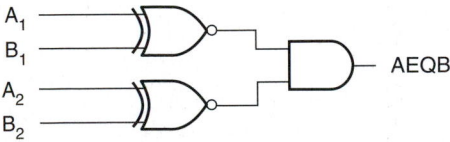

FIGURE 4.102 Problem 4.26: 2-Bit Equality Comparator

TABLE 4.4 Pin Assignments for Equality Comparator

	Pin		
Function	**UP-2**	**PLDT-2**	**eSOC**
A1	34	34	50
A2	33	33	51
B1	39	39	57
B2	41	41	58
AEQB	44	44	4

4.27   Use Quartus II to create a Block Diagram File for the half-adder circuit shown in Figure 4.103. The half adder adds two bits to generate a sum and a carry output. Save the file as *drive:*\qdesigns\textbook\ch04\problems\halfadd\halfadd.bdf. Save and compile the project. Create a symbol file for the half-adder design file. Do not assign pin numbers at this time.

FIGURE 4.103   Problem 4.27: Half Adder

4.28   Use Quartus II to create a Block Diagram File for the full-adder circuit shown in Figure 4.104. The full adder combines two bits *A* and *B,* plus an input carry from a previous stage to generate a sum and a carry output.

FIGURE 4.104   Problem 4.28: Full Adder

Save the file as *drive:*\qdesigns\textbook\ch04\problems\fulladd\fulladd.bdf. Assign pin numbers as shown in Table 4.5. Save and compile the project. Write a set of simulation criteria and use them to create a simulation that verifies the operation of the circuit.

TABLE 4.5   Pin Assignments for Full Adder

	Pin		
Function	**UP-2**	**PLDT-2**	**eSOC**
A	34	34	50
B	33	33	51
CARRY_IN	36	36	52
SUM	45	45	5
CARRY_OUT	44	44	4

4.8   Compiling and Simulating a Design in Quartus II

4.29   Describe a situation where it would be important to have a simulation time that shows input changes that are of the same order of magnitude as the input-to-output delay time of a CPLD.

4.9   Transferring a Design to a Target CPLD

4.30   Describe two methods that can be used to assign pin numbers to a project in Quartus II.

4.31   What do ISP and ICR stand for? Briefly state the difference between the two terms.

4.32   How many wires are defined for the JTAG interface? List their names and functions.

4.33   Under what condition is it necessary to set up programming hardware in Quartus II?

4.10 Using the Quartus II Block Editor to Create a Hierarchical Design

4.34   Examine the half-adder circuit in Figure 4.102 and the full adder circuit in Figure 4.103. You should find two half-adders in the full-adder circuit. Use the half-adder symbol you created in Problem 4.27 to create a full-adder as a hierarchical design, consisting of two half-adders and other logic. Save the file as *drive:* **qdesigns\textbook\ch04\problems\fulladd2\fulladd2.bdf.** Assign the pin numbers as in Table 4.5.

ANSWERS TO SECTION REVIEW PROBLEMS

4.1

4.1   The function of a programmable logic device is determined by the user and, in many cases, can be altered as necessary. The function of fixed-function logic is determined when it is manufactured and cannot be changed.

4.2

4.2   Sum-of-products (SOP).

4.3

4.3   PLD design software is used to assign fuse mapping in a modern PLD.

4.4

4.4   Exclusive OR (XOR).

4.5

4.5   A target device is the specific PLD for which a digital circuit is designed.

4.6

4.6   1. Schematic or block diagram (Block Diagram File).
        2. Text entry (Hardware Description Language).

4.7

4.7   1. Save a design file in a new folder, making sure that the box labeled **Create new project based on this file** is checked. Follow the prompts in the New Project Wizard.
        2. Select **New Project Wizard** from the **File** menu.

4.8

4.8   1. Understand correct operation of the design;
        2. Write a set of criteria for testing all possible correct operational and failure modes.
        3. Test all possibilities.

4.9

4.9   A CPLD is shipped blank and thus has no assigned input and output pins. In order to be able to tell which pins on the CPLD correspond to the inputs and outputs of our design, we must assign pin numbers and recompile the project before programming.

4.10

4.10   A user library path would not be required in a hierarchical design if all component design files were located in the same folder as the top-level design file of the project.

KEY TERMS

Architecture Body

Array

ASCII

Character

Concatenate

Concurrent Signal Assignment
  Statement

Concurrent Statement

Entity Declaration

IEEE Std. 1164 Multi-Value Logic

Library

Mode

Named Identifier

Package

Port

Range

Reserved Keyword

Selected Signal Assignment
  Statement

Signal

Signal Concurrency

String

Syntax

Type

VHDL

VHSIC

OUTLINE

5.1 VHDL Basics

5.2 Making a VHDL File in Quartus II

5.3 VHDL Syntax for Port, Mode, and Type

5.4 Signals in VHDL

CHAPTER OBJECTIVES

After successful completion of this chapter, you will be able to:

- State what VHDL stands for.
- State the functions of a VHDL entity declaration and architecture body.
- Write VHDL statements defining an entity and architecture, including ports, complete with mode and type.
- Define and use BIT, STD_LOGIC, and INTEGER types.
- Encode Boolean expressions in VHDL, using concurrent signal assignment statements.
- Encode truth tables in VHDL, using selected signal assignment statements.
- Use Quartus II software to enter, compile, and simulate a VHDL design.
- Use Quartus II software to assign pins to a CPLD and program the CPLD.

CHAPTER 5

Introduction to VHDL

VHDL CONSTRUCTS USED IN THIS CHAPTER

- Architecture body
- Array
- BIT and BIT_VECTOR types
- Comments
- Concatenation operator (&)
- Concurrent signal assignment statements
- Entity declaration
- INTEGER, NATURAL, and POSITIVE types
- LIBRARY declaration
- Mode
- Naming conventions
- `others` clause
- Port definitions
- Range (to and downto)
- Reserved keywords
- Selected signal assignment statement
- SIGNAL
- STD_LOGIC and STD_LOGIC_VECTOR types
- Type
- `use` clause
- VHDL package: ieee.std_logic_1164

VHDL stands for VHSIC Hardware Description Language. VHSIC means Very High Speed Integrated Circuit. VHDL is an industry-standard language for modeling and synthesizing digital hardware, particularly for programmable logic or Application Specific Integrated Circuits (ASICs). An ASIC is a custom-designed fixed-function device. An example of an ASIC would be a specially designed chip for a fax/modem.

VHDL was developed in the 1980s by contractors working in the U.S. defense industry and has since been standardized by the Institute of Electrical and Electronics Engineers (IEEE), first in 1987, then revised in 1993. It is currently defined by IEEE Standard 1076-1993. A related standard for certain data types in VHDL is IEEE Standard 1164-1993.

VHDL has enjoyed great popularity in the electronics design community due to its ability to easily migrate code from one system to another. Since a VHDL design is just a text file, it can be used with a wide variety of software tools without worrying about data compatibility. This feature is nearly impossible to achieve with graphic design formats, particularly because these formats tend to evolve over time as technology improves.

VHDL can be used to describe digital hardware at a number of levels of abstraction. It can be used simply as a tool to enter Boolean equations and truth tables, techniques that we will study in this chapter, or it can be used to design hardware by providing a description of the hardware's behavior. This latter technique will be studied in further chapters. ■

5.1 VHDL BASICS

■ KEY TERMS

VHDL VHSIC Hardware Description Language.

VHSIC (pronounced "VIZ-ick") Very High Speed Integrated Circuit.

Syntax The rules of construction, or "grammar," of a programming language.

Entity Declaration The part of a VHDL file that defines the inputs and outputs of the design.

Architecture Body The part of a VHDL design that defines the relationship between its inputs and outputs.

Port An input or output in a VHDL file.

Mode The property of a port, signal, or variable that specifies the direction of data flow.

Type The property of a port, signal, or variable that specifies the values it can have.

Concurrent Signal Assignment Statement A VHDL statement that assigns a value to a port or signal based on a Boolean expression or constant. The output port or signal is affected as soon as there is a change to the input port.

Concurrent Statement A VHDL statement that defines a relationship between inputs and outputs of a design entity such that the output responds immediately to a change of input.

Signal Concurrency A property of VHDL whereby all concurrent statements in a VHDL design are evaluated at the same time.

VHDL, or **VHSIC** Hardware Description Language, is a computer language used to model and synthesize digital hardware. As its title implies, VHDL is used to describe the structure or behavior of hardware, then determine how it should operate (modeling) or how it should be built (synthesis). A designer enters text according to the **syntax** of the language, then uses appropriate software to compile, simulate, and synthesize the design, much as was done in Chapter 4 with graphic design entry. VHDL designs can also be combined with graphic designs in the Quartus Graphic Editor by creating symbols based on the VHDL code.

In this text, we will concentrate on the use of VHDL for synthesis. We will not study all possible constructs in VHDL, just the most useful subset for fairly simple digital logic design. Since VHDL was originally conceived as a modeling tool, there are many parts of the language that have no physical meaning when it is used as a synthesis tool. For example, there are statements in VHDL that define time as a data type and specify circuit timing constraints. If we are using VHDL to synthesize circuits in particular CPLDs, these constraints are already part of the hardware and cannot be affected by statements in a hardware description language. These statements *are* useful when we wish to model the behavior of a CPLD that has not yet been designed. We will not look at any of these non-synthesis features of the language.

Entity and Architecture

The two basic constructs that are required by every VHDL file are an **entity declaration** and an **architecture body.** The entity declaration defines the inputs and outputs of the design. The architecture body defines the relationship between the inputs and outputs.

Figure 5.1a shows the logic circuit of a majority vote circuit, where the output is HIGH with two or more HIGH inputs. The Boolean expression $Y = AB + BC + AC$ shows the relationship between the circuit inputs and outputs and can thus represent a VHDL architecture. Figure 5.1b shows a graphic symbol that represents the VHDL file in a Quartus II Block Diagram File. This block, which can represent the VHDL entity, shows the VHDL file name (**majority_vote.vhd**), which is also the name of the entity. In addition, the block shows the circuit inputs, *a, b,* and *c* and output *y*.

The VHDL file that defines the majority vote circuit is as follows. Let us analyze it in detail.

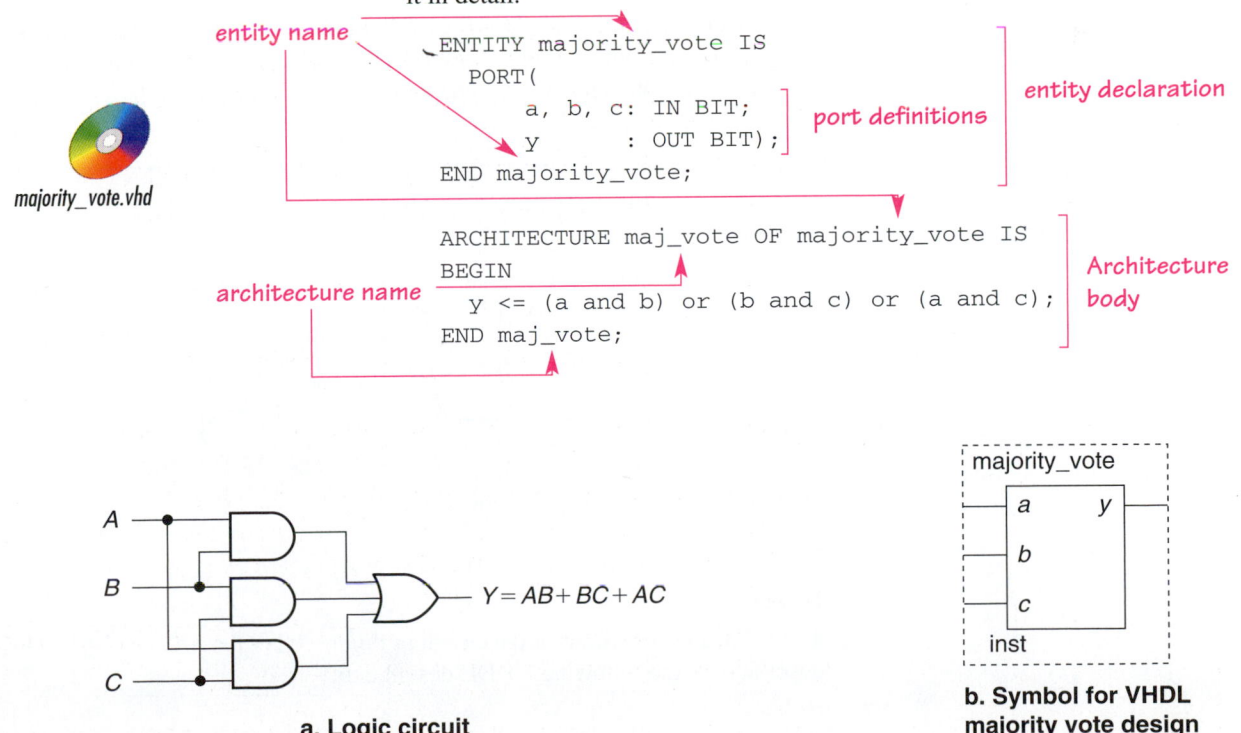

majority_vote.vhd

entity name

```
ENTITY majority_vote IS
    PORT(
        a, b, c: IN BIT;
        y      : OUT BIT);
END majority_vote;
```
port definitions — entity declaration

```
ARCHITECTURE maj_vote OF majority_vote IS
BEGIN
    y <= (a and b) or (b and c) or (a and c);
END maj_vote;
```
architecture name — Architecture body

$Y = AB + BC + AC$

a. Logic circuit

majority_vote

a   y
b
c
inst

b. Symbol for VHDL majority vote design

FIGURE 5.1   Majority Vote Circuit

The entity declaration consists of the first five lines of code. The first and last lines of this section define the name of the entity (in this case, `majority_vote`). Within the entity declaration is the PORT statement. This statement lists the port names, as well as the **mode** and **type** of each one. The mode of a port is the direction that data flows through it. The port's type defines the range of values that the port can have. For example, the statement `y : OUT BIT;` states that port **y** is of mode OUT and of type BIT. This means that data flows only out of the port and it can only have the values of '0' and '1'. We will examine other alternatives for mode and type in a later section of the chapter.

Statements in VHDL are terminated by semicolons. Thus, the PORT statement has the form `PORT();` Inside this statement, the various port definitions (the third and fourth lines of code) and are separated by semicolons. Because semicolons are used as separators within the statement, the last element in the list is not followed by a semicolon.

The architecture body consists of the last four lines of the example code. The first line has the form `ARCHITECTURE __architecture_name OF __entity name IS`, which basically says, "this is the architecture (called `maj_vote`) of the entity (called `majority_vote`) defined above." The actual architecture body is bracketed by a pair of statements: `BEGIN` and `END __architecture_name;`. The output is defined in relation to the inputs by the Boolean expression in the second last line of the code.

The operator <=, called an "assignment operator," is a compound symbol, consisting of the "less than" sign and the "equal" sign. It is meant to look like an arrow pointing to the left, indicating that whatever is on the left side of the statement is assigned the value of whatever is on the right. It is conventional to say "gets" when you see this operator, as in "y gets (a and b) or (b and c) or (a and c)".

The Boolean operators, AND, OR, NOT, NAND, NOR, XOR, and XNOR are simply represented by those words. VHDL has an equal order of precedence for these Boolean operators, so everything must be written explicitly with parentheses. This sometimes leads to difficult-to-read statements such as:

```
y <= (a and (not b)) or ((not a) and b and (not c));
```

for the Boolean expression $Y = A\bar{B} + \bar{A}B\bar{C}$. Fortunately, there are better ways to represent complex Boolean expressions in VHDL. We will examine some of these ways later in this chapter and in other chapters.

EXAMPLE 5.1

Write a VHDL file for the logic circuit in Figure 5.2, assuming all ports are of type BIT.

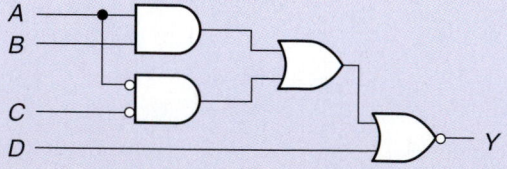

FIGURE 5.2   Example 5.1: Logic Circuit

■ **Solution**

The Boolean expression for the circuit in Figure 5.2 is $Y = \overline{AB + \bar{A}\,\bar{C} + D}$. This is described by the following VHDL design entity.

```
ENTITY logic_circuit IS
   PORT(
        a, b, c, d : IN BIT;
        y          : OUT BIT);
END logic_circuit;


ARCHITECTURE cct OF logic_circuit IS
BEGIN
   y <= not((a and b) or ((not a) and (not c)) or d);
END cct;
```

Signal Concurrency

The statements for these Boolean expressions are examples of **concurrent signal assignment statements.** These are specific examples of a larger class of VHDL statements, called **concurrent statements,** that exhibit a property called **signal concurrency.** The idea behind signal concurrency is that all concurrent statements are evaluated at the same time. (*Concurrent* means "simultaneous.")

If you have done any computer programming in a language such as BASIC, Pascal, or C, this idea might seem crazy. In those languages, statements are executed in the order in which they are written and the idea that a bunch of statements would be executed all at once does not appear logical. However, for a hardware description language, signal concurrency does make sense since its purpose is to synthesize hardware rather than run a sequence of program statements.

For example, look at the circuit in Figure 5.3, a digital arithmetic circuit called a half adder. The circuit adds two bits, *A* and *B,* producing a sum and carry. The circuit is defined by two equations: $SUM = A \oplus B$ and $CARRY\_OUT = A \cdot B$. Suppose both inputs are 1. Then $CARRY\_OUT = 1$ and $SUM = 0$. Further suppose that input *B* changes to a 0. Now $CARRY\_OUT = 0$ and $SUM = 1$.

Which changes first, *SUM* or *CARRY_OUT?* Since both outputs depend on the same two inputs, both outputs ideally change at the same time. (There may be differences in the times due to differing internal delays in the gate circuits. We will ignore these for now.) In VHDL, we would write these statements as follows:

```
sum       <= a xor b;
carry_out <= a and b;
```

If VHDL is to accurately represent the hardware in Figure 5.3 then it cannot matter which order those two statements are written in. Which is executed first? Neither; both are executed concurrently, because that's how the hardware behaves.

A
B
— SUM = $A \oplus B$

— CARRY_OUT = $A \cdot B$

FIGURE 5.3   Half Adder

TABLE 5.1   Example Truth Table

A	B	C	Y1	Y0
0	0	0	0	0
0	0	1	1	0
0	1	0	1	0
0	1	1	0	1
1	0	0	0	0
1	0	1	0	0
1	1	0	0	1
1	1	1	0	0

■ SECTION 5.1 REVIEW PROBLEM

5.1   Write the VHDL statements required to implement the truth table shown in Table 5.1.

5.2   MAKING A VHDL FILE IN QUARTUS II

VHDL files can be created using the Quartus II Text Editor. As an example, let us walk through the procedure required to enter, compile, and simulate a VHDL file that implements the majority vote circuit of Figure 5.1.

Design Entry

Before starting a new file, create a new folder in Windows. Let us call the folder *drive:***\qdesigns\ch05\majority_vote**. To start a new file, click the **New File** toolbar button or select **New** from the **File** menu. The dialog box shown in Figure 5.4 will appear. Select **VHDL File** from the **Device Design Files** tab.

The text editor window will open. Immediately save the file, using the **Save As** function from the **File** menu, as shown in Figure 5.5. The file name and path is

FIGURE 5.4   Quartus New File Dialog

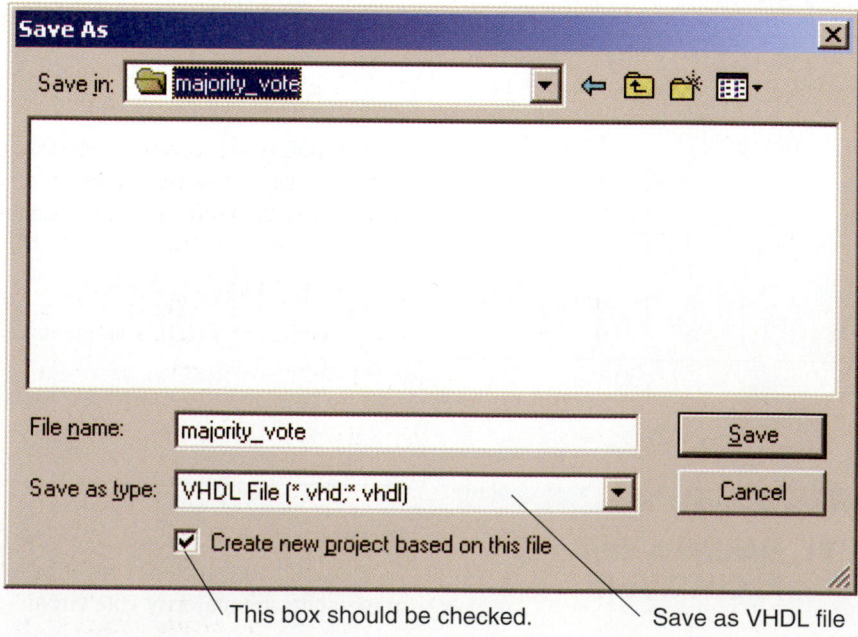

This box should be checked.   Save as VHDL file

FIGURE 5.5   "Save As" Dialog

drive:\qdesigns\ch05\majority_vote\majority_vote.vhd. Make sure the box labeled **Create new project based on this file** is checked. Click **Save.** Quartus II will respond with a dialog box, shown in Figure 5.6, asking if you want to create a new project with this file. Click **Yes** and follow the procedure to create the new project. This is the same series of steps outlined in Chapter 4.

Enter the VHDL text in the text editor window, as shown in Figure 5.7. The editor will number the lines of text for reference. When you have finished entering the text, save and compile the file, using the procedure outlined in Chapter 4.

FIGURE 5.6   New Project Query

```
 1  ENTITY majority_vote IS
 2      PORT(
 3          a, b, c: IN    BIT;
 4          y      : OUT   BIT);
 5  END majority_vote;
 6
 7  ARCHITECTURE maj_vote OF majority_vote IS
 8  BEGIN
 9      y <= (a and b) or (b and c) or (a and c);
10  END maj_vote;
```

FIGURE 5.7   VHDL Windows (Text Editor)

Simulation

In order to ensure the correctness of our design, we should simulate the VHDL design we have just created. To do so, we must create a Vector Waveform File to contain input waveforms that will be used to drive the simulation. Create the Vector Waveform File from the **New File** dialog, as shown in Figure 5.4. Select **Vector Waveform File** from the **Other Files** tab and click **OK.**

Use the **Node Finder** to enter inputs **a, b,** and **c** and output **y,** as outlined in Chapter 4. Set the end time of the simulation to 100 μs, again as outlined in Chapter 4. The resulting default waveforms are shown in Figure 5.8.

For convenience, we can group the input waveforms together to represent a multiple-bit binary value. Do this by highlighting the waveforms **a, b,** and **c,** as shown in Figure 5.9. (Click on waveform **a,** then drag down to waveform **c.**)

Right-click on the selected waveforms and select **Group** from the pop-up menu that appears. You will see the **Group** dialog box, as shown in Figure 5.10. For **Group Name,** type **inputs.** Select **Binary** from the drop-down box labeled as **Radix.** Click **OK.** The grouped inputs appear as shown in Figure 5.11.

Click on the **inputs** group to highlight it. We will apply an increasing binary count to the inputs that can be used as a test stimulus for the VHDL design. Click

FIGURE 5.8 Default Waveforms in Vector Waveform File

Click and drag to highlight a, b, and c.

FIGURE 5.9 Highlighted Waveform Group

FIGURE 5.10 "Group" Dialog

FIGURE 5.11 Waveforms Showing Grouped Inputs

FIGURE 5.12 "Count Value" Toolbar Button

FIGURE 5.13 Count Value Dialog (Counting)

the **Count Value** button on the toolbar (shown in Figure 5.12). The **Count Value** dialog box, shown in Figures 5.13 and 5.14, will appear. In the **Counting** tab (Figure 5.13), enter the values shown: **Radix: Binary; Start value: 000; Increment by: 1;** and **Count type: Binary.** In the **Timing** tab, enter the values shown in Figure 5.14: **Start time: 0 ps; End time: 100 us; Count every: 10.0 us; Multiplied by: 1.** Click **OK.** Select the **View menu** and **Fit in Window.** The waveforms shown in Figure 5.15 will appear.

The individual input waveforms can be viewed as well as the combined input group. Click the box with the + symbol beside the group name (**inputs**) to show the

FIGURE 5.14   Count Value Dialog (Timing)

FIGURE 5.15   Input Waveforms Showing Group

individual and group names, as in Figure 5.16. To collapse the group, click on the box with the – symbol.

Run the simulation by clicking on the **Simulation** toolbar button or else select **Run Simulation** from the Quartus II **Processing** menu. The simulation result will appear as shown in Figure 5.17.

When we examine the input and output waveforms, we note that the output, *y*, goes HIGH when two or more of inputs *a, b,* or *c* are HIGH.

FIGURE 5.16 Group and Individual Input Waveforms

FIGURE 5.17 Simulation Result

Pin Assignments

Pins can be assigned to a VHDL design exactly the same way as for a graphic design. Select the **Assignments** menu, then **Pin Assignments** to get a dialog box to perform the pin assignment function. Alternatively, you can assign pins using the **Assignment Editor,** also available in the **Assignments** menu. For further details, refer to the corresponding section of Chapter 4.

■ SECTION 5.2 REVIEW PROBLEM

5.2   Why should we create a simulation of a VHDL file?

5.3 VHDL SYNTAX FOR PORT, MODE, AND TYPE

> ### ■ KEY TERMS
>
> **Named Identifier** An object in VHDL, such as an entity, architecture, port, signal, or variable that is given a name to identify it within a VHDL program.
>
> **Reserved Keyword** A word that is part of the VHDL language and thus may not be used as a named identifier.
>
> **Type** A property of a port, signal, or variable that specifies its allowable values.
>
> **Character** A number, letter, or other printable symbol, as represented in a computer code, such as ASCII. Shown by enclosing the character in single quotes (e.g., `'0'` is a character.)
>
> **ASCII** American Standard Code for Information Interchange. A 7- or 8-bit binary code used to represent printable characters and control codes.
>
> **Array** A group of related objects of the same type, organized by numbering the various elements of the array.
>
> **Range** The property of a port, variable, or signal that defines the upper and lower bounds of the values that the object can have.
>
> **String** A one-dimensional array of characters. Shown by enclosing the string in double quotes (e.g., `"0110"` is a string of four characters.)
>
> **Selected Signal Assignment Statement** A VHDL statement that assigns a value to a port, signal, or variable, based on alternative values of another port, signal, or variable.
>
> **IEEE Std. 1164 Multi-Value Logic** A nine-valued type used in VHDL to describe logic hardware as a STD_LOGIC or STD_LOGIC_VECTOR type.
>
> **Library** A collection of VHDL design units that have been previously compiled.
>
> **Package** A group of uncompiled VHDL design elements that can be used by more than one VHDL file.

Having examined some simple VHDL design entities and a design entry procedure, let us more formally lay out some of the rules and options available to us when we create a design in VHDL.

Valid Names

A valid name for a port, signal, variable, entity name, architecture body, or similar object, called a **named identifier** in VHDL, consists of a letter followed by any number of letters or numbers, without spaces. VHDL is not case sensitive (i.e., the named identifiers **my_function** and **MY_FUNCTION** are treated as identical), but Altera has created a style guideline that can be accessed via the Quartus II Help menu. With some exceptions, according to the Altera style guidelines, VHDL keywords should be capitalized and user names written in lower case.

An underscore may be used within a name, but may not begin or end the name. Two consecutive underscores are not permitted. Also, there are a number of **reserved keywords** that may not be used as object names, as they are words that make up the VHDL language and thus have specific functions.

Examples:

Valid names:

```
decode8
just_in_time
What_4
```

Invalid names:	`8decode` (begins with a digit)
	`in__time` (two consecutive underscores)
	`_What_4` (begins with underscore)
	`my design` (space inside a name)
	`your_words?` (special character ? not allowed)
	`signal` (reserved keyword)

The list of reserved keywords for VHDL 1993 is as follows:

abs	file	nand	select
access	for	new	severity
after	function	next	signal
alias		nor	shared
all	generate	not	sla
and	generic	null	sll
architecture	group		sra
array	guarded	of	srl
assert		on	subtype
attribute	if	open	
	impure	or	then
begin	in	others	to
block	inertial	out	transport
body	inout		type
buffer	is	package	
bus		port	unaffected
	label	postponed	units
case	library	procedure	until
component	linkage	process	use
configuration	literal	pure	
constant	loop		variable
		range	
disconnect	map	record	wait
downto	mod	register	when
		reject	while
else		rem	with
elsif		report	
end		return	xnor
entity		rol	xor
exit		ror	

Comments

A comment is explanatory text that is ignored by the VHDL compiler. It is indicated by two consecutive hyphens.

```
Example: -- This is a comment.
```

Modes

The mode of a port defines its direction of data flow. The four modes we will use are IN, OUT, INOUT, and BUFFER. (Another mode called LINKAGE does not concern us.) These modes are illustrated in Figure 5.18. A port of mode IN only allows data to flow from an input pin to the CPLD logic. A port of mode OUT allows data to flow from the CPLD logic to the output port. A port of mode INOUT is bidirectional; it will allow data to flow in both directions.

BUFFER is a special case of OUT, which both allows the port value to be "updated" or modified by the CPLD logic and to be fed back into the CPLD logic

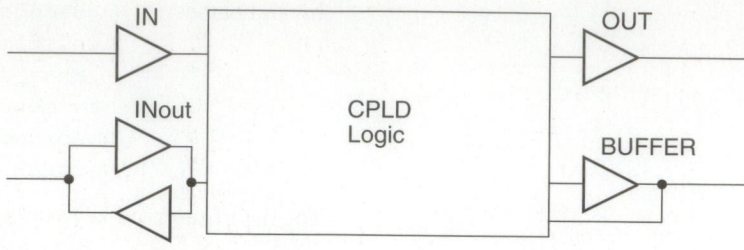

FIGURE 5.18   Modes of VHDL Ports

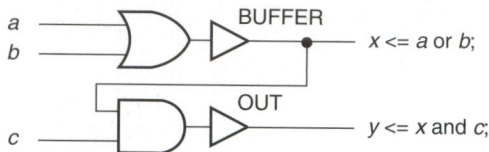

FIGURE 5.19   BUFFER and OUT Modes

to be reused by another function. This distinction is shown in Figure 5.19. The port associated with the OR function is of mode BUFFER and is assigned the function `x <= a or b;`. This function is fed back into the CPLD circuit and is reused by the AND function: `y <= x and c;`. The AND output does not need to be of mode BUFFER because `y` is not being reused or modified by any other part of the circuit.

■ SECTION 5.3A REVIEW PROBLEM

5.3   Which of the following is a valid name in VHDL? If a name is not valid, explain why it is not.

    a. MyFunction

    b. _a_and_b

Type, Array, and Range

A **type** in VHDL is a property applied to a port, signal, or variable that determines what values that object can have. Some of the most common types we will use in VHDL are BIT, STD_LOGIC, and INTEGER.

BIT and BIT_VECTOR

The BIT type is native to VHDL, in that its definition is stored in a standard library that is part of all versions of VHDL. BIT can have only two values: '0' and '1'. The values are placed in single quotes because VHDL treats them like **ASCII characters.** For many situations, BIT is good enough, as a large number of digital circuits use only these values.

If we want to assign a multiple-bit number to a series of related input or output ports, we can use the type BIT_VECTOR. A BIT_VECTOR is defined as a one-dimensional **array** of elements, each of type BIT. The **range** of the array is indicated by listing its upper and lower bounds. For example, we can define a 4-bit vector as follows:

```
d : IN BIT_VECTOR(3 downto 0);
```

Each of the elements of this vector can be individually addressed by indicating its position by a number in parentheses. This assignment is shown in Figure 5.20a. In this format, the most significant bit is positioned on the left-hand side of the vector and the elements number "downto" the least significant bit. ("Downto", written as a single word, is a VHDL reserved keyword.)

a) *d* (3 downto 0)

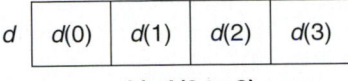

b) *d* (0 to 3)

FIGURE 5.20 Elements in a One-Dimensional Array (Vector)

We can also assign the vector with the elements going in the other direction. The statement d: IN BIT_VECTOR(0 to 3); will assign the elements from the LSB on the left to the MSB on the right, as shown in Figure 5.20b. The rule here is that VHDL always makes vector assignments from left to right and the contents of those vectors can be chosen by selecting how the elements are numbered.

Constant values assigned to BIT_VECTORs are written in double quotes, as they are treated like **strings,** or arrays of characters, by the VHDL compiler. For example, if a port called *d* is defined by d : IN BIT_VECTOR(3 downto 0);, then the following groups of statements are equivalent.

```
d(3) <= '0';        d <= "0101";
d(2) <= '1';
d(1) <= '0';
d(0) <= '1';
```

If *d* is defined by d : IN BIT_VECTOR(0 to 3);, then the following groups of statements are equivalent.

```
d(3) <= '0';        d <= "1010";
d(2) <= '1';
d(1) <= '0';
d(0) <= '1';
```

EXAMPLE 5.2

FIGURE 5.21 4-Bit AND Array

bitwise_and_4.vhd

Figure 5.21 shows a 4-bit AND array, that is, an array of four 2-bit AND gates. The following VHDL file describes the array.

```
-- 4-bit bitwise and function
-- y3 = a3 and b3; y2 = a2 and b2; etc.
ENTITY bitwise_and_4 IS
    PORT(
        a3, a2, a1, a0 : IN BIT;          Ports defined
        b3, b2, b1, b0 : IN BIT;          individually
        y3, y2, y1, y0 : OUT BIT);
END bitwise_and_4;


ARCHITECTURE and_gate OF bitwise_and_4 IS
BEGIN
    y3 <= a3 and b3;
    y2 <= a2 and b2;     Outputs assigned
    y1 <= a1 and b1;     individually
    y0 <= a0 and b0;
END and_gate;
```

Rewrite the file to make use of BIT_VECTOR rather than BIT types.

■ Solution

The file can be rewritten as follows.

bitwise_and_vec_4.vhd

```
-- 4-bit bitwise AND function
-- y = a and b;
ENTITY bitwise_and_vec_4 IS
    PORT(
        a, b : IN  BIT_VECTOR(3 downto 0);      Ports defined
        y    : OUT BIT_VECTOR(3 downto 0));     as vectors
END bitwise_and_vec_4;
```

(continued)

```
ARCHITECTURE and_gate OF bitwise_and_vec_4 IS
BEGIN
   y <= a and b; ] Outputs assigned as a vector
END and_gate;
```

Let us make some observations about the two files in Example 5.2.

- Both files describe the same hardware (Figure 5.21)
- The ports in the first file (*a3*, *a2*, etc.) are not numbered in parentheses because they are individual port names, not elements of a vector.
- In the second file, we can name more than one port in the same statement (e.g., *a, b*) if they are the same type and mode.
- We can use a single statement to apply the AND function to all elements of the *a* and *b* vectors simultaneously if they and the result vector *y* are all of the same type and length. The input vectors are automatically ANDed bit-by-bit in the same sequence they are defined.

Making a Graphic Symbol from a VHDL File   Quartus II has a utility that creates a symbol from a VHDL text file for use in a Block Diagram File. To use this utility, open the VHDL file and its associated project. Then selected **Create/Update** from the **File** menu, as shown in Figure 5.22. From the submenu that appears, select **Create Symbol Files for Current File,** as shown in Figure 5.23.

The resulting graphic symbol indicates whether the underlying VHDL code defines the inputs and outputs as separate ports, as in Figure 5.24, or as vectors, as in Figure 5.25. The single-bit ports in Figure 5.24 are shown as thin lines, whereas the multiple-bit ports in Figure 5.25 are shown as thick lines. When these symbols are used in Block Diagram Files, they may need to be connected differently. We will examine this feature in a later chapter.

■ SECTION 5.3B REVIEW PROBLEM

5.4a   Write a VHDL statement that declares an 8-bit output port, *y*, of type BIT_VECTOR, with the most significant bit on the left.

5.4b   Write a VHDL statement that declares an 8-bit output port, *y*, of type BIT_VECTOR, with the most significant bit on the right.

Selected Signal Assignment Statement   Suppose we wish to encode the following Boolean expression in a VHDL file:

$$Y = \bar{D}_3\bar{D}_2 D_1 D_0 + \bar{D}_3 D_2 D_1 \bar{D}_0 + D_3 \bar{D}_2 \bar{D}_1 D_0 + D_3 D_2 \bar{D}_1 \bar{D}_0$$

FIGURE 5.22   File Menu (Create/Update)

FIGURE 5.23   Creating a Symbol

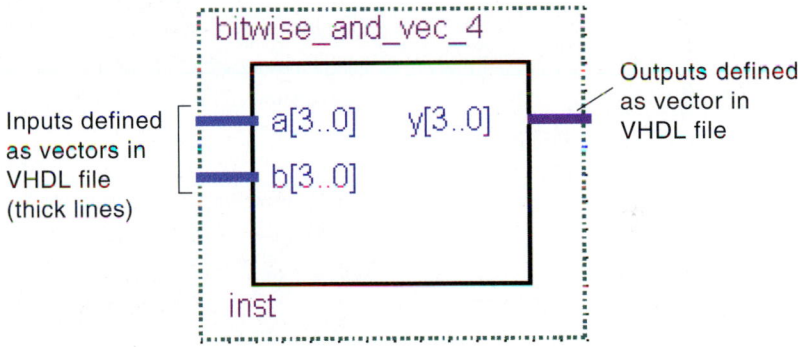

FIGURE 5.24 Symbol for 4-bit AND Array (Separate Inputs)

FIGURE 5.25 Symbol for 4-bit AND Array (Combined Inputs)

TABLE 5.2 Example Truth Table

D_2	D_1	D_0	Y
0	0	0	0
0	0	1	0
0	1	0	1
0	1	1	0
1	0	0	0
1	0	1	0
1	1	0	0
1	1	1	1

This expression is in its simplest sum-of-products (SOP) form. Since it has four 4-variable product terms, each of which has two NOT functions, the expression is awkward to type out in VHDL format, difficult to read, and prone to errors in the typing. A better way to encode this function is to use a **selected signal assignment statement.** This statement allows us to assign alternative values to an object, based on the status of a reference signal. The form of the statement is:

```
WITH __expression SELECT
    __signal <= __expression WHEN __constant_value,
                __expression WHEN __constant_value,
                __expression WHEN __constant_value,
                __expression WHEN __constant_value;
```

This type of statement can be used to encode an SOP Boolean expression, as represented by a truth table. For example, if we know that a circuit has the truth table shown in Table 5.2 we can represent the corresponding expression in VHDL as follows, assuming we have defined *d* as having type BIT_VECTOR (2 downto 0):

```
WITH d SELECT
y <= '1' WHEN "010",      ⟵ y is '1' when d is 010
     '1' WHEN "111",      ⟵ y is '1' when d is 111
     '0' WHEN others;     ⟵ otherwise y is '0'.
```

This statement can be interpreted as saying, "when the value of *d* is either "010" or "111", set the output *y* to '1'. Otherwise, set the value of *y* to '0'." The keyword others is used to specify all cases other than those explicitly selected.

EXAMPLE 5.3

Use a VHDL selected signal assignment statement to encode the Boolean expression given by: $Y = \bar{D}_3\bar{D}_2D_1D_0 + \bar{D}_3D_2D_1\bar{D}_0 + D_3\bar{D}_2\bar{D}_1D_0 + D_3D_2\bar{D}_1\bar{D}_0$

Also, create a simulation in Quartus II that verifies the correct operation of the design.

TABLE 5.3 Truth Table for Boolean Expression in Example 5.3

D_3	D_2	D_1	D_0	Y
0	0	0	0	0
0	0	0	1	0
0	0	1	0	0
0	0	1	1	1
0	1	0	0	0
0	1	0	1	0
0	1	1	0	1
0	1	1	1	0
1	0	0	0	0
1	0	0	1	1
1	0	1	0	0
1	0	1	1	0
1	1	0	0	1
1	1	0	1	0
1	1	1	0	0
1	1	1	1	0

■ Solution

The Boolean expression can be represented by the truth table shown in Table 5.3. For each line where $Y = 1$, we find a minterm that can be used as a term in a VHDL selected signal assignment statement.

The VHDL file for the truth table in Table 5.3 is as follows:

```
ENTITY select_example IS
    PORT(
            d : IN  BIT_VECTOR(3 downto 0);
            y : OUT BIT);
END select_example;

ARCHITECTURE cct OF select_example IS
BEGIN                                        ┌──── d(3)
    WITH d SELECT                        ┌─ d(0)
        y        <=    '1' WHEN "0011",
Select the value       '1' WHEN "0110",
of y based on          '1' WHEN "1001",
the value of d.        '1' WHEN "1100",
                Value  '0' WHEN others;   Default
    END cct;    of y                      case
```

➡ **LAB SOURCE**
*Lab #8,
Introduction to VHDL*

Figure 5.26 shows the simulation of the design, indicating that the outputs are HIGH where selected in the VHDL file and LOW elsewhere.

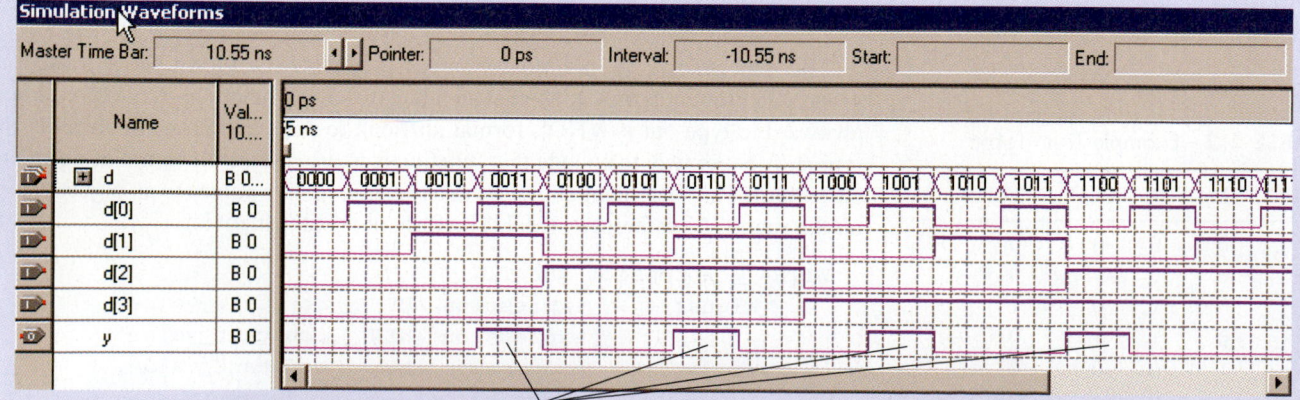

Output HIGH for selected values of d (0011, 0110, 1000, 1100), otherwise LOW.

FIGURE 5.26 Example 5.3: Simulation of VHDL File

STD_LOGIC and STD_LOGIC_VECTOR

The STD_LOGIC (standard logic) type, also called **IEEE Std. 1164 Multi-Valued Logic,** has been defined to give a broader range of output values than just '0' and '1'. Any port, signal, or variable of type STD_LOGIC or STD_LOGIC_VECTOR can have any of the following nine values:

```
'U',  -- Uninitialized
'X',  -- Forcing Unknown
'0',  -- Forcing 0
'1',  -- Forcing 1
'Z',  -- High Impedance
'W',  -- Weak     Unknown
'L',  -- Weak     0
'H',  -- Weak     1
'-'   -- Don't care
```

"Forcing" levels are deemed to be the equivalent of a gate output. "Weak" levels are specified by a pull-up or pull-down resistor. ("Weak" levels are usually used in circuit modeling, where it is important to distinguish between gate outputs and pull-up/down resistors. These levels will not be of importance to us.) The 'Z' state is used as the high-impedance state of a tristate buffer.

The majority of applications can be handled by 'X', '0', '1', and 'Z' values. If we use only '0' and '1' in a design, the statements have the identical effect to those we would make with BIT or BIT_VECTOR types.

To use STD_LOGIC in a VHDL file, you must include the following reference to the VHDL **library** called **ieee** and the **std_logic_1164 package** before the entity declaration:

```
LIBRARY ieee;
USE ieee.std_logic_1164.ALL;
```

The **std_logic_1164** package contains the type definitions of the STD_LOGIC types.

EXAMPLE 5.4

Rewrite the VHDL file for the 4-bit AND array of Example 5.2 using STD_LOGIC_VECTOR types.

■ **Solution**

```
LIBRARY ieee;                    ┐ These lines required when using
USE ieee.std_logic_1164.ALL;     │ STD_LOGIC or
                                 ┘ STD_LOGIC_VECTOR

ENTITY bitwise_and_std_4 IS
  PORT(
      a, b : IN  STD_LOGIC_VECTOR(3 downto 0);
      y    : OUT STD_LOGIC_VECTOR(3 downto 0));
END bitwise_and_std_4;


ARCHITECTURE and_gate OF bitwise_and_std_4 IS
BEGIN
   y <= a and b;
END and_gate;
```

Why use STD_LOGIC rather than BIT, if we only use '0' and '1' values? The usual reason is for compatibility with existing VHDL components that might be used in our design entities. For example, the Altera Library of Parameterized Modules (LPM) contains predesigned components that are written using STD_LOGIC types. To include these components in a VHDL design, the design must be written with STD_LOGIC types as well.

Also, STD_LOGIC is required if we have any tristate outputs in our design. For example, the quadruple tristate buffer shown in Figure 5.27 can be represented by the truth table in Table 5.4.

FIGURE 5.27 Quadruple Tristate Buffer

TABLE 5.4 Truth Table for a Quad Tristate Buffer

$\overline{G}$	Y1	Y2	Y3	Y4
0	A1	A2	A3	A4
1	'Z'	'Z'	'Z'	'Z'

EXAMPLE 5.5

Write a VHDL design file that describes the operation of the quadruple tristate buffer shown in Figure 5.27. Create a simulation in Quartus II that verifies the correct operation of the design.

■ **Solution**

*quad_tri.vhd
quad_tri.vwf*

```
LIBRARY ieee;
USE ieee.std_logic_1164.ALL;

ENTITY quad_tri IS
    PORT(
        a: IN   STD_LOGIC_VECTOR(3 downto 0);
        g: IN   STD_LOGIC;
        y: OUT STD_LOGIC_VECTOR(3 downto 0));
END quad_tri;


ARCHITECTURE quad_buff OF quad_tri IS
BEGIN
    WITH g SELECT
        y          <=  a WHEN '0',
                       "ZZZZ" WHEN others;
END quad_buff;
```

Both y and a are 4 bits →

g is only one bit

All 4 bits of y in "Hi-Z" state

We can interpret the selected signal assignment statement by saying, "Select a value of y based on the value of g. When g is 0, y = a; otherwise all four bits of y are in the high-impedance state."

Figure 5.28 shows a simulation of the 4-bit tristate buffer. As long as *g* is LOW, output *y* follows input *a,* first with a hexadecimal value of 3, then C. When *g* is HIGH, the output is in the high-impedance state, shown by Z.

y = a   y = high-impedance

FIGURE 5.28 Simulation of a 4-bit Tristate Buffer

◼ **SECTION 5.3C REVIEW PROBLEM**

5.5 State two reasons why we would use STD_LOGIC types rather than BIT in a VHDL design.

Integer, Natural, and Positive

Integers are positive and negative whole numbers. Thus, the VHDL INTEGER type consists of the range of positive and negative numbers that can be represented by the number of bits supported by the computer for which a given VHDL compiler is designed. In the case of Quartus II for a Windows/PC platform, this is the positive and negative range of 32-bit numbers: –2,147,483,648 to +2,147,483,647. There are two subtypes of INTEGER, called NATURAL and POSITIVE. NATURAL is the set of all integers greater than or equal to 0. POSITIVE is the set of all integers greater than or equal to 1. Constants in all these types are written in VHDL without quotes (e.g., `y <= 3;`).

We can define a multi-bit port of type INTEGER without specifying how many bits the port must have. The VHDL compiler will calculate the required number of bits, based on the range of the possible Boolean combinations of the port. For example, a 3-bit input port can be specified either by `d: IN BIT_VECTOR(3 downto 0);` or `d: IN INTEGER RANGE 0 to 7;` Care should be taken to define the range of an integer. If this is not done, a lot of unnecessary hardware might be generated.

EXAMPLE 5.6

A digital circuit is specified by the truth table shown in Table 5.5.

TABLE 5.5 Truth Table for Example 5.5

D_2	D_1	D_0	Y	D_2	D_1	D_0	Y
0	0	0	0	1	0	0	0
0	0	1	1	1	0	1	1
0	1	0	0	1	1	0	1
0	1	1	0	1	1	1	0

(continued)

The VHDL following code has been written to describe the operation of the circuit:

```
LIBRARY ieee;
USE ieee.std_logic_1164.ALL;

ENTITY truth_table IS
    PORT(
        d : IN  STD_LOGIC_VECTOR(3 downto 0);
        y : OUT STD_LOGIC);
END truth_table;

ARCHITECTURE a OF truth_table IS
BEGIN
    WITH d SELECT
        y <= '1' WHEN "001",
             '1' WHEN "101",
             '1' WHEN "110"
             '0' WHEN others;
END a;
```

Rewrite the code so that d is specified as type INTEGER.

■ Solution

```
LIBRARY ieee;
USE ieee.std_logic_1164.ALL;
ENTITY truth_table IS
    PORT(
        d: IN  INTEGER RANGE 0 to 7;
        y: OUT STD_LOGIC);
END truth_table;

ARCHITECTURE a OF truth_table IS
BEGIN
    WITH d SELECT              ◄—— d is type INTEGER
        y   <= '1' WHEN 1,
               '1' WHEN 5,
               '1' WHEN 6
               '0' WHEN others;
END a;                    ▲ —— y is type STD_LOGIC
```

Note that the output, *y*, which is of type STD_LOGIC, is assigned constant values in single quotes, whereas the values of *d*, of type INTEGER, are specified without quotes.

5.4 SIGNALS IN VHDL

■ KEY TERMS

Signal   An internal connection within a VHDL architecture that connects parts of the design together. A signal acts like a connecting wire inside the design.

Concatenate   Link together in sequence.

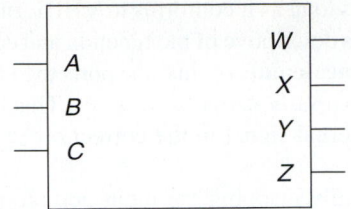

FIGURE 5.29 Digital Function Block

Multi-Bit Signals

Suppose we have a digital circuit, represented by the function block in Figure 5.29, that has the truth table shown in Table 5.6.

TABLE 5.6 Truth Table for the Circuit in Figure 5.29

A	B	C	W	X	Y	Z
0	0	0	1	0	0	0
0	0	1	0	1	0	0
0	1	0	0	1	1	0
0	1	1	1	0	0	1
1	0	0	0	1	1	0
1	0	1	0	0	0	1
1	1	0	1	0	0	1
1	1	1	0	0	1	0

The Boolean expressions for the four outputs are given by:

$$W = \overline{A}\ \overline{B}\ \overline{C} + \overline{A}\ B\ C + A\ B\ \overline{C}$$
$$X = \overline{A}\ \overline{B}\ C + \overline{A}\ B\ \overline{C} + A\ \overline{B}\ \overline{C}$$
$$Y = \overline{A}\ B\ \overline{C} + A\ \overline{B}\ \overline{C} + A\ B\ C$$
$$Z = \overline{A}\ B\ C + A\ \overline{B}\ C + A\ B\ \overline{C}$$

Since all of these equations are in their simplest possible SOP forms, it would probably be advantageous to code Table 5.6 as a selected signal assignment statement, rather than use four concurrent signal assignment statements. However, this form requires a single name for input and a single name for output. These names can represent multiple-bit objects of type BIT_VECTOR, STD_LOGIC_VECTOR, or INTEGER, but they must still have only one name. We cannot write something like:

```
-- Illegal syntax
WITH a, b, c SELECT
    w, x, y, z <= "1000" WHEN "000",
    etc.
```

So, to use the selected signal assignment, we need to find a way to treat all inputs as one object, even though they are not numerically related. We can do this by creating a **signal** that "bundles" together all inputs into a single group of type BIT_VECTOR, STD_LOGIC_VECTOR, or INTEGER. We also require a multi-bit signal for all outputs that we can use in a selected signal assignment statement and "peel off" the various bits and map them to single-bit outputs.

A signal is a VHDL construct used for connecting internal parts of a design, something like an internal connecting wire. Figure 5.30 shows the general idea. Ports *A*, *B*, and *C* come into the entity as separate ports. Inside the design they are bundled together like a 3-wire cable in a signal called *inputs*. ("Inputs" is a

FIGURE 5.30 Internal Signals for the Block in Figure 5.29

	inputs(2)	inputs(1)	inputs(0)
inputs	a	b	c

FIGURE 5.31 Input Ports Mapped to a 3-Bit Signal

user-chosen name; it could be any name at all, as long as it conforms to VHDL rules for valid names. We chose this name because it is descriptive of the function and easy to type.) To bundle the ports together, we can **concatenate,** or link, the ports together using the & operator. The VHDL statement is inputs <= a & b & c; This has the effect of mapping the input ports to the internal signal in the correct order, as shown in Figure 5.31.

We can achieve the same thing by individually mapping the input ports to the internal signal one bit at a time, using three separate statements, as follows:

```
inputs(2) <= a;
inputs(1) <= b;
inputs(0) <= c;
```

The output of the selected signal assignment statement is called *outputs,* and is designated as a 4-bit signal. Again, *outputs* is a user-chosen name, which could be any name that conforms to VHDL rules. The individual bits of this signal are peeled off, like wires in a cable, and assigned to the output ports *W, X, Y,* and *Z.* To achieve this, we assign each bit of the signal to an output port with a concurrent signal assignment statement, as follows:

```
w <= outputs(3);
x <= outputs(2);
y <= outputs(1);
z <= outputs(0);
```

The mapping for these above statements is shown in Figure 5.32.

	outputs(3)	outputs(2)	outputs(1)	outputs(0)
outputs	w	x	y	z

FIGURE 5.32 Signal Mapped to Output Ports

The complete VHDL file for the circuit in this example is given as follows. Signals are declared in between the first line of the architecture body and the BEGIN statement. Signals require a type to be assigned, but not a mode, as they are totally internal to the design. A signal is considered to be global, that is, valid throughout the whole architecture body.

```
LIBRARY ieee;
USE ieee.std_logic_1164.ALL;

ENTITY signal_ex IS
    PORT(
        a, b, c    : IN STD_LOGIC;
        w, x, y, z : OUT STD_LOGIC);
END signal_ex;

ARCHITECTURE sig OF signal_ex IS
    -- Declaration area
    -- Define signals here
    SIGNAL inputs : STD_LOGIC_VECTOR(2 downto 0);
    SIGNAL outputs: STD_LOGIC_VECTOR(3 downto 0);
BEGIN
    -- Concatenate input ports into 3-bit signal
    inputs <= a & b & c;
```

```
                              wxyz          abc
                              | | | |       | | |
WITH inputs SELECT
     outputs        <=    "1000" WHEN "000",
                          "0100" WHEN "001",
                          "0110" WHEN "010",
                          "1001" WHEN "011",
                          "0110" WHEN "100",
                          "0001" WHEN "101",
                          "1001" WHEN "110",
                          "0010" WHEN "111",
                          "0000" WHEN others;

-- Separate signal into individual ports
w <= outputs(3);
x <= outputs(2);
y <= outputs(1);
z <= outputs(0);
END sig;
```

signal_ex.vhd
signal_ex.vwf

The simulation for the example design, including internal signals, is shown in Figure 5.33.

■ SECTION 5.4 REVIEW PROBLEM

5.6　　What is the hardware equivalent of a signal in a VHDL file?

Single-Bit Signals

Signals need not be multi-bit objects. They can be used as single-bit objects any time it is useful to connect two parts of a design internally. Examine the logic diagram in Figure 5.34.

The Boolean expression for the circuit is given by:

$$Y = \overline{A}\,B + A\,\overline{B} + \overline{C}\,D$$

This can also be written as:

$$Y = (A \oplus B) + \overline{C}\,D$$

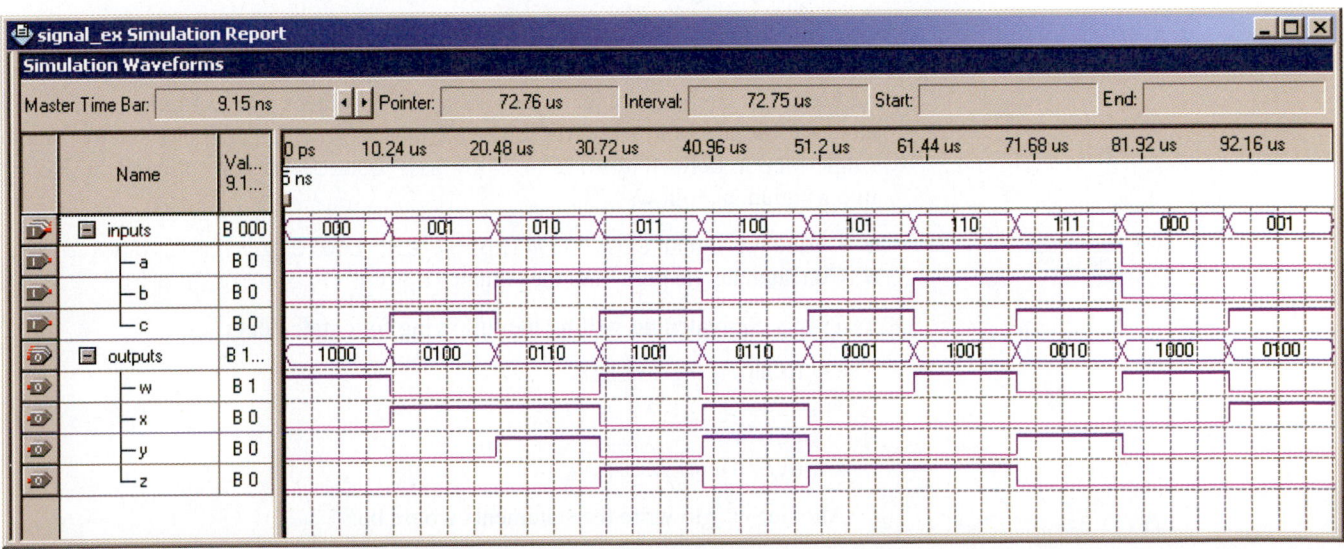

FIGURE 5.33　Simulation for Signal Example

FIGURE 5.34 Logic Diagram Showing a Signal Connection

The output of the top OR gate does not feed an output port pin and is thus entirely internal to the design. We can represent the connection between the OR gates as a signal, which we can call *a_xor_b*. A VHDL file that implements this circuit is as follows:

signal_ex2.vhd

```
LIBRARY ieee;
USE ieee.std_logic_1164.ALL;

ENTITY signal_ex2 IS
   PORT(
        a, b, c, d: IN STD_LOGIC;
        y          : OUT STD_LOGIC);
END signal_ex2;

ARCHITECTURE cct of signal_ex2 IS
   -- Declare signal
   SIGNAL a_xor_b : STD_LOGIC;
BEGIN
   -- Define signal in terms of ports a and b
   a_xor_b <= ((not a) and b) or (a and (not b));
   -- Combine signal with ports c and d
   y <= a_xor_b or ((not c) and d);
END cct;
```

Combining Single- and Multiple-Bit Signals

We have seen how we can combine several ports into a signal and also how we can create internal signals of single-bit width. It is also possible in VHDL to concatenate multiple and single-bit objects into one multi-bit object. For example, suppose we have a 3-bit port defined by `d: IN STD_LOGIC_VECTOR(2 downto 0);` and a single-bit port defined by `enable : IN STD_LOGIC;` We can combine these ports into a signal, as follows.

- Define the signal: `SIGNAL : inputs STD_LOGIC_VECTOR(3 downto 0);`
- Concatenate the ports into the signal: `inputs <= enable & d;`

This is equivalent to writing the following four statements:

```
inputs(3) <= enable;
inputs(2) <= d(2);
inputs(1) <= d(1);
inputs(0) <= d(0);
```

Another way to write the statements would be:

```
inputs(3) <= enable;
input(2 downto 0) <= d;
```

In this case we specify the elements of the *inputs* vector because we are assigning different parts of the signal in different ways. We do not need to specify the element numbers of *d* because the three elements are being assigned all together. We know that VHDL assigns elements of a vector left-to-right, so we know that *d* is assigned to *inputs* in the correct order.

SUMMARY

1. VHDL (VHSIC Hardware Description Language) is a text-based programming language used to model and synthesize digital circuits.

2. Every VHDL file requires an entity declaration, which describes the external aspects of the design (inputs and outputs), and an architecture body, which describes the relationship between the inputs and outputs.

3. The entity declaration defines ports (inputs and outputs), and the mode and type of each port.

4. The mode of a port defines the direction of data flow through the port.

5. The type of a port defines the range of values the port can have.

6. Some common types are BIT ('0' or '1'), STD_LOGIC (nine-valued standard logic), and INTEGER (whole numbers).

7. The STD_LOGIC type can take on any of the following values:

```
'U',  -- Uninitialized
'X',  -- Forcing Unknown
'0',  -- Forcing 0
'1',  -- Forcing 1
'Z',  -- High Impedance
'W',  -- Weak    Unknown
'L',  -- Weak    0
'H',  -- Weak    1
'-'   -- Don't care
```

8. STD_LOGIC is defined in a library called **ieee.** To use STD_LOGIC, include the following two statements at the beginning of a file

```
LIBRARY ieee;
USE ieee.std_logic_1164.ALL;
```

9. The simplest way to relate inputs and outputs in a VHDL design is with a concurrent signal assignment statement, which has the form: x <= (a and b) or c; The port or signal on the left side is assigned the value of the logic expression on the right side.

10. Logical operators such as **and, or,** and **not** all have equal precedence in VHDL, so the required order of operation must be explicitly indicated by parentheses.

11. All concurrent signal assignments in a VHDL file are evaluated simultaneously. This is required to emulate the behavior of combinational hardware.

12. VHDL design entry in Quartus II is done through the Quartus Text Editor.

13. A design should be simulated before programming a CPLD in order to ensure the correctness of the design logic.

14. Pins can be assigned directly to a VHDL design in Quartus II using the pin assignment utility.

15. VHDL designs can be embedded in a Block Diagram File as graphic symbols. These symbols are created using a utility in Quartus II.

16. A valid name in VHDL begins with a letter, followed by any number of letters and numbers. Underscores are permitted, but not two or more in a row. VHDL is not case sensitive.

17. There are a number of reserved keywords in VHDL, none of which can be used as named identifiers.

18. The Altera style guideline for VHDL indicates that in most cases, reserved keywords should be typed in uppercase letters and user names in lowercase.

19. VHDL comments are indicated by two hyphens at the beginning of a line.

20. VHDL ports can have one of the following modes: IN (input only), OUT (output only), INOUT (bidirectional), or BUFFER (output, with feedback).

21. A port, signal, or variable can have a multiple-bit construction of type BIT_VECTOR or STD_LOGIC_VECTOR. These structures are called vectors and can be referred to as separate elements (e.g., `y(3) <= a(3) and b(3);`) or as a group (e.g., `y <= "1000"`).

22. Vectors can be created from single-bit ports or signals by concatenation (e.g., `y <= d(1) & d(0);`) or by assigning the bits individually (e.g., `y(1) <= d(1); y(0) <= d(0);`).

23. Vectors in VHDL are assigned left-to-right. Thus, the statement `d: IN BIT_VECTOR (3 downto 0);` places `d(3)` on the left. The statement `d: IN BIT_VECTOR (0 to 3);` places `d(0)` on the left.

24. Entire vectors can be assigned to other vectors in VHDL with a single statement that is, if both *d* and *y* are assigned as `BIT_VECTOR(3 downto 0)`, then the statement `y <= d;` is the same as the following four statements:

```
y(3) <= d(3);
y(2) <= d(2);
y(1) <= d(1);
y(0) <= d(0);
```

25. Single-bit constants of BIT and STD_LOGIC types are treated as characters and placed in single quotes (e.g., '0', '1', 'Z'). Multiple-bit constants of type BIT_VECTOR and STD_LOGIC_VECTOR are treated as strings and placed in double quotes (e.g., "0110", "ZZZZ"). Constants of type INTEGER, NATURAL, and POSITIVE are written without quotes (e.g., 0, 34, 4096).

26. A signal in VHDL is an internal connection, like a wire. Signals are declared in the architecture body of a VHDL file, after the first line of the architecture and before the BEGIN statement.

27. Signals must have a declared type, but not a mode. There is no mode because the signal is not connected to the outside of the design.

28. A selected signal assignment statement can act as a truth table in VHDL. It assigns alternative values to one or more outputs, depending on the alternative values on one or more inputs.

29. Elements in signals can be assigned individually (`int(1) <= a; int(0) <= b;`), all at once (`int <= a & b;`), or in some combination (`int(2) <= x; int(1 downto 0) <= z;`)

GLOSSARY

Architecture Body   The part of a VHDL design that defines the relationship between its inputs and outputs.

Array   A group of related objects of the same type, organized by numbering the various elements of the array.

ASCII   American Standard Code for Information Interchange. A 7- or 8-bit binary code used to represent printable characters and control codes.

Character   A number, letter, or other printable symbol, as represented in a computer code, such as ASCII. Shown by enclosing the character in single quotes (e.g., '0' is a character).

Concatenate   Link together in sequence.

Concurrent Signal Assignment Statement   A VHDL statement that assigns a value to a port or signal based on a Boolean expression or constant. The output port or signal is affected as soon as there is a change to the input port.

Concurrent Statement   A VHDL statement that defines a relationship between inputs and outputs of a design entity such that the output responds immediately to a change of input.

Entity Declaration   The part of a VHDL file that defines the inputs and outputs of the design.

IEEE Std. 1164 Multi-Value Logic   A nine-valued type used in VHDL to describe logic hardware as a STD_LOGIC or STD_LOGIC_VECTOR type.

Library   A collection of VHDL design units that have been previously compiled.

Mode   The property of a port, signal, or variable that specifies the direction of data flow.

Named Identifier   An object in VHDL, such as an entity, architecture, port, signal, or variable that is given a name to identify it within a VHDL program.

Package   A group of uncompiled VHDL design elements that can be used by more than one VHDL file.

Port   An input or output in a VHDL file.

Range   The property of a port, variable, or signal that defines the upper and lower bounds of the values that the object can have.

Reserved Keyword   A word that is part of the VHDL language and thus may not be used as a named identifier.

Selected Signal Assignment Statement   A VHDL statement that assigns a value to a port, signal, or variable, based on alternative values of another port, signal, or variable.

Signal   An internal connection within a VHDL architecture that connects parts of the design together. A signal acts like a connecting wire inside the design.

Signal Concurrency   A property of VHDL whereby all concurrent statements in a VHDL design are evaluated at the same time.

String   A one-dimensional array of characters. Shown by enclosing the string in double quotes (e.g., "0110" is a string of four characters).

Syntax   The rules of construction, or "grammar", of a programming language.

Type   The property of a port, signal, or variable that specifies the values it can have.

VHDL   VHSIC Hardware Description Language.

VHSIC   (pronounced "VIZ-ick") Very High Speed Integrated Circuit.

PROBLEMS

Problem numbers set in color indicate more difficult problems.

5.1   VHDL Basics

5.1   State the full names for which VHDL and VHSIC are abbreviations.

5.2   State the minimum two structures required to write a VHDL file and briefly describe each one.

5.3   State the values that a port of type BIT can have.

5.4   **a.** Which logical operators (and, or, not, nand, not, xor, xnor) have precedence in VHDL? How is order of precedence specified for logical operators in VHDL?

   b. Write the following Boolean expressions as VHDL concurrent signal assignment statements:

   i. $Y = ABC + AD$

   ii. $Y = (\overline{A} + C)(B + \overline{D})$

   iii. $Y = A\overline{B}C + \overline{AD}$

5.5   Write a VHDL file that describes the circuit shown in Figure 5.35, assuming that all ports are of type BIT.

FIGURE 5.35   Problem 5.5: Logic Circuit

5.6   Why are semicolons used after all port definitions except the last one in a PORT statement?

5.7   Briefly explain the concept of signal concurrency and state why it is important in a hardware description language like VHDL.

5.3   VHDL Syntax for Port, Mode, and Type

5.8   Which of the following are valid names in VHDL? If a name is not valid, state the reason.

a.	and	**f.**	right*justified
b.	sum	**g.**	output enable
c.	_yellow_dog	**h.**	output_enable
d.	decode4	**i.**	output__enable
e.	4_decode	**j.**	output-enable

5.9   List four possible modes for a VHDL port and briefly describe each one.

5.10   Briefly state the difference between a port of mode OUT and a port of mode BUFFER.

5.11   **a.** Write a VHDL statement that defines an 8-bit input port, **addr,** with the most significant bit on the left. Assume the port is of type BIT or BIT_VECTOR.

   b. Write a VHDL statement that defines an 8-bit input port, **addr,** with the most significant bit on the right. Assume the port is of type BIT or BIT_VECTOR.

5.12   Write a VHDL file that implements the gates in Figure 5.36 using BIT_VECTORs.

5.13   Modify the VHDL file for Problem 5.12 so that it now implements an array of eight XOR gates.

5.14   An input port is defined as `x : IN BIT_VECTOR (5 downto 0);` It is assigned a value by the statement `x <= "100101";` Write the equivalent VHDL statements that will assign the port values separately for each bit.

5.15   Repeat Problem 5.14 for the case where the input is defined as `x : IN BIT_VECTOR (0 to 5);` and x is assigned by the statement `x <= "100101";`

5.16   State the difference between type BIT and type STD_LOGIC. Why does STD_LOGIC have so many values?

5.17   Use a selected signal assignment statement in VHDL file to encode the following truth table.

FIGURE 5.36 Problem 5.12: XOR Array

D_3	D_2	D_1	D_0	Y
0	0	0	0	0
0	0	0	1	0
0	0	1	0	0
0	0	1	1	0
0	1	0	0	0
0	1	0	1	0
0	1	1	0	0
0	1	1	1	1
1	0	0	0	0
1	0	0	1	0
1	0	1	0	0
1	0	1	1	1
1	1	0	0	0
1	1	0	1	1
1	1	1	0	1
1	1	1	1	0

FIGURE 5.37 Problem 5.18: 2-Bit Equality Comparator

5.18   Write a VHDL file that implements the circuit for the 2-bit equality comparator, shown in Figure 5.37. (The circuit sets $Y = 1$ when $A_2A_1 = B_2B_1$. The circuit is studied in detail in Chapter 6.)

5.19   **a.** Write the Boolean expression and truth table for the logic circuit shown in Figure 5.38.

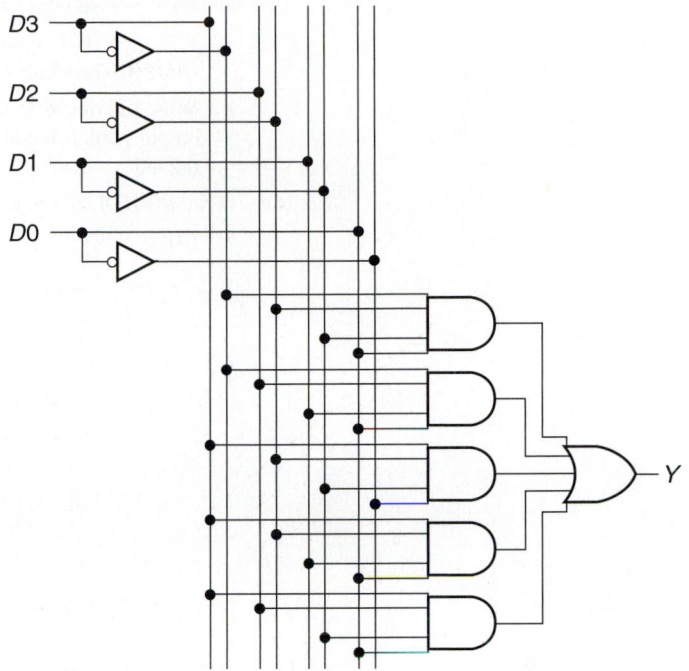

FIGURE 5.38   Problem 5.19: SOP Circuit

    b. Which would be simpler to use in a VHDL description of the circuit in Figure 5.38, a concurrent signal assignment statement or a selected signal assignment statement? Why?

    c. Write a VHDL file that describes the circuit in Figure 5.38, based on your choice in part b of this question.

5.20   **a.** Write the Boolean expression and truth table for the full adder circuit shown in Figure 5.38. (This circuit adds inputs A and B and a carry bit from a previous sum to get a sum bit and a carry bit. The circuit is studied in detail in Chapter 7.)

FIGURE 5.39   Problem 5.20: Full Adder

    b. Which would be simpler to use in a VHDL description of the circuit in Figure 5.39, a concurrent signal assignment statement or a selected signal assignment statement? Why?

    c. Write a VHDL file that describes the circuit in Figure 5.39, based on your choice in part b of this question.

5.21   Some circuits are easier to describe in VHDL with a concurrent signal assignment statement and others with a selected signal assignment statement. Based on your answers to Problems 5.19 and 5.20, write a general description of the characteristics of the circuit that is most suited to each type of statement.

5.22   Rewrite the VHDL file from Problem 5.19 using INTEGER types for the inputs.

5.4   Signals in VHDL

5.23   A VHDL design has five input ports labeled **enable** (MSB), **read, write, selector,** and **compare** (LSB) and three outputs labeled **address_latch** (MSB), **data_enable,** and **strobe** (LSB).

   a.   Write a VHDL statement that concatenates all input lines into a signal called **control.** The signal is written with the most significant bit on the left.

   b.   Write a series of VHDL statements that separates a signal called **status** into the output ports listed above. The ports are written with the most significant bit on the left.

5.24   Write a VHDL file that uses a selected signal assignment statement to encode the following truth table.

A	B	C	X	Y	Z
0	0	0	1	0	0
0	0	1	0	1	1
0	1	0	0	1	0
0	1	1	1	0	1
1	0	0	0	1	0
1	0	1	1	0	1
1	1	0	1	0	0
1	1	1	0	1	0

5.25   Write a VHDL file that encodes the following truth table for a 4-bit equality comparator. (The circuit compares two 4-bit inputs, A and B, and sets the output HIGH when they are equal.)

A_3	A_2	A_1	A_0	B_3	B_2	B_1	B_0	A_EQ_B
0	0	0	0	0	0	0	0	1
0	0	0	1	0	0	0	1	1
0	0	1	0	0	0	1	0	1
0	0	1	1	0	0	1	1	1
0	1	0	0	0	1	0	0	1
0	1	0	1	0	1	0	1	1
0	1	1	0	0	1	1	0	1
0	1	1	1	0	1	1	1	1
1	0	0	0	1	0	0	0	1
1	0	0	1	1	0	0	1	1
1	0	1	0	1	0	1	0	1
1	0	1	1	1	0	1	1	1
1	1	0	0	1	1	0	0	1
1	1	0	1	1	1	0	1	1
1	1	1	0	1	1	1	0	1
1	1	1	1	1	1	1	1	1
all other values								0

5.26   The circuit in Figure 5.25 is a 2-, 3-, and 4-bit equality comparator. It has outputs that determine if the last two, the last three, or all four bits of two 4-bit numbers are equal.

   a.   Write a VHDL file for the circuit in Figure 5.40, using two internal signals to connect the various parts of the circuit together.

   b.   Write a VHDL file for the circuit in Figure 5.40 that does not use internal signals. (*Hint:* Check the modes of the ports.)

FIGURE 5.40 Problem 5.26: 2-, 3-, and 4-Bit Comparator

ANSWERS TO SECTION REVIEW PROBLEMS

5.1

5.1

```
ENTITY srp5_1 IS
    PORT(
        a, b, c : IN BIT;
        y1, y0  : OUT BIT);
END srp5_1;

ARCHITECTURE cct OF srp5_1 IS
BEGIN
    y1 <= ((not a) and (not b) and (    c))
       or ((not a) and (    b) and (not c));

    y0 <= ((not a) and (    b) and (    c))
       or ((    a) and (    b) and (not c));
END cct;
```

5.2

5.2   A VHDL simulation allows us to verify the correctness of a design before we program our CPLD. This saves the time required for physically connecting and testing the design.

5.3A

5.3   Valid names in VHDL have the following criteria: Start with a letter; contain any number of letters or numbers; internal underscores permitted, but not two or more in a row; not case sensitive. By these criteria:

  a. MyFunction is a valid name.

  b. _a_and_b is not a valid name (does not begin with a letter).

5.3B

5.4a. `y : OUT BIT_VECTOR(7 downto 0);`

5.4b. `y : OUT BIT_VECTOR(0 to 7);`

5.3C

5.5   **1.** Compatibility with predefined components used in a VHDL file is required.

  2. Tristate buffers, which require the 'Z' state, are used in a design.

5.4

5.6   A signal is equivalent to an internal connecting wire or wires.

KEY TERMS

Address Decoder

BCD

Bit Multiplexing

Byte (or word) Multiplexing

CASE Statement

Clock

CMOS Analog Switch

Common Anode Display

Common Cathode Display

Conditional Signal Assignment
   Statement

Counter

Data Inputs

Decoder

Demultiplexer

Double-Subscript Notation

Encoder

Even Parity

GENERATE Statement

IF Statement

Index Variable

Magnitude Comparator

Multiplexer

Odd Parity

Parity

Parity Bit

Positive Edge

Priority Encoder

PROCESS

RBI

RBO

Response Waveforms

Ripple Blanking

Select Inputs

Selected Signal Assignment Statement

Sensitivity List

Seven-segment Display

Signal Attribute

Simulation

Stimulus Waveforms

Time Division Multiplexing (TDM)

Time Slot

Timing Diagram

OUTLINE

6.1   Decoders

6.2   Encoders

6.3   Multiplexers

6.4   Demultiplexers

6.5   Magnitude Comparators

6.6   Parity Generators and Checkers

CHAPTER OBJECTIVES

Upon successful completion of this chapter you will be able to:

- Design binary decoders using logic gates.

- Create decoder designs in Quartus II, using Block Diagram Files or VHDL.

- Create Quartus II simulation files to verify the operation of combinational circuits.

- Design BCD-to-seven-segment and hexadecimal-to-seven-segment decoders, including special features such as ripple blanking, using VHDL and Block Diagram Files in Quartus II.

- Use Quartus II Block Diagram Files and VHDL to generate the design for a 3-bit binary and a BCD priority encoder.

- Describe the circuit and operation of a simple multiplexer and program these functions in VHDL.

- Draw logic circuits for multiplexer applications, such as single-channel data selection, multibit data selection, waveform generation, and time-division multiplexing (TDM).

- Describe demultiplexer circuits and program them using VHDL.
- Define the operation of a CMOS analog switch and its use in multiplexers and demultiplexers.
- Define the operation of a magnitude comparator and program its function in VHDL.
- Explain the use of parity as an error-checking system and draw simple parity-generation and checking circuits.

Combinational Logic Functions

VHDL CONSTRUCTS USED IN THIS CHAPTER

- CASE statement
- Conditional signal assignment statement
- GENERATE statement
- IF statement
- PROCESS statement
- Selected signal assignment statement
- SIGNAL
- Signal attributes ('HIGH, 'LOW, 'LEFT, 'RIGHT)
- STD_LOGIC and STD_LOGIC_VECTOR types
- VHDL package: ieee.std_logic_1164
- VHDL package: ieee.std_logic_unsigned

A number of standard combinational logic functions have been developed that represent many of the useful tasks that can be performed with digital circuits. Decoders detect the presence of particular binary states and can activate other circuits based on their input values or can convert an input code to a different output code. Encoders generate a binary or binary coded decimal (BCD) code corresponding to an active input.

CHAPTER 6

Multiplexers and demultiplexers are used for data routing. They select a transmission path for incoming or outgoing data, based on a selection made by a set of binary-related inputs.

Magnitude comparators determine whether one binary number is less than, greater than, or equal to another binary number.

Parity generators and checkers are used to implement a system of checking for errors in groups of data. ◼︎

6.1   DECODERS

The general function of a **decoder** is to activate one or more circuit outputs upon detection of a particular digital state. The simplest decoder is a single logic gate, such as a NAND or AND, whose output activates when *all* its inputs are HIGH. When combined with one or more inverters, a NAND or AND can detect any unique combination of binary input values.

An extension of this type of decoder is a device containing several such gates, each of which responds to a different input state. Usually, for an *n*-bit input, there are 2^n logic gates, each of which decodes a different combination of input variables. A variation is a BCD device with 4 input variables and 10 outputs, each of which activates for a different BCD input.

Some types of decoders translate binary inputs to other forms, such as the decoders that drive seven-segment numerical displays, those familiar figure-8 arrangements of LED or LCD outputs ("segments"). The decoder has one output for every segment in the display. These segments illuminate in unique combinations for each input code.

Single-Gate Decoders

The simplest decoder is a single gate, sometimes in combination with one or more inverters, used to detect the presence of one particular binary value. Figure 6.1 shows two such decoders, both of which detect an input $D_3D_2D_1D_0 = 1111$.

The decoder in Figure 6.1a generates a logic HIGH when its input is 1111. The decoder in Figure 6.1b responds to the same input, but makes the output LOW instead.

In Figure 6.1, we designate D_3 as the most significant bit of the input and D_0 the least significant bit. We will continue this convention for multibit inputs.

In Boolean expressions, we will indicate the active levels of inputs and outputs separately. For example, in Figure 6.1, the inputs to both gates are the same, so we write $D_3D_2D_1D_0$ for the inputs of both gates. The gates in Figures 6.1a and Figure 6.1b have outputs with opposite active levels, so we write the output variables as complements (Y and $\overline{Y}$).

a. Active-HIGH indication

b. Active-LOW indication

FIGURE 6.1   Single-Gate Decoders

EXAMPLE 6.1

Figure 6.2 shows three single-gate decoders. For each one, state the output active level and the input code that activates the decoder. Also write the Boolean expression of each output.

a.   b.   c.

FIGURE 6.2   Single-Gate Decoders

■ **Solution**

Each decoder is a NAND or AND gate. For each of these gates, the output is *active* when *all inputs are HIGH*. Because of the inverters, each circuit has a different code that fulfils this requirement.

Figure 6.2a: Output: Active LOW

Input code: $D_3D_2D_1D_0 = 1001$

$$\overline{Y} = D_3\overline{D}_2\overline{D}_1D_0$$

Figure 6.2b: Output: Active LOW:

Input code: $D_2D_1D_0 = 001$

$$\overline{Y} = \overline{D}_2\overline{D}_1D_0$$

Figure 6.2c: Output: Active HIGH

Input code: $D_3D_2D_1D_0 = 1010$

$$Y = D_3\overline{D}_2D_1\overline{D}_0$$

Single-gate decoders are often used to activate other digital circuits under various operating conditions, particularly if there is a choice of circuits to activate. For example, single-gate decoders are used to enable peripheral devices in a personal computer (PC). A combination of binary values, called the address, specifies a unique set of conditions to enable a particular peripheral device.

EXAMPLE 6.2

A PC has two serial port cards called COM1 and COM2. Each card is activated when either one of two control inputs called $\overline{IOR}$ (Input/Output Read) and $\overline{IOW}$ (Input/Output Write) are active and a unique 10-bit address is present. $\overline{IOR}$ and $\overline{IOW}$ are active-LOW. The address is specified by bits $A_9A_8A_7A_6A_5A_4A_3A_2A_1A_0$, which can be represented by three hexadecimal digits. The decoder outputs, $\overline{COM1\_Enable}$ and $\overline{COM2\_Enable}$ are both active-LOW.

The card for COM1 activates when ($\overline{IOR}$ OR $\overline{IOW}$ is LOW) AND the address is between 3F8H and 3FFH.

The card for COM2 activates when ($\overline{IOR}$ OR $\overline{IOW}$ is LOW) AND the address is between 2F8H and 2FFH.

Create a Block Diagram File in Quartus II that implements the specified decoder.   (*continued*)

■ Solution

The lowest address that activates COM1 is

$$A_9A_8A_7A_6A_5A_4A_3A_2A_1A_0 = \text{3F8H} = 11\ 1111\ 1000$$

The highest COM1 address is

$$A_9A_8A_7A_6A_5A_4A_3A_2A_1A_0 = \text{3FFH} = 11\ 1111\ 1111$$

Since *any* address in this range is valid, we can represent the last three bits, $A_2A_1A_0$, as don't care states. Thus, for COM1, we should decode the address:

$$A_9A_8A_7A_6A_5A_4A_3A_2A_1A_0 = 11\ 1111\ 1XXX$$

Similarly, for COM2:

Low address: $A_9A_8A_7A_6A_5A_4A_3A_2A_1A_0 = \text{2F8H} = 10\ 1111\ 1000$
High address: $A_9A_8A_7A_6A_5A_4A_3A_2A_1A_0 = \text{2FFH} = 10\ 1111\ 1111$
Decode: $A_9A_8A_7A_6A_5A_4A_3A_2A_1A_0 = 10\ 1111\ 1XXX$

Figure 6.3 shows the **bdf** representation of the decoder circuit, including inputs for the control signals $\overline{IOR}$ and $\overline{IOW}$. The active-LOW inputs are indicated by a prefix, labeled *n* (e.g., *nIOR*), since we cannot draw the bar over the active-LOW input names.

FIGURE 6.3 Example 6.2: COM Port Decoders

■ SECTION 6.1A REVIEW PROBLEM

6.1 Draw a single-gate decoder that detects the input state $D_3D_2D_1D_0 = 1100$

   a. with active-HIGH indication

   b. with active-LOW indication

Multiple-Output Decoders

Decoder circuits often are constructed with multiple outputs. In effect, such a device is a collection of decoding gates controlled by the same inputs. A decoder circuit with

FIGURE 6.4   2-Line-to-4-Line Decoder

n inputs can activate up to $m = 2^n$ load circuits. Such a decoder is usually described an n-line-to-m-line decoder.

Figure 6.4 shows the logic circuit of a 2-line-to-4-line decoder. The circuit detects the presence of a particular state of the 2-bit input $D_1 D_0$, as shown by the truth table in Table 6.1. One and only one output is HIGH for any input combination. The active input of each line is shown in color. The subscript of the active output is the same as the value of the 2-bit input. For example, if $D_1 D_0 = 10$, output Y_2 is active since 10 (binary) = 2 (decimal).

TABLE 6.1   Truth Table of a 2-to-4 Decoder with Enable

D_1	D_0	Y_0	Y_1	Y_2	Y_3
0	0	1	0	0	0
0	1	0	1	0	0
1	0	0	0	1	0
1	1	0	0	0	1

If we are using the decoder to activate one of four output loads, it is possible that there are situations where we want no output to be active. For instance, in the COM ports decoders in Example 6.2, there will often be cases where no COM port is active. This is not possible in the circuit shown in Figure 6.4, as there is always one output active. A variation of the decoder that allows this type of control is shown in Figure 6.5.

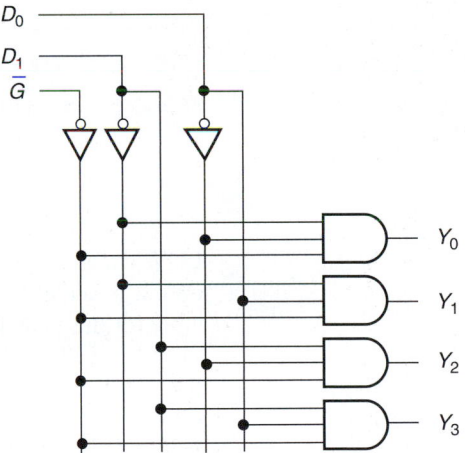

FIGURE 6.5   2-Line to-4-Line Decoder with Active-LOW Enable

The 2-line-to-4-line decoder shown in Figure 6.5 has an additional input called $\overline{G}$ that controls whether or not any input is active. If we make $\overline{G}$ LOW, the decoder acts the same as the one in Figure 6.4. If $\overline{G}$ is HIGH, then all outputs are deactivated (made LOW). The truth table for this decoder is shown in Table 6.2.

TABLE 6.2   Truth Table of a 2-Line-to-4-Line Decoder with Enable

$\overline{G}$	D_1	D_0	Y_0	Y_1	Y_2	Y_3
0	0	0	1	0	0	0
0	0	1	0	1	0	0
0	1	0	0	0	1	0
0	1	1	0	0	0	1
1	X	X	0	0	0	0

FIGURE 6.6 3-Line-to-8-Line Decoder with Enable

Figure 6.6 shows the circuit for a 3-line-to-8-line decoder, again with an active-LOW enable, $\overline{G}$. In this case, the decoder outputs are active LOW. One and only one output is active for any given combination of $D_2 D_1 D_0$. Table 6.3 shows the truth table for this decoder. Again if the enable line is HIGH, no output is active.

TABLE 6.3 Truth Table of a 3-to-8 Decoder with Enable

$\overline{G}$	D_2	D_1	D_0	$\overline{Y_0}$	$\overline{Y_1}$	$\overline{Y_2}$	$\overline{Y_3}$	$\overline{Y_4}$	$\overline{Y_5}$	$\overline{Y_6}$	$\overline{Y_7}$
0	0	0	0	0	1	1	1	1	1	1	1
0	0	0	1	1	0	1	1	1	1	1	1
0	0	1	0	1	1	0	1	1	1	1	1
0	0	1	1	1	1	1	0	1	1	1	1
0	1	0	0	1	1	1	1	0	1	1	1
0	1	0	1	1	1	1	1	1	0	1	1
0	1	1	0	1	1	1	1	1	1	0	1
0	1	1	1	1	1	1	1	1	1	1	0
1	X	X	X	1	1	1	1	1	1	1	1

74138

FIGURE 6.7 74138 Decoder

A 3-line-to-8-line decoder that is commercially available in a single chip is the 74138 decoder, shown in Figure 6.7. Depending on the logic family, this device will be designated 74LS138, 74HC138, or some other such number.

This decoder has active-LOW outputs and three enable inputs, one active-HIGH and two active-LOW, all of which must be active to enable any of the outputs. The truth table for this decoder is shown in Table 6.4.

TABLE 6.4 Truth Table for a 74138 Decoder

G1	$\overline{G2A}$	$\overline{G2B}$	C	B	A	$\overline{Y_0}$	$\overline{Y_1}$	$\overline{Y_2}$	$\overline{Y_3}$	$\overline{Y_4}$	$\overline{Y_5}$	$\overline{Y_6}$	$\overline{Y_7}$
0	X	X	X	X	X	1	1	1	1	1	1	1	1
1	1	X	X	X	X	1	1	1	1	1	1	1	1
1	X	1	X	X	X	1	1	1	1	1	1	1	1
1	0	0	0	0	0	0	1	1	1	1	1	1	1
1	0	0	0	0	1	1	0	1	1	1	1	1	1
1	0	0	0	1	0	1	1	0	1	1	1	1	1
1	0	0	0	1	1	1	1	1	0	1	1	1	1
1	0	0	1	0	0	1	1	1	1	0	1	1	1
1	0	0	1	0	1	1	1	1	1	1	0	1	1
1	0	0	1	1	0	1	1	1	1	1	1	0	1
1	0	0	1	1	1	1	1	1	1	1	1	1	0

Notice that the Y outputs are active-LOW and that they are selected by a combination of inputs CBA, where C is the most significant bit.

EXAMPLE 6.3

Figure 6.8 shows how a 74138 decoder can be used in a microcomputer memory system as an **address decoder**. Each block labeled **mem_8k** is a memory chip capable of holding 8192 (8K) bytes of data. Since there are eight such devices, the whole system can hold $8 \times 8192 = 65,536$ (64K) bytes. (Although this amount of memory may seem small by the standards of a desktop computer, it may be typical of a small stand-alone computer system (called an embedded system or a microcontroller) that is used in control applications.)

Each 8K block is enabled by a LOW at its nCS input $nCS = Chip Select$. Briefly explain the function of the decoder in the system.

■ Solution

Since only one decoder output is LOW at any one time, the decoder allows only one memory block to be active at any one time. The active block is chosen by inputs $ADDR_{15}ADDR_{14}ADDR_{13}$, which are connected to CBA on the decoder. The active memory block is the one connected to the y output whose subscript matches the binary value of these inputs. For example, when $ADDR_{15}ADDR_{14}ADDR_{13} = 110$, the block connected to Y_6 is active.

No outputs will be active, and therefore no memory block will be enabled, when $G_1 = 0$. (Note that Quartus II cannot represent an input or output with an inversion bar. Some conventions would represent an active-LOW terminal with an "n" prefix, indicating "NOT" (e.g., nCS). This is a matter of personal choice, but without such an indication it is not possible to tell the active level of an input or output from the Quartus II Block Diagram File.)

(continued)

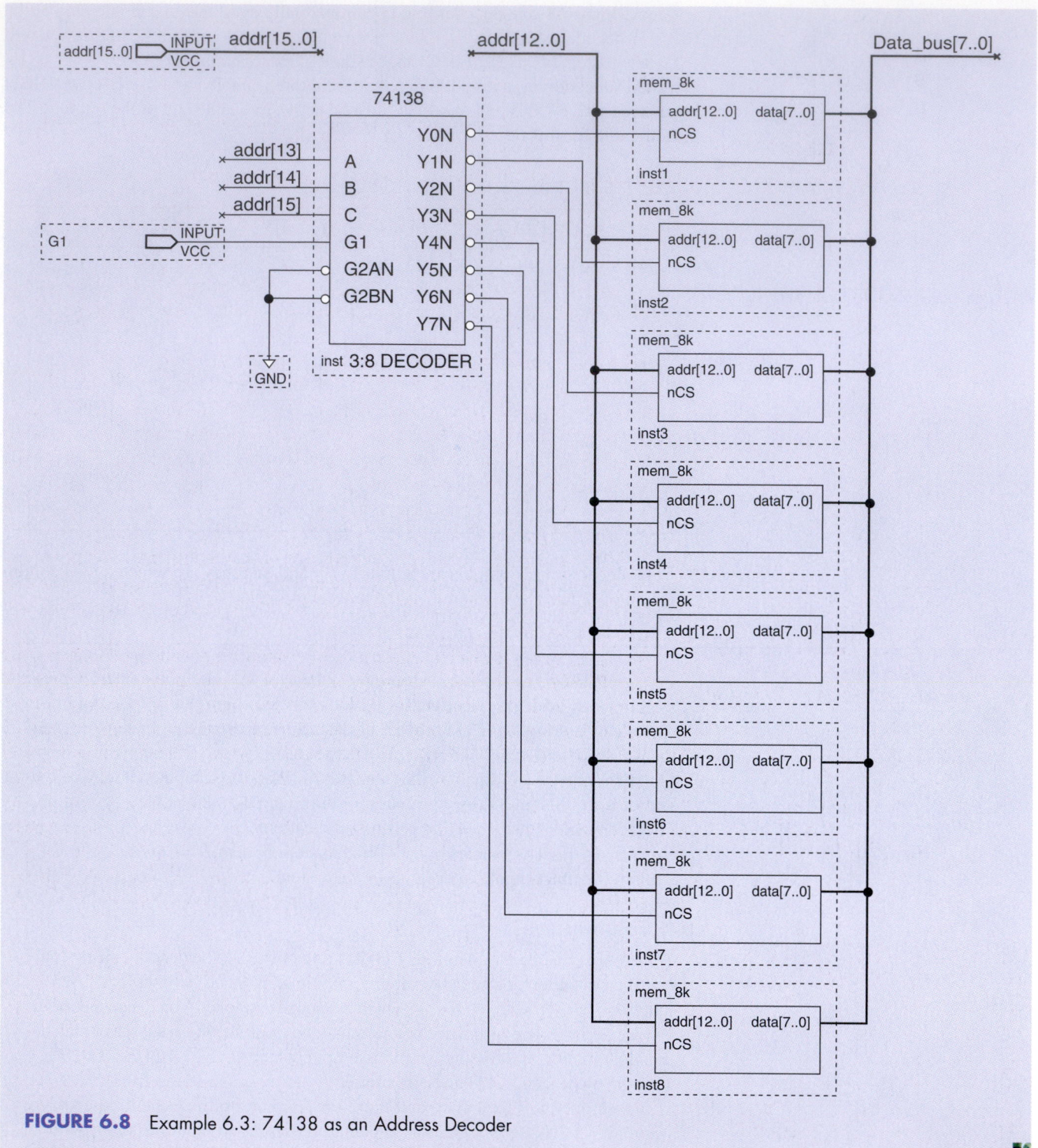

FIGURE 6.8 Example 6.3: 74138 as an Address Decoder

■ **SECTION 6.1B REVIEW PROBLEM**

6.2 How many inputs are required for a binary decoder with 16 outputs? How many inputs are required for a decoder with 32 outputs?

Simulation Criteria for *n*-Line-to-*m*-Line Decoders

➡ **LAB SOURCE**
*Lab #10 Binary Decoders
(Block Diagram File)*

> ■ **KEY TERMS**
>
> **Simulation**   The verification, using timing diagrams, of the logic of a digital design before programming it into a PLD.
>
> **Timing Diagram**   A diagram showing how two or more digital waveforms in a system relate to each other over time.
>
> **Stimulus Waveforms**   A set of user-defined input waveforms in a simulator file designed to imitate input conditions of a digital circuit.
>
> **Response Waveforms**   A set of output waveforms generated by a simulator for a particular digital design in response to a set of stimulus waveforms.

An important part of the CPLD design process is **simulation** of the design. A simulation tool allows us to see whether the output responses to a set of circuit inputs are what we expected in our initial design idea. The simulator works by creating a **timing diagram.** We specify a set of input (**stimulus**) waveforms. The simulator looks at the relationship between inputs and outputs, as defined by the design file, and generates a set of **response** outputs.

Figure 6.9 shows a set of simulation waveforms created for the 2-line-to-4-line decoder in Figure 6.5. The inputs *D1* and *D0* are combined as a single 2-bit value, to which an increasing binary count is applied as a stimulus. The decoder output waveforms are observed individually to determine the decoder's response. Once we have entered the design in the Quartus II Block Editor and compiled it, we can create the waveforms of Figure 6.9.

We have already learned the mechanics of creating a simulation in Chapters 4 and 5. The effectiveness of a simulation depends on our choosing input waveforms that will tell us everything we need to know about the operation of our design. For a decoder, this means applying input values that will show how the outputs activate in the correct order when the inputs are applied in all possible combinations. However, we must also test that the outputs do not activate when the enable input or inputs are not active. In other words, in order to use a simulation effectively, we must be sure to understand the correct operation of a design and test it for all possible working and failure modes.

Let us list some reasonable criteria for the simulation of a 2-line-to-4-line decoder with an active-LOW enable input and active-HIGH decoded outputs.

FIGURE 6.9   Simulation Waveforms for a 2-Line-to-4-Line Decoder with Active-LOW Enable

SIMULATION CRITERIA

- Each output must respond to its appropriate binary input by going HIGH when selected.
- Only one output must be HIGH at any time.
- If an ascending 2-bit binary count is applied to inputs $D1$ and $D0$, the outputs will go HIGH in the sequence $Y0$, $Y1$, $Y2$, and $Y3$, then repeat.
- If $\overline{G} = 0$, the decoder outputs can activate as described in the previous criteria.
- If $\overline{G} = 1$, no output will activate for any value of $D1$ and $D0$.

To test these criteria, we apply an ascending 2-bit count to inputs $D1$ and $D0$, with $\overline{G} = 0$, then apply the same count when $\overline{G} = 1$. As shown in Figure 6.9, the outputs activate in the correct order when the decoder is enabled and no outputs activate when the decoder is disabled.

EXAMPLE 6.4

The 2-line-to-4-line decoder in Figure 6.5 is tested with input waveforms meeting the simulation criteria previously discussed. The simulation result, shown in Figure 6.10, indicates an error condition in the decoder. What is likely to be the circuit fault?

FIGURE 6.10   Example 6.4: Simulation of a 2-Line-to-4-Line Decoder with Error in Enable/Disable Function

■ Solution

The simulation waveforms show the outputs behaving correctly when the $\overline{G}$ input is LOW, that is, when the decoder is enabled. When the decoder should be disabled (when $\overline{G}$ is HIGH), the $Y3$ output activates whenever $D1 = 1$ and $D0 = 1$. This is the correct decoding for this output, but it should not be decoding at all when the decoder is disabled. This could result from an improper connection from $\overline{G}$ to the gate that decodes $Y3$. The connection from $\overline{G}$ to $Y3$ is open-circuited or stuck HIGH.

EXAMPLE 6.5

The 2-line-to-4-line decoder of Figure 6.5 is tested with input waveforms meeting appropriate simulation criteria. The simulation result, shown in Figure 6.11, indicates an error condition in the decoder. What is likely to be the circuit fault?

FIGURE 6.11   Example 6.5: Simulation of 2-Line-to-4-Line Decoder with Improperly Assigned D Inputs.

■ **Solution**

The outputs are enabled and disabled when they are supposed to be, but, when enabled, the outputs activate in the wrong order. We can sort this out by comparing the actual outputs to the expected outputs, as shown in Table 6.5.

TABLE 6.5   Actual and Expected Decoder Outputs

Output	Should activate when $D =$	Actually activates when $D =$
$Y0$	00	00
$Y1$	01	10
$Y2$	10	01
$Y3$	11	11

Notice that the values in the actual and expected columns are mirror images of one another. This implies that when we entered the design in the Quartus II Block Editor, we mixed up the order of $D1$ and $D0$, as shown in Figure 6.12. Reconnecting (or renaming) the inputs will fix the problem.

(continued)

FIGURE 6.12   Example 6.5: Incorrectly Labeled *D* Inputs on a Decoder.

EXAMPLE 6.6

A 74138 3-line-to-8-line decoder is connected in a Quartus II Block Diagram File, as shown in Figure 6.13. Make a list of simulation criteria that will fully test the decoder and create a Quartus II simulation to verify the operation of the device.

FIGURE 6.13   Example 6.6: 74138 Decoder in Quartus II Block Diagram File

■ Solution

SIMULATION CRITERIA

- Each decoder output should respond to its appropriate binary input value by going LOW.
- Only one output should be LOW at any given time.
- An increasing 3-bit binary count on inputs *CBA* should make the outputs go LOW one at a time, starting with *Y*0, followed by *Y*1, *Y*2, and so on until *Y*7, and then repeat.
- The enable inputs must all be active for the outputs to activate: $G1 = 1$, $\overline{G2A} = 0$, and $\overline{G2B} = 0$
- If any enable input is inactive, no decoder output should be active, regardless of the values of the input *CBA*.

FIGURE 6.14 Example 6.6: Simulation of a 74138 Decoder

Figure 6.14 shows a simulation that meets these criteria. An increasing binary count on *CBA* (also shown grouped as *Decode inputs*), activates the outputs in sequence, as long as all three enables are active. Any one of the enables inactive prevents any output from activating. Notice that each of the enable inputs are inactive for the entire binary count range.

VHDL Binary Decoder

➡ **LAB SOURCE**
Lab #11 Binary and Seven-Segment Decoders in VHDL

> ■ **KEY TERMS**
>
> **Selected Signal Assignment Statement** A concurrent signal assignment in VHDL in which a value is assigned to a signal, depending on the alternative values of another signal or variable.
>
> **Conditional Signal Assignment Statement** A concurrent VHDL construct that assigns a value to a signal, depending on a sequence of conditions being true or false.

We can use several possible constructs to code a binary decoder in VHDL. We can use straight Boolean equations in a set of concurrent signal assignments. This becomes unwieldy when the decoder becomes fairly large. Alternatively, we can use a **selected signal assignment statement** or a **conditional signal assignment statement.**

The following VHDL code implements a 2-line-to-4-line decoder using a selected signal assignment statement.

decode3.vhd

```
LIBRARY ieee;
USE ieee.std_logic_1164.ALL;

ENTITY decode3 IS
    PORT(
        d : IN  STD_LOGIC_VECTOR (1 downto 0);
        y : OUT STD_LOGIC_VECTOR (3 downto 0));
END decode3;
```

```
ARCHITECTURE decoder OF decode3 IS
BEGIN
    WITH d SELECT
        y <=    "0001" WHEN "00",
                "0010" WHEN "01",
                "0100" WHEN "10",
                "1000" WHEN "11",
                "0000" WHEN others;
END decoder;
```

[annotations: d(1), d(0) pointing to the "00" value; — default case; y(0), y(3) pointing to "0000"]

The selected signal assignment statement evaluates input **d.** For every possible combination of the 2-bit input vector, **d,** a particular value is assigned to the 4-bit vector, **y.** (For example, for the case $d_1 d_0 = 10 (= 2_{10})$, the output y_2 is HIGH: $y_3 y_2 y_1 y_0 = 0100$.)

The default case (**"WHEN others"**) is required because of the multivalued logic type STD_LOGIC_VECTOR. Since a STD_LOGIC_VECTOR can have values other than '0' and '1', the values listed for **d** don't cover all possible cases. The default output (which will never occur if we only use '0' and '1' inputs) is chosen such that no output is active in the default case. The default case would not be required if we chose to use BIT_VECTOR, rather than STD_LOGIC_VECTOR, since the listed combinations of **d** cover all possible combinations of a BIT_VECTOR.

In order to include an enable input (**g**) in a decoder, we can increase the input vector size to include the **g** input, as shown in the following code.

decode 3a.vhd

```
LIBRARY ieee;
USE ieee.std_logic_1164.ALL;


ENTITY decode3a IS
    PORT(
        d  : IN  STD_LOGIC_VECTOR (1 downto 0);
        g  : IN  STD_LOGIC;
        y  : OUT STD_LOGIC_VECTOR (3 downto 0));
END decode3a;


ARCHITECTURE decoder OF decode3a IS
    SIGNAL inputs : STD LOGIC VECTOR (2 downto 0);
BEGIN
    inputs (2)           <= g;
    inputs (1 downto 0)  <= d;
    WITH inputs SELECT
        y <=    "0001" WHEN "000",
                "0010" WHEN "001",
                "0100" WHEN "010",
                "1000" WHEN "011",
                "0000" WHEN others;
END decoder;
```

[annotations: Concatenate g (1 bit) and d (2 bits) to get 3-bit vector; g, d(1), d(0) pointing to "000"; — default case; y(0), y(3) pointing to "0000"]

To include **g** and **d** in a single vector, we create a signal called **inputs,** a vector with three elements in the sequence **g, d(1), d(0).** When assigning the **d** to the last two elements of **inputs,** we must be explicit about which elements of **inputs** we want to use. Since **d** only contains two elements and we are assigning them to two elements of **inputs,** we don't need to list the elements of **d** explicitly.

We can use a selected signal assignment statement to evaluate all inputs, including **g,** and assign outputs accordingly. When **g** = '0', the decoder outputs are assigned the same as they were in the example without the enable input. The cases where **g** = '1' are covered by the **others** clause. In this default case, all decoder outputs are LOW (inactive).

Another way to include an enable input is to use a conditional signal assignment statement, which makes an assignment based on a Boolean expression. The template for the conditional signal assignment statement is:

```
__signal <= __expression WHEN __boolean_expression ELSE
            __expression WHEN __boolean_expression ELSE
            __expression;
```

The first Boolean expression in the statement is evaluated. If it is true, the corresponding expression is assigned to the signal. If false, the next Boolean expression is evaluated, and so on until a true Boolean expression is found. If none are true, the signal is assigned a default expression, listed last in the statement.

The VHDL code below implements the decoder with an active-LOW enable. If **g** is LOW, one decoder output activates, depending on the value of **d.** Note that the **d** inputs are defined as type INTEGER, rather than BIT_VECTOR or STD_LOGIC_VECTOR. In this situation, we don't need to specify the number of inputs; the compiler automatically defines the required inputs **d1** and **d0** when fitting the design to the selected CPLD. Also, since **d** is of type INTEGER, we write its value in the selected signal assignment statement directly, without quotes.

decode4g.vhd

```
—decode4g.vhd
LIBRARY ieee;
USE ieee.std_logic_1164.all;

ENTITY decode4g IS
   PORT (
            d : IN  INTEGER Range 0 to 3;
            g : IN  STD_LOGIC;
            y : OUT STD_LOGIC_VECTOR (0 to 3));
END decode4g;

ARCHITECTURE a OF decode4g IS                    Assign this value to y
BEGIN                                            under this condition
   y <=    "1000" WHEN (d=0 and g='0') ELSE
           "0100" WHEN (d=1 and g='0') ELSE
           "0010" WHEN (d=2 and g='0') ELSE
           "0001" WHEN (d=3 and g='0') ELSE
           "0000";                               STD_LOGIC
END a;                                           INTEGER
                          default case
```

Seven-Segment Decoders

■ KEY TERMS

Seven-Segment Display   An array of seven independently controlled light-emitting diode (LED) or liquid crystal display (LCD) elements, shaped like a figure-8, which can be used to display decimal digits and other characters by turning on the appropriate elements.

Common Anode Display   A seven-segment LED display where the anodes of all the LEDs are connected to the circuit supply voltage. Each segment is illuminated by a logic LOW at its cathode.

Common Cathode Display   A seven-segment display in which the cathodes of all LEDs are connected together and grounded. A logic HIGH illuminates a segment when applied to its anode.

Display

The **seven-segment display,** shown in Figure 6.15, is a numerical display device used to show digital circuit outputs as decimal digits (and sometimes hexadecimal digits or other alphabetic characters). It is called a seven-segment display because it consists of seven luminous segments, usually LEDs or liquid crystals, arranged in a figure-8. We can display any decimal digit by turning on the appropriate elements, designated by lowercase letters, *a* through *g.* It is conventional to designate the top segment as *a* and progress clockwise around the display, ending with *g* as the center element.

FIGURE 6.15 Seven-Segment Numerical Display

FIGURE 6.16 Convention for Displaying Decimal Digits

Figure 6.16 shows the usual convention for decimal digit display. Some variation from this convention is possible. For example, we could have drawn the digits 6 and 9 with "tails" (i.e., with segment *a* illuminated for 6 or segment *d* for 9). By convention, we display digit 1 by illuminating segments *b* and *c,* although segments *e* and *f* would also work.

The electrical requirements for an LED circuit are simple. Since an LED is a diode, it conducts when its anode is positive with respect to its cathode, as shown in Figure 6.17a. A decoder/driver for an LED display will illuminate an element by completing this circuit, either by supplying V_{CC} or ground. A series resistor limits the current to prevent the diode from burning out and to regulate its brightness. If the anode is +5 volts with respect to cathode, the resistor value should be in the range of 220 Ω to 470 Ω.

a. Circuit requirements for an illuminated LED

b. Common cathode

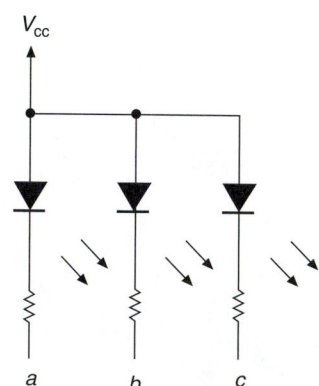

b. Common anode

FIGURE 6.17 Electrical Requirements for LED Displays

FIGURE 6.18 Physical Placement of LEDs in a Common Anode Display

Seven-segment displays are configured as **common anode** or **common cathode,** as shown in Figure 6.17b and Figure 6.17c. In a common cathode display, the cathodes of all LEDs are connected together and brought out to one or more pin connections on the display package. The cathode pins are wired externally to the circuit ground. We illuminate the segments by applying logic HIGHs to individual anodes.

Similarly, the common anode display has the anodes of the segments brought out to one or more common pins. These pins must be tied to the circuit power supply (V_{CC}). The segments illuminate when a decoder/driver makes their individual cathodes LOW. Figure 6.18 shows how the diodes could be physically laid out in a common anode display.

The two types of displays allow the use of either active HIGH or active LOW circuits to drive the LEDs, thus giving the designer some flexibility. However, it should be noted that the majority of seven-segment decoders are for common-anode displays.

EXAMPLE 6.7

Sketch the segment patterns required to display all 16 hexadecimal digits on a seven-segment display. What changes from the patterns in Figure 6.16 need to be made?

■ **Solution**

The segment patterns are shown in Figure 6.19.

FIGURE 6.19 Example 6.7: Hexadecimal Digit Display Format

Hex digits B and D must be displayed as lowercase letters, b and d, to avoid confusion between B and 8 and between D and 0. To make 6 distinct from b, 6 must be given a tail (segment *a*) and to make 6 and 9 symmetrical, 9 should also have a tail (segment *d*).

Decoder

■ **KEY TERM**

BCD Binary-coded decimal. A code in which each individual digit of a decimal number is represented by a 4-bit binary number (e.g., 905 (decimal) = 1001 0000 0101 (BCD)).

A BCD-to-seven-segment decoder is a circuit with a 4-bit input for a **BCD** digit and seven outputs for segment selection. To display a number, the decoder must translate the input bits to a combination of active outputs. For example, the input digit $D_3D_2D_1D_0 = 0000$ must illuminate segments *a, b, c, d, e,* and *f* to display the digit 0. We can make a truth table for each of the outputs, showing which must be active for every digit we wish to display. The truth table for a common-anode decoder (active LOW outputs) is given in Table 6.6.

TABLE 6.6 Truth Table for Common Anode BCD-to-Seven-Segment Decoder

Digit	D_3	D_2	D_1	D_0	a	b	c	d	e	f	g
0	0	0	0	0	0	0	0	0	0	0	1
1	0	0	0	1	1	0	0	1	1	1	1
2	0	0	1	0	0	0	1	0	0	1	0
3	0	0	1	1	0	0	0	0	1	1	0
4	0	1	0	0	1	0	0	1	1	0	0
5	0	1	0	1	0	1	0	0	1	0	0
6	0	1	1	0	1	1	0	0	0	0	0
7	0	1	1	1	0	0	0	1	1	1	1
8	1	0	0	0	0	0	0	0	0	0	0
9	1	0	0	1	0	0	0	1	1	0	0
	1	0	1	0	X	X	X	X	X	X	X
	1	0	1	1	X	X	X	X	X	X	X
Invalid Range	1	1	0	0	X	X	X	X	X	X	X
	1	1	0	1	X	X	X	X	X	X	X
	1	1	1	0	X	X	X	X	X	X	X
	1	1	1	1	X	X	X	X	X	X	X

The illumination of each segment is determined by a Boolean function of the input variables, $D_3D_2D_1D_0$. From the truth table, the function for segment *a* is

$$a = \bar{D}_3\bar{D}_2\bar{D}_1 D_0 + \bar{D}_3 D_2 \bar{D}_1 \bar{D}_0 + \bar{D}_3 D_2 D_1 \bar{D}_0$$

(Since the display is active-LOW, this means segment *a* is OFF for digits 1, 4, and 6.)

If we assume that inputs 1010 to 1111 are never going to be used ("don't care states," symbolized by X), we can make any of these states produce HIGH or LOW outputs, depending on which is most convenient for simplifying the segment functions. Figure 6.20a shows a Karnaugh map simplification for segment *a*. The resultant function is

$$a = \bar{D}_3\bar{D}_2\bar{D}_1 D_0 + D_2 \bar{D}_0$$

The corresponding partial decoder is shown in Figure 6.20b.

We could do a similar analysis for each of the other segments, but if we are programming the decoder function into a CPLD, it is just as simple to write the truth table directly into a selected signal assignment statement, as shown in the VHDL code that follows.

bcd_7seg.vhd

```
—bcd_7seg.vhd
—BCD-to-seven-segment decoder
LIBRARY ieee;
USE ieee.std_logic_1164.ALL;
```

Segment *a*

a. K=map

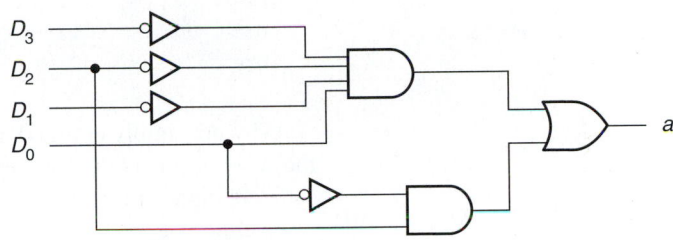

b. Decoder for segment *a* (common anode)

FIGURE 6.20   Decoding Segment *a*

```
ENTITY bcd_7seg IS
    PORT(
        d3, d2, d1, d0        : IN  STD_LOGIC;
        a, b, c, d, e, f, g : OUT STD_LOGIC);
END bcd_7seg;

ARCHITECTURE seven_segment OF bcd_7seg IS
    SIGNAL input  : STD_LOGIC_VECTOR (3 downto 0);
    SIGNAL output : STD_LOGIC_VECTOR (6 downto 0);
BEGIN
    input <= d3 & d2 & d1 & d0;                      d3            d0
    WITH input SELECT
        output <= "0000001" WHEN "0000", -- display 0
                  "1001111" WHEN "0001", -- display 1
                  "0010010" WHEN "0010", -- display 2
                  "0000110" WHEN "0011", -- display 3
                  "1001100" WHEN "0100", -- display 4
                  "0100100" WHEN "0101", -- display 5
                  "1100000" WHEN "0110", -- display 6
                  "0001111" WHEN "0111", -- display 7
                  "0000000" WHEN "1000", -- display 8
                  "0001100" WHEN "1001", -- display 9
                  "1111111" WHEN others;
                a              g
```

```
-- Separate the output vector to make individual pin outputs.
a <= output (6);
b <= output (5);
c <= output (4);
d <= output (3);
e <= output (2);
f <= output (1);
g <= output (0);

END seven_segment;
```

The inputs $D_3D_2D_1D_0$ are defined separately, then concatenated (linked in sequence) by the **&** operator to make a BIT_VECTOR called **input.** This is equivalent to the following four concurrent signal assignments:

```
input (3) <= d3;
input (2) <= d2;
input (1) <= d1;
input (0) <= d0;
```

Why not simply define **d** as a vector? If we wish to create a graphic symbol for the seven-segment decoder, the previous method creates a symbol shown with four separate inputs, rather than a single thick line for a 4-bit bus input. The design will work either way.

For each value of **input,** a signal assignment defines the output vector, each bit of which represents the value of one segment. For example, the first clause ("0000001" WHEN "0000") sets all segments ON except segment *g*, thus displaying the digit "0".

As a variation, we could define a signal called **d_inputs** of type INTEGER with RANGE 0 to 9. The WHEN clauses would evaluate the integer values 0 to 9, as follows.

```
WITH d_inputs SELECT
    output <= "0000001" WHEN 0,
              "1001111" WHEN 1,
              "0010010" WHEN 2,
              "0000110" WHEN 3,

              "1001100" WHEN 4,
              "0100100" WHEN 5,
              "0100000" WHEN 6,
              "0001111" WHEN 7,

              "0000000" WHEN 8,
              "0000100" WHEN 9,
              "1111111" WHEN others; -- blank
```

Ripple Blanking

■ KEY TERMS

Ripple Blanking   A technique used in a multiple-digit numerical display that suppresses leading or trailing zeros in the display, but allows internal zeros to be displayed.

$\overline{RBI}$ Ripple blanking input.

$\overline{RBO}$ Ripple blanking output.

IF Statement   A VHDL construct within a process that executes a series of statements, if a Boolean test condition is true.

CASE Statement   A VHDL construct in which there is a choice of statements to be executed, depending on the value of a signal or variable.

PROCESS   A VHDL construct that contains statements that are executed if there is a change in a signal in its sensitivity list.

Sensitivity List   A list of signals in a PROCESS statement that are monitored to determine whether the PROCESS should be executed.

A feature often included in seven-segment decoders is **ripple blanking.** The ripple blanking feature allows for suppression of leading or trailing zeros in a multiple digit display, while allowing zeros to be displayed in the middle of a number.

Each display decoder has a ripple blanking input ($\overline{RBI}$) and a ripple blanking output ($\overline{RBO}$), which are connected in cascade, as shown in Figure 6.21. If the decoder input $D_3D_2D_1D_0$ is 0000, it displays digit 0 if $\overline{RBI} = 1$ and shows a blank if $\overline{RBI} = 0$.

If $\overline{RBI} = 1$ OR $D_3D_2D_1D_0$ is (NOT 0000), then $\overline{RBO} = 1$. When we cascade two or more displays, these conditions suppress leading or trailing zeros (but not both) and still display internal zeros.

To suppress leading zeros in a display, ground the $\overline{RBI}$ of the most significant digit decoder and connect the $\overline{RBO}$ of each decoder to the $\overline{RBI}$ of the next least significant digit. Any zeros preceding the first nonzero digit (9 in this case) will be blanked, as $\overline{RBI} = 0$ AND $D_3D_2D_1D_0 = 0000$ for each of these decoders. The 0 inside the number 904 is displayed since its $\overline{RBI} = 1$.

Trailing zeros are suppressed by reversing the order of $\overline{RBI}$ and $\overline{RBO}$ from the previous example. $\overline{RBI}$ is grounded for the least significant digit and the $\overline{RBO}$ for each decoder cascades to the $\overline{RBI}$ of the next most significant digit.

We can implement the ripple blanking feature in VHDL by modifying a design entity for a BCD- or hexadecimal-to-seven-segment decoder to replace the selected signal assignment statement with an **IF statement** and a **CASE statement.** These are two VHDL statements that are each used for selecting an alternative set of statements to execute.

An IF statement executes one or more VHDL statements, depending on the state of a Boolean test condition. It has the following syntax.

```
IF __expression THEN
    __statement;
    __statement;
ELSIF __expression THEN
    __statement;
    __statement;
ELSE
    __statement;
    __statement;
END IF;
```

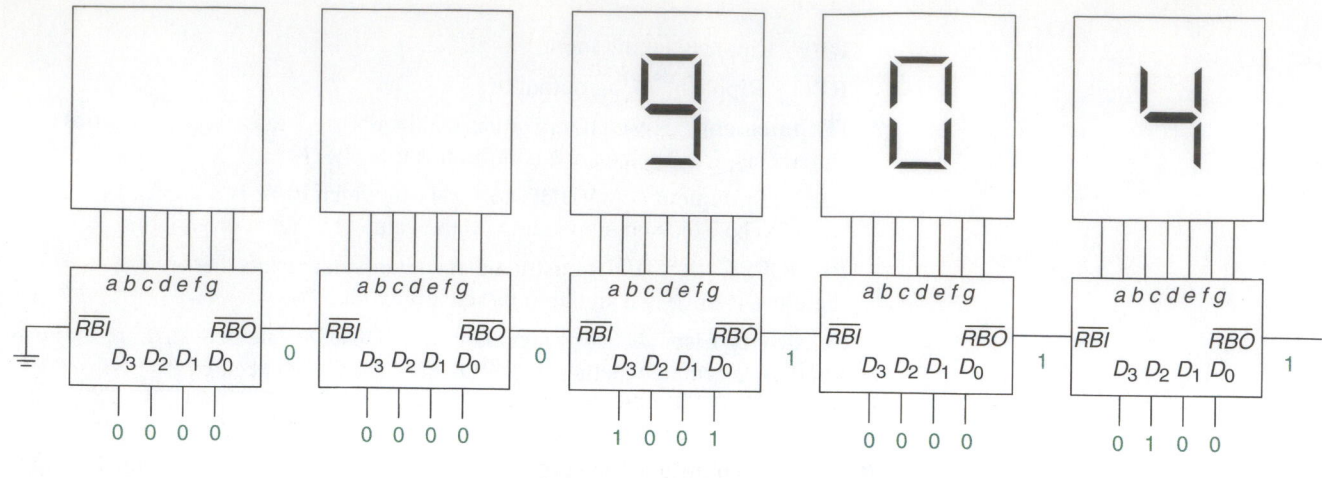

a. Suppression of leading 0s

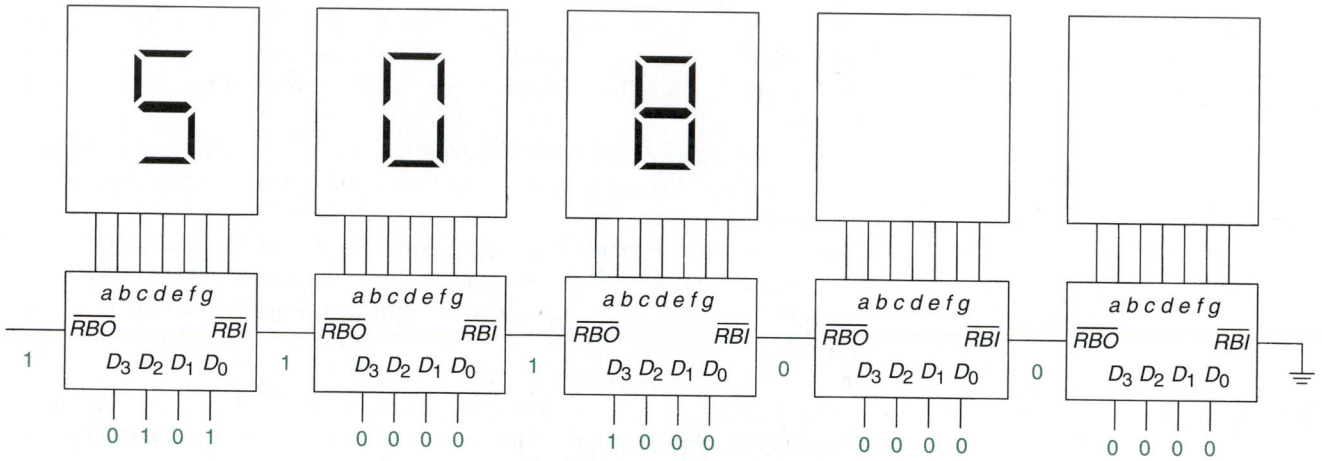

b. Suppression of trailing 0s

FIGURE 6.21   Zero Suppression in Seven-Segment Displays

The expressions in the various clauses of the IF statements are Boolean expressions, each yielding a true or false result. The IF clause executes one or more statements if the test condition in the first expression is true. If the test condition is false, we can test for other conditions with one or more optional ELSIF clauses. If none of the test conditions are true, then the ELSE clause is executed by default.

For example, if we want to blank out the display digit if the $\overline{RBI}$ input is LOW and all **D** inputs, defined as a 4-bit vector called **input,** are also all LOW, we can use the following construction:

```
IF (nRBI = '0' and input = "0000") THEN
    output <= "1111111";              -- 0 suppressed
    nRBO   <= '0';                    -- Next 0 suppressed
ELSE
    nRBO <= '1';                      -- Next 0 displayed
    all other cases display numeric output
END IF;
```

To select the other values of input, we could use a series of nested IF statements or an IF with a number of ELSIF clauses, but VHDL generates more efficient hardware if we use a CASE statement instead. A CASE statement can be used to select among several alternatives and takes the following form:

```
CASE __expression IS
    WHEN __constant_value =>
            __statement;
            __statement;
    WHEN constant_value =>
            __statement;
            __statement;
    WHEN OTHERS =>
            __statement;
            __statement;
END CASE;
```

The expression in the CASE statement is usually a port or signal whose value we wish to test. For example, we can use a CASE statement to select one of ten alternative output patterns for the ten allowable input combinations, as follows.

```
CASE input IS
    WHEN "0000" =>
            output <= "0000001"; -- 0 displayed
    WHEN "0001"    =>
            output <= "1001111"; -- 1
    WHEN "0010"    =>
            output <= "0010010"; -- 2
    WHEN "0011"    =>
            output <= "0000110"; -- 3
    WHEN "0100"    =>
            output <= "1001100"; -- 4
    WHEN "0101"    =>
            output <= "0100100"; -- 5
    WHEN "0110"    =>
            output <= "1100000"; -- 6
    WHEN "0111"    =>
            output <= "0001111"; -- 7
    WHEN "1000"    =>
            output <= "0000000"; -- 8
    WHEN "1001"    =>
            output <= "0001100"; -- 9
    WHEN others    =>
            output <= "1111111"; -- blank
END CASE;
```

VHDL syntax requires an IF statement or a CASE statement to be contained within a **PROCESS.** A PROCESS is a construct containing statements that are executed if a signal in the **sensitivity list** of the PROCESS changes. The general form of a PROCESS is:

```
[label:] PROCESS (sensitivity list)
BEGIN
    statements;
END PROCESS;
```

For example, in the PROCESS:

```
PROCESS (nRBI, input)
BEGIN
    statements;
END PROCESS;
```

the statements after the BEGIN clause will execute whenever there is a change in **nRBI** or **input.**

Combining all these constructs, we can write the VHDL code for the seven-segment decoder with ripple-blanking as follows:

```
-- sevsegrb.vhd
-- BCD-to-seven-segment decoder with ripple blanking
-- Leading or trailing zeroes can be suppressed in a
-- multi-digit display.
-- A zero is displayed IF input=0000 and nRBI=1.
-- A zero is suppressed (display blank) if input=0000 and nRBI=0.
-- nRBO = 0 for nrBI = 0 and input = 0000. Otherwise nRBO = 1.

ENTITY sevsegrb IS
    PORT (
        -- Use separate I/Os, not bus
        nRBI, d3, d2, d1, d0     : IN  BIT;
        a, b, c, d, e, f, g, nRBO: OUT BIT);
END sevsegrb;

ARCHITECTURE seven_segment OF sevsegrb IS
    SIGNAL input: BIT_VECTOR (3 DOWNTO 0);  -- Bit vectors for internal
    SIGNAL output: BIT_VECTOR (6 DOWNTO 0); -- use in decoder statements
BEGIN
    -- Concatenate inputs to make bit vector
    input  <=  d3 & d2 & d1 & d0;

    -- Process Statement
    PROCESS (input, nRBI)
    BEGIN
    IF(nRBI = '0' and input = "0000" THEN
        output  <= "1111111"; -- 0 suppressed
        nRBO    <= '0';       -- Next 0 suppressed
    ELSE
        nRBO    <= '1';       -- Next 0 displayed
        CASE input IS
            WHEN "0000"   =>
                output <= "0000001"; -- 0 displayed
            WHEN "0001"   =>
                output <= "1001111"; -- 1
            WHEN "0010"   =>
                output <= "0010010"; -- 2
            WHEN "0011"   =>
                output <= "0000110"; -- 3
            WHEN "0100"   =>
                output <= "1001100"; -- 4
            WHEN "0101"   =>
                output <= "0100100"; -- 5
            WHEN "0110"   =>
                output <= "1100000"; -- 6
            WHEN "0111"   =>
                output <= "0001111"; -- 7
            WHEN "1000"   =>
                output <= "0000000"; -- 8
            WHEN "1001"   =>
                output <= "0001100"; -- 9
```

sevsegrb.vhd

```
          WHEN others    =>
                output <= "1111111"; -- blank
          END CASE;
        END IF;
    END PROCESS assign_out;

-- Separate the output vector to make individual pin outputs.
        a   <= output (6);
        b   <= output (5);
        c   <= output (4);
        d   <= output (3);
        e   <= output (2);
        f   <= output (1);
        g   <= output (0);

    END seven_segment;
```

◼ **SECTION 6.1C REVIEW PROBLEM**

6.3   When would it be logical to suppress trailing zeros in a multiple-digit display and when should trailing zeros be displayed?

6.2   ENCODERS

◼ **KEY TERMS**

Encoder   A circuit that generates a binary code at its outputs in response to one or more active input lines.

Priority Encoder   An encoder that generates a binary or BCD output corresponding to the subscript of the active input having the highest priority. This is usually defined as the active input with the largest subscript value.

The function of a digital **encoder** is complementary to that of a digital decoder. A decoder activates a specified output for a unique digital input code. An encoder operates in the reverse direction, producing a particular digital code (e.g., a binary or BCD number) at its outputs when a specific input is activated.

Figure 6.22 shows a 3-bit binary encoder. The circuit generates a unique 3-bit binary output for every active input provided *only one input* is active at a time.

The encoder has only 8 permitted input states out of a possible 256. Table 6.7 shows the allowable input states, which yield the Boolean equations used to design the encoder. These Boolean equations are:

$$Q_2 = D_7 + D_6 + D_5 + D_4$$
$$Q_1 = D_7 + D_6 + D_3 + D_2$$
$$Q_0 = D_7 + D_5 + D_3 + D_1$$

The D_0 input is not connected to any of the encoding gates, as all outputs are in their LOW (inactive) state when the 000 code is selected.

Priority Encoder

The shortcoming of the encoder circuit shown in Figure 6.22 is that it can generate wrong codes if more than one input is active at the same time. For example, if we

FIGURE 6.22 3-Bit Encoder (No Input Priority)

TABLE 6.7 Partial Truth Table for a 3-Bit Encoder

D_7	D_6	D_5	D_4	D_3	D_2	D_1	Q_2	Q_1	Q_0
0	0	0	0	0	0	0	0	0	0
0	0	0	0	0	0	1	0	0	1
0	0	0	0	0	1	0	0	1	0
0	0	0	0	1	0	0	0	1	1
0	0	0	1	0	0	0	1	0	0
0	0	1	0	0	0	0	1	0	1
0	1	0	0	0	0	0	1	1	0
1	0	0	0	0	0	0	1	1	1

➡ **LAB SOURCE**
Lab #12
Priority Encoders

make D_3 and D_5 HIGH at the same time, the output is neither 011 or 101, but 111; the output code does not correspond to either active input.

One solution to this problem is to assign a priority level to each input and, if two or more are active, make the output code correspond to the highest-priority input. This is called a **priority encoder.** Highest priority is assigned to the input whose subscript has the largest numerical value.

EXAMPLE 6.8

Figures 6.23 shows a priority encoder with three different combinations of inputs. Determine the resultant output code for each figure. Inputs and outputs are active HIGH.

FIGURE 6.23 Example 6.8: Priority Encoder Inputs

Solution

Figure 6.23a: The highest-priority active input is D_5. D_4 and D_1 are ignored. $Q_2Q_1Q_0 = 101$.

Figure 6.23b: The highest-priority active input is D_4. D_1 is ignored. $Q_2Q_1Q_0 = 100$.

Figure 6.23c: The highest-priority active input is D_7. All other inputs ignored. $Q_2Q_1Q_0 = 111$.

NOTE . . .

The encoding principle of a prority encoder is that a low-priority input must not change the code resulting from a higher-priority input.

For example, if inputs D_3 and D_5 are both active, the correct output code is $Q_2Q_1Q_0 = 101$. The code for D_3 would be $Q_2Q_1Q_0 = 011$. Thus, D_3 must not make $Q_1 = 1$. The Boolean expressions for Q_2Q_1 and Q covering only these two codes are:

$$Q_2 = D_5 \qquad \text{(HIGH if } D_5 \text{ is active.)}$$
$$Q_1 = D_3\bar{D}_5 \qquad \text{(HIGH if } D_3 \text{ is active AND } D_5 \text{ is NOT active.)}$$
$$Q_0 = D_3 + D_5 \qquad \text{(HIGH if } D_3 \text{ OR } D_5 \text{ is active.)}$$

The truth table of a 3-bit priority encoder is shown in Table 6.8.

TABLE 6.8 Truth Table for a 3-Bit Priority Encoder

D_7	D_6	D_5	D_4	D_3	D_2	D_1	Q_2	Q_1	Q_0
0	0	0	0	0	0	0	0	0	0
0	0	0	0	0	0	1	0	0	1
0	0	0	0	0	1	X	0	1	0
0	0	0	0	1	X	X	0	1	1
0	0	0	1	X	X	X	1	0	0
0	0	1	X	X	X	X	1	0	1
0	1	X	X	X	X	X	1	1	0
1	X	X	X	X	X	X	1	1	1

Restating the encoding principle, a bit goes HIGH if it is part of the code for an active input AND it is NOT kept LOW by an input with a higher priority. We can use this principle to develop a mechanical method for generating the Boolean equations of the outputs.

1. Write the codes in order from highest to lowest priority, as in Table 6.9.

TABLE 6.9 Binary Outputs and Corresponding Decimal Values

Q_2	Q_1	Q_0	Code Value
1	1	1	7
1	1	0	6
1	0	1	5
1	0	0	4
0	1	1	3
0	1	0	2
0	0	1	1
0	0	0	0

2. Examine each code. For a code with value n, add a D_n term to each Q equation where there is a 1. For example, for code 111, add the term D_7 to the equations for Q_2, Q_1, and Q_0. For code 110, add the term D_6 to the equations for Q_2 and Q_1. (Steps 1 and 2 generate the nonpriority encoder equations listed earlier.)

3. Modify any D_n terms to ensure correct priority. Every time you write a D_n term, look at the previous lines in the table. For each previous code with a 0 in the same column as the 1 that generates D_n, use an AND function to combine D_n with a corresponding $\bar{D}$. For example, code 101 generates a D_5 term in the equations for Q_2 and Q_0. The term in the Q_2 equation need not be modified because there are no previous codes with a 0 in the same column. The term in the Q_0 equation must be modified since there is a 0 in the Q_0 column for code 110. This generates the term $\bar{D}_6 D_5$. There is no D_5 term in the equation for Q_1.

The equations from the 3-bit encoder of Figure 6.22 are modified by the priority encoding principle as follows:

$$Q_2 = D_7 + D_6 + D_5 + D_4$$
$$Q_1 = D_7 + D_6 + \bar{D}_5\bar{D}_4\,D_3 + \bar{D}_5\,\bar{D}_4 D_2$$
$$Q_0 = D_7 + \bar{D}_6 D_5 + \bar{D}_6\bar{D}_4 D_3 + \bar{D}_6\bar{D}_4\,\bar{D}_2 D_1$$

VHDL Priority Encoder

The most obvious way to program a priority encoder in VHDL is to use the equations derived in the previous section in a set of concurrent signal assignment statements, as follows.

hi_pri8a.vhd

```
-- hi_pri8a.vhd
ENTITY hi_pri8a IS

  PORT(
      d  : IN  BIT_VECTOR(7 downto 0);
      q  : OUT BIT_VECTOR (2 downto 0));
  END hi_pri8a;

ARCHITECTURE a OF hi_pri8a IS
BEGIN
  -- Concurrent Signal Assignments
  q(2) <= d(7) or d(6) or d(5) or d(4);

  q(1) <= d(7) or d(6)
       or ((not d(5)) and (not d(4)) and d(3))
       or ((not d(5)) and (not d(4)) and d(2));

  q(0) <= d(7) or ((not d(6)) and d(5))
       or ((not d(6)) and (not d(4)) and d(3))
       or ((not d(6)) and (not d(4)) and (not d(2)) and d(1));
  END a;
```

Although this code works, it is not terribly elegant, nor does it give any insight into the operation of the encoder circuit. Also, if we expand our encoder output by one or more bits, the equations become more cumbersome with each new bit and soon become impractically large and susceptible to typing errors. A VHDL conditional signal assignment statement is an ideal alternative for use in a priority encoder circuit. A section of VHDL code using this format is as follows:

Quartus II:
hi_pri8b.vhd
hi_pri8b.vwf
MAX+PLUS II:
hi_pri8b.vhd
hi_pri8b.scf

```
-- hi_pri8b.vhd
ENTITY hi_pri8b IS
  PORT(
      d  : IN  BIT_VECTOR (7 downto 0);
      q  : OUT INTEGER RANGE 0 to 7);
  END hi_pri8b;
```

```
ARCHITECTURE a OF hi_pri8b IS
BEGIN
 --  Conditional Signal Assignment
encoder:
 q  <=  7 WHEN d(7)='1' ELSE
         6 WHEN d(6)='1' ELSE
         5 WHEN d(5)='1' ELSE
         4 WHEN d(4)='1' ELSE
         3 WHEN d(3)='1' ELSE
         2 WHEN d(2)='1' ELSE
         1 WHEN d(1)='1' ELSE
         0;
 END a;
```

Output **q** is defined as type INTEGER. Since it ranges from 0 to 7, the VHDL compiler will automatically assign three outputs: Q_2, Q_1, and Q_0. The conditional signal assignment statement evaluates the first WHEN clause to determine if its condition (d(7) = '1') is true. If so, it assigns **q** the value of 7 ($Q_2Q_1Q_0 = 111$). If the first condition is false, the next WHEN clause is evaluated, assigning **q** the value 6 ($Q_2Q_1Q_0 = 110$) if true, and so on until all WHEN clauses have been evaluated. If no clause is true, then the default value (0: $Q_2Q_1Q_0 = 000$) is assigned to the output.

In the conditional signal assignment, the highest-priority condition is examined first. If it is true, the output is assigned according to that condition and no further conditions are evaluated. If the first condition is false, the condition of next priority is evaluated, and so on until the end. Thus, a low-priority input cannot alter the code resulting from an input of higher priority, as required by the priority encoding principle.

The effect is similar to that of an IF statement, where a sequence of conditions is evaluated, but only one output assignment is made. However, an IF statement must be used within a PROCESS statement, if we choose to use it. The IF statement for a priority encoder is as follows:

```
PROCESS (d)
BEGIN
 IF (d(7) = '1') THEN
         q <= 7;
     ELSIF (d(6) = '1') THEN
         q <= 6;
     .
     .
     .
     ELSIF (d(1) = '1' THEN
         q <= 1;
     ELSE
         q <= 0;
     END IF;
END PROCESS;
```

EXAMPLE 6.9

Write a set of simulation criteria for the 3-bit priority encoder using a conditional signal assignment statement. Use these criteria to create a simulation in Quartus II.

■ Solution

SIMULATION CRITERIA

■ With all input bits HIGH, d(7) is highest priority of active inputs. Expected output is 111_2 (= 7_{10}).   (*continued*)

■ With d(7) LOW and all other inputs HIGH, d(6) is now highest priority active input. Expected output is 110_2 (= 6_{10}).

■ With d(7) and d(6) LOW and remaining bits HIGH, highest priority is d(5). Expected output is 101_2 (= 5_{10}).

■ As each bit goes LOW in order, output code should count down by one with each step.

Figure 6.24 shows the simulation of the three bit priority encoder, based on the previous criteria. The **d** inputs are shown separately, so that we can easily determine which inputs are active. The **q** outputs are grouped so as to show the output code.

FIGURE 6.24   Example 6.9: Simulation Waveforms for a 3-Bit Priority Encoder

BCD Priority Encoder

A BCD priority encoder, illustrated in Figure 6.25, accepts ten inputs and generates a BCD code (0000 to 1001), corresponding to the highest-priority active input. The truth table for this circuit is shown in Table 6.10, with a simulation of the circuit shown in Figure 6.26.

Derivation of the BCD priority encoder equations and development of a VHDL description of the circuit are left as exercises in the end-of-chapter problems.

■ SECTION 6.2 REVIEW PROBLEM

6.4   State the main limitation of the 3-bit binary encoder shown in Figure 6.22. How can the encoder be modified to overcome this limitation?

HIPR/BCD

FIGURE 6.25 BCD Priority Encoder

TABLE 6.10 Truth Table of a BCD Priority Encoder

D_9	D_8	D_7	D_6	D_5	D_4	D_3	D_2	D_1	Q_3	Q_2	Q_1	Q_0
0	0	0	0	0	0	0	0	0	0	0	0	0
0	0	0	0	0	0	0	0	1	0	0	0	1
0	0	0	0	0	0	0	1	X	0	0	1	0
0	0	0	0	0	0	1	X	X	0	0	1	1
0	0	0	0	0	1	X	X	X	0	1	0	0
0	0	0	0	1	X	X	X	X	0	1	0	1
0	0	0	1	X	X	X	X	X	0	1	1	0
0	0	1	X	X	X	X	X	X	0	1	1	1
0	1	X	X	X	X	X	X	X	1	0	0	0
1	X	X	X	X	X	X	X	X	1	0	0	1

FIGURE 6.26 Simulation Waveforms for a BCD Priority Encoder

6.3 MULTIPLEXERS

■ **KEY TERMS**

Multiplexer A circuit that directs one of several digital signals to a single output, depending on the states of several select inputs.

Data Inputs The multiplexer inputs that feed a digital signal to the output when selected.

Select Inputs The multiplexer inputs that select a digital input channel.

Double-Subscript Notation A naming convention where two or more numerically related groups of signals are named using two subscript numerals. Generally, the first digit refers to a group of signals and the second to an element of a group. (e.g., X_{03} represents element 3 of group 0 for a set of signal groups, X.)

FIGURE 6.27 4-to-1 Multiplexer

A **multiplexer** (abbreviated MUX) is a device for switching one of several digital signals to an output, under the control of another set of binary inputs. The inputs to be switched are called the **data inputs;** those that determine which signal is directed to the output are called the **select inputs.**

Figure 6.27 shows the logic circuit for a 4-to-1 multiplexer, with data inputs labeled D_0 to D_3 and the select inputs labeled S_0 and S_1. By examining the circuit, we can see that the 4-to-1 MUX is described by the following Boolean equation:

$$Y = D_0 \overline{S}_1 \overline{S}_0 + D_1 \overline{S}_1 S_0 + D_2 S_1 \overline{S}_0 + D_3 S_1 S_0$$

For any given combination of $S_1 S_0$, only one of the previous four product terms will be enabled. For example, when $S_1 S_0 = 10$, the equation evaluates to:

$$Y = (D_0 \cdot 0) + (D_1 \cdot 0) + (D_2 \cdot 1) + (D_3 \cdot 0) = D_2$$

TABLE 6.11 4-to-1 MUX Truth Table

S_1	S_0	Y
0	0	D_0
0	1	D_1
1	0	D_2
1	1	D_3

The MUX equation can be described by a truth table as in Table 6.11. The subscript of the selected data input is the decimal equivalent of the binary combination $S_1 S_0$.

Figure 6.28 shows a symbol used for a 4-to-1 multiplexer.

In general, a multiplexer with n select inputs will have $m = 2^n$ data inputs. Thus, other common multiplexer sizes are 8-to-1 (for 3 select inputs) and 16-to-1 (for 4 select inputs). Data inputs can also be multiple-bit busses, as in Figure 6.29. The slash through a thick data line and the number 4 above the line indicate that it represents four related data signals. In this device, the select inputs switch groups of data inputs, as shown in the truth table in Table 6.12.

FIGURE 6.28 4-to-1 MUX Symbol Showing Individual Lines

FIGURE 6.29 4-to-1 4-Bit Bus Multiplexer

TABLE 6.12 Truth Table for a 4-to-1 4-bit Bus MUX

S_1	S_0	Y_3	Y_2	Y_1	Y_0
0	0	D_{03}	D_{02}	D_{01}	D_{00}
0	1	D_{13}	D_{12}	D_{11}	D_{10}
1	0	D_{23}	D_{22}	D_{21}	D_{20}
1	1	D_{33}	D_{32}	D_{31}	D_{30}

LAB SOURCE
*Lab #13
Multiplexers*

The naming convention shown in Table 6.12, known as **double-subscript notation,** is used frequently for identifying variables that are bundled in numerically related groups, the elements of which are themselves numbered. The first subscript identifies the group that a variable belongs to; the second subscript indicates which element of the group a variable represents.

Multiplexing of Time-Varying Signals

We can observe the function of a multiplexer by using time-varying waveforms, such as a series of digital pulses. If we apply a different digital signal to each data input, and step the select inputs through an increasing binary sequence, we can see the different input waveforms appear at the output in a predictable sequence. This can be used as a basis for creating a simulation of a multiplexer.

EXAMPLE 6.10

Derive a set of simulation criteria that will fully test the operation of the 4-to-1 multiplexer shown in Figure 6.27. Create a simulation in Quartus II, based on the criteria you derive.

■ Solution

SIMULATION CRITERIA

■ Each data input channel of the multiplexer will be selected in an ascending sequence by applying a binary count to the combined select inputs.

■ Each data input should be easily recognizable by having a "signature" waveform applied to it. Each channel should be selected for a period no less than about two or three cycles of the signature waveform.

■ The output waveform should display a series of unique signature waveforms, indicating the selection of the data channels in the correct sequence.

FIGURE 6.30   Simulation Waveforms for a 4-to-1 MUX

The simulation in Figure 6.30 is created in Quartus II by assigning a count value to the inputs as follows:

End Time:	100 μs
S[1..0]	25 μs
D3:	1 μs
D2:	2 μs
D1:	0.5 μs
D0:	4 μs

(continued)

In Figure 6.30, we initially see the D_0 waveform appearing at the Y output when $S_1S_0 = 00$, followed in sequence by the D_1, D_2, and D_3 waveforms when $S_1S_0 = 01, 10,$ and 11, respectively. The frequencies shown in the simulation were chosen to make as great a contrast as possible between adjacent inputs so that the different selected inputs could easily be seen. The select input count waveforms are set to allow three cycles of the longest waveform (D_0) to appear at Y when selected.

VHDL Implementation of Multiplexers

A multiplexer can be represented in Quartus II as a Block Diagram File, similar to the diagram of Figure 6.27, or in a hardware description language such as VHDL.

Several different VHDL constructs can be used to define a multiplexer. We can use a concurrent signal assignment statement, a selected signal assignment statement, or a CASE statement within a PROCESS. We will briefly look at each form for a 4-to-1 multiplexer. Later, you will be required to extend these constructs to larger multiplexer circuits.

Concurrent Signal Assignment

Recall that the concurrent signal assignment statement takes the form:

```
__signal <= __expression;
```

We can use this to encode the Boolean expression that describes a 4- to-1 MUX. The VHDL file that incorporates this statement is as follows.

mux4.vhd
mux4.vwf

```
-- mux4.vhd
-- 4-to-1 multiplexer
-- Directs one of four input signals (d0 to d3) to output,
--   depending on status of select bits (s1, s0).
ENTITY mux4 IS
PORT(
    d0, d1, d2, d3 : IN  BIT;
    s              : IN  BIT_VECTOR (1 downto 0);
    y              : OUT BIT);
END mux4;

ARCHITECTURE mux4to1 OF mux4 IS
BEGIN
-- Concurrent Signal Assignment
y <=  ((not s(1)) and (not s(0)) and d0)
   or ((not s(1)) and (    s(0)) and d1)
   or ((    s(1)) and (not s(0)) and d2)
   or ((    s(1)) and (    s(0)) and d3);
END mux4to1;
```

Though the concurrent signal assignment is fairly easy to derive, it becomes cumbersome for larger multiplexers, such as 8-to-1 or greater.

The entity declaration will be identical for the other VHDL examples except for the entity name. The only other change we will make will be to replace the concurrent signal assignment in the architecture body with some other VHDL construct.

Selected Signal Assignment Statement

This construct has the following form (the label is optional):

```
__label:
WITH __expression SELECT
__signal <= __expression WHEN __constant_value,
            __expression WHEN __constant_value,
            __expression WHEN __constant_value,
            __expression WHEN __constant_value;
```

The 4-to-1 MUX can be described in VHDL as follows, using a selected signal assignment:

mux4sel.vhd

```
ENTITY mux4sel IS
 PORT(
     d0, d1, d2, d3 : IN  BIT;
     s              : IN  BIT_VECTOR (1 downto 0);
     y              : OUT BIT);
END mux4sel;

ARCHITECTURE mux4to1 OF mux4sel IS
BEGIN
M: WITH s SELECT
    y  <=  d0 WHEN "00",
           d1 WHEN "01",
           d2 WHEN "10",
           d3 WHEN "11";
END mux4to1;
```

The selected signal assignment evaluates the expression in the WITH clause (in this case, the 2-bit vector, s) and, depending on its value, selects an expression to assign to y. Thus, if $s_1s_0 = 00$, $y = d_0$. If $s_1s_0 = 01$, then $y = d_1$, and so on for the remaining values of s_1s_0.

CASE Statement within a PROCESS

A CASE statement within a PROCESS has the following form (the label is optional):

```
__process_label:
PROCESS (sensitivity list)
BEGIN
  CASE __expression IS
        WHEN __constant_value =>
             __statement;
             __statement;
        WHEN __constant_value =>
             __statement;
             __statement;
        WHEN OTHERS =>
             __statement;
             __statement;

     END CASE;
  END PROCESS __process_label;
```

In our MUX example, we could use a CASE statement as follows:

mux4case.vhd

```
-- Define inputs and outputs
ENTITY mux4case IS
    PORT(
        d0, d1, d2, d3 : IN  BIT;                       -- data inputs
        s              : IN  BIT_VECTOR (1 downto 0); -- select inputs
        y              : OUT  BIT);
END mux4case;
```

```
ARCHITECTURE mux4to1 OF mux4case IS
BEGIN
    -- Monitor select inputs and execute if they change
    PROCESS(s)
    BEGIN
        CASE s IS
            WHEN "00" =>
                y   <= d0;
            WHEN "01" =>
                y   <= d1;
            WHEN "10" =>
                y   <= d2;
            WHEN "11" =>
                y   <= d3;
            WHEN others =>
                y   <= '0';
        END CASE;
    END PROCESS;
END mux4to1;
```

If the select inputs change, the PROCESS statements are executed. The CASE statement evaluates the select input vector, *s,* and chooses a signal assignment based on its value. It is good design practice to include a default case (the "others" clause) even when there are no obvious other cases. A default case is essential when using STD_LOGIC types rather than BIT types, as '0' and '1' values do not cover all possible cases for STD_LOGIC signals. (Recall from Chapter 5 that STD_LOGIC is a nine-valued logic type, incorporating such things as "Don't Care" ('-'), "Unknown" ('X'), and "High-Impedance" ('Z'), as well as '0' and '1'.)

Multiplexer Applications

→ LAB SOURCE
Lab #14
Multiplexer Applications

Multiplexers are used for a variety of applications, including selection of one data stream out of several choices, switching multiple-bit data from several channels to one multiple-bit output, sharing data on one output over time, and generating bit patterns or waveforms.

Single-Channel Data Selection

The simplest way to use a multiplexer is to switch the select inputs manually in order to direct one data source to the MUX output. Example 6.11 shows a pair of single-pole single-throw (SPST) switches supplying the select input logic for this type of application.

EXAMPLE 6.11

Figure 6.31 shows a digital audio switching system. The system shown can select a signal from one of four sources (compact disc [CD] players, labeled CD_0 to CD_3) and direct it to a digital signal processor (DSP) at its output. We assume we have direct access to the audio signals in digital form.

Make a table listing which digital audio source in Figure 6.31 is routed to the DSP for each combination of the multiplexer select inputs, S_1 and S_0.

FIGURE 6.31 Example 6.11: Single-Channel Data Selection

■ **Solution**

TABLE 6.13 Sources Selected by a 4-to-1 MUX in Figure 6.31

S_1	S_0	Selected Input	Selected Source
0	0	D_0	CD_0
0	1	D_1	CD_1
1	0	D_2	CD_2
1	1	D_3	CD_3

Multi-Channel Data Selection

Example 6.11 assumes that the output of a multiplexer is a single bit or stream of bits. Some applications require several bits to be selected in parallel, such as when data would be represented on a numerical display.

Figure 6.32 shows a circuit, based on a quadruple (4-channel) 2-to-1 multiplexer, that will direct one of two BCD digits to a seven-segment display. The bits $D_{03}D_{02}D_{01}D_{00}$ act as a 4-bit group input since the first digit of all four subscripts is 0. When the MUX select input (S) is 0, these inputs are all connected to the outputs $Y_3Y_2Y_1Y_0$. Similarly, when the select input is 1, inputs $D_{13}D_{12}D_{11}D_{10}$ are connected to the Y outputs.

The seven-segment display in Figure 6.32 will display "4" if $S = 0$ (D_0 inputs selected) and "9" if $S = 1$ (D_1 inputs selected).

FIGURE 6.32   Quadruple 2-to-1 MUX as a Digital Output Selector

EXAMPLE 6.12

Draw the symbol for a multiplexer that will select one of four 4-bit channels and direct it to a 4-bit output. Create a VHDL file that implements this function and derive simulation criteria showing the operation of the device. Create the simulation waveforms in Quartus II.

■ Solution

Figure 6.33 shows the symbol for the 4-channel, 4-bit multiplexer. This symbol is shown with the data inputs and outputs in bus form. The data inputs are labeled in groups D_0 to D_3, which contain the individual inputs. $D_0[3..0]$ to $D_3[3..0]$

A VHDL file describing this function is as follows:

FIGURE 6.33   Example 6.12:
4-Channel 4-Bit MUX

quad4to1.vhd
quad4to1.vwf

```
-- quad4to1.vhd

ENTITY quad4to1 IS
 PORT(
      s  : IN   INTEGER RANGE 0 to 3;
      d0 : IN   BIT_VECTOR (3 downto 0);
      d1 : IN   BIT_VECTOR (3 downto 0);
      d2 : IN   BIT_VECTOR (3 downto 0);
      d3 : IN   BIT_VECTOR (3 downto 0);
      y  : OUT BIT_VECTOR (3 downto 0));
END quad4to1;

ARCHITECTURE mux4 OF quad4to1 IS
BEGIN
 --  Selected Signal Assignment
MUX4:   WITH s SELECT
      y  <=  d0 WHEN 0,
             d1 WHEN 1,
             d2 WHEN 2,
             d3 WHEN 3;
END mux4;
```

SIMULATION CRITERIA

- An increasing binary count applied to the select inputs should send each of the 4-bit data inputs to the output in the sequence $d0[3..0]$ to $d3[3..0]$.
- Each data input should have a recognizable 4-bit value. Constant hexadecimal values on the data inputs allow us to easily see the data change.

Figure 6.34 shows a set of simulation waveforms for the multiplexer. The D inputs are shown in groups of four, the value of each shown as a steady hexadecimal value. The select inputs are grouped, showing an increasing 2-bit binary count. As the S inputs select each group of D inputs, their combined value is directed to the Y output group.

FIGURE 6.34   Example 6.12: Simulation Waveforms for a Quadruple 4-to-1 MUX

Time-Dependent Multiplexer Applications

■ KEY TERMS

Counter   A digital circuit whose output produces a fixed sequence of binary states when an input called the clock receives a series of pulses. The output advances by one for each clock pulse (e.g., the output state of a 4-bit binary counter progresses in order from 0000 to 1111, then repeats).

Clock   A signal that controls the operation of a sequential digital circuit, such as a counter, by advancing the counter's outputs to the next state when it receives a clock pulse.

Positive Edge   The point on a digital waveform where the logic level of the waveform makes a LOW-to-HIGH transition.

A time-dependent multiplexer application is one that uses the MUX input channels one after the other in a repeating time sequence. We can create such an application by applying a set of changing binary signals to the MUX select inputs. For this function, we can use a 3-bit binary **counter** to generate a binary sequence that goes from 000 to 111 (8 states) and repeats indefinitely, the outputs advancing by one with every pulse applied to the **clock** input of the counter.

Figure 6.35 shows the timing diagram of a 3-bit counter. The outputs $Q_2Q_1Q_0$ change every time the clock signal makes a transition from LOW to HIGH. If you read the Q waveforms from bottom to top, you will see that they generate a repeating binary sequence (000, 001, 010, 011, 100, 101, 110, 111, 000 . . .).

FIGURE 6.35 Timing Diagram of a 3-Bit Counter

FIGURE 6.36 Time-Dependent Selection of Eight Multiplexer Channels

If we connect the counter outputs $Q_2Q_1Q_0$ to the select inputs of an 8-to-1 MUX, as in Figure 6.36, we will select the channels in sequence, one after the other. The counter is labelled CTR DIV 8 because its most significant bit output has a frequency equal to the clock frequency divided by eight. The triangle on the clock input indicates that it is active when the clock waveform makes a transition from one logic level to another. Since there is no inverting bubble on the clock input, we know that the active clock transition is from LOW to HIGH (i.e., a **positive edge**). We say that the counter is "positive-edge triggered."

Waveform Generation   A multiplexer and counter can be used as a programmable waveform generator. The output waveform can be programmed to any pattern by switching the logic levels on the data inputs. This is an easy way to generate an asymmetrical waveform, a task which is more complicated using other digital circuits. The circuit can also generate symmetrical waveforms by alternating the logic levels of consecutive groups of inputs.

EXAMPLE 6.13

Draw a circuit that uses an 8-to-1 multiplexer to generate a programmable 8-bit repeating pattern. Draw the timing diagram of the select inputs and the output waveform for the following pattern of data inputs.

D_7	D_6	D_5	D_4	D_3	D_2	D_1	D_0
1	0	1	0	0	1	1	0

■ **Solution**

Figure 6.37a shows the waveform generator circuit. The output waveform with respect to the counter inputs is shown in Figure 6.37b. This pattern is relatively difficult to generate by other means since it has several unequal HIGH and LOW sequences in one period. Note that the waveform comes out "backwards" from the defined input value, since D_0 is output first and D_7 is output last.

The waveform can easily be changed by changing the settings of the switches applied to the D inputs.

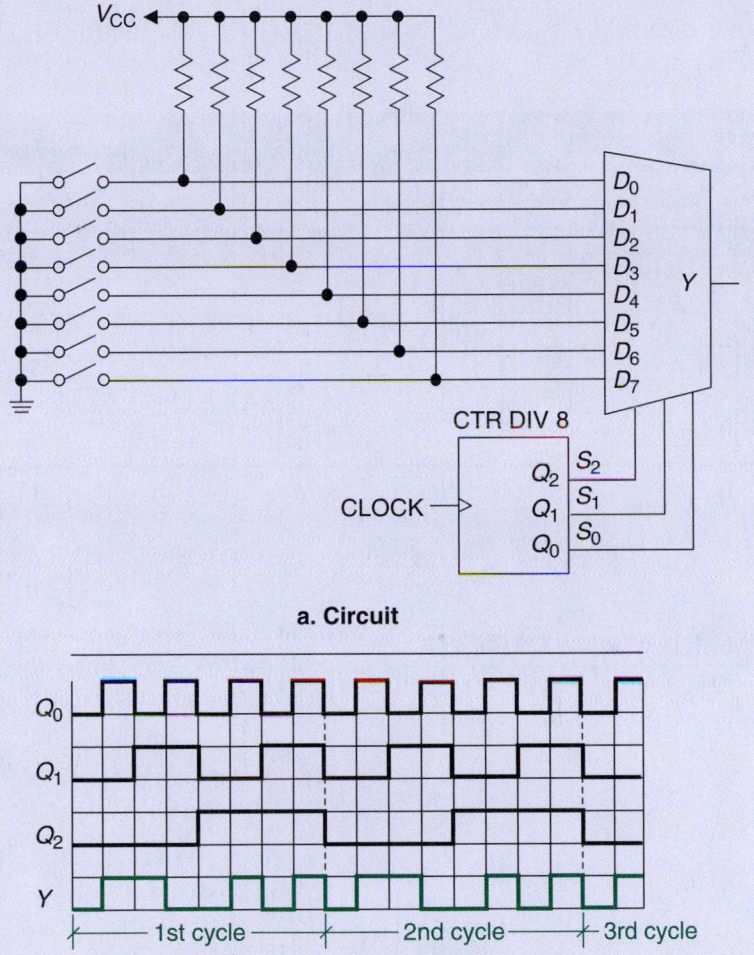

a. Circuit

b. Timing diagram

FIGURE 6.37　Example 6.13: Programmable Waveform Generator

EXAMPLE 6.14

Use the circuit and timing diagrams in Figure 6.37 as a basis for a set of simulation criteria for the MUX-based programmable waveform generator of Example 6.13. Use Quartus II to create a simulation based on these criteria.

■ **Solution**

Figure 6.38a shows a circuit drawn in the Quartus II Block Editor for the programmable waveform generator. Switches connected to the **d** inputs would be external to this design.

FIGURE 6.38A   Example 6.14: Block Diagram File Showing a MUX-Based Waveform Generator

FIGURE 6.38B   Example 6.14: Simulation Waveforms for a MUX-Based Waveform Generator (Arbitrary Waveform)

SIMULATION CRITERIA

- An increasing binary count applied to the select inputs should send each of the constant 1-bit data inputs to the output in the sequence $d0$ to $d7$. The count can be generated automatically by applying a series of clock pulses to the counter.

- Expected output: series of 1s and 0s as determined by the reverse order of $d[7..0]$, each lasting for the period of one clock cycle.
- Simulator settings:

  End Time: 100 μs
  Clock period: 4 μs
  d inputs: 10100110

waveform_generator.bdf
waveform_generator.vwf

Figure 6.38b shows a simulation of the waveform generator with three cycles of the output waveform. The **q** outputs of the counter are shown in hexadecimal for easy reference.

EXAMPLE 6.15

The programmable waveform generator in Figure 6.37 generates a symmetrical pulse waveform having a frequency of 125 kHz when the data inputs are set as follows.

D_7	D_6	D_5	D_4	D_3	D_2	D_1	D_0
1	1	1	1	0	0	0	0

How should the switches be set to generate a symmetrical 250 kHz waveform? A symmetrical 500 kHz waveform? Show the results of all three sets of input on the simulation created for example 6.14.

■ **Solution**

Pattern for 250 kHz:

D_7	D_6	D_5	D_4	D_3	D_2	D_1	D_0
1	1	0	0	1	1	0	0

Pattern for 500 kHz:

D_7	D_6	D_5	D_4	D_3	D_2	D_1	D_0
1	0	1	0	1	0	1	0

Figure 6.39 shows the simulation for the three sets of waveforms. The settings of the vector waveform file are:

End time: 25 μs
Clock frequency: 1 MHz

a. f = 125 kHz

a. f = 250 kHz

FIGURE 6.39 Example 6.15: Simulation of MUX-Based Programmable Waveform Generator

(continued)

a. f = 500 kHz

FIGURE 6.39   Example 6.15 *(continued)*

Time Division Multiplexing

> ### ■ KEY TERMS
>
> **Time Division Multiplexing (TDM)**   A technique of using one transmission line to send many signals simultaneously by making them share the line for equal fractions of time.
>
> **Time Slot**   A period of time during which a transmitted data element has sole access to a transmission path.
>
> **Bit Multiplexing**   A TDM technique in which one bit is sent from each channel during the channel's assigned time slot.
>
> **Byte (or Word) Multiplexing**   A TDM technique in which a byte (or word) is sent from each channel during its assigned time slot. (A byte is eight bits; a word is a group of bits whose size varies with the particular system.)

Time division multiplexing (TDM) is a method of improving the efficiency of a transmission system by sharing one transmission path among many signals. For example, if we wish to send four 4-bit numbers over a single transmission line, we can transmit the bits one after the other, as shown in Figure 6.40.

FIGURE 6.40 4×4 Data Stream (Bit Multiplexing)

FIGURE 6.41   4 × 4 Data Stream (Word Multiplexing)

In Figure 6.40, we see the least significant bit of the 4-bit word $p0$ transmitted, followed by the LSB of $p1$, $p2$, then $p3$. After that, the second bit of each word is transmitted in sequence, then all the third bits, and finally, all MSBs in sequence. Each bit is assigned a **time slot** in the sequence. During that time, the bit has sole access to the transmission line. When its time elapses, the next bit is sent and so on in sequence, until the channel assignment returns to the original location. This technique, known as **bit multiplexing,** can be implemented by a circuit similar to the waveform generator shown in Figure 6.37. Rather than fixed switch inputs, the data inputs would be some data source, such as a digitized audio signal.

We can also arrange our circuit so that one byte (8 bits) or one word (a group of bits) is sent through a selected channel. In this case, we must keep the channel selected for enough clock pulses to transmit the byte or word, then move to the next one. This technique is called **byte (or word) multiplexing.** Figure 6.41 shows a data stream of four 4-bit words that are word-multiplexed down a data transmission path.

Telephone companies use TDM to maximize the use of their phone lines. Speech or data is digitally encoded for transmission. Each speech or data channel becomes a multiplexer data input which shares time with all other channels on a single phone line. A counter on the MUX selects the speech channels one after the other in a continuous sequence. The counter must switch the channels fast enough so that there is no apparent interruption of the transmitted conversation or data stream.

EXAMPLE 6.16

Draw a diagram of a circuit that uses an 8-to-1 multiplexer to share one telephone line among eight digitized speech channels.

Write a VHDL file for the multiplexer and create a simulation to show its operation.

■ **Solution**

Figure 6.42 shows the required multiplexer circuit. Each channel is connected to a data input and a 3-bit binary counter is connected to the select inputs.

(continued)

FIGURE 6.42 Example 6.16: Time-Division Multiplexing of Telephone Channels

The VHDL code for the multiplexer is:

```
ENTITY mux_8ch IS
  PORT(
    sel : IN  BIT_VECTOR (2 downto 0);
    d   : IN  BIT_VECTOR (7 downto 0);
    y   : OUT BIT);
END mux_8ch;

ARCHITECTURE a OF mux_8ch IS
BEGIN
-- Selected Signal Assignment
MUX8:   WITH sel SELECT
    y  <=  d(0) WHEN "000",
           d(1) WHEN "001",
           d(2) WHEN "010",
           d(3) WHEN "011",
           d(4) WHEN "100",
           d(5) WHEN "101",
           d(6) WHEN "110",
           d(7) WHEN "111";
END a;
```

Quartus II:
mux_8ch.vhd
mux_8ch.vwf
Max+PLUS II:
mux_8ch.vhd
mux_8ch.scf

The simulation is shown in Figure 6.43. The output shows the channel data in sequence, starting with channel 0.

FIGURE 6.43 Example 6.16: Simulation of an 8-Channel Time Division Multiplexer

■ **SECTION 6.3 REVIEW PROBLEM**

6.5 What defines whether a multiplexer application is time-dependent or not? What additional component can be added to make a MUX application time-dependent?

6.4 DEMULTIPLEXERS

> ■ **KEY TERM**
>
> **Demultiplexer** A circuit that uses a binary decoder to direct a digital signal from a single source to one of several destinations.

A **demultiplexer** performs the reverse function of a multiplexer. A multiplexer (MUX) directs one of several input signals to a single output; a demultiplexer (DMUX) directs a single input signal to one of several outputs. In both cases, the selected input or output is chosen by the state of an internal decoder.

Figure 6.44 shows the logic circuit for a 1-to-4 demultiplexer. Compare this to Figure 6.5, a 4-output decoder. These circuits are the same except that the active-LOW enable input has been changed to an active-HIGH data input. The circuit in Figure 6.44 could still be used as a decoder, except that its enable input would be active-HIGH.

Each AND gate in the demultiplexer enables or inhibits the signal output according to the state of the select inputs, thus directing the data to one of the output lines. For instance, $S_1 S_0 = 10$ directs incoming digital data to output Y_2.

Figure 6.45 illustrates the use of a single device as either a decoder or a demultiplexer. In Figure 6.45a, input D is tied HIGH. When an output is selected by S_1 and S_0, it goes HIGH, acting as a decoder with active-HIGH outputs. In Figure 6.45b, D acts as a demultiplexer data input. The data are directed to the output selected by S_1 and S_0.

FIGURE 6.44   4-Bit Decoder/Demultiplexer

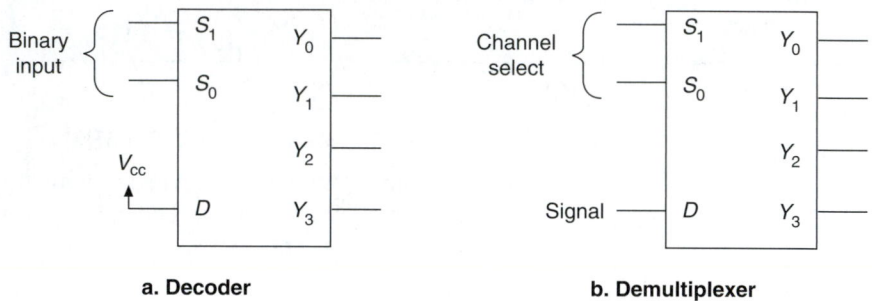

a. Decoder   **b. Demultiplexer**

FIGURE 6.45   Same Device Used as a Decoder or Demultiplexer

➡ **LAB SOURCE**
Lab #15
Demultiplexers

> **NOTE . . .**
>
> Since a single device can be used either way, this implies that any of the
> VHDL binary decoder designs used in this chapter can also be used as demul-
> tiplexers, provided they include an enable input.

A decoder/demultiplexer can have active-LOW outputs, but only if the D input
is also active-LOW. This is important because the demultiplexer data must be
inverted twice to retain its original logic values.

Demultiplexing a TDM Signal

In Example 6.16, we saw how a multiplexer could be used to send 8 digital channels across a single line, multiplexed over time. Obviously, such a system is not
of much value if the signals cannot be sorted out at the receiving end. The
received digital data must be demultiplexed and sent to their appropriate
destinations.

The process is the reverse of multiplexing; data are sent to an output selected by
a counter at the DMUX select inputs. (We assume that the counters at the MUX and
DMUX select inputs are somehow synchronized or possibly, if located close together,
are the same counter.)

EXAMPLE 6.17

Draw a demultiplexing circuit that will take the multiplexed output of the circuit in Figure 6.42 and distribute it to 8 different local telephone circuits. Write a VHDL file for the demultiplexer circuit and specify a set of simulation criteria that would fully test its operation. Use the criteria to make a simulation in Quartus II. Use active-LOW outputs for the demultiplexer. How does this affect the outputs when they are not transmitting data?

■ Solution

Figure 6.46 shows the original multiplexing circuit connecting to the new demultiplexing circuit. The diagram indicates that the two sides of the circuit are separated by some distance. The clock is shared between both sides of the circuit, but is generated on the MUX side. Both sides share a common ground. Each side of the circuit has its own 3-bit counter.

FIGURE 6.46 Example 6.17: Time-Division Multiplexing and Demultiplexing

The VHDL code for the demultiplexer is as follows. (This is the same implementation as a 3-line-to-8-line decoder with an enable input.)

Quartus II:
dmux8.vhd
dmux8.vwf
MAX+PLUS II:
dmux8.vhd
dmux8.scf

```
-- dmux8.vhd
-- 1-to-8 demultiplexer/decoder
-- Decoder: set d to '0'; outputs are activated
-- by binary combination of s.
-- Demultiplexer: apply data stream to d; data directed to
-- y output with subscript same as value of s.
-- Outputs and d are active-LOW. DMUX data are inverted
-- twice to keep them true.
```

(continued)

```
LIBRARY ieee;
USE ieee.std_logic_1164.ALL;

ENTITY dmux8 IS
    PORT (
            s  : IN  STD_LOGIC_VECTOR(2 downto 0);
            d  : IN  STD_LOGIC;
            y  : OUT STD_LOGIC_VECTOR(0 to 7));
END dmux8;

ARCHITECTURE a OF dmux8 IS
    SIGNAL inputs : STD_LOGIC_VECTOR(3 downto 0);
BEGIN
    inputs <= d & s;

    WITH inputs SELECT
        y   <=  "01111111" WHEN "0000",
                "10111111" WHEN "0001",
                "11011111" WHEN "0010",
                "11101111" WHEN "0011",
                "11110111" WHEN "0100",
                "11111011" WHEN "0101",
                "11111101" WHEN "0110",
                "11111110" WHEN "0111",
                "11111111" WHEN others;

End a;
```

SIMULATION CRITERIA

- An input data stream will be split into 8 output data streams. Each channel is selected by a specified value on the select inputs of the demultiplexer.
- Each output data stream should have a recognizable signature waveform to distinguish it from other streams. This implies that the signature of each stream should be built in to the input stream.
- The input stream can be derived from the output of the 8-channel multiplexer of Example 6.16.
- An increasing count on the select inputs will cause the output channels to be selected in sequence from 0 to 7.

FIGURE 6.47   Example 6.17: Simulation Waveforms for an 8-Channel Demultiplexer

The simulation, shown in Figure 6.47, has as its input data the output of the original MUX simulation in Figure 6.43. Data are distributed to the outputs in sequence. Compare the DMUX output data to the MUX input data in Figure 6.43. Note that idle channels sit HIGH. This is opposite from the status of the idle MUX lines and may affect circuit operation. If so, a DMUX with active-HIGH outputs and active-HIGH enable should be used.

CMOS Analog Multiplexer/Demultiplexer

> ■ **KEY TERM**
>
> **CMOS Analog Switch**   A CMOS device that will pass an analog or digital signal in either direction, when enabled. Also called a transmission gate. There is no TTL equivalent.

An interesting device used in some CMOS medium-scale integration multiplexers and demultiplexers, as well as other applications, is the **CMOS analog switch,** or transmission gate. This device has the property of allowing signals to pass in two directions, instead of only one, thus allowing both positive and negative voltages and currents to pass. It also has no requirement that the voltages be of a specific value such as +5 volts. These properties make the device suitable for passing analog signals.

Figure 6.48 shows several symbols, indicating the development of the transmission gate concept. Figure 6.48a and Figure 6.48b show amplifiers whose output and input are clearly defined by the direction of the triangular amplifier symbol. A signal has one possible direction of flow. Figure 6.48 includes an active-LOW gating input, which can turn the signal on and off.

Figure 6.48c shows two opposite-direction overlapping amplifier symbols, with a gating input to enable or inhibit the bidirectional signal flow. The signal through the transmission gate may be either analog or digital.

Analog switches are available in packages of four switches with part numbers such as 4066B (standard CMOS) or 74HC4066 (high-speed CMOS).

Several available CMOS MUX/DMUX chips use analog switches to send signals in either direction. Figure 6.49 illustrates the design principle as applied to a 4-channel MUX/DMUX.

If four signals are to be multiplexed, they are connected to inputs D_0 to D_3. The decoder, activated by S_1 and S_0, selects which one of the four switches is enabled. Figure 6.49 shows Channel 2 active ($S_1S_0 = 10$).

Since all analog switch outputs are connected together, any selected channel connects to Y, resulting in a multiplexed output. To use the circuit in Figure 6.49 as a demultiplexer, the inputs and outputs are merely reversed.

FIGURE 6.48   Line Drivers

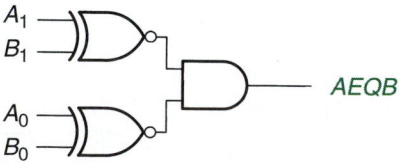

FIGURE 6.49   4-Channel CMOS MUX/DMUX

Some analog MUX/DMUX devices in high-speed CMOS include: 74HC4051 8-channel MUX/DMUX, 74HC4052 dual 4-channel MUX/DMUX, and 74HC4053 triple 2-channel MUX/DMUX.

■ SECTION 6.4 REVIEW PROBLEM

6.6   Refer to the symbol for the 74138 3-line-to-8-line enable in Figure 6.7. Show how to connect the inputs of this device so that it can be used as a 1-to-8 demultiplexer.

6.5 MAGNITUDE COMPARATORS

FIGURE 6.50   Exclusive NOR Gate

TABLE 6.14   XNOR Truth Table

A	B	$AEQB$
0	0	1
0	1	0
1	0	0
1	1	1

■ KEY TERM

Magnitude Comparator   A circuit that compares two n-bit binary numbers, indicates whether or not the numbers are equal, and, if not, which one is larger.

If we are interested in finding out whether or not two binary numbers are the same, we can use a **magnitude comparator.** The simplest comparison circuit is the Exclusive NOR gate, whose circuit symbol is shown in Figure 6.50 and whose truth table is given in Table 6.14.

The output of the XNOR gate is 1 if its inputs are the same ($A = B$, symbolized $AEQB$) and 0 if they are different. For this reason, the XNOR gate is sometimes called a coincidence gate.

We can use several XNORs to compare each bit of two multi-bit binary numbers. Figure 6.51 shows a 2-bit comparator with one output that goes HIGH if all bits of A and B are identical.

FIGURE 6.51   2-Bit Magnitude Comparator

If the most significant bit (MSB) of A equals the MSB of B, the output of the upper XNOR is HIGH. If the least significant bits (LSBs) are the same, the output of the lower XNOR is HIGH. If both these conditions are satisfied, then

FIGURE 6.52 2-Bit Comparator with *AEQB*, *AGTB*, and *ALTB* Outputs

$A = B$, which is indicated by a HIGH at the AND output. This general principle applies to any number of bits:

$$AEQB = (\overline{A_{n-1} \oplus B_{n-1}}) \cdot (\overline{A_{n-1} \oplus B_{n-2}}) \dots (\overline{A_1 \oplus B_1}) \cdot (\overline{A_0 \oplus B_0})$$

for two n-bit numbers, A and B.

Some magnitude comparators also include an output that activates if A is greater than B (symbolized $A > B$ or $AGTB$) and another that is active when A is less than B (symbolized $A < B$ or $ALTB$). Figure 6.52 shows the comparator of Figure 6.51 expanded to include the "greater than" and "less than" functions.

Let us analyze the $AGTB$ circuit. The $AGTB$ function has two AND-shaped gates that compare A and B bit-by-bit to see which is larger.

1. The 2-input gate examines the MSBs of A and B. If $A_1 = 1$ AND $B_1 = 0$, then we know that $A > B$. (This implies one of the following inequalities: $10 > 00$; $10 > 01$; $11 > 00$; or $11 > 01$.)

2. If $A_1 = B_1$, then we don't know whether or not $A > B$ until we compare the next most significant bits, A_0 and B_0. The 3-input gate makes this comparison. Since this gate is enabled by the XNOR, which compares the two MSBs, it is only active when $A_1 = B_1$. This yields the term $(\overline{A_1 \oplus B_1})A_0\overline{B_0}$ in the Boolean expression for the $AGTB$ function.

3. If $A_1 = B_1$ AND $A_0 = 1$ AND $B_0 = 0$, then the 3-input gate has a HIGH output, telling us, via the OR gate, that $A > B$. (The only possibilities are $(01 > 00)$ and $(11 > 10)$.)

Similar logic works in the $ALTB$ circuit, except that inversion is on the A, rather than the B bits. Alternatively, we can simplify either the $AGTB$ or the $ALTB$ function by using a NOR function. For instance, if we have developed a circuit to indicate $AEQB$ and $ALTB$, we can make the $AGTB$ function from the other two, as follows:

$$AGTB = \overline{AEQB + ALTB}$$

This Boolean expression implies that if A is not equal to or less than B, then it must be greater than B.

Figure 6.53 shows a 4-bit comparator with $AEQB$, $ALTB$, and $AGTB$ outputs.

FIGURE 6.53 4-Bit Magnitude Comparator

The Boolean expressions for the outputs are:

$$AEQB = \overline{(A_3 \oplus B_3)}\,\overline{(A_2 \oplus B_2)}\,\overline{(A_1 \oplus B_1)}\,\overline{(A_0 \oplus B_0)}$$

$$ALTB = \overline{A_3}B_3 + \overline{(A_3 \oplus B_3)}\overline{A_2}\,B_2 + \overline{(A_3 \oplus B_3)}\,\overline{(A_2 \oplus B_2)}\overline{A_1}B_1 + \overline{(A_3 \oplus B_3)}$$
$$\overline{(A_2 \oplus B_2)}\,\overline{(A_1 \oplus B_1)}\overline{A_0}B_0$$

$$AGTB = \overline{AEQB + ALTB}$$

This comparison technique can be expanded to as many bits as necessary. A 4-bit comparator requires four AND-shaped gates for its *ALTB* function. We can interpret the Boolean expression for this function as follows.

$A < B$ if:

1. The MSB of A is less than the MSB of B, OR
2. The MSBs are equal, but the second bit of A is less than the second bit of B, OR
3. The first two bits are equal, but the third bit of A is less than the third bit of B, OR
4. The first three bits are equal, but the LSB of A is less than the LSB of B

Expansion to more bits would use the same principle of comparing bits one at a time, beginning with the MSBs.

EXAMPLE 6.18

A digital thermometer has two input probes. A circuit in the thermometer converts the measured temperature at each probe to an 8-bit number, as shown by the block in Figure 6.54.

FIGURE 6.54 Example 6.18: Two-Channel Digital Thermometer

In addition to measuring the temperature at each input, the thermometer has a comparison function that indicates whether the temperature at one input is greater than, equal to, or less than the temperature at the other input.

Draw a logic diagram showing how a magnitude comparator could be connected to light a green LED for *AGTB*, an amber LED for *AEQB*, and a red LED for *ALTB*.

■ **Solution**

Figure 6.55 shows the logic diagram of the magnitude comparator connected to the thermometer's digital output.

FIGURE 6.55 Example 6.18: Temperature Comparator Block Diagram

When one of the comparator outputs goes HIGH, it sets the output of the corresponding inverter LOW. This provides a current path to ground for the indicator LED for that output, causing it to illuminate.

VHDL Magnitude Comparators

The most obvious way to create a VHDL representation of a magnitude comparator is to use a concurrent signal assignment statement for each comparing function. For example, the following VHDL code can represent the 2-bit magnitude comparator of Figure 6.51:

```
-- compare2.vhd
ENTITY compare2 IS
 PORT(
     a, b               : IN  BIT_VECTOR (1 downto 0);
     agtb, aeqb, altb   : OUT BIT);
END compare2;

ARCHITECTURE a OF compare2 IS
BEGIN
     altb <= (not (a(1)) and b(1))
          or ((not (a(1) xor b(1))) and (not (a(0) and b(0)));
     aeqb <= (not (a(1) xor b(1))) and (not (a(0) xor b(0)));
     agtb <= (a(1) and not (b(1)))
          or ((not (a(1) xor b(1))) and (a(0) and not (b(0))));
END a;
```

A simulation for this file is shown in Figure 6.56. The comparison outputs go HIGH to indicate $A = B$, $A < B$, or $A > B$.

Although this approach works, it is not a very good one. Due to the complexity of the Boolean equations for *ALTB* and *AGTB*, it is difficult to type them without making errors. (Try it!) The difficulty increases greatly with the number of required inputs.

The following code for a 4-bit comparator illustrates a much more efficient method. Since VHDL allows inputs to be represented as integers, we can define the required size of inputs A and B and compare them using IF statements. For every comparison, we assign an output vector consisting of bits for *ALTB*, *AEQB*, and *AGTB* one of the values 110, 101, or 011, for active-LOW outputs. For example, if $A = 12$ and $B = 9$, then the output vector would be 011 (i.e., $A > B$). An active-LOW output will illuminate a LOW-sense LED, such as those on the Altera UP-2 board. For boards with active-HIGH outputs, such as the RSR PLDT-2 the vectors will have opposite values: 001, 010, or 100.

FIGURE 6.56   Simulation Waveforms for a 2-Bit Magnitude Comparator

```
-- compare4.vhd
LIBRARY ieee;
USE ieee.std_logic_1164.ALL;

ENTITY compare4 IS
    PORT(
          a, b                : IN  INTEGER RANGE 0 TO 15;
          agtb, aeqb, altb : OUT STD_LOGIC);
END compare4;

ARCHITECTURE a OF compare4 IS
    SIGNAL compare : STD_LOGIC_VECTOR (2 downto 0);
BEGIN
    PROCESS (a,b)
    BEGIN
        IF a < b THEN
            compare    <=  "110";
        ELSIF a=b THEN
            compare    <=  "101";
        ELSIF a > b THEN
            compare    <=  "011";
        ELSE
            compare    <=  "111";
        END IF;
        agtb   <=  compare(2);
        aeqb   <=  compare(1);
        altb   <=  compare(0);
    END PROCESS;
END a;
```

Quartus II:
compare4.vhd
compare4.vwf
MAX+PLUS II:
compare4.vhd
compare4.scf

The beauty of this method is that the number of input bits can be changed by modifying one number: the range of the INTEGER-type input. For example, a 12-bit comparator is identical to the 4-bit comparator in the previous VHDL code, except that the inputs have a range of 0 to 4095 (= $2^{12} - 1$). Using this method, we can program an EPM7128S CPLD with a comparator up to 28 bits wide (range of 0 to 268,435,455). If we do, however, there is no room for anything else.

An alternative way to code a multibit comparator in VHDL is to use STD_LOGIC types for *A* and *B*. However, in order to compare two values of type STD_LOGIC, we must treat them as if they were unsigned integer values. This requires us to include the **std_logic_unsigned** package, which is included in the **ieee** library. The VHDL code for this variation of the 4-bit comparator is as follows:

```
LIBRARY ieee;
USE ieee.std_logic_1164.ALL;
USE ieee.std_logic_unsigned.ALL;

ENTITY compare4a IS
    PORT(
          a, b                : IN  STD_LOGIC_VECTOR(3 downto 0);
          agtb, aeqb, altb : OUT STD_LOGIC);
END compare4a;

ARCHITECTURE a OF compare4a IS
    SIGNAL compare    : STD_LOGIC_VECTOR(2 downto 0);
BEGIN
    PROCESS (a,b)
    BEGIN
        IF a<b THEN
            compare <=  "110";
```

compare4a.vhd
compare4a.vwf

```
        ELSIF a=b THEN
                compare   <=   "101";
        ELSIF a>b THEN
                compare   <=   "011";
        ELSE
                compare   <=   "111";
        END IF;
        agtb  <=  compare (2);
        aeqb  <=  compare (1);
        altb  <=  compare (0);
    END PROCESS;
END a;
```

The advantage of using this format is that, for large comparators, we do not need to calculate·the power of two representing the maximum size of the comparator inputs. For example, instead of writing a, b : IN INTEGER RANGE 0 to 268435455; we can write a,b : IN STD_LOGIC_VECTOR (27 downto 0); In all likelihood, we will not have the power-of-two value at our fingertips for a large number such as 2^{28}, so it is easier and more transparent to use the STD_LOGIC types.

Even for smaller comparators, it may be sensible to use the STD_LOGIC format, as we are often more interested in the number of input bits than the binary values they can represent.

EXAMPLE 6.19

Write a VHDL file that uses IF statements to compare two 8-bit numbers *A* and *B*. The design should have outputs for *AEQB*, *ALTB*, and *AGTB*. Use INTEGER types for *A* and *B*.

■ Solution

compare8.vhd

```
-- compare8.vhd
LIBRARY ieee;
USE ieee.std_logic_1164.ALL;

ENTITY compare8 IS
   PORT(
       a, b              : IN  INTEGER RANGE 0 TO 255;
       agtb, aeqb, altb : OUT STD_LOGIC);
END compare8;

ARCHITECTURE a OF compare8 IS
   SIGNAL compare : STD_LOGIC_VECTOR (2 downto 0);
BEGIN
   PROCESS (a,b)
   BEGIN
     IF a<b THEN
         compare     <=   "110";
     ELSIF a=b THEN
         compare     <=   "101";
     ELSIF a>b THEN
         compare     <=   "011";
     ELSE
         compare     <=   "111";
     END IF;
     agtb  <= compare(2);
     aeqb  <= compare(1);
     altb  <= compare(0);
   END PROCESS;
END a;
```

6.7 Write the Boolean equations for the *AEQB*, *ALTB*, and *AGTB* outputs of a 3-bit magnitude comparator.

6.6 PARITY GENERATORS AND CHECKERS

■ **KEY TERMS**

Parity A system that checks for errors in a multi-bit binary number by counting the number of 1s.

Parity Bit A bit appended to a binary number to make the number of 1s even or odd, depending on the type of parity.

Even Parity An error-checking system that requires a binary number to have an even number of 1s.

Odd Parity An error-checking system that requires a binary number to have an odd number of 1s.

When data are transmitted from one device to another, it is necessary to have a system of checking for errors in transmission. These errors, which appear as incorrect bits, occur as a result of electrical limitations such as line capacitance or induced noise.

Parity error checking is a way of encoding information about the correctness of data before they are transmitted. The data can then be verified at the system's receiving end. Figure 6.57 shows a block diagram of a parity error-checking system.

The parity generator in Figure 6.57 examines the outgoing data and adds a bit called the **parity bit** that makes the number of 1s in the transmitted data odd or even, depending on the type of parity. Data with **EVEN parity** have an even number of 1s, including the parity bit, and data with **ODD parity** have an odd number of 1s.

The data receiver "knows" whether to expect EVEN or ODD parity. If the incoming number of 1s matches the expected parity, the parity checker responds by indicating that correct data have been received. Otherwise, the parity checker indicates an error.

FIGURE 6.57 Parity Error Checking

EXAMPLE 6.20

Data are transmitted from a PC serial port to a modem in groups of 7 data bits plus a parity bit. What should the parity bit, *P*, be for each of the following data if the parity is EVEN? If the parity is ODD?

a. 0110110
b. 1000000
c. 0010101

■ **Solution**

a. 0110110   Four 1s in data. (4 is an even number.)
             EVEN parity: $P = 0$
             ODD parity: $P = 1$
b. 1000000   One 1 in data. (1 is an odd number.)
             EVEN parity: $P = 1$
             ODD parity: $P = 0$
c. 0010101   Three 1s in data. (3 is an odd number.)
             EVEN parity: $P = 1$
             ODD parity: $P = 0$

FIGURE 6.58 Exclusive OR Gate

TABLE 6.15 Exclusive OR Truth Table

A	*B*	$A \oplus B$
0	0	0
0	1	1
1	0	1
1	1	0

An Exclusive OR gate can be used as a parity generator or a parity checker. Figure 6.58 shows the gate, and Table 6.15 is the XOR truth table. Notice that each line of the XOR truth table has an even number of 1s if we include the output column.

Figure 6.59 shows the block diagram of a circuit that will generate an EVEN parity bit from 2 data bits, *A* and *B,* and transmit the three bits one after the other, that is, serially, to a data receiver.

Figure 6.60 shows a parity checker for the parity generator in Figure 6.59. Data are received serially, but read in parallel. The parity bit is re-created from the received values of *A* and *B*, and then compared to the received value of *P* to give an error indication, *P'*. If *P* and $A \oplus B$ are the same, then $P' = 0$ and the transmission is correct. If *P* and $A \oplus B$ are different, then $P' = 1$ and there has been an error in transmission.

FIGURE 6.59 Even Parity Generation

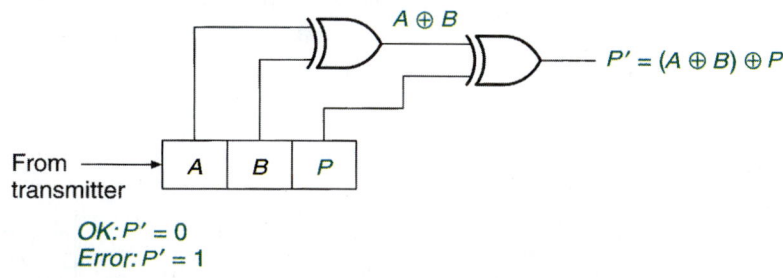

FIGURE 6.60 Even Parity Checking

EXAMPLE 6.21

The following data and parity bits are transmitted four times: $ABP = 101$. Data and parity are checked by the circuit in Figure 6.60.

1. State the type of parity used.
2. The transmission line over which the data are transmitted is particularly noisy and the data arrive differently each time as follows:
   a. $ABP = 101$
   b. $ABP = 100$
   c. $ABP = 111$
   d. $ABP = 110$

Indicate the output P' of the parity checker in Figure 6.60 for each case and state what the output means.

■ Solution

1. The system is using EVEN parity.
2. The parity checker produces the following responses:
   a. $ABP = 101$
   $A \oplus B = 1 \oplus 0 = 1$
   $\quad P' = (A \oplus B) \oplus P = 1 \oplus 1 = 0$ Data received correctly.
   b. $ABP = 100$
   $A \oplus B = 1 \oplus 0 = 1$
   $\quad P' = (A \oplus B) \oplus P = 1 \oplus 0 = 1$ Transmission error. (Parity bit incorrect.)
   c. $ABP = 111$
   $A \oplus B = 1 \oplus 1 = 0$
   $\quad P' = (A \oplus B) \oplus P = 0 \oplus 1 = 1$ Transmission error. (Data bit B incorrect.)
   d. $ABP = 110$
   $A \oplus B = 1 \oplus 1 = 0$
   $\quad P' = (A \oplus B) \oplus P = 0 \oplus 0 = 0$ Transmission error undetected. (B and P incorrectly received.)

The second and third cases in Example 6.21 show that parity error-detection cannot tell which bit is incorrect.

The fourth case points out the major flaw of parity error detection: An even number of errors cannot be detected. This is true whether the parity is EVEN or ODD. If a group of bits has an even number of 1s, a single error will change that to an odd number of 1s, but a double error will change it back to even. (Try a few examples to convince yourself that this is true.)

An ODD parity generator and checker can be made using an Exclusive NOR, rather than an Exclusive OR, gate. If a set of transmitted data bits require a 1 for EVEN parity, it follows that they require a 0 for ODD parity. This implies that EVEN and ODD parity generators must have opposite-sense outputs.

EXAMPLE 6.22

Modify the circuits in Figures 6.59 and 6.60 to operate with ODD parity. Verify their operation with the data bits $AB = 11$ transmitted twice and received once as $AB = 11$ and once as $AB = 01$.

■ **Solution**

Figure 6.61a shows an ODD parity generator and Figure 6.61b shows an ODD parity checker. The checker circuit still has an Exclusive OR output since it presents the same error codes as an EVEN parity checker. The parity bit is re-created at the receive end of the transmission path and compared with the received parity bit. If they are the same, $P' = 0$ (correct transmission). If they are different, $P' = 1$ (transmission error).

FIGURE 6.61   Example 6.22: ODD Parity Generator and Checker

Verification:

Generator:
  Data: $AB = 11$ Parity: $P = \overline{A \oplus B} = \overline{1 \oplus 1} = 1$

Checker:
  Received data: $AB = 11$
  $P' = (A \oplus B) \oplus P = (1 \oplus 1) \oplus 1 = 1 \oplus 1 = 0$ (Correct transmission)

Generator:
  Data: $AB = 11$ Parity: $P = \overline{A \oplus B} = \overline{1 \oplus 1} = 1$

Checker:
  Received data: $AB = 01$
  $P' = (A \oplus B) \oplus P = (0 \oplus 1) \oplus 1 = 0 \oplus 1 = 1$ (Incorrect transmission)

Parity generators and checkers can be expanded to any number of bits by using an XOR gate for each pair of bits and combining the gate outputs in further stages of 2-input XOR gates. The true form of the generated parity bit is P_E, the EVEN parity bit. The complement form of the bit is P_O, the ODD parity bit.

Table 6.16 shows the XOR truth table for 4 data bits and the ODD and EVEN parity bits. The EVEN parity bit P_E is given by $(A \oplus B) \oplus (C \oplus D)$. The ODD parity bit P_O is given by $\overline{P}_E = \overline{(A \oplus B) \oplus (C \oplus D)}$. For every line in Table 6.16, the bit combination $ABCDP_E$ has an even number of 1s and the group $ABCDP_O$ has an odd number of 1s.

TABLE 6.16 Even and Odd Parity Bits for 4-bit Data

A	B	C	D	$A \oplus B$	$C \oplus D$	P_E	P_O
0	0	0	0	0	0	0	1
0	0	0	1	0	1	1	0
0	0	1	0	0	1	1	0
0	0	1	1	0	0	0	1
0	1	0	0	1	0	1	0
0	1	0	1	1	1	0	1
0	1	1	0	1	1	0	1
0	1	1	1	1	0	1	0
1	0	0	0	1	0	1	0
1	0	0	1	1	1	0	1
1	0	1	0	1	1	0	1
1	0	1	1	1	0	1	0
1	1	0	0	0	0	0	1
1	1	0	1	0	1	1	0
1	1	1	0	0	1	1	0
1	1	1	1	0	0	0	1

EXAMPLE 6.23

Use Table 6.16 to draw a 4-bit parity generator and a 4-bit parity checker that can generate and check either EVEN or ODD parity, depending on the state of one select input.

■ Solution

Figure 6.62 shows the circuit for a 4-bit parity generator. The XOR gate at the output is configured as a programmable inverter to give P_E or P_O. When $\overline{\text{EVEN}}/\text{ODD}$ = 0, the parity output is not inverted and the circuit generates P_E. When $\overline{\text{EVEN}}/$ ODD = 1, the XOR inverts the parity bit, giving P_O.

FIGURE 6.62 Example 6.23: 4-Bit Parity Generator

The 4-bit parity checker, shown in Figure 6.63, is the same circuit, with an additional XOR gate to compare the parity bit re-created from data and the previously encoded parity bit.

FIGURE 6.63 Example 6.23: 4-Bit Parity Checker

EXAMPLE 6.24

Draw the circuit for an 8-bit *EVEN/ODD* parity generator.

■ **Solution**

An 8-bit parity generator is an expanded version of the 4-bit generator in the previous example. The circuit is shown in Figure 6.64.

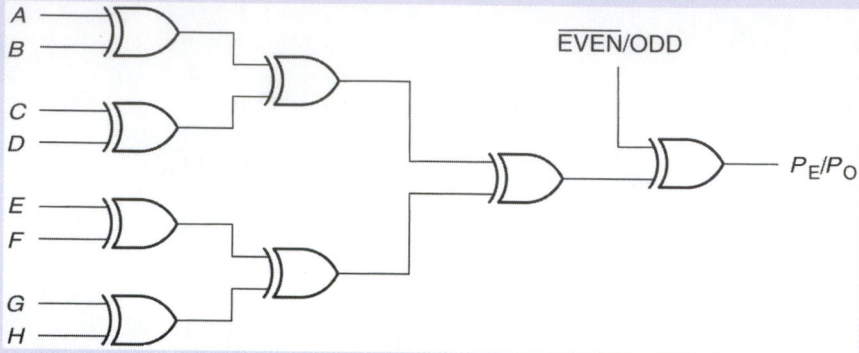

FIGURE 6.64 Example 6.24: 8-Bit Parity Generator

Parity Circuits in VHDL Using the GENERATE Statement

■ **KEY TERMS**

GENERATE Statement A VHDL statement that allows the creation of repetitive hardware structures.

Index Variable A variable that references one instance of a repetitive hardware structure in a statement like the GENERATE statement in VHDL.

Signal Attribute A property of a signal that allows it to be examined for size or monitored for a type of activity.

A simple way to make a small (e.g., 4-bit) parity generator in VHDL is to encode the Boolean expression as a concurrent signal assignment statement. For a 4-bit parity generator, we can define a set of inputs as **a, b, c, d** and write the equation as:

```
pe <= (a xor b) xor (c or d);
```

This gets more complicated for a larger generator. For example, for an 8-bit circuit, we might write:

```
pe <= ((a xor b) xor (c xor d)) xor ((e xor f) xor (g xor h));
```

This is still manageable, but VHDL allows us to write statements that can easily be scaled to different sizes without rewriting large parts of the file. We can start building this kind of description by examining the 4-bit parity generator shown in Figure 6.65.

This circuit can be broken up into several pieces that can be described by the following concurrent signal assignment statements:

```
p(1) <= d(1) xor d(0);
p(2) <= p(1) xor d(2);
p(3) <= p(2) xor d(3);
pe  <= p(3);
```

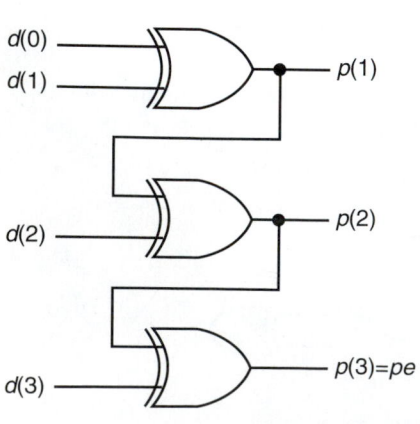

FIGURE 6.65 4-Bit Parity Generator

The statements for **p(2)** and **p(3)** can be written more generally as:

```
p(i) <= p(i - 1) xor d(i);
```

where **i** is an **index variable** that shows the relationship between a specific instance of **p** and other hardware constructs in the design. For repetitive hardware such as this circuit, we can use a VHDL **GENERATE statement** that has the general form:

```
__generate_label:
FOR __index variable IN __range GENERATE
 __statement;
 __statement;
END GENERATE;
```

For example, the equations for **p(2)** and **p(3)** can be created with the following GENERATE statement:

```
parity_generate:
FOR i IN 2 to 3 GENERATE
   P(i) <= p(i-1) xor d(i);
END GENERATE;
```

The complete VHDL file for this design is shown below.

parity_gen.vhd
parity_gen.vwf

```
-- 4-bit parity generator using GENERATE statement.
LIBRARY ieee;
USE ieee.std_logic_1164.ALL;

ENTITY parity4_gen IS
    PORT(
        d  : IN  STD_LOGIC_VECTOR(0 TO 3);
        pe : OUT STD_LOGIC);
END parity4_gen;

ARCHITECTURE parity OF parity4_gen IS
    SIGNAL p : STD_LOGIC_VECTOR(1 to 3);
BEGIN
    p(1) <= d(0) xor d(1);

    parity_generate:
    FOR i IN 2 to 3 GENERATE
        p(i) <= p(i-1) xor d(i);
    END GENERATE;

    pe <= p(3);
END parity;
```

EXAMPLE 6.25

Expand the circuit of the 4-bit parity generator in Figure 6.65 to make an 8-bit parity generator. Write a VHDL file that uses a GENERATE statement to implement the design.

■ Solution

Figure 6.66 shows the circuit for the 8-bit parity generator. The VHDL file for this design follows. The file is the same as for the 4-bit parity generator, except that the vector sizes and index variable are changed.

(continued)

FIGURE 6.66 Example 6.25: 8-Bit Parity Generator

parity8_gen.vhd

```
--8-bit parity generator using GENERATE statement
LIBRARY ieee;
USE ieee.std_logic_1164.ALL;

ENTITY parity8_gen IS
   PORT(
      d  : IN  STD_LOGIC_VECTOR(0 to 7);
      pe : OUT STD_LOGIC);
END parity8_gen;

ARCHITECTURE parity OF parity8_gen IS
   SIGNAL p : STD_LOGIC_VECTOR(1 to 7);
BEGIN
   p(1) <= d(0) xor d(1);

   parity generate:
   FOR i IN 2 to 7 GENERATE
      p(i) <= p (i - 1) xor d(i);
   END GENERATE;

   pe <= p(7);
END parity;
```

Using Signal Attributes

In Example 6.25, we had to make four changes to convert the parity generator from a 4-bit to an 8-bit circuit: the vector sizes of **d** and **p,** the index variable in the GEN-ERATE statement, and the element value in the statement that assigns the last element of **p** to **pe.** If we forget to make any of these changes, we will end up with a circuit that incorrectly calculates the parity value of **pe.** If we forget to change the vector sizes, the VHDL project will probably not even compile.

We can write the file in a more general way that allows for easy resizing by using **signal attributes.** VHDL has a number of predefined signal attributes that allow us to refer to certain properties of a signal. An attribute is indicated by an apostrophe following the signal name and preceding the attribute.

For example, suppose we have a signal defined as p : STD_LOGIC_VECTOR (1 to 7); The attribute p'HIGH (pronounced "p tick high") returns the value 7, since 7 is the high bound of the vector range. Table 6.17 shows the values of several other attributes of **p.**

TABLE 6.17 Attributes of Signal **p** [1..7]

Attribute	Description	Value
p' HIGH	Upper bound of **p**	7
p' LOW	Lower bound of **p**	1
p' LEFT	Left bound of **p**	1
p' RIGHT	Right bound of **p**	7

Note that if **p** is redefined as STD_LOGIC_VECTOR(7 downto 1) that the upper and lower bounds remain the same, but the left and right bounds are now reversed, as shown in Table 6.18.

TABLE 6.18 Attributes of Signal **p**[7..1]

Attribute	Description	Value
p' HIGH	Upper bound of **p**	7
p' LOW	Lower bound of **p**	1
p' LEFT	Left bound of **p**	7
p' RIGHT	Right bound of **p**	1

EXAMPLE 6.26

Rewrite the VHDL, code in example 6.25 to redefine the size of signal **p** and the index variable of the GENERATE statement in terms of the upper bound of input **d.**

■ Solution

The modified code is as follows. To rescale this design, only one value must be changed: the size of input **d.**

```
-- n-bit parity generator using GENERATE statement
LIBRARY ieee;
USE ieee.std_logic_1164.ALL;

ENTITY parity_genx IS
    PORT(
        d  : IN  STD_LOGIC_VECTOR(0 to 7);
        pe : OUT STD_LOGIC);
END parity_genx;
```

(continued)

parity_genx.vhd
parity_genx.vwf

```
ARCHITECTURE parity OF parity_genx IS
    SIGNAL p : STD_LOGIC_VECTOR(1 to d'HIGH);
BEGIN
   p(1) <= d(0) xor d(1);

   parity_generate:
   FOR i IN 2 to d'HIGH GENERATE
       p(i) <= p(i-1) xor d (i);
   END GENERATE;

   pe <= p(d'HIGH);
END parity;
```

■ SECTION 6.6 REVIEW PROBLEM

6.8 Data (including a parity bit) are detected at a receiver configured for checking ODD parity. Which of the following data do we know are incorrect? Could there be errors in the remaining data? Explain.

    a. 010010
    b. 011010
    c. 1110111
    d. 1010111
    e. 1000101

SUMMARY

1. A decoder detects the presence of a particular binary code. The simplest decoder is an AND or NAND gate, which can detect a binary code when combined with the right combination of input inverters.

2. Multiple-output decoders are implemented by a series of single-gate decoders, each of which responds to a different input code.

3. For an n-input decoder, there can be as many as 2^n unique outputs.

4. Some multiple-input decoders have an enable input that allows all decoder outputs to be deactivated when the enable input is inactive.

5. The effectiveness of the simulation of a CPLD design in software depends on selecting good criteria for a simulation. Ideally, good simulation criteria must test the design in all working modes and all potential failure modes.

6. Selected signal assignment statements and conditional signal assignment statements allow us to encode the behavior of a decoder in VHDL.

7. A seven-segment display is an array of seven luminous segments (usually LED or LCD), arranged in a figure-8 pattern, used to display numerical digits.

8. The segments in a seven-segment display are designated by lowercase letters a through g. The sequence of labels goes clockwise, starting with segment a at the top and ending with g in the center.

9. Seven-segment displays are configured as common anode (active-LOW inputs) or common cathode (active-HIGH segments).

10. A seven-segment decoder can be described with a truth table or Boolean equation for each segment function. Since the segment functions do not simplify very much, it is often easier to program a CPLD with a VHDL truth table, in the form of a selected signal assignment statement, rather than with the Boolean equations of the decoder.

11. Ripple blanking is a technique that allows leading or trailing zeros in a multiple-digit numerical display to be suppressed, while allowing internal zeros to be displayed. The

technique works by cascading a ripple blanking output ($\overline{RBO}$) to the ripple blanking input ($\overline{RBI}$) of the next decoder. $\overline{RBO}$ is LOW only if the corresponding decoder input is 0000 AND $\overline{RBI}$ on the same decoder is also LOW. A zero is suppressed if the decoder input is 0000 and $\overline{RBI} = 0$. A zero is displayed if the decoder input is 0000 and $\overline{RBI} = 1$.

12. An encoder is the complementary device to a decoder. It generates a binary code corresponding to the number of an active input. Without introducing priority circuitry, only one input can be active at a time, or erroneous codes can be generated.

13. A priority encoder allows more than one input to be active at a time. The output binary code corresponds to the active input with the highest priority, usually the one with the highest number.

14. A low-priority input must not change to code resulting from a higher-priority input. This is done by using OR gates to create the required output codes and AND gates to block certain code combinations.

15. A VHDL conditional signal assignment statement is ideally suited for priority encoding, because it checks for presence of the highest-priority condition, then proceeds through lower-priority conditions until the correct condition is found.

16. A multiplexer (MUX) is a circuit that directs a signal or group of signals (called the data inputs) to an output, based on the status of a set of select inputs.

17. Generally, for n select inputs in a multiplexer, there are $m = 2^n$ data inputs. Such a multiplexer is referred to as an m-to-1 multiplexer.

18. The selected data input in a MUX is usually denoted by a subscript that is the decimal equivalent of the combined binary value of the select inputs. For example, if the select inputs in an 8-to-1 MUX are set to $S_2S_1S_0 = 100$, data input D_4 is selected since 100 (binary) = 4 (decimal).

19. A MUX can be designed to switch groups of signals to a multibit output. The inputs can be denoted by double subscript notation, where the first subscript indicates the number of the signal group and the second subscript the element in the group. For example, a MUX can have a 4-bit set of inputs called $D_{03}D_{02}D_{01}D_{00}$ and another 4-bit input group called $D_{13}D_{12}D_{11}D_{10}$, each of which can be switched to a 4-bit output called $Y_3Y_2Y_1Y_0$ by the state of one select input.

20. A MUX can be created in VHDL using a selected signal assignment statement or a CASE statement to select one input of several to assign to the output. Inputs and outputs can be single- or multiple-bit.

21. A multiplexer can be used in time-dependent applications if a binary counter is applied to its select inputs.

22. Some examples of time-dependent MUX applications are waveform or bit pattern generation and time-division multiplexing (TDM).

23. In time division multiplexing, several digital signals share a single transmission path by allotting a time slot for every signal, during which that signal has sole access to the transmission path.

24. TDM can be configured for bit multiplexing, in which a channel transmits one bit each time it is selected, or byte (or word) multiplexing, in which a channel transmits and entire byte or word each time it is selected.

25. A demultiplexer (DMUX) receives data from a single source and directs the data to one of several outputs, which is selected by the status of a set of select inputs.

26. A decoder with an enable input can also act as a demultiplexer if the enable input of the decoder is used as a data input for a demultiplexer.

27. Any VHDL design for a decoder with an enable input can be also used as a demultiplexer. If the outputs are active-LOW, the enable input must also be active-LOW to keep the data true.

28. A TDM signal can be demultiplexed by applying a binary count to the DMUX's select inputs at the same rate as the count is applied to the select input of the multiplexer that originally sent the data.

29. A CMOS analog multiplexer or demultiplexer works by using a decoder to enable a set of analog data transmission switches. It can be used in either direction.

30. A magnitude comparator determines whether two binary numbers are equal and, if not, which one is greater.

31. The simplest equality comparator is an XNOR gate, whose output is HIGH if both inputs are the same.

32. A pair of multiple-bit numbers can be compared by a set of XNOR gates whose outputs are ANDed. The circuit compares the two numbers bit-by-bit.

33. Given two numbers A and B, the Boolean function $\overline{A}_n B_n$, if true, indicates that the nth bit of A is less than the nth bit of B.

34. Given two numbers A and B, the Boolean function $A_n \overline{B}_n$, if true, indicates that the nth bit of A is greater than the nth bit of B.

35. The less-than and greater-than functions can be combined with an equality comparator to determine, bit-by-bit, how two numbers compare in magnitude to one another.

36. A magnitude comparator can be easily implemented in VHDL by using INTEGER types for the inputs and using IF statements to compare their respective magnitudes.

37. A VHDL magnitude comparator can also be created using STD_LOGIC_VECTORs for the inputs and IF statements to compare the magnitudes provided that the **std_logic_unsigned** package is included in the file. This allows easy comparison of two multiple-bit values without needing to know the maximum integer value of either input.

38. Parity checking is a system of error detection that works by counting the number of 1s in a group of bits.

39. Even parity requires a group of bits to have an even number of 1s. Odd parity requires a group of bits to have an odd number of 1s. This is achieved by appending a parity bit to the data whose value depends on the number of 1s in the data bits.

40. An XOR gate is the simplest even parity generator. Each line in its truth table has an even number of 1s, if the output column is included.

41. An XNOR gate can be used to generate an odd parity bit from two data bits.

42. A parity checker consists of a parity generator on the receive end of a transmission system and a comparator to determine if the locally generated parity bit is the same as the transmitted parity bit.

43. Parity generators and checkers can be expanded to any number of bits by using an XOR gate for each pair of bits and combining the gate outputs in further stages of 2-input XOR gates.

44. A GENERATE statement in VHDL can be used to create repetitive instances of hardware, such as the multiple instances of an XOR gate used in a parity generator/checker. An index variable in the GENERATE statement counts the number of instances required.

45. A signal attribute in VHDL is used to refer to a property of a signal, such as its upper or lower bound. For example, for a port defined as d : IN STD_LOGIC_VECTOR (7 downto 0); some attributes of **d** and their values are: d'LOW (= 0), d'HIGH (=7), d'RIGHT (= 0), and d'LEFT (= 7). A signal attribute can be used in place of a constant to make a VHDL description more general and therefore easier to use in more than one design.

GLOSSARY

Address Decoder   A device that selects memory or peripheral devices for a block of addresses in a computer system.

BCD   Binary coded decimal. A code in which each individual digit of a decimal number is represented by a 4-bit binary number. (e.g., 905 [decimal] = 1001 0000 0101 [BCD]).

Bit Multiplexing   A TDM technique in which one bit is sent from each channel during its assigned time slot.

Byte (or Word) Multiplexing   A TDM technique in which a byte (or word) is sent from each channel during its assigned time slot. (A byte is eight bits; a word is a group of bits whose size varies with the particular system.)

CASE Statement   A VHDL construct in which there is a choice of statements to be executed, depending on the value of a signal or variable.

Clock   A signal that controls the operation of a sequential digital circuit, such as a counter, by advancing the counter's outputs to the next state when it receives a clock pulse.

CMOS Analog Switch A CMOS device that will pass an analog or digital signal in either direction, when enabled. Also called a transmission gate. There is no TTL equivalent.

Common Anode Display A seven-segment LED display where the anodes of all the LEDs are connected to the circuit supply voltage. Each segment is illuminated by a logic LOW at its cathode.

Common Cathode Display A seven-segment display in which the cathodes of all LEDs are connected together and grounded. A logic HIGH illuminates a segment when applied to its anode.

Conditional Signal Assignment Statement A concurrent VHDL construct that assigns a value to a signal, depending on a sequence of conditions being true or false.

Counter A digital circuit whose output produces a fixed sequence of binary states when an input called the clock receives a series of pulses. The output advances by one for each clock pulse (e.g., the output state of a 4-bit binary counter progresses in order from 0000 to 1111, then repeats).

Data Inputs The multiplexer inputs that feed a digital signal to the output when selected.

Decoder A digital circuit designed to detect the presence of a particular digital state.

Demultiplexer A circuit that uses a binary decoder to direct a digital signal from a single source to one of several destinations.

Double-Subscript Notation A naming convention where two or more numerically related groups of signals are named using two subscript numerals. Generally, the first digit refers to a group of signals and the second to an element of a group. (e.g., X_{03} represents element 3 of group 0 for a set of signal groups, X.)

Encoder A circuit that generates a digital code at its outputs in response to one or more active input lines.

Even Parity An error-checking system that requires a binary number to have an even number of 1s.

GENERATE statement A VHDL statement that allows the creation of repetitive hardware structures.

IF Statement A VHDL construct within a process that executes a series of statements, if a Boolean test condition is true.

Index Variable A variable that references one instance of a repetitive hardware structure in a statement like the GENERATE statement in VHDL.

Magnitude Comparator A circuit that compares two n-bit binary numbers, indicates whether or not the numbers are equal, and, if not, which one is larger.

Multiplexer A circuit that directs one of several digital signals to a single output, depending on the states of several select inputs.

Odd Parity An error-checking system that requires a binary number to have an odd number of 1s.

Parity A system that checks for errors in a multi-bit binary number by counting the number of 1s.

Parity Bit A bit appended to a binary number to make the number of 1s even or odd, depending on the type of parity.

Positive Edge The point on a digital waveform where the logic level of the waveform makes a LOW-to-HIGH transition.

Priority Encoder An encoder that generates a binary or BCD output corresponding to the subscript of the active input having the highest priority. This is usually defined as the active input with the largest subscript value.

PROCESS A VHDL construct that contains statements that are executed if there is a change in a signal in its sensitivity list.

$\overline{\text{RBI}}$ Ripple Blanking Input.

$\overline{\text{RBO}}$ Ripple Blanking Output.

Response Waveforms A set of output waveforms generated by a simulator tool for a particular digital design in response to a set of stimulus waveforms.

Ripple Blanking A technique used in a multiple-digit numerical display that suppresses leading or trailing zeros in the display, but allows internal zeros to be displayed.

Select Inputs The multiplexer inputs which select a digital input channel.

Selected Signal Assignment Statement A concurrent signal assignment in VHDL in which a value is assigned to a signal, depending on the alternative values of another signal or variable.

Sensitivity List A list of signals in a PROCESS statement that are monitored to determine whether the PROCESS should be executed.

Seven-Segment Display An array of seven independently controlled light-emitting diode (LED) or liquid crystal display (LCD) elements, shaped like a figure-8, which can be used to display decimal digits and other characters by turning on the appropriate elements.

Simulation The verification of the logic of a digital design before programming it into a PLD.

Signal Attribute A property of a signal that allows it to be examined for size or monitored for a type of activity.

Stimulus Waveforms A set of user-defined input waveforms on a simulator file designed to imitate input conditions of a digital circuit.

Time Division Multiplexing (TDM) A technique of using one transmission line to send many signals simultaneously by making them share the line for equal fractions of time.

Time Slot A period of time during which a transmitted data element has sole access to a transmission path.

Timing Diagram A diagram showing how two or more digital waveforms in a system relate to each other over time.

PROBLEMS

Problem numbers set in color indicate more difficult problems.

6.1 Decoders

6.1 When a HIGH is on the outputs of each of the decoding circuits shown in Figure 6.67, what is the binary code appearing at the inputs? Write the Boolean expression for each decoder output.

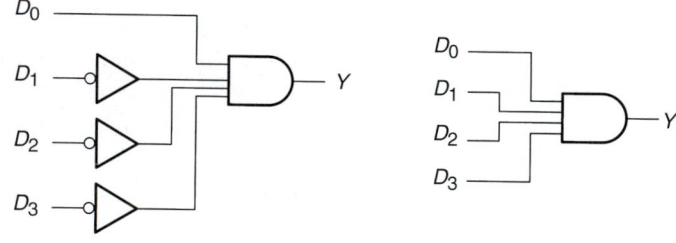

FIGURE 6.67 Problem 6.1: Decoding Circuits

6.2 Draw the decoding circuit for each of the following Boolean expressions:

a. $\bar{Y} = \bar{D}_3 D_2 \bar{D}_1 D_0$

b. $\bar{Y} = \bar{D}_3 \bar{D}_2 D_1 D_0$

c. $Y = \bar{D}_3 \bar{D}_2 D_1 D_0$

d. $\bar{Y} = D_3 D_2 \bar{D}_1 \bar{D}_0$

e. $Y = D_3 D_2 \bar{D}_1 D_0$

6.3 a. Create a Block Diagram File in Quartus II for a 2-line-to-4-line decoder with active-HIGH outputs and an active-LOW enable input.

b. Write a list of simulation criteria for the decoder.

c. Use the simulation criteria to create a set of simulation waveforms for the decoder using the Quartus II simulation tool.

6.4   **a.** Create a Block Diagram File in Quartus II for a 3-line-to-8-line decoder with active-HIGH outputs and an active-LOW enable input.

       b. Write a list of simulation criteria for the decoder.

       c. Use the simulation criteria to create a set of simulation waveforms for the decoder using the Quartus II simulation tool.

6.5   For a generalized n-line-to-m-line decoder, state the value of m if n is:

       a. 5

       b. 6

       c. 8

    Write the equation giving the general relation between n and m.

6.6   A microcomputer system has a RAM capacity of 128 megabytes (MB), split into 16 MB portions. Each RAM device is enabled by a low at an nCs input. Draw a logic diagram showing how a binary decoder can select one particular RAM device.

6.7   **a.** Write a VHDL file for a 3-line-to-8-line decoder with active-LOW outputs and no enable. Use a selected signal assignment statement.

       b. Write a list of simulation criteria for the decoder.

       c. Use the simulation criteria to create a set of simulation waveforms for the decoder using the Quartus II simulation tool.

6.8   **a.** Write a truth table for a binary-coded decimal (BCD) decoder. The decoder should decode a 4-bit input with a binary value between 0000 and 1001 by making one of ten outputs LOW. The outputs are labeled $\overline{Y}_0$ to $\overline{Y}_9$. Unused codes should disable all decoder outputs.

       b. Write a VHDL file that implements the decoder described in part a of this problem.

       c. Write a list of simulation criteria for the decoder.

       d. Use the simulation criteria to create a set of simulation waveforms for the decoder using the Quartus II simulation tool.

6.9   **a.** Create a VHDL file for a 3-line-to-8-line decoder with the same output and enable characteristics as a 74138 decoder.

       b. Write a list of simulation criteria for the decoder.

       c. Use the simulation criteria to create a set of simulation waveforms for the decoder using the Quartus II simulation tool.

6.10   Write a truth table for a hexadecimal-to-seven-segment decoder for a common anode display. Use the digit patterns of Figure 6.19 as a model.

6.11   Use the truth table derived in Problem 6.10 to derive the Boolean equations for each segment driver. Simplify the equations as much as possible, using any convenient method.

6.12   Write a VHDL file for the hexadecimal-to-seven-segment decoder described in Problem 6.10.

6.13   Modify the VHDL file for the hexadecimal-to-seven-segment decoder from Problem 6.12 to add a ripple-blanking feature.

6.14   Draw a diagram consisting of four seven-segment displays, each driven by a BCD-to-seven-segment decoder with ripple blanking. The circuit should be configured to suppress all leading zeros. Show the displayed digits and $\overline{RBO}/\overline{RBI}$ logic levels for each of the following displayed values: 100, 217, 1024.

6.2   Encoders

6.15   Figure 6.68 shows a BCD priority encoder with three different sets of inputs. Determine the resulting output code for each input combination. Inputs and outputs are active HIGH.

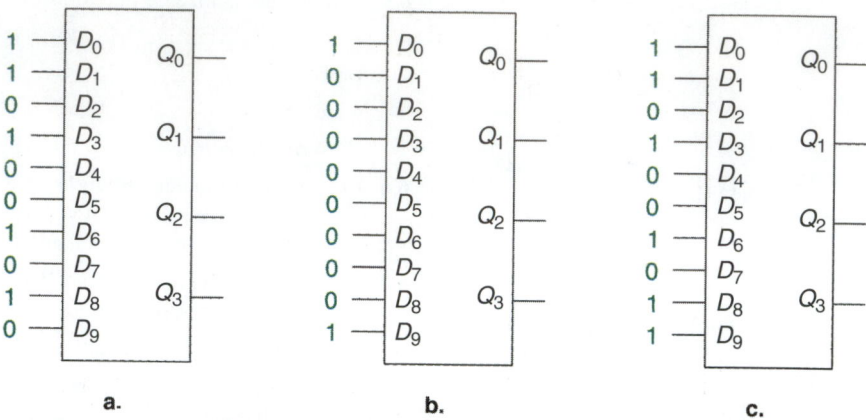

a.   **b.**   **c.**

FIGURE 6.68   Problem 6.15: BCD Priority Encoder

6.16   Derive the Boolean equations for the outputs of a BCD priority encoder, based on the encoding principle stated in Section 6.2. Show all work.

6.17   **a.** Create a Quartus II Block Diagram File for a BCD priority encoder, based on the equations derived in Problem 6.16.

   b. Write a list of simulation criteria for the encoder.

   c. Use the simulation criteria to create a set of simulation waveforms for the encoder using the Quartus II simulation tool.

6.18   **a.** Create a VHDL file that implements the functions of a BCD priority encoder.

   b. Write a list of simulation criteria for the encoder.

   c. Use the simulation criteria to create a set of simulation waveforms for the encoder using the Quartus II simulation tool.

6.19   **a.** Create a VHDL file that implements the functions of a 4-bit priority encoder.

   b. Write a list of simulation criteria for the encoder.

   c. Use the simulation criteria to create a set of simulation waveforms for the encoder using the Quartus II simulation tool.

6.3   Multiplexers

6.20   Make a table listing which digital audio source in Figure 6.69 is routed to output Y for each combination of the multiplexer select inputs. (CD = compact disc; DAT = digital audio tape.)

FIGURE 6.69 Problem 6.20: Digital Audio Multiplexer

6.21 Draw symbols for an 8-to-1 and a 16-to-1 multiplexer. Write the truth table for each multiplexer, showing which data input is selected for every binary combination of the select inputs.

6.22 **a.** Create a Block Diagram File in Quartus II for an 8-to-1 multiplexer circuit.

b. Write a list of simulation criteria for the multiplexer.

c. Use the simulation criteria to create a set of simulation waveforms for the multiplexer using the Quartus II simulation tool.

6.23 Write the Boolean expression describing an 8-to-1 multiplexer. Evaluate the equation for the case where input D_5 is selected.

6.24 Draw the symbol for a quadruple 8-to-1 multiplexer (i.e., a MUX with eight switched groups of 4 bits each). Write the truth table for this device, showing which data inputs are selected for every binary combination of the select inputs. Use double-subscript notation.

6.25 **a.** Create a VHDL file for the quadruple 8-to-1 multiplexer in Problem 6.24.

b. Write a list of simulation criteria for the multiplexer.

c. Use the simulation criteria to create a set of simulation waveforms for the multiplexer using the Quartus II simulation tool.

6.26 Draw the symbol for an octal 4-to-1 multiplexer (i.e., a MUX with four switched groups of 8 bits each). Write the truth table for this device, showing which data inputs are selected for every binary combination of the select inputs. Use double-subscript notation.

6.27 **a.** Create a VHDL file for the octal 4-to-1 multiplexer in Problem 6.26.

b. Write a list of simulation criteria for the multiplexer.

c. Use the simulation criteria to create a set of simulation waveforms for the multiplexer using the Quartus II simulation tool.

6.28   Write a VHDL file for an 8-to-1 multiplexer using a concurrent signal assignment statement to encode the multiplexer's Boolean equation directly. Would this be a good method for encoding a larger device, such as a 16-to-1 multiplexer? Explain your answer.

6.29   Write a VHDL file for an 8-to-1 multiplexer using a selected signal assignment statement. Would this be a good method for encoding a larger device, such as a 16-to-1 multiplexer? Explain your answer.

6.30   Write a VHDL file for a 16-to-1 multiplexer using the method you believe to be most efficient.

6.31   Draw the circuit of a programmable waveform generator based on an 8-to-1 multiplexer. Draw a timing diagram of this circuit for the following input data:

   a. $D_7D_6D_5D_4D_3D_2D_1D_0 = 01100101$

   b. $D_7D_6D_5D_4D_3D_2D_1D_0 = 01010101$

6.32   The data pattern in Problem 6.31b generates a symmetrical 12 kHz waveform. Write the data patterns required to produce a 6 kHz waveform and a 3 kHz waveform at the output of a MUX-based programmable waveform generator.

6.4   Demultiplexers

6.33   **a.**   Create a Block Diagram File in Quartus II for a 1-to-4 demultiplexer circuit with active-LOW outputs and an active-LOW data input.

   b.   Write a list of simulation criteria for the demultiplexer that shows how the circuit can be used as a demultiplexer or a decoder.

   c.   Use the simulation criteria to create a set of simulation waveforms for the decoder/demultiplexer using the Quartus II simulation tool.

6.34   **a.**   Create a Block Diagram File in Quartus II for a 1-to-8 demultiplexer circuit with active-HIGH outputs.

   b.   Write a list of simulation criteria for the demultiplexer.

   c.   Use the simulation criteria to create a set of simulation waveforms for the demultiplexer using the Quartus II simulation tool.

6.35   Write a VHDL file that implements the function of a 1-to-16 demultiplexer.

6.36   Briefly state what characteristics of an analog switch make it suitable for transmitting analog signals.

6.37   Draw a diagram showing how eight analog switches can be connected to a decoder to form an 8-channel MUX/DMUX circuit. Briefly explain why the same circuit can be used as a multiplexer or as a demultiplexer.

6.5   Magnitude Comparators

6.38   **a.**   Create a Block Diagram File in Quartus II for a 3-bit magnitude comparator that has outputs for functions *AEQB*, *AGTB*, and *ALTB*.

   b.   Write a set of simulation criteria for the circuit.

   c.   Based on the simulation criteria, create a set of simulation waveforms in Quartus II for the 3-bit magnitude comparator.

6.39   Create a Block Diagram File in Quartus II for the *ALTB* portion of a 4-bit magnitude comparator.

6.40   Write the VHDL file for a 12-bit magnitude comparator with active-HIGH outputs. Use STD_LOGIC types for all ports and signals.

6.41   Write the Boolean expressions for the *AEQB*, *ALTB*, and *AGTB* outputs of a 6-bit magnitude comparator.

6.42   Write a VHDL file that implements the functions $A < B$, $A = B$ and $A > B$ for two 16-bit numbers.

6.43   Write a VHDL file that implements the following six comparison functions in a single device for two 4-bit inputs A and B: $A < B$, $A \leq B$, $A = B$, $A \neq B$, $A \geq B$, and $A > B$. Make the outputs indicate active-LOW.

6.44 Create a simulation that verifies the operation of the six-function comparator in Problem 6.43, including a set of simulation criteria.

6.6 Parity Generators and Checkers

6.45 What parity bit, P, should be added to the following data if the parity is EVEN? If the parity is ODD?

a. 1111100

b. 1010110

c. 0001101

6.46 The following data are transmitted in a serial communication system (P is the parity bit). What parity is being used in each case?

a. $ABCDEFGHP = 010000101$

b. $ABCDEFGHP = 011000101$

c. $ABCDP = 01101$

d. $ABCDEP = 101011$

e. $ABCDEP = 111011$

6.47 The data $ABCDEFGHP = 110001100$ are transmitted in a serial communication system. Give the output P' of a receiver parity checker for the following received data. State the meaning of the output P' for each case.

a. $ABCDEFGHP = 110101100$

b. $ABCDEFGHP = 110001101$

c. $ABCDEFGHP = 110001100$

d. $ABCDEFGHP = 110010100$

6.48 **a.** Create a Block Diagram File in Quartus II for a 5-bit parity generator with a switchable EVEN/ODD output.

b. Write a set of simulation criteria for the circuit.

c. Based on the simulation criteria, create a set of simulation waveforms in Quartus II for the parity generator.

6.49 **a.** Create a Block Diagram File in Quartus II for 5-bit parity checker corresponding to the parity generator in Problem 6.48.

b. Write a set of simulation criteria for the circuit.

c. Based on the simulation criteria, create a set of simulation waveforms in Quartus II for the parity checker.

6.50 Create a set of simulation criteria for the 4-bit parity generator in Figure 6.65. Make a simulation in Quartus II to check the correctness of the design.

6.51 **a.** Write the VHDL code for a 4-bit parity generator that can select either EVEN or ODD parity.

b. Write a set of simulation criteria for the parity generator in part a.

c. Make a simulation in Quartus II to check the correctness of the design.

6.52 **a.** Write the VHDL code for a 4-bit parity checker that can select either EVEN or ODD parity.

b. Write a set of simulation criteria for the parity checker in part a.

c. Make a simulation in Quartus II to check the correctness of the design.

6.53 Write the VHDL code for an 8-bit parity checker that can select either EVEN or ODD parity. Use the signal attribute ′HIGH in the code, so that the width of the circuit is specified only once.

6.54 Write the VHDL code for an 8-bit EVEN parity checker that uses the signal attribute ′RIGHT to indicate the width of the circuit.

ANSWERS TO SECTION REVIEW PROBLEMS

6.1A

6.1 The decoders are shown in Figure 6.70.

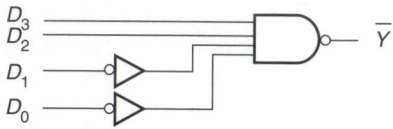

FIGURE 6.70 Decoders

6.1B

6.2 A decoder with 16 outputs requires 4 inputs. A decoder with 32 outputs requires 5 inputs.

6.1C

6.3 Trailing zeros could logically be suppressed after a decimal point or if there are digits displaying a power-of-ten exponent (e.g., 455. or 4.55 02), that is, if the zeros are nonsignificant. The zeros should be displayed if they set the location of the decimal point (e.g., 450).

6.2

6.4 The encoder in Figure 6.22 can have only one input active at any time. If more than one input is active, it may generate incorrect output codes. The circuit can be modified according to the priority encoding principle, as expressed by the Boolean equations for the 3-bit priority encoder, to ensure that a low-priority input is not able to modify the code generated by a higher-priority input.

6.3

6.5 A multiplexer application is time-dependent if its channels are selected in a repeating sequence. This can be accomplished by connecting a binary counter to the select inputs of the multiplexer.

6.4

6.6 The 74138 should be connected as shown in Figure 6.71. The $G1$ input is held HIGH and the $\overline{G2B}$ input LOW to enable the circuit. Input data are applied to the $\overline{G2A}$ input. The output channel is selected by a combination of inputs CBA, where C is the most significant bit.

FIGURE 6.71 Section Review Problem 6.6: 74138 as a Demultiplexer

6.5

6.7 The equations for the 3-bit magnitude comparator are:

$$AEQB = (\overline{A_2 \oplus B_2})(\overline{A_1 \oplus B_1})(\overline{A_0 \oplus B_0})$$
$$ALTB = \overline{A}_2 B_2 + (\overline{A_2 \oplus B_2})\,\overline{A}_1 B_1 + (\overline{A_2 \oplus B_2})(\overline{A_1 \oplus B_1})\overline{A}_0 B_0$$
$$AGTB = A_2 \overline{B}_2 + (\overline{A_2 \oplus B_2})A_1 \overline{B}_1 + (\overline{A_2 \oplus B_2})(\overline{A_1 \oplus B_1})A_0 \overline{B}_0$$

6.6

6.8 Parts a and c are certainly incorrect because each has an even number of 1s. Items b, d, and e could have an even number of errors, which is undetectable by parity checking.

KEY TERMS

1's Complement
2's Complement
8421 Code
9's Complement
Addend
Alphanumeric Code
ASCII
Augend
BCD Adder
Binary-Coded Decimal (BCD)
Borrow
Carry
Carry Bit
Cascade
Case Shift
Component
Component Declaration Statement
Component Instantiation Statement
Difference
End-Around Carry
Excess-3 Code
Fast Carry (or Look-Ahead Carry)
Full Adder
GENERATE Statement
Gray Code
Half Adder
Hierarchy
Instantiate
Magnitude Bits
Minuend
Operand
Overflow
Parallel Binary Adder
Port
Ripple Carry
Self-Complementing

Sign Bit
Sign Extension
Signed Binary Arithmetic
Signed Binary Number
Speed Grade
Subtrahend
Sum
Sum Bit (Single-Bit Addition)
True-Magnitude Form
Unsigned Binary Number

OUTLINE

7.1   Digital Arithmetic

7.2   Representing Signed Binary Numbers

7.3   Signed Binary Arithmetic

7.4   Hexadecimal Arithmetic

7.5   Numeric and Alphanumeric Codes

7.6   Binary Adders and Subtractors

7.7   BCD Adders

7.8   Carry Generation in Quartus II

CHAPTER OBJECTIVES

Upon successful completion of this chapter, you will be able to:

■ Add or subtract two unsigned binary numbers.

■ Write a signed binary number in true-magnitude, 1's complement, or 2's complement form.

■ Add or subtract two signed binary numbers.

■ Explain the concept of overflow.

■ Calculate the maximum sum or difference of two signed binary numbers that will not result in an overflow.

CHAPTER 7

Digital Arithmetic and Arithmetic Circuits

- Add or subtract two hexadecimal numbers.
- Write decimal numbers in BCD codes, such as 8421 (Natural BCD) and Excess-3 code.
- Construct a Gray code sequence.
- Use the ASCII table to convert alphanumeric characters to hexadecimal or binary numbers and vice versa.
- Derive the logic gate circuits for full and half adders, given their truth tables.
- Demonstrate the use of full and half adder circuits in arithmetic and other applications.
- Add and subtract *n*-bit binary numbers, using parallel binary adders and logic gates.
- Explain the difference between ripple carry and parallel carry.
- Design a circuit to detect sign-bit overflow in a parallel adder.
- Draw circuits to perform BCD arithmetic and explain their operation.
- Use VHDL to program CPLD devices to perform various arithmetic functions, such as parallel adders, overflow detectors, and 1's complementers.

VHDL CONSTRUCTS USED IN THIS CHAPTER

- CASE statement
- Component declaration statement
- Component instantiation statement, including GENERIC MAP and PORT MAP
- Explicit port mapping
- GENERATE statement
- GENERIC clause (entity declaration)
- IF statement
- Implicit port mapping
- PROCESS statement
- SIGNAL declaration
- VHDL package: ieee.std_logic_unsigned

There are two ways of performing binary integer arithmetic: with unsigned binary numbers or with signed binary numbers. Signed binary numbers incorporate a bit defining the sign of a number; unsigned binary numbers do not. Several ways of writing signed binary numbers are true-magnitude form, which maintains the magnitude of the number in binary value, and 1's complement and 2's complement forms, which modify the magnitude but are more suited to digital circuitry.

Hexadecimal arithmetic is used for calculations that would be awkward for a human to read in binary due to the large number of bits involved. Important applications of hexadecimal arithmetic are found in microcomputer systems.

In addition to positional number systems, binary numbers can be used in a variety of nonpositional number codes, which can represent numbers, letters, and computer control codes. Binary coded decimal (BCD) codes represent decimal digits as individually encoded groups of bits. Gray code is a binary code used in special applications. American Standard Code for Information Interchange (ASCII) represents alphanumeric and control code characters in a 7- or 8-bit format.

There are a number of different digital circuits for performing digital arithmetic, most of which are based on the parallel binary adder, which in turn is based on the full adder and half adder circuits. The half adder adds two bits and produces a sum and a carry. The full adder also allows for an input carry from a previous adder stage. Parallel adders have many full adders in cascade, with carry bits connected between the stages.

Specialized adder circuits are used for adding and subtracting binary numbers, generating logic functions, and adding numbers in binary-coded decimal (BCD) form. ■

7.1   DIGITAL ARITHMETIC

> ■ **KEY TERMS**
>
> **Signed Binary Number**   A binary number of fixed length whose sign is represented by one bit, usually the most significant bit, and whose magnitude is represented by the remaining bits.
>
> **Unsigned Binary Number**   A binary number whose sign is not specified by a sign bit. A positive sign is assumed unless explicitly stated otherwise.

Digital arithmetic usually means binary arithmetic, or perhaps BCD arithmetic. Binary arithmetic can be performed using **signed binary numbers,** in which the MSB of each number indicates a positive or negative sign, or **unsigned binary numbers,** in which the sign is presumed to be positive.

The usual arithmetic operations of addition and subtraction can be performed using signed or unsigned binary numbers. Signed binary arithmetic is often used in digital circuits for two reasons:

1. Calculations involving real-world quantities require us to use both positive and negative numbers.
2. It is easier to build circuits to perform some arithmetic operations, such as subtraction, with certain types of signed numbers than with unsigned numbers.

Unsigned Binary Arithmetic

> ### ■ KEY TERMS
>
> **Sum** The result of an addition operation.
>
> **Carry** A digit that is "carried over" to the next most significant position when the sum of two single digits is too large to be expressed as a single digit.
>
> **Augend** The number in an addition operation to which another number is added.
>
> **Addend** The number in an addition operation that is added to another.
>
> **Operand** A number upon which an arithmetic function operates (e.g., in the expression $x + y = z$, x and y are the operands).
>
> **Sum Bit (Single-Bit Addition)** The least significant bit of the sum of two 1-bit binary numbers.
>
> **Carry Bit** A bit that holds the value of a carry (0 or 1) resulting from the sum of two binary numbers.

Addition

When we add two numbers, they combine to yield a result called the **sum.** If the sum is larger than can be contained in one digit, the operation generates a second digit, called the **carry.** The two numbers being added are called the **augend** and the **addend,** or more generally, the **operands.**

For example, in the decimal addition $9 + 6 = 15$, 9 is the augend, 6 is the addend, and 15 is the sum. Since the sum cannot fit into a single digit, a carry is generated into a second digit place.

Four binary sums give us all of the possibilities for adding two n-bit binary numbers:

$$0 + 0 = 00$$
$$1 + 0 = 01$$
$$1 + 1 = 10 \qquad (1_{10} + 1_{10} = 2_{10})$$
$$1 + 1 + 1 = 11 \qquad (1_{10} + 1_{10} + 1_{10} = 3_{10})$$

Each of these results consists of a **sum bit** and a **carry bit.** For the first two results shown, the carry bit is 0. The final sum in the table is the result of adding a carry bit from a sum in a less significant position.

When we add two 1-bit binary numbers in a logic circuit, the result *always* consists of a sum bit and a carry bit, even when the carry is 0, since each bit corresponds to a measurable voltage at a specific circuit location. Just because the value of the carry is 0 does not mean it has ceased to exist.

EXAMPLE 7.1

Calculate the sum $10010 + 1010$.

■ Solution

```
         ┌──── (Carry from sum of 2nd LSBs)
        1
      10010
   +   1010
      11100
```

EXAMPLE 7.2

Calculate the sum 10111 + 10010.

■ **Solution**

$$
\begin{array}{r}
\text{(Carry bits)}\\
1\;\;11\\
10111\\
+\;\;\underline{10010}\\
101001
\end{array}
$$

■ **SECTION 7.1A REVIEW PROBLEMS**

7.1   Add 11111 + 1001.

7.2   Add 10011 +1101.

Subtraction

■ **KEY TERMS**

Subtrahend   The number in a subtraction operation that is subtracted from another number.

Minuend   The number in a subtraction operation from which another number is subtracted.

Difference   The result of a subtraction operation.

Borrow   A digit brought back from a more significant position when the subtrahend digit is larger than the minuend digit.

In unsigned binary subtraction, two operands, called the **subtrahend** and the **minuend,** are subtracted to yield a result called the **difference.** In the operation $x = a - b,$ x is the difference, a is the minuend, and b is the subtrahend. To remember which comes first, think of the minuend as the number that is di*min*ished (i.e., something is taken away from it).

Unsigned binary subtraction is based on the following four operations:

$$
\begin{array}{l}
0 - 0 = 0\\
1 - 0 = 1\\
1 - 1 = 0\\
10 - 1 = 1 \qquad (2_{10} - 1_{10} = 1_{10})
\end{array}
$$

The last operation shows how to obtain a positive result when subtracting a 1 from a 0: **borrow** 1 from the next most significant bit.

Borrowing Rules:

1. If you are borrowing from a position that contains a 1, leave behind a 0 in the borrowed-from position.
2. If you are borrowing from a position that already contains a 0, you must borrow from a more significant digit that contains a 1. All 0s up to that point become 1s, and the last borrowed-from digit becomes a 0.

EXAMPLE 7.3

Subtract 1110 – 1001.

■ Solution

(New 2nd LSB) ─────┐ ┌───── (Bit borrowed from 2nd LSB)

```
        01
      11\0
    – 1001
      0101
```

EXAMPLE 7.4

Subtract 10000 – 101.

■ Solution

```
                            1111
  10000  (original        1\0\0\0\0  (After borrowing
–   101   problem)      –   101    from higher-order bits)
                           1011
```

■ SECTION 7.1B REVIEW PROBLEMS

7.3 Subtract 10101 – 10010.

7.4 Subtract 10000 – 1111.

7.2 REPRESENTING SIGNED BINARY NUMBERS

■ KEY TERMS

Sign Bit A bit, usually the MSB, that indicates whether a signed binary number is positive or negative.

Magnitude Bits The bits of a signed binary number that tell us how large the number is (i.e., its magnitude).

True-Magnitude Form A form of signed binary number whose magnitude is represented in true binary.

1's Complement A form of signed binary notation in which negative numbers are created by complementing all bits of a number, including the sign bit.

2's Complement A form of signed binary notation in which negative numbers are created by adding 1 to the 1's complement form of the number.

Binary arithmetic operations are performed by digital circuits that are designed for a fixed number of bits, since each bit has a physical location within a circuit. It is useful to have a way of representing binary numbers within this framework that accounts not only for the magnitude of the number, but for the sign as well.

This can be accomplished by designating one bit of a binary number, usually the most significant bit, as the **sign bit** and the rest as **magnitude bits.** When the number is negative, the sign bit is 1, and when the number is positive, the sign bit is 0.

There are several ways of writing the magnitude bits, each having its particular advantages. **True-magnitude form** represents the magnitude in straight binary form, which is relatively easy for a human operator to read. Complement forms, such as **1's complement** and **2's complement,** modify the magnitude so that it is more suited to digital circuitry.

> **NOTE...**
> Positive numbers are the same in all three notations.

True-Magnitude Form

In true-magnitude form, the magnitude of a number is translated into its true binary value. The sign is represented by the MSB, 0 for positive and 1 for negative.

EXAMPLE 7.5

Write the following numbers in 6-bit true-magnitude form:

a. $+25_{10}$ **b.** -25_{10} **c.** $+12_{10}$ **d.** -12_{10}

■ **Solution**

Translate the magnitudes of each number into 5-bit binary, padding with leading zeros as required, and set the sign bit to 0 for a positive number and 1 for a negative number.

a. 011001   **b.** 111001   **c.** 001100   **d.** 101100

1's Complement Form

True-magnitude and 1's complement forms of binary numbers are the same for positive numbers—the magnitude is represented by the true binary value and the sign bit is 0. We can generate a negative number in one of two ways:

1. Write the positive number of the same magnitude as the desired negative number. Complement each bit, including the sign bit; or
2. Subtract the n-bit positive number from a binary number consisting of n 1s.

EXAMPLE 7.6

Convert the following numbers to 8-bit 1's complement form:

a. $+57_{10}$ **b.** -57_{10} **c.** $+72_{10}$ **d.** -72_{10}

■ **Solution**

Positive numbers are the same as numbers in true-magnitude form. Negative numbers are the bitwise complements of the corresponding positive number.

a. $+57_{10} = 00111001$ **b.** $-57_{10} = 11000110$
c. $+72_{10} = 01001000$ **d.** $-72_{10} = 10110111$

We can also generate an 8-bit 1's complement negative number by subtracting its positive magnitude from 11111111 (eight 1s). For example, for part b:

$$
\begin{array}{r}
11111111 \\
-\ \underline{00111001} \quad (+57_{10}) \\
11000110 \quad (-57_{10})
\end{array}
$$

2's Complement Form

Positive numbers in 2's complement form are the same as in true-magnitude and 1's complement forms. We create a negative number by adding 1 to the 1's complement form of the number.

EXAMPLE 7.7

Convert the following numbers to 8-bit 2's complement form:

a. $+57_{10}$ **b.** -57_{10} **c.** $+72_{10}$ **d.** -72_{10}

■ **Solution**

a. $+57 = 00111001$
b. $-57 = 11000110$ (1's complement)
$$\underline{1}$$
11000111 (2's complement)
c. $+72 = 01001000$
d. $-72 = 10110111$ (1's complement)
$$\underline{1}$$
10111000 (2's complement)

A negative number in 2's complement form can be made positive by 2's complementing it again. Try it with the negative numbers in Example 7.7.

7.3 SIGNED BINARY ARITHMETIC

■ **KEY TERM**

Signed Binary Arithmetic   Arithmetic operations performed using signed binary numbers.

Signed Addition

Signed addition is done in the same way as unsigned addition. The only difference is that both operands *must* have the same number of magnitude bits, and each has a sign bit.

EXAMPLE 7.8

Add $+30_{10}$ and $+75_{10}$. Write the operands and the sum as 8-bit signed binary numbers.

■ **Solution**

$$
\begin{array}{rl}
+30 & 00011110 \\
+75 & +01001011 \\
\hline
+105 & 01101001
\end{array}
$$

(Magnitude bits)
(Sign bit)

Subtraction

The real advantage of using complement notation for **signed binary arithmetic** becomes evident when we subtract signed binary numbers. In complement notation, we add a negative number instead of subtracting a positive number. We thus have only one kind of operation—addition—and can use the same circuitry for both addition and subtraction.

This idea does not work for true-magnitude numbers. In the complement forms, the magnitude bits change depending on the sign of the number. In true-magnitude form, the magnitude bits are the same regardless of the sign of the number.

Let us subtract $80_{10} - 65_{10} = 15_{10}$ using 8-bit 1's complement and 2's complement addition. We will also show that the method of adding a negative number to perform subtraction is not valid for true-magnitude signed numbers.

1's Complement Method

> ### ■ KEY TERM
>
> **End-Around Carry**   An operation in 1's complement subtraction where the carry bit resulting from a sum of two 1's complement numbers is added to that sum.

Add the 8-bit 1's complement values of 80 and –65. If the sum results in a carry beyond the sign bit, perform an **end-around carry.** That is, add the carry to the sum.

$$+80_{10} = 01010000$$
$$+65_{10} = 01000001$$
$$-65_{10} = 10111110 \quad \text{(1's complement)}$$

$$
\begin{array}{rl}
(+80) & 01010000 \\
+ (-65) & +\ \underline{10111110} \\
& 1\ 00001110 \\
& \quad\ \ \boxed{\longrightarrow\ 1} \quad \text{(End-around carry)} \\
(+15) & 00001111
\end{array}
$$

2's Complement Method

Add the 8-bit 2's complement values of 80 and –65. If the sum results in a carry into the 9th bit, discard it. (We discard this bit because in a defined 8-bit size, there is no place to hold it, much as the leading digit in a car's odometer is not seen when the mileage rolls over from 99999 to 100000. The "1" is not seen because there is no place to show it.)

$$+80_{10} = 01010000$$
$$+65_{10} = 01000001$$
$$-65_{10} = 10111110 \quad \text{(1's complement)}$$
$$+\ \underline{\qquad 1}$$
$$10111111 \quad \text{(2's complement)}$$

$$
\begin{array}{rl}
(+80) & 01010000 \\
+(-65) & +\ \underline{10111111} \\
(+15) & 1\ 00001111 \\
& \underline{\qquad\qquad\qquad} \text{(Discard carry)}
\end{array}
$$

True-Magnitude Method

$$+80_{10} = 01010000$$
$$+65_{10} = 01000001$$
$$-65_{10} = 11000001$$

$$
\begin{array}{rl}
(+80) & 01010000 \\
+(-65) & +\ \underline{11000001} \\
(?) & 1\ 00010001
\end{array}
$$

If we perform an end-around carry, the result is $00010010 = 18_{10}$. If we discard the carry, the result is $00010001 = 17_{10}$. Neither answer is correct. Thus, adding a negative true-magnitude number is not equivalent to subtraction.

Negative Sum or Difference

All examples to this point have given positive-valued results. When a 2's complement addition or subtraction yields a negative sum or difference, we can't just read the magnitude from the result since a 2's complement operation modifies the bits of a negative number. We must calculate the 2's complement of the sum or difference, which will give us the positive number that has the same magnitude. That is, $-(-x) = +x$.

EXAMPLE 7.9

Subtract $65_{10} - 80_{10}$ in 8-bit 2's complement form.

■ **Solution**

$$
\begin{array}{lll}
+65_{10} = & 01000001 & \\
+80_{10} = & 01010000 & \\
-80_{10} = & 10101111 & \text{(1's complement)} \\
+ & \underline{\qquad 1} & \\
& 10110000 & \text{(2's complement)}
\end{array}
$$

$$
\begin{array}{ll}
(+65) & 01000001 \\
+ (-80) + & \underline{10110000} \\
& 11110001
\end{array}
$$

Take the 2's complement of the difference to find the positive number with the same magnitude.

$$
\begin{array}{lll}
11110001 & & (-15) \\
00001110 & \text{(1's complement)} & \\
+ \underline{\qquad 1} & & \\
00001111 & \text{(2's complement)} & (+15)
\end{array}
$$

$00001111 = +15_{10}$. We generated this number by complementing 11110001. Thus, $11110001 = -15_{10}$.

Range of Signed Numbers

The largest positive number in 2's complement notation is a 0 followed by n 1s for a number with n magnitude bits. For instance, the largest positive 4-bit number is $0111 = +7_{10}$. The negative number with the largest magnitude is *not* the 2's complement of the largest positive number. We can find the largest negative number by extension of a sequence of 2's complement numbers.

The 4-bit 2's complement form of -7_{10} is $1000 + 1 = 1001$. The positive and negative numbers with the next largest magnitudes are $0110 (= +6_{10})$ and $1010 (= -6_{10})$. If we continue this process, we will get the list of numbers in Table 7.1.

We have generated the 4-bit negative numbers from -1_{10} (1111) through -7_{10} (1001) by writing the 2's complement forms of the positive numbers 1 through 7. Notice that these numbers count down in binary sequence. The next 4-bit number in the sequence (which is the only binary number we have left) is 1000. By extension, $1000 = -8_{10}$. This number is its own 2's complement. (Try it.) It exemplifies a general rule for the largest-magnitude numbers with n magnitude bits.

NOTE . . .

A 2's complement number consisting of a 1 followed by n 0s is equal to -2^n. Therefore, the range of a signed number, x, is $-2^n \leq x \leq 2^n - 1$ for a number with n magnitude bits.

TABLE 7.1   4-Bit 2's Complement Numbers

Decimal	2's Complement
+7	0111
+6	0110
+5	0101
+4	0100
+3	0011
+2	0010
+1	0001
0	0000
−1	1111
−2	1110
−3	1101
−4	1100
−5	1011
−6	1010
−7	1001
−8	1000

EXAMPLE 7.10

Write the largest positive and negative numbers for an 8-bit signed number in 2's complement and decimal notation.

■ **Solution**

$$01111111 = +127 \quad (\text{7 magnitude bits: } 2^7 - 1 = 127)$$
$$10000000 = -128 \quad (\text{1 followed by seven 0s: } -2^7 = -128)$$

EXAMPLE 7.11

Write -16_{10}:

a. As an 8-bit 2's complement number
b. As a 5-bit 2's complement number

(8-bit numbers are more common than 5-bit numbers in digital systems, but it is useful to see how we must write the same number differently with different numbers of bits.)

■ **Solution**

a. An 8-bit number has 7 magnitude bits and 1 sign bit.

$$
\begin{array}{rl}
+16 = 00010000 & \\
-16 = 11101111 & (\text{1's complement}) \\
+ \underline{1} & \\
11110000 & (\text{2's complement})
\end{array}
$$

b. A 5-bit number has 4 magnitude bits and 1 sign bit. Four magnitude bits are not enough to represent +16. However, a 1 followed by n 0s is equal to -2^n. For a 1 and four 0s, $-2^n = -2^4 = -16$. Thus, $10000 = -16_{10}$.

The last five bits of the binary equivalent of –16 are the same in both the 5-bit and 8-bit numbers.

Sign Extension

> ### ■ KEY TERM
>
> **Sign Extension**   The process of fitting a number into a fixed size of 2's complement number by padding the number with leading 0s for a positive number and leading 1s for a negative number.

Example 7.11 implies that the magnitude bits of a 2's complement number remain the same, regardless of the bit size of the number, but the number may be padded out with leading 1s or 0s, depending on its sign. This is called **sign extension.** Sign extension can apply to any positive or negative number in 2's complement format. If we wish to write a given positive number with a larger number of bits, we pad the number with leading 0s (positive sign bits). If we wish to extend the number of bits in a negative number, we pad it with leading 1s (negative sign bits).

For example, the smallest bit size required to write +25 is six bits: a sign bit and five magnitude bits. The number is written: 011001. The 6-bit negative value in 2's complement form is 100111. In 8-bit and 12-bit form, these numbers are simply padded out with additional leading sign bits, as follows:

$$8\text{-bit: } +25 = 00011001$$
$$-25 = 11100111$$
$$12\text{-bit: } +25 = 000000011001$$
$$-25 = 111111100111$$

■ SECTION 7.3 REVIEW PROBLEMS

7.5   Write –32 as an 8-bit 2's complement number.

7.6   Write –32 as a 6-bit 2's complement number.

Sign Bit Overflow

> ### ■ KEY TERM
>
> **Overflow**   An erroneous carry into the sign bit of a signed binary number that results from a sum or difference larger than can be represented by the number of magnitude bits.

Signed addition of positive numbers is performed in the same way as unsigned addition. The only problem occurs when the number of bits in the sum of two numbers exceeds the number of magnitude bits and **overflows** into the sign bit. This causes the number to appear to be negative when it is not. For example, the sum $75 + 96 = 171$ causes an overflow in 8-bit signed addition. In unsigned addition the binary equivalent is:

$$\begin{array}{r} 1001011 \\ + \underline{1100000} \\ 10101011 \end{array}$$

In signed addition, the sum is the same, but has a different meaning.

$$\begin{array}{r} 0\ 1001011 \\ + \underline{0\ 1100000} \\ 1\ 0101011 \end{array}$$

(Sign bit) ⌐   └——┴——┴—(Magnitude bits)

The sign bit is 1, indicating a negative number, which cannot be true, since the sum of two positive numbers is always positive.

> **NOTE . . .**
>
> A sum of positive signed binary numbers must not exceed $2^n - 1$ for numbers having n magnitude bits. Otherwise, there will be an overflow into the sign bit.

Overflow in Negative Sums

Overflow can also occur with large negative numbers. For example, the 8-bit addition of -80_{10} and -65_{10} should produce the result:

$$-80_{10} + (-65_{10}) = -145_{10}$$

In 8-bit 2's complement notation, we get:

$$
\begin{array}{rll}
+80_{10} = & 01010000 & \\
-80_{10} = & 10101111 & \text{(1's complement)} \\
+ & \underline{\qquad 1} & \\
& 10110000 & \text{(2's complement)} \\
+65_{10} = & 01000001 & \\
-65_{10} = & 10111110 & \text{(1's complement)} \\
+ & \underline{\qquad 1} & \\
& 10111111 & \text{(2's complement)}
\end{array}
$$

$$
\begin{array}{rl}
-80 & 10110000 \\
+ (-65) & + \underline{10111111} \\
? & 1\ 01101111
\end{array}
$$

(Incorrect magnitude = 111_{10})
(Erroneous sign bit = 0)
(Discard carry)

This result shows a positive sum of two negative numbers—clearly incorrect. We can extend the statement we made earlier about permissible magnitudes of sums to include negative as well as positive numbers.

> **NOTE . . .**
>
> A sum of signed binary numbers must be within the range of $-2^n \leq \text{sum} \leq 2^n - 1$ for numbers having n magnitude bits. Otherwise, there will be an overflow into the sign bit.

For an 8-bit signed number in 2's complement form, the permissible range of sums is $10000000 \leq \text{sum} \leq 01111111$. In decimal, this range is $-128 \leq \text{sum} \leq +127$.

> **NOTE . . .**
>
> A sum of two positive numbers is always positive. A sum of two negative numbers is always negative. Any 2's complement addition or subtraction operation that appears to contradict these rules has produced an overflow into the sign bit.

EXAMPLE 7.12

Which of the following sums will produce a sign bit overflow in 8-bit 2's complement notation? How can you tell?

a. $67_{10} + 33_{10}$
b. $67_{10} + 63_{10}$
c. $-96_{10} - 22_{10}$
d. $-96_{10} - 42_{10}$

■ Solution

A sign bit overflow is generated if the sum of two positive numbers appears to produce a negative result or the sum of two negative numbers appears to produce a positive result. In other words, overflow occurs if the operand sign bits are both 1 and the sum sign bit is 0 or vice versa. We know this will happen if an 8-bit sum is outside the range ($-128 \leq$ sum $\leq + 127$).

a.
(+67)	01000011	(no overflow;
+(+33)	+00100001	sum of positive numbers
(+100)	01100100	is positive.)

b.
(+67)	01000011	(Overflow; sum of
+(+63)	00111111	positive numbers is negative.
+(130)	10000010	Sum > +127; out of range.)

c.
(+96) =	01100000	
(−96) =	10011111	(1's complement)
+	1	
	10100000	(2's complement)

(+22) =	00010110	
(−22)	11101001	(1's complement)
+	1	
	11101010	(2's complement)

(−96)	10100000	
+(−22)	11101010	
(−118)	1 10001010	

————— (Magnitude bits)
————— (Sign bit)
————— (Discard carry)

(No overflow; sum of two negative numbers is negative.)

d.
(+96) =	01100000	
(−96) =	10011111	(1's complement)
+	1	
	10100000	(2's complement)

(+42) =	00101010	
(−42)	11010101	(1's complement)
+	1	
	11010110	(2's complement)

(−96)	10100000	
+(−42)	11010110	
(−138)	1 01110110	

————— (Magnitude bits)
————— (Sign bit)
————— (Discard carry)

(Overflow; sum of two negative numbers is positive. Sum < −128; out of range.)

> **NOTE . . .**
>
> The carry bit generated in 1's and 2's complement operations is not the same as an overflow bit. (See Example 7.12, parts c and d.) An overflow is a change in the sign bit, which leads us to believe that the number is opposite in sign from its true value. A carry is the result of an operation carrying beyond the physical limits of an n-bit number. It is similar to the idea of an odometer rolling over from 99999.9 to 1 00000.0. There are not enough places to hold the new number, so it goes back to the beginning and starts over.

7.4   HEXADECIMAL ARITHMETIC

(This section may be omitted without loss of continuity.)

The main reason to be familiar with addition and subtraction in the hexadecimal system is that it is useful for calculations related to microcomputer and memory systems. Microcomputer systems often use binary numbers of 8, 16, 20, 32, or 64 bits. Rather than write out all these bits, we use hex numbers as shorthand. Binary numbers having 8, 16, 20, 32, or 64 bits can be represented by 2, 4, 5, 8, or 16 hex digits, respectively.

Hex Addition

Hex addition is very much like decimal addition, except that we must remember how to deal with the hex digits A to F. A few sums are helpful:

$$F + 1 = 10$$
$$F + F = 1E$$
$$F + F + 1 = 1F$$

The positional multipliers for the hexadecimal system are powers of 16. Thus, the most significant digit of the first sum is in the 16's column. The equivalent sum in decimal is:

$$15_{10} + 1_{10} = 16_{10} = 10H$$

The second sum is the largest possible sum of two hex digits; the carry to the next position is 1. This shows that the sum of two hex digits will never produce a carry larger than 1. The second sum can be calculated as follows:

$$FH + FH = 15_{10} + 15_{10}$$
$$= 30_{10}$$
$$= 16_{10} + 14_{10}$$
$$= 10H + EH$$
$$= 1EH$$

The third sum shows that if there is a carry from a previous sum, the carry to the next bit will still be 1.

> **NOTE . . .**
>
> It is useful to think of any digits larger than 9 as their decimal equivalents. For any number greater than 15_{10} (FH), subtract 16_{10}, convert the difference to its hex equivalent, and carry 1 to the next digit position.

EXAMPLE 7.13

Add 6B3H + A9CH.

■ **Solution**

Hex	Decimal Equivalents
6B3	(6) (11) (3)
+A9C	+ (10) (9) (12)
	(16) (20) (15)

For sums greater than 15, subtract 16 and carry 1 to the next position:

	Hex		Decimal Equivalents
(Carry) ————	11		(1) (1)
	6B3		(6) (11) (3)
+	A9C	+	(10) (9) (12)
	114F		(1) (1) (4) (15)

Sum: 6B3H + A9CH = 114FH.

Hex Subtraction

There are two ways to subtract hex numbers. The first reverses the addition process in the previous section. The second is a complement form of subtraction.

EXAMPLE 7.14

Subtract 6B3H – 49CH.

■ **Solution**

Hex	Decimal Equivalent
6B3	(6) (11) (3)
– 49C	– (4) (9) (12)

To subtract the least significant digits, we must borrow 10H (16_{10}) from the previous position. This leaves the subtraction looking like this:

	Hex		Decimal Equivalent
(Borrow) ————	1		
	6A3		(6) (10) (16 + 3)
–	49C	–	(4) (9) (12)
	217		(2) (1) (7)

The second subtraction method is a complement method, where, as in 2's complement subtractions, we add a negative number to subtract a positive number.

Calculate the 15's complement of a hex number by subtracting it from a number having the same number of digits, all Fs. Calculate the 16's complement by adding 1 to this number. This is the negated value of the number.

EXAMPLE 7.15

Negate the hex number 15AC by calculating its 16's complement.

■ **Solution**

$$
\begin{array}{r}
\text{FFFF} \\
-\ \underline{\text{15AC}} \\
\text{EA53} \qquad \text{(15's complement)} \\
+\ \underline{\hspace{0.6cm}1} \\
\text{EA54} \qquad \text{(16's complement)}
\end{array}
$$

The original value, 15AC, can be restored by calculating the 16's complement of EA54. Try it.

EXAMPLE 7.16

Subtract 8B63 – 55D7 using the complement method.

■ **Solution**

Find the 16's complement of 55D7.

$$
\begin{array}{r}
\text{FFFF} \\
-\ \underline{\text{55D7}} \\
\text{AA28} \qquad \text{(15's complement)} \\
+\ \underline{\hspace{0.6cm}1} \\
\text{AA29} \qquad \text{(16's complement)}
\end{array}
$$

Therefore, –55D7 = AA29.

$$
\begin{array}{r}
1 \\
8\text{B}63 \\
+\ \underline{\text{AA29}} \\
1\ \ 358\text{C}
\end{array}
$$

(Discard ——⌐
carry)

Difference: 8B63 – 55D7 = 358C.

■ **SECTION 7.4 REVIEW PROBLEM**

7.7   Perform the following hexadecimal calculations:

   a. A25F + 74A2

   b. 7380 – 5FFF

7.5 NUMERIC AND ALPHANUMERIC CODES

BCD Codes

■ **KEY TERM**

Binary-Coded Decimal (BCD).   A code that represents each digit of a decimal number by a binary value.

BCD stands for **binary-coded decimal.** As the name implies, BCD is a system of writing decimal numbers with binary digits. There is more than one way to do this, as BCD is a *code,* not a positional number system. That is, the various positions of the bits do not necessarily represent increasing powers of a specified number base.

Two commonly used BCD codes are 8421 code, where the bits for *each decimal digit* are weighted, and Excess-3 code, where each decimal digit is represented by a binary number that is 3 larger than the true binary value of the digit.

8421 Code

TABLE 7.2 Decimal Digits and Their 8421 BCD Equivalents

Decimal Digit	BCD (8421)
0	0000
1	0001
2	0010
3	0011
4	0100
5	0101
6	0110
7	0111
8	1000
9	1001

■ **KEY TERM**

8421 code   A BCD code that represents each digit of a decimal number by its 4-bit true binary value.

The most straightforward BCD code is the **8421 code,** also called Natural BCD. Each decimal digit is represented by its 4-bit true binary value. When we talk about BCD code, this is usually what we mean.

This code is called 8421 because these are the positional weights of each digit. Table 7.2 shows the decimal digits and their BCD equivalents.

8421 BCD is not a positional number system, because each decimal digit is encoded separately as a 4-bit number.

EXAMPLE 7.17

Write 4987_{10} in both binary and 8421 BCD.

■ **Solution**

The binary value of 4987_{10} can be calculated by repeated division by 2:

$$4987_{10} = 1\ 0011\ 0111\ 1011_2$$

The BCD digits are the binary values of each decimal digit, encoded separately. We can break bits into groups of 4 for easier reading. Note that the first and last BCD digits each have a leading zero to make them 4 bits long.

$$4987_{10} = 0100\ 1001\ 1000\ 0111_{BCD}$$

Excess-3 Code

■ **KEY TERMS**

Excess-3 Code   A BCD code that represents each digit of a decimal number by a binary number derived by adding 3 to its 4-bit true binary value.

Self-Complementing   A code that automatically generates a negative equivalent (e.g., 9's complement for a decimal code) when all its bits are inverted.

9's Complement   A way of writing decimal numbers where a number is made negative by subtracting each of its digits from 9 (e.g., $-726 = 999 - 726 = 273$ in 9's complement).

TABLE 7.3 Decimal Digits and Their 8421 and Excess-3 Equivalents

Decimal Digit	8421	Excess-3
0	0000	0011
1	0001	0100
2	0010	0101
3	0011	0110
4	0100	0111
5	0101	1000
6	0110	1001
7	0111	1010
8	1000	1011
9	1001	1100

Excess-3 code is a type of BCD code that is generated by adding $11_2\ (3_{10})$ to the 8421 BCD codes. Table 7.3 shows the Excess-3 codes and their 8421 and decimal equivalents.

The advantage of this code is that it is **self-complementing.** If the bits of the Excess-3 digit are inverted, they yield the **9's complement** of the decimal equivalent.

We can generate the 9's complement of an *n*-digit number by subtracting it from a number made up of *n* 9s. Thus, the 9's complement of 632 is $999 - 632 = 367$.

The Excess-3 equivalent of 632 is 1001 0110 0101. If we invert all the bits, we get 0110 1001 1010. The decimal equivalent of this Excess-3 number is 367, the 9's complement of 632.

This property is useful for performing decimal arithmetic digitally.

Gray Code

> ■ **KEY TERM**
>
> **Gray Code**   A binary code that progresses such that only one bit changes between two successive codes.

TABLE 7.4   4-Bit Gray Code

Decimal	True Binary	Gray Code
0	0000	0000
1	0001	0001
2	0010	0011
3	0011	0010
4	0100	0110
5	0101	0111
6	0110	0101
7	0111	0100
8	1000	1100
9	1001	1101
10	1010	1111
11	1011	1110
12	1100	1010
13	1101	1011
14	1110	1001
15	1111	1000

Table 7.4 shows a 4-bit **Gray code** compared to decimal and binary values. Any two adjacent Gray codes differ by exactly one bit.

Gray code can be extended indefinitely if you understand the relationship between the binary and Gray digits. Let us name the binary digits $b_3 b_2 b_1 b_0$, with b_3 as the most significant bit, and the Gray code digits $g_3 g_2 g_1 g_0$ for a 4-bit code. The Gray code bits are defined as follows:

$$g_3 = b_3$$
$$g_2 = b_3 \oplus b_2$$
$$g_1 = b_2 \oplus b_1$$
$$g_0 = b_1 \oplus b_0$$

For an n-bit code, the MSBs are the same in Gray and binary ($g_n = b_n$). The other Gray digits are generated by the Exclusive OR function of the binary digits in the same position and the next most significant position.

Another way to generate a Gray code sequence is to recognize the inherent symmetry in the code. For example, a 2-bit Gray code sequence is given by:

$$00$$
$$01$$
$$11$$
$$10$$

(Note that this is the same sequence as the numbering of cells in a Karnaugh map.) To generate a 3-bit Gray code, write the 2-bit sequence, then write it again in reverse order.

$$00$$
$$01$$
$$11$$
$$10$$
$$10$$
$$11$$
$$01$$
$$00$$

Add an MSB of 0 to the first four codes and an MSB of 1 to the last four codes. The sequence followed by the last two bits of all codes is symmetrical about the center of the sequence.

$$000$$
$$001$$
$$011$$

010

110

111

101

100

We can apply a similar process to generate a 4-bit Gray code. Write the 3-bit sequence, then again in reverse order. Add an MSB of 0 to the first half of the table and an MSB of 1 to the second half. This procedure yields the code in Table 7.4.

We will see further applications of Gray code in a later chapter.

ASCII Code

> ### ■ KEY TERMS
>
> **Alphanumeric Code**   A code used to represent letters of the alphabet and numerical characters.
>
> **ASCII**   American Standard Code for Information Interchange. A 7-bit code for representing alphanumeric and control characters.
>
> **Case Shift**   Changing letters from capitals (UPPERCASE) to small letters (lowercase) or vice versa.

Digital systems and computers could operate perfectly well using only binary numbers. However, if there is any need for a human operator to understand the input and output data of a digital system, it is necessary to have a system of communication that is understandable to both a human operator and the digital circuit.

A code that represents letters (alphabetic characters) and numbers (numeric characters) as binary numbers is called an **alphanumeric code.** The most commonly used alphanumeric code is **ASCII** ("askey"), which stands for American Standard Code for Information Interchange. ASCII code represents letters, numbers, and other printable characters in 7 bits. In addition, ASCII has a repertoire of "control characters," codes that are used to send control instructions to and from devices such as video display terminals, printers, and modems.

Table 7.5 shows the ASCII code in both binary and hexadecimal forms. The code for any character consists of the bits in the column heading, then those in the row heading. For example, the ASCII code for "A" is 1000001_2 or 41H. The code for "a" is 1100001_2 or 61H. The codes for capital (uppercase) and lowercase letters differ only by the second most significant bit, for all letters. Thus, we can make an alphabetic **case shift,** like using the Shift key on a typewriter or computer keyboard, by switching just one bit.

Numeric characters are listed in column 3, with the least significant digit of the ASCII code being the same as the represented number value. For example, the numeric character "0" is equivalent to 30H in ASCII. The character "9" is represented as 39H.

The codes in columns 0 and 1 are control characters. They cannot be displayed on any kind of output device, such as a printer or video monitor, although they may be used to control the device. For instance, if the codes 0AH (Line Feed) and 0DH (Carriage Return) are sent to a printer, the paper will advance by one line and the print head will return to the beginning of the line.

The displayable characters begin at 20H ("space") and continue to 7EH ("tilde"). Spaces are considered ASCII characters.

TABLE 7.5 ASCII Code

LSBs	000 (0)	001 (1)	010 (2)	011 (3)	100 (4)	101 (5)	110 (6)	111 (7)
0000 (0)	NUL	DLE	SP	0	@	P	`	p
0001 (1)	SOH	DC1	!	1	A	Q	a	q
0010 (2)	STX	DC2	"	2	B	R	b	r
0011 (3)	ETX	DC3	#	3	C	S	c	s
0100 (4)	EOT	DC4	$	4	D	T	d	t
0101 (5)	ENQ	NAK	%	5	E	U	e	u
0110 (6)	ACK	SYN	&	6	F	V	f	v
0111 (7)	BEL	ETB	'	7	G	W	g	w
1000 (8)	BS	CAN	(	8	H	X	h	x
1001 (9)	HT	EM	)	9	I	Y	i	y
1010 (A)	LF	SUB	*	:	J	Z	j	z
1011 (B)	VT	ESC	+	;	K	[	k	{
1100 (C)	FF	FS	,	<	L	\	l	|
1101 (D)	CR	GS	-	=	M	]	m	}
1110 (E)	SO	RS	.	>	N	^	n	~
1111 (F)	SI	US	/	?	O	—	o	DEL

Control Characters:

NUL—NULL
SOH—Start of Header
STX—Start Text
ETX—End Text
EOT—End of Transmission
ENQ—Enquiry
ACK—Acknowledge
BEL—Bell
BS—Backspace
HT—Horizontal Tabulation
LF—Line Feed
VT—Vertical Tabulation
FF—Form Feed
CR—Carriage Return
SO—Shift Out
SI—Shift In
SP—Space

DLE—Data Link Escape
DC1—Device Control 1
DC2—Device Control 2
DC3—Device Control 3
DC4—Device Control 4
NAK—No Acknowledgment
SYN—Synchronous Idle
ETB—End of Transmission Block
CAN—Cancel
EM—End of Medium
SUB—Substitute
ESC—Escape
FS—Form Separator
GS—Group Separator
RS—Record Separator
US—Unit Separator
DEL—Delete

EXAMPLE 7.18

Encode the following string of characters into ASCII (hexadecimal form). Do not include quotation marks.

"Total system cost: $4,000,000. @ 10%"

■ **Solution**

Each character, including spaces, is represented by two hex digits as follows:

54	6F	74	61	6C	20	73	79	73	74	65	6D	20	63	6F	73	74	3A	20
T	o	t	a	l	SP	s	y	s	t	e	m	SP	c	o	s	t	:	SP

24	34	2C	30	30	30	2C	30	30	30	2E	20	40	20	31	30	25
$	4	,	0	0	0	,	0	0	0	.	SP	@	SP	1	0	%

■ **SECTION 7.5 REVIEW PROBLEM**

7.8 Decode the following sequence of hexadecimal ASCII codes.

54 72 75 65 20 6F 72 20 46 61 6C 73 65 3A 20 31
2F 34 20 3C 20 31 2F 32

7.6 BINARY ADDERS AND SUBTRACTORS

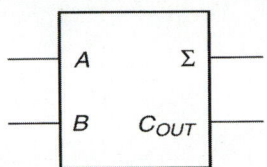

FIGURE 7.1 Half Adder

TABLE 7.6 Half Adder Truth Table

A	B	C_{OUT}	Σ
0	0	0	0
0	1	0	1
1	0	0	1
1	1	1	0

FIGURE 7.2 Half Adder Circuit

FIGURE 7.3 Full Adder

TABLE 7.7 Full Adder Truth Table

A	B	C_{IN}	C_{OUT}	Σ
0	0	0	0	0
0	0	1	0	1
0	1	0	0	1
0	1	1	1	0
1	0	0	0	1
1	0	1	1	0
1	1	0	1	0
1	1	1	1	1

Half and Full Adders

> ■ **KEY TERMS**
>
> **Half Adder**   A circuit that will add two bits and produce a sum bit and a carry bit.
>
> **Full Adder**   A circuit that will add a carry bit from another full or half adder and two operand bits to produce a sum bit and a carry bit.

There are only three possible sums of two 1-bit binary numbers:

$$0 + 0 = 00$$
$$0 + 1 = 01$$
$$1 + 1 = 10$$

We can build a simple combinational logic circuit to produce these sums. Let us designate the bits on the left side of these equalities as inputs to the circuit and the bits on the right side as outputs. Let us call the LSB of the output the sum bit, symbolized by Σ, and the MSB of the output the carry bit, designated C_{OUT}.

Figure 7.1 shows the logic symbol of the circuit, which is called a **half adder.** Its truth table is given in Table 7.6. Since addition is subject to the commutative property ($A + B = B + A$), the second and third lines of the truth table are the same.

The Boolean functions of the two outputs, derived from the truth table, are:

> **NOTE . . .**
>
> $$C_{OUT} = AB$$
> $$\Sigma = \overline{A}B + A\overline{B} = A \oplus B$$

The corresponding logic circuit is shown in Figure 7.2.

The half adder circuit cannot account for an *input* carry, that is, a carry from a lower-order 1-bit addition. A **full adder,** shown in Figure 7.3, can add two 1-bit numbers *and* accept a carry bit from a previous adder stage. Operation of the full adder is based on the following sums:

$$0 + 0 + 0 = 00$$
$$0 + 0 + 1 = 01$$
$$0 + 1 + 1 = 10$$
$$1 + 1 + 1 = 11$$

Designating the left side of the above equalities as circuit inputs A, B, and C_{IN} and the right side as outputs C_{OUT} and Σ, we can make the truth table in Table 7.7. (The second and third of the above sums each account for three lines in the full adder truth table.)

The unsimplified Boolean expressions for the outputs are:

$$C_{OUT} = \overline{A} B C_{IN} + A \overline{B} C_{IN} + A B \overline{C}_{IN} + A B C_{IN}$$
$$\Sigma = \overline{A} \overline{B} C_{IN} + \overline{A} B \overline{C}_{IN} + A \overline{B} \overline{C}_{IN} + A B C_{IN}$$

There are a couple of ways to simplify these expressions.

FIGURE 7.4   K-Maps for a Full Adder

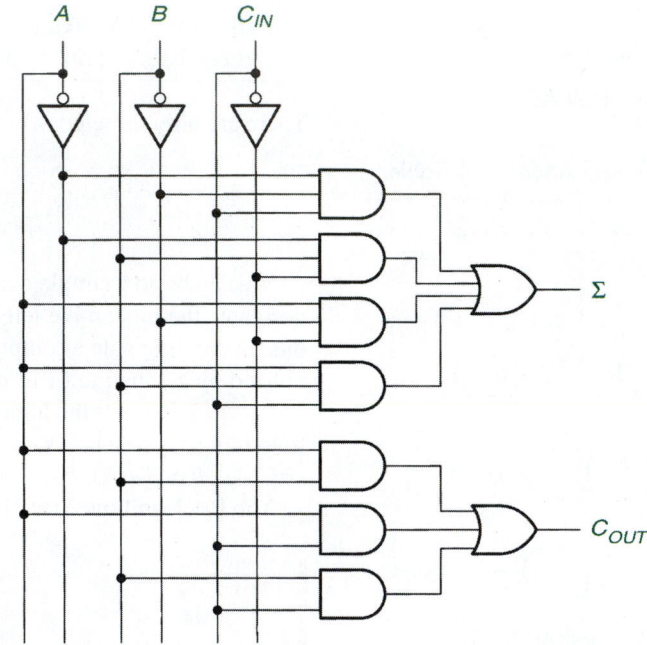

FIGURE 7.5   Full Adder from K-Map Simplification

Karnaugh Map Method

Since we have expressions for Σ and C_{OUT} in sum-of-products form, let us try to use the Karnaugh maps in Figure 7.4 to simplify them. The expression for Σ doesn't reduce at all. The simplified expression for C_{OUT} is:

$$C_{OUT} = A\,B + A\,C_{IN} + B\,C_{IN}$$

The corresponding logic circuits for Σ and C_{OUT}, shown in Figure 7.5, don't give us much of a simplification.

Boolean Algebra Method

The simplest circuit for C_{OUT} and Σ involves the Exclusive OR function, which we cannot derive from K-map groupings. This can be shown by Boolean algebra, as follows:

$$C_{OUT} = \bar{A}\,B\,C_{IN} + A\,\bar{B}\,C_{IN} + A\,B\,\bar{C}_{IN} + A\,B\,C_{IN}$$
$$= (\bar{A}\,B + A\,\bar{B})C_{IN} + A\,B\,(\bar{C}_{IN} + C_{IN})$$
$$= (A \oplus B)\,C_{IN} + A\,B$$

$$\Sigma = (\bar{A}\,\bar{B} + AB)\,C_{IN} + (\bar{A}\,B + A\,\bar{B})\,\bar{C}_{IN}$$
$$= (\overline{A \oplus B})\,C_{IN} + (A \oplus B)\,\bar{C}_{IN} \qquad \text{Let } x = A \oplus B$$
$$= \bar{x}\,C_{IN} + x\,\bar{C}_{IN}$$
$$= x \oplus C_{IN}$$
$$= (A \oplus B) \oplus C_{IN}$$

The simplified expressions are as follows:

> **NOTE . . .**
>
> $$C_{OUT} = (A \oplus B)\,C_{IN} + A\,B$$
> $$\Sigma = (A \oplus B) \oplus C_{IN}$$

Figure 7.6 shows the logic circuit derived from these equations. If you refer back to the half adder circuit in Figure 7.2, you will see that the full adder can be constructed from two half adders and an OR gate, as shown in Figure 7.7. The carry output is HIGH whenever there is a carry from either half adder, or both.

FIGURE 7.6 Full Adder from Logic Gates

FIGURE 7.7 Full Adder from Two Half Adders

 EXAMPLE 7.19

Evaluate the Boolean expression for Σ and C_{OUT} of the full adder in Figure 7.8 for the following input values. What is the binary value of the outputs in each case?

a. $A = 0, B = 0, C_{IN} = 1$
b. $A = 1, B = 0, C_{IN} = 0$
c. $A = 1, B = 0, C_{IN} = 1$
d. $A = 1, B = 1, C_{IN} = 0$

(continued)

FIGURE 7.8 Example 7.19: Full Adder

■ Solution

The output of a full adder for any set of inputs is simply given by $C_{OUT} \Sigma = A + B + C_{IN}$. For each of the stated sets of inputs:

a. $C_{OUT} \Sigma = A + B + C_{IN} = 0 + 0 + 1 = 01$
b. $C_{OUT} \Sigma = A + B + C_{IN} = 1 + 0 + 0 = 01$
c. $C_{OUT} \Sigma = A + B + C_{IN} = 1 + 0 + 1 = 10$
d. $C_{OUT} \Sigma = A + B + C_{IN} = 1 + 1 + 0 = 10$

We can verify each of these sums algebraically by plugging the specified inputs into the full adder Boolean equations:

$$C_{OUT} = (A \oplus B) C_{IN} + A B$$
$$\Sigma = (A \oplus B) \oplus C_{IN}$$

a. $C_{OUT} = (0 \oplus 0) \cdot 1 + 0 \cdot 0$
$= 0 \cdot 1 + 0$
$= 0 + 0 = 0$
$\Sigma = (0 \oplus 0) \oplus 1$
$= 0 \oplus 1 = 1$ (Binary equivalent: $C_{OUT} \Sigma = 01$)

b. $C_{OUT} = (1 \oplus 0) \cdot 0 + 1 \cdot 0$
$= 1 \cdot 0 + 0$
$= 0 + 0 = 0$
$\Sigma = (1 \oplus 0) \oplus 0$
$= 1 \oplus 0 = 1$ (Binary equivalent: $C_{OUT} \Sigma = 01$)

c. $C_{OUT} = (1 \oplus 0) \cdot 1 + 1 \cdot 0$
$= 1 \cdot 1 + 0$
$= 1 + 0 = 1$
$\Sigma = (1 \oplus 0) \oplus 1$
$= 1 \oplus 1 = 0$ (Binary equivalent: $C_{OUT} \Sigma = 10$)

d. $C_{OUT} = (1 \oplus 1) \cdot 0 + 1 \cdot 1$
$= 0 \cdot 0 + 1$
$= 0 + 1 = 1$
$\Sigma = (1 \oplus 1) \oplus 0$
$= 0 \oplus 0 = 0$ (Binary equivalent: $C_{OUT} \Sigma = 10$)

In each case, the binary equivalent is the same as the number of HIGH inputs, regardless of which inputs they are.

EXAMPLE 7.20

Combine a half adder and a full adder to make a circuit that will add two 2-bit numbers. Check that the circuit will work by adding the following numbers and writing the binary equivalents of the inputs and outputs:

a. $A_2 A_1 = 01$, $B_2 B_1 = 01$
b. $A_2 A_1 = 11$, $B_2 B_1 = 10$

■ Solution

The 2-bit adder is shown in Figure 7.9. The half adder combines A_1 and B_1; A_2, B_2, and C_1 are added in the full adder. The carry output, C_1, of the half adder is connected to the carry input of the full adder. (A half adder can be used only in the LSB of a multiple-bit addition.)

FIGURE 7.9 Example 7.20: 2-Bit Adder

Sums
a. $01 + 01 = 010$
$A_1 = 1, B_1 = 1$ $C_1 = 1, \Sigma_1 = 0$

$A_2 = 0, B_2 = 0, C_1 = 1$ $C_2 = 0, \Sigma_2 = 1$
(Binary equivalent: $A_2 A_1 + B_2 B_1 = C_2 \Sigma_2 \Sigma_1 = 010$)
b. $11 + 10 = 101$
$A_1 = 1, B_1 = 0$ $C_1 = 0, \Sigma_1 = 1$

$A_2 = 1, B_2 = 1, C_1 = 0$ $C_2 = 1, \Sigma_2 = 0$
(Binary equivalent: $A_2 A_1 + B_2 B_1 = C_2 \Sigma_2 \Sigma_1 = 101$)

Parallel Binary Adder/Subtractor

> ■ **KEY TERMS**
>
> **Cascade** To connect an output of one device to an input of another, often for the purpose of expanding the number of bits available for a particular function.
>
> **Parallel Binary Adder** A circuit, consisting of n full adders, that will add two n-bit binary numbers. The output consists of n sum bits and a carry bit.
>
> **Ripple Carry** A method of passing carry bits from one stage of a parallel adder to the next by connecting C_{OUT} of one full adder to C_{IN} of the following stage.

As Example 7.20 implies, a binary adder can be expanded to any number of bits by using a full adder for each bit addition and connecting their carry inputs and outputs in **cascade.** Figure 7.10 shows four full adders connected as a 4-bit **parallel binary adder.**

FIGURE 7.10 4-Bit Parallel Binary Adder

The first stage (LSB) can be either a full adder with its carry input forced to logic 0 or a half adder, as there is no previous stage to provide a carry. The addition is done one bit at a time, with the carry from each adder propagating to the next stage.

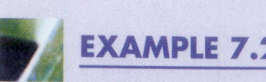 **EXAMPLE 7.21**

Verify the summing operation of the circuit in Figure 7.10 by calculating the output for the following sets of inputs:

a. $A_4 A_3 A_2 A_1 = 0101$, $B_4 B_3 B_2 B_1 = 1001$
b. $A_4 A_3 A_2 A_1 = 1111$, $B_4 B_3 B_2 B_1 = 0001$

■ **Solution**

At each stage, $A + B + C_{IN} = C_{OUT} \Sigma$.

a. $0101 + 1001 = 1110$
$(5_{10} + 9_{10} = 14_{10})$
$A_1 = 1, B_1 = 1, C_0 = 0; C_1 = 1, \Sigma_1 = 0$
$A_2 = 0, B_2 = 0, C_1 = 1; C_2 = 0, \Sigma_2 = 1$
$A_3 = 1, B_3 = 0, C_2 = 0; C_3 = 0, \Sigma_3 = 1$
$A_4 = 0, B_4 = 1, C_3 = 0; C_4 = 0, \Sigma_4 = 1$
(Binary equivalent: $C_4 \Sigma_4 \Sigma_3 \Sigma_2 \Sigma_1 = 01110$)
b. $1111 + 0001 = 10000$
$(15_{10} + 1_{10} = 16_{10})$
$A_1 = 1, B_1 = 1, C_0 = 0; C_1 = 1, \Sigma_1 = 0$
$A_2 = 1, B_2 = 0, C_1 = 1; C_2 = 1, \Sigma_2 = 0$
$A_3 = 1, B_3 = 0, C_2 = 1; C_3 = 1, \Sigma_3 = 0$
$A_4 = 1, B_4 = 0, C_3 = 1; C_4 = 1, \Sigma_4 = 0$
(Binary equivalent: $C_4 \Sigma_4 \Sigma_3 \Sigma_2 \Sigma_1 = 10000$)

The internal carries in the parallel binary adder in Figure 7.10 are achieved by a system called **ripple carry.** The carry output of one full adder cascades directly to the carry input of the next. Every time a carry bit changes, it "ripples" through some or all of the following stages. A sum is not complete until the carry from another stage has arrived. The equivalent circuit of a 4-bit ripple carry is shown in Figure 7.11.

A potential problem with this design is that the adder circuitry does not switch instantaneously. A carry propagating through a ripple adder adds delays to the summation time and, more importantly, can introduce unwanted intermediate states.

Examine the sum $(1111 + 0001 = 10000)$. For a parallel adder having a ripple carry, the output goes through the following series of changes as the carry bit propagates through the circuit:

$$C_4 \Sigma_4 \Sigma_3 \Sigma_2 \Sigma_1 = 01111$$
$$01110$$
$$01100$$
$$01000$$
$$10000$$

FIGURE 7.11   4-Bit Ripple Carry Chain

If the output of the full adder is being used to drive another circuit, these unwanted intermediate states may cause erroneous operation of the load circuit.

Fast Carry

An alternative carry circuit is called **fast carry** or **look-ahead carry.** The idea behind fast carry is that the circuit will examine all the *A* and *B* bits simultaneously and produce an output carry that uses fewer levels of gating than a ripple carry circuit. Also, since there is a carry bit gate network for each internal stage, the propagation delay is the same for each full adder, regardless of the input operands.

The algebraic relation between operand bits and fast carry output is presented here, without proof. It can be developed from the fast carry circuit of Figure 7.12 by tracing the logic of the gates in the circuit.

$$C_4 = A_4 B_4 + A_3 B_3 (A_4 + B_4) + A_2 B_2 (A_4 + B_4)(A_3 + B_3)$$
$$+ A_1 B_1 (A_4 + B_4)(A_3 + B_3)(A_2 + B_2)$$
$$+ C_0 (A_4 + B_4)(A_3 + B_3)(A_2 + B_2)(A_1 + B_1)$$

We can make some intuitive sense of this expression by examining it one term at one time. The first term says if the MSBs of both operands are 1, there will be a carry (e.g., $1000 + 1000 = 10000$; carry generated).

The second term says if both second bits are 1 AND at least one MSB is 1, there will be a carry (e.g., $0100 + 1100 = 10000$, or $1100 + 1100 = 11000$; carry generated in either case). This pattern can be followed logically through all the terms.

The internal carry bits are generated by similar circuits that drive the carry input of each full adder stage in the parallel adder. In general, we can generate each internal carry by expanding the following expression:

$$C_n = A_n B_n + C_{n-1} (A_n + B_n)$$

The algebraic expressions for the remaining carry bits are:

$$C_1 = A_1 B_1 + C_0 (A_1 + B_1)$$
$$C_2 = A_2 B_2 + A_1 B_1 (A_2 + B_2) + C_0 (A_2 + B_2)(A_1 + B_1)$$
$$C_3 = A_3 B_3 + A_2 B_2 (A_3 + B_3) + A_1 B_1 (A_3 + B_3)(A_2 + B_2)$$
$$+ C_0 (A_3 + B_3)(A_2 + B_2)(A_1 + B_1)$$

■ **SECTION 7.6A REVIEW PROBLEM**

7.9   Refer to the logic diagrams for the ripple carry and fast carry circuits (Figures 7.11 and 7.12). How many gates must a carry bit propagate through in each device if the effect of the carry input ripples through to the C_4 bit?

Using VHDL Components to Implement a Parallel Adder

FIGURE 7.12 4-Bit Fast Carry Circuit

LAB SOURCE
Lab # 18
Full Adder and Parallel
Binary Adder

Component   A complete VHDL design entity that can be used as a part of a higher-level file in a hierarchical design.

Component Declaration Statement   A statement that defines the input and output port names of a component used in a VHDL design entity.

Component Instantiation Statement   A statement that maps port names of a VHDL component to the port names, internal signals, or variables of a higher-level VHDL design entity.

Port   An input or output of a VHDL design entity or component.

Instantiate   To use an instance of a component.

VHDL designs can be created using a **hierarchy** of design entities. Certain functions, such as full adders, decoders, and so on, can be created once and used in many designs or multiple times in a single design.

We can create a parallel adder in VHDL by using multiple instances of a full adder **component** in the top-level file of a VHDL design hierarchy. Figure 7.13 shows a graphical illustration of this concept. Each full adder shown is an instance of a component written in VHDL, as shown in the following.

Full_add.vhd

```
-- full_add.vhd
-- Full adder: adds two bits, a and b, plus input carry
-- to yield sum bit and output carry.
LIBRARY ieee;
USE ieee.std_logic_1164. ALL;

ENTITY full_add IS
PORT (
    a, b, c_in : IN  STD_LOGIC;
    c_out, sum : OUT STD_LOGIC);
END full_add;

ARCHITECTURE adder OF full_add IS
BEGIN
    c_out <= ((a xor b) and c_in) or (a and b) ;
    sum   <= (a xor b) xor c_in;
END adder;
```

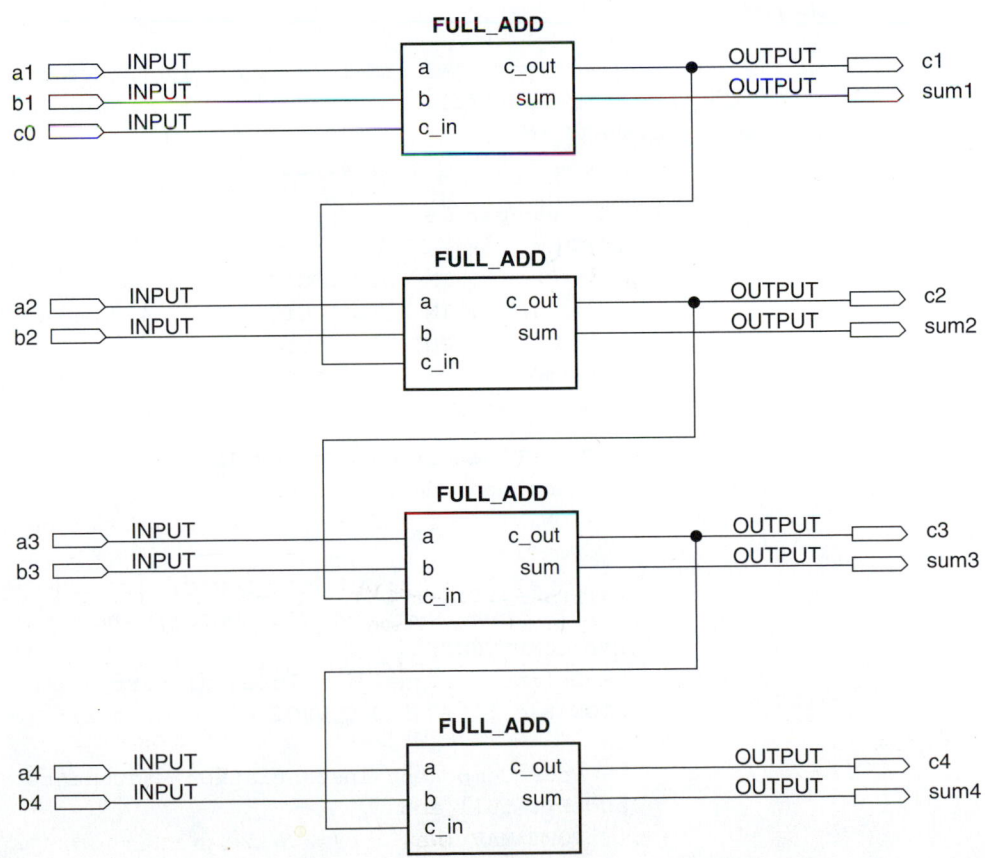

FIGURE 7.13 4-Bit Parallel Adder with Ripple Carry

We can create the same design as in Figure 7.13 using VHDL only. To make this hierarchical design we require:

1. A separate component file for a full adder (**full_add.vhd**), saved in a folder where the compiler can find it (i.e., on a library path)
2. A **component declaration statement** in the top-level file of the design hierarchy
3. A **component instantiation statement** for each instance of the full adder component

The general form of a design entity using components is:

```
ENTITY entity_name IS
    PORT ( input and output definitions);
END entity_name;

ARCHITECTURE arch_name OF entity_name IS
    component declaration(s);
    signal declaration(s);
BEGIN
    Component instantiation(s);
    Other statements;
END arch_name;
```

add4par.vhd

The VHDL file for a 4-bit parallel adder using full adder components is shown next.

```
-- add4par.vhd
-- 4-bit parallel adder, using 4 instances of the
-- component full_add
LIBRARY ieee;
USE ieee.std_logic_1164.ALL;

ENTITY add4par IS
    PORT(
        c0      : IN  STD_LOGIC;
        a, b    : IN  STD_LOGIC_VECTOR(4 downto 1);
        c4      : OUT STD_LOGIC;
        sum     : OUT STD_LOGIC_VECTOR(4 downto 1));
END add4par;

ARCHITECTURE adder OF add4par IS
    -- Component declaration
    COMPONENT full_add          Full_add function defined in separate file.
        PORT(
            a, b, c_in: IN  STD_LOGIC;        Ports must be the same as
            c_out, sum: OUT STD_LOGIC);       those in component file.
    END COMPONENT;
    -- Define a signal for internal carry bits
    SIGNAL c : STD_LOGIC_VECTOR(3 downto 1);
BEGIN
    -- Four Component Instantiation Statements
    adder1: full_add
        PORT MAP ( a       => a(1),
                   b       => b(1),
                   c_in    => c0
```

```
                    c_out    => c(1),
                    sum      => sum (1));
        adder2: full_add
           PORT MAP ( a       => a(2),
                      b       => b(2),
                      c_in    => c(1),
                      c_out   => c(2),
                      sum     => sum(2));
        adder3: full_add
           PORT MAP ( a       => a(3),
                      b       => b(3),
                      c_in    => c(2),
                      c_out   => c(3),
                      sum     => sum (3));
        adder4: full_add
           PORT MAP ( a       => a(4),
                      b       => b(4),
                      c_in    => c(3),
                      c_out   => c4,
                      sum     => sum (4));
        END adder;
```

Signal c(1) connects carry output of adder 1 to carry input of adder 2.

Ports of each full_add component are mapped explicitly.

User port and signal names
Component port names

The component declaration statement defines the **ports** of the component with the same names as in the **full_add.vhd.** Note that the form of the component declaration statement is almost the same as that of the component's entity declaration. In effect, we are redefining the component entity in the top-level file of the design hierarchy.

The component instantiation statement is of the following form:

```
__instance_name: __component_name
   GENERIC MAP (__parameter_name => __parameter_value ,
                __parameter_name => __parameter_value)
   PORT MAP (__component_port => __connect_port,
             __component_port => __connect_port);
```

In the generic map, a generalized parameter name can be mapped to a specific value when the component is **instantiated.** (To *instantiate* is to create an instance of something.) For example, a parameter name can be given a value that specifies the number of component output bits.

In the port map, component ports are the names of the ports used in the component file and connect ports are the names of the ports, variables, or signals used in the higher-level design entity. For example, the component ports of the full adder component are **a, b, c_in, c_out,** and **sum.** The connect ports for the instance **adder1** are **a(1), b(1), c0, c(1),** and **sum(1).** The ripple carry from **adder1** to **adder2** is achieved by mapping the port **c_in** of **adder2** to **c(1),** which is also mapped to the port **c_out** of **adder1.**

We can write the component instantiation statements more efficiently if we decide to use all ports of the component in the order they are defined. In this case, we can simply list the connect ports in the port map in the correct order, as follows:

```
adder1: full_add PORT MAP (a(1),b(1),c0,  c(1),sum(1));
adder2: full_add PORT MAP (a(2),b(2),c(1),c(2),sum(2));
adder3: full_add PORT MAP (a(3),b(3),c(2),c(3),sum(3));
adder4: full_add PORT MAP (a(4),b(4),c(3),c4, sum(4));
```

This is called implicit port mapping. If we only wish to use some of the component ports or use them in a different order than the order in which they were originally defined, we must use the previous form of port map (i.e., a => a(1), etc.). This form is called explicit port mapping.

GENERATE Statements

> ### ■ KEY TERM
>
> **GENERATE Statement**   A VHDL construct that is used to create repetitive portions of hardware.

The four component instantiation statements shown previously can be written in a more general form:

```
adder: full_add PORT MAP (a(i), b(i), c(i-1), c(i), sum(i));
```

A statement that can be written in this indexed form can be implemented using a **GENERATE statement,** which has the form:

```
label:
FOR index IN range GENERATE
    statements;
END GENERATE;
```

The VHDL code that follows shows how to use the statement to create a 4-bit adder.

add4gen.vhd

```
-- add4gen.vhd
-- 4-bit parallel adder, using a generate statement and
-- components
LIBRARY ieee;
USE ieee.std_logic_1164.ALL;

ENTITY add4gen IS
    PORT(
        c0    : IN  STD_LOGIC;
        a, b  : IN  STD_LOGIC_VECTOR(4 downto 1);
        c4    : OUT STD_LOGIC;
        sum   : OUT STD_LOGIC_VECTOR(4 downto 1));
END add4gen;

ARCHITECTURE adder OF add4gen IS
    -- Component declaration
    COMPONENT full_add
        PORT(
            a, b, c_in : IN  STD_LOGIC;
            c_out, sum : OUT STD_LOGIC);
    END COMPONENT;
    -- Define a signal for internal carry bits
    SIGNAL c : STD_LOGIC_VECTOR (4 downto 0);
BEGIN
    c(0)   <=  c0;——— Input port c0 mapped to internal signal c(0).

    adders:
    FOR i IN 1 to 4 GENERATE
        adder: full_add PORT MAP (a(i), b(i),    ⎤ Implicit port
                c(i-1), c(i), sum(i));           ⎦ mapping
    END GENERATE;

    c4 <= c(4); ———Output port c4 mapped to internal signal c(4).
END adder;
```

The GENERATE statement will create hardware that corresponds to the range of the index variable, i. In this case i goes from 1 to 4, so the statement creates four instances of the full adder. Since we have an input carry, an output carry and three internal carries, we must use a 5-bit signal (`STD_LOGIC_VECTOR (4 downto 0)`) if we are to include all carry bits in indexed form. The input carry, c0, defined in the entity declaration, is assigned to the vector element c(0). Similarly, the output, c4, is assigned the value of the element c(4).

We can create an adder of unspecified width by including a GENERIC clause in the entity declaration of the VHDL file that defines the adder. This clause is of the form `GENERIC(parameter : INTEGER := default_value);` The VHDL compiler requires the parameter to be "locally resolvable." That is, the value of the parameter must be known if the component is compiled on its own. Thus, a default value of the parameter is required.

To change the width (number of bits), we use this adder as a component in a higher-level file and specify the width of the component in a GENERIC MAP. The VHDL code for the *n*-bit adder is as follows. The default width is 8 bits.

addxgen.vhd

```
-- addxgen.vhd
-- n-bit parallel adder, using a GENERATE statement and
   components
-- Default width is 8 bits.
LIBRARY ieee;
USE ieee.std_logic_1164.ALL;

ENTITY addxgen IS
      GENERIC (width : INTEGER := 8);
      PORT(
            c0    : IN  STD LOGIC;
            a, b  : IN  STD_LOGIC_VECTOR(width downto 1);
            c_max : OUT STD_LOGIC;
            sum   : OUT STD_LOGIC_VECTOR(width downto 1));
END addxgen;

ARCHITECTURE adder OF addxgen IS
      -- Component declaration
      COMPONENT full_add
            PORT(
                  a, b, c_in : IN  STD_LOGIC;
                  c_out, sum : OUT STD_LOGIC);
      END COMPONENT;
      -- Define a signal for internal carry bits
      SIGNAL c : STD_LOGIC_VECTOR (width downto 0);
BEGIN
      c(0)  <=    c0;

      adders:
      FOR i IN 1 to width GENERATE
      adder: full_add PORT MAP (a(i), b(i), c(i-1), c(i),
            sum(i));
END GENERATE;

      c_max <=    c(width);
END adder;
```

Default value required for parameter.

If we want to create a 16-bit adder, we can use the adder component created previously and instantiate it as a 16-bit value. The code is as follows. In order to use the code in Quartus II, all components (**addxgen** and **full_add**) must be specified as part of the current project.

```
-- add16gen.vhd
-- 16-bit adder using the component addxgen.vhd
-- Also requires full_add.vhd for hierarchy
LIBRARY ieee;
USE ieee.std_logic_1164.ALL;

ENTITY add16gen IS
        PORT(
                c0    : IN   STD_LOGIC;
                a, b  : IN   STD_LOGIC_VECTOR(16 downto 1);
                c16   : OUT  STD_LOGIC;
                sum   : OUT  STD_LOGIC_VECTOR(16 downto 1));
END add16gen;

ARCHITECTURE adder of add16gen IS
        COMPONENT addxgen
                GENERIC (width : INTEGER);
                PORT(
                        c0    : IN   STD_LOGIC;
                        a, b  : IN   STD_LOGIC_VECTOR(width downto 1);
                        c_max : OUT  STD_LOGIC;
                        sum   : OUT  STD_LOGIC_VECTOR(width downto 1));
        END COMPONENT;
BEGIN
        add16 : addxgen
                        GENERIC MAP(width => 16)
                        PORT MAP(c0, a, b, c16, sum);
END adder;
```

Required component width is specified in entity declaration.

Default value of "width" not required in component declaration.

Parameter "width" specified in GENERIC MAP.

add16gen.vhd

2's Complement Subtractor

Recall the technique for subtracting binary numbers in 2's complement notation. For example, to find the difference $0101 - 0011$ by 2's complement subtraction:

1. Find the 2's complement of 0011:

$$
\begin{array}{ll}
0011 & \\
1100 & \text{(1's complement)} \\
\underline{+1} & \\
1101 & \text{(2's complement)}
\end{array}
$$

2. Add the 2's complement of the subtrahend to the minuend:

$$
\begin{array}{ll}
0101 & (+5) \\
+ \underline{1101} & (-3) \\
1\,0010 & (+2)
\end{array}
$$

(Discard carry)

We can easily build a circuit to perform 2's complement subtraction, using a parallel binary adder and an inverter for each bit of one of the operands. The circuit shown in Figure 7.14 performs the operation $(A - B)$.

The four inverters generate the 1's complement of B. The parallel adder generates the 2's complement by adding the carry bit (held at logic 1) to the 1's complement at the B inputs. Algebraically, this is expressed as:

$$A - B = A + (-B) = A + \bar{B} + 1$$

where $\bar{B}$ is the 1's complement of B, and $(\bar{B} + 1)$ is the 2's complement of B.

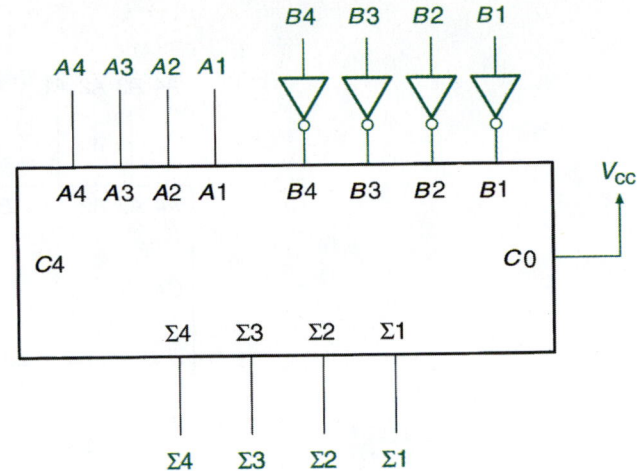

FIGURE 7.14 2's Complement Subtractor

EXAMPLE 7.22

Verify the operation of the 2's complement subtractor in Figure 7.14 by subtracting:

a. $1001 - 0011$ (unsigned)
b. $0100 - 0111$ (signed)

■ **Solution**

Let $\overline{B}$ be the 1's complement of B.

a. Inverter inputs (B):   0011
    Inverter outputs ($\overline{B}$):   1100
    Sum ($A + \overline{B} + 1$):   1001   (9)
                         1100   + (−3)
                  +   1   ———
                  1 0110   (6)

    (Discard carry)———┘

b. Inverter inputs (B):   0111
    Inverter outputs ($\overline{B}$):   1000
    Sum ($A + \overline{B} + 1$):   0100   (+4)
                           1000   + (−7)
                        +   1   ———
    Negative result: 1101   (−3)◄─┐
    1's complement of 1101:   0010   │
                        +   1   ———   │
    2's complement of 1101:   0011   (+3)——┘

Parallel Binary Adder/Subtractor

Figure 7.15 shows a parallel binary adder configured as a programmable adder/subtractor. The Exclusive OR gates work as programmable inverters to pass B to the parallel adder in either true or complement form, as shown in Figure 7.16.

FIGURE 7.15 2's Complement Adder/Subtractor

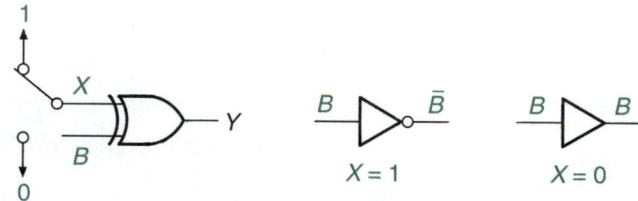

FIGURE 7.16 XOR as a Programmable Inverter

The $\overline{ADD/SUB}$ input is tied to the XOR inverter/buffers and to the carry input of the parallel adder. When $\overline{ADD/SUB} = 1$, B is complemented and the 1 from the carry input is added to the complement sum. The effect is to subtract $(A - B)$. When $\overline{ADD/SUB} = 0$, the B inputs are presented to the adder in true form and the carry input is 0. This produces an output equivalent to $(A + B)$.

This circuit can add or subtract 4-bit signed or unsigned binary numbers.

EXAMPLE 7.23

Write a VHDL file to implement the 4-bit adder/subtractor shown in Figure 7.15. Write a set of simulation criteria to test the functionality of the adder/subtractor. Create a simulation in Quartus II based on the criteria.

■ **Solution**

The VHDL file is as follows:

addsub4g.vhd

```
-- addsub4g.vhd
ENTITY addsub4g IS
    PORT (
        sub  : IN  BIT;
        a, b : IN  BIT_VECTOR (4 downto 1);
        c4   : OUT BIT;
        sum  : OUT BIT_VECTOR (4 downto 1));
END addsub4g;
```

```
ARCHITECTURE adder OF addsub4g IS
   -- Component declaration
   COMPONENT full_add
      PORT (
            a, b, c_in : IN BIT;
            c_out, sum : OUT BIT);
   END COMPONENT;
   -- Define a signal for internal carry bits
   SIGNAL c      : BIT_VECTOR (4 downto 0);
   SIGNAL b_comp : BIT_VECTOR (4 downto 1);
BEGIN
   -- add/subtract select to carry input (sub=1 for subtract)
   c(0)   <= sub;
   adders:
   FOR i IN 1 to 4 GENERATE
      -- invert b for subtract (b(i) xor 1),
      -- do not invert for add (b(i) xor 0)
      b_comp(i) <= b(i) xor sub;
      adder: full_add PORT MAP (a(i), b_comp(i), c(i-1), c(i),
sum(i));
   END GENERATE;
   c4   <= c(4);
END adder;
```

The VHDL code for the adder/subtractor is the same as that for the 4-bit adder created using a GENERATE statement, except that there is an input to select the add or subtract function. This input (**sub**) is tied to **c(0)** and to a set of XOR functions that invert **b** for subtraction. Input **b** is transferred through the XOR functions without inversion for the add function.

SIMULATION CRITERIA

The most thorough way to test the adder/subtractor is to apply 256 different combinations of input for the Add state and repeat the combinations for the Subtract state. However, a smaller set of input combinations will allow full testing of the circuit by checking all combinations of input for each full adder, even though not all combinations of the entire adder/subtractor are tested. This can be done as follows:

- set all inputs to 0 and **Sub** = 0 (Add function).
- set the inputs HIGH in sequence, starting at the LSB of *B* and going to MSB of *A*.
- set the inputs LOW in sequence, starting at the LSB of *B* and going to MSB of *A*.
- Repeat the input sequence for **Sub** = 1 (Subtract function).

Table 7.8 and Table 7.9 summarize the operation for the Add and Subtract functions, respectively. The simulation waveforms are shown in Figure 7.17.

(continued)

TABLE 7.8   Simulation Data for Add Function

A	B	Carry	Sum	Comment
0000	0000	0	0000	
0000	0001	0	0001	Sum follows B
0000	0011	0	0011	inputs, since
0000	0111	0	0111	$0 + B = B.$
0000	1111	0	1111	
0001	1111	1	0000	Sum fills with
0011	1111	1	0010	1s from 2nd
0111	1111	1	0110	LSB.
1111	1111	1	1110	
1111	1110	1	1101	A 0 "walks
1111	1100	1	1011	through" sum
1111	1000	1	0111	from 2nd LSB.
1111	0000	0	1111	
1110	0000	0	1110	Sum follows A
1100	0000	0	1100	inputs, since
1000	0000	0	1000	$A + 0 = A.$
0000	0000	0	0000	

TABLE 7.9   Simulation Data for Subtract Function

A	B	Carry	Sum	Comment
0000	0000	1	0000	
0000	0001	0	1111	Sum is 2's
0000	0011	0	1101	complement of
0000	0111	0	1001	B inputs, since
0000	1111	0	0001	$0 - B = -B.$
0001	1111	0	0010	A 1 "walks
0011	1111	0	0100	through" sum
0111	1111	0	1000	from 2nd LSB.
1111	1111	1	0000	
1111	1110	1	0001	Sum fills with
1111	1100	1	0011	1s.
1111	1000	1	0111	
1111	0000	1	1111	
1110	0000	1	1110	Sum follows A
1100	0000	1	1100	inputs, since
1000	0000	1	1000	$A - 0 = A.$
0000	0000	1	0000	

SECTION 7.6 ▰▰▰▰▰▰ **Binary Adders and Subtractors** **385**

FIGURE 7.17 Example 7.23: Simulation Waveforms for a 4-bit Adder/Subtractor

Note that the Carry output for the Subtract function is not perhaps what one might expect. (For example, why does the Carry = 1 when $A - B = 0000 - 0000$?) The Subtract function carry can be explained by manipulating the full adder carry equation, accounting for the case where $SUB = 1$:

$$C_{OUT} = (A \oplus (B \oplus SUB))C_{IN} + A(B \oplus SUB)$$
$$= (A \oplus \overline{B})C_{IN} + A\overline{B}$$
$$= (\overline{A}\overline{B} + AB)C_{IN} + A\overline{B}$$

A full adder will carry when C_{IN} is HIGH and A and B are both LOW or both HIGH, or when A is HIGH and B is LOW. These cases can all be found for the MSB adder in Table 7.9 when the carry output is HIGH.

The final carry resulting from this equation (C_4) is an unintended artifact of the system, which is really only designed to do 2's complement addition and subtraction. However, the internal carries (C_1, C_2, and C_3) resulting from the equation are necessary to generate a correct 2's complement output.

Overflow Detection

We will examine two methods for detecting overflow in a binary adder/subtractor: one that requires access to the sign bits of the operands and result and another that requires access to the internal carry bits of the circuit.

Recall from Example 7.12 the condition for detecting a sign bit overflow in a sum of two binary numbers.

> **NOTE...**
>
> If the sign bits of both operands are the same and the sign bit of the sum is different from the operand sign bits, an overflow has occurred.

This implies that overflow is not possible if the sign bits of the operands are different from each other. This is true because the sum of two opposite-sign numbers will always be smaller in magnitude than the larger of the two operands.

Here are two examples:

1. $(+15) + (-7) = (+8)$; $+8$ has a smaller magnitude than $+15$.
2. $(-13) + (+9) = (-4)$; -4 has a smaller magnitude than -13.

No carry into the sign bit will be generated in either case.

An 8-bit parallel binary adder will add two signed binary numbers as follows:

$$\begin{array}{ll} S_A A_7 A_6 A_5 A_4 A_3 A_2 A_1 & (S_A = \text{Sign bit of } A) \\ \underline{S_B B_7 B_6 B_5 B_4 B_3 B_2 B_1} & (S_B = \text{Sign bit of } B) \\ S_\Sigma \Sigma_7 \Sigma_6 \Sigma_5 \Sigma_4 \Sigma_3 \Sigma_2 \Sigma_1 & (S_\Sigma = \text{Sign bit of sum}) \end{array}$$

From our condition for overflow detection, we can make a truth table for an overflow variable, V, in terms of S_A, S_B, and S_Σ. Let us specify that $V = 1$ when there is an overflow condition. This condition occurs when $(S_A = S_B) \neq S_\Sigma$. Table 7.10 shows the truth table for the overflow detector function.

The SOP Boolean expression for the overflow detector is:

$$V = S_A S_B \overline{S}_\Sigma + \overline{S}_A \, \overline{S}_B \, S_\Sigma$$

Figure 7.18 shows a logic circuit that will detect a sign bit overflow in a parallel binary adder. The inputs S_A, S_B, and S_Σ are the MSBs (sign bits) of the adder A and B inputs and Σ outputs, respectively.

TABLE 7.10 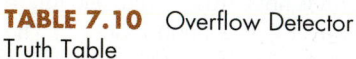 Overflow Detector Truth Table

S_A	S_B	S_Σ	V
0	0	0	0
0	0	1	1
0	1	0	0
0	1	1	0
1	0	0	0
1	0	1	0
1	1	0	1
1	1	1	0

FIGURE 7.18 Overflow Detector

EXAMPLE 7.24

Combine two instances of the 4-bit adder/subtractor shown in Figure 7.15 and other logic to make an 8-bit adder/subtractor that includes a circuit to detect sign bit overflow.

■ **Solution**

Figure 7.19 represents the 8-bit adder/subtractor with an overflow detector of the type shown in Figure 7.18.

FIGURE 7.19 Example 7.24: 8-Bit Adder with Overflow Detector

A second method of overflow detection generates an overflow indication by examining the carry bits into and out of the MSB of a 2's complement adder/subtractor.

Consider the following 8-bit 2's complement sums. We will use our previous knowledge of overflow to see whether overflow occurs and then compare the carry bits into and out of the MSB.

a. 80H + 80H
b. 7FH + 01H
c. 7FH + 80H
d. 7FH + C0H

a. 80H = 10000000

$$
\begin{array}{r}
10000000 \\
+\ 10000000 \\
\hline
1\ 00000000
\end{array}
\quad \text{(Sign bit overflow; } V = 1\text{)}
$$

Carry into MSB = 0
Carry out of MSB = 1

(continued)

b. 7FH = 01111111   01111111

    01H = 00000001   + 00000001

                        0 10000000   (Sign bit overflow; $V = 1$)

Carry into MSB = 1

Carry out of MSB = 0

---

c. 7FH = 01111111   01111111

    80H = 10000000   + 10000000

                        0 11111111   (No sign bit overflow; $V = 0$)

Carry into MSB = 0

Carry out of MSB = 0

---

d. 7FH = 01111111   01111111

    C0H = 11000000   + 11000000

                        1 00111111   (No sign bit overflow; $V = 0$)

Carry into MSB = 1

Carry out of MSB = 1

These examples suggest that a 2's complement sum has overflowed if there is a carry into or out of the MSB, but not both. For an 8-bit adder/subtractor, we can write the Boolean equation for this condition as $V = C_8 \oplus C_7$. More generally, for an n-bit adder/subtractor, $V = C_n \oplus C_{n-1}$.

Figure 7.20 shows a circuit that can implement the overflow detection from the carry into and out of the MSB of an 8-bit adder.

FIGURE 7.20   Example 7.24: Overflow Detector Using Internal Carry Bits

■ SECTION 7.6B REVIEW PROBLEM

7.10   What is the permissible range of values of a sum or difference, x, in a 12-bit parallel binary adder if it is written as:

    a. A signed binary number?

    b. An unsigned binary number?

7.7   BCD ADDERS

■ KEY TERM

BCD Adder   A parallel adder whose output is in groups of 4 bits, each group representing a BCD digit.

It is sometimes convenient to have the output of an adder circuit available as a BCD number, particularly if the result is to be displayed numerically. The problem is that most parallel adders have binary outputs, and 6 of the 16 possible 4-bit binary sums—1010 to 1111—are not within the range of the BCD code.

BCD numbers range from 0000 to 1001, or 0 to 9 in decimal. The unsigned binary sum of any two BCD numbers plus an input carry can range from 00000 (= 0000 + 0000 + 0) to 10011 (= 1001 + 1001 + 1 = 19_{10}).

For any sum up to 1001, the BCD and binary values are the same. Any sum greater than 1001 must be modified, since it requires a second BCD digit. For example, the binary value of 19_{10} is 10011_2. The BCD value of 19_{10} is 0001 1001$_{BCD}$. (The most significant digit of a sum of two BCD digits and a carry will never be larger than 1, as the largest such sum is 19_{10}.)

Table 7.11 shows the complete list of possible binary sums of two BCD digits (*A* and *B*) and a carry *(C)*, their decimal equivalents, and their corrected BCD values. The MSD of the BCD sum is shown only as a carry bit, with leading zeros suppressed.

TABLE 7.11 Binary Sums of Two BCD Digits and a Carry Bit

Binary Sum $(A + B + C)$	Decimal	Corrected BCD (Carry + BCD)
00000	0	0 + 0000
00001	1	0 + 0001
00010	2	0 + 0010
00011	3	0 + 0011
00100	4	0 + 0100
00101	5	0 + 0101
00110	6	0 + 0110
00111	7	0 + 0111
01000	8	0 + 1000
01001	9	0 + 1001
01010	10	1 + 0000
01011	11	1 + 0001
01100	12	1 + 0010
01101	13	1 + 0011
01110	14	1 + 0100
01111	15	1 + 0101
10000	16	1 + 0110
10001	17	1 + 0111
10010	18	1 + 1000
10011	19	1 + 1001

Figure 7.21 shows how we can add two BCD digits and get a corrected output. The **BCD adder** circuit consists of a standard 4-bit parallel adder to get the binary sum and a code converter to translate it into BCD.

The Binary-to-BCD code converter operates on the binary inputs as follows:

1. A carry output is generated if the binary sum is in the range 01010 ≤ sum ≤ 10011 (BCD equivalent: 1 0000 ≤ sum ≤ 1 1001).
2. If the binary sum is less than or equal to 01001, the output of the code converter is the same as the input.
3. If the sum is in the range 01010 ≤ sum ≤ 10011, the four LSBs of the input must be corrected to a BCD value. This can be done by adding 0110 to the four LSBs of the input and discarding any resulting carry. We add 0110_2 (6_{10}) because we must account for six unused codes.

Let's look at how each of these requirements can be implemented by a digital circuit.

Carry Output

The carry output will be automatically 0 for any uncorrected sum from 00000 to 01001 and automatically 1 for any sum from 10000 to 10011. Thus, if the binary adder's carry output, which we will call C_4', is 1, the BCD adder's carry output, C_4, will also be 1.

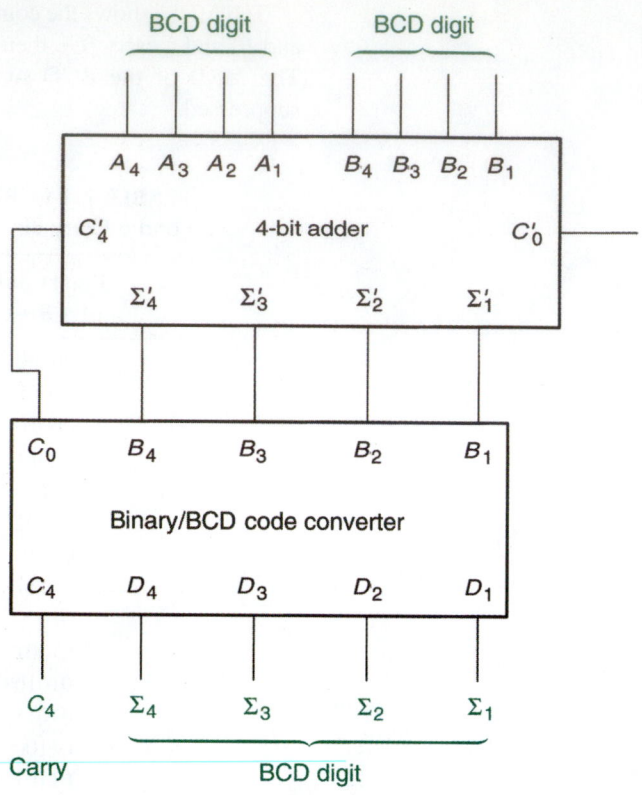

FIGURE 7.21   BCD Adder ($1\frac{1}{2}$ Digit Output)

Any sum falling between these ranges, that is, between 01010 and 01111, must have its MSB modified. This modifying condition is a function, designated C_4'', of the binary adder's sum outputs when its carry output is 0. This function can be simplified by a Karnaugh map, as shown in Figure 7.22, resulting in the following Boolean expression.

$$C_4'' = \Sigma_4' \Sigma_3' + \Sigma_4' \Sigma_2'$$

The BCD carry output C_4 is given by:

$$\begin{aligned} C_4 &= C_4' + C_4'' \\ &= C_4' + \Sigma_4' \Sigma_3' + \Sigma_4' \Sigma_2' \end{aligned}$$

The BCD carry circuit is shown in Figure 7.23.

Sum Correction

The four LSBs of the binary adder output need to be corrected if the sum is 01010 or greater and need not be corrected if the binary sum is 01001 or less. This condition

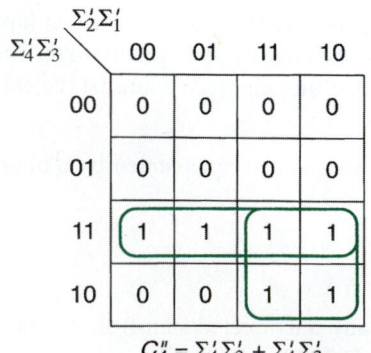

FIGURE 7.22   Carry as a Function of Sum Bits When $C_4' = 0$.

FIGURE 7.23   BCD Carry Circuit

FIGURE 7.24 BCD Adder

is indicated by the BCD carry. Let us designate the binary sum outputs as $\Sigma'_4 \, \Sigma'_3 \, \Sigma'_2 \, \Sigma'_1$ and the BCD sum outputs as $\Sigma_4 \, \Sigma_3 \, \Sigma_2 \, \Sigma_1$.

$$\text{If } C_4 = 0, \Sigma_4 \, \Sigma_3 \, \Sigma_2 \, \Sigma_1 = \Sigma'_4 \, \Sigma'_3 \, \Sigma'_2 \, \Sigma'_1 + 0000;$$
$$\text{If } C_4 = 1, \Sigma_4 \, \Sigma_3 \, \Sigma_2 \, \Sigma_1 = \Sigma'_4 \, \Sigma'_3 \, \Sigma'_2 \, \Sigma'_1 + 0110.$$

Figure 7.24 shows a BCD adder, complete with a binary adder, BCD carry, and sum correction. A second parallel adder is used for sum correction. The B inputs are the uncorrected binary sum inputs. The A inputs are either 0000 or 0110, depending on the value of the BCD carry.

EXAMPLE 7.25

Write a VHDL file for a BCD adder, using two parallel adder components, such as in the logic diagram in Figure 7.24.

■ **Solution**

bcd_add.vhd

```
--bcd_add.vhd
--BCD adder, using 2 instances of the component add4par
--See Figure 7.24 for component interconnections.

LIBRARY ieee;
USE ieee.std_logic_1164.ALL;
USE ieee.std_logic_unsigned.ALL;

ENTITY bcd_add IS
    PORT(
        c0        : IN  STD_LOGIC;
        a, b      : IN  STD_LOGIC_VECTOR(4 downto 1);
        c4        : OUT STD_LOGIC;
        sum       : OUT STD_LOGIC_VECTOR(4 downto 1));
END bcd_add;
```

(continued)

```vhdl
ARCHITECTURE adder OF bcd_add IS
   -- Component declaration
   COMPONENT add4par
   PORT(
      c0      : IN  STD_LOGIC;
      a, b    : IN  STD_LOGIC_VECTOR(4 downto 1);
      c4      : OUT STD_LOGIC;
      sum     : OUT STD_LOGIC_VECTOR(4 downto 1));
   END COMPONENT;
   SIGNAL c4_bin : STD_LOGIC;
   SIGNAL c4_bcd : STD_LOGIC;
   SIGNAL sum_bin: STD_LOGIC_VECTOR(4 downto 1);
   SIGNAL binary : STD_LOGIC_VECTOR(5 downto 1);
   SIGNAL a_bcd  : STD_LOGIC_VECTOR(4 downto 1);
   SIGNAL c0_bcd : STD_LOGIC;
BEGIN
   -- Instantiate 4-bit adder (binary sum)
   add_bin: add4par
      PORT MAP ( c0  => c0,
                 a   => a,
                 b   => b,
                 c4  => c4_bin,
                 sum => sum_bin);

   -- Instantiate 4-bit adder (binary-BCD converter)
   converter: add4par
      PORT MAP ( c0  => c0_bcd,
                 a   => a_bcd,
                 b   => sum_bin,
                 sum => sum);

   -- BCD carry input (code converter has no input carry)
   c0_bcd     <= '0';

   -- Concatenate binary carry and sum for comparison functions
   binary <= c4_bin & sum bin;
-- Generate a BCD carry for number from 10 to 19
   -- If there is a BCD carry, add 6 to sum
   -- If no BCD carry, add 0
   PROCESS (binary)
   BEGIN
      -- This comparison requires std_logic_unsigned package
      IF (binary >= "01010" and binary <= "10011") THEN
         c4_bcd  <= '1';
         a_bcd   <= "0110";
      ELSE
         c4_bcd  <= '0';
         a_bcd   <= "0000";
      END IF;
   END PROCESS;

   -- Map c4 port to internal BCD carry
   c4 <= c4_bcd;

END adder;
```

FIGURE 7.25 $4\frac{1}{2}$-Digit BCD Adder

Multiple-Digit BCD Adders

Several BCD adders can be cascaded to add multidigit BCD numbers. Figure 7.25 shows a $4\frac{1}{2}$-digit BCD adder. The carry output of the most significant digit is considered to be a half-digit since it can only be 0 or 1. The output range of the $4\frac{1}{2}$-digit BCD adder is 00000 to 19999.

BCD adders are cascaded by connecting the code converter carry output of one stage to the binary adder carry input of the next most significant stage. Each BCD output digit represents a decade, designated as the units, tens, hundreds, thousands, and ten thousands digits.

■ SECTION 7.7 REVIEW PROBLEM

7.11 What is the maximum BCD sum of two 3-digit numbers with no carry input? How many digits are required to display this result on a numerical output?

7.8 CARRY GENERATION IN QUARTUS II

■ KEY TERM

Speed Grade A specification that indicates the internal delay time that can be expected of a CPLD.

The VHDL adder circuits implemented in Section 7.6 were all defined using a ripple carry format. Is it necessary to design a fast carry circuit when compiling one of these adder designs in Quartus II? Probably not.

Recall that the design strategy behind the fast carry circuit was to flatten the gate network, that is, to replace a long network (many gates for the carry bit to pass through) with a wide one (fewer levels of gating). Also recall that any combinational logic function can be implemented as a sum-of-products (SOP) network, which inherently is a very flat network form.

FIGURE 7.26 Quartus II Compilaton Report Menu

The internal circuit of a MAX 7000S CPLD is a programmable SOP network. In order to program such a device, the Quartus II compiler must analyze the design entity, break it into product terms and reassemble it as an SOP network. (This is an oversimplification. Sometimes SOP outputs are fed back into the circuit to be reused by other parts of the circuit, thus lengthening the logic path.)

We can determine the circuit delays for a parallel adder by examining the Timing Analyses that Quartus II automatically generates when a project is compiled. Figure 7.26 shows a menu generated by Quartus II to allow the user to view different parts of the project's Compilation Report.

If we select **tpd (Pin to Pin Delays),** we will see the window in Figure 7.27, which shows a timing analysis for the 4-bit adder, **add4gen.vhd.** The VHDL code for this adder uses a GENERATE statement to create a 4-bit parallel adder with ripple carry, fitted to an EPM7128SLC84-7 CPLD. But does the VHDL compiler actually synthesize a ripple carry design? We can find out by examining the timing analysis.

The values in the Destination column are the delays from logic level changes on the inputs specified by the Source column. For example, a change on input **a[1]** reaches the output **sum[1]** in 7.5 ns and **sum[4]** in 7.5 ns to 10.4 ns. Many of the entries in the table have two values (e.g., 7.5 ns and 10.4 ns), indicating a range of delay from input to output. The actual delay will depend on the logic level change that takes place on the source line and thus which logic path is taken. The delay time of 7.5 ns is about the lowest value possible in the EPM7128SLC84–7 device. (The "–7" tells us that the chip has a **speed grade** of minus 7, meaning an internal delay of about 7 ns.)

The delay to the **c4** output is the same from all bits of inputs **a** and **b,** indicating that a fast carry circuit has been synthesized. The delay from **c0** to **c4** is 7.5 ns in both the shortest and longest delay, indicating that perhaps it bypasses one more gate layer than **a** and **b.** Given the equality of delay times through all pathways, we find that we do not need to explicitly design an adder circuit to have a fast carry function. The carry circuit will be optimized by the compiler to give the shortest possible delay.

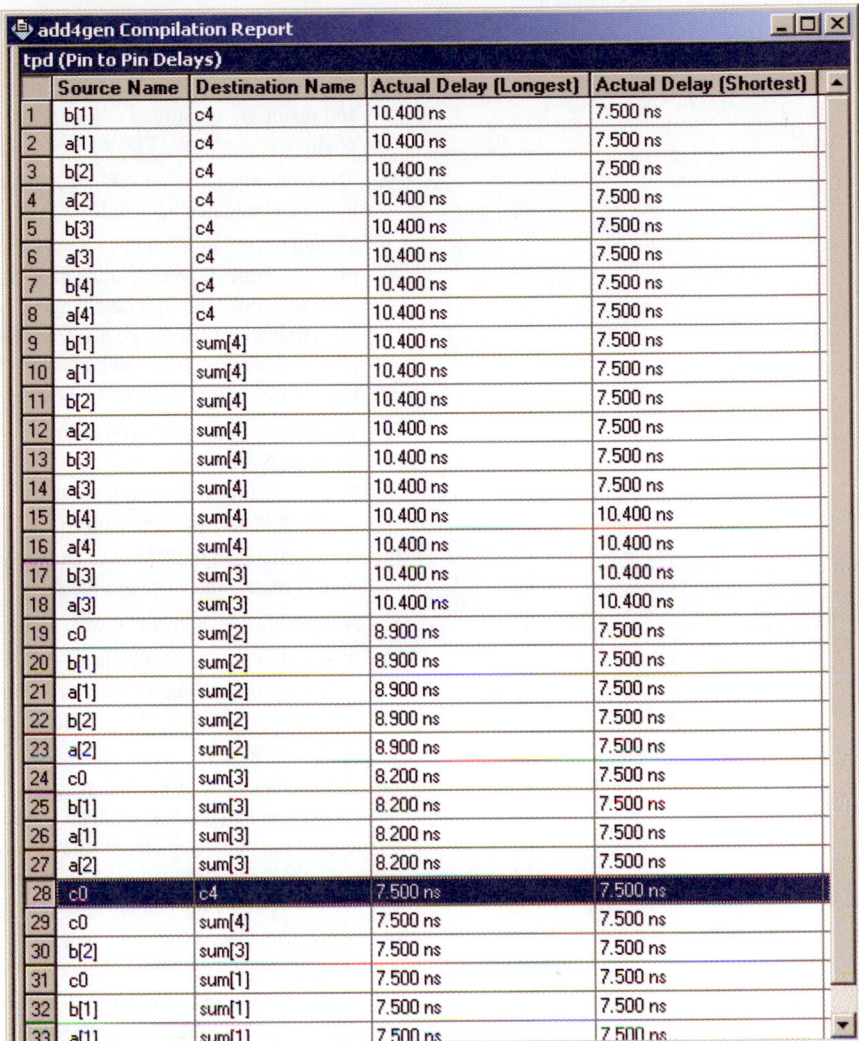

FIGURE 7.27 Quartus II Pin-to-Pin Delay Window

SUMMARY

1. Addition combines an addend (x) and an augend (y) to get a sum $(z = x + y)$.

2. Binary addition is based on four sums:

$$0 + 0 = 00$$
$$0 + 1 = 01$$
$$1 + 1 = 10$$
$$1 + 1 + 1 = 11$$

3. A sum of two bits generates a sum bit and a carry bit. (For the first two sums just shown, the carry bit is 0; the last two sums have a carry of 1. The last sum includes a carry from a lower-order bit.)

4. Subtraction combines a minuend (x) and a subtrahend (y) to get a difference $(z = x - y)$.

5. Binary subtraction is based on the following four differences:

$$0 - 0 = 0$$
$$1 - 0 = 1$$
$$1 - 1 = 0$$
$$10 - 1 = 1$$

6. If the subtrahend bit is larger than the minuend bit, as in the fourth difference above, a 1 must be borrowed from the next higher-order bit.

7. Binary addition or subtraction can be unsigned, where the magnitudes of the operands and result are presumed to be positive, or signed, where the operands and result can be positive or negative. The sign is indicated by a sign bit.

8. The sign bit (usually MSB) of a binary number indicates that the number is positive if it is 0 and negative if it is 1.

9. Signed binary numbers can be written in true-magnitude, 1's complement, or 2's complement form. True magnitude has the same binary value for positive and negative numbers, with only the sign bit changed. A 1's complement negative number is generated by inverting all bits of the positive number of the same magnitude. A 2's complement negative number is generated by adding 1 to the equivalent 1's complement number. Positive numbers are the same in all three forms.

10. 1's complement or 2's complement binary numbers are used in signed addition or subtraction. Subtraction is performed by adding a negative number in complement form to another number in complement form (i.e., $x - y = x + (-y)$). This technique does not work for true-magnitude form.

11. A negative sum or difference in 2's complement subtraction must be converted to a positive form to read its magnitude (i.e., $-(-x) = +x$).

12. A signed binary number, x, with n magnitude bits has a valid range of $-2^n \leq x \leq + (2^n - 1)$.

13. A negative number with a power-of-2 magnitude (i.e., -2^n) is written in 2's complement form as n 0s preceded by all 1s to fill the defined size of the number (e.g., in 8-bit 2's complement form, $-128 = 10000000$ (1 followed by seven 0s; $128 = 2^7$); in 8-bit 2's complement form, $-8 = 11111000$ (all 1s, followed by three 0s; $8 = 2^3$).

14. If the sum of two n-bit 2's complement numbers carries beyond the nth bit, discard the carry. The carry is not retained because in an n-bit number there is no $(n+1)^{th}$ position to hold the carry (e.g., the is no 9th bit in an 8-bit number). Analogy: a car odometer does not hold the leading value of "1" when the mileage goes from 99999 to 100000. The "1" is not shown because there is no place to hold it.

15. A 2's complement number can be expanded to a larger bit size by the process of sign extension. To expand a positive number to a larger bit size, pad the number with leading 0s. To expand a negative number, pad it with leading 1s. For example, +5 is written 0101 as a 4-bit number and 00000101 as an 8-bit number. The negative value −5 is written 1011 as a 4-bit number and 11111011 as an 8-bit number.

16. If a sum or difference falls outside the permissible range of magnitudes for a 2's complement number, it generates an overflow into the sign bit of the number. The result is that the sum of two positive numbers appears to be negative (e.g., 01111111 + 00000010 = 10000001) or the sum of two negative numbers appears to be positive (e.g., 11111111 + 10000000 = 01111111, where the carry beyond the 8th place is discarded).

17. When adding two hexadecimal digits, any digit sum greater than 15 (F) can be converted to a hexadecimal value by subtracting 16 and carrying a 1 to the next digit position.

18. Hexadecimal numbers can be subtracted conventionally or by a complement method. For the complement method, first obtain the 15's complement by subtracting the positive hexadecimal value from another hexadecimal number with the same number of digits, all of which are Fs. Add 1 to the 15's complement to obtain the 16's complement. Perform the subtraction by adding the complement value to the other hex number.

19. Binary numbers can be used in nonpositional codes to represent numbers or alphanumeric characters.

20. Binary coded decimal (BCD) codes represent decimal numbers as a series of 4-bit groups of numbers. Natural BCD or 8421 code does this as a positionally weighted code for each digit (e.g., 158 = 0001 0101 1000 [NBCD]). Other codes, such as Excess-3, are not positionally weighted.

21. Gray code is a binary code that has a difference of one bit between adjacent codes. It can be generated by a set of XOR functions or by recognizing the symmetry inherent in the code. In any Gray code sequence, the MSB is 0 for the first half of the sequence and 1 for the second half. The remaining bits are symmetrical about the halfway point of the sequence.

22. ASCII code represents alphanumeric characters and computer control codes as a 7-bit group of binary numbers. Alpha characters are listed in uppercase in columns 4 and 5 of the ASCII table. Lowercase alpha characters are in columns 6 and 7. Numeric characters are in column 3.

23. A half adder combines two bits to generate a sum and a carry. It can be represented by the following truth table:

A	B	C_{OUT}	Σ
0	0	0	0
0	1	0	1
1	0	0	1
1	1	1	0

24. From the half adder truth table, we can derive two equations:

$$C_{OUT} = AB$$
$$\Sigma = A \oplus B$$

25. A full adder can accept an input carry from a lower-order adder and combine the input carry with two operands to generate a sum and output carry. Its operation can be summarized in the following truth table:

A	B	C_{IN}	C_{OUT}	Σ
0	0	0	0	0
0	0	1	0	1
0	1	0	0	1
0	1	1	1	0
1	0	0	0	1
1	0	1	1	0
1	1	0	1	0
1	1	1	1	1

26. The following Boolean equations for a full adder can be derived from the truth table and Boolean algebra:

$$C_{OUT} = (A \oplus B) C_{IN} + AB$$
$$\Sigma = (A \oplus B) \oplus C_{IN}$$

27. Two half adders can be combined to make a full adder. Operands *A* and *B* go to the first half adder. The sum output of the first half adder and the carry input go to the inputs of the second half adder. The carry outputs of both half adders are combined in an OR gate.

28. Multiple full adders can be cascaded to make a parallel binary adder. Operands A_1 and B_1 are applied to the first full adder. Carry bit C_0 is grounded. A_2 and B_2 go to the second adder stage, and so on. The carry output of one stage is cascaded to the carry input of the following stage. This connection is called ripple carry.

29. Ripple carry has the disadvantage of increasing the time required to generate an output result as more stages are added. Fast carry, or look-ahead carry, examines all adder inputs simultaneously and generates each internal and output carry with a separate circuit. This makes the carry circuit wider, but flatter, thus reducing the delay time of the circuit.

30. A parallel adder can be implemented in VHDL by creating a design entity for a full adder, then using multiple instances of the full adder as components in the parallel adder.

31. To use a component in a VHDL design hierarchy, we require a design entity that defines the component, a component declaration in the design entity that uses the component, and a component instantiation statement for every instance of the component in the higher-level design entity.

32. The general form of a design entity using components is:

```
ENTITY entity_name IS
    PORT ( input and output definitions);
END entity_name;
ARCHITECTURE arch_name OF entity_name IS
    component declaration(s);
    signal declaration(s);
BEGIN
    Component instantiation(s);
    Other statements;
END arch_name;
```

33. The port map of a component maps the port names defined in the component to the port, signal, or variable names defined in the design entity that uses the component.

34. If all ports of a component are to be used in the same order as in the component definition in the original component design entity, the port map can simply contain the user names in the same order. This is called implicit port mapping. For example:

```
adder1: full_add PORT MAP (a(1), b(1), c0, c(1), sum(1));
```

35. If only a portion of the component ports are to be used or they are not used in the same sequence as they are declared, the port map must use explicit port mapping. For example:

```
adder1: full_add
    PORT MAP ( b       => b(1),
               a       => a(1),
               c_in    => c0,
               sum     => sum(1),
               c_out   => c(1));
```

36. A GENERATE statement can be used to instantiate multiple instances of a component. The GENERATE statement has the form:

```
label:
FOR index IN range GENERATE
    statements;
END GENERATE;
```

37. A parallel adder of unspecified width can be created in VHDL by using a parameter in a GENERIC clause in the entity declaration. The GENERIC clause appears before the PORT clause and has the form `GENERIC(parameter : INTEGER := default_value)`. When the component is instantiated, the parameter in the GENERIC clause is given a value in a GENERIC MAP.

38. The Quartus II compiler will flatten the carry network of a parallel adder so that a fast carry need not be explicitly designed.

39. A parallel binary adder can be made into a 2's complement subtractor by inverting one set of inputs and tying the input carry to a logic HIGH.

40. A parallel binary adder can be made into a 2's complement adder/subtractor by using a set of XOR gates as programmable inverters and connecting the XOR control line to the carry input of the adder.

41. One method of detecting a sign bit overflow in a 2's complement adder/subtractor is to compare the sign bits of the operands to the sign bit of the result. If the sign bits of the operands are the same as each other, but different from the sign bit of the result, there has been an overflow. The Boolean equation for this detector is given by $V = \overline{S_\Sigma} \cdot S_A \cdot S_B + S_\Sigma \cdot \overline{S_A} \cdot \overline{S_B}$.

42. Another method of overflow detection compares the carry out of the MSB of the adder/subtractor to the carry into the MSB. An overflow occurs if there is a carry out of or into the MSB, but not both. The Boolean equation for this detector is given by $V = C_n \oplus C_{n-1}$, for an n-bit adder/subtractor.

43. A BCD adder adds two binary-coded decimal (BCD) digits and generates a BCD digit and a carry bit.

44. Since BCD is a 4-bit code, BCD addition can be done with a 4-bit binary adder and a code converter. The code converter can be synthesized from another 4-bit binary adder and a circuit to generate a carry.

GLOSSARY

1's Complement   A form of signed binary notation in which negative numbers are created by complementing all bits of a number, including the sign bit.

2's Complement   A form of signed binary notation in which negative numbers are created by adding 1 to the 1's complement form of the number.

8421 Code (or NBCD; Natural Binary Coded Decimal)   A BCD code that represents each digit of a decimal number by its 4-bit true binary value.

9's Complement   A way of writing decimal numbers where a number is made negative by subtracting each of its digits from 9 (e.g., $-726 = 999 - 726 = 273$ in 9's complement).

Addend   The number in an addition operation that is added to another.

Alphanumeric Code   A code used to represent letters of the alphabet and numerical characters.

ASCII   American Standard Code for Information Interchange. A 7-bit code for representing alphanumeric and control characters.

Augend   The number in an addition operation to which another number is added.

BCD Adder   A parallel adder whose output is in groups of 4 bits, each group representing a BCD digit.

Binary-Coded Decimal (BCD)   A code that represents each digit of a decimal number by a 4-bit binary value.

Borrow   A digit brought back from a more significant position when the subtrahend digit is larger than the minuend digit.

Carry   A digit which is "carried over" to the next most significant position when the sum of two single digits is too large to be expressed as a single digit.

Carry Bit   A bit that holds the value of a carry (0 or 1) resulting from the sum of two binary numbers.

Cascade   To connect an output of one device to a input of another, often for the purpose of expanding the number of bits available for a particular function.

Case Shift   Changing letters from capitals (UPPERCASE) to small letters (lowercase), or vice versa.

Component   A complete VHDL design entity that can be used as a part of a higher-level file in a hierarchical design.

Component Declaration Statement   A statement that defines the input and output port names of a component used in a VHDL design entity.

Component Instantiation Statement   A statement that maps port names of a VHDL component to the port names, internal signals, or variables of a higher-level VHDL design entity.

Difference   The result of a subtraction operation.

End-Around Carry   An operation in 1's complement subtraction where the carry bit resulting from a sum of two 1's complement numbers is added to that sum.

Excess-3 Code   A BCD code that represents each digit of a decimal number by a binary number derived by adding 3 to its 4-bit true binary value. Excess-3 code has the advantage of being "self-complementing."

Fast Carry (or Look-Ahead Carry)   A gate network which generates a carry bit directly from *all* incoming operand bits, independent of the operation of each full adder stage.

Full Adder   A circuit that will add a carry bit from another full or half adder and two operand bits to produce a sum bit and a carry bit.

GENERATE Statement A VHDL construct that is used to create repetitive portions of hardware.

Gray Code A binary code which progresses such that only one bit changes between two successive codes.

Half Adder A circuit that will add two bits and produce a sum bit and a carry bit.

Hierarchy A group of design entities associated in a series of levels or layers in which complete designs form portions of another, more general design entity. The more general design is considered to be the higher level of the hierarchy.

Instantiate To use an instance of a component.

Magnitude Bits The part of a signed binary number that tell us how large the number is (i.e., its magnitude).

Minuend The number in a subtraction operation from which another number is subtracted.

Operand A number upon which an arithmetic function operates (e.g., in the expression $x + y = z$, x and y are the operands).

Overflow An erroneous carry into the sign bit of a signed binary number which results from a sum larger than can be represented by the number of magnitude bits.

Parallel Binary Adder A circuit, consisting of n full adders, which will add two n-bit binary numbers. The output consists of n sum bits and a carry bit.

Port An input or output of a VHDL design entity or component.

Ripple Carry A method of passing carry bits from one stage of a parallel adder to the next by connecting C_{OUT} of one full adder to C_{IN} of the following stage.

Self-Complementing A code that automatically generates a negative-equivalent (e.g., 9's complement for a decimal code) when all its bits are inverted.

Sign Bit A bit, usually the MSB, that indicates whether a signed binary number is positive or negative.

Sign Extension The process of fitting a number into a fixed size of 2's complement number by padding the number with leading 0s for a positive number and leading 1s for a negative number.

Signed Binary Arithmetic Arithmetic operations performed using signed binary numbers.

Signed Binary Number A binary number of fixed length whose sign is represented by one bit, usually the most significant bit, and whose magnitude is represented by the remaining bits.

Speed Grade A specification that indicates the internal delay time that can be expected of a CPLD.

Subtrahend The number in a subtraction operation that is subtracted from another number.

Sum The result of an addition operation.

Sum Bit (Single-Bit Addition) The least significant bit of the sum of two 1-bit binary numbers.

True-Magnitude Form A form of signed binary number whose magnitude is represented in true binary.

Unsigned Binary Number A binary number whose sign is not indicated by a sign bit. A positive sign is assumed unless explicitly stated otherwise.

PROBLEMS

For maximum learning benefit, do not use a calculator for the problems in this chapter, except to check your work. Problem numbers set in color indicate more difficult problems.

7.1 Digital Arithmetic

7.1 Add the following unsigned binary numbers.

a. $10101 + 1010$

b. $10101 + 1011$

c. $1111 + 1111$

d. $11100 + 1110$

e. $11001 + 10011$

f. $111011 + 101001$

7.2 Subtract the following unsigned binary numbers.

   a. 1100 – 100

   b. 10001 – 1001

   c. 10101 – 1100

   d. 10110 – 1010

   e. 10110 – 1001

   f. 10001 – 1111

   g. 100010 – 10111

   h. 1100011 – 100111

7.2   Representing Signed Binary Numbers

7.3 Write the following decimal numbers in 8-bit true-magnitude, 1's complement, and 2's complement forms.

   a. –110

   b. 67

   c. –54

   d. –93

   e. 0

   f. –1

   g. 127

   h. –127

7.3   Signed Binary Arithmetic

7.4 Perform the following arithmetic operations in the true-magnitude (addition only), 1's complement, and 2's complement systems. Use 8-bit numbers consisting of a sign bit and 7 magnitude bits. (The numbers shown are in the decimal system.)

   Convert the results back to decimal to prove the correctness of each operation. Also demonstrate that the idea of adding a negative number to perform subtraction is not valid for the true-magnitude form.

   a. 37 + 25

   b. 85 + 40

   c. 95 – 63

   d. 63 – 95

   e. –23 – 50

   f. 120 – 73

   g. 73 – 120

7.5 What are the largest positive and negative numbers, expressed in 2's complement notation, that can be represented by an 8-bit signed binary number?

7.6 Perform the following *signed* binary operations, using 2's complement notation where required. State whether or not sign bit overflow occurs. Give the signed decimal equivalent values of the sums in which overflow does *not* occur.

   a. 01101 + 00110

   b. 01101 + 10110

   c. 01110 – 01001

   d. 11110 + 00010

   e. 11110 – 00010

7.7 Without doing any binary complement arithmetic, indicate which of the following operations will result in 2's complement overflow. (Assume 8-bit representation consisting of a sign bit and 7 magnitude bits.) Explain the reasons for each choice. (All numbers are shown in the decimal system.)

    a. $-109 + 36$

    b. $109 + 36$

    c. $65 + 72$

    d. $-110 - 29$

    e. $117 + 11$

    f. $117 - 11$

7.8 Explain how you can know, by examining sign or magnitude bits of the numbers involved, when overflow has occurred in 2's complement addition or subtraction.

7.9 **a.** Write +19 as 2's complement value with the smallest number of bits possible.

    b. Write +19 as an 8-bit 2's complement number.

    c. Write +19 as a 12-bit 2's complement number.

7.10 **a.** Write −19 as a 2's complement value with the smallest number of bits possible.

    b. Write −19 as an 8-bit 2's complement number.

    c. Write −19 as a 12-bit 2's complement number.

7.4 Hexadecimal Arithmetic

7.11 Add the following hexadecimal numbers.

    a. 27H + 16H

    b. 87H + 99H

    c. A55H + C5H

    d. C7FH + 380H

    e. 1FFFH + A80H

7.12 Subtract the following hexadecimal numbers.

    a. F86H − 614H

    b. E72H − 229H

    c. 37FFH − 137FH

    d. 5764H − ACBH

    e. 7D30H − 5D33H

    f. 5D33H − 7D30H

    g. 813AH − A318H

7.5 Numeric and Alphanumeric Codes

7.13 Convert the following decimal numbers to true binary, 8421 BCD code, and Excess 3 code.

    a. 709_{10}

    b. 1889_{10}

    c. 2395_{10}

    d. 1259_{10}

    e. 3972_{10}

    f. 7730_{10}

7.14 Make a table showing the equivalent Gray codes corresponding to the range from 0_{10} to 31_{10}.

7.15 Write your name in ASCII code.

7.16 Encode the following text into ASCII code: "10% off purchases over $50. (Monday only)"

7.17 Decode the following sequence of ASCII characters.

43	41	55	54	49	4F	4E	21	20	45	72	61	73	
69	6E	67	20	61	6C	6C	20	64	61	74	61	21	20
41	72	65	20	79	6F	75	20	73	75	72	65	3F	

7.6 Binary Adders and Subtractors

7.18 Write the truth table for a half adder, and from the table derive the Boolean expressions for both C_o (carry output) and Σ (sum output) in terms of inputs A and B. Draw the half adder circuit.

7.19 Write the truth table for a full adder, and from the table derive the simplest possible Boolean expressions for C_{OUT} and Σ in terms of A, B, and C_{IN}.

7.20 From the equations in Problems 7.18 and 7.19, draw a circuit showing a full adder constructed from two half adders.

7.21 Evaluate the Boolean expression for Σ and C_{OUT} of the full adder in Figure 7.7 for the following input values. What is the binary value of the outputs in each case?

  a. $A = 0$, $B = 0$, $C_{IN} = 0$

  b. $A = 0$, $B = 1$, $C_{IN} = 0$

  c. $A = 0$, $B = 1$, $C_{IN} = 1$

  d. $A = 1$, $B = 1$, $C_{IN} = 1$

7.22 Verify the summing operation of the circuit in Figure 7.10, as follows. Determine the output of each full adder based on the inputs shown below. Calculate each sum manually and compare it to the 5-bit output ($C_4 \Sigma_4 \Sigma_3 \Sigma_2 \Sigma_1$) of the parallel adder circuit.

  a. $A_4 A_3 A_2 A_1 = 0100$, $B_4 B_3 B_2 B_1 = 1001$

  b. $A_4 A_3 A_2 A_1 = 1010$, $B_4 B_3 B_2 B_1 = 0110$

  c. $A_4 A_3 A_2 A_1 = 0101$, $B_4 B_3 B_2 B_1 = 1101$

  d. $A_4 A_3 A_2 A_1 = 1111$, $B_4 B_3 B_2 B_1 = 0111$

7.23 Briefly describe the differences in the underlying design strategies of the ripple carry adder and the fast carry adder (i.e., what makes the fast carry faster than the ripple carry?). What is the main limitation for the fast carry circuit?

7.24 Write the general form of the fast carry equation. Use it to generate Boolean expression for C_1, C_2, and C_3 for a fast carry adder.

7.25 The following equation describes the carry output function for a parallel binary adder:

$$\begin{aligned}
C_{OUT} = {} & A_4 B_4 + A_3 B_3 (A_4 + B_4) \\
& + A_2 B_2 (A_4 + B_4)(A_3 + B_3) \\
& + A_1 B_1 (A_4 + B_4)(A_3 + B_3)(A_2 + B_2) \\
& + C_{IN} (A_4 + B_4)(A_3 + B_3)(A_2 + B_2)(A_1 + B_1)
\end{aligned}$$

Briefly explain how to interpret the third term of this equation.

7.26 Write a VHDL file for an 8-bit parallel adder, using eight instances of a full adder component.

7.27 Create a simulation for the 8-bit adder of Problem 7.26, showing a representative sample of sums. How many different sums would be required to show all possible combinations of inputs?

7.28 Write a VHDL file that creates a 12-bit adder using a GENERATE statement.

7.29 Create a simulation file in Quartus II for the 12-bit adder of Problem 7.28 showing only one transition, as follows. Set input **a** to 000 from 0 to 500 ns, then 001 from 500 ns to 1 μs. Set input **b** to FFF from 0 to 1 μs. From the simulation, determine the internal delays from **a1** to each of the **sum** bits. Confirm your observations with the pin-to-pin delay window from a timing analysis.

7.30 Use Quartus II to create a Block Diagram File for a 2's complement subtractor based on a 4-bit parallel binary adder. Explain how the circuit generates the 2's complement of B for the subtraction $A - B$.

7.31 Use Quartus II to create a Block Diagram File for a 2's complement adder/subtractor based on a 4-bit parallel binary adder. Explain how the circuit is programmed to add or subtract and how it produces the 2's complement of B for the subtraction $A - B$.

7.32 Use Quartus II to draw a circuit that will detect an overflow condition in a 4-bit 2's complement adder/subtractor. The detector output should go HIGH upon overflow detection. Draw the circuit truth table, explain what all input and output variables are, and show any Boolean equations you need to complete the circuit design.

7.33   Modify the 4-bit adder/subtractor drawn in Figure 7.15 to include an overflow detection circuit.

7.34   Create a simulation for the 4-bit adder/subtractor with overflow detection (Problem 7.33), using the following representative hexadecimal input values: $F + 1 = 10$ (carry, but no overflow); $7 + 1 = 8$ (overflow, but no carry); $8 + 8 = 10$ (carry and overflow); $0 - 1 = F$ (result $= -1$).

7.35   Modify the VHDL file for the 4-bit parallel binary adder/subtractor (**addsub4g.vhd**) to include an overflow detection circuit. Use two different methods.

7.36   What is the permissible range of values that a sum or difference, x, can have in a 16-bit parallel binary adder if it is written as:

   a. A signed binary number

   b. An unsigned binary number

7.7   BCD Adders

7.37   What is the maximum BCD sum of two 3-digit BCD numbers plus an input carry? How many digits are needed to display the result?

7.38   What is the maximum BCD sum of two 4-digit BCD numbers plus an input carry? How many digits are needed to display the result?

7.39   Based on the answers to Problems 7.37 and 7.38, formulate a general rule to calculate the maximum BCD sum of two n-digit BCD numbers plus a carry bit.

7.40   Derive the Boolean expression for a BCD carry output as a function of the sum of two BCD digits.

7.41   Draw the circuit for a binary-to-BCD code converter.

7.42   Write a VHDL file to implement a binary-to-BCD code converter for a BCD adder. Use a selected signal assignment or CASE statement.

7.43   Write a VHDL file that uses the binary-to-BCD code converter of Problem 7.42 and a 4-bit parallel binary adder as components in a BCD adder.

7.44   Write a VHDL file that uses a code converter and parallel adder as components in a design that will add two 2-digit BCD numbers and produce a $2\frac{1}{2}$ digit result.

7.45   Draw the block diagram of a circuit that will add two 3-digit BCD numbers and display the result as a series of decimal digits. How many digits will the output display?

ANSWERS TO SECTION REVIEW PROBLEMS

7.1A

7.1	101000
7.2	100000

7.1B

7.3	11
7.4	1

7.3

7.5	11100000
7.6	100000

7.4

7.7a	11701H
7.7b	1381H

7.5

7.8 "True or False: 1/4 < 1/2"

7.6A

7.9 Figures 7.28 and 7.29 show the propagation paths for the carry bits.

Fast carry: 2 gates

Ripple carry: 8 gates

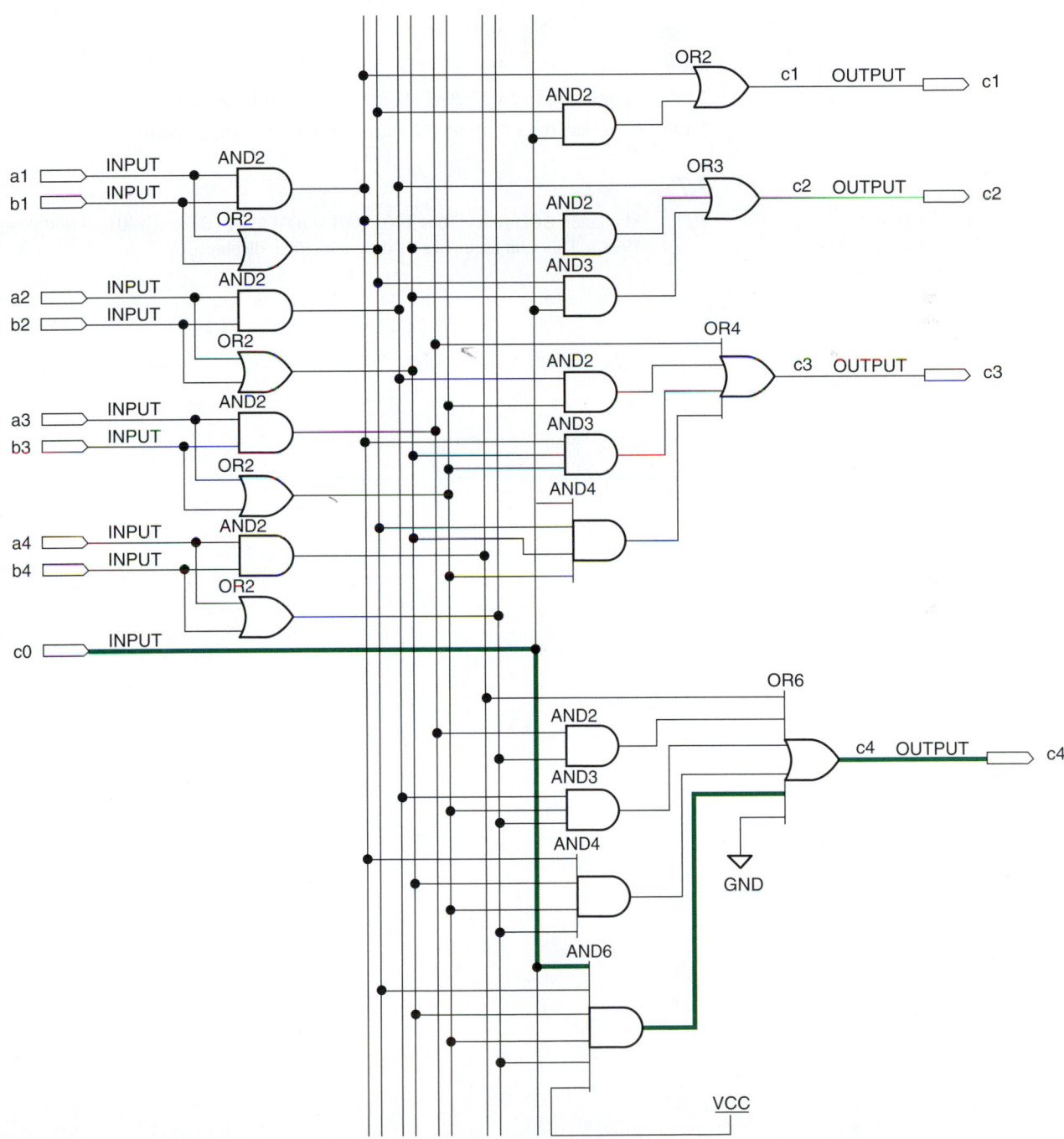

FIGURE 7.28 Fast Carry from C0 to C4.

FIGURE 7.29 Ripple Carry from C_0 to C_4

7.6B

7.10a Signed: $-2048 \leq x \leq +2047$ (11 magnitude bits, 1 sign bit)

7.10b Unsigned: $0 \leq x \leq +4095$ (12 magnitude bits, no sign bit: positive implied)

7.7

7.11 Maximum BCD sum = 1001 1001 1001 + 1001 1001 1001 = 1 1001 1001 1000$_{BCD}$ = 1998_{10}. This sum requires a $3\frac{1}{2}$-digit numerical display.

KEY TERMS

Architecture Cell
Asynchronous
Asynchronous Inputs
Buried Logic
Carry Chain
Cascade Chain
Clear
CLOCK
CPLD
Edge
Edge Detector
Edge-Sensitive
Edge-Triggered
Embedded Array Block (EAB)
Flip-Flop
Gated SR Latch
Generic Array Logic (GAL)
Generic Map
Global Clock
In-System Programmability (ISP)
I/O Control Block
JTAG Port
Latch
Level-Sensitive
Library of Parameterized Modules (LPM)
Logic Array Block (LAB)
Logic Element (LE)
Look-Up Table (LUT)
Master Reset
MegaWizard Plug-In Manager
One-Time Programmable (OTP)
Output Logic Macrocell (OLMC)
Parallel Logic Expanders

Parameter (in an LPM Component)
Port Map
Preset
Programmable Interconnect Array (PIA)
Register
Registered Output
RESET
Sequential Circuit
SET
Shared Logic Expanders
Steering Gates
Synchronous
Synchronous Inputs
T (Toggle) Flip-Flop
Toggle
Transparent Latch (Gated D Latch)

OUTLINE

8.1   Latches

8.2   NAND/NOR Latches

8.3   Gated Latches

8.4   Edge-Triggered D Flip-Flops

8.5   Edge-Triggered JK Flip-Flops

8.6   Edge-Triggered T Flip-Flops

8.7   Flip-Flops in PLDs (Registered Outputs)

8.8   Generic Array Logic (GAL)

8.9   MAX 7000S CPLD

8.10   FLEX 10K CPLD

CHAPTER 8

Introduction to Sequential Logic

CHAPTER OBJECTIVES

Upon successful completion of this chapter, you will be able to:

- Explain the difference between combinational and sequential circuits.
- Define the set and reset functions of an SR latch.
- Draw circuits, function tables, and timing diagrams of NAND and NOR latches.
- Explain the effect of each possible input combination to a NAND and a NOR latch, including set, reset, and no change functions, as well as the ambiguous or forbidden input condition.
- Design circuit applications that employ NAND and NOR latches.
- Describe the use of the *ENABLE* input of a gated SR or D latch as an enable/inhibit function and as a synchronizing function.
- Outline the problems involved with using a level-sensitive *ENABLE* input on a gated SR or D latch.
- Explain the concept of edge-triggering and why it is an improvement over level-sensitive enabling.
- Draw circuits, function tables, and timing diagrams of edge-triggered D, JK, and T flip-flops.
- Describe the toggle function of a JK flip-flop and a T flip-flop.
- Describe the operation of the asynchronous preset and clear functions of D, JK, and T flip-flops and be able to draw timing diagrams showing their functions.
- Use Quartus II to create simple circuits and simulations with D latches and D, JK, and T flip-flops.
- Create simple latch and flip-flop designs using VHDL.
- Describe the structure of registered PAL outputs.
- Determine the number and type of outputs from a PAL/GAL part number.
- Explain the structure of an output logic macrocell (OLMC).
- State differences between generic array logic (GAL) and standard PAL.
- Interpret the logic diagrams of GAL devices to determine the number of outputs and product terms and the type of control signals available in a device.

- Interpret block diagrams to determine the basic structure of an Altera MAX 7000S CPLD, including macrocell configuration, logic array blocks (LABs), control signals, and product term expanders.
- State the differences between PLDs based on sum-of-products (SOP) architecture versus look-up table (LUT) architecture.
- Interpret block diagrams to determine the basic structure of a logic element in an Altera FLEX 10K CPLD, including look-up tables, cascade chains, carry chains, and control signals.
- Interpret block diagrams to determine how a logic element in a FLEX 10K device relates to the overall structure of the device.
- Interpret block diagrams to determine how logic array blocks and embedded array blocks relate to the overall structure of a FLEX 10K CPLD.

VHDL CONSTRUCTS USED IN THIS CHAPTER

- CASE statement
- Component declaration statement
- Component instantiation statement, including GENERIC MAP and PORT MAP
- GENERATE statement
- IF statement
- Library of Parameterized Modules (LPM)
- PROCESS statement
- Selected signal assignment statement
- SIGNAL
- STD_LOGIC and STD_LOGIC_VECTOR types
- VHDL package: altera.maxplus2
- VHDL package: ieee.std_logic_1164
- VHDL package: ieee.std_logic_unsigned
- VHDL package: lpm.lpm_components

The digital circuits studied to this point have all been combinational circuits, that is, circuits whose outputs are functions only of their present inputs. A particular set of input states will always produce the same output state in a combinational circuit.

This chapter will introduce a new category of digital circuitry: the sequential circuit. The output of a sequential circuit is a function both of the present input conditions and the previous conditions of the inputs and/or outputs. The output depends on the sequence in which the inputs are applied.

We will begin our study of sequential circuits by examining the two most basic sequential circuit elements: the latch and the flip-flop, both of which are part of the general class of circuits called bistable multivibrators. These are similar devices, each being used to store a single bit of information indefinitely. The difference between a latch and a flip-flop is the condition under which the stored bit is allowed to change.

Latches and flip-flops are also used as integral parts of more complex devices, such as programmable logic devices (PLDs), usually when an input or output state must be stored.

A flip-flop is found in a PLD either as a registered output, where the output of a product term array is loaded to a flip-flop, which then connects to an output pin, or as part of an output logic macrocell (OLMC), which can be configured as an output with or without the flip-flop. We will examine registered outputs in PAL-type devices and macrocells in GAL22V10 (GAL = generic array logic) and MAX 7000S devices.

Some higher-density CPLDs, such as the Altera FLEX 10K series, are constructed using SRAM-based technology to implement a look-up table (LUT) structure. In a look-up table, the truth table of a logic function is stored in a 16-bit memory. The output is generated by looking up the function value for any given input combination. ■

8.1 LATCHES

■ **KEY TERMS**

Sequential Circuit   A digital circuit whose output depends not only on the present combination of inputs, but also on the history of the circuit.

Latch   A sequential circuit with two inputs called *SET* and *RESET,* which make the latch store a logic 0 (reset) or 1 (set) until actively changed.

SET   1. The stored HIGH state of a latch circuit.
             2. A latch input that makes the latch store a logic 1.

RESET   1. The stored LOW state of a latch circuit.
                2. A latch input that makes the latch store a logic 0.

All of the circuits we have seen up to this point have been combinational circuits. That is, their present outputs depend only on their present inputs. The output state of a combinational circuit results only from a combination of input logic states.

The other major class of digital circuits is the **sequential circuit.** The present outputs of a sequential circuit depend not only on its present inputs, but also on its past input states.

The simplest sequential circuit is the SR **latch,** whose logic symbol is shown in Figure 8.1a. The latch has two inputs, *SET (S)* and *RESET (R),* and two complementary outputs, Q and $\overline{Q}$. If the latch is operating normally, the outputs are always in opposite logic states.

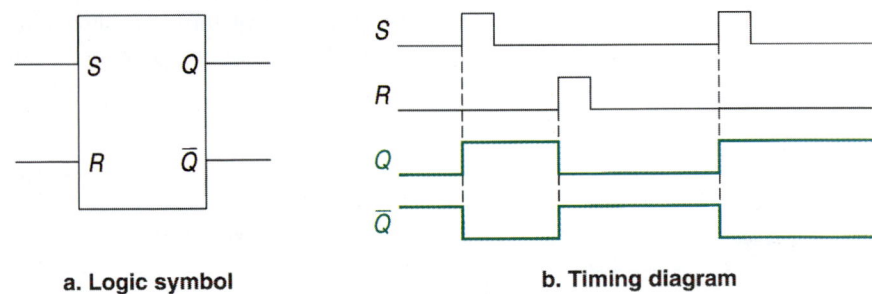

a. Logic symbol   b. Timing diagram

FIGURE 8.1   SR Latch (Active-HIGH Inputs)

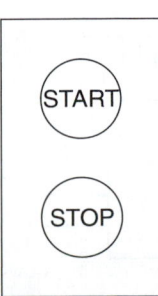

FIGURE 8.2   Industrial Pushbutton (e.g., Motor Starter)

The latch operates like a momentary-contact pushbutton with START and STOP functions, shown in Figure 8.2. A momentary-contact switch operates only when it is held down. When released, a spring returns the switch to its rest position.

Suppose the switch in Figure 8.2 is used to control a motor starter. When you push the START button, the motor begins to run. Releasing the START switch does not turn the motor off; that can be done only by pressing the STOP button. If the motor is running, pressing the START button again has no effect, except continuing to let the motor run. If the motor is not running, pressing the STOP switch has no effect, since the motor is already stopped.

There is a conflict if we press both switches simultaneously. In such a case we are trying to start and stop the motor at the same time. We will come back to this point later.

The latch *SET* input is like the START button in Figure 8.2. The *RESET* input is like the STOP button.

> **NOTE . . .**
>
> By definition:
> A latch is set when $Q = 1$ and $\bar{Q} = 0$.
> A latch is reset when $Q = 0$ and $\bar{Q} = 1$.

The latch in Figure 8.1 has active-HIGH *SET* and *RESET* inputs. To set the latch, make $R = 0$ and make $S = 1$. This makes $Q = 1$ until the latch is actively reset, as shown in the timing diagram in Figure 8.1b. To activate the reset function, make $S = 0$ and make $R = 1$. The latch is now reset ($Q = 0$) until the set function is next activated.

Combinational circuits produce an output by *combining inputs*. In sequential circuits, it is more accurate to think in terms of *activating functions*. In the latch described, S and R are not *combined* by a Boolean function to produce a particular result at the output. Rather, the set function is *activated* by making $S = 1$, and the reset function is *activated* by making $R = 1$, much as we would activate the START or STOP function of a motor starter by pressing the appropriate pushbutton.

The timing diagram in Figure 8.1b shows that the inputs need not remain active after the set or reset functions have been selected. In fact, the S or R input *must* be inactive before the opposite function can be applied, in order to avoid conflict between the two functions.

 EXAMPLE 8.1

Latches can have active-HIGH or active-LOW inputs, but in each case $Q = 1$ after the set function is applied and $Q = 0$ after reset. For each latch shown in Figure 8.3, complete the timing diagram shown. Q is initially LOW in both cases. (The state of Q before the first active *SET* or *RESET* is unknown unless specified, since the present state depends on previous history of the circuit.)

■ Solution

The Q and $\bar{Q}$ waveforms are shown in Figure 8.3. Note that the outputs respond only to the first set or reset command in a sequence of several pulses.

a. Latch with active-HIGH inputs   b. Latch with active-LOW inputs

FIGURE 8.3 Example 8.1: SR Latch

EXAMPLE 8.2

Figure 8.4 shows a latching HOLD circuit for an electronic telephone. When HIGH, the *HOLD* output allows you to replace the handset without disconnecting a call in progress.

FIGURE 8.4 Example 8.2: Latching HOLD Button

The two-position switch is the telephone's hook switch (the switch the handset pushes down when you hang up), shown in the off-hook (in-use) position. The normally closed pushbutton is a momentary-contact switch used as a *HOLD* button. The circuit is such that the *HOLD* button does not need to be held down to keep the *HOLD* active. The latch "remembers" that the switch was pressed, until told to "forget" by the reset function.

Describe the sequence of events that will place a caller on hold and return the call from hold. Also draw timing diagrams showing the waveforms at the *HOLD* input, hook switch inputs, S input, and *HOLD* output for one hold-and-return sequence. (*HOLD* output = 1 means the call is on hold.)

■ Solution

To place a call on hold, we must set the latch. We can do so if we press and hold the *HOLD* switch, then the hook switch. This combines two HIGHs—one from the *HOLD* switch and one from the on-hook position of the hook switch—into the AND gate, making $S = 1$ and $R = 0$. Note the sequence of events: press *HOLD*, hang up, release *HOLD*. The S input is HIGH only as long as the *HOLD* button is pressed. The handset can be kept on-hook and the *HOLD* button released. The latch stays set, as $S = R = 0$ (neither *SET* nor *RESET* active) as long as the handset is on-hook.

To restore a call, lift the handset. This places the hook switch into the off-hook position and now $S = 0$ and $R = 1$, which resets the latch and turns off the *HOLD* condition.

Figure 8.5 shows the timing diagram for the sequence described.

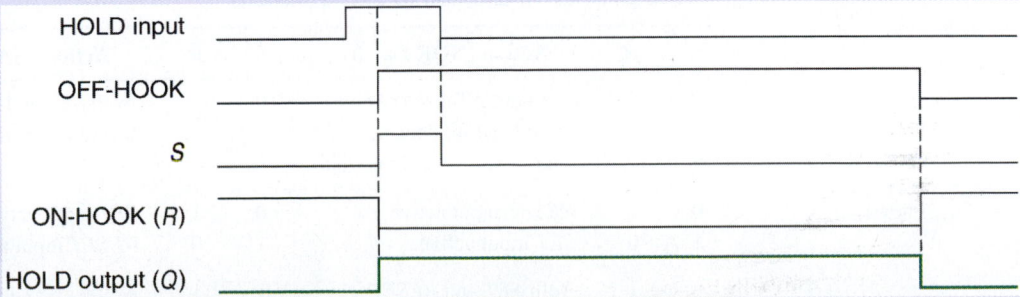

FIGURE 8.5 Example 8.2: HOLD Timing Diagram

■ **SECTION 8.1 REVIEW PROBLEM**

8.1   A latch with active-HIGH *S* and *R* inputs is initially set. *R* is pulsed HIGH three times, with *S* = 0. Describe how the latch responds.

8.2   NAND/NOR LATCHES

An SR latch is easy to build with logic gates. Figure 8.6 shows two such circuits, one made from NOR gates and one from NANDs. The NAND gates in the second circuit are drawn in DeMorgan equivalent form.

a. NOR latch   **b. NAND latch**

FIGURE 8.6   SR Latch Circuits

The two circuits both have the following three features:

1. OR-shaped gates
2. Logic level inversion between the gate input and output
3. Feedback from the output of one gate to an input of the opposite gate

During our examination of the NAND and NOR latches, we will discover why these features are important.

A significant difference between the NAND and NOR latches is the placement of *SET* and *RESET* inputs with respect to the Q and $\bar{Q}$ outputs. Once we define which output is Q and which is $\bar{Q}$, the locations of the *SET* and *RESET* inputs are automatically defined.

In a NOR latch, the gates have active-HIGH inputs and active-LOW outputs. When the input to the Q gate is HIGH, $Q = 0$, since either input HIGH makes the output LOW. Therefore, this input must be the *RESET* input. By default, the other is the *SET* input.

In a NAND latch, the gate inputs are active LOW (in DeMorgan equivalent form) and the outputs are active HIGH. A LOW input on the Q gate makes $Q = 1$. This, therefore, is the *SET* input, and the other gate input is *RESET*.

Since the NAND and NOR latch circuits have two binary inputs, there are four possible input states. Table 8.1 summarizes the action of each latch for each input

TABLE 8.1   NOR and NAND Latch Functions

S	*R*	Action (NOR Latch)	$\bar{S}$	$\bar{R}$	Action (NAND Latch)
0	0	Neither *SET* nor *RESET* active; output does not change from previous state	0	0	Both *SET* and *RESET* active; forbidden condition
0	1	*RESET* input active	0	1	*SET* input active
1	0	*SET* input active	1	0	*RESET* input active
1	1	Both *SET* and *RESET* active; forbidden condition	1	1	Neither *SET* nor *RESET* active; output does not change from previous state

a. Set   b. Reset

FIGURE 8.7 NAND Latch Stable States

combination. The functions are the same for each circuit, but they are activated by opposite logic levels.

We will examine the NAND latch circuit for each of the input conditions in Table 8.1. The analysis of a NOR latch is similar and will be left as an exercise.

NAND Latch Operation

Figure 8.7 shows a NAND latch in its two possible stable states. In each case the inputs $\overline{S}$ and $\overline{R}$ are both HIGH (inactive).

Figure 8.7a shows a latch stable in its *SET* condition ($Q = 1$). Note the following characteristics of this state:

■ The upper gate in Figure 8.7a has a LOW on its "inner" input. Since, for a NAND gate, either input LOW makes output HIGH, the output Q is HIGH.
■ The HIGH at Q feeds back to the "inner" input of the lower gate. Both inputs of this gate are HIGH, making the output $\overline{Q}$ LOW.
■ The LOW at $\overline{Q}$ feeds back to the "inner" input of the upper gate. This returns us to our starting point with the same logic level as we had to begin with. Therefore, we find a consistent path of logic levels with no tendency for the latch outputs to change.

The stable *RESET* state ($Q = 0$), shown in Figure 8.7b, is a mirror image of the *SET* condition. It has the following characteristics:

■ The lower gate in Figure 8.7b has a LOW on its "inner" input. Since either input LOW makes output HIGH, this makes $\overline{Q}$ HIGH.
■ The HIGH on $\overline{Q}$ feeds back to the upper gate. Since both inputs of this gate are HIGH, output Q is LOW.
■ The LOW on Q feeds back to the "inner" input of the lower gate. Since this is the same logic level as our starting point, we find that the logic levels are consistent throughout the path and the latch has no tendency to change its output states.

The stability of the latch depends on the feedback connections between the two gates, which supply the logic levels to the "inner" inputs of the latch gates. Notice that in the two stable states, only one of the four inputs to the two NAND gates has a 0. The difference between *SET* and *RESET* states has to do with the placement of the logic 0 on one of these "inner" inputs. Whichever gate has the 0 input will have the HIGH output; the other gate will have a LOW output. When we change the state of the latch, what we are doing is moving this 0 from one side of the latch to the other. Thus, since the output state of the latch can be determined solely by these "inner" inputs, it is possible for the same input values of $\overline{S} = 1$ and $\overline{R} = 1$ to yield two different sets of output values for Q and $\overline{Q}$.

What function do $\overline{S}$ and $\overline{R}$ have then? They are used to change the state of the latch.

**a) Stable in the RESET condition.
Set and Reset inputs inactive.**

b) Set input activates.

**c) Change propagates through upper gate.
(Either input LOW makes output HIGH.)**

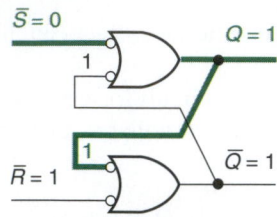

**d) HIGH transfers across feedback line to lower
gate, removing active input condition.**

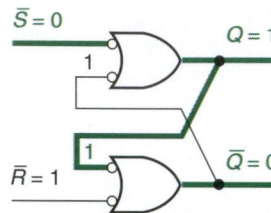

**e) Change propagates through lower gate.
(Both inputs HIGH, therefore output LOW.)**

**f) Feedback transfers LOW to upper gate,
completing change to new state.**

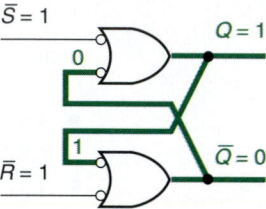

**g) S input goes back to inactive state. SET state
held by LOW at inner input of upper gate.**

FIGURE 8.8   RESET-to-SET Transition

Figure 8.8 shows the transition of the latch from the *RESET* state to the *SET* state. The following actions occur in the transition:

- The latch begins in the stable *RESET* state ($Q = 0$), as shown in Figure 8.7b and Figure 8.8a. In this state, $\overline{S} = 1$ and $\overline{R} = 1$, the no change condition.
- To set the latch, we make $\overline{S} = 0$ (Figure 8.8b).
- This change propagates through the upper gate of the latch circuit. Since either input LOW makes output HIGH, $Q = 1$ (Figure 8.8c).
- The HIGH on Q transfers across to the lower gate via the feedback line, removing its active input condition (Figure 8.8d).
- The lower gate changes state. Both inputs are now HIGH, making $\overline{Q} = 0$ (Figure 8.8e).
- The feedback line carries the LOW to the inner input of the upper gate. At this point the *RESET*-to-*SET* transition is complete (Figure 8.8f).
- Since only one LOW input is required to hold the output of the upper gate in the HIGH state, we can remove the LOW on input $\overline{S}$. The latch is now stable in the *SET* condition (Figure 8.8g).

A similar action occurs when the latch makes a transition from the *SET* state to the *RESET* state. This is shown in Figure 8.9.

- The latch begins in the stable *SET* condition ($Q = 1$), as shown in Figures 8.7a and 8.9a. $\overline{S} = 1$ and $\overline{R} = 1$ (no change).
- To reset the latch, we make $\overline{R} = 0$ (Figure 8.9b).

a) Stable in the SET condition.
Set and Reset inputs inactive.

b) Reset input activates.

c) Change propagates through lower gate.
(Either input LOW makes output HIGH.)

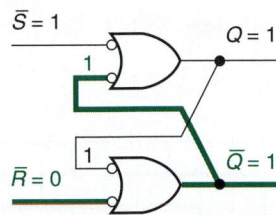

d) HIGH transfers across feedback line to upper
gate, removing active input condition.

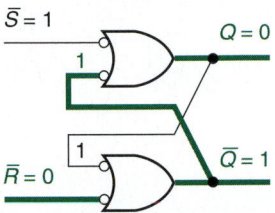

e) Change propagates through upper gate.
(Both inputs HIGH, therefore output LOW.)

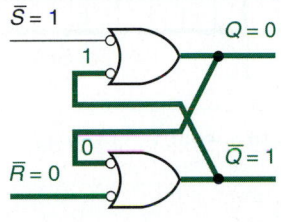

f) Feedback line transfers LOW to lower gate,
completing transition to RESET state.

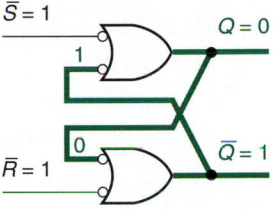

g) R goes back to inactive level. Latch is stable in
new state, held by the 0 on inner input of lower gate.

FIGURE 8.9 SET-to-RESET Transition

- This change propagates through the lower gate of the circuit. Since either input LOW makes output HIGH, $\overline{Q} = 1$ (Figure 8.9c).
- The HIGH on $\overline{Q}$ transfers across the feedback line to the inner input of the upper gate, removing its active input condition (Figure 8.9d).
- Since both inputs on the upper gate are now HIGH, output $Q = 0$ (Figure 8.9e).
- The 0 on Q transfers across the feedback line to the inner input of the lower gate, thus completing the transition from *RESET* to *SET* (Figure 8.9f).
- Since only one LOW input on the lower gate is required to maintain the *RESET* state, we can remove the 0 from the $\overline{R}$ input. The latch is now stable in the *RESET* state (Figure 8.9g).

Note that for each of these cases, the "outer" inputs of the circuit (i.e., $\overline{S}$ and $\overline{R}$) are used to *change* the state of the latch, whereas the "inner" inputs (i.e., the feedback connections) are used to *maintain* the present state of the latch.

Also note that the transition between states is not complete until the change initiated at $\overline{S}$ or $\overline{R}$ propagates through both gates; the circuit is not stable until the 0 transfers to the inner input of the opposite gate. If the set or reset pulse is shorter than the time required for the change to propagate through the gates, the latch output will oscillate between states. In practice this is not a huge problem since the total delay through the latch is only on the order of 12 to 20 ns. Any manual input to the latch, such as a pushbutton, will be far longer than this. Electronic inputs, such as logic gate outputs, will have to account for this delay, but as they, too, are subject to their own delay times, this seldom presents a practical problem.

FIGURE 8.10   NAND Latch
Forbidden State

FIGURE 8.11   Transition from Forbidden State to RESET State

FIGURE 8.12   Transition from Forbidden State to SET State

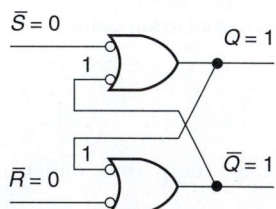

a) Both Set and Reset inputs active.
Either input LOW makes output HIGH.
Therefore, both outputs HIGH.

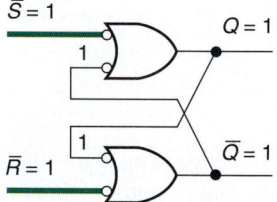

b) Set and Reset inputs
deactivate simultaneously.

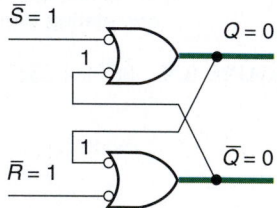

c) Change propagate through gates
simultaneously.

d) New output levels cross circuit via
feedback lines.

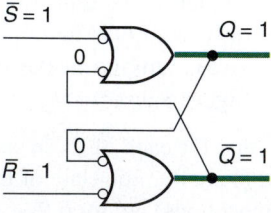

e) Either input LOW makes output HIGH.
Changes on feedback lines propagate
through both gates simultaneously.

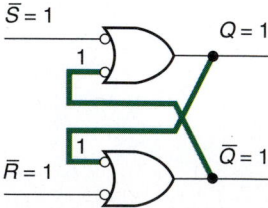

f) Output logic levels cross circuit via
feedback lines. Cycle repeats and
circuit oscillates.

FIGURE 8.13   Transition from Forbidden State to Oscillation

Figure 8.10 shows a NAND latch with $\overline{S} = \overline{R} = 0$. This implies that both *SET* and *RESET* functions are active. Since a NAND gate requires at least one input LOW to make the output HIGH, both outputs respond by going HIGH. This condition is not unstable in and of itself, but instability can result when the inputs change.

There are three possible results when the outputs go back to the HIGH state.

1. The *SET* input goes HIGH before the *RESET* input. In this case the latch resets, as *RESET* is the last input active. This is shown in the timing diagram in Figure 8.11.
2. The *RESET* input goes HIGH before *SET*. In this case, the latch sets, as shown in Figure 8.12.
3. The *SET* and *RESET* inputs go HIGH at the same time. This is an unstable case. Figure 8.13 shows how the latch will oscillate under this condition. When the inputs $\overline{S}$ and $\overline{R}$ are both LOW (Figure 8.13a), both latch outputs are HIGH. When $\overline{S}$ and $\overline{R}$ go HIGH (Figure 8.13b), all gate inputs are HIGH. This makes both outputs LOW (Figure 8.13c). The LOWs transfer across the latch to the opposite gates (Figure 8.13d) and, after a delay, make both outputs HIGH (Figure 8.13e). At this point, oscillations will be sustained until the latch is *SET* or *RESET*. The waveforms in Figure 8.14 show the oscillatory condition of the latch outputs.

In practice, the oscillatory condition of Figure 8.14 is unlikely to be sustained for very long. One of the two gates is likely to be slightly faster than the other, which will allow one state or the other to dominate.

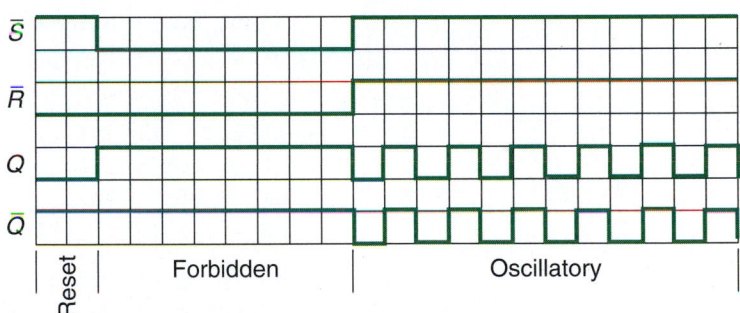

FIGURE 8.14 Transition from Forbidden State to Oscillation

TABLE 8.2 NAND Latch Function Table

$\overline{S}$	$\overline{R}$	Q_{t+1}	$\overline{Q}_{t+1}$	Function
0	0	1	1	Forbidden
0	1	1	0	Set
1	0	0	1	Reset
1	1	Q_t	$\overline{Q}_t$	No change

The operation of the NAND latch can be summarized in a function table, shown in Table 8.2. The notation Q_{t+1} indicates that the column shows the value of Q *after* the specified input is applied. Q_t indicates the present state of the Q input.* Thus, the entry for the no change state indicates that after the inputs $\overline{S} = 0$ and $\overline{R} = 0$ are applied, the next state of the output is the same as its present state.

Table 8.3 shows the function table for the NOR latch.

TABLE 8.3 NOR Latch Function Table

S	R	Q_{t+1}	$\overline{Q}_{t+1}$	Function
0	0	Q_t	$\overline{Q}_t$	No change
0	1	0	1	Reset
1	0	1	0	Set
1	1	0	0	Forbidden

*Many sources (such as datasheets) use the notation Q_0 to refer to the previous state of Q. We will use the notation indicated (Q_t for present state and Q_{t+1} for next state) so as to be able to reserve Q_0 for the least significant bit of a circuit requiring multiple Q outputs.

FIGURE 8.15   Block Diagram File Representation of a NAND Latch

Practical Synthesis of a NAND Latch in Quartus II

(This section may be omitted without loss of continuity.)

Figure 8.15 shows a Block Diagram File of a NAND latch created in Quartus II. The gate components for the negative-NANDs are called BOR2, for "Bubbled-OR, 2-inputs." Inputs are labeled *nS* and *nR*. Outputs are *Q* and *nQ*. The *n* prefix takes the place of a logic inversion bar, which cannot be typed in the Quartus Block Editor.

Quartus II does not synthesize the latch exactly as shown in Figure 8.15. In order to fit the circuit more efficiently into the selected CPLD, Quartus II analyzes the Boolean equations of the original latch and reformats the circuit to fit the target device. When we compile the design, the **Fitter Equations,** found in the **Fitter** section of the compilation report, contains the following text.

Quartus II:
nand_latch.bdf
nand_latch.vwf
MAX+PLUS II:
nd_latch.gdf

```
--A1L2 is inst~23 at LC3
A1L2_p1_out = A1L2 & nR;
A1L2_or_out = A1L2_p1_out # !nS;
A1L2 = A1L2_or_out;

--A1L1 is inst1~9 at LC5
A1L1_p1_out = A1L2 & nR;
A1L1_or_out = A1L1_p1_out;
A1L1 = !(A1L1_or_out);

--nS is nS at Pin_33
--operation mode is input

nS = INPUT();

--nR is nR at Pin_81
--operation mode is input

nR = INPUT();

--Q is Q at Pin_12
--operation mode is output
Q = OUTPUT(A1L2);

--nQ is nQ at Pin_11
--operation mode is output

nQ = OUTPUT(A1L1);
```

The Fitter Equations section of the Compilation Report uses the following Boolean operators:

&—AND
#—OR
!—NOT
$—XOR (not shown in this example)

The Boolean equations for the latch circuit can be determined by substituting pin names into the equation text. This translates to the following two equations:

$$Q = Q \cdot nR + \overline{nS}$$
$$nQ = \overline{Q \cdot nR}$$

The circuit derived from these equations is shown in Figure 8.16.

The latching in the circuit occurs in the feedback loop from Q to the upper AND gate. The lower AND and NOT gates are used solely to provide a complementary output, nQ. In this circuit the complementary output is not required to maintain or change the latching condition of either state of the latch; it is only there because the original circuit specified a $\overline{Q}$ output. A major advantage of this circuit is that it has an unambiguous transition from the forbidden state to the no change state.

Figure 8.17 shows tables from the Quartus II Compilation Report for the latch circuit indicating the pin to pin delays from input to output. The tables (found under **Timing Analyzer** in the Compilation Report) indicate that the delays are 7.5 ns from either input to Q and 7.5 ns or 12.8 ns to nQ, depending on the input-to-output path

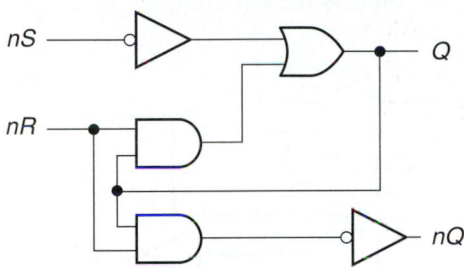

FIGURE 8.16   NAND Latch as Synthesized by Quartus II

nand_latch Compilation Report

tpd

	Slack	Required P2P Time	Actual P2P Time	Source Name	Destination Name	
1	N/A	None	12.800 ns	nR	nQ	
2	N/A	None	12.800 ns	nS	nQ	
3	N/A	None	7.500 ns	nR	Q	
4	N/A	None	7.500 ns	nS	Q	

FIGURE 8.17A   Pin-to-Pin Delays (t_{pd}) for the Latch in Figures 8.15 and 8.16

nand_latch Compilation Report

Minimum tpd

	Minimum Slack	Required P2P Time	Actual P2P Time	Source Name	Destination Name	
1	N/A	None	7.500 ns	nS	Q	
2	N/A	None	7.500 ns	nR	Q	
3	N/A	None	7.500 ns	nR	nQ	
4	N/A	None	12.800 ns	nS	nQ	

FIGURE 8.17B   Pin-to-Pin Delays (Minimum t_{pd}) for the Latch in Figures 8.15 and 8.16

a) Stable in RESET state.

b) Set input activates.

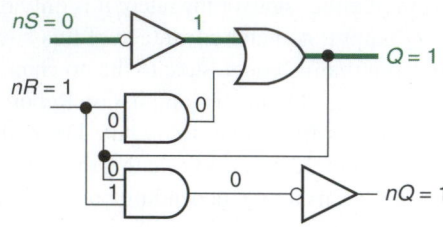

c) Change propagates through
OR gate and sets $Q = 1$.

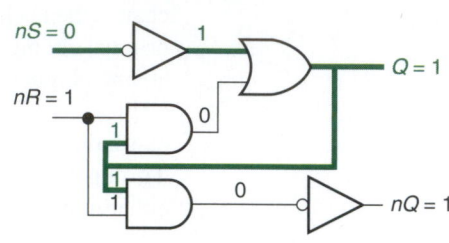

d) Feedback takes HIGH to
AND gates.

e) HIGH propagates through upper AND
gate, completing latch transition.

f) Change reaches nQ.

g) HIGH supplied to OR gate from AND gate keeps
latch SET. LOW at *nS* can be removed.

FIGURE 8.18   RESET-to-SET Transition for a Latch Synthesized by Quartus II

taken for a particular transition. We will see how these delays are expressed when we analyze the operation of the latch.

Figure 8.18 shows the *RESET*-to-*SET* transition of the latch in Figure 8.16. The transition goes through the following steps.

■ The latch begins in the *RESET* state. *nS* and *nR* are both HIGH. Since the *Q* output is LOW, feedback ensures that both inputs to the OR gate are LOW. Thus, $Q = 0$ (Figure 8.18a).
■ The *nS* input goes LOW (Figure 8.18b).
■ The change at *nS* propagates through the NOT gate and the OR gate, making $Q = 1$ (Figure 8.18c).
■ The HIGH at *Q* transfers to the AND inputs via the feedback line (Figure 8.18d).
■ Since *nR* = 1, the HIGH from the feedback line propagates through both AND gates. The upper gate completes the latching sequence, since it applies a HIGH

FIGURE 8.19 Waveforms for RESET-to-SET Transition of a NAND Latch Synthesized by Quartus II

to the OR gate input, which only requires a single HIGH for a HIGH output. The lower AND gate sends the HIGH to the NOT gate that drives nQ (Figure 8.18e).

◼ The change propagates through the NOT gate and reaches nQ (Figure 8.18f).

◼ The latch is stable in the *SET* condition. The LOW at nS can be removed (Figure 8.18g).

Figure 8.19 shows the Quartus II simulation waveforms for the *RESET*-to-*SET* transition of the latch. Each grid space is 5 ns. The Q output goes HIGH 7.5 ns after nS goes LOW. The nQ output goes LOW after 12.8 ns. (We know these times from the pin-to-pin delays shown in Figure 8.17.) The nQ transition takes longer than Q because the nS-to-nQ path is four gates long and the nS-to-Q path is only two gates.

Figure 8.20 shows the transition of the latch from *SET* to *RESET,* as follows.

◼ The latch begins in the stable *SET* state (Figure 8.20a).

◼ Input nR goes LOW (Figure 8.20b).

◼ The LOW at nR changes both AND outputs to LOW (Figure 8.20c).

◼ The upper AND gate makes the OR output LOW, since neither input is now HIGH. This makes $Q = 0$. The lower AND gate applies a LOW to the lower NOT gate, making $nQ = 1$ (Figure 8.20d).

◼ The feedback line transfers the LOW at Q to the input of the upper AND gate, completing the latch transition (Figure 8.20e).

◼ The latch is now stable in the *RESET* condition. The LOW at nR can be removed (Figure 8.20f).

The simulation waveforms for the *SET*-to-*RESET* transition are shown in Figure 8.21. In this case, Q and nQ change simultaneously, 7.5 ns after nR goes LOW, because both outputs are the same number of gates away from nR.

The transition from the forbidden state of the latch to the no change state is shown in Figure 8.22. Recall that the forbidden state occurs when both inputs are active, indicating that the latch is setting and resetting at the same time. In the original latch circuit, this conflict was not predictably resolvable if the latch inputs changed simultaneously to the inactive state. In this circuit, however, the latch will always drop to the *SET* state when both inputs go from 0 to 1 at the same time.

The transition takes the following steps:

◼ The latch begins in the forbidden state. Both inputs are LOW. The LOW at nS is inverted and applies a HIGH to one input of the OR gate, making $Q = 1$. The LOW at nR ensures that both AND gates have LOW outputs. The LOW applied to the NOT gate that drives nQ makes $nQ = 1$ (Figure 8.22a).

a) Stable in SET state.

b) Reset input activates.

c) LOW propagates through AND gates.

d) LOW propagates through OR and NOT gates, changing states of Q and nQ.

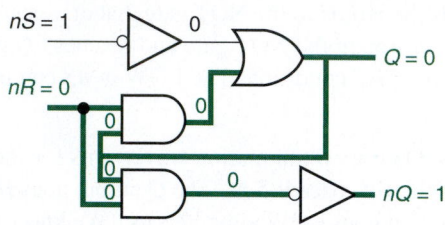

e) Feedback transfers LOW to AND gate inputs, completing latch transition.

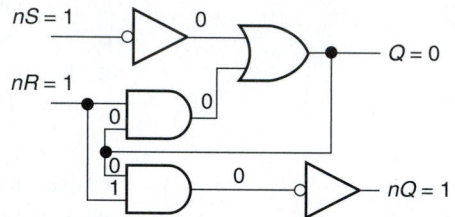

f) LOW at AND input ensures LOW at OR input. LOW on *nR* can be removed. Stable in RESET state.

FIGURE 8.20 SET-to-RESET Transition for a Latch Synthesized by Quartus II

nand_latch Simulation Report

Simulation Waveforms

| Master Time Bar: | 100.0 ns | ◀ ▶ Pointer: | 17.49 ns | Interval: | -82.51 ns | Start: | | End: | |

Name	Value at 100.0 ns	20.0 ns	30.0 ns	40.0 ns	50.0 ns	60.0 ns
nS	B 1					
nR	B 1					
Q	B 0					
nQ	B 1					

FIGURE 8.21 Waveforms for SET-to-RESET Transition of a NAND Latch Synthesized by Quartus II

a) Latch in forbidden state.

b) Inputs revert to no change state.

c) Changes propagate through AND and NOT gates.

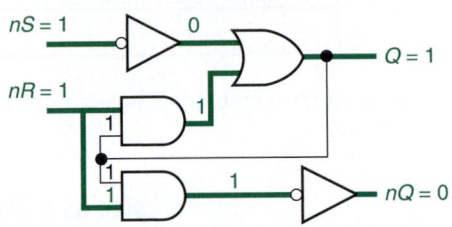

d) Changes propagate through OR and NOT gates to Q and nQ.

e) Feedback loop closes. Latch is stable in SET condition.

FIGURE 8.22 Forbidden State-to-No Change State Transition for a Latch Synthesized by Quartus II

- Both inputs go HIGH at the same time (Figure 8.22b).
- The changes propagate through the first layer of gates. The HIGH at nS goes through the NOT gate and applies a LOW to the OR input. The HIGH at nR is applied to the upper AND gate and applies a HIGH to the OR input. Thus the OR output remains unchanged since it requires only a single HIGH to make its output HIGH. Meanwhile, the HIGH at nR propagates through the lower AND gate towards nQ (Figure 8.22c).
- The HIGH at the OR input keeps Q HIGH. The lower NOT gate inverts the HIGH from nR and makes nQ LOW (Figure 8.22d).
- The HIGH at Q transfers to the AND input via the feedback line, completing the transition. The latch is now stable in the *SET* condition (Figure 8.22e).

Figure 8.23 shows the waveforms for the forbidden state transition. The simulation begins in the *RESET* state. The inputs then go to the forbidden state, followed by a transition to the no change condition, shown as a *SET* state. During the forbidden state, both outputs are HIGH. When the latch drops into the *SET* state, the nQ output goes LOW 7.5 ns after the inputs both go HIGH.

nand_latch Simulation Report

Simulation Waveforms

| Master Time Bar: | 100.0 ns | ◀ ▶ Pointer: | 87.65 ns | Interval: | -12.35 ns | Start: | | End: | |

	Name	Value at 100.0 ns	0 ps	20.0 ns	40.0 ns	60.0 ns	80.0 ns	100.0 ns
	nS	B 1						
	nR	B 1						
	Q	B 1						
	nQ	B 0						

FIGURE 8.23   Waveforms for a Forbidden State-to-No Change State of a NAND Latch Synthesized by Quartus II

Latch as a Switch Debouncer

Pushbutton or toggle switches are sometimes used to generate pulses for digital circuit inputs, as illustrated in Figure 8.24. However, when a switch is operated and contact is made on a new terminal, the contact, being mechanical, will bounce a few times before settling into the new position. Figure 8.24d shows the effect of contact bounce on the waveform for a pushbutton switch. The contact bounce is shown only on the terminal where contact is being made, not broken.

Contact bounce can be a serious problem, particularly when a switch is used as an input to a digital circuit that responds to individual pulses. If the circuit expects to receive one pulse, but gets several from a bouncy switch, it will behave unpredictably.

A latch can be used as a switch debouncer, as shown in Figure 8.25a. When the pushbutton is in the position shown, the latch is set, since $\overline{S} = 0$ and $\overline{R} = 1$. (Recall that the NAND latch inputs are active LOW.) When the pushbutton is pressed, the $\overline{R}$ contact bounces a few times, as shown in Figure 8.25b. However, on the first bounce, the latch is reset. Any further bounces are ignored, since the resulting input state is either $\overline{S} = \overline{R} = 1$ (no change) or $\overline{S} = 1$, $\overline{R} = 0$ (reset).

Similarly, when the pushbutton is released, the $\overline{S}$ input bounces a few times, setting the latch on the first bounce. The latch ignores any further bounces, as they either do not change the latch output ($\overline{S} = \overline{R} = 1$) or set it again ($\overline{S} = 0$, $\overline{R} = 1$). The resulting waveforms at Q and $\overline{Q}$ are free of contact bounce and can be used reliably as inputs to digital sequential circuits.

a. Pushbutton

b. Toggle

c. Ideal waveform

d. Effect of contact bounce

FIGURE 8.24   Switches as Pulse Generators

a. Switch debouncer

b. Timing diagram

FIGURE 8.25 NAND Latch as a Switch Debouncer

EXAMPLE 8.3

A NOR latch can be used as a switch debouncer, but not in the same way as a NAND latch. Figure 8.26 shows two NOR latch circuits, only one of which works as a switch debouncer. Draw a timing diagram for each circuit, showing R, S, Q, and $\overline{Q}$, to prove that the circuit in Figure 8.26b eliminates switch contact bounce but the circuit in Figure 8.26a does not.

a.   b.

FIGURE 8.26 Example 8.3: NOR Latch Circuits

■ **Solution**

Figure 8.27 shows the timing diagrams of the two NOR latch circuits. In the circuit in Figure 8.26a, contact bounce causes the latch to oscillate in and out of the forbidden state of the latch ($S = R = 1$). This causes one of the two outputs to bounce for each contact closure. (Use the function table of the NOR latch to examine each part of the timing diagram to see that this is so.)

(continued)

a. Timing diagram for circuit of Figure 6.13a

b. Timing diagram for circuit of Figure 6.13b

FIGURE 8.27   Example 8.3: NOR Latch Circuits

By making the resistors pull down rather than pull up, as in Figure 8.26b, the latch oscillates in and out of the no change state ($S = R = 0$) as a result of contact bounce. The first bounce on the *SET* terminal sets the latch, and other oscillations are disregarded. The first bounce on the *RESET* input resets the latch, and further pulses on this input are ignored.

The principle illustrated here is that a closed switch must present the active input level to the latch, since switch bounce is only a problem on contact closure. Thus, a closed switch must make the input of a NOR latch HIGH or the input of a NAND latch LOW to debounce the switch waveform.

NOTE...

The NOR latch is seldom used in practice as a switch debouncer. The pull-down resistors need to be about 500 Ω or less to guarantee a logic LOW at the input of a TTL NOR gate. In such a case, a constant current of about 10 mA flows through the resistor connected to the normally closed portion of the switch. This value is unacceptably high in most circuits, as it draws too much idle current from the power supply. For this reason, the NAND latch, which uses higher-value pull-up resistors (about 1 kΩ or larger) and therefore draws less idle current, is preferred for a switch debouncer.

■ **SECTION 8.2 REVIEW PROBLEM**

8.2   Why is the input state $S = R = 1$ considered forbidden in the NOR latch? Why is the same state in the NAND latch the no change condition?

8.3   GATED LATCHES

Gated SR Latch

It is not always desirable to allow a latch to change states at random times. The circuit shown in Figure 8.28, called a **gated SR latch,** regulates the times when a latch is allowed to change state.

a. Circuit

b. ENABLE used as an ON/OFF signal

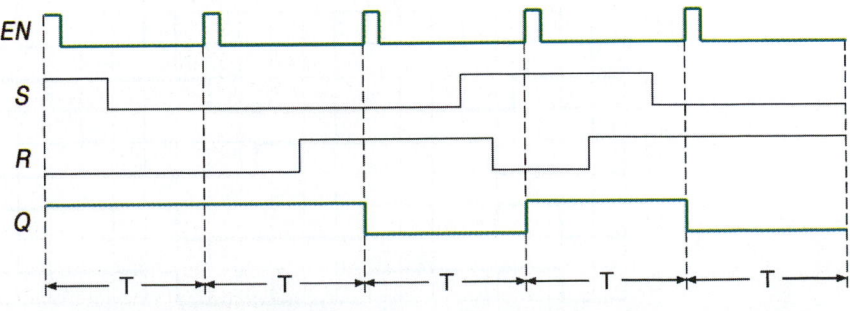

T = equal time interval

c. ENABLE used as a synchronizing signal

FIGURE 8.28   Gated SR Latch

The gated SR latch has two distinct subcircuits. One pair of gates is connected as an SR latch. A second pair, called the **steering gates,** can be enabled or inhibited by a control signal, called *ENABLE,* allowing one or the other of these gates to pass a *SET* or *RESET* signal to the latch gates.

The *ENABLE* input can be used in two principal ways: (1) as an ON/OFF signal, and (2) as a synchronizing signal. This latter technique is also called "strobing."

Figure 8.28b shows the *ENABLE* input functioning as an ON/OFF signal. When *ENABLE* = 1, the circuit acts as an active-HIGH latch. The upper gate converts a HIGH at *S* to a LOW at $\bar{S}$, setting the latch. The lower gate converts a HIGH at *R* to a LOW at $\bar{R}$, thus resetting the latch.

When *ENABLE* = 0, the steering gates are inhibited and do not allow *SET* or *RESET* signals to reach the latch gate inputs. In this condition, the latch outputs cannot change.

Figure 8.28c shows the *ENABLE* input as a synchronizing or "strobe" signal. A periodic pulse waveform is present on the *ENABLE* line. The *S* and *R* inputs are free to change at random, but the latch outputs will change only when the *ENABLE* input is active. Since the *ENABLE* pulses are equally spaced in time, changes to the latch output can occur only at fixed intervals. The outputs can change out of synchronization if *S* or *R* change when *ENABLE* is HIGH. We can minimize this possibility by making the *ENABLE* pulses as short as possible.

Table 8.4 represents the function table for a gated SR latch.

TABLE 8.4   Gated SR Latch Function Table

EN	S	R	Q_{t+1}	$\bar{Q}_{t+1}$	Function
1	0	0	Q_t	$\bar{Q}_t$	No change
1	0	1	0	1	Reset
1	1	0	1	0	Set
1	1	1	1	1	Forbidden
0	X	X	Q_t	$\bar{Q}_t$	Inhibited

EXAMPLE 8.4

Figure 8.29 shows two gated latches with the same *S* and *R* input waveforms but different *ENABLE* waveforms. EN_1 has a 50% duty cycle. EN_2 has a duty cycle of 16.67%.

Draw the output waveforms, Q_1 and Q_2. Describe how the length of the *ENABLE* pulse affects the output of each latch, assuming that the intent of each circuit is to synchronize the output changes to the beginning of the *ENABLE* pulse.

FIGURE 8.29   Example 8.4: Effect of ENABLE Pulse Width

■ **Solution**

Figure 8.29 shows the completed timing diagram. The longer *ENABLE* pulse at latch 1 allows the output to switch too soon during pulses 1 and 4. ("Too soon" means before the beginning of the next *ENABLE* pulse.) In each of these cases, the *S* and *R* inputs change while the *ENABLE* input is HIGH. This premature switching is eliminated in latch 2 because the *S* and *R* inputs change after the shorter *ENABLE* pulse is finished. A shorter pulse gives less chance for synchronization error since the time for possible output changes is minimized.

Transparent Latch (Gated D Latch)

Figure 8.30 shows the equivalent circuit of a gated D ("data") latch, or **transparent latch**. This circuit has two modes. When the *ENABLE* input is HIGH, the latch is *transparent* because the output *Q* goes to the level of the data input, *D*. (We say, "*Q* follows *D*.") When the *ENABLE* input is LOW, the latch *stores* the data that was present at *D* when *ENABLE* was last HIGH. In this way, the latch acts as a simple memory circuit.

The latch in Figure 8.30 is a modification of the gated SR latch, configured so that the *S* and *R* inputs are always opposite. Under these conditions, the states $S = R = 0$ (no change) and $S = R = 1$ (forbidden) can never occur. However, the equivalent of the no change state happens when the *ENABLE* input is LOW, when the latch steering gates are inhibited.

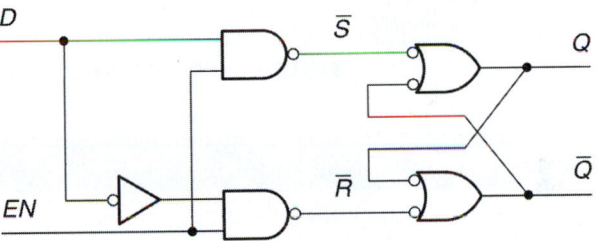

FIGURE 8.30   Transparent Latch

Figure 8.31 shows the operation of the transparent latch in the inhibit (no change), set, and reset states. When the latch is inhibited, the steering gates block any LOW pulses to the latch gates; the latch does not change states, regardless of the logic level at *D*.

If *EN* = 1, *Q* follows *D*. When *D* = 1, the upper steering gate transmits a LOW to the *SET* input of the latch and *Q* = 1. When *D* = 0, the lower steering gate transmits a LOW to the *RESET* input of the output latch and *Q* = 0.

Table 8.5 shows the function table for a transparent latch.

TABLE 8.5   Function Table of a Transparent Latch

EN	*D*	Q_{t+1}	$\bar{Q}_{t+1}$	**Function**	**Comment**
0	X	Q_t	$\bar{Q}_t$	No change	Store
1	0	0	1	Reset	Transparent
1	1	1	0	Set	

a. No change state   b. Set condition

c. Reset condition

FIGURE 8.31 Operation of Transparent Latch

Quartus II:
d_latch.bdf
d_latch.vwf
MAX+PLUS II:
d_latch.gdf
d_latch.scf

FIGURE 8.32 D Latch in a Quartus II Block Diagram

![d_latch Simulation Report]

Name	Value at 0 ps
ENA	B 1
D	B 0
Q	B 0

FIGURE 8.33 Simulation for a D Latch

Implementing D Latches in Quartus II

A D latch can be implemented in Quartus II as a primitive in a Block Diagram File or in a VHDL design entity. It can also be created with a behavioral or structural description in a VHDL file.

Figure 8.32 shows a D latch primitive in a Quartus II Block Diagram File. Figure 8.33 shows a simulation of the latch. From 0 to 500 ns, *ENABLE* is HIGH and the latch is in the transparent mode (Q follows D). When *ENABLE* goes LOW, the last value of D (0) is stored until *ENABLE* goes high again, just before 800 ns. When *ENABLE* goes LOW again, a new value of D (HIGH) is stored until the end of the simulation.

In VHDL, a PROCESS statement is *concurrent*, but the statements inside the PROCESS are *sequential*. In other words, anything described by a PROCESS acts like a separate component in a design entity. However, the interior of the component so described acts as a sequential circuit. Since the behavior of a D latch is sequential, its description can be created inside a PROCESS. (You can pull a latch out of a bin of parts and connect it in a circuit, but the inside of the part is sequential.) The following VHDL code describes a D latch.

Quartus II:
d_latch_vhdl.vhd
d_latch_vhdl.vwf
MAX+PLUS II:
d_lch.vhd
d_lch.scf

```
-- d_latch_vhdl.vhd
-- D latch with active-HIGH level-sensitive enable

ENTITY d_latch_vhdl IS
    PORT(
        d, ena : IN  BIT;
        q      : OUT BIT);
END d_latch_vhdl;

ARCHITECTURE a OF d_latch_vhdl IS
BEGIN
    PROCESS (d, ena)
    BEGIN
        IF (ena = '1') THEN
            q  <= d;
        END IF;
    END PROCESS;
END a;
```

Another method of creating a latch in VHDL is to instantiate a LATCH primitive in a VHDL file. The primitive is contained in the **altera** library, in a package called **maxplus2.** The component declaration for this primitive is:

```
COMPONENT LATCH
    PORT (d   : IN  STD_LOGIC;
          ena : IN  STD_LOGIC;
          q   : OUT STD_LOGIC);
END COMPONENT;
```

Since the component declaration is in the **maxplus2** package, you do not have to declare it in the file in which you are using it. A VHDL file that uses the latch primitive is listed next. The component declaration uses STD_LOGIC types, so we must include the type definitions in the **ieee** library (**std_logic_1164** package).

Quartus:
latch_primitive.vhd
latch_primitive.vwf
MAX+PLUS II:
lch_prim.vhd

```
-- latch_primitive.vhd
-- D latch with active-HIGH level-sensitive enable

LIBRARY ieee;                      ⎤ Required for STD_LOGIC types
USE ieee.std_logic_1164.ALL;       ⎦
LIBRARY altera;                    ⎤ Required for latch component
USE altera.maxplus2.ALL;           ⎦

ENTITY latch_primitive IS
    PORT(
        d_in, enable : IN  STD_LOGIC;
        q_out        : OUT STD_LOGIC);
END latch_primitive;

ARCHITECTURE a OF latch_primitive IS
```

```
BEGIN
    -- Instantiate a latch from a primitive
    latch_prim: latch
        PORT MAP (d   => d_in,
                  ena => enable,
                  q   => q_out);
END a;
```

User names (defined in entity declaration)

Component names (defined in maxplus2 package)

More information about Quartus II packages and primitives can be found in the Quartus II Help menu. In the **Help** menu, select **Index.** For information on packages, type in **Using Quartus II Packages.** Double click the index item to see the help for this topic. For information on available primitives, type **primitives** in the **Index** tab. Select the entry called **list of** under **primitives.** By clicking on the name of a particular primitive, you can determine whether it can be instantiated in a VHDL file and what its component declaration is, if available.

EXAMPLE 8.5

A system for monitoring automobile traffic is set up at an intersection, with four sensors, placed as shown in Figure 8.34. Each sensor monitors traffic for a particular direction. When a car travels over a sensor, it produces a logic HIGH. The status of the sensor system is captured for later analysis by a set of D latches, as shown in Figure 8.35. A timing pulse enables the latches once every 5 seconds and thus stores the system status as a "snapshot" of the traffic pattern. This technique of data capture is known as strobing.

FIGURE 8.34 Example 8.5: Sensor Placement in a Traffic Intersection

FIGURE 8.35 Example 8.5: D Latch Collection of Data

Figure 8.36 shows the timing diagram of a typical traffic pattern at the intersection. The *D* inputs show the cars passing through the intersection in the various lanes. Complete this timing diagram by drawing the *Q* outputs of the latches. How should we interpret the *Q* output waveforms?

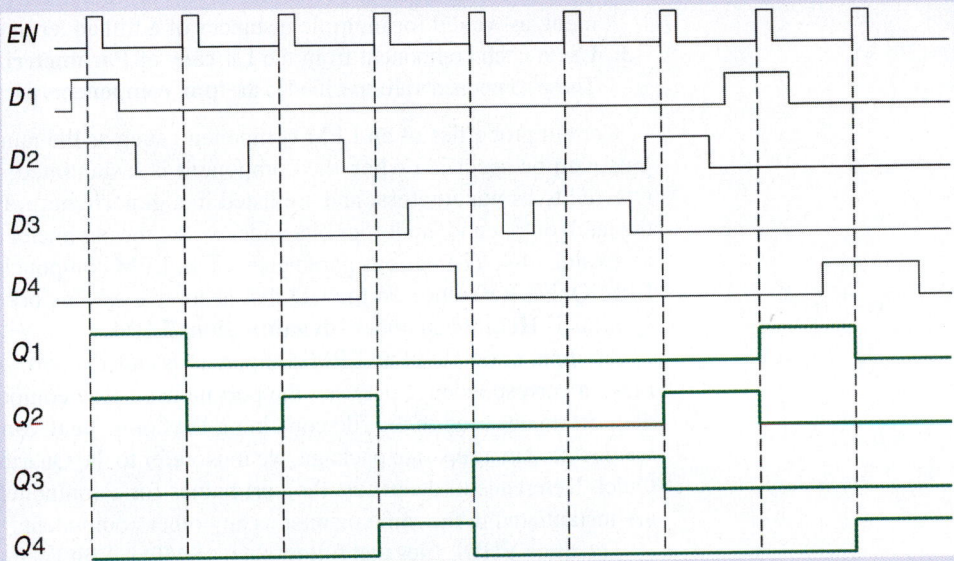

FIGURE 8.36 Example 8.5: Latch Timing Diagram

■ **Solution**

Figure 8.36 shows the completed timing diagram. The *ENABLE* input synchronizes the random sensor pattern to a 5-second standard interval. A *HIGH* on any *Q* output indicates a car over a sensor at the beginning of the interval. For example, at the beginning of the first interval, there is a car in the northbound lane (*Q*1) and one in the southbound lane (*Q*2). Similar interpretations can be made for each interval.

Multibit Latches in VHDL

■ **KEY TERMS**

Library of Parameterized Modules (LPM) A standardized set of components for which certain properties can be specified when the component is instantiated.

Parameter (in an LPM Component) A property of a component that can be specified when the component is instantiated.

Generic Map A VHDL construct that maps one or more parameters of a component to a value for that instance of the component.

Port Map A VHDL construct that maps the name of a port in a component to the name of a port, variable, or signal in a design entity that uses the component.

We can easily use VHDL to implement latches with multiple *D* inputs and *Q* outputs, but with a common *ENABLE* line, as in Figure 8.35. Three approaches are:

1. Use a behavioral description, as we did earlier for a single latch (**d_latch_vhdl.vhd**). Use STD_LOGIC_VECTOR types for *D* and *Q*, rather than STD_LOGIC.
2. We can use multiple LATCH primitives, instantiated by a GENERATE statement, as we did for multiple instances of a full adder in Chapter 7.
3. Use a latch component from the **Library of Parameterized Modules (LPM).** These components are specified in the **lpm_components** package in the **lpm** library.

Certain properties of an LPM component, such as the number of inputs or outputs, can be specified when the component is instantiated. These properties are referred to as **parameters,** and are listed in a **generic map.** For example, to make the latch output and input four bits wide, we set the parameter called LPM_WIDTH to a value of 4. The various parameters of an LPM component can be found in the LPM Quick Reference on the CD that accompanies this book (lpm.pdf) or in the Quartus II **Help** menu under **Megafunctions/LPM.**

An input or output of an LPM component is called a port. A **port map** is used to make a correspondence between the port names in the component declaration and the port names used in the file containing the component. Since LPM components are declared in a separate package, we must refer to the Quartus II Help or the LPM Quick Reference to determine the port names for a component. LPM components are instantiated in the same manner as any other component.

The three VHDL files that follow each specify a 4-bit latch with common enable, each using one of the previously mentioned methods.

Behavioral Description

Quartus II:
latch4_behavioral.vhd
MAX+PLUS II:
ltch4bhv.vhd

```
-- latch4_behavioral.vhd
-- D latch with active-HIGH level-sensitive enable

LIBRARY ieee;
USE ieee.std_logic_1164.ALL;

ENTITY latch4_behavioral IS
    PORT(d      : IN  STD_LOGIC_VECTOR (3 downto 0);
         enable : IN  STD_LOGIC;
         q      : OUT STD_LOGIC_VECTOR (3 downto 0));
END latch4_behavioral;

ARCHITECTURE a OF latch4_behavioral IS
BEGIN
    PROCESS (enable, d)
    BEGIN
        IF (enable = "1") THEN
            q <= d;
        END IF;
    END PROCESS;
END a;
```

4 LATCH Primitives and a GENERATE Statement

Quartus II:
latch4_primitive.vhd
MAX+PLUS II:
ltch4prm.vhd

```
-- latch4_primitive.vhd
-- D latch with active-HIGH level-sensitive enable

LIBRARY ieee;
USE ieee.std_logic_1164.ALL;
LIBRARY altera;
USE altera.maxplus2.ALL;
```

```
ENTITY latch4_primitive IS
    PORT(d_in : IN  STD_LOGIC_VECTOR (3 downto 0);
        enable : IN  STD_LOGIC;
        q_out  : OUT STD_LOGIC_VECTOR (3 downto 0));
END latch4_primitive

ARCHITECTURE a OF latchl4_primitive IS
BEGIN
    -- Instantiate a latch primitive from maxplus2 package
    latch4:
    FOR i IN 3 downto 0 GENERATE
    latch_primitive: latch
        PORT MAP (d => d_in(i), ena => enable, q => q_out(i));
    END GENERATE;
END a;
```

LPM Latch

Quartus II:
latch4_lpm
latch4_lpm.scf
MAX+PLUS II
latch4_lpm.vhd
latch4_lpm.scf

```
-- latch4_lpm.vhd
-- 4-BIT D latch with active-HIGH level-sensitive enable
-- Uses a latch component from
-- the Library of Parameterized Modules (LPM)

LIBRARY ieee;
USE ieee.std_logic_1164.ALL;        ] Required for STD_LOGIC types

LIBRARY lpm;
USE lpm.lpm_components.ALL;          ] Required for LPM components

ENTITY latch4_lpm IS
    PORT(d_in : IN  STD_LOGIC_VECTOR (3 downto 0);
        enable : IN  STD_LOGIC;
        q_out  : OUT STD_LOGIC_VECTOR (3 downto 0));
END latch4_lpm;

ARCHITECTURE a OF latch4_lpm IS
BEGIN
    -- Instantiate latch from an LPM component
    latch4: lpm_latch                        Parameters assigned with
        GENERIC MAP (LPM_WIDTH => 4) —— GENERIC MAP
        PORT MAP ( data    => d_in,
                   gate    => enable,    Ports assigned with
                   q       => q_out);    PORT MAP
END a;
```

Note that in the LPM_LATCH component, the **enable** input is called **gate.** In digital parlance, to "gate" an output is to enable or inhibit it.

All three files can be tested with a simulation having the following criteria.

SIMULATION CRITERIA

- No change on **q_out** when **enable** = 0.
- **q_out** follows **d_in** when **enable** = 1.
- Since the latch is four bits wide, each bit of **q_out** should be checked for a $0 \rightarrow 1$ and a $1 \rightarrow 0$ transition.

FIGURE 8.37 Simulation of a 4-Bit D Latch

The resultant simulation waveforms are shown in Figure 8.37. The inputs, **d_in,** represent a 4-bit group of signals, as do the outputs, **q_out.** An increasing count, from 5 to C (0101 to 1100) is applied to **d_in.** This count contains both states (0 and 1) for each input bit. For each applied input state, the output bus, **q_out,** does not change until the enable line goes HIGH.

Multibit Latches in the Quartus II Block Editor

> ■ **KEY TERM**
>
> **MegaWizard Plug-In Manager**   A tool provided with Quartus II (and MAX+PLUS II) that allows the user to automatically set the ports and parameter values of an LPM component for use in an HDL or Block Diagram File.

In addition to the VHDL applications of the previous section, LPM components can also be instantiated in a graphic format in the Block Editor of Quartus II. We can do this in two different ways:

1. We can enter the symbol using the **MegaWizard Plug-In Manager,** a tool that sets the ports and parameters of the component by directing the user through a sequence of setup screens. This method also generates the required VHDL code for the component which can be used if you want to instantiate it in a VHDL file.
2. We can enter the symbol manually and set the ports and parameters in a **Properties** dialog box.

MegaWizard Plug-In Manager

To enter the latch using the MegaWizard Plug-In Manager, first create a Block Diagram File in Quartus II, by selecting **New** from the **File** menu. Save the file and create a new project. To enter a new symbol, double click somewhere in the Block Editor desktop space. The **Symbol** dialog box, shown in Figure 8.38, will appear. In the **Name** box, type **lpm_latch.** The dialog box will change to the form shown in Figure 8.39. The box labeled **Launch MegaWizard Plug-In** will be automatically checked. Click **OK.**

The first screen of the MegaWizard Plug-In Manager will appear, as shown in Figure 8.40. Select **VHDL** as the output file type. Use the **Browse** button to select the folder and file name for the component output files or select the default file name given in the box. This folder should be the same as the Quartus II project folder. The file name is the name of the instance of the component, in this case **lpm_latch1.** Click **Next.**

FIGURE 8.38 Default Symbol Dialog Box

Default component symbol

Make sure this box is checked.

LPM component name

FIGURE 8.39 Symbol Dialog Box (LPM_LATCH)

Select VHDL output

FIGURE 8.40   MegaWizard Plug-in Manager (Selecting File Name and Type)

4-bit latch

These boxes should not be checked for this example.

FIGURE 8.41   MegaWizard Plug-In Manager (Specifying Component Ports and Parameters)

The next screen, shown in Figure 8.41, allows you to specify the width of the latch inputs and outputs. Type in "4" or select it from the drop-down box. Leave the other two boxes unchecked for this example. Click **Next.**

The final screen of the MegaWizard Plug-In Manager, shown in Figure 8.42, lists the file and path names created by the wizard. Click **Finish.** The modified LPM_LATCH symbol appears, as in Figure 8.43, with only the selected ports of the component showing. (Recall that the **gate** input is the same as an **enable** input.) The component can be used in a Block Diagram File, as shown in Figure 8.44.

FIGURE 8.42 MegaWizard Plug-In Manager (File Names)

FIGURE 8.43 LPM_LATCH Symbol

FIGURE 8.44 Block Diagram File Containing LPM_LATCH

The MegaWizard Plug-In Manager also generates VHDL code for the component. This code is as follows:

```
-- megafunction wizard: %LPM_LATCH%
-- GENERATION: STANDARD
-- VERSION: WM1.0
-- MODULE: lpm_latch
-- ========================================================
-- File Name: lpm latch1.vhd
-- Megafunction Name(s):
                    lpm_latch
-- ========================================================
-- *******************************************************
-- THIS IS A WIZARD-GENERATED FILE. DO NOT EDIT THIS FILE!
-- *******************************************************


--Copyright (C) 1991-2003 Altera Corporation
--Any megafunction design, and related netlist (encrypted or
--decrypted), support information, device programming or
--simulation file, and any other associated documentation or
--information provided by Altera or a partner under Altera's
--Megafunction Partnership Program may be used only to program
--PLD devices (but not masked PLD devices) from Altera. Any
--other use of such megafunction design, netlist, support
--information, device programming or simulation file, or any
--other related documentation or information is prohibited for
--any other purpose, including, but not limited to modification,
--reverse engineering, de-compiling, or use with any other
--silicon devices, unless such use is explicitly licensed under
--a separate agreement with Altera or a megafunction partner.
--Title to the intellectual property, including patents,
--copyrights, trademarks, trade secrets, or maskworks, embodied
--in any such megafunction design, netlist, support information,
--device programming or simulation file, or any other related
--documentation or information provided by Altera or a
--megafunction partner, remains with Altera, the megafunction
--partner, or their respective licensors. No other licenses,
--including any licenses needed under any third party's
--intellectual property, are provided herein.

LIBRARY ieee;
USE ieee.std_logic_1164.all;

LIBRARY lpm;
USE lpm.lpm_components.all;

ENTITY lpm_latch1 IS
      PORT
      (
              data        : IN  STD_LOGIC_VECTOR (3 DOWNTO 0);
              gate        : IN  STD_LOGIC ;
              q           : OUT STD_LOGIC_VECTOR (3 DOWNTO 0)
      );
END lpm_latch1;
```

```
ARCHITECTURE SYN OF lpm_latch1 IS

        SIGNAL sub_wire0 : STD_LOGIC_VECTOR (3 DOWNTO 0);


        COMPONENT lpm_latch
        GENERIC (
                lpm_width        : NATURAL;
                lpm_type         : STRING
        );
        PORT (
                        q    : OUT STD_LOGIC_VECTOR (3 DOWNTO 0);
                        data : IN STD_LOGIC_VECTOR (3 DOWNTO 0);
                        gate : IN STD_LOGIC
        );
        END COMPONENT;

BEGIN
        q   <= sub_wire0(3 DOWNTO 0);

        lpm_latch_component : lpm_latch
        GENERIC MAP (
                lpm_width => 4,
                lpm_type => "LPM_LATCH"
        )
        PORT MAP (
                data => data,
                gate => gate,
                q => sub_wire0
        );

END SYN;


-- =======================================================
-- CNX file retrieval info
-- =======================================================
-- Retrieval info: PRIVATE: nBit NUMERIC "4"
-- Retrieval info: PRIVATE: aclr NUMERIC "0"
-- Retrieval info: PRIVATE: aset NUMERIC "0"
-- Retrieval info: CONSTANT: LPM_WIDTH NUMERIC "4"
-- Retrieval info: CONSTANT: LPM_TYPE STRING "LPM_LATCH"
-- Retrieval info: USED_PORT: data 0 0 4 0 INPUT NODEFVAL data[3..0]
-- Retrieval info: USED_PORT: q 0 0 4 0 OUTPUT NODEFVAL q[3..0]
-- Retrieval info: USED_PORT: gate 0 0 0 0 INPUT NODEFVAL gate
-- Retrieval info: CONNECT: @data 0 0 4 0 data 0 0 4 0
-- Retrieval info: CONNECT: q 0 0 4 0 @q 0 0 4 0
-- Retrieval info: CONNECT: @gate 0 0 0 0 gate 0 0 0 0
-- Retrieval info: LIBRARY: lpm lpm.lpm_components.all
```

Manual Entry of an LPM Component

To enter an LPM component manually, follow the same procedure as shown in Figures 8.38 and 8.39, but make sure that the box labeled **Launch MegaWizard Plug-In** is *unchecked*. The default component symbol, shown in Figure 8.45, will appear. Place the component on the desktop of the Block Editor by clicking somewhere in the desktop space. In addition to the presently selected ports of the component, the symbol also shows a box at the top right corner that shows the present parameter values.

Parameter	Value
LPM_WIDTH	
LPM_AVALUE	

LPM_LATCH

aset
data[]   q[]
gate
aconst
aclr

inst

FIGURE 8.45   Default Symbol for an LPM_LATCH

Symbol Properties

General | Ports | Parameters | Format

Port

Name: aclr   Status: Used

Type: INPUT   Alias: aclr

Inversion
- None
- All
- Pattern: [   ] Binary

☐ Hide alias

Change

Existing ports:

Name	Alias	Inversion	Status	Direction	Hide Alias
aclr	aclr	None	Used	INPUT	No
aconst	aconst	None	Used	INPUT	No
aset	aset	None	Used	INPUT	No
data[LPM_WIDTH-1..0]	data[]	None	Used	INPUT	No
gate	gate	None	Used	INPUT	No
q[LPM_WIDTH-1..0]	q[]	None	Used	OUTPUT	No

OK   Cancel

All ports used (default status)

FIGURE 8.46   Default Port Settings for an LPM_LATCH

Click on the component to highlight it. Right-click the highlighted component and select **Properties** from the pop-up menu. Figure 8.46 shows the **Ports** tab of the resultant dialog box with all port settings in their default condition. Figure 8.47 shows the default setting of the **Parameters** tab.

Change the port settings to the required values, as shown in Figure 8.48. The first three ports, **aclr, aconst,** and **aset,** are not used in this example. For each one, click on the port name in the **Existing ports** box, select **Unused** from the drop-down menu in the **Status** box, and click **Change.** After this is done, go to the **Parameters** tab, shown in Figure 8.49. Select the **LPM_WIDTH** parameter and change the setting to **4,** either by typing or by selecting the value from the drop-down menu in the **Setting** box. Click **Change,** then click **OK.**

The symbol for the modified LPM latch is shown in Figure 8.50. This component is ready to be used in a Block Diagram File.

No parameter values specified (default)

FIGURE 8.47   Default Parameter Settings for an LPM_LATCH

Selecting aclr as unused

Ports shown as used or unused

FIGURE 8.48   Modified Port Settings for a 4-Bit LPM_LATCH

Specifying value of parameter LPM_width

FIGURE 8.49   Modified Parameter Settings for a 4-Bit LPM_LATCH

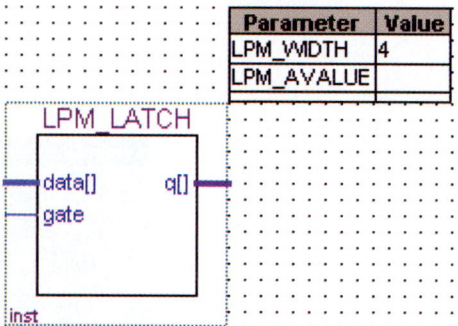

FIGURE 8.50   Modified Symbol for a 4-Bit LPM_LATCH

LPM Components with Active-LOW Ports

If your CPLD application has active-LOW inputs or outputs, such as the active-LOW LED indicators on the Altera UP-1 and UP-2 boards, you can alter the active level of a component port in the **Ports** tab of the component **Properties** dialog box. To change the **q** outputs of the LPM_LATCH component to active-LOW, open the **Properties** dialog box, shown in Figure 8.51, and select the **Ports** tab. Select **q[LPM_WIDTH-1..0]** from the **Existing ports** box. In the **Inversion** box, select **All** and click **Change.** Click **OK.**

The modified component symbol is shown in Figure 8.52. Note that the **q[ ]** port has a bubble now, indicating an active-LOW port.

latch4_lpm_graphic2.bdf
latch4_lpm_graphic2.vwf

■ **SECTION 8.3 REVIEW PROBLEM**

8.3   Write the VHDL code for a 16-bit latch with common active-HIGH enable, using MAX+PLUS II **latch** primitives.

q is active-Low

FIGURE 8.51 Port Settings for an LPM Latch with Active-LOW Outputs

FIGURE 8.52 Symbol for an LPM Latch with Active-LOW Outputs

8.4 EDGE-TRIGGERED D FLIP-FLOPS

➡ LAB SOURCE
Lab #20
NAND Latches, D Latches,
and D Flip-Flops

■ KEY TERMS

CLOCK An enabling input to a sequential circuit that is sensitive to the positive- or negative-going edge of a waveform.

Edge The HIGH-to-LOW (negative edge) or LOW-to-HIGH (positive edge) transition of a pulse waveform.

Edge-Triggered Enabled by the positive or negative edge of a digital waveform.

Edge-Sensitive Edge-triggered.

Level-Sensitive Enabled by a logic HIGH or LOW level.

CLK ⎯⎯⎯⎯⎯⎯⎯⎿   CLK ⎯⎯⎯⎯⎯⎯⏋⎯⎯

a. LOW-to-HIGH transition   **b. HIGH-to-LOW transition**
(positive edge)   **(negative edge)**

FIGURE 8.53 Edges of a *CLOCK* Waveform

In Example 8.4, we saw how a shorter pulse width at the *ENABLE* input of a gated latch increased the chance of the output being synchronized to the *ENABLE* pulse waveform. This is because a shorter *ENABLE* pulse gives less chance for the *SET* and *RESET* inputs to change during the time the latch is enabled.

A logical extension of this idea is to enable the latch for such a small time that the width of the *ENABLE* pulse is almost zero. The best approximation we can make to this is to allow changes to the circuit output only when an enabling, or **CLOCK,** input receives the **edge** of an input waveform. An edge is the part of a waveform that is in transition from LOW to HIGH (positive edge) or HIGH to LOW (negative edge), as shown in Figure 8.53. We can say that a device enabled by an edge is **edge-triggered** or **edge-sensitive.**

Since the *CLOCK* input enables a circuit only while in transition, we can refer to it as a "dynamic" input. This is in contrast to the *ENABLE* input of a gated latch, which is **level-sensitive** or "static," and will enable a circuit for the entire time it is at its active level.

Latches Versus Flip-Flops

> ■ **KEY TERMS**
>
> **Flip-Flop** A sequential circuit based on a latch whose output changes when its *CLOCK* input receives an edge.
>
> **Edge Detector** A circuit in an edge-triggered flip-flop that converts the active edge of a *CLOCK* input to an active-level pulse at the internal latch's *SET* and *RESET* inputs.

A gated latch with a clock input is called a **flip-flop.** Although the distinction is not always understood, we will define a *latch* as a circuit with a *level-sensitive enable* (e.g., gated D latch) or *no enable* (e.g., NAND latch) and a *flip-flop* as a circuit with an *edge-triggered clock* (e.g., D flip-flop). A NAND or NOR latch is sometimes called an SR flip-flop. By our definition this is not correct, as neither of these circuits has a clock input. (An SR flip-flop would be like the gated SR latch of Figure 8.28 with a clock instead of an enable input.)

The symbol for the D, or data, flip-flop is shown in Figure 8.54. The D flip-flop has the same behavior as a gated D latch, except that the outputs change only on the positive edge of the clock waveform, as opposed to the HIGH state of the enable input. The triangle on the *CLK* (clock) input of the flip-flop indicates that the device is edge-triggered.

Table 8.6 shows the function table of a positive edge-triggered D flip-flop.

FIGURE 8.54 D Flip-Flop Logic Symbol

TABLE 8.6 Function Table for a Positive Edge-Triggered D Flip-Flop

CLK	D	Q_{t+1}	$\bar{Q}_t + 1$	Function
↑	0	0	1	Reset
↑	1	1	0	Set
0	X	Q_t	$\bar{Q}_t$	Inhibited
1	X	Q_t	$\bar{Q}_t$	Inhibited
↓	X	Q_t	$\bar{Q}_t$	Inhibited

FIGURE 8.55 D Flip-Flop Equivalent Circuit

Figure 8.55 shows the equivalent circuit of a positive edge-triggered D flip-flop. The circuit is the same as the transparent latch of Figure 8.30, except that the enable input (called *CLK* in the flip-flop) passes through an **edge detector,** a circuit that converts a positive edge to a brief positive-going pulse. (A negative edge detector converts a negative edge to a positive-going pulse.)

Figure 8.56 shows a circuit that acts as a simplified positive edge detector. Edge detection depends on the fact that a gate output does not switch immediately when its input switches. There is a delay of about 3 to 10 ns from input change to output change, called propagation delay.

When input *x*, shown in the timing diagram of Figure 8.56, goes from LOW to HIGH, the inverter output, $\bar{x}$, goes from HIGH to LOW after a short delay. This delay causes both *x* and $\bar{x}$ to be HIGH for a short time, producing a high-going pulse at the circuit output immediately following the positive edge at *x*.

When *x* returns to LOW, $\bar{x}$ goes HIGH after a delay. However, there is no time in this sequence when both AND inputs are HIGH. Therefore, the circuit output stays LOW after the negative edge of the input waveform.

Figure 8.57 shows how the D flip-flop circuit operates. When $D = 0$ and the edge detector senses a positive edge at the *CLK* input, the output of the lower NAND gate steers a low-going pulse to the *RESET* input of the latch, thus storing a 0 at *Q*. When $D = 1$, the upper NAND gate is enabled. The edge detector sends a high-going pulse to the upper steering gate, which transmits a low-going *SET* pulse to the output latch. This action stores a 1 at *Q*.

a. Simplified circuit

b. Waveforms

FIGURE 8.56 Positive Edge Detector

a. Reset action

b. Set action

FIGURE 8.57 Operation of a D Flip-Flop

EXAMPLE 8.6

Figure 8.58 shows a Quartus II Block Diagram File with a D latch and a D flip-flop connected to the same data input and clock. Create a set of simulation criteria that illustrate the difference between the behavior of a latch (level-sensitive enable) and a flip-flop (edge-triggered clock). Use the simulation criteria to create a simulation that verifies the operation of the circuit.

FIGURE 8.58   Example 8.6: D Latch and D Flip-Flop

■ Solution

SIMULATION CRITERIA

- Check that *Q_latch* follows *D* when *CLK* = 1 and that *Q_flip_flop* follows *D* when there is a positive edge on *CLK*.
- If the outputs are LOW and *D* is at a constant HIGH level, both outputs should go HIGH when *CLK* goes HIGH.
- If the outputs are HIGH and *D* is at a constant LOW level, both outputs should go LOW when *CLK* goes HIGH.
- If the outputs are LOW and *D* changes from LOW to HIGH while *CLK* = 1, *Q_latch* will go HIGH immediately, but *Q_flip_flop* will wait until the next positive clock edge.
- If the outputs are HIGH and *D* changes from HIGH to LOW while *CLK* = 1, *Q_latch* will go LOW immediately, but *Q_flip_flop* will wait until the next positive clock edge.

The simulation, shown in Figure 8.59, has a 160 ns grid. Several points on the waveform indicate the similarities and differences between the latch and flip-flop operation.

FIGURE 8.59   Simulation Showing the Difference between a D Latch and a D Flip-Flop

1. At 500 ns, D goes HIGH. The latch output *(Q_latch)* and the flip-flop output *(Q_flip_flop)* both go HIGH at 960 ns since the beginning of the enable HIGH state and the positive edge of the *CLK* both correspond to this time.
2. D goes LOW at 2 μs. Both Q outputs go LOW at 2.24 μs since the positive edge of the *CLK* and its HIGH level occur at the same time.
3. The D input goes HIGH at about 5 μs, in the middle of a *CLK* pulse. Since the *CLK* line is HIGH, *Q_latch* changes immediately. *Q_flip_flop* does not change until the next positive edge, at about 6 μs.
4. D goes LOW at about 7.5 μs. *Q_latch* also changes at this time, since *CLK* is HIGH. *Q_flip_flop* changes on the next positive edge, at 8.64 μs.

Note that the latch and flip-flop outputs are LOW, even before the first *CLK* pulse. This is because Altera CPLDs have power-on reset circuitry that ensures that registered outputs in a CPLD are LOW immediately after power is applied to the device. The Quartus II simulator accounts for this condition.

EXAMPLE 8.7

Two positive edge-triggered D flip-flops are connected as shown in Figure 8.60a. Inputs D_0 and *CLK* are shown in the timing diagram. Complete the timing diagram by drawing the waveforms for Q_0 and Q_1, assuming that both flip-flops are initially reset.

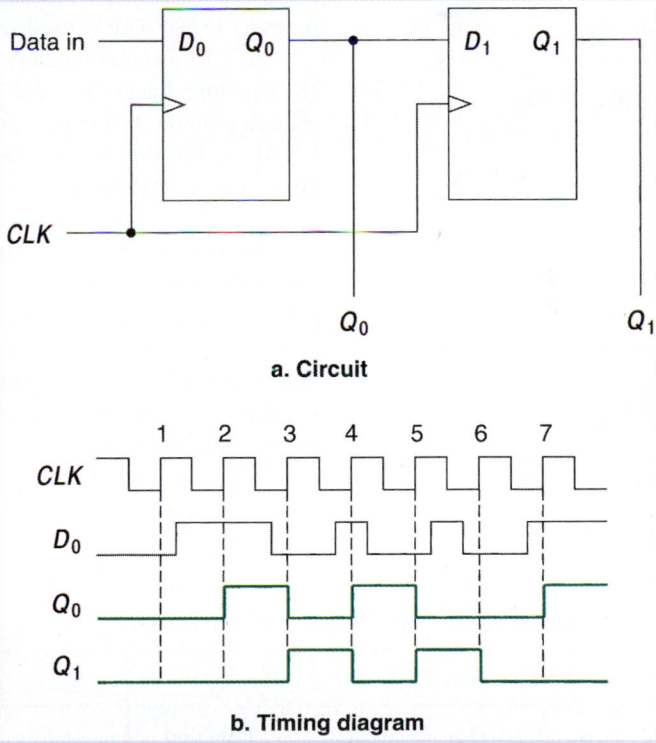

FIGURE 8.60   Example 8.7: Circuit and Timing Diagram

■ **Solution**

Figure 8.60b shows the output waveforms. Q_0 follows D_0 at each point where the clock input has a positive edge. One result of this is that the HIGH pulse on D_0 between clock pulses 5 and 6 is ignored, since $D_0 = 0$ on positive edges 5 and 6.

(continued)

Since $D_1 = Q_0$ and Q_1 follows D_1, the waveform at Q_1 is the same as at Q_0, but delayed by one clock cycle. If Q_0 changes due to *CLK*, we assume that the value of D_1 is the same as Q_0 just *before* the clock pulse. This is because delays within the circuitry of the flip-flops ensure that their outputs will not change for several nanoseconds after an applied clock pulse. Therefore, the level at D_1 remains constant long enough for it to be clocked into the second flip-flop.

The data entering the circuit at D_0 are moved, or shifted, from one flip-flop to the next. This type of data movement, called "serial shifting," is frequently used in data communication and digital arithmetic circuits.

■ SECTION 8.4 REVIEW PROBLEM

8.4   Which part of a D flip-flop accounts for the difference in operation between a D flip-flop and a D latch? How does it work?

8.5   EDGE-TRIGGERED JK FLIP-FLOPS

a. Positive edge-triggered

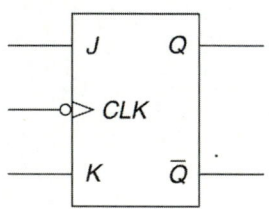

b. Negative edge-triggered

FIGURE 8.61   Edge-Triggered JK Flip-Flops

■ KEY TERM

Toggle   Alternate between opposite binary states with each applied clock pulse.

A versatile and widely used sequential circuit is the JK flip-flop.

Figure 8.61 shows the logic symbols of a positive- and a negative-edge triggered JK flip-flop. *J* acts as a *SET* input and *K* acts as a *RESET* input, with the output changing on the active clock edge in response to *J* and *K*. When *J* and *K* are both HIGH, the flip-flop will **toggle** between opposite logic states with each applied clock pulse. The function tables of the devices in Figure 8.61 are shown in Table 8.7.

Figure 8.62 shows the simplified circuit of a negative-edge triggered JK flip-flop. The circuit is like that of a gated SR latch with an edge detector (an SR flip-flop), except that there are two extra feedback lines from the latch outputs to the steering gate inputs. This extra feedback is responsible for the flip-flop's toggling action.

Figure 8.63 illustrates how the additional two lines cause the flip-flop to toggle. The cross-feedback from Q to K and from $\bar{Q}$ to J enables one, but not both, of the steering gates. The edge detector just after the *CLK* input produces a short positive-going pulse upon detecting a negative edge on the *CLK* waveform. The enabled steering gate complements and transmits this pulse to the latch, activating either the

TABLE 8.7   Function Tables for Edge-Triggered JK Flip-Flops

Positive Edge-Triggered						Negative Edge-Triggered					
CLK	**J**	**K**	Q_{t+1}	$\bar{Q}_{t+1}$	**Function**	**CLK**	**J**	**K**	Q_{t+1}	$\bar{Q}_{t+1}$	**Function**
↑	0	0	Q_t	$\bar{Q}_t$	No change	↓	0	0	Q_t	$\bar{Q}_t$	No change
↑	0	1	0	1	Reset	↓	0	1	0	1	Reset
↑	1	0	1	0	Set	↓	1	0	1	0	Set
↑	1	1	$\bar{Q}_t$	Q_t	Toggle	↓	1	1	$\bar{Q}_t$	Q_t	Toggle
0	X	X	Q_t	$\bar{Q}_t$	Inhibited	0	X	X	Q_t	$\bar{Q}_t$	Inhibited
1	X	X	Q_t	$\bar{Q}_t$	Inhibited	1	X	X	Q_t	$\bar{Q}_t$	Inhibited
↓	X	X	Q_t	$\bar{Q}_t$	Inhibited	↑	X	X	Q_t	$\bar{Q}_t$	Inhibited

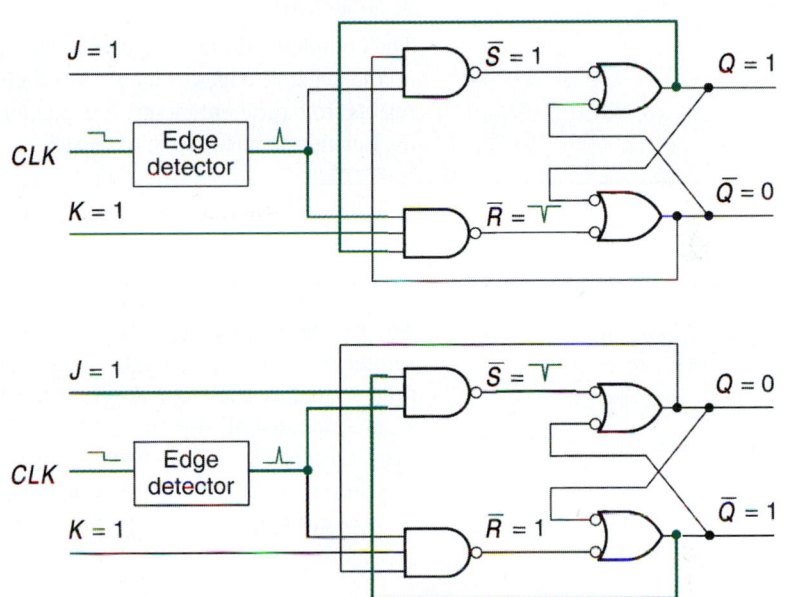

FIGURE 8.62 JK Flip-Flop Circuit (Simplified)

FIGURE 8.63 Toggle Action of a JK Flip-Flop (shown after CLK pulse, but before Q changes)

set or reset function. This in turn changes the latch state and enables the opposite steering gate.

Since all inputs of the steering gates must be HIGH to enable one of the latch functions, J and K must both be HIGH to sustain a repeated toggling action. Under these conditions, $\bar{Q}$ and Q alternately enable one of the steering gates.

EXAMPLE 8.8

The J, K, and CLK inputs of a negative edge-triggered JK flip-flop are as shown in the timing diagram in Figure 8.64. Complete the timing diagram by drawing the waveforms for Q and $\bar{Q}$. Indicate which function (no change, set, reset, or toggle) is performed at each clock pulse. The flip-flop is initially reset.

FIGURE 8.64 Example 8.8: Timing Diagram (Negative-Edge-Triggered JK Flip-Flop)

■ Solution

The completed timing diagram is shown in Figure 8.64. The outputs change only on the negative edges of the *CLK* waveform. Note that the same output sometimes results from different inputs. For example, the function at clock pulse 4 is reset and the function at pulses 5 and 6 is no change, but the Q waveform is LOW in each case.

EXAMPLE 8.9

The toggle function of a JK flip-flop is often used to generate a desired output sequence from a series of flip-flops. The circuit shown in Figure 8.65 is configured so that all flip-flops are permanently in toggle mode.

Assume that all flip-flops are initially reset. Draw a timing diagram showing the *CLK*, Q_0, Q_1, and Q_2 waveforms when eight clock pulses are applied. Make a table showing each combination of Q_2, Q_1, and Q_0. What pattern do the outputs form over the period shown on the timing diagram?

```
1 ──┤ J₀    Q₀ ├──   1 ──┤ J₁    Q₁ ├──   1 ──┤ J₂    Q₂ ├──
    ─○▷ CLK              ─○▷ CLK              ─○▷ CLK
1 ──┤ K₀      ├──   1 ──┤ K₁      ├──   1 ──┤ K₂      ├──
```

FIGURE 8.65 Example 8.9: Flip-Flop Circuit

■ Solution

The circuit timing diagram is shown in Figure 8.66. All flip-flops are in toggle mode. Each time a negative clock edge is applied to the flip-flop *CLK* input, the Q output will change to the opposite state.

For flip-flop 0, this happens with every clock pulse, as it is clocked directly by the *CLK* waveform. Each of the other flip-flops is clocked by the Q output waveform of the previous stage. Flip-flop 1 is clocked by the negative edge of the Q_0 waveform. Flip-flop 2 toggles when Q_1 goes from HIGH to LOW.

TABLE 8.8 Sequence of Outputs for Circuit in Figure 8.65

Clock Pulse	Q_2	Q_1	Q_0
0	0	0	0
1	0	0	1
2	0	1	0
3	0	1	1
4	1	0	0
5	1	0	1
6	1	1	0
7	1	1	1
8	0	0	0

FIGURE 8.66 Example 8.9: Timing Diagram

Table 8.8 shows the flip-flop outputs after each clock pulse. The outputs form a 3-bit number, in the order $Q_2\ Q_1\ Q_0$, that counts from 000 to 111 in binary sequence, then returns to 000 and repeats.

This flip-flop circuit is called a 3-bit asynchronous counter.

Synchronous Versus Asynchronous Circuits

■ KEY TERMS

Asynchronous Not synchronized to the system clock

Synchronous Synchronized to the system clock.

Quartus II:
asynch3.bdf
asynch3.vwf
MAX+PLUS II:
asynch3.gdf
asynch3.scf

The **asynchronous** counter in Figure 8.65 has the advantage of being simple to construct and analyze. However, because it is asynchronous (that is, not synchronized to a single clock), it is seldom used in modern digital designs. The main problem with this and other asynchronous circuits is that their outputs do not change at the same time, due to delays in the flip-flops. This yields intermediate states that are not part of the desired output sequence.

Figure 8.67 shows a simulation of a circuit similar to that in Figure 8.65. The outputs are shown separately, and also as a group labeled **Q[2..0]** that shows the combined binary value of the outputs.

FIGURE 8.67 Simulation for a 3-Bit Asynchronous Counter

FIGURE 8.68   Detail of Simulation for a 3-Bit Asynchronous Counter

FIGURE 8.69   3-Bit Synchronous Counter

Quartus II:
sync3.bdf
sync3.vwf
MAX+PLUS II:
sync3.gdf
sync3.scf

Figure 8.68 shows a detail of the simulation at the point where the output goes from 7 to 0 (111 to 000). At 640 ns, the circuit output is 111. A negative clock edge, applied to flip-flop 0, makes Q_0 toggle after a short delay. The output is now 110 ($=6_{10}$). The resulting negative edge on Q_0 clocks flip-flop 1, making it toggle, and yields a new output of 100 ($=4_{10}$). The negative edge on Q_1 clocks flip-flop 2, making the output equal to 000 after a short delay.

Thus, the output goes through two short intermediate states that are not in the desired output sequence. Instead of going directly from 111 to 000, as in Figure 8.66, the output goes in the sequence 111-110-100-000. We see in Figure 8.67 that the counter output goes through one or more intermediate transitions after each negative edge of the Q_0 waveform. In other words, intermediate states arise whenever a change propagates through more than one flip-flop. This happens because the flip-flops are clocked from different sources.

Figure 8.69 shows the circuit of a 3-bit **synchronous** counter. Unlike the circuit in Figure 8.65, the flip-flops in this circuit are clocked from a common source. Therefore, flip-flop delays do not add up through the circuit, and all the outputs change at the same time. Figure 8.70 shows a simulation of the circuit of Figure 8.69. Note that the outputs progress in a binary sequence, and there are no intermediate states.

The circuit works as follows:

1. Flip-flop 0 is configured for toggle mode ($J_0K_0 = 11$). Since the flip-flops in Figure 8.69 are positive edge-triggered, Q_0 toggles on each positive clock edge.
2. Q_0 is connected to inputs J_1 and K_1. Since these inputs are tied together, only two states are possible: no change ($JK = 00$) or toggle ($JK = 11$). If $Q_0 = 1$,

FIGURE 8.70 Simulation for a 3-Bit Synchronous Counter

Q_1 toggles. Otherwise, it does not change. This results in a Q_1 waveform that toggles at half the rate of Q_0.

3. J_2 and K_2 are both tied to the output of an AND gate. The AND gate output is HIGH if *both* Q_1 and Q_0 are HIGH. This makes Q_2 toggle, since $J_2K_2 = 11$. In all other cases, there is no change on Q_2. The result of this is that Q_2 toggles every fourth clock pulse, the only times when Q_1 and Q_0 are both HIGH.

Asynchronous Inputs (Preset and Clear)

■ KEY TERMS

Synchronous Inputs The inputs of a flip-flop that do not affect the flip-flop's Q outputs unless a clock pulse is applied. Examples include D, J, and K inputs.

Asynchronous Inputs The inputs of a flip-flop that change the flip-flop's Q outputs immediately, without waiting for a pulse at the CLK input. Examples include preset and clear inputs.

Preset An asynchronous set function.

Clear An asynchronous reset function.

The *D, J,* and *K* inputs of the flip-flops examined so far are called **synchronous inputs.** This is because any effect they have on the flip-flop outputs is synchronized to the *CLK* input.

Another class of inputs is also provided on many flip-flops. These inputs, called **asynchronous inputs,** do not need to wait for a clock pulse to make a change at the output. The two functions usually provided are **preset,** an asynchronous set function, and **clear,** an asynchronous reset function. These functions are generally active LOW, and are abbreviated $\overline{PRE}$ and $\overline{CLR}$.

Figure 8.71 shows a modification to the JK flip-flop of Figure 8.62. The $\overline{PRE}$ and $\overline{CLR}$ inputs have direct access to the latch gates of the flip-flop and thus are not affected by the *CLK* input. They act exactly the same as the *SET* and *RESET* inputs of an SR latch and will override any synchronous input functions currently active.

FIGURE 8.71 $\overline{PRE}$ and $\overline{CLR}$ Inputs

EXAMPLE 8.10

The waveforms for the *CLK*, *J*, *K*, $\overline{PRE}$ and $\overline{CLR}$ inputs of a negative edge-triggered JK flip-flop are shown in the timing diagram of Figure 8.72. Complete the diagram by drawing the waveform for output *Q*. Assume that *Q* is initially HIGH.

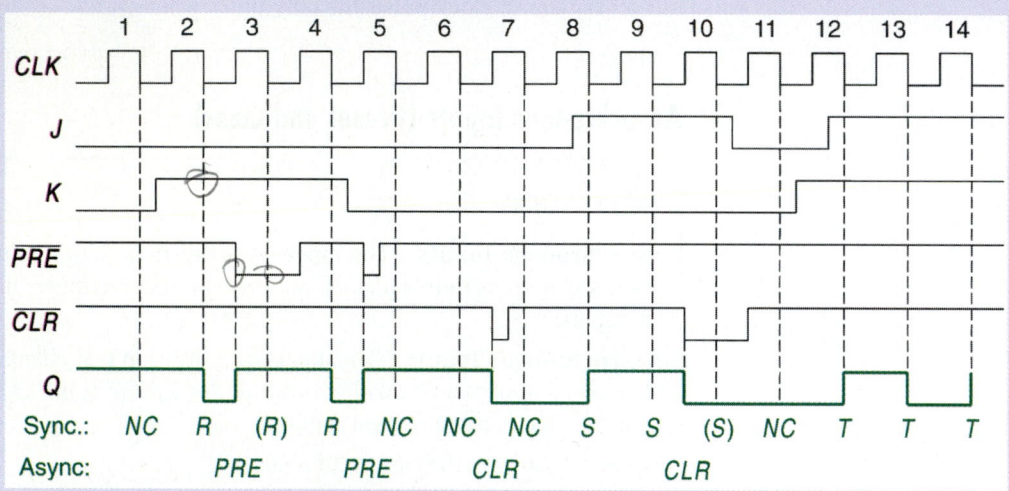

FIGURE 8.72 Example 8.10: Flip-Flop Waveforms Showing Synchronous and Asynchronous Operation

■ Solution

The *Q* waveform is shown in Figure 8.72. The asynchronous inputs cause an immediate change in *Q*, whereas the synchronous inputs must wait for the next negative clock edge. If asynchronous and synchronous inputs are simultaneously active, the asynchronous inputs have priority. This occurs in two places: pulse 3 (*K*, $\overline{PRE}$) and pulse 10 (*J*, $\overline{CLR}$).

The diagram shows the synchronous function (no change, reset, set, and toggle) at each clock pulse and the asynchronous functions (preset and clear) at the corresponding transition points.

The function table of a negative edge-triggered JK flip-flop with preset and clear functions is shown in Table 8.9.

TABLE 8.9 Function Table of a Negative Edge-Triggered JK Flip-Flop with Preset and Clear Functions

	$\overline{PRE}$	$\overline{CLR}$	CLK	J	K	Q_{t+1}	$\overline{Q}_t+1$	Function
Synchronous Functions	1	1	↓	0	0	Q_t	$\overline{Q}_t$	No change
	1	1	↓	0	1	0	1	Reset
	1	1	↓	1	0	1	0	Set
	1	1	↓	1	1	$\overline{Q}_t$	Q_t	Toggle
Asynchronous Functions	0	1	X	X	X	1	0	Preset
	1	0	X	X	X	0	1	Clear
	0	0	X	X	X	1	1	Forbidden
	1	1	0	X	X	Q_t	$\overline{Q}_t$	Inhibited
	1	1	1	X	X	Q_t	$\overline{Q}_t$	Inhibited
	1	1	↑	X	X	Q_t	$\overline{Q}_t$	Inhibited

X = Don't care   ↓ = HIGH-to-LOW transition
Q_t = Present state of Q ↑ = LOW-to-HIGH transition
Q_{t+1} = Next state of Q

NOTE . . .

If preset and clear functions are not used, they should be disabled by connecting them to logic HIGH (for active-LOW inputs). This prevents them from being activated inadvertently by circuit noise. The synchronous functions of some flip-flops will not operate properly unless $\overline{PRE}$ and $\overline{CLR}$ are HIGH. In Quartus II, the asynchronous inputs of all flip-flop primitives are set to a default level of HIGH.

Using Asynchronous Reset in a Synchronous Circuit

■ KEY TERM

Master Reset An asynchronous reset input used to set a sequential circuit to a known initial state.

Figure 8.73 shows an application of asynchronous clear inputs in a 3-bit synchronous counter. An input called *RESET* is tied to the asynchronous $\overline{CLR}$ inputs of all flip-flops. The counter output is set to 000 when the *RESET* line goes LOW.

FIGURE 8.73 Synchronous Counter with Asynchronous Reset

FIGURE 8.74 Simulation of a 3-Bit Synchronous Counter with Asynchronous Reset

Quartus II:
synch3r.bdf
synch3r.vwf
MAX+PLUS II:
synch3r.gdf
synch3r.scf

Figure 8.74 shows a set of simulation waveforms that illustrate the asynchronous clear function. When *RESET* is HIGH, the count proceeds normally. The positive clock edge at 635 ns drives the counter to state 011. The reset pulse at 650 ns sets the counter to 000 as soon as it goes LOW. On the next clock edge, the count proceeds from 000.

The function that sets all flip-flops in a circuit to a known initial state is sometimes called **Master Reset.**

■ **SECTION 8.5 REVIEW PROBLEM**

8.5 What is the main difference between synchronous and asynchronous circuits, such as the two counters in Figures 8.65 and 8.69? What disadvantage is there to an asynchronous circuit?

8.6 EDGE-TRIGGERED T FLIP-FLOPS

➡ *LAB SOURCE*
Lab #21
JK and T Flip-Flops

■ **KEY TERM**

T (Toggle) Flip-Flop A flip-flop whose output toggles between HIGH and LOW states on each applied clock pulse when a synchronous input, called *T*, is active.

In the section on the JK flip-flop, we saw how that device can be set to toggle between HIGH and LOW output states. Other types of flip-flops can perform this function, as well. For example, Figure 8.75 shows a D flip-flop configured for toggle operation. Since Q follows D and $D = \bar{Q}$ in this circuit, then the flip-flop output must change to its opposite state with each clock pulse. Figure 8.76 shows a Quartus II simulation of this circuit.

It is seldom useful for flip-flops in synchronous circuits to be permanently configured in toggle mode. What made the JK flip-flops suitable elements for the synchronous counter in Figure 8.69 was the fact that sometimes they toggled and sometimes they didn't, depending on the current point in the output sequence of the counter. Figure 8.77 shows a D flip-flop configured for a switchable toggle function.

Quartus II:
notq2d.bdf
notq2d.vwf
MAX+PLUS II:
notq2d.gdf
notq2d.scf

FIGURE 8.75 D Flip-Flop Configured for Toggle Function

FIGURE 8.76 Simulation of a D Flip-Flop in Toggle Mode

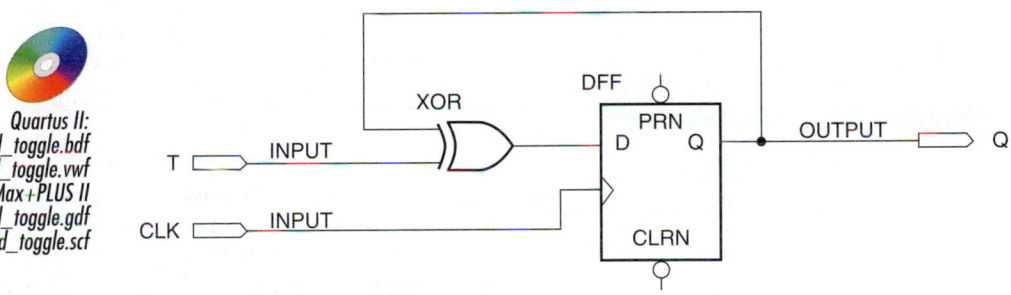

Quartus II:
d_toggle.bdf
d_toggle.vwf
Max+PLUS II
d_toggle.gdf
d_toggle.scf

FIGURE 8.77 Switchable Toggle Function for a D Flip-Flop

FIGURE 8.78 T Flip-Flop

Quartus II:
t_flipflop.bdf
t_flipflop.vwf
MAX+PLUS II:
t_flipflop.gdf
t_flipflop.scf

The XOR gate acts as an inverter when the T input is HIGH and as a noninverting buffer when T is LOW. Thus, when T is LOW, the Q output is circulated back to the D input of the flip-flop and the current value of Q is reloaded on the next clock pulse. This is the no change state of the flip-flop. When T is HIGH, the circuit acts like that of Figure 8.75 and toggles.

A **T flip-flop** has this equivalent function. Figure 8.78 shows the symbol of a T flip-flop in a Quartus II Block Diagram File. A Quartus II simulation in Figure 8.79

FIGURE 8.79 Simulation of a T Flip-Flop

TABLE 8.10 Function Table for a T Flip-Flop

CLK	T	Q_{t+1}	Function
$\uparrow$	0	$\bar{Q}_t$	No change
$\uparrow$	1	$\bar{Q}_t$	Toggle
0	X	Q_t	Inhibited
1	X	Q_t	Inhibited
$\downarrow$	X	Q_t	Inhibited

shows the operation of this device. The Q output toggles on each clock pulse when T is HIGH; otherwise Q retains its last value. A function table for the T flip-flop is shown in Table 8.10.

■ SECTION 8.6 REVIEW PROBLEM

8.6 Draw a circuit showing how the JK flip-flops in Figure 8.69 can be replaced by T flip-flops.

8.7 FLIP-FLOPS IN PLDs (REGISTERED OUTPUTS)

Before reading this section, you may wish to briefly review the material on programmable logic devices in Sections 4.2–4.4 of Chapter 4.

■ KEY TERMS

Registered Output   An output of a programmable logic device (PLD) having a flip-flop (usually D-type) that stores the output state.

Register   A digital circuit such as a flip-flop or array of flip-flops that stores one or more bits of digital information.

Flip-flops are generally found in programmable logic devices as **registered outputs**. A **register** is one or more flip-flops used to store data. Registered outputs in programmable array logic (PAL) devices can be used for the same functions as individual flip-flops.

Figure 8.80 shows the logic diagram of a PAL device with eight registered outputs: a PAL16R8. The fuse matrix is identical to that of the PAL16L8 device that we examined in Chapter 4: the differences between the two devices are the registered outputs, a dedicated clock input (pin 1), and a pin for enabling all registered outputs (pin 11).

With Registered PAL, the number of outputs shown in the part number indicates the number of registered outputs. For example, a PAL16R4 device has four registered outputs and four combinational I/O pins, a PAL16R6 device has six registered outputs and two combinational I/O pins, and a PAL16R8 has eight registered outputs.

FIGURE 8.80 PAL16R8 Logic Diagram

EXAMPLE 8.11

A common data operation is that of "rotation." Figure 8.81 illustrates how a 4-bit number can be rotated to the right by 0, 1, 2, or 3 places by a circuit called a "barrel shifter," commonly used in digital signal processing applications to speed up computations. To rotate the data, move all bits the required number of places to the right. As data reach the rightmost position, more them to the beginning so that they are transferred in a closed loop.

This operation is usually performed by serially shifting the data the required number of places and feeding back the last output to the first input of a serial shift register.

(continued)

FIGURE 8.81   Example 8.11: Rotation to the Right (4-Bit Data)

Rotation can be accomplished more quickly by a parallel transfer operation. We can load the bits of the input into four D flip-flops in the order determined by two select inputs, S_1 and S_0. Assume that the binary number $S_1 S_0$ is the same as the rotation number in Figure 8.81. Table 8.11 summarizes the contents of the circuit after one clock pulse is applied.

TABLE 8.11   Rotation to the Right by a Selectable Number of Bits

S_1	S_0	Q_A	Q_B	Q_C	Q_D	Rotation
0	0	A	B	C	D	0
0	1	D	A	B	C	1
1	0	C	D	A	B	2
1	1	B	C	D	A	3

Sketch a circuit, using gates and flip-flops, that can accomplish this rotation as a parallel transfer function. Briefly explain its operation.

Write the Boolean expression(s) for the circuit.

Show how the circuit can be implemented by a PAL16R4 device by drawing fuses on its logic diagram.

■ Solution

Figure 8.82 shows a parallel transfer circuit (barrel shifter) that will perform the specified rotation. The circuit works by enabling one AND gate in each group of four for each combination of S_1 and S_0. For example, when $S_1 S_0 = 00$, the rotation is 0 and the leftmost AND gate of each group is enabled, transferring the parallel data into the flip-flops so that $D_A = A$, $D_B = B$, $D_C = C$, and $D_D = D$. After one clock pulse, $Q_A Q_B Q_C Q_D = ABCD$.

Similarly, if $S_1 S_0 = 10$, we select a rotation of 2. The third AND gate from the left is selected in each group of four. This makes the data $D_A = C$, $D_B = D$, $D_C = A$, and $D_D = B$ appear at the flip-flop inputs. After one clock pulse, $Q_A Q_B Q_C Q_D = CDAB$.

FIGURE 8.82 Example 8.11: Rotation by Parallel Transfer (Barrel Shifter)

The same principle governs the circuit operation for the other two select codes. The Boolean equations for the circuit are:

$$Q_A = \overline{S}_1\,\overline{S}_0\,A + \overline{S}_1\,S_0\,D + S_1\,\overline{S}_0\,C + S_1\,S_0\,B$$
$$Q_B = \overline{S}_1\,\overline{S}_0\,B + \overline{S}_1\,S_0\,A + S_1\,\overline{S}_0\,D + S_1\,S_0\,C$$
$$Q_C = \overline{S}_1\,\overline{S}_0\,C + \overline{S}_1\,S_0\,B + S_1\,\overline{S}_0\,A + S_1\,S_0\,D$$
$$Q_D = \overline{S}_1\,\overline{S}_0\,D + \overline{S}_1\,S_0\,C + S_1\,\overline{S}_0\,B + S_1\,S_0\,A$$

(continued)

FIGURE 8.83   Example 8.11: Programmed PLD for Selectable Bit Rotation

These equations imply that each registered output requires us to use four product lines, one for each product term. The programmed logic diagram is shown in Figure 8.83.

8.8 GENERIC ARRAY LOGIC (GAL)

There are several limitations of standard low density PALs. First, these devices are **one-time programmable (OTP).** Since the AND matrix of a PAL is programmable by blowing metal fuse links, programming is permanent; there is no opportunity to correct or update a design. In development of a new design, where many modifications must be made to the original design, this can be particularly wasteful. Second, standard PAL outputs are permanently configured either as combinational or registered. A given PAL has a certain number of each type of output, which may not be optimum for the design. Third, a standard PAL cannot be programmed while it is installed in a circuit.

A number of low-density PLDs have been developed to address these concerns. Devices such as the GAL16V8 and GAL22V10 **generic array logic (GAL)** (Lattice Semiconductor) are based on sum-of-products fuse matrices, just as the earlier version PALs. However, these devices are based on electrically erasable read-only memory (EEPROM or E^{2}PROM) cells, rather than fuses, which allow them to be erased and reprogrammed about 10,000 times. A programmed device will hold its data for about 20 years.

GALs also have programmable input/output configurations. An I/O pin can be configured as a registered output, a combinational output, or a dedicated input, as required. Additionally, an output can be specified as active-HIGH or active-LOW.

Devices such as the ispGAL22V10 or the Altera MAX 7000S series can be programmed while installed in a circuit via a standard four-wire interface called a **JTAG port.** This property is known as **in-system programmability (ISP).**

GAL22V10

Figure 8.84 shows one I/O pin and its associated circuitry for a GAL22V10 generic array logic device. (The "V" stands for "variable" or "versatile" architecture.) It consists of a programmable SOP array with 8 product terms and an **output logic macrocell (OLMC),** or just "macrocell," which determines the I/O configuration for

FIGURE 8.84   Typical Macrocell for a GAL22V10

that pin. Other macrocells in the GAL22V10 have other numbers of product lines, as described later in this section. The various configuration options are selected by a pair of multiplexers that are programmed by two **architecture cells,** S_1 and S_0, that set the MUX select inputs HIGH or LOW, as required. S_1 determines whether the output is combinational or registered. S_0 determines whether it is active-HIGH or active-LOW.

Figure 8.85 shows the four different output configurations of the GAL22V10 macrocell. Table 8.12 lists the configuration options. The PLD design and programming software determines the required status of S_1 and S_0 for each macrocell and sets them accordingly when the GAL is programmed.

TABLE 8.12   GAL22V10 Macrocell Configurations

S_1	S_0	Configuration
0	0	Registered, Active-LOW
0	1	Registered, Active-HIGH
1	0	Combinational Active-LOW
1	1	Combinational, Active-HIGH

If there are registered outputs, the clock input (pin 1) provides a **global clock** function. That is, all registered outputs are clocked simultaneously by this signal. (Some other PLDs provide an option to clock a registered output from a product term in the AND matrix, allowing several clock functions in one chip.) If there are no registered outputs used in the PLD, pin 1 can be used as an input.

a. Registered/active LOW   b. Registered/active HIGH

c. Combinatorial/active LOW   d. Combinatorial/active HIGH

FIGURE 8.85 GAL22V10 OLMC Configurations

Figure 8.86 shows the logic diagram of a GAL22V10. This device has a number of features that make it superior to low-density PALs, such as the PAL16L8.

1. There are more outputs (10 as opposed to 8 for the 16L8).
2. There are more inputs (11 dedicated inputs, plus any I/O lines used as inputs).
3. The output logic macrocells are of different sizes, allowing expressions with larger numbers of product terms in some macrocells than others. There are two macrocells with each of the following numbers of product lines: 8, 10, 12, 14, and 16. This allows more flexibility in design, while minimizing the number of product lines.
4. Macrocells are easy to configure in a GAL22V10. Two architecture cells per macrocell, S_0 and S_1, select the output type, as shown in Figures 8.84 and 8.85.
5. There are product lines for Synchronous Preset *(SP)* and Asynchronous Reset *(AR)*. The *SP* line sets all flip-flops HIGH on the first clock pulse after it becomes active. The *AR* line sets all flip-flops LOW as soon as it activates, without waiting for the clock pulse. (Note that these lines set or reset the *Q* output of each flip-flop. An active-LOW registered output inverts this state at the output pin.)

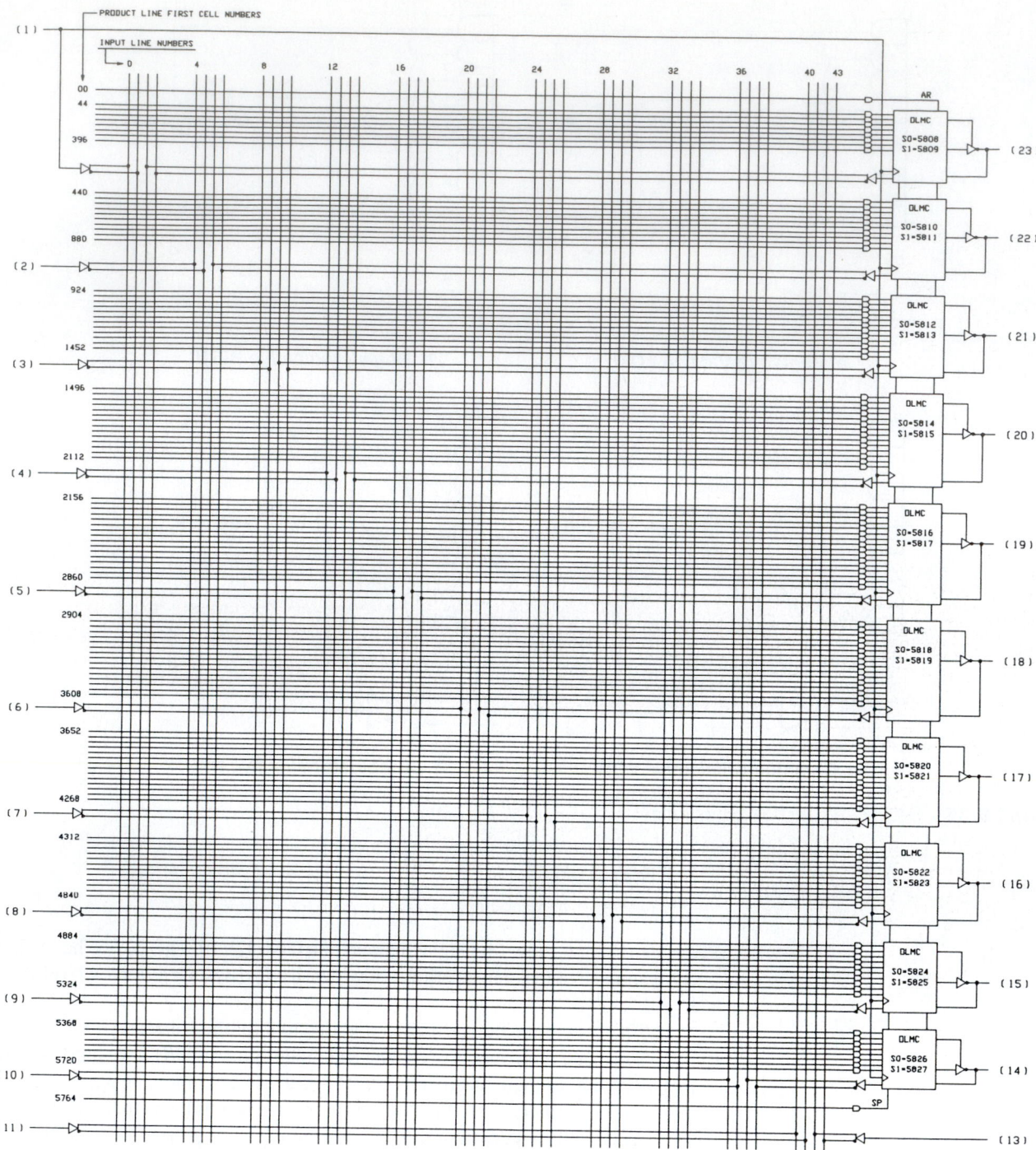

FIGURE 8.86 GAL22V10 Logic Diagram

8.9   MAX 7000S CPLD

Figure 8.87 shows the block diagram of an Altera MAX 7000S **Complex PLD (CPLD),** A device of this type—the EPM7128SLC84—is one of the two devices installed on the Altera UP-1 and UP-2 University Program boards, so we will use it as a specific example of the MAX 7000S family of devices.

FIGURE 8.87   MAX 7000E and MAX 7000S Device Block Diagram
(*Source:* Altera Corporation)

The part number breaks up as follows:

EPM7	MAX 7000(S) family
128	number of macrocells
S	in-system programmable
LC84	84-pin PLCC package

The main structure of the MAX 7000S is a series of **Logic Array Blocks (LABs)**, linked by a **Programmable Interconnect Array (PIA)**. Each LAB is a group of 16 macrocells that can share common product terms and lend or borrow unused product terms among each other. A single LAB has similar I/O and programming capability to a low-density PLD, so a CPLD like the MAX 7000S can be thought of as an array of interconnected PALs or GALs on a single chip.

An EPM7128S has 8 LABs, for a total of $8 \times 16 = 128$ macrocells. However, these are not all available to the user as I/Os; the number of available I/O pins depends on the device package. Figure 8.87 indicates that each LAB in a MAX 7000S device has from 6 to 16 I/O pins. For an EPM7128S in a 160-pin PQFP package, there are 12 I/Os per LAB, for a total of 96 available pins. For the same device in an 84-pin PLCC package, there are only 8 I/Os per LAB, for a total of 64 pins.

In practice, if an EPM7128SLC84 is to be programmed in-circuit (i.e., while installed on a circuit board), there are only 60 I/Os available, as four pins required for the programming interface. The macrocells that are not connected to user I/O pins can only be used for **buried logic,** or logic that is internal to the chip only.

As implied in Figure 8.87, all I/O pins connect to and from their associated LAB via an **I/O Control Block** (a circuit that controls the tristate switching of signals at an I/O pin). The I/O pin signals also connect directly to the PIA, where they are available for use in other LABs. Sixteen lines connect the macrocell outputs of each LAB to the PIA, again for use throughout the device. The PIA communicates to each LAB via 36 product lines to provide connections from other LABs.

The MAX 7000S family has four pins that can be configured as control signals or inputs. *GCLK1* is a global clock that is common to all macrocells in the device and can be used to synchronously clock all registers. *OE1* is an output enable that can globally activate or disable the tristate outputs of the device macrocells. *GCLRn* is an active-LOW global clear function. The fourth control pin can be configured as an input, as can the other three pins, or as a second global clock *(GCLK2)* or output enable *(OE2)*. If the control functions are not used, these pins add four inputs to the available total. These assignments can be made by the Quartus II software during the design process.

Table 8.13 summarizes the pin allocations for the EPM7128S in an 84-pin PLCC package.

Figure 8.88 shows a macrocell from a MAX 7000S device. The macrocell is similar to that of a GAL in that it provides a sum-of-products function with active-HIGH or -LOW options and the choice of registered or combinational output. Registered outputs can be clocked with one of two global clocks or by a product term from the AND matrix. The register can be cleared globally or by a product term and preset with a product term.

The macrocell has five dedicated product terms, which is fewer than found in the PAL and GAL matrices we examined earlier. This is generally sufficient to implement most logic functions. If more terms are required, they can be supplied by a set of **shared logic expanders** or **parallel logic expanders.**

Shared logic expanders do not add more product terms to a given macrocell. They do make the programming of the entire LAB more efficient by allowing a product term to be programmed once and used in several macrocells of the same LAB. One product term per macrocell is inverted and fed back into the shared expander pool of product terms. Since there are 16 macrocells per LAB, the shared logic expander pool has up to 16 product terms.

TABLE 8.13 Pin Allocations for an EPM7128SLC84 CPLD

Function	Pins
V_{cc}	8
Ground	8
JTAG port	4
GCLK1	1
OE1	1
GCLRn	1
GCLK2/OE2	1
User I/Os	60
Total pins	**84**

FIGURE 8.88 MAX 7000E and MAX 7000S Device Macrocell
(*Source:* Altera Corporation)

Parallel logic expanders allow a macrocell to borrow up to 15 product terms from its three lower-numbered neighbors (5 product terms per neighboring macrocell). For example, macrocell 4 can borrow up to 5 terms each from macrocells 3, 2, and 1. By using its 5 dedicated product terms and the maximum number of parallel expanders, a macrocell can have up to 20 product terms at its disposal. These borrowed terms are not usable by the macrocell from which they were borrowed. The parallel expanders are set up so that a lower-number cell lends product terms to a higher-number cell, so the number of available terms depends on how close to the end of a chain a macrocell is. Expander assignments are done automatically by Quartus II at compile time.

8.10   FLEX 10K CPLD

▪ KEY TERMS

Look-Up Table (LUT)   A circuit that implements a combinational logic function by storing a list of output values that correspond to all possible input combinations.

Logic Element (LE)   A circuit internal to a CPLD used to implement a logic function as a look-up table.

Cascade Chain   A circuit in a CPLD that allows the input width of a Boolean function to expand beyond the width of one logic element.

> **Carry Chain**   A circuit in a CPLD that is optimized for efficient operation of carry functions between logic elements.
>
> **Embedded Array Block (EAB)**   A relatively large block of storage elements in a CPLD (2048 bits in a FLEX 10K device), used for implementing complex logic functions in look-up table format.

All programmable logic devices we have seen until now have been based on sum-of-products arrays. Another major type of PLD is based on **look-up table (LUT)** architecture. In this architecture, a number of storage elements are used to synthesize logic functions by storing each function as a truth table. To illustrate the look-up table concept, let us use the truth table of a 2-bit equality comparator, shown in Table 8.14.

TABLE 8.14   Truth Table for a 2-Bit Equality Comparator

A_1	A_0	B_1	B_0	Decimal	$AEQB$
0	0	0	0	0	1
0	0	0	1	1	0
0	0	1	0	2	0
0	0	1	1	3	0
0	1	0	0	4	0
0	1	0	1	5	1
0	1	1	0	6	0
0	1	1	1	7	0
1	0	0	0	8	0
1	0	0	1	9	0
1	0	1	0	10	1
1	0	1	1	11	0
1	1	0	0	12	0
1	1	0	1	13	0
1	1	1	0	14	0
1	1	1	1	15	1

The comparator examines inputs. A_1A_0 and B_1B_0 and makes output $AEQB$ equal to logic 1 if $A_1A_0 = B_1B_0$. If we were to implement the circuit as an SOP array, we would first find the Boolean expression by combining the four product terms from the truth table and then program the appropriate cells in a CPLD AND matrix. The look-up table implementation of this function is based on a totally different concept.

Figure 8.89 shows the structural concept of a 4-bit look-up table circuit. An array of 16 flip-flops (Q_0 through Q_{15}) contain data for all possible combinations of $A_1A_0B_1B_0$, one flip-flop per combination. The LUT inputs $A_1A_0B_1B_0$ are decoded by an internal address decoder. Each decoder output activates a tristate buffer that passes or blocks the output of one flip-flop. The active buffer passes the contents of the flip-flop to $AEQB$; all other buffers are in the high-impedance state, blocking the data from the other flip-flops.

The contents of the flip-flops are loaded when the look-up table is configured (programmed) with the required function. After that the flip-flops retain their information until they are reconfigured. For our comparator example, flip-flops 0, 5, 10, and 15 are all set ($Q = 1$). All other flip-flops are reset ($Q = 0$). Examine Table 8.14 to confirm that this is true.

a. 2-bit comparator look-up table

b. Stuctural concept of a look-up table

FIGURE 8.89   Look-Up Table

The 16-bit storage element in Figure 8.89, combined with switching to choose a combinational or registered output and to interconnect with other parts of the chip, is called a **logic element (LE).** A logic element performs a function similar to that of a macrocell in SOP-type PLDs.

Figure 8.90 shows the structure of a logic element in an Altera FLEX 10K CPLD. In addition to the LUT, the LE has circuitry to select various control functions, such as clock and reset, a flip-flop for registered output, some expansion circuitry (cascade and carry), and interconnections to local and global busses.

The **cascade chain** circuit, shown in Figure 8.91 allows the user to program Boolean functions with more than four inputs, thus requiring more than one LUT. The cascade chain can be AND- or OR-type, depending on what DeMorgan equivalent form is most appropriate.

The **carry chain,** shown in Figure 8.92 allows for efficient fast-carry implementation of adders, comparators, and other circuits that depend on the combination of low-order bits to define high-order functions (i.e., circuits whose inputs become wider with higher-order bits). Figure 8.92 shows the carry chain as implemented by an n-bit adder.

FIGURE 8.90   FLEX 10K Logic Element (*Source:* Altera Corporation)

FIGURE 8.91   Cascade Chain Operation (*Source:* Altera Corporation)

FIGURE 8.92 Carry Chain Operation (*n*-Bit Full Adder)
(*Source:* Altera Corporation)

FIGURE 8.93   FLEX 10K LAB (*Source:* Altera Corporation)

A Logic Array Block (LAB), shown in Figure 8.93 consists of eight logic elements and a local interconnect. The LAB is connected to the rest of the device by a series of row and column interconnects, which Altera calls a FastTrack Interconnect. Figure 8.94 shows the overall structure of a FLEX 10K device, with several LABs and a number of **Embedded Array Blocks (EABs).** An EAB is an array of 2048 storage elements that can be used to efficiently implement complex logic functions.

The FLEX 10K device found on the Altera UP-1 board—the EPF10K20RC240-4—has an array of 6 rows by 24 columns of LABs, which gives a total of 144 LABs ($= 8 \times 144 = 1152$ logic elements). The device also has 6 EABs ($6 \times 2048 = 12288$ bits of EAB storage). The FLEX 10K device on the UP-2 board (EPF10K70RC240-4) has a total of 468 LABs (3744 Logic Elements) and 9 EABs (18432 bits).

The FLEX 10K series of CPLDs (and LUT-based devices generally) are based on static random access memory (SRAM) technology. The advantage of this configuration is that it can be manufactured with a very high density of storage cells and it programs quickly compared to an EEPROM-based SOP device. The disadvantage is that SRAM cells are volatile; that is, they do not retain their data when power is removed from the circuit. An SRAM-based device must be reconfigured every time it is powered up.

Embedded Array Block (EAB)

I/O Element (IOE)

Column Interconnect

EAB

Logic Array

Logic Array Block (LAB)

Row Interconnect

EAB

Logic Element (LE)

Local Interconnect

Logic Array

Embedded Array

FIGURE 8.94 FLEX 10K Device Block Diagram (*Source:* Altera Corporation)

SUMMARY

1. A combinational circuit combines inputs to generate a particular output logic level that is always the same, regardless of the order in which the inputs are applied. A sequential circuit can generate different outputs for the same inputs, depending on the sequence in which the inputs were applied.

2. An SR latch is a sequential circuit with *SET (S)* and *RESET (R)* inputs and complementary outputs (Q and $\overline{Q}$). By definition, a latch is set when $Q = 1$ and reset when $Q = 0$.

3. A latch sets when its S input activates. When S returns to the inactive state, the latch remains in the set condition until explicitly reset by activating its R input.

4. A latch can have active-HIGH inputs (designated S and R) or active-LOW inputs (designated $\overline{S}$ and $\overline{R}$).

5. Two basic SR latch circuits are the NAND latch and the NOR latch, each consisting of two gates with cross-coupled feedback. In the NAND form, we draw the gates in their DeMorgan equivalent form so that each circuit has OR-shaped gates, inversion from input to output, and feedback to the opposite gate.

6. A NOR latch has active-HIGH inputs. It is described by the following function table:

S	R	Q_{t+1}	$\overline{Q}_{t+1}$	*Function*
0	0	Q_t	$\overline{Q}_t$	No change
0	1	0	1	Reset
1	0	1	0	Set
1	1	0	0	Forbidden

7. A NAND latch has active-LOW inputs and is described by the following function table:

$\bar{S}$	$\bar{R}$	Q_{t+1}	$\bar{Q}_{t+1}$	Function
0	0	1	1	Forbidden
0	1	1	0	Set
1	0	0	1	Reset
1	1	Q_t	$\bar{Q}_t$	No change

8. The transition from the forbidden state of a NAND or NOR latch to the no change state is not always predictable. If the latch inputs do not change at the same time, the latch will take the state represented by the last input to change. If both inputs change simultaneously, one of the latch gates will be slightly faster than the other, causing the latch to drop into the Set or Reset state. However, it cannot be determined beforehand which state will prevail.

9. Quartus II does not synthesize the latch circuit exactly as entered in the Quartus Block Editor or Text Editor. The circuit will be optimized to fit the target CPLD, resulting in a circuit that has a single resolution to the forbidden state-to-no change transition.

10. A NAND latch can be used as a switch debouncer for a switch with a grounded common terminal, a normally open, and a normally closed contact. When the switch operates, one contact closes, resetting the latch on the first bounce. Further bounces are ignored. When the switch returns to its normal position, it sets the latch on the first bounce and further bounces are ignored.

11. A NOR latch can also be used as a debouncer, but the logic switch to be debounced must use pull-down resistors, rather than pull-up, and the common terminal must be connected to V_{CC}, not ground. This configuration is seldom used because of its tendency to draw unacceptably large amounts of idle current from the power supply.

12. A gated SR latch controls the times when a latch can switch. The circuit consists of a pair of latch gates and a pair of steering gates. The steering gates are enabled or inhibited by a control signal called *ENABLE*. When the steering gates are enabled, they can direct a set or reset pulse to the latch gates. When inhibited, the steering gates block any set or reset pulses to the latch gates so the latch output cannot change.

13. A gated D ("data") latch can be constructed by connecting opposite logic levels to the *S* and *R* inputs of an SR latch. Since *S* and *R* are always opposite, the D latch has no forbidden state. The no change state is provided by the inhibit property of the *ENABLE* input.

14. In a gated D latch (or transparent latch), *Q* follows *D* when *ENABLE* is active. This is the transparent mode of the latch. When *ENABLE* is inactive, the latch stores the last value of *D*.

15. A D latch can be described in VHDL by an IF statement within a PROCESS. The PROCESS statement in VHDL is concurrent, but the statements inside the PROCESS are sequential.

16. A D latch can also be implemented in VHDL by instantiating a LATCH primitive as a component in a VHDL design entity or by instantiating a component called **lpm_latch** from the Library of Parameterized Modules (LPM).

17. An LPM component is a standard component with certain properties, called parameters, that can be specified when the component is instantiated. The inputs and outputs of an LPM component are called ports. Parameter values are assigned in the generic map of a component instantiation statement. Component port names are associated with user port names in the port map of a component instantiation statement.

18. Multibit latches can be instantiated in VHDL by:

    a. using a behavioral description and STD_LOGIC_VECTOR or BIT_VECTOR types;

    b. instantiating multiple LATCH primitives in a GENERATE statement; or

    c. instantiating an LPM component with the LPM_WIDTH parameter used to specify the number of bits in the latch.

19. Multibit latches can be created in the Quartus II Block Editor by:

    a. customizing an LPM_LATCH component with the MegaWizard Plug-In Manager, a tool that allows the user to specify ports and parameters of a component via a sequence of setup screens; or

    b. manually entering an LPM component and modifying its ports and parameters in a **Properties** dialog box.

20. A flip-flop is like a gated latch that responds to the edge of a pulse applied to an enable input called *CLOCK*. A flip-flop output will change only when the input makes a transition from LOW to HIGH (for a positive edge-triggered device) or HIGH to LOW (for a negative edge-triggered device).

21. In a positive edge-triggered D flip-flop. Q follows D when there is a positive edge on the clock input.

22. A JK flip-flop has two synchronous inputs, called *J* and *K, J* acts as an active HIGH set input. *K* acts as an active-HIGH reset function. When both inputs are asserted, the flip-flop toggles between 0 and 1 with each applied clock pulse.

23. The toggle function in a JK flip-flop is implemented with additional cross-coupled feedback from the latch gate outputs to the steering gate inputs.

24. A chain of JK flip-flops can implement an asynchronous binary counter if the Q of each flip-flop is connected to the clock input of the next and each flip-flop is configured to toggle. Although this is an easy way to create a counter, it is seldom used because internal flip-flop delays result in unwanted intermediate states in the count sequence.

25. JK flip-flops can be combined with a network of logic gates to make a synchronous binary counter. The gates are connected in such a way that each flip-flop toggles when all previous bits are HIGH: otherwise the flip-flops are in a no change state. Although more complex than an asynchronous counter, a synchronous counter is free of unwanted intermediate states.

26. Many flip flops are provided with asynchronous preset (set) and clear (reset) functions. Since these functions are connected directly to the latch gates of a flip-flop, they act immediately, without waiting for the clock. In most cases, these functions are active-LOW.

27. Asynchronous inputs, such as preset and clear, are usually designed so that they will override the synchronous inputs, such as *D* or *JK*.

28. Unused asynchronous inputs should be disabled by tying them to a logic HIGH (for an active-LOW input). Flip-flop primitives in Quartus II automatically have their asynchronous inputs connected to HIGH unless otherwise specified by a design entry file.

29. The outputs of a T (toggle) flip-flop toggle with each clock pulse when the *T* input is HIGH and do not change when *T* is LOW.

30. A registered PLD output consists of a flip-flop (usually D-type) on the output of an SOP matrix.

31. A PAL part number indicates the number of registered outputs (e.g., a PAL16R8 has eight registered outputs).

32. Early-version standard PALs are limited in that they are one-time programmable (OTP), their outputs are permanently configured as combinational or registered, and they cannot be programmed in-system. Later-version PLDs, such as GAL22V10s (GAL = generic array logic) overcome these limitations.

33. PLDs with configurable architecture have outputs that can be combinational or registered, with various input or feedback options.

34. Configurable output circuits in a PLD are called output logic macrocells (OLMCs), or just macrocells.

35. Macrocells are configured by programming architecture cells. Global architecture cells affect all macrocells in a device. A local architecture cell affects only the macrocell in which it is found.

36. GALs have global control signals, such as clock, clear, and output enable, that can be applied to all macrocells in the device.

37. A GAL22V10 has ten macrocells, a global clock that can be used as a combinational input for non-clocked designs, and eleven dedicated inputs.

38. The GAL22V10 macrocells are not all the same size. The are two macrocells with each of the following numbers of product terms: 8, 10, 12, 14, 16.

39. PLDs that can be programmed while installed in a circuit are called in-system programmable (ISP). There are programmed by a 4-wire interface that complies to a standard published by the Joint Test Action Group (JTAG) and the IEEE (Std. 1149.1).

40. An Altera MAX 7000S CPLD consists of groups of 16 macrocells, called Logic Array Blocks (LABs), that are interconnected by an internal bus called a Programmable Interconnect Array (PIA).

41. The number of macrocell outputs in an LAB that are connected to I/O pins depends on the CPLD package type. Macrocells that do not have external connections can still be used for buried logic functions.

42. MAX 7000S devices have four programmable control pins: global clock *(GCLK1)*, Global Output Enable *(OE1)*, Global Clear *(GCLRn)*, and a pin that can be configured as a second global clock *(GCLK2)* or as a second global output enable *(OE2)*. If these functions are not used, the associated pins can be used as standard I/Os.

43. If the ISP capability of a MAX 7000S CPLD is to be used, there are four fewer pins available on the CPLD for user I/O.

44. Each MAX 7000S macrocell has five dedicated product lines and the capability to borrow or share additional product terms with neighboring macrocells in the same LAB.

45. Shared logic expanders allow one product term per macrocell to be shared with other macrocells in the LAB, totaling 16 product terms per LAB. The expander inverts the product term and feeds it back into the LAB AND matrix.

46. Parallel logic expanders allow a macrocell to borrow product lines from neighboring macrocells in the same LAB. These borrowed product lines are not available to the macrocell from which they are borrowed.

47. Expander assignments are done automatically by Quartus II at compile time.

48. MAX 7000S are based on EEPROM cells and are thus nonvolatile.

49. The Altera FLEX 10K series of CPLDs is based on a look-up table (LUT) architecture. A look-up table consists of a 16-bit array of storage elements that are selected by four logic inputs.

50. An LUT, combined with switching, configuration, and expansion circuitry, comprise a logic element (LE), whose function is equivalent to a macrocell in an SOP-type device.

51. Eight logic elements and a local interconnect make up a Logic Array Block (LAB).

52. LABs in a FLEX 10K device are interconnected by global row and column busses.

53. The number of inputs in a logic function can be expanded beyond the capacity of one logic element by using cascade chains.

54. Carry chains can be used to more efficiently implement carry functions in adders, counters, and comparators.

55. FLEX 10K devices are based on SRAM technology and are therefore volatile; they must be reconfigured each time power is applied to the circuit.

GLOSSARY

Architecture Cell A programmable cell that, in combination with other architecture cells, sets the configuration of a macrocell.

Asynchronous Not synchronized to the system clock.

Asynchronous Inputs The inputs of a flip-flop that change the flip-flop's *Q* outputs immediately, without waiting for a pulse at the *CLK* input. Examples include preset and clear inputs.

Buried Logic Logic circuitry in a PLD that has no connection to the input or output pins of the PLD, but is used solely as internal logic.

Carry Chain A circuit in a CPLD that is optimized for efficient operation of carry functions between logic elements.

Cascade Chain A circuit in a CPLD that allows the input width of a Boolean function to expand beyond the width of one logic element.

Clear An asynchronous reset function.

CLOCK A enabling input to a sequential circuit that is sensitive to the positive- or negative-going edge of a waveform.

CPLD Complex programmable logic device. A programmable logic device consisting of several interconnected programmable blocks.

Edge The HIGH-to-LOW (negative edge) or LOW-to-HIGH (positive edge) transition of a pulse waveform.

Edge Detector A circuit in an edge-triggered flip-flop that converts the active edge of a CLOCK input to a short active-level pulse at the internal latch's Set and Reset inputs.

Edge-Sensitive Edge-triggered.

Edge-Triggered Enabled by the positive or negative edge of a digital waveform.

Embedded Array Block (EAB) A relatively large block of storage elements in a CPLD (2048 bits in a FLEX 10K device), used for implementing complex logic functions in look-up table format.

Flip-Flop A sequential circuit based on a latch whose output changes when its *CLOCK* input receives a pulse.

Gated SR Latch An SR latch whose ability to change states is controlled by an extra input called the *ENABLE* input.

Generic Array Logic (GAL) A type of programmable logic device whose outputs can be configured as combinational or registered and whose programming matrix is based on electrically erasable logic cells.

Generic Map A VHDL construct that maps one or more parameters of a component to a value for that instance of the component.

Global Clock A clock signal in a PLD that clocks all registered outputs in the device.

In-System Programmability (ISP) The ability of a PLD to be programmed through a standard four-wire interface while installed in a circuit.

I/O Control Block A circuit in an Altera CPLD that controls the type of tristate switching used in a macrocell output.

JTAG Port A four-wire interface specified by the Joint Test Action Group (JTAG) used for loading test data or programming data into a PLD installed in a circuit.

Latch A sequential circuit with two inputs called *SET* and *RESET*, which make the latch store a logic 0 (reset) or 1 (set) until actively changed.

Level-Sensitive Enabled by a logic HIGH or LOW level.

Library of Parameterized Modules (LPM) A standardized set of components for which certain properties can be specified when the component is instantiated.

Logic Array Block (LAB) A group of macrocells that share common resources in a CPLD.

Logic Element (LE) A circuit internal to a CPLD used to implement a logic function as a look-up table.

Look-Up Table (LUT) A circuit that implements a combinational logic function by storing a list of output values that correspond to all possible input combinations.

Master Reset An asynchronous reset input used to set a synchronous circuit to a known initial state.

Mega Wizard Plug-In Manager A tool provided with Quartus II (and MAX+PLUS II) that allows the user to automatically set the ports and parameter values of an LPM component for use in an HDL or Block Diagram File.

One-Time-Programmable (OTP) A property of some PLDs that allows them to be programmed, but not erased.

Output Logic Macrocell (OLMC) An input/output circuit that can be programmed for a variety of input or output configurations, such as active-HIGH or active-LOW, combinational or registered. Often just called a **macrocell.**

Parallel Logic Expanders Product terms that are borrowed from neighboring macrocells in the same LAB.

Parameter (in an LPM Component) A property of a component that can be specified when the component is instantiated.

Port Map A VHDL construct that maps the name of a port in a component to the name of a port, variable, or signal in a design entity that uses the component.

Preset An asynchronous set function.

Programmable Interconnect Array (PIA) An internal bus with programmable connections that link together the Logic Array Blocks of a CPLD.

Register A digital circuit such as a flip-flop that stores one or more bits of digital information.

Registered Output An output of a Programmable Array Logic (PAL) device having a flip-flop (usually D-type) which stored the output state.

RESET 1. The stored LOW state of a latch circuit.

2. A latch input that makes the latch store a logic 0.

Sequential Circuit A digital circuit whose output depends not only on the present combination of inputs, but also history of the circuit.

SET 1. The stored HIGH state of a latch circuit.

2. A latch input that makes the latch store a logic 1.

Shared Logic Expanders Product terms that are inverted and fed back into the programmable AND matrix of an LAB for use by any other macrocell in the LAB.

Steering Gates Logic gates, controlled by the *ENABLE* input of a gated latch, that steer a *SET* or *RESET* pulse to the correct input of an SR latch circuit.

Synchronous Synchronized to the system clock.

Synchronous Inputs The inputs of a flip-flop that do not affect the flip-flop's *Q* outputs unless a clock pulse is applied. Examples include *D*, *J*, and *K* inputs.

T (Toggle) Flip-Flop A flip-flop whose output toggles between HIGH and LOW states on each applied clock pulse when a synchronous input, called *T*, is active.

Toggle To alternate between binary states with each applied clock pulse.

Transparent Latch (Gated D Latch) A latch whose output follows its data input when its *ENABLE* input is active.

PROBLEMS

Problem numbers set in color indicate more difficult problems.

8.1 Latches

8.1 Complete the timing diagram in Figure 8.95 for the active-HIGH latch shown. The latch is initially set.

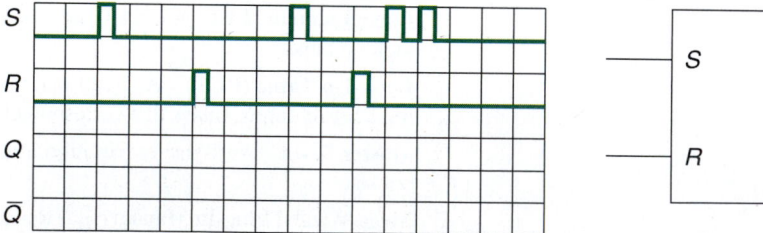

FIGURE 8.95 Problem 8.1: Timing Diagram

8.2   Repeat Problem 8.1 for the timing diagram shown in Figure 8.96.

FIGURE 8.96   Problem 8.2: Timing Diagram

8.3   Complete the timing diagram in Figure 8.97 for the active-LOW latch shown.

FIGURE 8.97   Problem 8.3: Timing Diagram

8.4   Figure 8.98 shows an active-LOW latch used to control a motor starter. The motor runs when $Q = 1$ and stops when $Q = 0$.

FIGURE 8.98   Problem 8.4: Latch for Motor Starter

The motor is housed in a safety enclosure that has an access hatch for service. A safety interlock prevents the motor from running when the hatch is open. The *HATCH* switch opens when the hatch opens, supplying a logic *HIGH* to the circuit. The *START* switch is a normally open momentary-contract pushbutton (LOW when pressed). The *STOP* switch is a normally closed momentary-contact pushbutton (HIGH when pressed).

Draw the timing diagram of the circuit, showing *START, STOP, HATCH, $\bar{S}$, $\bar{R}$, and Q* for the following sequence of events:

a. *START* is pressed and released.

b. The hatch cover is opened.

c. *START* is pressed and released.

d. The hatch cover is closed.

    e. *START* is pressed and released.

    f. *STOP* is pressed and released.

Briefly describe the functions of the three switches and how they affect the motor operation.

8.2   NAND/NOR Latches

8.5   Draw a NAND latch, correctly labeling the inputs and outputs. Describe the operation of a NAND latch for all four possible combinations of $\overline{S}$ and $\overline{R}$.

8.6   Draw a NOR latch, correctly labeling the inputs and outputs. Describe the operation of a NOR latch for all four possible combinations of S and R.

8.7   The timing diagram in Figure 8.99 shows the input waveforms of a NAND latch. Complete the diagram by showing the output waveforms.

FIGURE 8.99   Problem 8.7: Timing Diagram

8.8   Figure 8.100 shows the input waveforms to a NOR latch. Draw the corresponding output waveforms.

FIGURE 8.100   Problem 8.8: Input Waveforms to a NOR Latch

FIGURE 8.101   Problem 8.9: Input Waveforms to a Latch

8.9 Figure 8.101 represents two input waveforms to a latch circuit.

    a. Draw the outputs Q and $\bar{Q}$ if the latch is a NAND latch.

    b. Draw the output waveforms if the latch is a NOR latch.

(Note that in each case, the waveforms will produce the forbidden state at some point. Even under this condition, it is still possible to produce unambiguous output waveforms. Refer to Figures 8.11 and 8.12 for guidance.)

8.10 **a.** Draw a timing diagram for a NAND latch showing each of the following sequences of events:

      i. $\bar{S}$ and $\bar{R}$ are both LOW; $\bar{S}$ goes HIGH before $\bar{R}$.

      ii. $\bar{S}$ and $\bar{R}$ are both LOW; $\bar{R}$ goes HIGH before $\bar{S}$.

      iii. $\bar{S}$ and $\bar{R}$ are both LOW; $\bar{S}$ and $\bar{R}$ go HIGH simultaneously.

    b. State why $\bar{S} = \bar{R} = 0$ is a forbidden state for the NAND latch.

    c. Briefly explain what the final result is for each of the above transitions.

8.11 **a.** Draw a timing diagram for a NOR latch showing each of the following sequences of events:

      i. S and R are both HIGH; S goes LOW before R.

      ii. S and R are both HIGH, R goes LOW before S.

      iii. S and R are both HIGH, S and R go LOW simultaneously.

    b. Briefly explain what the final result is for each of the transitions listed in part a of this question.

    c. State why $S = R = 1$ is a forbidden state for the NOR latch.

8.12 **a.** Use the Quartus II Block Editor to draw a NOR latch like the one in Figure 8.6a. Call the inputs R and S. Call the outputs Q and nQ. Compile the project and examine the fitter equations in the Quartus II Compilation Report. Write the Boolean equations for Q and nQ from the report, in terms of Q, R, and S.

    b. Use the equations from part a of this problem to draw the logic diagram of the NOR latch, as synthesized by Quartus II.

8.13 Write a brief description of the operation of the NOR latch synthesized by Quartus II in Problem 8.12. Include set-to-reset, reset-to-set, and forbidden state transitions.

8.14 Figure 8.102 shows the effect of mechanical bounce on the switching waveforms of a single-pole double-throw (SPDT) switch.

    a. Briefly explain how this effect arises.

    b. Draw a NAND latch circuit that can be used to eliminate this mechanical bounce, and briefly explain how it does so.

FIGURE 8.102 Problem 8.14: Effect of Mechanical Bounce on a SPDT Switch

8.3   Gated Latches

8.15   Complete the timing diagram for the gated latch show in Figure 8.103.

Q is initially LOW

FIGURE 8.103   Problem 8.15: Gated Latch

8.16   Complete the timing diagram for the gated latch shown in Figure 8.104.

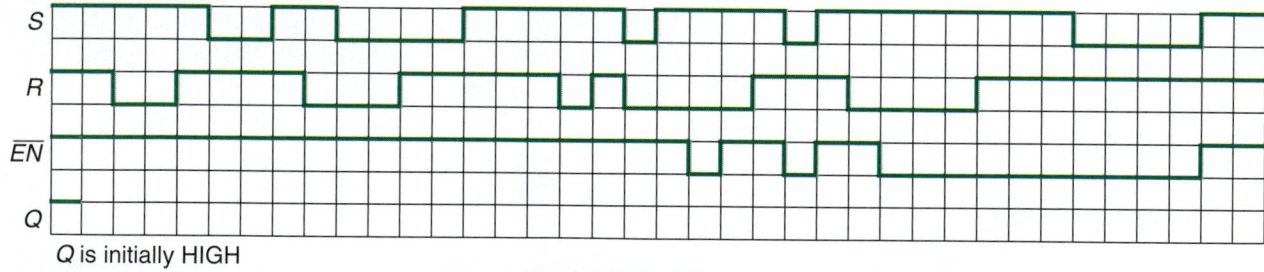

Q is initially HIGH

FIGURE 8.104   Problem 8.16: Gated Latch

8.17   A pump motor can be started at two different locations with momentary-contact pushbuttons S_1 and S_2. It can be stopped by momentary-contact pushbuttons ST_1 and ST_2. As in Problem 8.4, a *RUN* input on the motor controller must be kept HIGH to keep the motor running. After the motor is stopped, a timer prevents the motor from starting for 5 minutes.

Draw a circuit block diagram showing how an SR latch and some additional gating logic can be used in such an application. The timer can be shown as a block activated by the *STOP* function. Assume that the timer output goes HIGH for 5 minutes when activated.

8.18 The *S* and *R* waveforms in Figure 8.105 are applied to two different gated latches. The *ENABLE* waveforms for the latches are shown as *EN₁* and *EN₂*. Draw the output waveforms Q_1 and Q_2, assuming that *S*, *R*, and *EN* are all active HIGH. Which output is least prone to synchronization errors? Why?

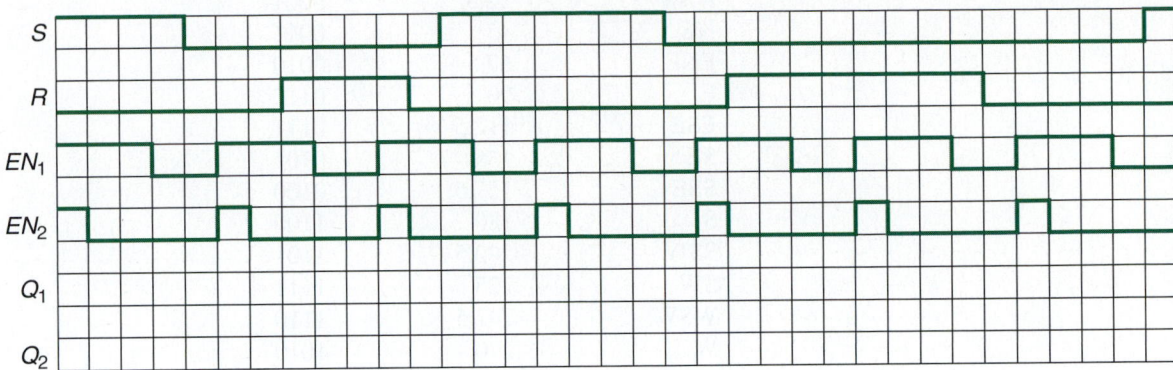

FIGURE 8.105 Problem 8.18: Waveforms

8.19 Figure 8.106 represents the waveforms of the *EN* and *D* inputs of a 4-bit transparent latch. Complete the timing diagram by drawing the waveforms for Q_1 to Q_4.

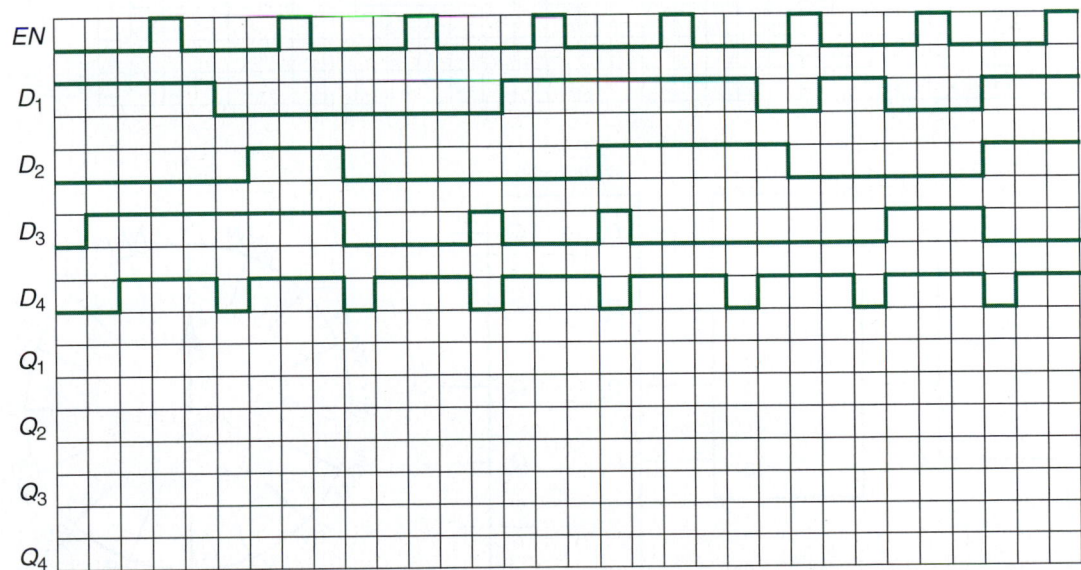

FIGURE 8.106 Problem 8.19: Waveforms

8.20 An electronic direction finder aboard an aircraft uses a 4-bit number to distinguish 16 different compass points as follows:

Direction	Degrees	Gray Code
N	0/360	0000
NNE	22.5	0001
NE	45	0011
ENE	67.5	0010
E	90	0110
ESE	112.5	0111
SE	135	0101
SSE	157.5	0100
S	180	1100
SSW	202.5	1101
SW	225	1111
WSW	247.5	1110
W	270	1010
WNW	295.5	1011
NW	315	1001
NNW	337.5	1000

FIGURE 8.107 Problem 8.20: Direction Finder and Sample Output

The output of the direction finder is stored in a 4-bit latch so that the aircraft flight path can be logged by a computer. The latch is periodically updated by a continuous pulse on the latch enable line.

Figure 8.107 shows a sample reading of the direction finder's output as presented to the latch.

a. Complete the timing diagram by filling in the data for the Q outputs.

b. Based on the completed timing diagram of Figure 8.107 make a rough sketch of the aircraft's flight path for the monitored time.

8.21   **a.** Write a VHDL file for an 8-bit latch, using LATCH primitives.

b. Write a set of simulation criteria that would verify the behavior of the latch designed in part a of this problem.

c. Use the criteria of part b to create a simulation of the 8-bit latch in Quartus II.

8.22   **a.** Write a VHDL file for an 8-bit latch, using a component from the Library of Parameterized Modules (LPM).

b. Use the simulation criteria of Problem 8.21 to create a simulation for the 8-bit latch in part a of this problem.

8.23   **a.** Use the Quartus II MegaWizard Plug-in Manager to create a 12-bit latch (no asynchronous inputs). Instantiate the latch in a Quartus II Block Diagram File.

b. Write a set of simulation criteria that verify the function of the latch of part a.

c. Create a simulation in Quartus II, using the criteria of part b.

8.24   **a.** Use the Quartus II Block Editor to manually enter a 12-bit latch based on the LPM component **lpm_latch.** Use only the following ports: **data[], gate,** and **q[].**

b. Use the simulation criteria written for Problem 8.23 to create a simulation that verifies the function of the LPM latch in part a of this problem.

8.4   Edge-Triggered D Flip-Flops

8.25   The waveforms in Figure 8.108 are applied to the inputs of a positive edge-triggered D flip-flop and a gated D latch. Complete the timing diagram where Q_1 is the output of the flip-flop and Q_2 is the output of the gated latch. Account for any differences between the Q_1 and Q_2 waveforms.

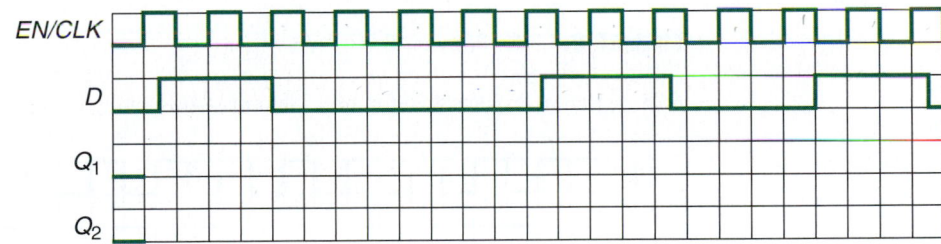

FIGURE 8.108   Problem 8.25: Waveforms

8.26   Complete the timing diagram for a positive edge-triggered D flip-flop if the waveforms shown in Figure 8.109 are applied to the flip-flop inputs.

FIGURE 8.109   Problem 8.26: Waveforms

8.27   Repeat Problem 8.26 for the waveforms shown in Figure 8.110.

FIGURE 8.110   Problem 8.27: Waveforms

8.28   Repeat Problem 8.26 for the waveforms shown in Figure 8.111.

FIGURE 8.111   Problem 8.28: Waveforms

8.29   Draw a logic diagram of a D flip-flop configured for toggle mode. (Hint: The D input must always be the opposite of the Q output.)

8.30   Write a VHDL file that defines a 12-bit D flip-flop with a clock common to all flip-flops, using primitives from the **maxplus2** package. The component declaration for the DFF component is as follows:

```
COMPONENT DFF
    PORT (d : IN STD_LOGIC;
        clk  : IN STD_LOGIC;
        clrn : IN STD_LOGIC;
        prn  : IN STD_LOGIC;
        q    : OUT STD_LOGIC;
END COMPONENT;
```

Disregard the **clrn** (active-LOW clear) and **prn** (active-LOW preset) ports for this problem. (Hint: you may have to use a component declaration in your file that only declares the ports **d, clk,** and **q.**)

8.31 Write a VHDL file that creates a 12-bit D flip-flop, using the LPM component **lpm_ff.** (This component is instantiated as a D flip-flop by default. The required LPM component port names are: **data, clock,** and **q.**)

8.5 Edge-Triggered JK Flip-Flops

8.32 The waveforms in Figure 8.112 are applied to a negative edge-triggered JK flip-flop. Complete the timing diagram by drawing the Q waveform.

FIGURE 8.112 Problem 8.32: Waveforms

8.33 Repeat Problem 8.32 for the waveforms in Figure 8.113.

FIGURE 8.113 Problem 8.33: Waveforms

8.34 Given the inputs *x, y,* and *z* to the circuit in Figure 8.114, draw the waveform for output Q.

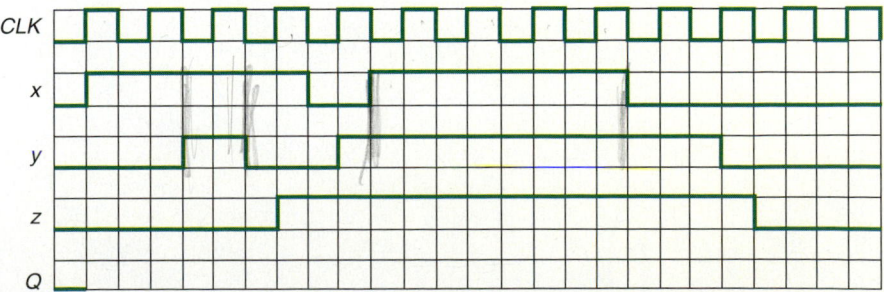

FIGURE 8.114 Problem 8.34: Inputs to Circuit

8.35   Assume that all flip-flops in Figure 8.115 are initially set. Draw a timing diagram showing the *CLK*, Q_0, Q_1, and Q_2 waveforms when eight clock pulses are applied. Make a table showing each combination of Q_2, Q_1, and Q_0. What pattern do the outputs form over the period shown on the timing diagram?

FIGURE 8.115   Problem 8.35: Flip-Flops

8.36   Refer to the JK flip-flop circuit in Figure 8.116. Is the circuit synchronous or asynchronous? Explain your answer.

FIGURE 8.116   Problem 8.36: Flip-Flop Circuit

8.37   Assume all flip-flops in the circuit in Figure 8.116 are reset. Analyze the operation of the circuit when sixteen clock pulses are applied by making a table showing the sequence of states of $Q_3Q_2Q_1Q_0$, beginning at 0000.

8.38   Draw a timing diagram showing the sequence of states from the table derived in Problem 8.37.

8.39 The waveforms shown in Figure 8.117 are applied to a negative edge-triggered JK flip-flop. The flip-flop's Preset and Clear inputs are active LOW. Complete the timing diagram by drawing the output waveforms.

FIGURE 8.117 Problem 8.39: Waveforms

8.40 Repeat Problem 8.39 for the waveforms in Figure 8.118.

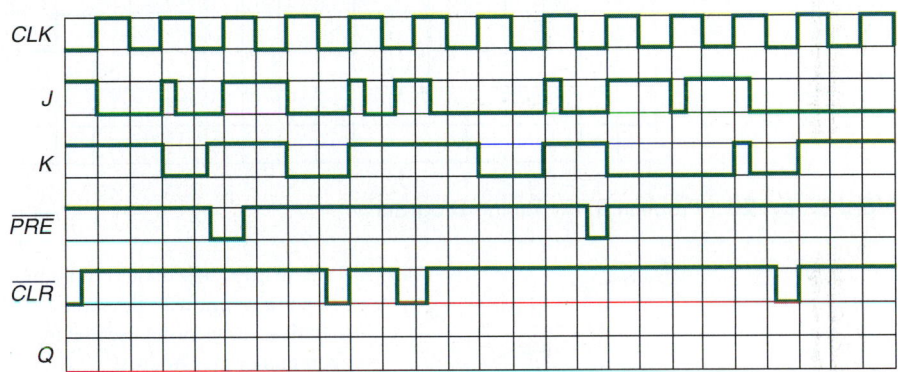

FIGURE 8.118 Problem 8.40: Waveforms

8.41 Create a Quartus II Block Diagram File for the synchronous circuit in Figure 8.116. Modify the circuit to add an asynchronous Master Reset function. Create a simulation file to verify the circuit operation.

8.42 Modify the **bdf** created in Problem 8.41 to include a Master Reset function *and* an asynchronous preset function that will set the state of the circuit to $Q_3Q_2Q_1Q_0 = 1010$ when activated. Create a simulation file to verify the circuit operation.

8.43 The term *asynchronous* is sometimes used to refer to the configuration of a circuit (e.g., a 3-bit asynchronous counter) and sometimes to a type of input to a device (e.g., an asynchronous clear input). Briefly explain how these two usages are similar and how they are different.

8.44 Write a VHDL file for a 12-bit D flip-flop that uses **maxplus2** DFF primitives, similar to that in Problem 8.30. Include active-LOW asynchronous clear (CLRN) and preset (PRN) inputs. Create a simulation file to verify the operation of your design.

8.45 Write a VHDL file for a 12-bit D flip-flop with asynchronous preset and clear, using the LPM component **lpm_ff**, similar to that in Problem 8.31. Required ports: **data, clock, aclr** (asynchronous clear), **aset** (asynchronous set), and **q**. Ports **aset** and **aclr** are active-HIGH. Add two signals to the VHDL design to make them active-LOW. Create a simulation file to verify the operation of your design.

8.6   Edge-Triggered T Flip-Flops

8.46   The *T* and *CLK* waveforms for a positive edge triggered T flip-flop are shown in Figure 8.119. Complete the timing diagram.

FIGURE 8.119   Problem 8.46: Timing Diagram

8.47   The *T* and *CLK* waveforms for a positive-edge triggered T flip-flop are shown in Figure 8.120. Complete the timing diagram.

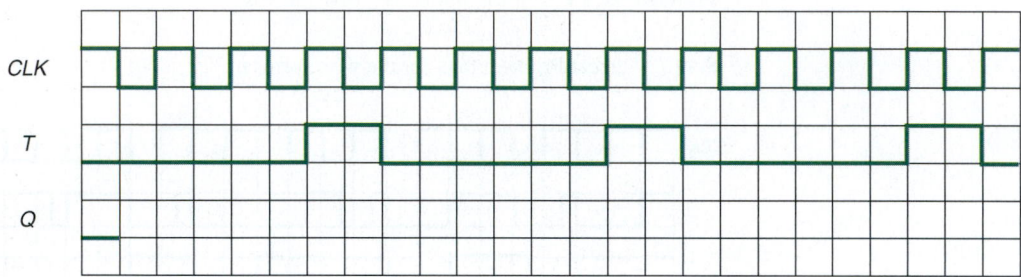

FIGURE 8.120   Problem 8.47: Timing Diagram

8.48   Refer to the synchronous circuit in Figure 8.116. Create a Quartus II Block Diagram File for a circuit with the same function using T flip-flops rather than JK flip-flops. Include an asynchronous reset input in the circuit. Create a simulation file to test the operation of the circuit.

8.49   Write a VHDL file that implements the circuit you drew in Problem 8.48. Use TFF primitives in the design.

8.7   Flip-Flops in PLDs (Registered Outputs)

8.50   What is a registered output?

8.51   State the number of registered outputs for each of the following PAL devices:

    a. PAL16R4

    b. PAL16R6

    c. PAL16R8

8.8   Generic Array Logic (GAL)

8.52   How many macrocells are there in a GAL22V10? How many product lines do these macrocells have?

8.53   State the four configurations possible with a macrocell in a GAL22V10.

8.54   Is there a global output enable function available for a GAL22V10?

8.55   Is there a global clear function available in the GAL22V10?

8.56   Can the registered outputs of a GAL22V10 be clocked by a product term function from the GAL AND matrix?

8.57   Are the Asynchronous Reset *(AR)* and Synchronous Preset *(SP)* functions in a GAL22V10 global or local? Explain your answer in one sentence.

8.9   MAX 7000S CPLD

8.58   State one way in which a Complex PLD, such as an Altera MAX 7000S, differs from a low-density PAL or GAL.

8.59   State how many macrocells are available in the following CPLDs:

   a.   EPM7032

   b.   EPM7064

   c.   EPM7128S

   d.   EPM7160S

8.60   Which of the CPLDs listed in Problem 8.59 are in-system programmable? What does it mean when a device is in-system programmable?

8.61   How many logic array blocks (LABs) are there in an Altera MAX 7000S CPLD?

8.62   How many user I/O pins are there in an EPM7128SLC84 CPLD? How many pins per LAB does this represent?

8.63   What can be done with the macrocells in an LAB that do not connect to I/O pins?

8.64   State the possible clock configurations of a MAX 7000S macrocell.

8.65   State the possible reset configurations of a MAX 7000S macrocell.

8.66   State the possible preset configurations of a MAX 7000S macrocell.

8.67   How many dedicated product terms are available in a MAX 7000S macrocell? How can this number of product terms be supplemented? What is the maximum number of product terms available to a macrocell?

8.68   How many shared logic expanders are available in an LAB?

8.10   FLEX 10K CPLD

8.69   Briefly state the difference between CPLDs having sum-of-products architecture and look-up table architecture.

8.70   How many inputs can a look-up table accept in an Altera FLEX 10K logic element? How can this be expanded?

8.71   What is the purpose of the carry chain in a FLEX 10K CPLD?

8.72   How many logic elements are there in a FLEX 10K LAB?

8.73   How many bits of storage are there in an Embedded Array Block in a FLEX 10K CPLD?

ANSWERS TO SECTION REVIEW PROBLEMS

8.1

8.1   The latch resets (i.e., Q goes LOW) upon receiving the first reset pulse. At that point, the latch is already reset, so further pulses are ignored.

8.2

8.2   The NOR latch has active-HIGH inputs. If you make both inputs HIGH, you are attempting to set and reset the latch at the same time, which is a contradictory action. A NAND latch has active-LOW inputs. Therefore, if both inputs are HIGH, neither the set nor reset function activates and there is no change on the latch output.

8.3

8.3

```
LIBRARY ieee;
USE ieee.std_logic_1164.ALL;
LIBRARY altera;
USE altera.maxplus2.ALL;
```

```
ENTITY lch16prm IS
    PORT(d_in   : IN  STD_LOGIC_VECTOR (15 downto 0);
          enable : IN  STD_LOGIC;
          q_out  : OUT STD_LOGIC_VECTOR (15 downto 0) );
END lch16prm;

ARCHITECTURE a OF lch16prm IS
BEGIN
    -- Instantiate a latch from a maxplus2 primitive
    latch4:
    FOR i IN 15 downto 0 GENERATE
    latch_primitive: latch
        PORT MAP (d => d_in(i),
          ena => enable, q => q_out(i) );
    END GENERATE;
END a;
```

8.4

8.4 The edge detector in the clock circuit accounts for the operational difference between a D flip-flop and a D latch. It works by using the difference in internal delay times between the gates that comprise the flip-flop's clock input circuit.

8.5

8.5 The flip-flops in asynchronous circuits are not all clocked at the same time; they are asynchronous with respect to the system clock. The flip-flops in a synchronous circuit have a common clock connection, which makes them synchronous to the system clock. The disadvantage to asynchronous circuits is that the internal delays of flip-flops can lead to unwanted intermediate states, since the flip-flops do not all change at the same time.

8.6

8.6 The circuit is shown in Figure 8.121.

FIGURE 8.121 Solution to Section Review Problem 8.6

KEY TERMS

Attribute
Behavioral Design
Bidirectional Counter
Bidirectional Shift Register
Binary Counter
Clear
Command Lines
Control Section
Count Enable
Count Sequence
Counter
Count-Sequence Table
Dataflow Design
DOWN Counter
Excitation Table
Full-Sequence Counter
IF Statement
Johnson Counter
Left Shift
Maximum Modulus (m_{max})
Memory Section
Modulo Arithmetic
Modulo-n (or mod-n) Counter
Modulus
Next State
Output Decoding
Parallel Load
Parallel-Load Shift Register
Parallel Transfer
Present State
Presettable Counter
Recycle
Right Shift
Ring Counter
Ripple Carry Out or Ripple Clock Out (RCO)

Rotation
Serial Shifting
Shift Register
SRGn
State Diagram
State Machine
Status Lines
Structural Design
Synchronous Counter
Terminal Count
Truncated-Sequence Counter
Universal Shift Register
UP Counter

OUTLINE

9.1   Basic Concepts of Digital Counters
9.2   Synchronous Counters
9.3   Design of Synchronous Counters
9.4   Programming Binary Counters for CPLDs
9.5   Control Options for Synchronous Counters
9.6   Programming Presettable and Bidirectional Counters for CPLDs
9.7   Shift Registers
9.8   Programming Shift Registers in VHDL
9.9   Shift Register Counters

CHAPTER OBJECTIVES

Upon successful completion of this chapter you will be able to:

- Determine the modulus of a counter.
- Determine the number of outputs required by a counter for a given modulus.
- Determine the maximum modulus of a counter, given the number of circuit outputs.

CHAPTER 9

Counters and Shift Registers

- Draw the count sequence table, state diagram, and timing diagram of a counter.
- Determine the recycle point of a counter's sequence.
- Calculate the frequencies of each counter output, given the input clock frequency.
- Draw a circuit for any full-sequence synchronous counter.
- Determine the count sequence, state diagram, timing diagram, and modulus of any synchronous counter.
- Complete the state diagram of a synchronous counter to account for unused states.
- Design the circuit of a truncated-sequence synchronous counter, using flip-flops and logic gates.
- Use Quartus II to create a Block Diagram File for any synchronous counter circuit.
- Use behavioral descriptions in VHDL to design synchronous counters of any modulus.
- Use a parameterized counter from the Library of Parameterized Modules in a Block Diagram File and a VHDL file.
- Use the Quartus II simulation tool to verify the operation of synchronous counters.
- Implement various counter control functions, such as parallel load, clear, count enable, and count direction, both in Block Diagram Files and in VHDL.
- Design a circuit to decode the output of the counter, both in a Quartus II Block Diagram File and in VHDL.
- Draw a logic circuit of a serial shift register and determine its contents over time given any input data.
- Draw a timing diagram showing the operation of a serial shift register.
- Draw the logic circuit of a general parallel-load shift register.
- Draw a timing diagram showing the operation of a parallel-load shift register.
- Draw the general logic circuit of a bidirectional shift register and explain the concepts of right-shift and left-shift.

- Use timing diagrams to explain the operation of a bidirectional shift register.
- Describe the operation of a universal shift register.
- Design shift registers, ring counters, and Johnson counters with the Quartus II Block Editor or VHDL.
- Verify the operation of shift registers, ring counters, and Johnson counters using the Quartus II simulation tool.
- Design a decoder for a Johnson counter.
- Use a ring counter or a Johnson counter as an event sequencer.
- Compare binary, ring, and Johnson counters in terms of the modulus and the required decoding for each circuit.

VHDL CONSTRUCTS USED IN THIS CHAPTER

- Clearing of vector using (others => '0')
- Component declaration statement
- Component instantiation statement, including GENERIC MAP and PORT MAP
- GENERIC clause in entity declaration
- IF statement
- Library of Parameterized Modules (LPM)
- Port mode BUFFER
- PROCESS statement
- Selected signal assignment statement
- Signal assignment operator <=
- Signal attribute ('EVENT)
- SIGNAL declaration
- STD_LOGIC and STD_LOGIC_VECTOR types
- Type conversion function CONV_STD_LOGIC
- Variable assignment operator :=
- VARIABLE declaration
- VHDL package: altera.maxplus2
- VHDL package: ieee.std_logic_1164
- VHDL package: ieee.std_logic_arith
- VHDL package: ieee.std_logic_unsigned
- VHDL package: lpm.lpm_components

Counters and shift registers are two important classes of sequential circuits. In the simplest terms, a counter is a circuit that counts pulses. As such, it is used in many circuit applications, such as event counting and sequencing, timing, frequency division, and control. A basic counter can be enhanced to incorporate functions such as synchronous or asynchronous parallel loading, synchronous or asynchronous clear, count enable, directional control, and output decoding. In this chapter, we will design counters using schematic entry, VHDL, and counters from the Library of Parameterized Modules and verify their operation using the Quartus II simulator.

Shift registers are circuits that store and move data. They can be used in serial data transfer, serial/parallel conversion, arithmetic functions, and delay elements. As with counters, many shift registers have additional functions such as parallel load, clear, and directional control. We can implement these circuits using schematic entry, VHDL, and LPM components.

9.1 BASIC CONCEPTS OF DIGITAL COUNTERS

Counter A sequential digital circuit whose output progresses in a predictable repeating pattern, advancing by one state for each clock pulse.

Recycle To make a transition from the last state of the count sequence to the first state.

Count Sequence The specific series of output states through which a counter progresses.

Modulus The number of states through which a counter sequences before repeating.

UP Counter A counter with an ascending sequence.

DOWN Counter A counter with a descending sequence.

State Diagram A diagram showing the progression of states of a sequential circuit.

Modulo-*n* (or mod-*n*) Counter A counter with a modulus of *n*.

Modulo Arithmetic A closed system of counting and adding, whereby a sum greater than the largest number in a sequence "rolls over" and starts from the beginning. For example, on a clock face, four hours after 10 A.M. is 2 P.M., so in a mod-12 system, $10 + 4 = 2$.

The simplest definition of a **counter** is "a circuit that counts pulses." Knowing only this, let us look at an example of how we might use a counter circuit.

 EXAMPLE 9.1

Figure 9.1 shows a 10-bit binary counter that can be used to count the number of people passing by an optical sensor. Every time the sensor detects a person passing by, it produces a pulse. Briefly describe the counter's operation. What is the maximum number of people it can count? What happens if this number is exceeded?

FIGURE 9.1 Example 9.1: 10-Bit Counter

■ **Solution**

The counter has a 10-bit output, allowing a binary number from 00 0000 0000 to 11 1111 1111 (0 to 1023) to appear at its output. The sensor causes the counter to advance by one binary number for every pulse applied to the counter's clock (*CLK*) input. If the counter is allowed to register *no people* (i.e., 00 0000 0000), then the circuit can count 1023 people since there are 1024 unique binary combinations of a 10-bit number, including 0. (This is because $2^{10} = 1024$.) When the 1024th pulse is applied to the clock input, the counter rolls over to 0 (or **recycles**) and starts counting again. (After this point, the counter would not accurately reflect the number of people counted.)

The counter is labeled CTR DIV 1024 to indicate that one full cycle of the counter requires 1024 clock pulses (i.e., the frequency of the MSB output signal, Q_9, is the clock frequency divided by 1024).

A counter is a digital circuit that has a number of binary outputs whose states progress through a fixed sequence. This **count sequence** can be ascending, descending, or nonlinear. The output sequence of a counter is usually defined by its **modulus,** that is, the number of states through which the counter progresses. An **UP counter** with a modulus of 12 counts through 12 states from 0000 up to 1011 (0 to 11 in decimal), recycles to 0000, and continues. A **DOWN counter** with a modulus of 12 counts from 1011 down to 0000, recycles to 1011, and continues downward. Both types of counter are called **modulo-12,** or just **mod-12** counters, since they both have sequences of 12 states.

State Diagram

The states of a counter can be represented by a **state diagram.** Figure 9.2 compares the state diagram of a mod-12 UP counter to an analog clock face. Each counter state is illustrated in the state diagram by a circle containing its binary value. The progression is shown by a series of directional arrows. With each clock pulse, the counter progresses by one state, from its present position in the state diagram to the next state in the sequence.

Both the clock face and the state diagram represent a closed system of counting. In each case, when we reach the end of the count sequence, we start over from the beginning of the cycle.

For instance, if it is 10:00 A.M. and we want to meet a friend in four hours, we know we should turn up for the appointment at 2:00 P.M. We arrive at this figure by starting at 10 on the clock face and counting 4 digits forward in a "clockwise" circle. This takes us two digits past 12, the "recycle point" of the clock face.

Similarly, if we want to know the 8th state after 0111 in a mod-12 UP counter, we start at state 0111 and count 8 positions in the direction of the arrows. This brings us to state 0000 (the recycle point) in 5 counts and then on to state 0011 in another 3 counts. This closed system of counting and adding is known as **modulo arithmetic.**

Number of Bits and Maximum Modulus

■ KEY TERMS

Maximum Modulus (m_{max})   The largest number of counter states that can be represented by n bits ($m_{max} = 2^n$).

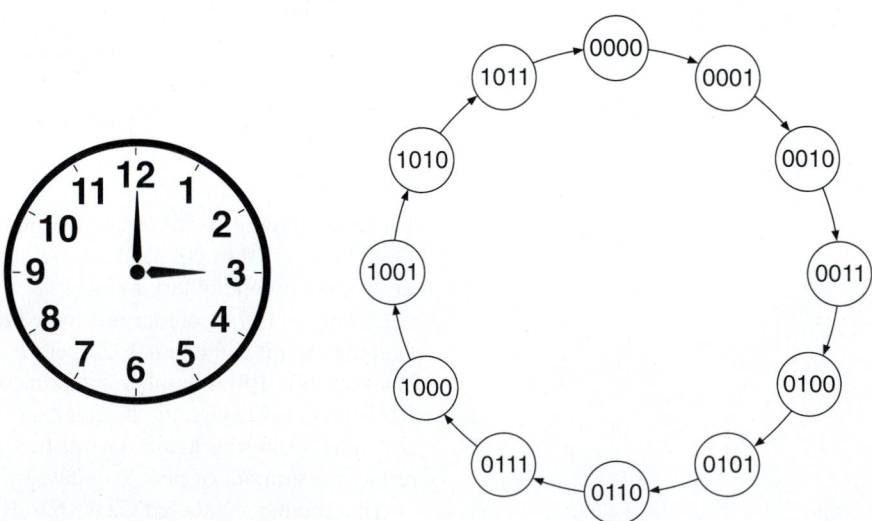

FIGURE 9.2 Mod-12 State Diagram and Analog Clock Face

> **Full-Sequence Counter** A counter whose modulus is the same as its maximum modulus ($m = 2^n$ for an n-bit counter).
>
> **Binary Counter** A counter that generates a binary count sequence.
>
> **Truncated-Sequence Counter** A counter whose modulus is less than its maximum modulus ($m < 2^n$ for an n-bit counter).

The state diagram of Figure 9.2 represents the states of a mod-12 counter as a series of 4-bit numbers. Counter states are always written with a fixed number of bits, since each bit represents the logic level of a physical location in the counter circuit. A mod-12 counter requires four bits because its highest count value is a 4-bit number: 1011.

The **maximum modulus** of a 4-bit counter is 16 (= 2^4). The count sequence of a mod-16 UP counter is from 0000 to 1111 (0 to 15 in decimal), as illustrated in the state diagram of Figure 9.3.

In general, an n-bit counter has a maximum modulus of 2^n and a count sequence from 0 to ($2^n - 1$) (i.e., all 0s to all 1s). Since a mod-16 counter has a modulus of 2^n (= m_{max}), we say that it is a **full-sequence counter.** We can also call this a **binary counter** if it generates the sequence in binary order. A counter, such as a mod-12 counter, whose modulus is less than 2^n, is called a **truncated sequence counter.**

Count-Sequence Table and Timing Diagram

> ■ **KEY TERM**
>
> **Count-Sequence Table** A list of counter states in the order of the count sequence.

Two ways to represent a count sequence other than a state diagram are by a **count-sequence table** and by a timing diagram. The count sequence table is simply a list of counter states in the same order as the count sequence. Table 9.1 and Table 9.2 show the count-sequence tables of a mod-16 UP counter and a mod-12 UP counter, respectively.

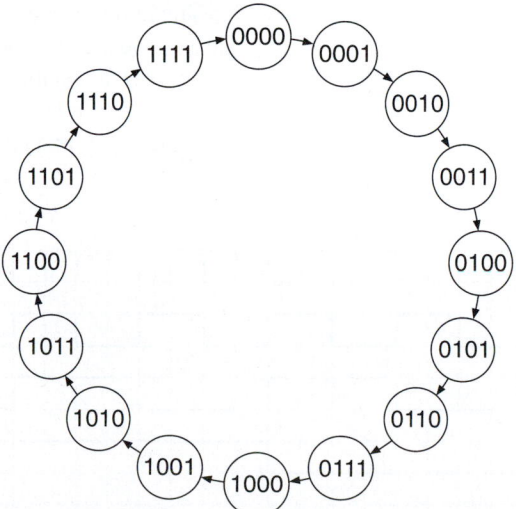

FIGURE 9.3 State Diagram of a Mod-16 Counter

TABLE 9.1 Mod-16 Count-Sequence Table

Q_3	Q_2	Q_1	Q_0
0	0	0	0
0	0	0	1
0	0	1	0
0	0	1	1
0	1	0	0
0	1	0	1
0	1	1	0
0	1	1	1
1	0	0	0
1	0	0	1
1	0	1	0
1	0	1	1
1	1	0	0
1	1	0	1
1	1	1	0
1	1	1	1

TABLE 9.2 Mod-12 Count-Sequence Table

Q_3	Q_2	Q_1	Q_0
0	0	0	0
0	0	0	1
0	0	1	0
0	0	1	1
0	1	0	0
0	1	0	1
0	1	1	0
0	1	1	1
1	0	0	0
1	0	0	1
1	0	1	0
1	0	1	1

We can derive timing diagrams from each of these tables. We know that each counter advances by one state with each applied clock pulse. The mod-16 count sequence shows us that the Q_0 waveform changes state with each clock pulse. Q_1 changes with every two clock pulses, Q_2 with every four, and Q_3 with every eight. Figure 9.4 shows this pattern for the mod-16 UP counter, assuming the counter is a positive edge-triggered device.

A divide-by-two ratio relates the frequencies of adjacent outputs of a binary counter. For example, if the clock frequency is $f_c = 16$ MHz, the frequencies of the output waveforms are: 8 MHz ($f_0 = f_c/2$); 4 MHz ($f_1 = f_c/4$); 2 MHz ($f_2 = f_c/8$); 1 MHz ($f_3 = f_c/16$).

We can construct a similar timing diagram, illustrated in Figure 9.5, for a mod-12 UP counter. The changes of state can be monitored by noting where Q_0 (the least significant bit) changes. This occurs on each positive edge of the *CLK* waveform. The sequence progresses by 1 with each *CLK* pulse until the outputs all go to 0 on the first *CLK* pulse after state $Q_3Q_2Q_1Q_0 = 1011$.

The output waveform frequencies of a truncated sequence counter do not necessarily have a simple relationship to one another as do binary counters. For the mod-12 counter the relationships between clock frequency, f_c, and output frequencies are: $f_0 = f_c/2$; $f_1 = f_c/4$; $f_2 = f_c/12$; $f_3 = f_c/12$. Note that both Q_2 and Q_3 have the same frequencies (f_2 and f_3), but are out of phase with one another.

FIGURE 9.4 Mod-16 Timing Diagram

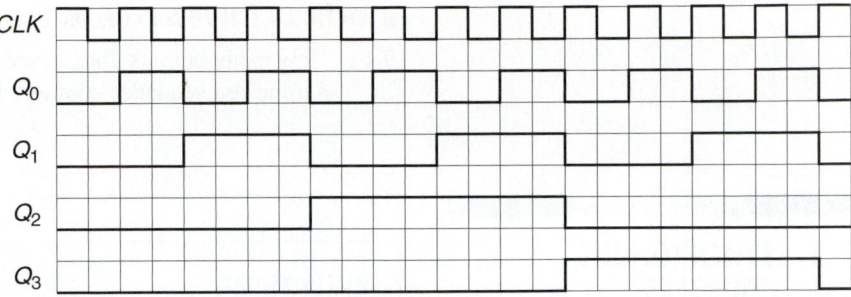

FIGURE 9.5 Mod-12 Timing Diagram

EXAMPLE 9.2

Draw the state diagram, count-sequence table, and timing diagram for a mod-12 DOWN counter.

■ **Solution**

Figure 9.6 shows the state diagram for the mod-12 DOWN counter. The states are identical to those of a mod-12 UP counter, but progress in the opposite direction. Table 9.3 shows the count sequence table of this circuit.

TABLE 9.3 Count-Sequence Table for a Mod-12 DOWN Counter

Q_3	Q_2	Q_1	Q_0
1	0	1	1
1	0	1	0
1	0	0	1
1	0	0	0
0	1	1	1
0	1	1	0
0	1	0	1
0	1	0	0
0	0	1	1
0	0	1	0
0	0	0	1
0	0	0	0

FIGURE 9.6 Example 9.2: State Diagram of a Mod-12 DOWN Counter

The timing diagram of this counter is illustrated in Figure 9.7. The output starts in state $Q_3Q_2Q_1Q_0 = 1011$ and counts DOWN until it reaches 0000. On the next pulse, it recycles to 1011 and starts over.

FIGURE 9.7 Example 9.2: Timing Diagram of a Mod-12 DOWN Counter

■ **SECTION 9.1 REVIEW PROBLEM**

9.1   How many outputs does a mod-24 counter require? Is this a full-sequence or a truncated sequence counter? Explain your answer.

9.2 SYNCHRONOUS COUNTERS

> ■ **KEY TERMS**
>
> **Synchronous Counter**   A counter whose flip-flops are all clocked by the same source and thus change in synchronization with each other.
>
> **Memory Section**   A set of flip-flops in a synchronous circuit that holds its present state.
>
> **Present State**   The current state of flip-flop outputs in a synchronous sequential circuit.
>
> **Control Section**   The combinational logic portion of a synchronous circuit that determines the next state of the circuit.
>
> **Next State**   The desired future state of flip-flop outputs in a synchronous sequential circuit after the next clock pulse is applied.
>
> **Status Lines**   Signals that communicate the present state of a synchronous circuit from its memory section to its control section.
>
> **Command Lines**   Signals that connect the control section of a synchronous circuit to its memory section and direct the circuit from its present to its next state.

In Chapter 8, we briefly examined the circuits of a 3-bit and a 4-bit **synchronous counter** (Figures 8.69 and 8.116, respectively). A synchronous counter is a circuit consisting of flip-flops and control logic, whose outputs progress through a regular predictable sequence, driven by a clock signal. The counter is synchronous because all flip-flops are clocked at the same time.

Figure 9.8 shows the block diagram of a synchronous counter, which consists of a **memory section** to keep track of the **present state** of the counter and a **control section** to direct the counter to its **next state.** The memory section is a sequential circuit (flip-flops) and the control section is combinational (gates). They communicate through a set of **status lines** that go from the Q outputs of the flip-flops to the control gate inputs and **command lines** that connect the control gate outputs to the synchronous inputs ($J, K, D,$ or T) of the flip-flops. Outputs can be tied directly to the status lines or can be decoded to give a sequence other than that of the flip-flop output states. The circuit might have inputs to implement one or more control functions, such as changing the count direction, clearing the counter, or presetting the counter to a specific value.

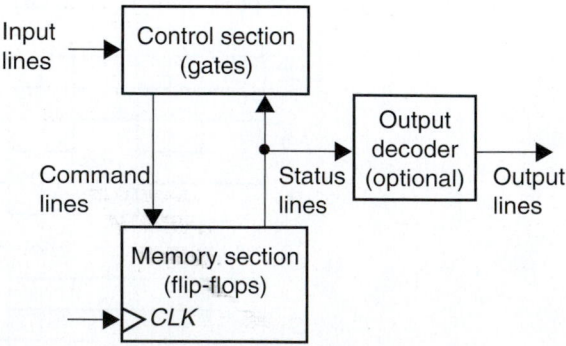

FIGURE 9.8 Synchronous Counter Block Diagram

FIGURE 9.9 3-Bit Synchronous Binary Counter

Analysis of Synchronous Counters

A 3-bit synchronous binary counter based on JK flip-flops is shown in Figure 9.9. Let us analyze its count sequence in detail so that we can see how the J and K inputs are affected by the Q outputs and how transitions between states are made. Later we will look at the function of truncated-sequence counter circuits and counters that are made from flip-flops other than JK.

The synchronous input equations are given by:

$$J_2 = K_2 = Q_1 \cdot Q_0$$
$$J_1 = K_1 = Q_0$$
$$J_0 = K_0 = 1$$

For reference, the JK flip-flop function table is shown in Table 9.4:

TABLE 9.4 Function Table of a JK Flip-Flop

J	K	Q_{t+1}	Function
0	0	Q_t	No change
0	1	0	Reset
1	0	1	Set
1	1	$\overline{Q_t}$	Toggle

Q_t indicates the state of Q before a clock pulse is applied (i.e, the present state). Q_{t+1} indicates the state of Q after the clock pulse (i.e., the next state).

Assume that the counter output is initially $Q_2Q_1Q_1 = 000$. Before any clock pulses are applied, the J and K inputs are at the following states:

$$J_2 = K_2 = Q_1 \cdot Q_0 = 0 \cdot 0 = 0 \quad \text{(No change)}$$
$$J_1 = K_1 = Q_0 = 0 \quad \text{(No change)}$$
$$J_0 = K_0 = 1 \text{ (Constant)} \quad \text{(Toggle)}$$

The transitions of the outputs after the clock pulse are:

$$Q_2: 0 \rightarrow 0 \quad \text{(No change)}$$
$$Q_1: 0 \rightarrow 0 \quad \text{(No change)}$$
$$Q_0: 0 \rightarrow 1 \quad \text{(Toggle)}$$

The output goes from $Q_2Q_1Q_1 = 000$ to $Q_2Q_1Q_1 = 001$ (see Figure 9.10). The transition is defined by the values of J and K *before* the clock pulse, since the propagation delays of the flip-flops prevent the new output conditions from changing the J and K values until after the transition.

The new conditions of the J and K inputs are:

$$J_2 = K_2 = Q_1 \cdot Q_0 = 0 \cdot 1 = 0 \quad \text{(No change)}$$
$$J_1 = K_1 = Q_0 = 1 \qquad\qquad\quad \text{(Toggle)}$$
$$J_0 = K_0 = 1 \text{ (Constant)} \qquad\quad \text{(Toggle)}$$

The transitions of the outputs generated by the second clock pulse are:

$$Q_2: 0 \rightarrow 0 \quad \text{(No change)}$$
$$Q_1: 0 \rightarrow 1 \quad \text{(Toggle)}$$
$$Q_0: 1 \rightarrow 0 \quad \text{(Toggle)}$$

The new output is $Q_2Q_1Q_0 = 010$, since both Q_0 and Q_1 change and Q_2 stays the same. The J and K conditions are now:

$$J_2 = K_2 = Q_1 \cdot Q_0 = 1 \cdot 0 = 0 \quad \text{(No change)}$$
$$J_1 = K_1 = Q_0 = 0 \qquad\qquad\quad \text{(No change)}$$
$$J_0 = K_0 = 1 \text{ (Constant)} \qquad\quad \text{(Toggle)}$$

The output transitions are:

$$Q_2: 0 \rightarrow 0 \quad \text{(No change)}$$
$$Q_1: 1 \rightarrow 1 \quad \text{(No change)}$$
$$Q_0: 0 \rightarrow 1 \quad \text{(Toggle)}$$

The output is now $Q_2Q_1Q_0 = 011$, which results in the JK conditions:

$$J_2 = K_2 = Q_1 \cdot Q_0 = 1 \cdot 1 = 1 \quad \text{(Toggle)}$$
$$J_1 = K_1 = Q_0 = 1 \qquad\qquad\quad \text{(Toggle)}$$
$$J_0 = K_0 = 1 \text{ (Constant)} \qquad\quad \text{(Toggle)}$$

The above conditions result in output transitions:

$$Q_2: 0 \rightarrow 1 \quad \text{(Toggle)}$$
$$Q_1: 1 \rightarrow 0 \quad \text{(Toggle)}$$
$$Q_0: 1 \rightarrow 0 \quad \text{(Toggle)}$$

All of the outputs toggle and the new output state is $Q_2Q_1Q_0 = 100$. The J and K values repeat this pattern in the second half of the counter cycle (states 100 to 111). Go through the exercise of calculating the J, K, and Q values for the rest of the cycle. Compare the result with the timing diagram in Figure 9.10.

FIGURE 9.10 Timing Diagram for a Synchronous 3-Bit Binary Counter

In the counter we have just analyzed, the combinational circuit generates either a toggle ($JK = 11$) or a no change ($JK = 00$) state at each point through the count sequence. We could use any combination of JK modes (no change, reset, set, or toggle) to make the transitions from one state to the next. For instance, instead of using only the no change and toggle modes, the $000 \rightarrow 001$ transition could also be done by making Q_0 set ($J_0 = 1$, $K_0 = 0$) and Q_1 and Q_2 reset ($J_1 = 0$, $K_1 = 1$ and $J_2 = 0$, $K_2 = 1$). To do so we would need a different set of combinational logic in the circuit.

The simplest synchronous counter design uses only the no change ($JK = 00$) or toggle ($JK = 11$) modes since the J and K inputs of each flip-flop can be connected together. The no change and toggle modes allow us to make any transition (i.e., not just in a linear sequence), even though for truncated sequence and nonbinary counters this is not usually the most efficient design.

There is a simple progression of algebraic expressions for the J and K inputs of a synchronous binary (full-sequence) counter, which uses only the no change and toggle states:

$$J_0 = K_0 = 1$$
$$J_1 = K_1 = Q_0$$
$$J_2 = K_2 = Q_1 \cdot Q_0$$
$$J_3 = K_3 = Q_2 \cdot Q_1 \cdot Q_0$$
$$J_4 = K_4 = Q_3 \cdot Q_2 \cdot Q_1 \cdot Q_0 \text{ and so on}$$

The J and K inputs of each stage are the ANDed outputs of all previous stages. This implies that a flip-flop toggles only when the outputs of *all* previous stages are HIGH. For example, Q_2 doesn't change unless *both* Q_1 AND Q_0 are HIGH (and therefore $J_2 = K_2 = 1$) before the clock pulse. In a 3-bit counter, this occurs only at states 011 and 111, after which Q_2 will toggle, along with Q_1 and Q_0, giving transitions to states 100 and 000 respectively. Look at the timing diagram of Figure 9.10 to confirm this.

Determining the Modulus of a Synchronous Counter

We can use a more formal technique to analyze any synchronous counter, as follows.

1. Determine the equations for the synchronous inputs (JK, D, or T) in terms of the Q outputs for all flip-flops. (For counters other than straight binary full sequence types, the equations will *not* be the same as the algebraic progressions previously listed.)
2. Lay out a table with headings for the Present State of the counter (Q outputs before *CLK* pulse), each Synchronous Input before *CLK* pulse, and Next State of the counter (Q outputs after the clock pulse).
3. Choose a starting point for the count sequence, usually 0, and enter the starting point in the Present State column.
4. Substitute the Q values of the initial present state into the synchronous input equations and enter the results under the appropriate columns.
5. Determine the action of each flip-flop on the next *CLK* pulse (e.g., for a JK flip-flop, the output either will not change ($JK = 00$), or will reset ($JK = 01$), set ($JK = 10$), or toggle ($JK = 11$)).
6. Look at the Q values for every flip-flop. Change them according to the function determined in Step 5 and enter them in the column for the counter's next state.
7. Enter the result from Step 6 on the next line of the column for the counter's present state (i.e., this line's next state is the next line's present state).
8. Repeat this process until the result in the next state column is the same as the initial state.

EXAMPLE 9.3

Find the count sequence of the synchronous counter shown in Figure 9.11 and, from the count sequence table, draw the timing diagram and state diagram. What is the modulus of the counter?

FIGURE 9.11 Example 9.3: Synchronous Counter of Unknown Modulus

■ **Solution**

J and *K* equations are:

$$J_2 = Q_1 \cdot Q_0 \qquad J_1 = Q_0 \qquad J_0 = \bar{Q}_2$$
$$K_2 = 1 \qquad K_1 = Q_0 \qquad K_0 = 1$$

The output transitions can be determined from the values of the *J* and *K* functions before each clock pulse, as shown in Table 9.5.

TABLE 9.5 State Table for Figure 9.11

Present State			Synchronous Inputs									Next State		
Q_2	Q_1	Q_0	J_2	K_2		J_1	K_1		J_0	K_0		Q_2	Q_1	Q_0
0	0	0	0	1	(R)	0	0	(NC)	1	1	(T)	0	0	1
0	0	1	0	1	(R)	1	1	(T)	1	1	(T)	0	1	0
0	1	0	0	1	(R)	0	0	(NC)	1	1	(T)	0	1	1
0	1	1	1	1	(T)	1	1	(T)	1	1	(T)	1	0	0
1	0	0	0	1	(R)	0	0	(NC)	0	1	(R)	0	0	0

Since there are five unique output states, the counter's modulus is 5.

The timing diagram and state diagram are shown in Figure 9.12. Since this circuit produces one pulse on Q_2 for every 5 clock pulses, we can use it as a divide-by-5 circuit.

a. Timing diagram   **b. State diagram**

FIGURE 9.12   Example 9.3: Timing Diagram and State Diagram of a Mod-5 Counter

The analysis in Example 9.3 did not account for the fact that the counter uses only 5 of a possible 8 output states. In any truncated sequence counter, it is good practice to determine the next state for each unused state to ensure that if the counter powers up in one of these unused states, it will eventually enter the main sequence.

An unused state might be problematic because a flip-flop is not necessarily guaranteed to power up in a particular state. The worst possible situation would be one such as where the counter flip-flops would power up in state 110, then make a transition to 111, then back to 110. The counter would be stuck between these two states and never enter the main count sequence. Therefore, we should check that the counter's combinational logic never lets it get stuck in one or more unused states.

EXAMPLE 9.4

Extend the analysis of the counter in Example 9.3 to include its unused states. Redraw the counter's state diagram to show how these unused states enter the main sequence (if they do).

■ **Solution**

The synchronous input equations are:

$$J_2 = Q_1 \cdot Q_0 \qquad J_1 = Q_0 \qquad J_0 = \bar{Q}_2$$
$$K_2 = 1 \qquad K_1 = Q_0 \qquad K_0 = 1$$

The unused states are $Q_2Q_1Q_0 = 101$, 110, and 111. Table 9.6 shows the transitions made by the unused states. Figure 9.13 shows the completed state diagram.

TABLE 9.6   State Table for Mod-5 Counter Including Unused States

Present State			Synchronous Inputs									Next State		
Q_2	Q_1	Q_0	J_2	K_2		J_1	K_1		J_0	K_0		Q_2	Q_1	Q_0
0	0	0	0	1	(R)	0	0	(NC)	1	1	(T)	0	0	1
0	0	1	0	1	(R)	1	1	(T)	1	1	(T)	0	1	0
0	1	0	0	1	(R)	0	0	(NC)	1	1	(T)	0	1	1
0	1	1	1	1	(T)	1	1	(T)	1	1	(T)	1	0	0
1	0	0	0	1	(R)	0	0	(NC)	0	1	(R)	0	0	0
1	0	1	0	1	(R)	1	1	(T)	0	1	(R)	0	1	0
1	1	0	0	1	(R)	0	0	(NC)	0	1	(R)	0	1	0
1	1	1	1	1	(T)	1	1	(T)	0	1	(R)	0	0	0

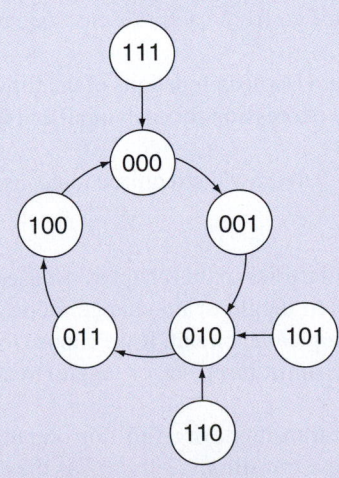

FIGURE 9.13   Example 9.4: Complete State Diagram

9.2   A 4-bit synchronous counter based on JK flip-flops is described by the following set of equations:

$$J_3 = Q_2Q_1Q_0 \qquad J_2 = Q_1Q_0 \qquad J_1 = \overline{Q}_3Q_0 \qquad J_0 = 1$$
$$K_3 = Q_0 \qquad K_2 = Q_1Q_0 \qquad K_1 = Q_0 \qquad K_0 = 1$$

Assume that the counter output is at 1000 in the count sequence. What will the output be after one clock pulse? After two clock pulses?

9.3 DESIGN OF SYNCHRONOUS COUNTERS

■ **KEY TERMS**

State Machine   A synchronous sequential circuit.

Excitation Table   A table showing the required input conditions for every possible transition of a flip-flop output.

A synchronous counter can be designed using established techniques that involve the derivation of Boolean equations for the counter's next state logic. Alternatively, several VHDL structures can be used to define counters; we can use a behavioral description of the counter, or we can use a **state machine** definition in VHDL that specifies each present and next state explicitly.

In addition to the classical counter design techniques, we will examine the design of a counter through a behavioral description in VHDL. We will leave the state machine design for the following chapter.

Classical Design Technique

There are several steps involved in the classical design of a synchronous counter.

1. Define the problem. Before you can begin design of a circuit, you have to know what its purpose is and what it should do under all possible conditions.
2. Draw a state diagram showing the progression of states under various input conditions and what outputs the circuit should produce, if any.
3. Make a state table which lists all possible Present States and the Next State for each one. *List the present states in **binary order**.*
4. Use flip-flop **excitation tables** to determine at what states the flip-flop synchronous inputs must be to make the circuit go from each Present State to its Next State.
5. The logic levels of the synchronous inputs are Boolean functions of the flip-flop outputs and the control inputs. Simplify the expression for each input and write the simplified Boolean expression.
6. Use the Boolean expressions found in step 5 to draw the required logic circuit.

Flip-Flop Excitation Tables

In the synchronous counter circuits we examined earlier in this chapter, we used JK flip-flops that were configured to operate only in toggle or no change mode. We can use any type of flip-flop for a synchronous sequential circuit. If we choose to use JK flip-flops, we can use any of the modes (no change, reset, set, or toggle) to make transitions from one state to another.

A flip-flop excitation table shows all possible transitions of a flip-flop output and the synchronous input levels needed to effect these transitions. Table 9.7 is the excitation table of a JK flip-flop.

If we want a flip-flop to make a transition from 0 to 1, we can use either the toggle function ($JK = 11$) or the set function ($JK = 10$). It doesn't matter what K is, as

TABLE 9.7 JK Flip-Flop Excitation Table

Transition	Function	JK	
$0 \rightarrow 0$	No change or reset	00 01	0X
$0 \rightarrow 1$	Toggle or set	11 10	1X
$1 \rightarrow 0$	Toggle or reset	11 01	X1
$1 \rightarrow 1$	No change or set	00 10	X0

TABLE 9.8 Condensed Excitation Table for a JK Flip-Flop

Transition	JK
$0 \rightarrow 0$	0X
$0 \rightarrow 1$	1X
$1 \rightarrow 0$	X1
$1 \rightarrow 1$	X0

long as $J = 1$. This is reflected by the variable pair ($JK = 1X$) beside the $0 \rightarrow 1$ entry in Table 9.7. The X is a don't care state, a 0 or 1 depending on which is more convenient for the simplification of the Boolean function of the J or K input affected.

Table 9.8 shows a condensed version of the JK flip-flop excitation table.

Design of a Synchronous Mod-12 Counter

We will follow the procedure previously outlined to design a synchronous mod-12 counter circuit, using JK flip-flops. The aim is to derive the Boolean equations of all J and K inputs and to draw the counter circuit.

1. *Define the problem.* The circuit must count in binary sequence from 0000 to 1011 and repeat. The output progresses by 1 for each applied clock pulse. Since the outputs are 4-bit numbers, we require 4 flip-flops.
2. *Draw a state diagram.* The state diagram for this problem is shown in Figure 9.14.

FIGURE 9.14 State Diagram for a Mod-12 Counter

3. *Make a state table showing each present state and the corresponding next state.*
4. *Use flip-flop excitation tables to fill in the J and K entries in the state table.* Table 9.9 shows the combined result of steps 3 and 4. Note that all present states are in binary order.

TABLE 9.9 State Table for a Mod-12 Counter

Present State				Next State				Synchronous Inputs							
Q_3	Q_2	Q_1	Q_0	Q_3	Q_2	Q_1	Q_0	J_3	K_3	J_2	K_2	J_1	K_1	J_0	K_0
0	0	0	0	0	0	0	1	0	X	0	X	0	X	1	X
0	0	0	1	0	0	1	0	0	X	0	X	1	X	X	1
0	0	1	0	0	0	1	1	0	X	0	X	X	0	1	X
0	0	1	1	0	1	0	0	0	X	1	X	X	1	X	1
0	1	0	0	0	1	0	1	0	X	X	0	0	X	1	X
0	1	0	1	0	1	1	0	0	X	X	0	1	X	X	1
0	1	1	0	0	1	1	1	0	X	X	0	X	0	1	X
0	1	1	1	1	0	0	0	1	X	X	1	X	1	X	1
1	0	0	0	1	0	0	1	X	0	0	X	0	X	1	X
1	0	0	1	1	0	1	0	X	0	0	X	1	X	X	1
1	0	1	0	1	0	1	1	X	0	0	X	X	0	1	X
1	0	1	1	0	0	0	0	X	1	0	X	X	1	X	1
1	1	0	0	X	X	X	X	X	X	X	X	X	X	X	X
1	1	0	1	X	X	X	X	X	X	X	X	X	X	X	X
1	1	1	0	X	X	X	X	X	X	X	X	X	X	X	X
1	1	1	1	X	X	X	X	X	X	X	X	X	X	X	X

We assume for now that states 1100 to 1111 never occur. If we assign their corresponding next states to be don't care states, they can be used to simplify the *J* and *K* expressions we derive from the state table.

Let us examine one transition to show how the table is completed. The transition from $Q_3Q_2Q_1Q_0 = 0101$ to $Q_3Q_2Q_1Q_0 = 0110$ consists of the following individual flip-flop transitions.

$Q_3: 0 \rightarrow 0$	(No change or reset;	$J_3K_3 = 0X$)
$Q_2: 1 \rightarrow 1$	(No change or set;	$J_2K_2 = X0$)
$Q_1: 0 \rightarrow 1$	(Toggle or set;	$J_1K_1 = 1X$)
$Q_0: 1 \rightarrow 0$	(Toggle or reset;	$J_0K_0 = X1$)

The other lines of the table are similarly completed.

5. *Simplify the Boolean expression for each input.* Table 9.9 can be treated as eight truth tables, one for each *J* or *K* input. We can simplify each function by Boolean algebra or by using a Karnaugh map.

Figure 9.15 shows K-map simplification for all eight synchronous inputs. These maps yield the following simplified Boolean expressions.

$$J_0 = 1$$
$$K_0 = 1$$
$$J_1 = Q_0$$
$$K_1 = Q_0$$
$$J_2 = \bar{Q}_3Q_1Q_0$$
$$K_2 = Q_1Q_0$$
$$J_3 = Q_2Q_1Q_0$$
$$K_3 = Q_1Q_0$$

6. *Draw the required logic circuit.* Figure 9.16 shows the circuit corresponding to the above Boolean expressions.

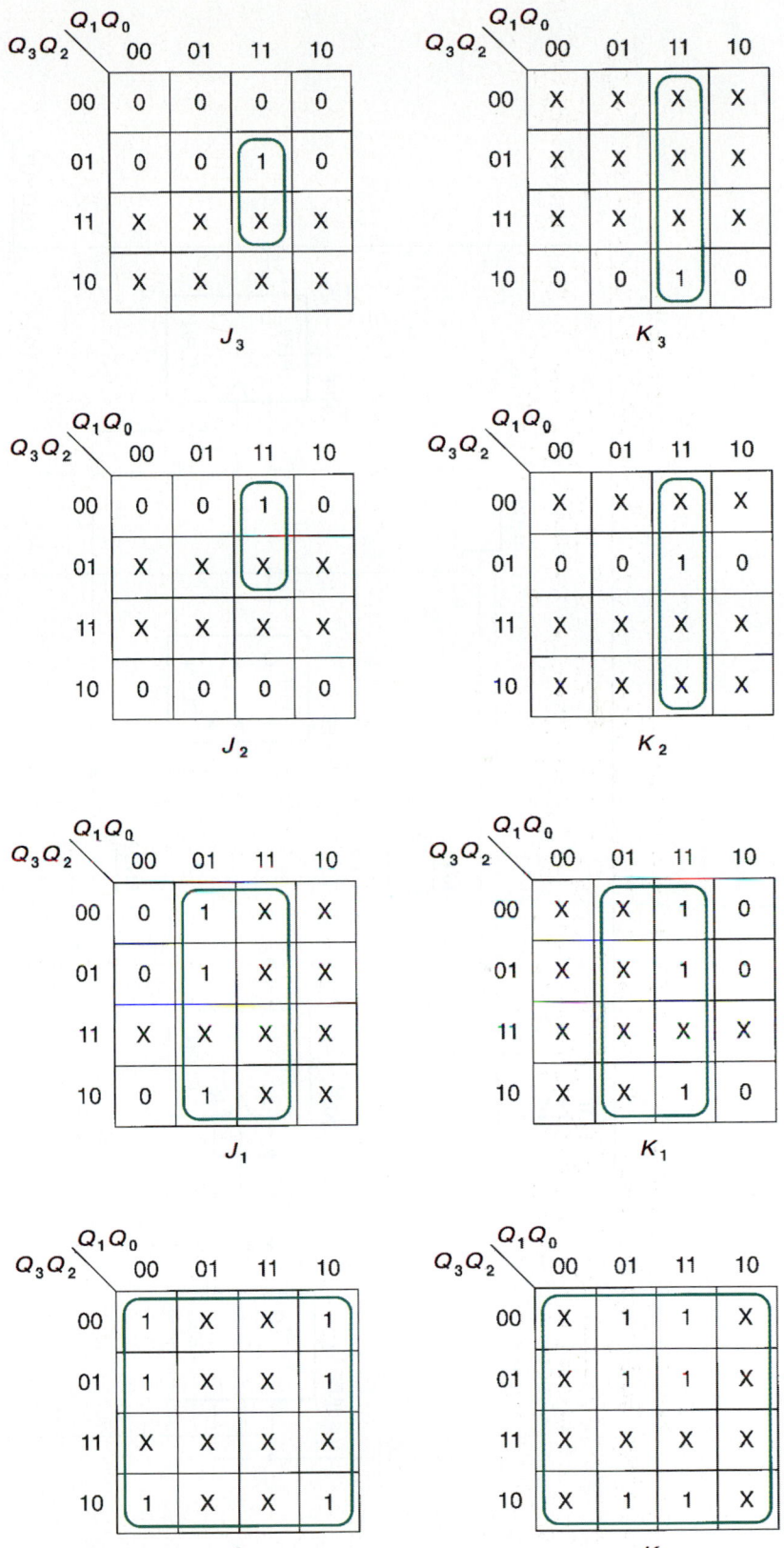

FIGURE 9.15 K-Map Simplification of Table 9.9

FIGURE 9.16 Synchronous Mod-12 Counter

We have assumed that states 1100 to 1111 will never occur in the operation of the mod-12 counter. This is normally the case, but when the circuit is powered up, there is no guarantee that the flip-flops will be in any particular state.

If a counter powers up in an unused state, the circuit should enter the main sequence after one or more clock pulses. To test whether or not this happens, let us make a state table, applying each unused state to the J and K equations as implemented, to see what the Next State is for each case. This analysis is shown in Table 9.10.

TABLE 9.10   Unused States in a Mod-12 Counter

Present State				Synchronous Inputs								Next State			
Q_3	Q_2	Q_1	Q_0	J_3	K_3	J_2	K_2	J_1	K_1	J_0	K_0	Q_3	Q_2	Q_1	Q_0
1	1	0	0	0	0	0	0	0	0	1	1	1	1	0	1
1	1	0	1	0	0	0	0	1	1	1	1	1	1	1	0
1	1	1	0	0	0	0	0	0	0	1	1	1	1	1	1
1	1	1	1	1	1	0	1	1	1	1	1	0	0	0	0

Figure 9.17 shows the complete state diagram for the designed mod-12 counter. If the counter powers up in an unused state, it will enter the main sequence in no more than four clock pulses.

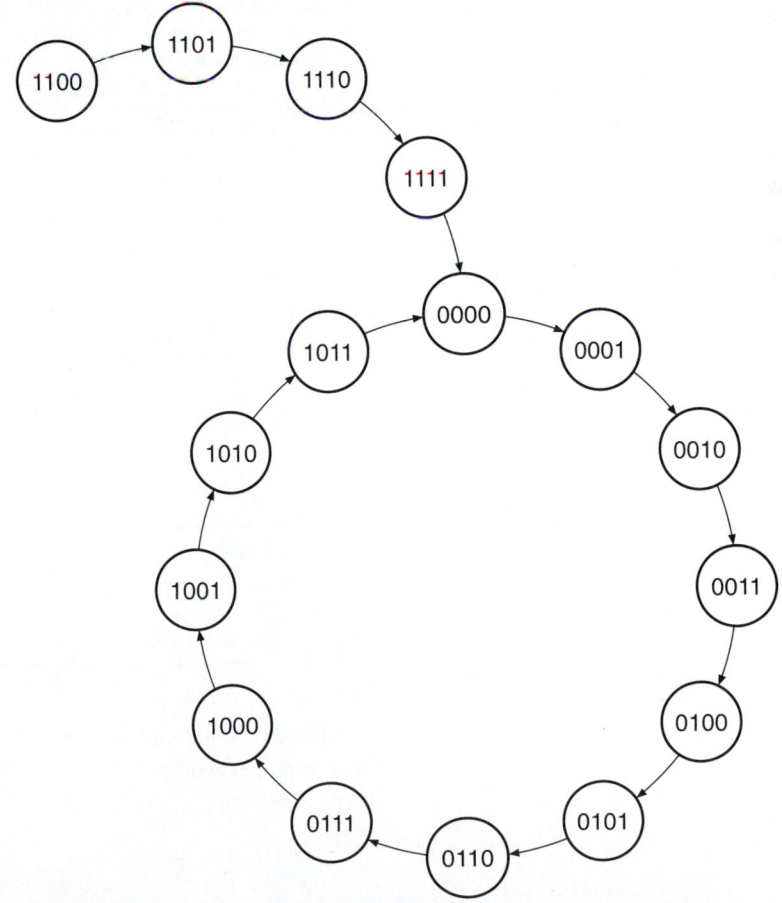

FIGURE 9.17   Complete State Diagram of Mod-12 Counter in Figure 9.16

If we want an unused state to make a transition directly to 0000 in one clock pulse, we have a couple of options:

1. We could reset the counter asynchronously and otherwise leave the design as is.
2. We could rewrite the state table to specify these transitions, rather than make the unused states don't cares.

Option 1 is the simplest and is considered perfectly acceptable as a design practice. Option 2 would yield a more complicated set of Boolean equations and hence a more complex circuit, but might be worthwhile if a direct synchronous transition to 0000 were required.

EXAMPLE 9.5

Derive the synchronous input equations of a 4-bit synchronous binary counter based on D flip-flops. Draw the corresponding counter circuit.

■ Solution

The first step in the counter design is to derive the excitation table of a D flip-flop. Recall that Q follows D when the flip-flop is clocked. Therefore the next state of Q is the same as the input D for any transition. This is illustrated in Table 9.11.

Next, we must construct a state table, shown in Table 9.12, with present and next states for all possible transitions. Note that the binary value of $D_3D_2D_1D_0$ is the same as the next state of the counter.

TABLE 9.11 Excitation Table of a D Flip-Flop

Transition	D
$0 \rightarrow 0$	0
$0 \rightarrow 1$	1
$1 \rightarrow 0$	0
$1 \rightarrow 1$	1

TABLE 9.12 State Table for a 4-Bit Binary Counter

Present State				Next State				Synchronous Inputs			
Q_3	Q_2	Q_1	Q_0	Q_3	Q_2	Q_1	Q_0	D_3	D_2	D_1	D_0
0	0	0	0	0	0	0	1	0	0	0	1
0	0	0	1	0	0	1	0	0	0	1	0
0	0	1	0	0	0	1	1	0	0	1	1
0	0	1	1	0	1	0	0	0	1	0	0
0	1	0	0	0	1	0	1	0	1	0	1
0	1	0	1	0	1	1	0	0	1	1	0
0	1	1	0	0	1	1	1	0	1	1	1
0	1	1	1	1	0	0	0	1	0	0	0
1	0	0	0	1	0	0	1	1	0	0	1
1	0	0	1	1	0	1	0	1	0	1	0
1	0	1	0	1	0	1	1	1	0	1	1
1	0	1	1	1	1	0	0	1	1	0	0
1	1	0	0	1	1	0	1	1	1	0	1
1	1	0	1	1	1	1	0	1	1	1	0
1	1	1	0	1	1	1	1	1	1	1	1
1	1	1	1	0	0	0	0	0	0	0	0

This state table yields four Boolean equations, for D_3 through D_0, in terms of the present state outputs. Figure 9.18 shows four Karnaugh maps used to simplify these functions.

FIGURE 9.18 Example 9.5: K-Maps for a 4-Bit Counter Based on D Flip-Flops

The simplified equations are:

$$D_3 = \bar{Q}_3 Q_2 Q_1 Q_0 + Q_3 \bar{Q}_2 + Q_3 \bar{Q}_1 + Q_3 \bar{Q}_0$$
$$D_2 = \bar{Q}_2 Q_1 Q_0 + Q_2 \bar{Q}_1 + Q_1 \bar{Q}_0$$
$$D_1 = \bar{Q}_1 Q_0 + Q_1 \bar{Q}_0$$
$$D_0 = \bar{Q}_0$$

These equations represent the maximum SOP simplifications of the input functions. However, we can rewrite them to make them more compact. For example the equation for D_3 can be rewritten using DeMorgan's theorem ($\bar{x} + \bar{y} + \bar{z} = \overline{xyz}$) and our knowledge of Exclusive OR (XOR) functions ($\bar{x}y + x\bar{y} = x \oplus y$).

$$D_3 = \bar{Q}_3 Q_2 Q_1 Q_0 + Q_3 \bar{Q}_2 + Q_3 \bar{Q}_1 + Q_3 \bar{Q}_0$$
$$= \bar{Q}_3 (Q_2 Q_1 Q_0) + Q_3 (\bar{Q}_2 + \bar{Q}_1 + \bar{Q}_0)$$
$$= \bar{Q}_3 (Q_2 Q_1 Q_0) + Q_3 (\overline{Q_2 Q_1 Q_0})$$
$$= Q_3 \oplus Q_2 Q_1 Q_0$$

We can write similar equations for the other D inputs as follows:

$$D_2 = Q_2 \oplus Q_1 Q_0$$
$$D_1 = Q_1 \oplus Q_0$$
$$D_0 = Q_0 \oplus 1$$

These equations follow a predictable pattern of expansion. Each equation for an input D_n is simply Q_n XORed with the logical product (AND) of all previous Qs.

Figure 9.19 shows the circuit for the 4-bit counter, including an asynchronous reset.

(continued)

FIGURE 9.19   Example 9.5: 4-bit Counter Using D Flip-Flops

In Section 8.6 (Edge-Triggered T Flip-Flops) of Chapter 8, we saw how a D flip-flop could be configured for a switchable toggle function (refer to Figure 8.75). The flip-flops in Figure 9.19 are similarly configured. Each flip-flop output, except Q_0, is fed back to its input through an Exclusive OR gate. The other input to the XOR controls whether this feedback is inverted (for toggle mode) or not (for no change mode). Recall that $x \oplus 0 = x$ and $x \oplus 1 = \bar{x}$.

For example, Q_3 is fed back to D_3 through an XOR gate. The feedback is inverted only if the 3-input AND gate has a HIGH output. Thus, the Q_3 output toggles only if all previous bits are HIGH ($Q_3Q_2Q_1Q_0 = 0111$ or 1111). The flip-flop toggle mode is therefore controlled by the states of the XOR and AND gates in the circuit.

■ **SECTION 9.3 REVIEW PROBLEM**

9.3   A 4-bit synchronous counter must make a transition from state $Q_3Q_2Q_1Q_0 = 1011$ to $Q_3Q_2Q_1Q_0 = 1100$. Write the required states of the synchronous inputs for a set of four JK flip-flops used to implement the counter. Write the required states of the synchronous inputs if the counter is made from D flip-flops.

9.4 PROGRAMMING BINARY COUNTERS FOR CPLDs

A binary counter can be designed for a CPLD by using a VHDL description or by entering a schematic in the Quartus II Block Editor. We could, in the latter case, draw the counter as a series of flip-flops and their associated logic. This, however, becomes unwieldy as the counter increases in size. Since predefined components, such as those in the Altera Library of Parameterized Modules, are available to us, it makes more sense to use them instead.

VHDL Counters

> ### ■ KEY TERMS
>
> **IF Statement**   A VHDL statement that is executed only if a specified Boolean condition is satisfied.
>
> **Attribute**   A property associated with a named identifier in VHDL. (For example, the attribute **EVENT,** when associated with the identifier **clk** (written **clk'EVENT**), indicates, when true, that a transition has occurred on the input called **clk.**)

When using VHDL to create a counter, we can take several approaches. We can encode the Boolean equations of the counter directly with concurrent signal assignment statements; we can use VHDL code to describe the behavior of the counter; we can use a CASE statement to implement the state diagram of the counter; or we can use a predefined counter, such as those found in the Altera Library of Parameterized Modules (LPM) and map its ports to the ports of a VHDL design entity.

If we chose to use concurrent signal assignments to encode the Boolean equations of a counter, we could derive the following equations for a 4-bit counter with D flip-flops.

➡ **LAB SOURCE**
Lab #22
Binary Counters

```
d(3)  <=  q(3)xor(q(2)and q(1)and q(0));,
d(2)  <=  q(2)xor(q(1)and q(0));
d(1)  <=  q(1)xor q(1);,
d(0)  <=  not q(0);,
```

In Chapter 6, we saw that using concurrent signal assignment statements is an inefficient way to code many digital functions. (For one thing, if we use this procedure, we must know what the equations are. Getting to that point requires a lot of work that can be done by the VHDL compiler.) Though acknowledging this as a possible option, we will not examine this method any further.

In this section, we will design counters using behavioral descriptions and using LPM counters. The design of a counter as a state machine will be examined in the next chapter.

Behavioral Description of Counters

The following VHDL code shows the behavioral description of a simple 8-bit counter **simple_int_counter.vhd** with asynchronous clear.

Quartus II:
simple_int_counter.vhd
MAX+PLUS II:
ct_simp.vhd

```
-- simple_int_counter.vhd
-- 8-bit synchronous counter with asynchronous clear.
-- Uses INTEGER type for counter output.
LIBRARY ieee;
USE ieee.std_logic_1164.ALL;
```

```
ENTITY simple_int_counter IS
   PORT(
        clock : IN  STD_LOGIC;
        reset : IN  STD_LOGIC;
        q     : OUT INTEGER RANGE 0 TO 255);
END simple_int_counter;

ARCHITECTURE counter OF simple_int_counter IS
   BEGIN
   PROCESS (clock, reset)
      VARIABLE count : INTEGER RANGE 0 TO 255;
   BEGIN
      IF (reset = '0') THEN
         count := 0;
      ELSE
         IF (clock'EVENT AND clock = '1') THEN
            count := count + 1;
         END IF;
      END IF;
      q <= count;
   END PROCESS;
END counter;
```

This VHDL code relies on three main constructs to perform the count function:

■ a PROCESS statement to monitor the two inputs, **clock** and **reset,** which control the state of the counter.
■ a variable, called **count,** to hold the present value of the counter.
■ an IF statement to evaluate the **clock** and **reset** inputs to determine whether the counter should increment or clear.

A PROCESS statement has the following syntax:

```
PROCESS (sensitivity list)
   [VARIABLE variable_name :type [range]; ]
BEGIN
   Process statements
END PROCESS;
```

Square brackets [ ] indicate an optional part of the code.

The purpose of the PROCESS statement is to determine when the counter should change state. When there is a change in an item in the sensitivity list (i.e., a change in state of **clock** or **reset**), the statements within the PROCESS are executed. For a synchronous counter, the list must include **clock,** since any action in a synchronous circuit depends on a clock transition. The reset function in this counter is asynchronous, and therefore independent of the clock, so the **reset** input must also be monitored for any changes.

To hold the accumulating output value of the counter, we define a variable called **count.** A variable is similar to a signal in that it is entirely internal to a VHDL design entity. Unlike a signal, which acts like a virtual wire, a variable is simply a segment of working memory used to hold a temporary value. A variable is always local to a process. That is, its value is defined only inside the process in which it is used. In order for a variable to have any effect external to the design, it must be mapped to a port in the design entity.

The variable **count** is presumed to have an initial value of 0, but is defined for the range of 0 to 255. This 8-bit value rolls over to 0 when the count exceeds 255. We update the value of **count** by an **IF statement,** with the form:

```
IF (condition) THEN
    Statement[s];
[ELSIF (condition) THEN
    statement[s];]
[ELSE
    statement[s];]
END IF;
```

An IF statement, which must always be contained within a PROCESS, executes one of several alternative sets of statements, based on a series of Boolean conditions. The IF statement examines the Boolean condition with the highest priority. If the condition is true, one or more statements are executed. If the condition is false, the IF statement uses the ELSIF clause to test the condition with the next-highest priority. If true, a different set of statements is executed. There may or may not be more ELSIF clauses, depending on the required conditions. When all conditions have been tested in order of priority and found false, the statements under the ELSE clause are executed by default.

It is important to note that only one of the alternative sets of statements under IF, ELSIF, or ELSE will be executed for each time the IF statement is executed. Also note that only the IF clause is required for correct syntax; depending on the conditions being tested, ELSIF and ELSE may not be used.

The clause (IF (reset='0') THEN) monitors the asynchronous reset function independently of the clock and executes the variable assignment that sets the output to 0 if the Boolean condition (reset='0') is true. Otherwise, the clock is monitored for a positive edge by the condition (clock'EVENT AND clock = '1'). EVENT is a predefined **attribute** that indicates if there has been activity on the signal of which it is an attribute. The clause clock'EVENT (pronounced "clock tick event") is true if there has just been a change on clock. The combination of this and the condition clock = '1' indicates that a positive edge has just occurred. If this is true, then **count** is incremented.

Since **reset** is tested first, it has the highest priority, which is as it should be. A reset should always have priority over a count function, as it is used to stop any action and bring the counter output to a known initial state.

As a final step, the accumulated count must be assigned to an output port. This is done in the concurrent signal assignment q <= count at the end of the process.

Note the difference in types of assignments. A *variable* is assigned by the := operator (e.g., count := count + 1;). A *signal* is assigned by the <= operator (e.g., q <= count).

We can also program a counter using only STD_LOGIC types. An example is as follows:

simple_std_counter.vhd

```
-- simple_std_counter.vhd
-- 8-bit synchronous counter with asynchronous clear.
-- Uses STD_LOGIC_VECTOR type for counter output.
LIBRARY ieee;
USE ieee.std_logic_1164.ALL;
USE ieee.std_logic_unsigned.ALL;

ENTITY simple_std_counter IS
    PORT(
        clock : IN     STD_LOGIC;
        reset : IN     STD_LOGIC;
        q     : BUFFER STD_LOGIC_VECTOR(7 downto 0));
END simple_std_counter;
```

```
ARCHITECTURE counter OF simple_std_counter IS
BEGIN
    PROCESS (clock, reset)
    BEGIN
        IF (reset = '0') THEN
            q   <= (others => '0');
        ELSE
            IF (clock'EVENT AND clock = '1') THEN
                    q <= q + 1;
            END IF;
        END IF;
    END PROCESS;
END counter;
```

In this case, we define the width of the counter, rather than its upper count range. To use this format, we must make the following changes to the original file:

- Output **q** is defined as type STD_LOGIC_VECTOR. The width of the vector can be defined explicitly or with a parameter in a GENERIC clause.
- Eliminate the variable, **count.** Operate on the port, **q,** directly.
- The statement q <= q + 1; indicates that **q** acts as both an output (left side of the assignment) and input (right side). To make this possible, **q** must be of mode BUFFER, rather than of mode OUT.
- To update the value of **q,** we add 1 to its present value, which means that we must be able to treat a port of type STD_LOGIC_VECTOR as an unsigned integer. You can't add anything to a STD_LOGIC_VECTOR, which is only a string of bits in a particular sequence. To allow it to be treated as an integer, you require the routines contained in the **std_logic_unsigned** package, which is part of the **ieee** library.
- For the reset function, we could write q <= "00000000"; for an 8-bit counter. It is more convenient to use the notation q <= (others => '0'); This statement can be applied to a counter of any width. It can be loosely translated as, "set each bit of **q** to '0', regardless of how many bits that actually is."

LPM Counters in VHDL

We can use a component **(lpm_counter)** from the Library of Parameterized Modules (LPM) to instantiate a counter in VHDL. When using an LPM counter, we don't need to describe the behavior of the counter, as this has been done for us in the module itself. All we need to do is map the ports and parameters of the LPM component to the ports of the VHDL design entity. We do this by using a generic map to specify the parameters we need and a port map to map the ports of the LPM device either to an external port or an internal signal. The code below shows the VHDL implementation of the same 8-bit counter as in the previous behavioral examples.

Quartus II:
simple_lpm_counter.vhd
MAX+PLUS II:
lpm_simp.vhd

```
-- simple_lpm_counter.vhd
-- Eight-bit binary counter based on a component
--   from the Library of Parameterized Modules (LPM)
-- Counter has an active-LOW asynchronous clear,

LIBRARY ieee;
USE ieee.std_logic_1164.ALL;
LIBRARY lpm;
USE lpm.lpm_components.ALL;

ENTITY simple_lpm_counter IS
    PORT(
        clk, clear : IN  STD_LOGIC;
        q_out      : OUT STD_LOGIC_VECTOR (7 downto 0));
END simple_lpm_counter;
```

```
ARCHITECTURE count OF simple_lpm_counter IS
    SIGNAL clrn : STD_LOGIC;────── Internal signal for active-LOW clear
BEGIN
-- Instantiate 8-bit counter
    count: lpm_counter
        GENERIC MAP (LPM_WIDTH => 8)
        PORT MAP ( clock       => clk,           Internal clear mapped
                   aclr        => clrn,          to asynch. clear port
                                                 of counter
                   q           => q_out(7 downto 0));
    clrn  <= not clear;──────  Input port inverted and mapped to
END count;                     internal clear signal
```

LPM components require us to use two packages: the **std_logic_1164** package in the **ieee** library to define STD_LOGIC types used in the LPM components and the **lpm_components** package in the **lpm** library to define the components themselves. Since LPM components are defined using STD_LOGIC and STD_LOGIC_VECTOR types, we should use these types for our other identifiers as well.

The entity declaration defines the inputs and outputs of our counter and need not correspond to the port names for the LPM counter. That correspondence is defined in the architecture body, where we instantiate the counter module. The counter is defined in a component instantiation statement, which takes the following form:

```
__instance_name: __component_name
    GENERIC MAP (__parameter_name => __parameter_value,
                 __parameter_name => __parameter_value)
    PORT MAP (__component_port => __connect_port,
              __component_port => __connect_port);
```

The component name is the name of the LPM component. Parameter names are those defined in the LPM component, such as LPM_WIDTH. Parameter values are those values assigned in the instance of the component. Component ports are the LPM port names. Connect ports are the names of identifiers declared in the entity or as ports, signals, or variables.

If we want to invert the active level of an LPM input port, we must use a signal assignment statement (e.g., clrn <= not clear;). We need to do this because a VHDL input port cannot be "updated" (modified); only an output can be assigned a new value as a result of a Boolean expression. Thus, we create a signal called **clrn** that maps to the **aclr** (asynchronous clear) port of the LPM counter. This is connected to the **clear** input of the counter circuit via an inverter. Figure 9.20 shows the graphic equivalent of this mapping.

Entering Simple LPM Counters with the Quartus II Block Editor

We can enter an LPM counter in the Quartus II Block Editor, either by using the MegaWizard Plug-In Manager or by manually entering the component. We will use each method to enter an 8-bit binary counter with active-LOW asynchronous clear.

MegaWizard Plug-In Manager

To use the MegaWizard Plug-In Manager, first create and save a new Block Diagram File, and use it to make a new project. Enter the component **lpm_counter,** as shown in **Symbol** dialog box, shown in Figure 9.21. Make sure the box labeled **Launch MegaWizard Plug-In** is *checked.* Click **OK.**

Figures 9.22 through 9.26 show the set-up screens used by the MegaWizard Plug-In Manager to create the LPM counter with the required ports and parameters. On the screen shown in Figure 9.22, choose **VHDL** and click **Next.**

FIGURE 9.20   Graphic Equivalent of an LPM Counter with Active-LOW Clear

FIGURE 9.21   Symbol Dialog (LPM_COUNTER)

- Synchronous or asynchronous **parallel load,** which allows the count to be set to any value whenever a *LOAD* input is asserted
- Synchronous or asynchronous **clear** (reset), which sets all of the counter outputs to zero
- **Count enable,** which allows the count sequence to progress when asserted and inhibits the count when deasserted
- **Bidirectional** control, which determines whether the counter counts up or down
- **Output decoding,** which activates one or more outputs when detecting particular states on the counter outputs
- **Ripple carry out or ripple clock out (RCO),** a special case of output decoding that produces a pulse upon detecting the **terminal count,** or last state, of a count sequence.

We will examine the implementation of these functions, first as Block Diagram Files in Quartus II, and then, in the next section, in VHDL, both as behavioral descriptions and as functions of LPM counters.

Parallel Loading

Figure 9.33 shows the symbol of a 4-bit **presettable counter** (i.e., a counter with a parallel load function). The parallel inputs, P_3 to P_0, have direct access to the flip-flops of the counter. When the *LOAD* input is asserted, the values at the P inputs are loaded directly into the counter and appear at the Q outputs.

FIGURE 9.33 4-Bit Counter with Parallel Load

> **NOTE . . .**
>
> Parallel loading requires at least two sets of inputs: the load *data* (P_3 to P_0) and the load *command (LOAD).* If the load function is synchronous, it also requires a clock input.

Parallel loading can be synchronous or asynchronous, as shown by the two counters in Figure 9.34. The Quartus II simulation in Figure 9.35 shows the difference. Two waveforms, QS[3..0] and QA[3..0], represent the outputs of two 4-bit counters with synchronous and asynchronous load, respectively. Both counters have the same clock, load, and P inputs. The count is already in progress at the beginning of the simulation window and shows both counters advancing with each clock pulse.

When *LOAD* goes HIGH, the value of P[3..0] (= AH) is loaded into the asynchronously loading counter (**qa**) immediately after a short propagation delay. The counter with synchronous load (**qs**) is not loaded until the next positive clock edge.

Synchronous Load

The logic diagram of Figure 9.36 shows the concept of synchronous parallel load. Depending on the status of the *LOAD* input, the flip-flop will either count according to its count logic (the next-state combinational circuit) or load an external value. The flip-flop shown is the most significant bit of a 4-bit binary counter, such as shown in Figure 9.19, but with the count logic represented only by an input pin. (For the fourth bit of a counter, the Boolean equation of the count logic is given by $D_3 = Q_3 \oplus Q_2 Q_1 Q_0$. It is left out in order to more clearly show the operation of the count/load function select circuit, which is essentially a 2-to-1 multiplexer.)

The *LOAD* input selects whether the flip-flop synchronous input will be fed by the count logic or by the parallel input P_3. When *LOAD* = 0, the upper AND gate steers the count logic to the flip-flop, and the count progresses with each clock pulse. When *LOAD* = 1, the lower AND gate loads the logic level at P_3 directly into the flip-flop on the next clock pulse.

FIGURE 9.34 Two 4-Bit Counters (Synchronous and Asynchronous Load)

FIGURE 9.35 Simulation of Two 4-Bit Counters

FIGURE 9.36 Count/Load Selection

FIGURE 9.37 Counter Element with Synchronous Load and Asynchronous Clear

FIGURE 9.38 Counter Element with Synchronous Load and Asynchronous Reset (sl_count)

Quartus II:
sl_count.bdf
4bit_sl.bdf
4bit_sl.vwf
MAX+PLUS II:
sl_count.gdf
4bit_sl.gdf
4bit_sl.scf

Figure 9.37 shows the same circuit, but includes the count logic. If we leave out the 3-input AND gate, as in Figure 9.38, we have a circuit that can be used as a general element (called **sl_count**) in a synchronous presettable counter. Figure 9.39 shows the logic diagram of a 4-bit synchronously presettable counter consisting of four instances of the counter element of Figure 9.38 and appropriate AND gates for a synchronous counter. This diagram implements a synchronous counter like that of Figure 9.19, but also incorporates a synchronous load function.

The counter in Figure 9.39 can be tested using the following criteria.

SIMULATION CRITERIA

- The counter should be clocked 16 times to observe a full count cycle from 0H to FH.
- Test the load function by assigning a constant value to the *P* inputs and making *LOAD* HIGH. Loading should happen on the first clock pulse after *LOAD* goes HIGH since the load function is synchronous.
- Test the asynchronous reset function. Make *RESET* LOW at any time other than just before a positive edge of the clock. *Q* should reset immediately, without waiting for a positive clock edge.

FIGURE 9.39 4-Bit Counter with Synchronous Load and Asynchronous Reset

Figure 9.40 shows a simulation of the counter in Figure 9.39. The first 18 clock pulses drive the counter through its normal 4-bit cycle from 0H to FH, then up to 1H. At this point, we set the *LOAD* input HIGH and the value at the *P* inputs (9H) is loaded into the counter on the rising edge of the next clock pulse. An asynchronous *RESET* pulse then drives the counter outputs to 0H, after which the count resumes.

Asynchronous Load

The asynchronous load function of a counter makes use of the asynchronous preset and clear inputs of the counter's flip-flops. Figure 9.41 shows the circuit implementation of the asynchronous load function, without any count logic.

When *ALOAD* (Asynchronous LOAD) is HIGH, both NAND gates in Figure 9.41 are enabled. If the *P* input is HIGH, the output of the upper NAND gate goes LOW, activating the flip-flop's asynchronous *PRESET* input, thus setting *Q* = 1. The lower NAND gate has a HIGH output, thus deactivating the flip-flop's *CLEAR* input.

FIGURE 9.40 Simulation of a 4-Bit Counter with Synchronous Load and Asynchronous Reset

FIGURE 9.41 Asynchronous Load Element

FIGURE 9.42 Asynchronous Load Element with Asynchronous Clear

If *P* is LOW the situation is reversed. The upper NAND output is HIGH and the lower NAND has a LOW output, activating the flip-flop's *CLEAR* input, resetting *Q*. Thus, *Q* will be the same value as *P* when the *ALOAD* input is asserted. When *ALOAD* is not asserted (= 0), both NAND outputs are HIGH and thus do not activate either the preset or clear function of the flip-flop.

Figure 9.42 shows the asynchronous load circuit with an asynchronous clear (reset) function added. The flip-flop can be cleared by a logic LOW either from the *P* input (via the lower NAND gate) or the *CLEAR* input pin. The clear function disables the upper NAND gate when it is LOW, preventing the flip-flop from being cleared and preset simultaneously. This extra connection also ensures that the clear function has priority over the load function.

EXAMPLE 9.6

Quartus II:
al_count.bdf
MAX+PLUS II:
al_count.gdf

Use Quartus II to redraw the circuit in Figure 9.42 to create a general element called **al_count** that can be used in a synchronous counter with asynchronous load and clear. (Refer to Figure 9.38 for a similar element with *synchronous* load.)

■ **Solution**

Figure 9.43 shows the modified circuit, which includes an XOR gate for part of the count logic. The remainder of the count logic must be supplied externally to this element for each bit of the counter.

FIGURE 9.43 Example 9.6 Counter Element with Asynchronous Load and Clear (al_count)

EXAMPLE 9.7

Quartus II:
4bit_al.bdf
4bit_al.vwf
MAX+PLUS II:
4bit_al.gdf
4 bit_al.scf

Draw a circuit with four instances of **al_count** (from Example 9.6) to make a 4-bit synchronous counter with asynchronous load and reset. Create a simulation that tests the function of the counter.

■ **Solution**

Figure 9.44 shows the circuit. (Compare this circuit to the counter with synchronous load in Figure 9.39. This difference between the two is in the load function, not the count logic.)

The Boolean function applied to the *COUNT* input of each instance of **al_count** consists of the logical product of all previous output bits. ($COUNT_3 = Q_2Q_1Q_0$, $COUNT_2 = Q_1Q_0$, $COUNT_1 = Q_0$, $COUNT_0 = 1$.) When combined with the XOR at the *COUNT* input of each element, this yields the Boolean equations for a binary counter based on D flip-flops, as derived in Example 9.5. The circuitry inside each instance of **al_count** also generates the asynchronous load and clear functions.

The counter in Figure 9.44 can be tested using the following criteria.

FIGURE 9.44   Example 9.7: 4-bit Counter with Asynchronous Load and Reset

SIMULATION CRITERIA

- The counter should be clocked 16 times to observe a full count cycle from 0H to FH.
- Test the load function by assigning a constant value to the *P* inputs and making *LOAD* HIGH. Loading should happen as soon as *LOAD* goes HIGH since the load function is asynchronous.
- Test the asynchronous reset function. Make *RESET* LOW at any time other than just before a positive edge of the clock. *Q* should reset immediately, without waiting for a positive clock edge.
- Test precedence of reset over load by making *LOAD* HIGH, then in the middle of the load pulse, make *RESET* LOW. The counter should load first, then reset, and load again when *RESET* goes HIGH. The count should resume when *LOAD* returns to LOW.

(continued)

FIGURE 9.45   Example 9.7: Simulation of a 4-bit Counter with Asynchronous Load and Reset

Figure 9.45 shows a Quartus II simulation of the counter. The counter cycles through its full range and continues. A pulse at about 700 ns loads the counter with the value 9H ($= 1001_2$), after which the count continues.

The reset pulse at about 900 ns clears the counter. The *LOAD* pulse starting at about 960 μs shows how the load function has precedence over the count function. When *LOAD* is asserted, 9H is loaded and the count does not increase until *LOAD* is deasserted. The *RESET* pulse overrides both load and count functions. When *RESET* is deasserted, 9H is asynchronously reloaded. Counting resumes from that value on the first positive clock edge after *LOAD* goes LOW.

Count Enable

The counter elements in Figures 9.38 (**sl_count**) and 9.43 (**al_count**) are just D flip-flops configured for switchable toggle operation with additional circuitry for load and clear functions. Normally, when these elements are used in synchronous counters, the count progresses when the input to the element's XOR gate goes HIGH. In other words, the count progresses when the counter element is switched from a no change to a toggle mode.

In order to arrest the count sequence, we must disable the count logic of the counter circuit. Figure 9.46 shows a simple modification to the 4-bit counter circuit of Figure 9.39 that can achieve this function. Each AND gate has an extra input which is used to enable or inhibit the count logic function to each flip-flop.

The counter in Figure 9.46 can be tested using the following criteria.

SIMULATION CRITERIA

- Observe the effect of the *COUNT_ENA* input. When *COUNT_ENA* = 1, the count should progress normally. When *COUNT_ENA* = 0, the count should be suspended without resetting.
- Test that the load function can still activate when the count is disabled.
- Test that the reset function can still activate when the count is disabled.

Figure 9.47 shows a simulation of the counter. Note that the count progresses normally when *COUNT_ENA* is HIGH and stops when *COUNT_ENA* is LOW, even though the clock pulses remain constant throughout the simulation.

Also note that the count enable has no effect on the synchronous load and asynchronous reset functions. In the latter part of the simulation, the count stops at AH ($Q_3Q_2Q_1Q_0 = 1010_2$), when *COUNT_ENA* goes LOW. At 760 ns, the synchronous

FIGURE 9.46 4-Bit Counter with Synchronous Load, Asynchronous Reset, and Count Enable

FIGURE 9.47 Simulation of 4-Bit Counter with Synchronous Load, Asynchronous Reset, and Count Enable load function loads the value of 9H into the counter. The counter stays at this value, even after LOAD is no longer active, since the count is still disabled. At 880 ns, an asynchronous reset pulse clears the counter. The count resumes on the first clock pulse after *COUNT_ENA* goes HIGH again.

Bidirectional Counters

Figure 9.48 shows the logic diagram of a 4-bit synchronous DOWN counter. Its count sequence starts at 1111 and counts backwards to 0000, then repeats. The Boolean equations for this circuit will not be derived at this time, but will be left for an exercise in an end-of-chapter problem.

We can intuitively analyze the operation of the counter if we understand that the upper three flip-flops will each toggle when their associated XOR gates have a HIGH input from the rest of the count logic.

Q_0 is set to toggle on each clock pulse. Q_1 toggles whenever Q_0 is LOW (every second clock pulse, at states 1110, 1100, 1010, 1000, 0110, 0100, 0010, and 0000). Q_2 toggles when Q_1 AND Q_0 are LOW (1100, 1000, 0100, and 0000). Q_3 toggles when Q_2 AND Q_1 AND Q_0 are LOW (1000 and 0000). The result of this analysis

Quartus II:
4bit_dn.bdf
element.bdf
MAX+PLUS II:
element.gdf

FIGURE 9.48   4-Bit Synchronous DOWN Counter

FIGURE 9.49 Synchronous DOWN Counter Simulation

FIGURE 9.50 Synchronous Counter Element (T Flip-Flop)

can be represented by a timing diagram, such as the simulation shown in Figure 9.49. As we expect, the counter will count down from 1111 (FH) to 0000 (0H) and repeat.

We can create a bidirectional counter by including a circuit to select count logic for an UP or DOWN sequence. Figure 9.50 shows a basic synchronous counter element that can be used to create a synchronous counter. The element is simply a D flip-flop configured for switchable toggle mode.

Four of these elements can be combined with selectable count logic to make a 4-bit bidirectional counter, as shown in Figure 9.51. Each counter element has a pair of AND-shaped gates and an OR gate to steer the count logic to the XOR in the element. When $DIR = 1$, the upper gate in each pair is enabled and the lower gates disabled, steering the UP count logic to the counter element. When $DIR = 0$, the lower gate in each pair is enabled, steering the DOWN count logic to the counter element. The directional function can also be combined with the load and count enable functions, as was shown for unidirectional UP counters.

The counter in Figure 9.51 can be tested using the following criteria.

SIMULATION CRITERIA

■ Observe the effect of the *DIR* input. When $DIR = 1$, the count should progress from 0H to FH. When $COUNT\_ENA = 0$, the count should go from FH to 0H.

Figure 9.52 shows a simulation of the bidirectional counter of Figure 9.51. The waveforms show the UP count when *DIR* is HIGH and the DOWN count when *DIR* is LOW.

FIGURE 9.51   4-Bit Bidirectional Counter

FIGURE 9.52   Simulation of 4-Bit Bidirectional Counter

Decoding the Output of a Counter

Figure 9.53 shows a Block Diagram File of a 4-bit bidirectional counter with an output decoder. The counter is the one shown in Figure 9.51, represented as a logic circuit symbol. The decoder component **decode16** is a module written in VHDL, as follows.

FIGURE 9.53 4-Bit Bidirectional Counter with Output Decoder

```
-- decode16.vhd

LIBRARY ieee;
USE ieee.std_logic_1164.ALL;

ENTITY decode16 IS
    PORT(
        sel : IN  INTEGER RANGE 0 to 15;
        y   : OUT STD_LOGIC_VECTOR (0 to 15));
END decode16;

ARCHITECTURE a OF decode16 IS
BEGIN
    WITH sel SELECT
        y <= x"7FFF" WHEN 0,
             x"BFFF" WHEN 1,
             x"DFFF" WHEN 2,
             x"EFFF" WHEN 3,
             x"F7FF" WHEN 4,
             x"FBFF" WHEN 5,
             x"FDFF" WHEN 6,
             x"FEFF" WHEN 7,
             x"FF7F" WHEN 8,
             x"FFBF" WHEN 9,
             x"FFDF" WHEN 10,
             x"FFEF" WHEN 11,
             x"FFF7" WHEN 12,
             x"FFFB" WHEN 13,
             x"FFFD" WHEN 14,
             x"FFFE" WHEN 15,
             x"FFFF" WHEN others;
    END a;
```

y values encoded in hexadecimal (e.g., x"7FFF" = "0111111111111111")

Quartus II:
decode16.vhd
4bit_dcd.bdf
4bit_dcd.vwf
MAX+PLUS II:
decode16.vhd
4bit_dcd.gdf
4bit_dcd.scf

The decoder has 16 outputs, one for each state of the counter. For each state, one and only one output will be low. (Refer to the section on binary decoders in Chapter 6 for a more detailed description of *n*-line-to-*m*-line binary decoders.)

The selected signal assignment state assigns the values of decoder output, **y,** as a 4-digit hexadecimal number. For example, the line in the assignment statement (x"FDFF" when 6,) assigns **y** the equivalent binary value of "1111110111111111" when the integer value of select input, **sel,** is 6. We use the hexadecimal assignment because it is easier to read than a large binary number.

The counter in Figure 9.53 can be tested using the following criteria.

SIMULATION CRITERIA

- The counter should be clocked at least 32 times to observe a full count cycle from 0H to FH and from FH to 0H.
- Test the directional function. Count should be up when $DIR = 1$ and down when $DIR = 0$.
- The decoder outputs should activate in the same sequence as the count value.

Figure 9.54 shows the simulation waveforms for the circuit in Figure 9.53. As the count progresses up or down, as shown by the waveform for Q[3..0], the decoder outputs respond by going LOW in sequence.

Output decoders for binary counters can also be configured to have active HIGH outputs. In this case, one and only one output would be HIGH for each output state of the counter.

FIGURE 9.54   Simulation of a 4-Bit Decoded Counter

Terminal Count and RCO

A special case of output decoding is a circuit that will detect the **terminal count,** or last state, of a count sequence and activate an output to indicate this state. The terminal count depends on the count sequence. A 4-bit binary UP counter has a terminal count of 1111; a 4-bit binary DOWN counter has a terminal count of 0000. A circuit to detect these conditions must detect the *maximum value* of an UP count and the *minimum value* of a DOWN count.

Quartus II:
term_dcd.bdf
MAX+PLUS II:
term_dcd.gdf

The decoder shown in Figure 9.55 fulfills both of these conditions. The directional input DIR enables the upper gate when HIGH and the lower gate when LOW. Thus, the upper gate generates a HIGH output when $DIR = 1$ AND $Q_3Q_2Q_1Q_0 = 1111$. The lower gate generates a HIGH when $DIR = 0$ AND $Q_3Q_2Q_1Q_0 = 0000$.

Figure 9.56 shows the terminal count decoder combined with a 4-bit bidirectional counter. The decoder is also used to enable a NAND gate output that gener-

FIGURE 9.55 Terminal Count Decoder for a 4-Bit Bidirectional Counter

FIGURE 9.56 4-Bit Bidirectional Counter with Terminal Count Detection

Quartus II:
4bit_rco.bdf
4bit_rco.vwf
MAX+PLUS II:
4bit_rco.gdf
4bit_rco.scf

ates an *RCO* signal. RCO stands for ripple carry out or ripple clock out. The purpose of *RCO* is to produce exactly one clock pulse upon terminal count and have the positive edge of *RCO* at the end of the counter cycle, for a counter that has a positive edge-triggered clock.

This function is generally found in counters with a fixed number of bits (i.e., fixed-function counter chips, not PLDs) and is used to asynchronously clock a further counter stage, as in Figure 9.57. This allows us to extend the width of the counter beyond the number of bits available in the fixed-function device. This is not necessary when designing synchronous counters in programmable logic, but is included for the sake of completeness.

The NAND gate in Figure 9.56 is enabled upon terminal count and passes the clock signal through to *RCO*. The NAND output sits HIGH when inhibited. The clock is inverted in the RCO circuit so that when the NAND gate inverts it again, the circuit generates a clock pulse in true form.

FIGURE 9.57   Counter Expansion Using RCO

The correctness of the counter design in Figure 9.56 can be verified with the following criteria.

SIMULATION CRITERIA

- The counter should be clocked for at least 32 clock cycles to observe a full up-count and a full down-count.
- When $DIR = 0$, the counter should count down from FH to 0H and recycle. When $DIR = 1$, the counter should count up from 0H to FH and recycle.
- When the count is DOWN, MAX_MIN should go HIGH when the count is 0H. RCO should reproduce one full cycle of the clock during this interval.
- When the count is UP, MAX_MIN should go HIGH when the count is FH. RCO should reproduce one full cycle of the clock during this interval.

Figure 9.58 shows the simulation of the circuit of Figure 9.56. In the first half of the simulation, the counter is counting DOWN. The terminal count decoder output, MAX_MIN, goes HIGH when $Q_3Q_2Q_1Q_0 = 0000$. RCO generates a pulse at that time. For the second half, the counter is counting UP. MAX_MIN is HIGH when $Q_3Q_2Q_1Q_0 = 1111$ and RCO generates a pulse at that time.

Note that the RCO pulse appears to be half the width of the MAX_MIN pulse. Although the NAND gate that generates RCO is enabled for the whole MAX_MIN pulse, the clock input is HIGH for the first half-period, which is the same as the RCO inhibit level.

FIGURE 9.58   Simulation of a 4-Bit Bidirectional Counter with Terminal Count Decoding

The positive edge of *RCO* is at the *end* of the pulse. The idea is to synchronize the positive edge of the clock with the positive edge of *RCO*. However, since the RCO decoder is combinational, a propagation delay of about 7 ns is introduced.

▪ SECTION 9.5 REVIEW PROBLEM

9.5 Figure 9.59 shows two presettable counters, one with asynchronous load and clear, the other with synchronous load and clear. The counter with asynchronous functions has a 4-bit output labeled *QA*. The synchronously loaded counter has a 4-bit output labeled *QS*. The load *and* reset inputs to both counters are active-LOW.

Figure 9.60 shows a partial timing diagram for the counters. Complete the diagram.

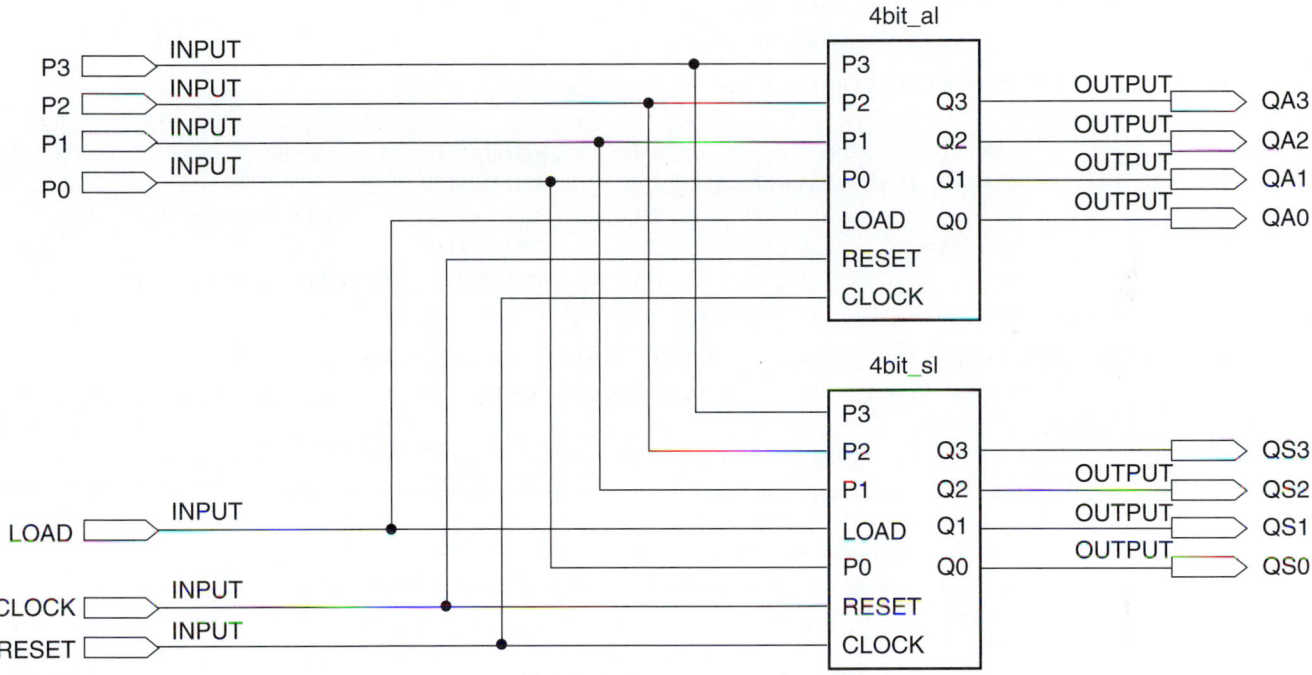

FIGURE 9.59 Section Review Problem 9.5: Two Presettable Counters

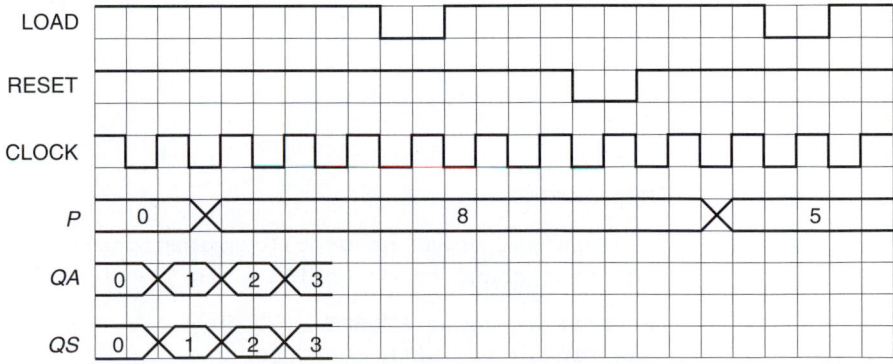

FIGURE 9.60 Timing Diagram for Counters in Figure 9.59

9.6 PROGRAMMING PRESETTABLE AND BIDIRECTIONAL COUNTERS FOR CPLDs

The presettable counters and bidirectional counters described in the previous section can be easily implemented in VHDL, either as behavioral descriptions or as LPM components. We will initially examine the behavioral descriptions of two counters, one with asynchronous load and clear and one with synchronous load and clear. We will then examine some options available in the module **lpm_counter** both as VHDL and graphical components.

➡ **LAB SOURCE**
Lab #23
Binary and BCD Counters
with Control Inputs

Behavioral Description

The following lists the VHDL code for an 8-bit bidirectional counter with count enable, terminal count decoding, and asynchronous load and clear:

```
-- presettable_8bit_counter_async_load.vhd

LIBRARY ieee;
USE ieee.std_logic_1164.ALL;
USE ieee.std_logic_unsigned.ALL;

ENTITY presettable_8bit_counter_async_load IS
    PORT(
        clk, count_ena        : IN      STD_LOGIC;
        clear, load, direction : IN      STD_LOGIC;
        p                     : IN      STD_LOGIC_VECTOR(7 downto 0);
        max_min               : OUT     STD_LOGIC;
        q                     : BUFFER  STD_LOGIC_VECTOR(7 downto 0));
END presettable_8bit_counter_async_load;

ARCHITECTURE a OF presettable_8bit_counter_async_load IS
    SIGNAL terminal_count : STD_LOGIC_VECTOR(8 downto 0);
BEGIN
    PROCESS (clk, clear, load, p)
    BEGIN
        IF (clear = '0') THEN                       -- Asynchronous clear
            q <= (others => '0');
        ELSIF (load = '1' and clear = '1') THEN -- Asynchronous load
            q <= p;
        ELSE
            IF (clk'EVENT AND clk = '1') THEN
                IF (count_ena = '1' and direction = '0') THEN
                    -- count down, if enabled
                    q <= q - 1;
                ELSIF (count_ena = '1' and direction = '1') THEN
                    -- count up, if enabled
                    q <= q + 1;
                END IF;
            END IF;
        END IF;
    END PROCESS;

    -- Terminal count decoder (combinational)
    terminal_count <= direction & q;

WITH terminal_count SELECT
    max_min <= '1' WHEN "000000000",
               '1' WHEN "111111111",
               '0' WHEN others;
END a;
```

Sensitivity list contains clk and all asynchronous controls — PROCESS (clk, clear, load, p)

Asynch. functions checked before clk

Synchronous functions checked as a condition of clk

presettable_8bit_counter_async_load.vhd

The load and clear functions of this counter are asynchronous, so these identifiers are part of the sensitivity list of the PROCESS statement; the statements in the process will execute if there is a change on the clear, load, **p,** or clock inputs. Load and clear are checked by IF statements, independently of the clock. Since load and clear are checked first, they have precedence over the clock. Clear has precedence over load since load can only activate if clear is not active.

If clear and load are not asserted, the clock status is checked by a clause in an IF statement: (IF (clk'EVENT and CLK = '1') THEN). If this condition is true, then the count is incremented or decremented, depending on the states of a count enable input and a directional control input. If the count enable input is not asserted, the count is neither incremented nor decremented.

After the PROCESS statement executes, **q** has several possible values. They are:

- **q** = 0 (if **clear** = 0),
- **q** = **p** (if **clear** = 1 and **load** = 1),
- **q** increments (if there is a positive clock edge, **count_ena** = 1, and **direction** = 1),
- **q** decrements (if there is a positive clock edge, **count_ena** = 1, and **direction** = 0), or
- **q** stays the same (if none of the conditions are met).

The terminal count is decoded by a selected signal assignment statement following the process that defines the counter. This statement evaluates the counter value and direction. If the count is maximum (11111111) and the direction is UP (**direction** = 1), the **max_min** output is HIGH. The **max_min** output is also HIGH if the count is at its minimum (00000000) and the direction is DOWN (**direction** = 0).

The code for the same 8-bit counter but with synchronous clear and load is shown next.

```
-- presettable_8bit_counter_sync_load
-- 8-bit presettable counter with synchronous clear and load
-- and terminal count decoding using STD_LOGIC types

LIBRARY ieee;
USE ieee.std_logic_1164.ALL;
USE ieee.std_logic_unsigned.ALL;

ENTITY presettable_8bit_counter_sync_load IS
   PORT(
        clk, count_ena          : IN        STD_LOGIC;
        clear, load, direction  : IN        STD_LOGIC;
        p                       : IN        STD_LOGIC_VECTOR(7 downto 0);
        max_min                 : OUT       STD_LOGIC;
        q                       : BUFFER    STD_LOGIC_VECTOR(7 downto 0));
END presettable_8bit_counter_sync_load;

ARCHITECTURE a OF presettable_8bit_counter_sync_load IS
   SIGNAL terminal_count : STD_LOGIC_VECTOR(8 downto 0);
BEGIN
   PROCESS (clk) —— Since all functions are synchronous, only clk is in sensitivity list
   BEGIN
      IF (clk'EVENT AND clk = '1') THEN
      IF (clear = '0') THEN -- Synchronous clear
         q <= (others => '0');
      ELSIF (load = '1') THEN -- Synchronous load
         q <= p;
      ELSIF (count_ena = '1' and direction = '0') THEN
         q <= q - 1;
```

*presettable_8bit_counter_
sync_load.vhd*

```
            ELSIF (count_ena = '1' and direction = '1') THEN
                q <= q + 1;
            END IF;
        END IF;
    END PROCESS;

    -- Terminal count decoder (combinational)
    terminal_count <= direction & q;

    WITH terminal_count SELECT
        max_min <= '1' WHEN "000000000",
                   '1' WHEN "111111111",
                   '0' WHEN others;

    END a;
```

The PROCESS statement in the synchronous counter has only one identifier in its sensitivity list—that of the clock input. Load and clear status are not evaluated until after the process checks for a positive clock edge. Otherwise the code is the same as for the asynchronously loading counter.

Figure 9.61 shows a detail of a simulation of the asynchronously loading counter. It shows the point where the count rolls over from FFH to 00H and activates the **max_min** output. The directional output changes shortly after this point and shows the terminal count decoding for a DOWN count, the point where the counter rolls over from 00H to FFH. In the UP count, **max_min** is HIGH when the counter output is FFH, but not 00H. In the DOWN count, **max_min** goes HIGH when the counter output is 00H, but not FFH.

Figure 9.62 shows the operation of the asynchronous load and clear functions. Figure 9.63 shows the synchronous load and clear. The inputs are identical for each simulation; each has two pairs of load pulses and a pair of clear pulses. The first pulse of each pair is arranged so that it immediately *follows* a positive clock edge; the second pulse of each pair immediately *precedes* a positive clock edge.

Quartus II:
*presettable_8bit_counter_*
async_load.vhd
MAX+PLUS II:
pre_ct8a.scf

FIGURE 9.61   Simulation Detail of 8-Bit Counter Showing Terminal Count Decode

In the counter with asynchronous load and clear, these functions are activated by the first pulse of each pair and again on the second pulse. For the counter with synchronous load and clear, only the second pulse of each pair has an effect, since the load and clear functions must be active *during* or *just prior to* an active clock edge, in order to satisfy setup-time requirements of the counter flip-flops. The end of the load and clear pulse can correspond to the positive clock edge, as the flip-flop hold time is zero. (See Chapter 11 for a more detailed description of setup and hold time.)

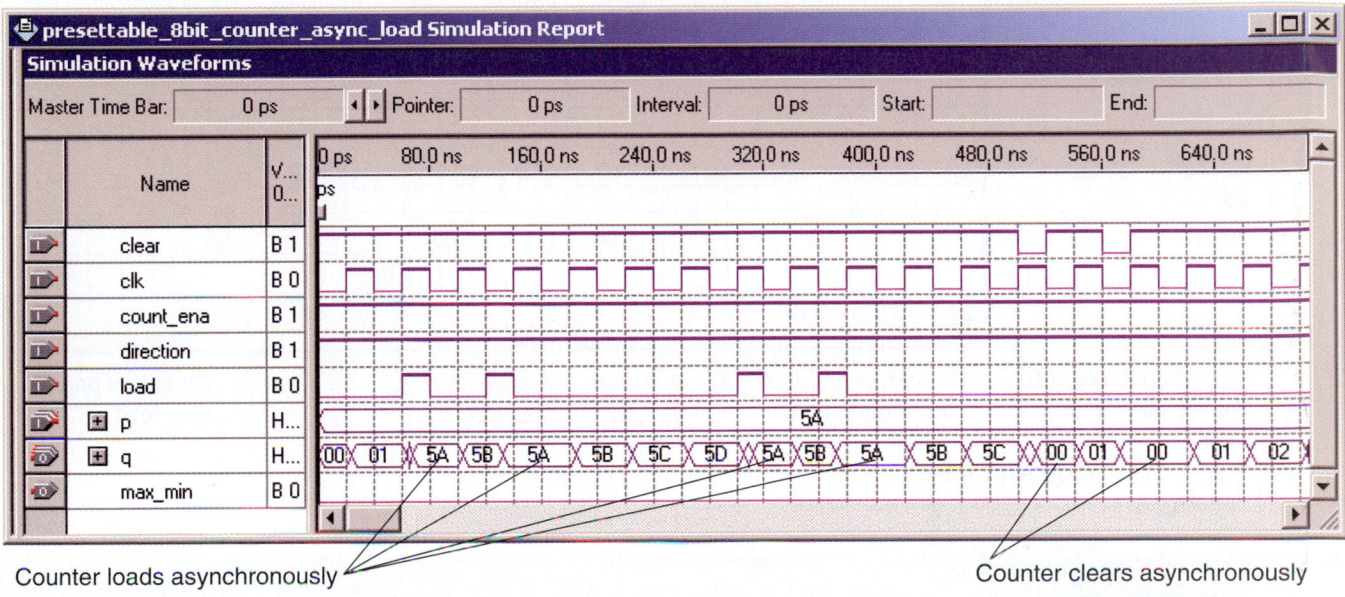

FIGURE 9.62 Simulation Detail of 8-Bit Counter Showing Asynchronous Load and Clear

FIGURE 9.63 Simulation Detail of 8-Bit Counter Showing Synchronous Load and Clear

FIGURE 9.64   Simulation Detail of 8-Bit Counter Showing Relative Priority of Count Enable, Asynchronous Load, and Asynchronous Clear

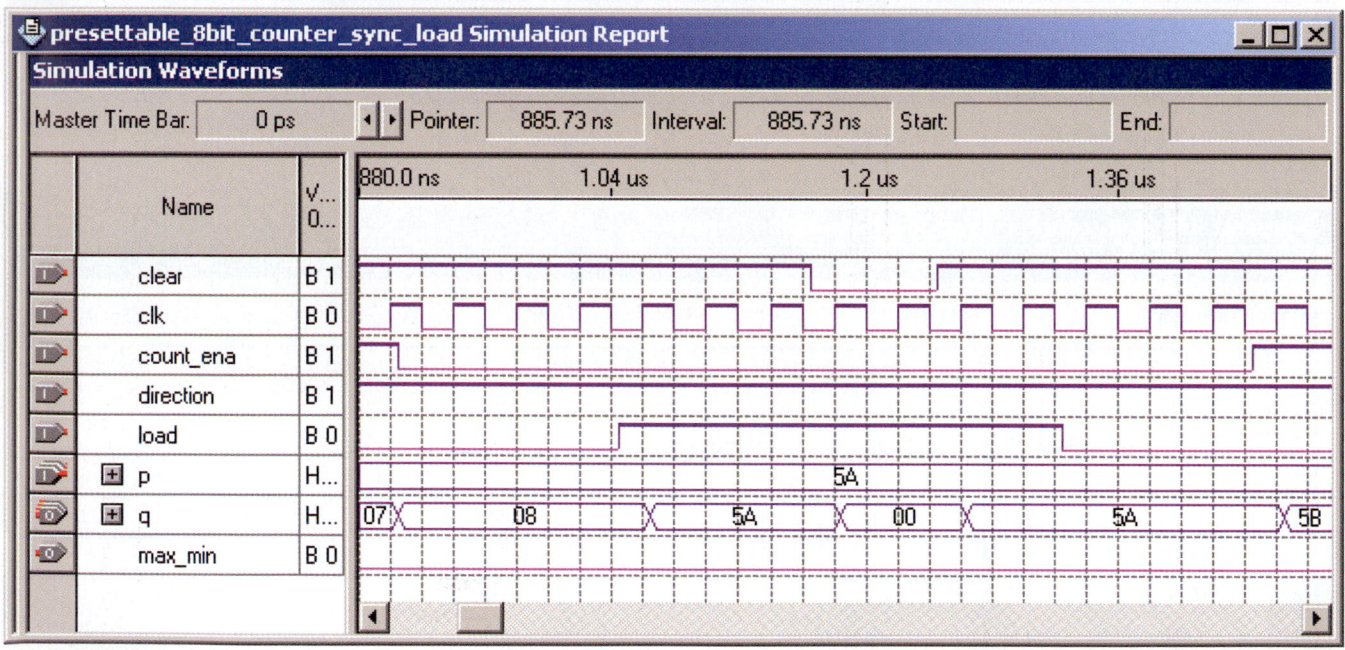

FIGURE 9.65   Simulation Detail of 8-Bit Counter Showing Relative Priority of Count Enable, Synchronous Load, and Synchronous Clear

Also note that the counter with synchronous load and clear has no intermediate glitch states on its outputs. (The simulation for the asynchronously loading counter shows glitch states on output q immediately following the first load pulse of each pair. Refer to the section on synchronous versus asynchronous circuits in Chapter 8 for further discussion of intermediate states in asynchronous circuits.)

Figure 9.64 and Figure 9.65 show further simulation details for our two VHDL counters. Both show the priority of the load, clear, and count enable functions. Both diagrams show that load and clear are independent of count enable and that clear has precedence over load. Note that the counter with synchronous load and clear waits for a clock pulse to act, whereas the asynchronous functions activate as soon as they are asserted.

Possible Design Errors

There are a number of mistakes that you can make when you are learning to program a counter function with a behavioral description. Several of the most common are described below.

1. Only one instance of the expression (`clk'EVENT and clk = '1'`) is allowed in a PROCESS statement. More than one instance implies more than one clock for the process, which is not allowed. The following code shows an example of illegal syntax.

```
-- Illegal syntax (more than one clock per process)
PROCESS(clk)
BEGIN
    IF(clk'EVENT and clk = '1') THEN
        IF(load = '1') THEN
            q <= p;
        END IF;
    END IF;
    IF(clk'EVENT and clk = '1') THEN
        IF(count_enable = '1') THEN
            q <= q+1;
        END IF;
    END IF;
END PROCESS;
```

The correct syntax for this function is shown below. In this case, the clock is evaluated once and all synchronous functions are tested inside the clause that evaluates the clock.

```
PROCESS(clk)
BEGIN
    IF (clk'EVENT and clk ='1') THEN
        IF (load = '1') THEN
            q <= p;
        ELSIF (count_enable = '1') THEN
            q <= q+1;
        END IF;
    END IF;
END PROCESS;
```

2. No other port, signal, or variable is allowed to be included with the expression that evaluates the clock. Any synchronous functions must be separately evaluated by IF statements that are nested within the IF that evaluates the clock. An example of code that violates this principle is shown below. The VHDL compiler will give an error for this code stating that the clocking for the process is too complex.

```
-- Illegal syntax (load evaluated in same statement as clk)
PROCESS(clk)
BEGIN
    IF(clk'EVENT and clk ='1' and load = '1') THEN
        q <= p;
    ELSE
        q <= q+1;
    END IF;
END PROCESS;
```

The correct syntax for this function is as follows. In this case, the clock is evaluated first and synchronous functions are evaluated separately within the clause that evaluates the clock.

```
PROCESS(clk)
BEGIN
    IF(clk'EVENT and clk = '1') THEN
        IF (load = '1') THEN
            q <= p;
        ELSE
            q <= q+1;
        END IF;
    END IF;
END PROCESS;
```

3. The statements in a process should be such that it is only possible to assign one value to a port, variable, or signal for each time that the process executes. This can be done by constructing the IF or CASE statements in the process so that all possible conditions that might change the value of an object are accounted for and the value of the object is assigned for each possible condition.

The following code violates this principle by having more than one opportunity to assign a value to **q**. If **count_enable, load,** and **clear** are all active, **q** will be assigned three different values in the process, one for each IF statement, finally taking on the value of the last assignment.

```
-- Ambiguous (but not illegal) syntax
-- (q assigned more than once in a process. May have an
-- unexpected result.)
PROCESS (clk)
BEGIN
    IF (clk'EVENT and clk = '1') THEN
        IF (count_enable = '1') THEN
            q <= q+1;
        END IF;
        IF (load = '1') THEN
            q <= p;
        END IF;
        IF (clear = '0') THEN
            q <= (others => '0');
        END IF;
    END IF;
END PROCESS;
```

The following code uses better syntax to assign the value of **q**. In this case, there is only one possible value of **q** for each execution of the process, since the inner IF statement has several clauses, each covering one of the possible conditions that assign a value to **q**.

```
PROCESS(clk)
BEGIN
    IF(clk'EVENT and clk = '1') THEN
        IF(clear = '0') THEN
            q <= (others => '0');
        ELSIF(load = '1') THEN     -- Only executes if clear = '1'
            q <= p;
        ELSIF(count_enable = '1') THEN -- Only executes if clear = '1'
            q <= q+1;                      -- and load = '0'
        END IF;
    END IF;
END PROCESS;
```

Using Generics in a VHDL Counter

In Chapter 7, we saw how a GENERIC clause in an entity declaration can be used to define a parameter whose value is specified when the design entity is instantiated as a component in another file. We used this property to define a parallel adder whose width was initially unspecified, but later determined when it was instantiated in a higher-level design entity.

We can also use this technique to specify various parameters of a counter. For example, we can define a parameter to determine the width of the counter and another parameter to specify a value to load into the counter when a control input is asserted. All parameters are defined by a GENERIC clause in the entity declaration and must be assigned a default value, using the form:

```
GENERIC ( parameter_1 : type := default_value;
          parameter_2 : type := default_value);
```

The following VHDL code defines two generics:

- the number of bits of the counter, specified by a parameter called **width**, which has a default value of 8, and
- the value to which the counter will synchronously set when a control input called **set** goes HIGH. The loaded value is specified by a parameter called **sync_set,** which has a default value of 127. Note that this is not the same as a load function. The loaded value in a counter with a parallel load function is determined by the value at a multiple-bit input port, which can be changed at any time. The synchronous set function has no parallel input for the value; the function always loads the same value, specified by the parameter **sync_set.**

generic_width_counter.vhd
generic_width_counter.vwf

```
-- generic_width_counter.vhd
-- Counter of generic width, based on a behavioral description
-- Counter sets to "sync_value" when set = '1' and there is a
-- positive edge on the clock

LIBRARY ieee;
-- Defines std_logic types
USE ieee.std_logic_1164.ALL;
-- Allows std_logic to be treated as integer in unsigned expressions
USE ieee.std_logic_unsigned.ALL;
-- Contains conversion functions (integer to std_logic)
USE ieee.std_logic_arith.ALL;

ENTITY generic_width_counter IS
    GENERIC (width    : INTEGER := 8;      -- Default width
             sync_set : INTEGER := 127);   -- Default set value
    PORT(
        clock, reset, set : IN STD_LOGIC;
        -- Output width declared generically
        q : BUFFER STD_LOGIC_VECTOR(width-1 downto 0));
END generic_width_counter;

ARCHITECTURE counter of generic_width_counter IS
BEGIN
    PROCESS(clock, reset)
    BEGIN
        IF (reset = '0') THEN
            -- Set all outputs LOW (unspecified number of bits)
            q <= (others => '0');
```

```
      ELSIF (clock'EVENT and clock = '1')THEN
          IF set = '1' THEN
              -- Set outputs to value of sync_set
              -- (Must convert from integer to std_logic)
              q <= CONV_STD_LOGIC_VECTOR(sync_set, width);
          ELSE
              -- Increment counter
              q <= q + 1;
          END IF;
      END IF;
    END PROCESS;
  END counter;
```

In this entity declaration, we do not specify the width of output **q,** but define it in terms of the width parameter. Any assignment to **q** in the architecture body must account for the fact that we don't know how many bits wide **q** is. Thus, when we reset the counter, we cannot assign the value in the form q <= "00000000", since this assumes that **q** is 8 bits wide, which is only true if we leave it at its default value.

There are two ways we can correctly make the assignment:

- We could use the construct q <= (others => '0'); to set all bits of the counter to '0', regardless of the number of bits.
- We could use a conversion function, q <= CONV_STD_LOGIC_VECTOR(0, width);, which converts an integer value of 0 to an array of STD_LOGIC values equivalent to 0, with the number of bits specified by **width.** The general form of the function is: sv <= CONV_STD_LOGIC_VECTOR(iv, bits);, where sv is the STD_LOGIC_VECTOR value, iv is the INTEGER value, and bits is the number of bits in the STD_LOGIC_VECTOR. This conversion function is contained in the **std_logic_arith** package of the **ieee** library.

We must use the conversion function when assigning **sync_set** to **q,** because **sync_set** is of type INTEGER and **q** is of type STD_LOGIC_VECTOR.

Since the counter parameters have been assigned default values, the counter can be compiled and simulated as a stand-alone device. The default counter can be tested using the following criteria.

SIMULATION CRITERIA

- Apply sufficient clock pulses to observe a full 8-bit count from 00H to FFH.
- Test the synchronous set function. When **set** = 1, **q** should go to 127_{10} (= 7FH) on the next positive clock edge.
- Test asynchronous clear function. Output **q** should go to 00H as soon as **reset** goes LOW.
- Test priority of **reset** over **set.** Activate **set** for several clock cycles. Activate reset for a short time in the middle of this period. The counter should set to its predetermined value, then reset, then set again.

Figure 9.66 shows a simulation of the default counter showing the end of the full count, as well as the synchronous set and asynchronous reset functions.

Since the default width of the counter is 8 bits, its full count is 255_{10} or FFH. The counter rolls over from FF to 00 near the beginning of the simulation. The synchronous set value is 127_{10} or 7FH. When **set** goes HIGH, the counter output goes to 7F on the next positive clock edge. The simulation also shows the priority of the functions. Synchronous set overrides the count function. Asynchronous reset overrides both count and set.

FIGURE 9.66   Simulation for a VHDL Counter with Generic Width and Set Value

If we choose to instantiate this counter into a higher-level file, we assign the parameter values in a GENERIC MAP. The following VHDL code instantiates the counter with a width of 5 and a synchronous set value of 25. The component file, **generic_width_counter.vhd,** must be included in the project for this design entity.

```
-- generic_ctr_instance.vhd
-- Counter of generic width, based on a behavioral description

LIBRARY ieee;
USE ieee.std_logic_1164.ALL;

ENTITY generic_ctr_instance IS
    PORT(
        clk, clear, sset : IN  STD_LOGIC;
        -- Must define output width explicitly for
        -- instantiated component
        q_out : BUFFER STD_LOGIC_VECTOR(4 downto 0));
END generic_ctr_instance;

ARCHITECTURE counter of generic_ctr_instance IS
        -- Declare component
        COMPONENT generic_width_counter
          GENERIC (width    : INTEGER;
                   sync_set  : INTEGER);
          PORT(
            clock, reset, set : IN  STD_LOGIC;
            q : BUFFER STD_LOGIC_VECTOR(width-1 downto 0));
        END COMPONENT;
    BEGIN
        -- Instantiate component
    count: generic_width_counter
        GENERIC MAP(width => 5, sync_set => 25)
        PORT MAP(clk, clear, sset, q_out);
END counter;
```

generic_ctr_instance.vhd
generic_ctr_instance.vwf

The width of the output of the higher-level design entity must be explicitly stated in its entity declaration. The instantiated component must match the entity in width.
The instantiated counter can be tested using the following criteria.

SIMULATION CRITERIA

- Apply sufficient clock pulses to observe a full 5-bit count from 00H to 1FH.
- Test the synchronous set function. When **set** = 1, **q** should go to 25_{10} (= 19H) on the next positive clock edge.
- Test asynchronous clear function. Output **q** should go to 00H as soon as **reset** goes LOW.
- Test priority of **reset** over **set**. Activate **set** for several clock cycles. Activate reset for a short time in the middle of this period. The counter should set to its predetermined value, then reset, then set again.

FIGURE 9.67   Simulation for an Instance of a Generic Counter

Figure 9.67 shows a simulation of the instantiated counter. The maximum value of a 5-bit counter is 31_{10} or 1FH. The hexadecimal equivalent of the synchronous set value is 19H. The simulation shows the counter rolling over from 1F to 00, then setting to 19 and clearing. Again, the simulation shows the priority of functions: clear, then set, then count.

→ **LAB SOURCE**
Lab #24
LPM Counters

LPM Counters

Earlier in this chapter, we saw how a parameterized counter from the Library of Parameterized Modules (LPM) could be used as a simple 8-bit counter. The component **lpm_counter** has a number of other functions that can be implemented using specific ports and parameters. These functions are indicated in Table 9.13.

The only ports that are required by an LPM counter are **clock,** and one of **q[]** (counter outputs) or **eq[]** (decoder outputs). The only required parameter is **LPM_WIDTH,** which specifies the number of counter output bits. All other ports and parameters are optional, although certain ones must be used together. (For instance, ports **sload** and **data[]** are optional, but both must be used for the synchronous load function.) If unused, a port or parameter will be held at a default logic level.

To use any of the functions of an LPM component in a VHDL file, we use a component instantiation statement and specify the required parameters in a generic map and the ports in a port map.

TABLE 9.13 Available Functions of an LPM counter

Function	Ports	Parameters	Description
Basic count operation	clock, q []	LPM_WIDTH	Output **q[]** increases by one with each positive clock edge. **LPM_WIDTH** is the number of output bits.
Synchronous load	sload, data []	none	When **sload** = 1, output **q[]** goes to the value at input **data[]** on the next positive clock edge. **data[]** has the same width as **q[]**.
Synchronous clear	sclr	none	When **sclr** = 1, output **q[]** goes to zero on positive clock edge.
Synchronous set	sset	LPM_SVALUE	When **sset** = 1, output goes to value of **LPM_SVALUE** on positive clock edge. If **LPM_SVALUE** is not specified, **q[]** goes to all 1s.
Asynchronous load	aload, data[]	none	Output goes to value at **data[]** when **aload** = 1.
Asynchronous clear	aclr	none	Output goes to zero when **aclr** = 1.
Asynchronous set	aset	LPM_AVALUE	Output goes to value of **LPM_AVALUE** when aset = 1. If **LPM_AVALUE** is not specified, outputs all go HIGH when **aset** = 1.
Directional control	updown	LPM_DIRECTION	Optional direction control. Default direction is UP. Only one of **updown** and **LPM_DIRECTION** can be used. **updown** = 1 for UP count, updown = 0 for DOWN count. **LPM_DIRECTION** = "UP", "DOWN", or "DEFAULT".
Count enable	cnt_en	none	When **cnt_en** = 1, count proceeds upon positive clock edges. No effect on other synchronous functions (**sload, sclr, sset**). Defaults to "enabled" when not specified.
Clock enable	clk_en	none	All synchronous functions are enabled when **clk_en** = 1. Defaults to "enabled" when not specified.
Modulus control	none	LPM_MODULUS	Modulus of counter is set to value of **LPM_MODULUS**.
Output decoding (BDF or AHDL only; not available in VHDL)	eq[15..0]	none	Sixteen active HIGH decoded outputs, one for each internal counter value from 0 to 15.

NOTE . . .

The following VHDL component declaration indicates that all parameters except LPM_WIDTH and LPM_MODULUS are defined as having type STRING, which requires the parameter value to be written in double quotes, even if numeric (e.g., LPM_AVALUE => "1001"). Since LPM_WIDTH is defined as type POSITIVE (i.e., any integer > 0) it must be written without quotes (e.g., `LPM_WIDTH => 8`). This is also true for LPM_MODULUS, which is defined as type NATURAL (e.g., LPM_MODULUS => 12). Default values of all ports and parameters are also included in the component declaration (e.g., `clk_en: IN STD_LOGIC := '1'`; the default value of the clock enable input is '1'). The LPM component declaration can also be found in the Quartus II Help menu (**Help; Megafunctions/LPM; lpm_counter**).

VHDL Component Declaration for lpm_counter:

```
COMPONENT lpm_counter
    GENERIC (LPM_WIDTH: POSITIVE;
        LPM_MODULUS: NATURAL := 0;
        LPM_AVALUE: STRING:= "UNUSED";
        LPM_SVALUE: STRING := "UNUSED";
        LPM_DIRECTION: STRING := "UNUSED";
        LPM_TYPE: STRING := "LPM_COUNTER";
        LPM_PVALUE: STRING := "UNUSED";
        LPM_HINT : STRING := "UNUSED");
    PORT (data: IN STD_LOGIC_VECTOR (LPM_WIDTH-1 DOWNTO 0) := (OTHERS => '0');
        clock: IN STD_LOGIC;
        cin: IN STD_LOGIC := '0';
        clk_en: IN STD_LOGIC := '1';
        cnt_en: IN STD_LOGIC := '1';
        updown: IN STD_LOGIC := '1'
        sload: IN STD_LOGIC := '0';
        sset: IN STD_LOGIC := '0';
        sclr: IN STD_LOGIC := '0';
        aload: IN STD_LOGIC := '0';
        aset: IN STD_LOGIC := '0';
        aclr: IN STD_LOGIC := '0';
        cout: OUT STD_LOGIC;
        q: OUT STD_LOGIC_VECTOR (LPM_WIDTH-1 DOWNTO 0) );
    END COMPONENT;
```

 EXAMPLE 9.8

Write a VHDL file for an 8-bit LPM counter with ports for the following functions: asynchronous load, asynchronous clear, directional control, and count enable.

■ Solution

The required VHDL file is as follows. Note that no behavioral descriptions are required for the functions, only a mapping from the defined port names to the entity inputs and outputs.

```
-- pre_lpm8
-- 8-bit presettable counter with asynchronous clear and load,
-- count enable, and a directional control port

LIBRARY ieee;
USE ieee.std_logic_1164.ALL;
LIBRARY lpm;
USE lpm.lpm_components.ALL;

ENTITY pre_lpm8 IS
    PORT (
        clk, count_ena        : IN  STD_LOGIC;
        clear, load, direction : IN  STD_LOGIC;
        p                     : IN  STD_LOGIC_VECTOR(7 downto 0);
        q_out                 : OUT STD_LOGIC_VECTOR(7 downto 0));
END pre_lpm8;
```

```
ARCHITECTURE a OF pre_lpm8 IS
BEGIN
    counter1: lpm_counter
        GENERIC MAP (LPM_WIDTH => 8)
        PORT MAP ( clock  => clk,
                   updown => direction,
                   cnt_en => count_ena,
                   data   => p,
                   aload  => load,
                   aclr   => clear,
                   q      => q_out);
END a;
```

EXAMPLE 9.9

Write a VHDL file that uses an LPM counter to generate a DOWN counter with a modulus of 500. Create a Quartus II simulation file to verify the counter's operation.

■ Solution

A mod-500 counter requires nine bits since $2^9 < 500 < 2^2$. Since the counter always counts DOWN, we can use the parameter LPM_DIRECTION to specify the DOWN counter rather than using an unnecessary port. The required VHDL code follows.

Note that the values of LPM_WIDTH and LPM_MODULUS are written without quotes since they are defined as types POSITIVE and NATURAL, respectively, in the component declaration. LPM_DIRECTION is written in double quotes since the component declaration defines it as type STRING.

Quartus II:
mod5c_lpm.vhd
mod5c_lpm.vwf
MAX+PLUS II:
mod5c_lpm.vhd
mod5c_lpm.scf

```
LIBRARY ieee;
USE ieee.std_logic_1164.ALL;
LIBRARY lpm;
Use lpm.lpm_components.ALL;

ENTITY mod5c_lpm IS
    PORT(
        clk : IN  STD_LOGIC;
        q   : OUT STD_LOGIC_VECTOR (8 downto 0) );
END mod5c_lpm;

ARCHITECTURE a OF mod5c_lpm IS
BEGIN
    counter1 : lpm_counter
        GENERIC MAP(LPM_WIDTH     => 9,
                    LPM_DIRECTION => "DOWN",
                    LPM_MODULUS   => 500)
        PORT MAP ( clock => clk,
                   q     => q);
END a;
```

Figure 9.68 shows a partial simulation of the counter, indicating the point at which the output rolls over from 0 to 499 (decimal).

FIGURE 9.68   Simulation of a Mod-500 LPM DOWN Counter

If we are designing a counter for the Altera UP-1 or UP-2 circuit board, we can simulate the on-board oscillator by choosing a clock period of 40 ns, which corresponds to a clock frequency of 25 MHz. The minimum value for the end of the simulation time is:

40 ns/clock period × 500 clock periods = 20000 ns = 20 μs.

If we wish to see a few clock cycles past the recycle point, we can set the simulation end time to 20.1 μs. Note that the waveform for **q** is shown in Figure 9.68 with a decimal radix.

EXAMPLE 9.10

Write a VHDL file that instantiates a 12-bit LPM counter with asynchronous clear and synchronous set functions. Design the counter to set to 2047 (decimal). Create a simulation to verify the counter operation.

■ **Solution**

The required VHDL file is:

```
-- sset_lpm.vhd
-- 12-bit LPM counter with sset and aclr

LIBRARY ieee;
USE ieee.std_logic_1164.ALL;
LIBRARY lpm;
USE lpm.lpm_components.ALL;

ENTITY sset_lpm IS
    PORT(
            clk         : IN  STD_LOGIC;
            clear, set  : IN  STD_LOGIC;
            q           : OUT STD_LOGIC_VECTOR (11 downto 0));
END sset_lpm;

ARCHITECTURE a OF sset_lpm IS
BEGIN
    counter1: lpm_counter
        GENERIC MAP      (LPM_WIDTH => 12,
                          LPM_SVALUE => "2047")
        PORT MAP ( clock => clk,
                   sset  => set,
                   aclr  => clear,
                   q     => q);
END a;
```

Quartus II:
sset_lpm.vhd
sset_lpm.vwf
MAX+PLUS II:
sset_lpm.vhd
sset_lpm.scf

FIGURE 9.69 Example 9.10: Simulation of a 12-Bit Counter with Synchronous Set to 2047 and Asynchronous Clear

Figure 9.69 shows the simulation file of the counter. The full count sequence would take over 160 μs, so we will assume the count portion of the design works properly. Only the set and clear functions are fully simulated. The count waveform is shown in decimal.

LPM Components in Quartus II Block Diagram Files

➡ **LAB SOURCE**
Lab #25
Digital Stopwatch
(Lab Project 1)

Earlier in the chapter, we learned to create LPM components in a graphical format, either by using the MegaWizard Plug-In Manager or manual entry of the component. Let us look at two further examples that exploit more of the features of the LPM counter than our previous examples did.

EXAMPLE 9.11

Use the MegaWizard Plug-In Manager to create an LPM counter with the following characteristics:

- modulus = 200_{10}
- active-HIGH synchronous load
- active-HIGH synchronous set to 50_{10}
- clock enable
- count enable
- active-LOW asynchronous clear

Instantiate this component in a Quartus II Block Diagram File. Write a set of simulation criteria to test the counter. Create a simulation in Quartus II, based on your criteria.

■ **Solution**

The following screens from the MegaWizard Plug-In Manager indicate the series of steps required to define the counter. In Figure 9.70, we select that the output file created by the MegaWizard Plug-In Manager should be a VHDL file. We accept the default filename (**lpm_counter0**).

(continued)

VHDL output   Default file name for component

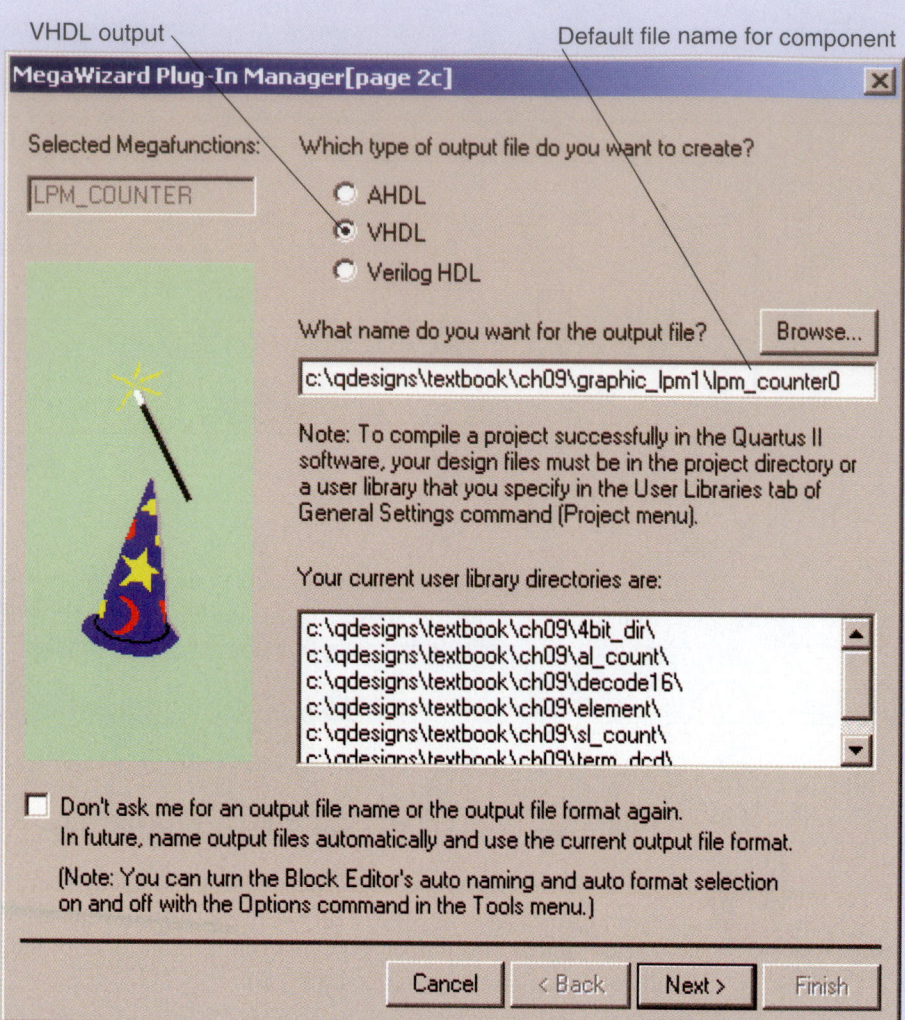

FIGURE 9.70   Example 9.11: MegaWizard Manager (Select VHDL Output)

In Figure 9.71, we determine that the counter should have eight bits, the minimum required for an output value of 199_{10}. (For a modulus of 200, the counter output ranges from 0 to 199.) We also specify that the counter will count up. Although this is not specified, it is a reasonable default option.

In Figure 9.72, we set the modulus of the counter to 200 and specify that the clock enable and count enable functions are required.

In Figure 9.73, we specify the required synchronous and asynchronous outputs. Load and set are synchronous, with the synchronous set value set to 50, as required. Clear is asynchronous.

After finishing the MegaWizard process, we must change the active level of the asynchronous clear to active-LOW. Click in the Block Editor window to place the component. Highlight the component and right-click. Choose **Properties** from the pop-up menu. In the **Symbol Properties** box, shown in Figure 9.74, click **aclr** to select the asynchronous clear input and select **All** in the box labeled **Inversion.** Click **OK.**

Figure 9.75 shows the completed symbol instantiated in a Quartus II Block Diagram File.

Define count direction   Set required width to 8 bits

FIGURE 9.71 Example 9.11: MegaWizard Manager (Number of Bits and Count Direction)

Define modulus

Specify both clock and count enable

FIGURE 9.72 Example 9.11: MegaWizard Manager (Modulus and Optional Ports)

Specify synchronous load   Specify asynchronous clear

Synchronous set value

FIGURE 9.73 Example 9.11: MegaWizard Manager (Synchronous and Asynchronous Inputs)

(continued)

Set aclr to Active-LOW

FIGURE 9.74 Example 9.11: Setting *aclr* to Active-LOW

FIGURE 9.75 Example 9.11: MegaWizard LPM Counter in a Block Diagram File

SIMULATION CRITERIA

- Count should proceed when **clear, count_enable,** and **clock_enable** are all HIGH and **load** and **set** are both LOW. The count should roll over from 199_{10} to 0.
- The counter outputs, **q[7..0],** should get the values of the parallel inputs, **p[7..0],** on the first positive clock edge after **load** = 1. (Load overrides count.)
- The counter outputs, **q[7..0],** should get the value 50_{10} on the first positive clock edge after **set** = 1. (Set overrides count.)
- The set function should have priority over the load function.
- The count should stop when either **count_enable** or **clock_enable** is LOW.
- When **count_enable** is LOW, it should not prevent **load, set,** or **clear** from acting. Only the count should stop.
- When **clock_enable** is LOW, it should prevent **load** and **set** from acting, since it affects all synchronous functions. It should not prevent the asynchronous clear from acting.
- **Clear** should override all other inputs.

Figures 9.76 through 9.79 show portions of the simulation of the LPM counter, based on the previously noted criteria. Figure 9.76 shows the counter rolling over from 199 to 0 as it counts up. (The count outputs and parallel inputs are shown as groups with an unsigned decimal radix.)

In Figure 9.77, the output goes to 50 on the first positive clock edge after **set** goes HIGH. When **load** goes HIGH, the output is loaded with 99 on the next positive clock edge. When both **load** and **set** are active, the set function predominates.

In Figure 9.78, the count stops when **count_enable** goes LOW. **Set** activates and loads the output with the value 50. While **set** is active, the counter asynchronously clears, then reloads 50 on the first clock after **clear** is deasserted. **Load** parallel-loads the value 99 into the counter. This value is asynchronously cleared,

FIGURE 9.76 Example 9.11: Partial Simulation of Graphic LPM Counter
(Recycle Point)

(continued)

FIGURE 9.77   Example 9.11: Partial Simulation of Graphic LPM Counter (Set and Load Functions)

FIGURE 9.78   Example 9.11: Partial Simulation of Graphic LPM Counter (Count Enable and Clear)

then reloads after **clear** deactivates. This shows that **count_enable** stops the count, but leaves other synchronous functions alone and that **clear** overrides them all.

Figure 9.79 shows the **clock_enable** going LOW, thus stopping the count. In this case the HIGH states on **load** and **set** are ignored because **clock_enable** affects all synchronous functions. Again, the asynchronous clear overrides all other inputs.

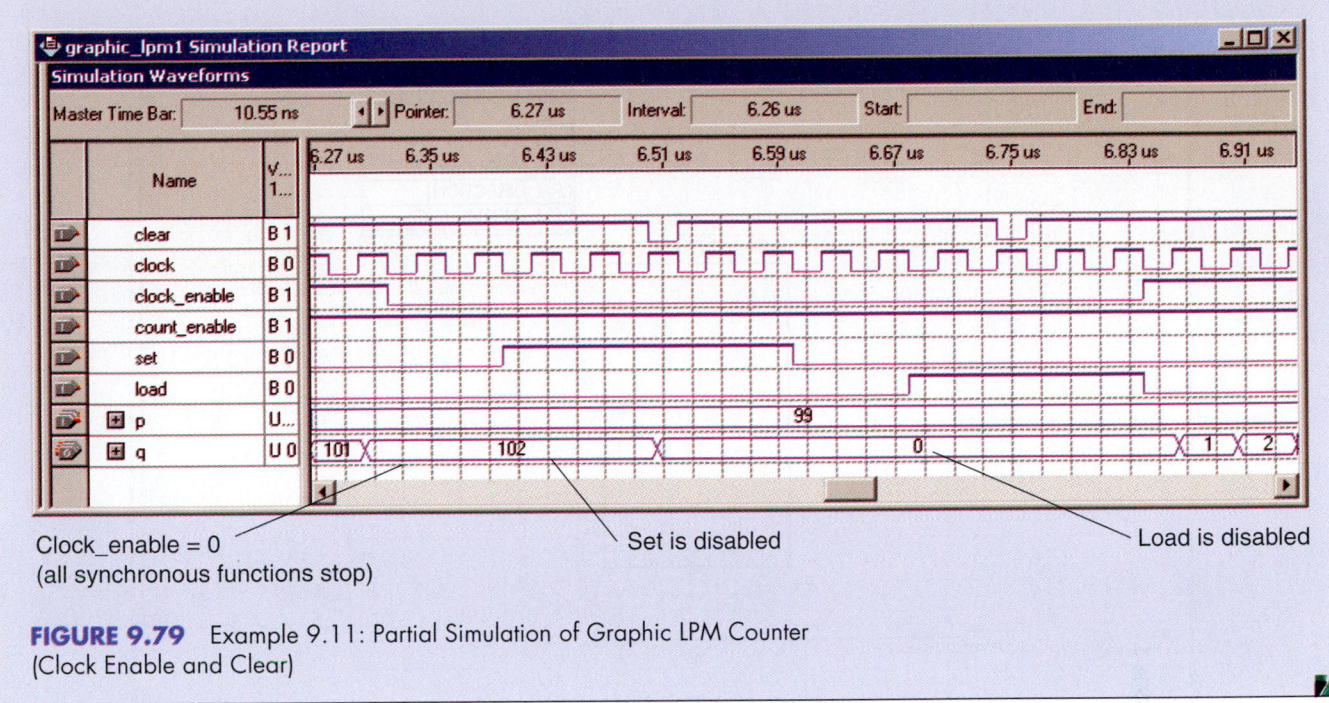

Clock_enable = 0
(all synchronous functions stop)

Set is disabled

Load is disabled

FIGURE 9.79 Example 9.11: Partial Simulation of Graphic LPM Counter (Clock Enable and Clear)

Decoded Outputs

One feature available for a graphical LPM counter, but not its VHDL version, is a set of 16 decoded outputs called **eq[15..0].** The way **eq** works is that a particular decoded output, **eq*n*,** goes HIGH when the value of counter outputs, **q,** *equals n.* For example, **eq0** goes HIGH when the count *equals* 0. Similarly, **eq1** goes HIGH when the count *equals* 1, and so on, up to **eq15,** which goes HIGH when the count *equals* 15. If the count is higher than 15, then no decoded outputs are HIGH. (**Eq** can also work even if there is no instantiated port for **q.** In such a case, **eq** decodes the internal value of the counter flip-flops.)

EXAMPLE 9.12

Create a Block Diagram File in the Quartus II Block Editor. Manually instantiate an LPM counter with a modulus of 20. Create the counter with the following ports: **clock, q[4..0], eq[15..0],** and active-LOW **aclr.** Create a Quartus II simulation that verifies the function of the design.

■ Solution

Figure 9.80 shows the Block Diagram File that contains the specified LPM counter. The simulation is shown in Figure 9.81, with the **q** and **eq** ports having an unsigned decimal radix. When the count is between 0 and 15, the **eq** port whose number is the same as the decimal count value is HIGH. When the count is between 16 and 19, no **eq** outputs are active. The count rolls over from 19 to 0 and continues.

(continued)

FIGURE 9.80 Example 9.12: LPM Counter with eq[ ] Port

FIGURE 9.81 Example 9.12: Simulation for LPM Counter with eq[ ] Port

9.6   The first part of a VHDL process statement includes a sensitivity list: `PROCESS (sensitivity list)`. How should this be written for a counter with asynchronous clear and for a counter with synchronous clear?

9.7   SHIFT REGISTERS

■ **KEY TERMS**

Shift Register   A synchronous sequential circuit that will store and move n-bit data, either serially or in parallel, in n flip-flops.

SRGn   Abbreviation for an n-bit shift register (e.g., SRG4 indicates a 4-bit shift register).

Serial Shifting   Movement of data from one end of a shift register to the other at a rate of one bit per clock pulse.

Parallel Transfer   Movement of data into all flip-flops of a shift register at the same time.

Rotation   Serial shifting of data with the output(s) of the last flip-flop connected to the synchronous input(s) of the first flip-flop. The result is continuous circulation of the same data.

Right Shift   A movement of data from the left to the right in a shift register. (Right is defined in Quartus II as toward the LSB.)

Left Shift   A movement of data from the right to the left in a shift register. (Left is defined in Quartus II as toward the MSB.)

Bidirectional Shift Register   A shift register that can serially shift bits left or right according to the state of a direction control input.

Parallel-Load Shift Register   A shift register that can be preset to any value by directly loading a binary number into its internal flip-flops.

Universal Shift Register   A shift register that can operate with any combination of serial and parallel inputs and outputs (i.e., serial in/serial out, serial in/parallel out, parallel in/serial out, parallel in/parallel out). A universal shift register is often bidirectional, as well.

A **shift Register** (abbreviated **SRGn** for an n-bit circuit) is a synchronous sequential circuit used to store or move data. It consists of several flip-flops, connected so that data are transferred into and out of the flip-flops in a standard pattern.

Figure 9.82 represents three types of data movement in three 4-bit shift registers. The circuits each contain four flip-flops, configured to move data in one of the ways shown.

Figure 9.82a shows the operation of **serial shifting.** The stored data are taken in one bit at a time from the input and moved one position toward the output with each applied clock pulse.

Parallel transfer is illustrated in Figure 9.82b. As with the synchronous parallel load function of a presettable counter, data move simultaneously into all flip-flops when a clock pulse is applied. The data are available in parallel at the register outputs.

Rotation, depicted in Figure 9.82c, is similar to serial shifting in that data are shifted one place to the right with each clock pulse. In this operation, however, data are continuously circulated in the shift register by moving the rightmost bit back to the leftmost flip-flop with each clock pulse.

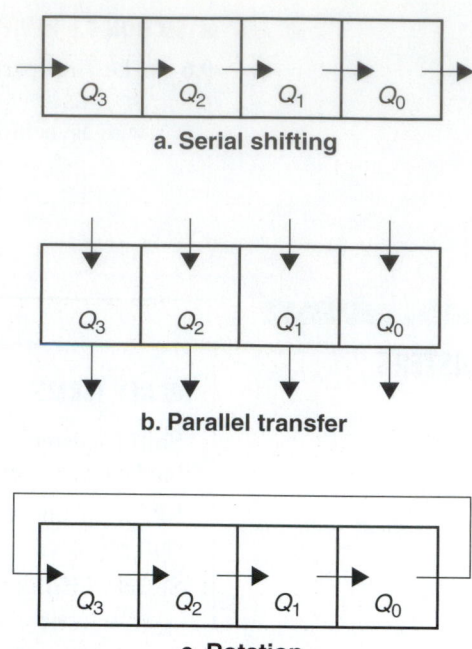

FIGURE 9.82 Data Movement in a 4-Bit Shift Register

Serial Shift Registers

Quartus II:
srg4_sr.bdf
srg4_sr.vwf
MAX+PLUS II:
srg4_sr.gdf
srg4_sr.scf

Figure 9.83 shows the most basic shift register circuit: the serial shift register, so called because data are shifted through the circuit in a linear or serial fashion. The circuit shown consists of four D flip-flops connected in cascade and clocked synchronously.

For a D flip-flop, Q follows D. The value of a bit stored in any flip-flop *after* a clock pulse is the same as the bit in the flip-flop to its left *before* the pulse. The result is that when a clock pulse is applied to the circuit, the contents of the flip-flops move one position to the right and the bit at the circuit input is shifted into Q_3. The bit stored in Q_0 is overwritten by the former value of Q_1 and is lost. Since the data move from left to right, we say that the shift register implements a **right shift** function. (Data movement in the other direction, requiring a different circuit connection, is called **left shift**.)

Let us track the progress of data through the circuit in two cases. All flip-flops are initially cleared in each case.

FIGURE 9.83 4-Bit Serial Shift Register Configured to Shift Right

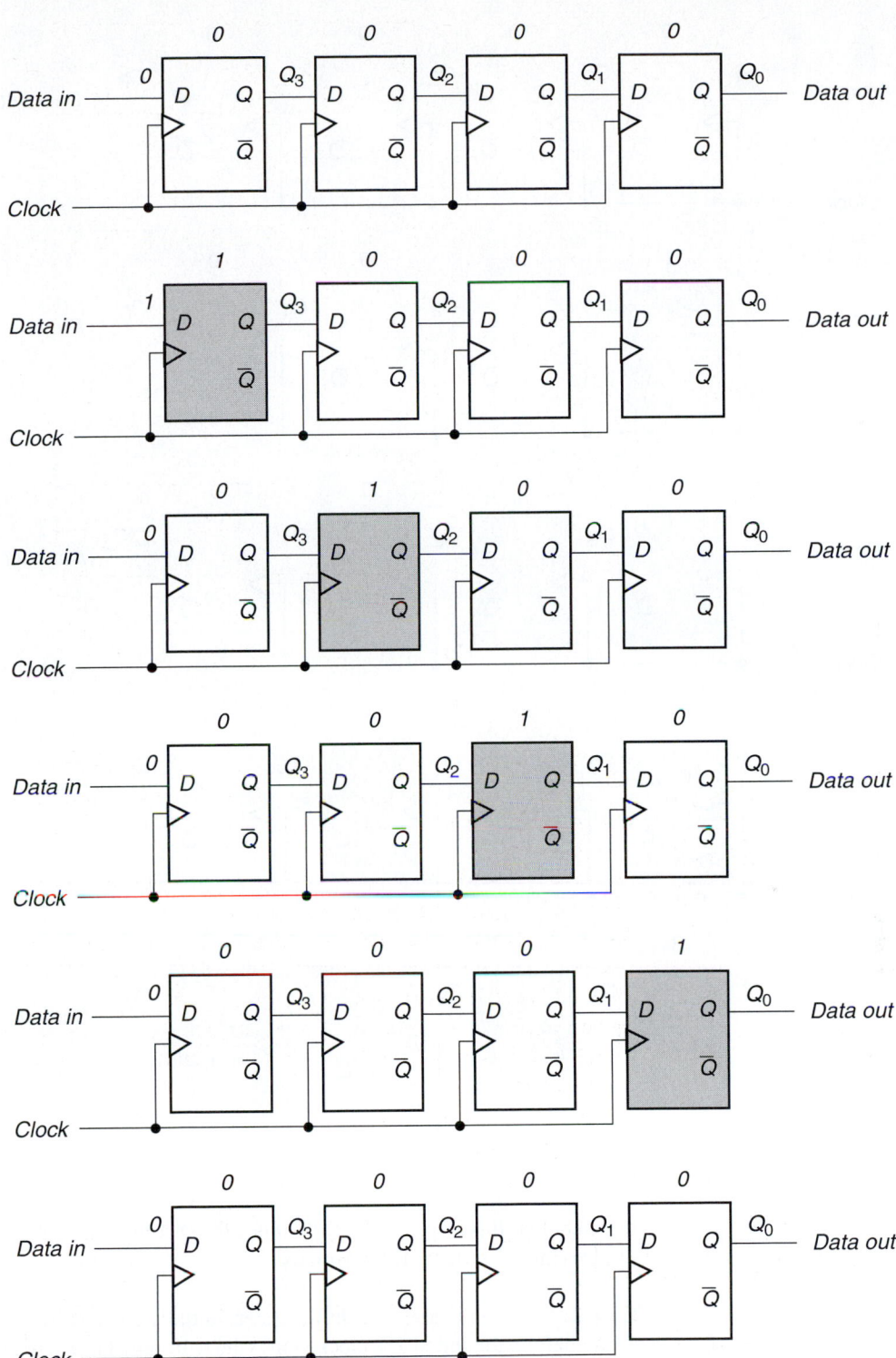

FIGURE 9.84 Shifting a "1" Through a Shift Register (Shift Right)

Case 1: A 1 is clocked into the shift register, followed by a string of 0s, as shown in Figure 9.84. The flip-flop containing the 1 is shaded.

Before the first clock pulse, all flip-flops are filled with 0s. Data In goes to a 1 and on the first clock pulse, the 1 is clocked into the first flip-flop. After that, the input goes to 0. The 1 moves one position right with each clock pulse, the register filling up with 0s behind it, fed by the 0 at Data In. After four clock pulses, the 1 reaches

FIGURE 9.85   Filling a Shift Register with "1"s (Shift Right)

the Data Out flip-flop. On the fifth pulse, the 0 coming behind overwrites the 1 at Q_0, leaving the register filled with 0s.

Case 2: Figure 9.85 shows a shift register, initially cleared, being filled with 1s.
   As before, the initial 1 is clocked into the shift register and reaches the Data Out line on the fourth clock pulse. This time, the register fills up with 1s, not 0s, because the Data input remains HIGH.

FIGURE 9.86 Simulation of a 4-Bit Shift Register (Shift Right)

Figure 9.86 shows a Quartus II simulation of the 4-bit serial shift register in Figures 9.83 through 9.85. The first half of the simulation shows the circuit operation for Case 1. The 1 enters the register at Q_3 on the first clock pulse after **serial_in** (Data In) goes HIGH. The 1 moves one position for each clock pulse, which is seen in the simulation as a pulse moving through the Q outputs.

Case 2 is shown in the second half of the simulation. Again, a 1 enters the register at Q_3. The 1 continues to be applied to **serial_in,** so all Q outputs stay HIGH after receiving the 1 from the previous flip-flop.

NOTE . . .

Conventions differ about whether the rightmost or leftmost bit in a shift register should be considered the most significant bit. The Altera Library of Parameterized Modules uses the convention of the leftmost bit being the MSB, so this is the convention we will follow. The convention has no physical meaning; the concept of right or left shift only makes sense on a logic diagram. The actual flip-flops may be laid out in any configuration at all in the physical circuit and still implement the right or left shift functions as defined on the logic diagram. (That is to say, wires, circuit board traces, and internal programmable logic connections can run wherever you want; left and right are defined on the logic diagram.)

EXAMPLE 9.13

Use the Quartus II Block Editor to create the logic diagram of a 4-bit serial shift register that shifts left, rather than right.

Quartus II:
srg4_sl.bdf
srg4_sl.vwf
MAX+PLUS II:
srg4_sl.gdf
srg4_sl.scf

■ **Solution**

Figure 9.87 shows the required logic diagram. The flip-flops are laid out the same way as in Figure 9.83, with the MSB (Q_3) on the left. The D input of each flip-flop is connected to the Q output of the flip-flop to its right, resulting in a looped-back connection. A bit at D_0 is clocked into the rightmost flip-flop. Data in the other flip-flops are moved one place to the left. The bit in Q_2 overwrites Q_3. The previous value of Q_3 is lost.

FIGURE 9.87 4-Bit Serial Shift Register Configured to Shift Left

EXAMPLE 9.14

Draw a diagram showing the movement of a single 1 through the register in Figure 9.87. Also draw a diagram showing how the register can be filled up with 1s.

■ **Solution**

Figures 9.88 and 9.89 show the required data movements.

FIGURE 9.88 Shifting a "1" Through a Shift Register (Shift Left)

(continued)

FIGURE 9.89   Filling a Shift Register with "1"s (Shift Left)

EXAMPLE 9.15

Use the Quartus II simulator to verify the operation of the shift-left serial shift register in Figure 9.87.

■ Solution

Figure 9.90 shows the simulation of the shift operations shown in Example 9.14. Compare this simulation to the one in Figure 9.86 to see how the opposite shift direction appears on a timing diagram.

FIGURE 9.90   Example 9.15: Simulation of a 4-Bit Shift Register (Shift Left)

Bidirectional Shift Registers

Quartus II:
srg4_bi.bdf
srg4_bi.vwf
MAX+PLUS II:
srg4_bi.gdf
srg4_bi.scf

Figure 9.91 shows the logic diagram of a **bidirectional shift register.** This circuit combines the properties of the right shift and left shift circuits, seen earlier in Figures 9.83 and 9.87. This circuit can serially move data right or left, depending on the state of a control input, called *DIRECTION.*

The shift direction is controlled by enabling or inhibiting four pairs of AND-OR circuit paths that direct the bits at the flip-flop outputs to other flip-flop inputs. When *DIRECTION* = 0, the right-hand AND gate in each pair is enabled and the flip-flop outputs are directed to the *D* inputs of the flip-flops one position left. Thus the enabled pathway is from *Left_Shift_In* to Q_0, then to Q_1, Q_2, and Q_3.

When *DIRECTION* = 1, the left-hand AND gate of each pair is enabled, directing the data from *Right_Shift_In* to Q_3, then to Q_2, Q_1, and Q_0. Thus, *DIRECTION* = 0 selects left shift and *DIRECTION* = 1 selects right-shift.

The design of the bidirectional shift register in Figure 9.91 can be tested using the following criteria.

SIMULATION CRITERIA

- When *DIRECTION* = 0, data applied to *Left_Shift_In* should transfer into Q_0 on the first clock pulse.
- Further clock pulses will move the data in the direction of Q_3, one bit position per clock pulse.
- Data applied to *Right_Shift_In* should be ignored when *DIRECTION* = 0.
- When *DIRECTION* = 1, data applied to *Right_Shift_In* should transfer into Q_3 on the first clock pulse.
- Further clock pulses will move the data in the direction of Q_0, one bit position per clock pulse.
- Data applied to *Left_Shift_In* should be ignored when *DIRECTION* = 1.

FIGURE 9.91 Bidirectional Shift Register

FIGURE 9.92 Simulation of a 4-Bit Bidirectional Shift Register

Figure 9.92 shows a Quartus II simulation of the bidirectional shift register in Figure 9.91. The simulation shows the left shift function from 0 to 500 ns and right shift after 500 ns. Both *Right_Shift_In* and *Left_Shift_In* are applied in both parts of the simulation, but the circuit responds only to one for each function.

For the left shift function, a 1 is applied to Q_0 at 140 ns and shifted left. The *Right_Shift_In* pulse is ignored. Similarly, for the right shift function, a 1 is applied to Q_3 at 540 ns and shifted right. *Left_Shift_In* is ignored.

Quartus II:
srg4_par.bdf
srg4_par.vwf
MAX+PLUS II:
srg4_par.gdf
srg4_par.scf

Shift Register with Parallel Load

Earlier in this chapter, we saw how a counter could be set to any value by synchronously loading a set of external inputs directly into the counter flip-flops. We can implement the same function in a **parallel-load shift register,** as shown in Figure 9.93.

The circuit is similar to that of the bidirectional shift register in Figure 9.91. The synchronous input of each flip-flop is fed by an AND-OR circuit that directs one of two signals to the flip-flop: the output of the previous flip-flop (shift function) or a parallel input (load function). The circuit is configured such that the shift function is enabled when *LOAD* = 0 and the load function is enabled when *LOAD* = 1.

Figure 9.94 shows a simulation of the parallel-load shift register circuit of Figure 9.93. In the first part of the simulation, the shift function is selected. This is tested by sending a 1 through the circuit in a right-shift pattern. Next, at 400 ns, *LOAD* goes HIGH, and the parallel input value AH (= 1010_2) is synchronously loaded into the circuit. The *LOAD* input goes LOW, thus causing the circuit to revert to the shift function. The data in the register are right-shifted out, followed by 0s. At 640 ns, the value FH (= 1111_2) is loaded into the circuit, then right-shifted out.

Universal Shift Register

Quartus II:
srg4_uni.bdf
srg4_uni.vwf
MAX+PLUS II:
srg4_uni.gdf
srg4_uni.scf

Figure 9.95 shows the logic circuit of a **universal shift register.** This circuit can implement any combination of serial and parallel inputs and outputs. It can also serially shift data left or right or hold data, depending on the states of S_1 and S_0, which form a 2-bit function select input.

Each AND-OR circuit acts as a multiplexer to direct one of several possible data sources to the synchronous inputs of each flip-flop. For instance, if we trace the paths through the corresponding AND-OR circuit, we find that the possible sources of data at D_2, the synchronous input of the second flip-flop, are Q_3 ($S_1S_0 = 01$), P_2 ($S_1S_0 = 11$), Q_1 ($S_1S_0 = 10$), and Q_2($S_1S_0 = 00$). These are the inputs required for the right-shift,

FIGURE 9.93 Serial Shift Register with Parallel Load

FIGURE 9.94 Simulation of a 4-Bit Serial Shift Register with Parallel Load

parallel load, left-shift, and hold functions, respectively. All functions are synchronous, including the parallel load and hold functions.

The hold function is a synchronous no change function, implemented by feeding back the Q output of a flip-flop to its synchronous (D) input. It is necessary to have this function, so that the flip-flops will not synchronously clear when none of the other functions is selected.

Table 9.14 summarizes the various possible inputs to each flip-flop as a function of S_1 and S_0.

TABLE 9.14 Flip-Flop Inputs as a Function of $S_1 S_0$ in a Universal Shift Register

S_1	S_0	Function	D_3	D_2	D_1	D_0
0	0	Hold	Q_3	Q_2	Q_1	Q_0
0	1	Shift Right	RSI^*	Q_3	Q_2	Q_1
1	0	Shift Left	Q_2	Q_1	Q_0	LSI^{**}
1	1	Load	P_3	P_2	P_1	P_0

*RSI = Right-shift input
**LSI = Left-shift input

EXAMPLE 9.16

Write a set of simulation criteria to verify the correctness of the universal shift register shown in Figure 9.95. Use these criteria to create a Quartus II simulation of the shift register.

■ **Solution**

The design of the universal shift register in Figure 9.95 can be tested using the following criteria.

SIMULATION CRITERIA

■ When $S_1 S_0 = 01$, data applied to RSI should transfer into Q_3 on the first clock pulse.
■ Further clock pulses will move the data in the direction of Q_0, one bit position per clock pulse.
■ Data applied to LSI should be ignored when $S_1 S_0 = 01$. *(continued)*

FIGURE 9.95   4-Bit Universal Shift Register

■ When $S_1S_0 = 10$, data applied to *LSI* should transfer into Q_0 on the first clock pulse.

■ Further clock pulses will move the data in the direction of Q_3, one bit position per clock pulse.

■ Data applied to *RSI* should be ignored when $S_1S_0 = 01$.

■ When $S_1S_0 = 11$, data applied to the parallel inputs should transfer into $Q_3Q_2Q_1Q_0$ on the first clock pulse.

■ When $S_1S_0 = 00$, the value of $Q_3Q_2Q_1Q_0$ should be retained from its previous value.

■ A LOW on *CLEAR* should immediately set the shift register contents to 0000.

Figure 9.96 shows a possible simulation. The following functions are tested: hold, right shift (*LSI* ignored), hold, left shift (*RSI* ignored), load FH, asynchronous clear, load FH, shift right for two clocks, shift left for three clocks.

FIGURE 9.96 Example 9.16: Simulation of a 4-Bit Universal Shift Register

■ **SECTION 9.7 REVIEW PROBLEM**

9.7 Can the D flip-flops in Figure 9.83 be replaced by JK flip-flops? If so, what modifications to the existing circuit are required?

9.8 PROGRAMMING SHIFT REGISTERS IN VHDL

➡ **LAB SOURCE**
Lab #26
Shift Registers

■ **KEY TERMS**

Structural Design A VHDL design technique that connects predesigned components using internal signals.

Dataflow Design A VHDL design technique that uses Boolean equations to define relationships between inputs and outputs.

Behavioral Design A VHDL design technique that uses descriptions of required behavior to create the design.

As with other circuit applications, we can take several approaches to programming shift registers in VHDL. Three basic design techniques are **structural, dataflow,** and **behavioral** descriptions. We will use each of these techniques to design a 4-bit shift register, such as the one shown in Figure 9.83.

Structural Design

Structural design is like taking components out of a bin and connecting them together to make a circuit. We can use the **DFF** component from the Quartus II primitives library and instantiate enough components to make a shift register, with connections made by internal signals. The code to make a 4-bit shift register using the structural design technique is shown here in the file **srg4strc.vhd.**

Quartus II:
srg4strc.vhd
srgstrc.vwf
MAX+PLUS II:
srg4strc.vhd
srg4strc.scf

```
-- srg4strc.vhd
-- Structural description of a 4-bit serial shift register
LIBRARY ieee;
USE ieee.std_logic_1164.ALL;
LIBRARY altera;
USE altera.maxplus2.ALL;

ENTITY srg4strc IS
    PORT(
        serial_in, clk : IN     STD_LOGIC;
        qo                  : BUFFER STD_LOGIC_VECTOR(3 downto 0));
END srg4strc;

ARCHITECTURE right_shift of srg4strc IS
    COMPONENT DFF
        PORT (d : IN  STD_LOGIC;
            clk : IN  STD_LOGIC;
             q  : OUT STD_LOGIC);
    END COMPONENT;
BEGIN
    flip_flop_3: dff
        PORT MAP (serial_in, clk, qo(3) );

    dffs:
    FOR i IN 2 downto 0 GENERATE
    flip_flops_2_to_0: dff
        PORT MAP (qo(i + 1), clk, qo(i) );
    END GENERATE;
END right_shift;
```

The design entity **srg4strc.vhd** instantiates four D flip-flops from the **altera .maxplus2** package and connects them by assigning common inputs and outputs to related components. A different way of writing the component instantiations would be as follows.

```
flip_flop_3: dff
    PORT MAP (serial_in, clk, qo(3) );
flip_flop_2: dff
    PORT MAP(qo(3), clk, qo(2) );
flip_flop_1: dff
    PORT MAP(qo(2), clk, qo(1) );
flip_flop_0: dff
    PORT MAP(qo(1), clk, qo(0) );
```

Since the component ports are in the sequence (*D, clk, Q*), these component instantiations imply that the *D* input of a flip-flop is fed by the *Q* of the previous flip-flop.

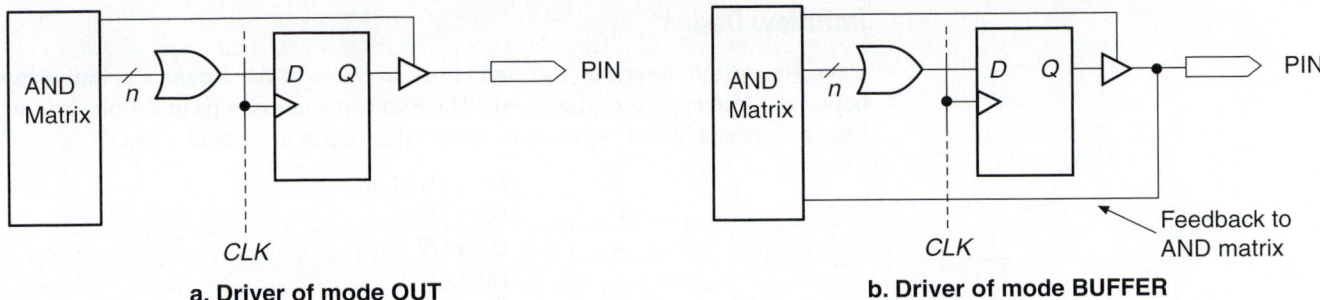

a. Driver of mode OUT   b. Driver of mode BUFFER

FIGURE 9.97 OUT Versus BUFFER

The port identifier **qo** is defined as mode BUFFER, not as OUT, because it is sometimes used as an input and sometimes as an output. A port of mode OUT can only be used as an output. A port of mode BUFFER has a feedback connection so that the output can be reused in the programmed AND matrix of the CPLD macrocell. Figure 9.97 illustrates the difference between these modes.

Rather than defining connections in the component instantiations, we would also be able to use an internal signal to connect the flip-flops. This method allows us to use a port of mode OUT, rather than BUFFER. The file **srg4str2.vhd** shows this alternative way.

Quartus II:
srg4str2.vhd
srg4str2.vwf
MAX+PLUS II:
srg4str2.vhd
srg4str2.scf

```
-- srg4str2.vhd
-- Structural description of a 4-bit serial shift register
LIBRARY ieee;
USE ieee.std_logic_1164.ALL;
LIBRARY altera;
USE altera.maxplus2.ALL;

ENTITY srg4str2 IS
    PORT (
        serial_in, clk : IN  STD_LOGIC;
        qo             : OUT STD_LOGIC_VECTOR(3 downto 0));
END srg4str2;

ARCHITECTURE right_shift of srg4str2 IS
    COMPONENT DFF
        PORT (d   : IN  STD_LOGIC;
              clk : IN  STD_LOGIC;
              q   : OUT STD_LOGIC);
    END COMPONENT;
    SIGNAL connect : STD_LOGIC_VECTOR(3 downto 0);
BEGIN
    flip_flop_3: dff
        PORT MAP (serial_in, clk, connect(3));
    dffs:
    FOR i IN 2 downto 0 GENERATE
    flip_flops_2_to_0: dff
        PORT MAP (connect(i + 1), clk, connect(i));
    END GENERATE;

    qo <= connect;
END right_shift;
```

In this case, the internal signal **connect** is used to tie the flip-flops together. The circuit output derives from a signal assignment statement at the end of the file. Since the internal signal **connect** is used to fulfil the flip-flop input/output functions, **qo** can be defined solely as an output.

Dataflow Design

Dataflow design describes a design entity in terms of the Boolean relationships between different parts of the circuit. The Boolean relationships in a 4-bit shift register are defined by the expressions for the flip-flop synchronous inputs:

$$D3 = serial\_in$$
$$D2 = Q3$$
$$D1 = Q2$$
$$D0 = Q1$$

The design entity **srg4dflw.vhd** illustrates the use of the dataflow design method for a 4-bit serial shift register.

Quartus II:
srg4dflw.vhd
srg4dflw.vwf
MAX+PLUS II:
srg4dflw.vhd
srg4dflw.scf

```
-- srg4dflw.vhd
-- Dataflow description of a 4-bit serial shift register
LIBRARY ieee;
USE ieee.std_logic_1164.ALL;

ENTITY srg4dflw IS
   PORT (
       serial_in, clk : IN      STD_LOGIC;
       q                 : BUFFER STD_LOGIC_VECTOR(3 downto 0) );
END srg4dflw;

ARCHITECTURE right_shift of srg4dflw IS
   SIGNAL d : STD_LOGIC_VECTOR(3 downto 0);
BEGIN
   PROCESS (clk)
   BEGIN
     -- Define a 4-bit D flip-flop
     IF clk'EVENT and clk = '1' THEN
        q <= d;
     END IF;
   END PROCESS;

   d <= serial_in & q(3 downto 1);
END right_shift;
```

Before the flip-flops can be connected, they must be defined in a PROCESS statement. The statements inside the process are sequential, as they must be to define a flip-flop, but the process itself is a concurrent statement. Signals are applied concurrently (simultaneously) to the construct implied by the process (the flip-flops) and all other concurrent constructs in the design entity (the connections between **q** and **d** and the serial input).

A signal assignment statement implements the Boolean equations for the shift register. It is written as a single statement for efficiency, but could also be written as four separate assignment statements, as follows:

```
d(3) <= serial_in;
d(2) <= q(3);
d(1) <= q(2);
d(0) <= q(1);
```

We must define **q** as mode BUFFER, since we are using it as both input and output.

Behavioral Design

We can create a VHDL design entity from the description of its desired behavior. In the case of a shift register, we know that after a clock pulse all data move over one position and the first flip-flop in the chain accepts a bit from a serial input, as

TABLE 9.15 Next States of Flip-Flops in a Serial Shift Register

Q_3	Q_2	Q_1	Q_0
serial_in	Q_3	Q_2	Q_1

Quartus II:
srg4behv.vhd
srg4behv.vwf
MAX+PLUS II:
srg4behv.vhd
srg4behv.scf

indicated in Table 9.15. We can use this behavioral description to implement a serial shift register, as shown in the VHDL file **srg4behv.vhd.**

```
-- srg4behv.vhd
-- Behavioral description of a 4-bit serial shift register
LIBRARY ieee;
USE ieee.std_logic_1164.ALL;

ENTITY srg4behv IS
    PORT (
        serial_in, clk : IN     STD_LOGIC;
        q                : BUFFER STD_LOGIC_VECTOR (3 downto 0) );
END srg4behv;

ARCHITECTURE right_shift of srg4behv IS
BEGIN
    PROCESS (clk)
    BEGIN
        IF (clk'EVENT and clk = '1') THEN
            q <= serial_in & q(3 downto 1);
        END IF;
    END PROCESS;
END right_shift;
```

In the behavioral design, we are not concerned with the flip-flop inputs or other internal connections; the behavioral description is sufficient for the VHDL compiler to synthesize the required hardware. Compare this to the dataflow description, where we created a set of flip-flops, then assigned Boolean functions to the *D* inputs. In this case, the behavioral design method combines these two steps into one.

EXAMPLE 9.17

Write the code for a VHDL design entity that implements a 4-bit bidirectional shift register with asynchronous clear. Create a simulation that verifies the design function.

■ **Solution**

The next states of the shift register can be described as follows:

Function	Q_3	Q_2	Q_1	Q_0	VHDL Signal Assignment
Right shift	*RSI*	Q_3	Q_2	Q_1	q <= rsi & q(3 downto 1);
Left shift	Q_2	Q_1	Q_0	*LSI*	q <= q(2 downto 0) & lsi;

The VHDL code for the bidirectional shift register, **srg4bidi.vhd,** follows. A CASE statement monitors the directional control of the shift register. We require the **others** clause of the CASE statement since the identifier **direction** is of type STD_LOGIC; the cases '0' and '1' do not cover all possible values of STD_LOGIC. We want no action to be taken in the default case, so we use the keyword **NULL.**

```
LIBRARY ieee;
USE ieee.std_logic_1164.ALL;

ENTITY srg4bidi IS
    PORT (
        clk, clear : IN STD_LOGIC;
        rsi, lsi   : IN STD_LOGIC;
```

(continued)

Quartus II:
srg4bidi.vhd
srg4bidi.vwf
MAX+PLUS II:
srg4bidi.vhd
srg4bidi.scf

```
    direction   : IN STD_LOGIC;
    q           : BUFFER STD_LOGIC_VECTOR (3 downto 0) );
END srg4bidi;

ARCHITECTURE bidirectional_shift of srg4bidi IS
BEGIN
    PROCESS (clk, clear)
    BEGIN
        IF clear = '0' THEN
            q <= (others => '0'); -- asynchronous clear
        ELSIF (clk'EVENT and clk = '1') THEN
            CASE direction IS
                WHEN '0' =>
                    q <= q(2 downto 0) & lsi; -- left shift
                WHEN '1' =>
                    q <= rsi & q(3 downto 1); -- right shift
                WHEN others =>
                    NULL;
            END CASE;
        END IF;
    END PROCESS;
END bidirectional_shift;
```

Figure 9.98 shows the simulation of the shift register, with the left shift function in the first half of the simulation and the right shift function in the second half.

FIGURE 9.98 Example 9.17: Simulation of a 4-Bit Bidirectional Shift Register

Shift Registers of Generic Width

We can use a GENERIC clause in an entity declaration to specify parameters for a shift register, much as we have done for other components. The design entity **srt_bhv.vhd** below behaviorally defines an *n*-bit right-shift register, with a default width of four bits given by the statement (GENERIC (width : POSITIVE := 4);).

```
-- srt_bhv.vhd
-- Behavioral description of an n-bit shift register
LIBRARY ieee;
USE ieee.std_logic_1164.ALL;
```

```
ENTITY srt_bhv IS
    GENERIC (width : POSITIVE := 4);
    PORT (
        serial_in, clk : IN      STD_LOGIC;
        q                 : BUFPER STD_LOGIC_VECTOR (width-1 downto 0));
END srt_bhv;

ARCHITECTURE right_shift of srt_bhv IS
BEGIN
    PROCESS (clk)
    BEGIN
        IF (clk'EVENT and clk = '1') THEN
            q(width-1 downto 0) <= serial_in & q(width-1 downto 1);
    END IF;
END PROCESS;
END right_shift;
```

Quartus II:
srt_bhv.vhd
srt8_bhv.vhd
srt8_bhv.vwf
MAX + PLUS II:
srt_bhv.vhd
srt8_bhv.vhd
srt8_bhv.vwf

The entity **srt8_bhv.vhd** instantiates the *n*-bit shift register as an 8-bit circuit by specifying the bit width in a GENERIC MAP. If no value is specified, the component is presumed to be have a default width of four, as defined in the component's entity declaration.

```
-- srt8_bhv.vhd
-- 8-bit shift register that instantiates srt_bhv
LIBRARY ieee;
USE ieee.std_logic_1164.ALL;

ENTITY srt8_bhv IS
    PORT (
        data_in, clock : IN      STD_LOGIC;
        qo                 : BUFFER STD_LOGIC_VECTOR(7 downto 0));
END srt8_bhv;

ARCHITECTURE right_shift of srt8_bhv IS
COMPONENT srt_bhv
    GENERIC (width : POSITIVE);
    PORT (
        serial_in, clk : IN  STD_LOGIC;
        q                 : OUT STD_LOGIC_VECTOR(7 downto 0));
END COMPONENT;
BEGIN
    Shift_right_8: srt_bhv
        GENERIC MAP (width=> 8)
        PORT MAP (serial_in => data_in,
                  clk        => clock,
                  q          => qo);
END right_shift;
```

EXAMPLE 9.18

Write the code for a VHDL design entity that defines a universal shift register with a generic width. (The default width is eight bits.) Instantiate this entity as a component in a file for a 16-bit universal shift register.

■ **Solution**

```
-- srg_univ.vhd
-- Universal shift register with generic width
-- Default width = 8 bits
```

(continued)

```
LIBRARY ieee;
USE ieee.std_logic_1164.ALL;
USE ieee.std_logic_arith.ALL;

ENTITY srg_univ IS
   GENERIC (width : POSITIVE := 8);
   PORT (
       clk, clear      : IN      STD_LOGIC;
       rsi, lsi        : IN      STD_LOGIC;
       function_select : IN      STD_LOGIC_VECTOR (1 downto 0);
       p               : IN      STD_LOGIC_VECTOR (width-1 downto 0);
       q               : BUFFER STD_LOGIC_VECTOR (width-1 downto 0) );
END srg_univ;

ARCHITECTURE universal_shift of srg_univ IS
BEGIN
   PROCESS (clk, clear)
   BEGIN
     IF clear = '0' THEN
         -- Conversion function to convert integer 0 to vector
         -- of any width. Requires ieee.std_logic_arith package.
         q <= CONV_STD_LOGIC_VECTOR (0, width);
       ELSIF (clk'EVENT and clk = '1') THEN
         CASE function_select IS
             WHEN "00" =>
                 q <= q; -- Hold
             WHEN "01" =>
                 q <= rsi & q(width-1 downto 1);-- Shift right
             WHEN "10" =>
                 q <= q(width-2 downto 0) & lsi;-- Shift left
             WHEN "11" =>
                 q <= p; --  Load
             WHEN OTHERS =>
                 NULL;
         END CASE;
       END IF;
   END PROCESS;
END universal_shift;
```

The following VHDL code instantiates the generic shift register as a 16-bit device.

```
-- srg16uni.vhd
-- 16-bit universal shift register (instantiates srg_univ)
LIBRARY ieee;
USE ieee.std_logic_1164.ALL;

ENTITY srg16uni IS
   PORT (
       clock, clr  : IN      STD_LOGIC;
       rsi, lsi    : IN      STD_LOGIC;
       s           : IN      STD_LOGIC_VECTOR (1 downto 0);
       parallel_in : IN      STD_LOGIC_VECTOR (15 downto 0);
       qo          : BUFFER STD_LOGIC_VECTOR (15 downto 0) );
END srg16uni;
```

srg_univ.vhd
srg16uni.vhd
srg16uni.vwf

```
            ARCHITECTURE universal_shift of srg16uni IS
               COMPONENT srg_univ
                  GENERIC (width : POSITIVE) ;
                  PORT (
                       clk, clear      : IN    STD_LOGIC;
                       rsi, lsi        : IN    STD_LOGIC;
                       function_select : IN    STD_LOGIC_VECTOR(1 downto 0);
                       p               : IN    STD_LOGIC_VECTOR(width-1 downto 0);
                       q               : BUFFER STD_LOGIC_VECTOR(width-1 downto 0));
               END COMPONENT;
            BEGIN
               Shift_universal_16: srg_univ
                  GENERIC MAP (width=> 16)
                  PORT MAP (clk            => clock,
                            clear          => clr,
                            rsi            => rsi,
                            lsi            => lsi,
                            function_select => s,
                            p              => parallel_in,
                            q              => qo);
            END universal_shift;
```

When we are designing the clear function in **srg_univ.vhd,** we must account for the fact that we must set all bits of a vector of unknown width to '0'. To get around this problem, we use a conversion function that changes an INTEGER value of 0 to a STD_LOGIC_VECTOR of **width** bits and assigns the value to the output. The required conversion function, CONV_STD_LOGIC_VECTOR (*value, number_of_bits*), is found in the **std_logic_arith** package in the **ieee** library. We could also use the construct

```
            q <= (others => '0');
```

which states that the default case is to set all bits of **q** to 0 when **clear** is 0. Since there is no other case specified, all bits of **q** are cleared.

LPM Shift Registers

The Library of Parameterized Modules contains a shift register component, **lpm_shiftreg,** that we can instantiate in a VHDL design entity. The various functions of **lpm_shiftreg** are listed in Table 9.16.

The following VHDL code instantiates **lpm_shiftreg** as an 8-bit shift register with serial input and serial output. By default the register direction is LEFT (i.e., toward the MSB).

```
            -- srg8_lpm.vhd
            -- 8-bit serial shift register (shift left by default)
            LIBRARY ieee;
            USE ieee.std_logic_1164.ALL;
            LIBRARY lpm;
            USE lpm.lpm_components.ALL;

            ENTITY srg8_lpm IS
               PORT(
                    clk        : IN  STD_LOGIC;
                    serial_in  : IN  STD_LOGIC;
                    serial_out : OUT STD_LOGIC);
            END srg8_lpm;
```

```
ARCHITECTURE lpm_shift of srg8_lpm IS
BEGIN
    Shift_8: lpm_shiftreg
        GENERIC MAP (LPM_WIDTH=> 8)
        PORT MAP (clock      => clk,
                  shiftin    => serial_in,
                  shiftout   => serial_out);
END lpm_shift;
```

TABLE 9.16 Available Functions for **lpm_shiftreg**

Function	Ports	Parameters	Description
Basic serial operation	clock, shiftin, shiftout, q[]	LPM_WIDTH	Data moves serially from **shiftin** to **shiftout**. Parallel outputs appear at **q[]**.
Load	sload, data[]	none	When **sload** = 1, **q[]** goes to the value at input **data[]** on the next positive clock edge. **Data[]** has the same width as **LPM_WIDTH.**
Synchronous clear	sclr	none	When **sclr** = 1, **q[]** goes to zero on positive clock edge.
Synchronous set	sset	LPM_SVALUE	When **sset** = 1, output goes to value of **LPM_SVALUE** on positive clock edge. If **LPM_SVALUE** is not specified **q[]** goes to all 1s.
Asynchronous clear	aclr	none	Output goes to zero when **aclr** = 1.
Asynchronous set	aset	LPM_AVALUE	Output goes to value of **LPM_AVALUE** when **aset** = 1. If **LPM_AVALUE** is not specified, outputs all go HIGH when **aset** = 1.
Directional control	none	LPM_DIRECTION	Optional direction control. Default direction is LEFT. **LPM_DIRECTION** = "LEFT" or "RIGHT". If **shiftin** and **shiftout** are used, the serial shift always goes through the entire shift register, in the direction given by **LPM_DIRECTION.**
Clock	enable	none	All synchronous functions are enabled when **enable = 1**. Defaults "enabled" when not specified.

Figure 9.99 shows a simulation of the shift register, with the data shifting from right to left (LSB to MSB). Since there is no parallel output (**q[]**) instantiated in our design, we would not normally be able to monitor the progress of bits from flip-flop to flip-flop; we would only see **shiftin,** and then, eight clock cycles later, **shiftout.** However, we are able to monitor the flip-flop states as buried nodes (**lpm_shiftreg: Shift_8|dffs[7..0]**). These buried nodes are the last eight lines in the simulation.

EXAMPLE 9.19

Modify the VHDL code just shown to make the serial shift register shift right, rather than left. Create a simulation to verify the circuit function. How do the positions of **shiftin** and **shiftout** ports relate to the internal flip-flops for the right-shift and left-shift implementations?

■ **Solution**

The modified VHDL code is shown next as design entity **srg8_lpm2.** The only difference is the addition of the parameter LPM_DIRECTION to the component instantiation statement.

FIGURE 9.99 Simulation of an 8-Bit LPM Shift Register (Shift Left)

Quartus II:
srg8_lpm2.vhd
srg8_lpm2.vwf
MAX+PLUS II:
srg8_lpm2.vhd
srg8_lpm2.scf

```
-- srg8_lpm2.vhd
-- 8-bit serial shift register (shift right)
LIBRARY ieee;
USE ieee.std_logic_1164.ALL;
LIBRARY lpm;
USE lpm.lpm_components.ALL;

ENTITY srg8_lpm2 IS
    PORT (
        clk        : IN  STD_LOGIC;
        serial_in  : IN  STD_LOGIC;
        serial_out : OUT STD_LOGIC);
END srg8_lpm2;

ARCHITECTURE lpm_shift of srg8_lpm2 IS
BEGIN
    Shift_8: lpm_shiftreg
        GENERIC MAP (LPM_WIDTH=> 8, LPM_DIRECTION => "RIGHT")
        PORT MAP ( clock    => clk,
                   shiftin  => serial_in,
                   shiftout => serial_out);
END lpm_shift;
```

The simulation for the right-shift register is shown in Figure 9.100. The inputs and outputs are identical to those of Figure 9.99, but the internal shift direction is opposite. The LPM component configures the serial shift input and output such that they allow data to go through the entire register, regardless of shift direction. For left-shift, **serial_in (shiftin)** is applied to **D0,** and is shifted toward **Q7.** For right-shift, the same **serial_in** is applied to **D7** and shifted toward **Q0.** Thus, there is no right shift input or left shift input in this component, and also no bidirectional shift that can be controlled by an input port. Shift direction can only be set by the value of a parameter and is therefore fixed when a component is instantiated.

(continued)

FIGURE 9.100   Example 9.19: Simulation of an 8-Bit LPM Shift Register (Shift Right)

EXAMPLE 9.20

Write the VHDL code for an 8-bit LPM shift register with both parallel and serial outputs, parallel load, and asynchronous clear. Create a simulation to verify the design operation.

■ Solution

The VHDL code for the parallel load shift register follows as **srg8lpm3.**

Quartus II:
srg8lpm3.vhd
srg8lpm3.vwf
MAX+PLUS II:
srg8lpm3.vhd
srg8lpm.scf

```
LIBRARY ieee;
USE ieee.std_logic_1164.ALL;
LIBRARY lpm;
USE lpm.lpm_components.ALL;

ENTITY srg8_lpm3 IS
    PORT (
        clk, ld, clr : IN  STD_LOGIC;
        p            : IN  STD_LOGIC_VECTOR (7 downto 0);
        q_out        : OUT STD_LOGIC_VECTOR (7 downto 0);
        serial_out   : OUT STD_LOGIC);
END srg8_lpm3;

ARCHITECTURE lpm_shift of srg8_lpm3 IS
BEGIN
    Shift_8: lpm_shiftreg
        GENERIC MAP (LPM_WIDTH => 8)
        PORT MAP (clock    => clk,
                  load     => ld,
                  aclr     => clr,
                  data     => p,
                  q        => q_out,
                  shiftout => serial_out);
END lpm_shift;
```

The simulation for **srg8_lpm3** is shown in Figure 9.101. The load input is initially HIGH, causing the shift register to load 55H (= 01010101) on the first clock pulse. Since we have not instantiated the serial input **shiftin,** the serial input reverts to a default value of '1', causing the register to be filled with 1s. If we did not want this to be the case, we would have to instantiate the **shiftin** port and set it to '0'.

FIGURE 9.101 Example 9.20: Simulation of an 8-Bit LPM Shift Register with Parallel Load

→ **LAB SOURCE**
Lab #28
Time Division Multiplexing
(Lab Project 2)

■ **SECTION 9.8 REVIEW PROBLEM**

9.8 When a shift register is encoded in VHDL, why are its outputs defined as BUFFER, not OUT?

9.9 SHIFT REGISTER COUNTERS

■ **KEY TERMS**

Ring Counter A serial shift register with feedback from the output of the last flip-flop to the input of the first.

Johnson Counter A serial shift register with complemented feedback from the output of the last flip-flop to the input of the first. Also called a twisted ring counter.

By introducing feedback into a serial shift register, we can create a class of synchronous counters based on continuous circulation, or rotation, of data.

If we feed back the output of a serial shift register to its input without inversion, we create a circuit called a **ring counter.** If we introduce inversion into the feedback loop, we have a circuit called a **Johnson counter.** These circuits can be decoded more easily than binary counters of similar size and are particularly useful for event sequencing.

FIGURE 9.102 *4-Bit Ring Counter*

Ring Counters

Figure 9.102 shows a 4-bit ring counter made from D flip-flops. This circuit could also be constructed from SR or JK flip-flops, as can any serial shift register.

A ring counter circulates the same data in a continuous loop. This assumes that the data have somehow been placed into the circuit upon initialization, usually by synchronous or asynchronous preset and clear inputs, which are not shown.

Figure 9.103 shows the circulation of a logic 1 through a 4-bit ring counter. If we assume that the circuit is initialized to the state $Q_3Q_2Q_1Q_0 = 1000$, it is easy to see that the 1 is shifted one place right with each clock pulse. The feedback connection from Q_0 to D_3 ensures that the input of flip-flop 3 will be filled by the contents of Q_0, thus recirculating the initial data. The final transition in the sequence shows the 1 recirculated to Q_3.

A ring counter is not restricted to circulating a logic 1. We can program the counter to circulate any data pattern we happen to find convenient.

Figure 9.104 shows a ring counter circulating a 0 by starting with an initial state of $Q_3Q_2Q_1Q_0 = 0111$. The circuit is the same as before; only the initial state has changed. Figure 9.105 shows the timing diagrams for the circuit in Figures 9.103 and 9.104.

Ring Counter Modulus and Decoding

The maximum modulus of a ring counter is the maximum number of unique states in its count sequence. In Figures 9.103 and 9.104, the ring counters each had a maximum modulus of 4. We say that 4 is the *maximum* modulus of the ring counters shown, since we can change the modulus of a ring counter by loading different data at initialization.

For example, if we load a 4-bit ring counter with the data $Q_3Q_2Q_1Q_0 = 1000$, the following unique states are possible: 1000, 0100, 0010, and 0001. If we load the same circuit with the data $Q_3Q_2Q_1Q_0 = 1010$, there are only two unique states: 1010 and 0101. Depending on which data are loaded, the modulus is 4 or 2.

Most input data in this circuit will yield a modulus of 4. Try a few combinations.

> **NOTE...**
>
> The maximum modulus of a ring counter is the same as the number of bits in its output.

A ring counter requires more flip-flops than a binary counter to produce the same number of unique states. Specifically, for n flip-flops, a binary counter has 2^n unique states and a ring counter has n.

This is offset by the fact that a ring counter requires no decoding. A binary counter used to sequence eight events requires three flip-flops and eight 3-input decoding gates. To perform the same task, a ring counter requires eight flip-flops and no decoding gates.

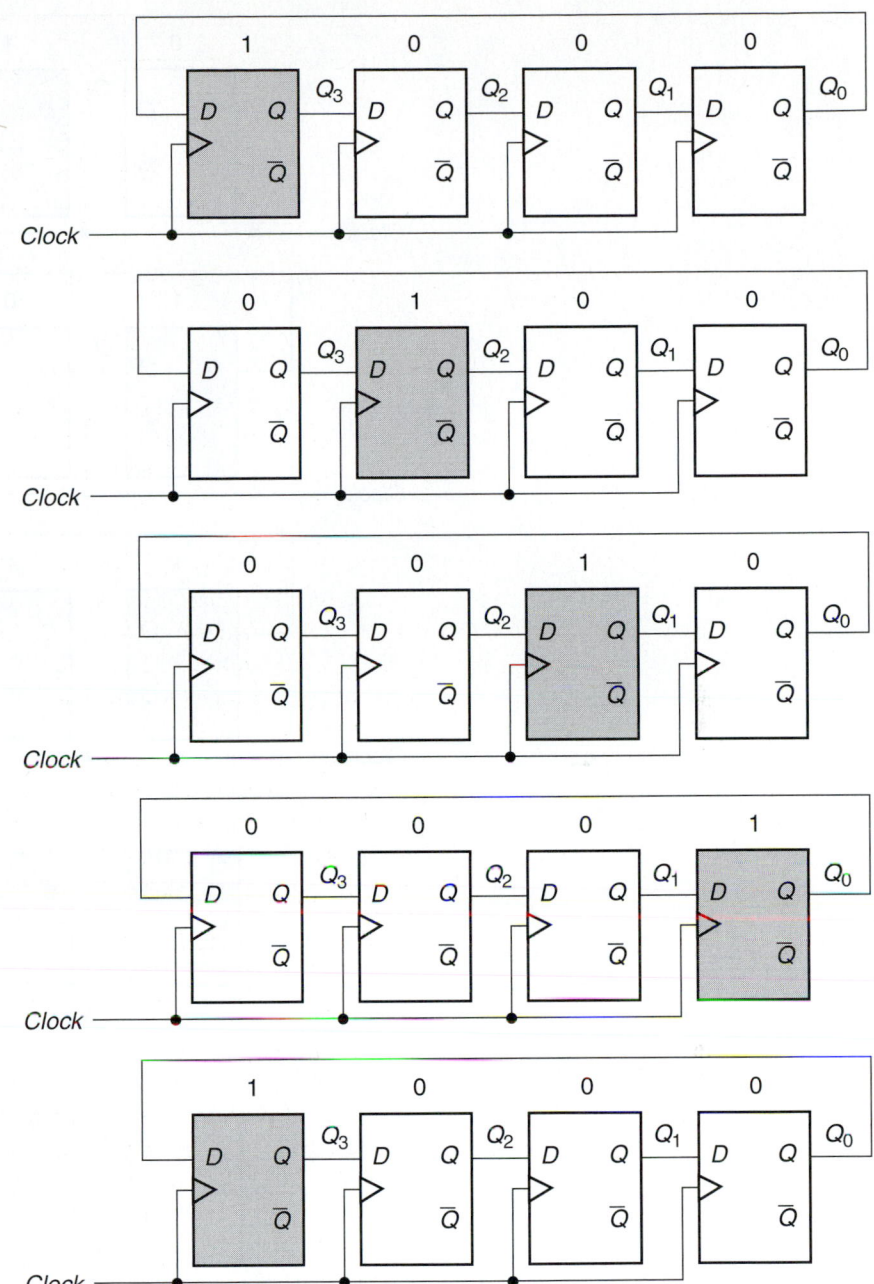

FIGURE 9.103   Circulating a 1 in a Ring Counter

As the number of output states of an event sequencer increases, the complexity of the decoder for the binary counter also increases. A circuit requiring 16 output states can be implemented with a 4-bit binary counter and sixteen 4-input decoding gates. If you need 18 output states, you must have a 5-bit counter ($2^4 < 18 < 2^5$) and eighteen 5-input decoding gates.

The only required modification to the ring counter is one more flip-flop for each additional state. A 16-state ring counter needs 16 flip-flops and an 18-state ring counter must have 18 flip-flops. No decoding is required for either circuit.

FIGURE 9.104   Circulating a 0 in a Ring Counter

Johnson Counters

Figure 9.106 shows a 4-bit Johnson counter constructed from D flip-flops. It is the same as a ring counter except for the inversion in the feedback loop where $\overline{Q}_0$ is connected to D_3. The circuit output is taken from flip-flop outputs Q_3 through Q_0. Since the feedback introduces a "twist" into the recirculating data, a Johnson counter is also called a "twisted ring counter."

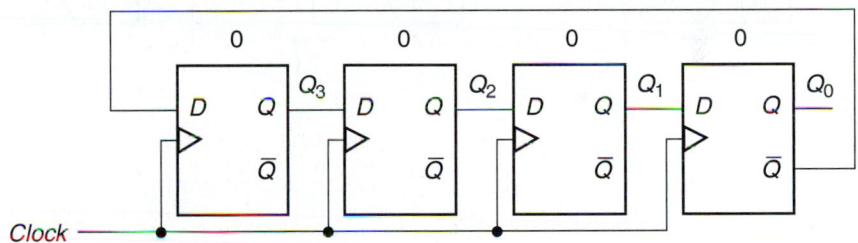

FIGURE 9.105 Timing Diagrams for Figures 9.103 and 9.104

FIGURE 9.106 4-Bit Johnson Counter

Figure 9.107 shows the progress of data through a Johnson counter that starts cleared ($Q_3Q_2Q_1Q_0 = 0000$). The shaded flip-flops represent 1s and the unshaded flip-flops are 0s. Every 0 at Q_0 is fed back to D_3 as a 1 and every 1 is fed back as a 0. The count sequence for this circuit is given in Table 9.17. There are eight unique states in the count sequence table.

TABLE 9.17 Count Sequence of a 4-Bit Johnson Counter

Q_3	Q_2	Q_1	Q_0
0	0	0	0
1	0	0	0
1	1	0	0
1	1	1	0
1	1	1	1
0	1	1	1
0	0	1	1
0	0	0	1

FIGURE 9.107 Data Circulation in a 4-Bit Johnson Counter

EXAMPLE 9.21

Write the VHDL code for a Johnson counter of generic width and instantiate it as an 8-bit counter. List the sequence of states in a table, assuming the counter is initially cleared, and create a simulation to verify the circuit's operation. Include a clear input (synchronous).

■ **Solution**

The VHDL, design entities for the generic-width component follows.

```
-- jnsn_ct.vhd
-- Johnson counter of generic width
LIBRARY ieee;
USE ieee.std_logic_1164.ALL;

ENTITY jnsn_ct IS
   GENERIC (width : POSITIVE := 4);
   PORT (
        clk, clr : IN     STD_LOGIC;
        q        : BUFFER STD_LOGIC_VECTOR (width-1 downto 0) );
END jnsn_ct;

ARCHITECTURE johnson_counter of jnsn_ct IS
BEGIN
   PROCESS (clk)
   BEGIN
      IF (clk'EVENT and clk = '1') THEN
         IF clr = '0' THEN
            q <= (others => '0'); -- n-bit clear function (n = width)
         ELSE
            q(width-1 downto 0) <= (not q(0) ) & q(width-1 downto 1);
         END IF;
      END IF;
   END PROCESS;
END johnson_counter;
```

Quartus II:
jnsn_ct.vhd
jnsn_ct8.vhd
jnsn_ct8.vwf
MAX+PLUS II:
jnsn_ct8.scf

The VHDL code for the generic Johnson counter instantiated as an 8-bit circuit is as follows.

```
-- jnsn_ct8.vhd
-- 8-bit Johnson counter using component jnsn_ct
LIBRARY ieee;
USE ieee.std_logic_1164.ALL;

ENTITY jnsn_ct8 IS
   PORT(
        clock, clear : IN     STD_LOGIC;
        qo           : BUFFER STD_LOGIC_VECTOR(7 downto 0));
END jnsn_ct8;

ARCHITECTURE johnson_counter of jnsn_ct8 IS
COMPONENT jnsn_ct GENERIC (width : POSITIVE);
   PORT(
        clk, clr : IN     STD_LOGIC;
        q        : BUFFER STD_LOGIC_VECTOR(7 downto 0) );
END COMPONENT;
BEGIN
   johnson: jnsn_ct
      GENERIC MAP (width=> 8)
         PORT MAP (clk => clock
                   clr => clear,
                   q   => qo);
END johnson_counter;
```

(continued)

Note that in the component file (**jnsn_ct.vhd**), the counter is cleared synchronously by the statement (q <= (others => '0');). Recall that the clause (others => '0') can be used to set all bits of a signal aggregate to the value '0'. This is a simple way to clear a vector of unknown width without using a conversion function.

Table 9.18 shows the count sequence for the 8-bit Johnson counter.

The simulation of the Johnson counter, including one full cycle and a clear, is shown in Figure 9.108.

TABLE 9.18   Count Sequence of an 8-Bit Johnson Counter

Q_7	Q_6	Q_5	Q_4	Q_3	Q_2	Q_1	Q_0
0	0	0	0	0	0	0	0
1	0	0	0	0	0	0	0
1	1	0	0	0	0	0	0
1	1	1	0	0	0	0	0
1	1	1	1	0	0	0	0
1	1	1	1	1	0	0	0
1	1	1	1	1	1	0	0
1	1	1	1	1	1	1	0
1	1	1	1	1	1	1	1
0	1	1	1	1	1	1	1
0	0	1	1	1	1	1	1
0	0	0	1	1	1	1	1
0	0	0	0	1	1	1	1
0	0	0	0	0	1	1	1
0	0	0	0	0	0	1	1
0	0	0	0	0	0	0	1

FIGURE 9.108   Example 9.21: Simulation of an 8-Bit Johnson Counter

Johnson Counter Modulus and Decoding

> **NOTE...**
>
> The maximum modulus of a Johnson counter is $2n$ for a circuit with n flip-flops.

The Johnson counter represents a compromise between binary and ring counters, whose maximum moduli are, respectively, 2^n and n for an n-bit counter.

If it is used for event sequencing, a Johnson counter must be decoded, unlike a ring counter. Its output states are such that each state can be decoded uniquely by a 2-input AND or NAND gate, depending on whether you need active-HIGH or active-LOW indication. This yields a simpler decoder than is required for a binary counter.

Table 9.19 shows the decoding of a 4-bit Johnson counter.

TABLE 9.19 Decoding a 4-Bit Johnson Counter

Q_3	Q_2	Q_1	Q_0	Decoder Outputs	Comment
0	0	0	0	$\bar{Q}_3\bar{Q}_0$	MSB = LSB = 0
1	0	0	0	$Q_3\bar{Q}_2$	"1/0"
1	1	0	0	$Q_2\bar{Q}_1$	Pairs
1	1	1	0	$Q_1\bar{Q}_0$	
1	1	1	1	Q_3Q_0	MSB = LSB = 1
0	1	1	1	$\bar{Q}_3Q_2$	"0/1"
0	0	1	1	$\bar{Q}_2Q_1$	Pairs
0	0	0	1	$\bar{Q}_1Q_0$	

Decoding a sequential circuit depends on the decoder responding uniquely to every possible state of the circuit outputs. If we want to use only 2-input gates in our decoder, it must recognize two variables for every state that are *both* active *only* in that state.

A Johnson counter decoder exploits what might be called the "1/0 interface" of the count sequence table. Careful examination of Table 9.17 and Table 9.18 reveals that for every state except where the outputs are all 1s or all 0s, there is a side-by-side 10 or 01 pair which exists only in that state.

Each of these pairs can be decoded to give unique indication of a particular state. For example, the pair $Q_3\bar{Q}_2$ uniquely indicates the second state since $Q_3 = 1$ AND $Q_2 = 0$ *only* in the second line of the count sequence table. (This is true for any size of Johnson counter; compare the second lines of Tables 9.17 and 9.18. In the second line of both tables, the MSB is 1 and the 2nd MSB is 0.)

FIGURE 9.109   4-Bit Johnson Counter with Output Decoding

For the states where the outputs are all 1s or all 0s, the most significant AND least significant bits can be decoded uniquely, these being the only states where MSB = LSB.

Figure 9.109 shows the decoder circuit for a 4-bit Johnson counter.

The output decoder of a Johnson counter does not increase in complexity as the modulus of the counter increases. The decoder will always consist of $2n$ 2-input AND or NAND gates for an n-bit counter. (For example, for an 8-bit Johnson counter, the decoder will consist of sixteen 2-input AND or NAND gates.)

EXAMPLE 9.22

Draw the timing diagram of the Johnson counter decoder of Figure 9.109, assuming that the counter is initially cleared.

■ **Solution**

Figure 9.110 shows the timing diagram of the Johnson counter and its decoder outputs.

FIGURE 9.110   Example 9.22: Johnson Counter Decoder Outputs

■ SECTION 9.9 REVIEW PROBLEM

9.9   How many flip-flops are required to produce 24 unique states in each of the following types of counters; binary counter, ring counter, Johnson counter? How many and what type of decoding gates are required to produce an active-LOW decoder for each type of counter?

SUMMARY

1. A counter is a circuit that progresses in a defined sequence at the rate of one state per clock pulse.

2. The modulus of a counter is the number of states through which the counter output progresses before repeating.

3. A counter with an ascending sequence of states is called an UP counter. A counter with a descending sequence of states is called a DOWN counter.

4. In general, the maximum modulus of a counter is given by 2^n for an n-bit counter.

5. A counter whose modulus is 2^n is called a full-sequence counter. The count progresses from 0 to $2^n - 1$, which corresponds to a binary output of all 0s to all 1s.

6. A counter whose output is less than 2^n is called a truncated sequence counter.

7. The adjacent outputs of a full-sequence binary counter have a frequency ratio of 2:1. The less significant of the two bits has the higher frequency.

8. The outputs of a truncated sequence counter do not necessarily have a simple frequency relationship.

9. A synchronous counter consists of a series of flip-flops, all clocked from the same source, that stores the present state of the counter and a combinational circuit that monitors the counter's present state and determines its next state.

10. A synchronous counter can be analyzed by a formal procedure that includes the following steps;

    a. Write the Boolean equations for the synchronous inputs of the counter flip-flops in terms of the present state of the flip-flip outputs.

    b. Evaluate each Boolean equation for an initial state to find the states of the synchronous inputs.

    c. Use flip-flop function tables to determine each flip-flop next state.

    d. Set the next state to the new present state.

    e. Continue until the sequence repeats.

11. The analysis procedure just noted should be applied to any unused states of the counter to ensure that they will enter the count sequence properly.

12. A synchronous counter can be designed using a formal method that relies on the excitation tables of the flip-flops used in the counter. An excitation table indicates the required logic levels on the flip-flop inputs to effect a particular transition.

13. The synchronous counter design procedure is based on the following steps:

    a. Draw the state diagram of the counter and use it to list the relationship between the counter's present and next states. The table should list the counter's present states in binary order.

    b. For the initial design, unused states can be set to a known destination, such as 0, or treated as don't care states.

    c. Use the flip-flop excitation table to determine the synchronous input levels for each present-to-next state transition.

    d. Use Boolean algebra or Karnaugh maps to find the simplest equations for the flip-flop inputs (*JK, D,* or *T*) in terms of *Q*.

    e. Unused states should be analyzed by substituting their values into the Boolean equations of the counter. This will verify whether or not an unused state will enter the count sequence properly.

14. If a counter must reset to 0 from an unused state, the flip-flops can be reset asynchronously to their initial states or the counter can be designed with the unused states always having 0 as their next state.

15. A counter can be designed in VHDL by using a behavioral description or a structural design that uses a component from the Library of Parameterized Modules (LPM).

16. A behavioral counter design requires a PROCESS statement that lists the clock signal and any asynchronous inputs in its sensitivity list. An IF statement inside the PROCESS can monitor the active clock edge by using the predefined EVENT attribute (e.g., clk'EVENT) and increment a count variable.

17. A variable is local to a PROCESS and is assigned with the : = operator. A signal is global to the VHDL design entity and is assigned wth the <= operator. (Recall that a signal is like an internal connecting wire and a variable is a piece of working memory.)

18. A binary counter can also be designed in VHDL using STD_LOGIC types only. The package **std_logic_unsigned** must be included to allow a STD_LOGIC_VECTOR to be incremented like an INTEGER. The output port must be of type BUFFER to allow it to be used as both input and output (i.e., q <= q + 1;). If a reset function is included, the output can be cleared with the construct q <= (others => '0');, which does not require a specific width of the output port.

19. A structural counter design can use an LPM component (**lpm_counter**) and instantiate the component in a component instantiation statement. The statement's generic map specifies the component parameters, and its port map indicates the correspondence between the component port names and the user port, signal, or variable names.

20. An LPM counter can be instantiated in a Quartus II Block Diagram File in two ways: (1) by using the MegaWizard Plug-In Manager, and (2) by manually instantiating the component and adjusting its ports and parameters, as required. The parameters of a MegaWizard component cannot be adjusted after the component has been created.

21. Some of the most common control features available in synchronous counters include:

    a. Synchronous or asynchronous parallel load, which allows the count to be set to any value whenever a *LOAD* input is asserted

    b. Synchronous or asynchronous clear (reset), which sets all of the counter outputs to zero

    c. Count enable, which allows the count sequence to progress when asserted and inhibits the count when deasserted

    d. Bidirectional control, which determines whether the counter counts up or down

    e. Output decoding, which activates one or more outputs when detecting particular states on the counter outputs

    f. Ripple carry out or ripple clock out (RCO), a special case of output decoding that produces a pulse upon detecting the terminal count, or last state, of a count sequence.

22. The parallel load function of a counter requires load *data* (the parallel input values) and a load *command* input, such as *LOAD,* that transfer the parallel data when asserted. If the load function is synchronous, a clock pulse is also required.

23. Synchronous load transfers data to the counter outputs on an active clock edge. Asynchronous load operates as soon as the load input activates, without waiting for the clock.

24. Synchronous load is implemented by a function select circuit that selects either the count logic or the direct parallel input to be applied to the synchronous input(s) of a flip-flop.

25. Asynchronous load can be implemented by enabling or inhibiting a pair of NAND gates, one of which asserts a flip-flop clear input and the other which asserts a preset input for the same flip-flop.

26. The count enable function enables or disables the count logic of a counter without affecting other functions, such as clock or clear. This can be done by ANDing the count logic with the count enable input signal.

27. A flip-flop in an UP counter toggles when all previous bits are HIGH. A flip-flop in a DOWN counter toggles when all previous bits are LOW. A circuit that selects one of these two conditions (a pair of AND-shape gates, combined in an OR gate; essentially a 2-to-1 multiplexer) can implement a bidirectional count.

28. An output decoder asserts one output for each counter state. A special case is a terminal count decoder that detects the last state of a count sequence.

29. RCO (ripple clock out) generates one clock pulse upon terminal count, with its positive edge at the end of the count cycle.

30. Asynchronous inputs to a behaviorally defined counter in VHDL must be included in the sensitivity list of the process defining the counter. Asynchronous inputs must be checked inside the process before the clock is checked for an active edge.

31. Synchronous inputs to a behaviorally defined counter should not be included in the sensitivity list of the process defining the counter. Synchronous inputs must be checked inside the IF statement that checks the clock edge.

32. A VHDL component can be created with parameters (such as width) that are specified when the component is instantiated. The parameters are listed in a GENERIC clause in the component's entity declaration. Each parameter must be given a default value. The parameters are specified in a generic map in the design entity that instantiates the component.

33. The conversion function CONV_STD_LOGIC is required when assigning an INTEGER value to a port of type STD_LOGIC_VECTOR. The function is contained in the **std_logic_arith** package of the **ieee** library and is of the form CONV_STD_LOGIC (integer_value, bits), where **bits** is the number of bits in the equivalent STD_LOGIC_VECTOR.

34. A shift register is a circuit for storing and moving data. Three basic movements in a shift register are serial (from one flip-flop to another), parallel (into all flip-flops at once), and

rotation (serial shift with a connection from the last flip-flop output to the first flip-flop input).

35. Serial shifting can be left (toward the MSB) or right (away from the MSB). This is the convention used by Quartus II. Some data sheets indicate the opposite relationship between right/left and LSB/MSB.

36. A function select circuit can implement several shift register variations: bidirectional serial shift, parallel load with serial shift, and universal shift (parallel/serial in/out and bidirectional in one device). The function select circuit directs data to the **D** inputs of each flip-flop from one of several sources, such as from the flip-flop immediately to the left or right or from an external parallel input.

37. A shift register can be created in VHDL by the structural, dataflow, or behavioral method.

38. A structural design instantiates components, such as D flip-flops, and connects them with internal signals.

39. A dataflow design uses internal Boolean relationships between inputs and outputs. It is similar to a structural model, except that it must contain a process statement to create the flip-flops.

40. A behavioral design method uses a description of the shift register function to generate the required hardware.

41. A ring counter is a serial shift register with the serial output fed back to the serial input so that the internal data is continuously circulated. The initial value is generally set by asynchronous preset and clear functions.

42. The maximum modulus of a ring counter is n for a circuit with n flip-flops, as compared to 2^n for a binary counter. A ring counter output is self-decoding, whereas a binary counter requires $m \leq 2^n$ AND or NAND gates with n inputs each.

43. A Johnson counter is a ring counter where the feedback is complemented. A Johnson counter has $2n$ states for an n-bit counter which can be uniquely decoded by $2n$ 2-input AND or NAND gates.

GLOSSARY

Attribute   A property associated with a named identifier in VHDL (e.g., the attribute EVENT, when associated with the identifier clk (written clk'EVENT) indicates whether a transition has occurred on the input called clk).

Behavioral Design   A VHDL design technique that uses descriptions of required behavior to create the design.

Bidirectional Counter   A counter that can count up or down, depending on the state of a control input.

Bidirectional Shift Register   A shift register that can serially shift bits left or right according to the state of a direction control input.

Binary Counter   A counter that generates a binary count sequence.

Clear   Reset (synchronous or asynchronous).

Command Lines   Signals that connect the control section of a synchronous circuit to its memory section and direct the circuit from its present to its next state.

Control Section   The combination logic portion of a synchronous circuit that determines the next state of the circuit.

Count Enable   A control function that allows a counter to progress through its count sequence when active and disables the counter when inactive.

Count Sequence   The specific series of output states through which a counter progresses.

Counter   A sequential digital circuit whose output progresses in a predictable repeating pattern, advancing by one state for each clock pulse.

Count-Sequence Table   A list of counter states in the order of the count sequence.

Dataflow Design   A VHDL design technique that uses Boolean equations to define relationships between inputs and outputs.

DOWN Counter   A counter with a descending sequence.

Excitation Table   A table showing the required input conditions for every possible transition of a flip-flop output.

Full-Sequence Counter   A counter whose modulus is the same as its maximum modulus ($m = 2^n$ for an n-bit counter.)

IF Statement   A VHDL statement that is executed only if a specified Boolean condition is satisfied.

Johnson Counter   A serial shift register with complemented feedback from the output of the last flip-flop to the input of the first. Also called a twisted ring counter.

Left Shift   A movement of data from the right to the left in a shift register. (Left is defined in Quartus II as toward the MSB.)

Maximum Modulus (m_{max})   The largest number of counter states that can be represented by n bits ($m_{max} = 2^n$).

Memory Section   A set of flip-flops in a synchronous circuit that hold its present state.

Modulo Arithmetic   A closed system of counting and adding, whereby a sum greater than the largest number in a sequence "rolls over" and starts from the beginning. For example, on a clock face, four hours after 10 A.M. is 2 P.M., so in a mod-12 system, $10 + 4 = 2$.

Modulo-n (or mod-n) counter   A counter with a modulus of n.

Modulus   The number of states through which a counter sequences before repeating.

Next State   The desired future state of flip-flop outputs in a synchronous sequential circuit after the next clock pulse is applied.

Output Decoding   A feature in which one or more outputs activate when a particular counter state is detected.

Parallel Load   A function that allows simultaneous loading of binary values into all flip-flops of a synchronous circuit. Parallel loading can be synchronous or asynchronous.

Parallel-Load Shift Register   A shift register that can be preset to any value by directly loading a binary number into its internal flip-flops.

Parallel Transfer   Movement of data into all flip-flops of a shift register at the same time.

Present State   The current state of flip-flop outputs in a synchronous sequential circuit.

Presettable Counter   A counter with a parallel load function.

Recycle   To make a transition from the last state of the count sequence to the first state.

Right Shift   A movement of data from the left to the right in a shift register. (Right is defined in Quartus II as toward the LSB).

Ring Counter   A serial shift register with feedback from the output of the last flip-flop to the input of the first.

Ripple Carry Out or Ripple Clock Out (RCO)   An output that produces one pulse with the same period as the clock upon terminal count.

Rotation   Serial shifting of data with the output(s) of the last flip-flop connected to the synchronous input(s) of the first flip-flop. The result is continuous circulation of the same data.

Serial Shifting   Movement of data from one end of a shift register to the other at a rate of one bit per clock pulse.

Shift Register   A synchronous sequential circuit that will store and move n-bit data, either serially or in parallel, in n flip-flops.

SRGn   Symbol for an n-bit shift register (e.g., SRG4 indicates a 4-bit shift register).

State Diagram   A diagram showing the progression of states of a sequential circuit.

State Machine   A synchronous sequential circuit.

Status Lines   Signals that communicate the present state of a synchronous circuit from its memory section to its control section.

Structural Design   A VHDL design technique that connects predesigned components using internal signals.

Synchronous Counter   A counter whose flip-flops are all clocked by the same source and thus change in synchronization with each other.

Terminal Count   The last state in a count sequence before the sequence repeats (e.g., 1111 is the terminal count of a 4-bit binary UP counter; 0000 is the terminal count of a 4-bit binary DOWN counter).

Truncated-Sequence Counter A counter whose modulus is less than its maximum modulus ($m < 2^n$ for an n-bit counter).

Universal Shift Register A shift register that can operate with any combination of serial and parallel inputs and outputs (i.e., serial in/serial out, serial in/parallel out, parallel in/serial out, parallel in/parallel out). A universal shift register is often bidirectional, as well.

UP Counter A counter with an ascending sequence.

PROBLEMS

Problem numbers set in color indicate more difficult problems.

9.1 Basic Concepts of Digital Counters

9.1 A parking lot at a football stadium is monitored before a game to determine whether or not there is available space for more cars. When a car enters the lot, the driver takes a ticket from a dispenser which also produces a pulse for each ticket taken.

The parking lot has space for 4095 cars. Draw a block diagram which shows how you can use a digital counter to light a LOT FULL sign after 4095 cars have entered. (Assume that no cars leave the lot until after the game, so you don't need to keep track of cars leaving the lot.) How many bits should the counter have?

9.2 Figure 9.111 shows a mod-16 which controls the operation of two digital sequential circuits, labeled Circuit 1 and Circuit 2. Circuit 1 is positive edge-triggered and clocked by counter output Q_1. Circuit 2 is negative edge-triggered and clocked by Q_3. (Q_3 is the MSB output of the counter.)

FIGURE 9.111 Problem 9.2: Mod-16 Counter Driving Two Sequential Circuits

a. Draw the timing diagram for one complete cycle of the circuit operation. Draw arrows on the active edges of the waveforms that activate Circuit 1 and Circuit 2.

b. State how many times Circuit 1 is clocked for each time that Circuit 2 is clocked.

9.3 Draw the timing diagram for one complete cycle of a mod-8 counter, including waveforms for *CLK*, Q_0, Q_1, and Q_2, where Q_0 is the LSB.

9.4 How many bits are required to make a counter with a modulus of 64? Why? What is the maximum count of such a counter?

9.5 **a.** Draw the state diagram of a mod-10 UP counter.

b. Use the state diagram drawn in part **a** to answer the following questions:

i. The counter is at state 0111. What is the count after 7 clock pulses are applied?

ii. After 5 clock pulses, the counter output is at 0001. What was the counter state prior to the clock pulses?

iii. The counter output is at 1000 after 15 clock pulses. What was the original output state?

9.6 What is the maximum modulus of a 6-bit counter? A 7-bit? 8-bit?

9.7 Draw the count sequence table and timing diagram of a mod-10 UP counter.

9.8 Draw the state diagram, count sequence table, and timing diagram of a mod-10 DOWN counter.

9.9 A mod-16 counter is clocked by a waveform having a frequency of 48 kHz. What is the frequency of each of the waveforms at Q_0, Q_1, Q_2, and Q_3?

9.10 A mod-10 counter is clocked by a waveform having a frequency of 48 kHz. What is the frequency of the Q_3 output waveform? The Q_0 waveform? Why is it difficult to determine the frequencies of Q_1 and Q_2?

9.2 Synchronous Counters

9.11 Draw the circuit for a synchronous mod-16 UP counter made from negative edge-triggered JK flip-flops.

9.12 Write the Boolean equations required to extend the counter drawn in Problem 9.11 to a mod-64 counter.

9.13 Write the J and K equations for the MSB of a synchronous mod-256 (8-bit) UP counter.

9.14 Analyze the operation of the synchronous counter in Figure 9.112 drawing a state table showing all transitions, including unused states. Use this state table to draw a state diagram and a timing diagram. What is the counter's modulus?

FIGURE 9.112 Problem 9.14: Synchronous Counter

9.15 **a.** Write the equations for the J and K inputs of each flip-flop of the synchronous counter represented in Figure 9.113.

FIGURE 9.113 Problem 9.15: Synchronous Counter

b. Assume that $Q_3Q_2Q_1Q_0 = 1010$ at some point in the count sequence. Use the equations from part **a** to predict the circuit outputs after each of three clock pulses.

9.16 Analyze the operation of the counter shown in Figure 9.114. Predict the count sequence by determining the J and K inputs and resulting transitions for each counter output state. Draw the state diagram and the timing diagram. Assume that all flip-flop outputs are initially 0.

FIGURE 9.114 Problem 9.16: Counter

9.3   Design of Synchronous Counters

9.17 Draw the timing diagram and state diagram of a synchronous mod-10 counter with a positive edge-triggered clock.

9.18 Design a synchronous mod-10 counter, using positive edge-triggered JK flip-flops. Check that unused states properly enter the main sequence. Draw a state diagram showing the unused states.

9.19 Design a synchronous mod-10 counter, using positive edge-triggered D flip-flops. Check that unused states properly enter the main sequence. Draw a state diagram showing the unused states.

9.20 Design a synchronous 3-bit binary counter using T flip-flops.

9.21 Table 9.20 shows the count sequence for a **biquinary sequence** counter. The sequence has ten states, but does not progress in binary order. The advantage of the sequence is that its most significant bit has a divide-by-10 ratio, relative to a clock input, and a 50% duty cycle. Design the synchronous counter circuit for this sequence, using D flip-flops. *Hint:* When making the state table, list all *present states* in binary order. The next states *will not* be in binary order.

9.4   Programming Binary Counters for CPLDs

9.22 Write the VHDL code for a behavioral description of a 6-bit binary counter with asynchronous clear.

9.23 Create a simulation file in Quartus II to verify the operation of the counter in Problem 9.22. (Use a 40 ns clock, which approximates the clock period of the oscillator on the Altera UP-1 and UP-2 boards.) **Note:** *To make a useful simulation, you must include the recycle point, which may be beyond the default end time of the simulation.*

9.24 Write a VHDL file that instantiates a counter from the Library of Parameterized Modules to make a 12-bit binary counter. Create a Quartus II simulation to verify the operation of the counter. (Refer to the note after Problem 9.23.)

TABLE 9.20 Biquinary Sequence

Q_3	Q_2	Q_1	Q_0
0	0	0	0
0	0	0	1
0	0	1	0
0	0	1	1
0	1	0	0
1	0	0	0
1	0	0	1
1	0	1	0
1	0	1	1
1	1	0	0

9.25 **a** Use the MegaWizard Plug-In Manager in Quartus II to create a 4-bit LPM counter, with active-LOW asynchronous clear.

b. Create a Quartus II simulation of the counter in part **a** of this problem to verify its operation.

9.26 **a.** Use the Quartus II Block Editor to manually enter a 5-bit LPM counter with active-LOW asynchronous clear.

b. Create a Quartus II simulation of the counter in part **a** of this problem to verify its operation.

9.5 Control Options for Synchronous Counters

9.27 Briefly explain the difference between asynchronous and synchronous parallel load in a synchronous counter. Draw a partial timing diagram that illustrates both functions for a 4-bit counter.

9.28 Refer to the 4-bit counter of Figure 9.39 (p. 538). The block diagram files for the counter are found on the CD accompanying this text as **4bit_sl.gdf** and **sl_count.gdf** in the folder *drive:*\qdesigns\textbook\ch09. Copy these files to a new folder and use the Quartus II block editor to expand the counter of Figure 9.39 to a 5-bit counter with synchronous load and asynchronous reset. Save and compile the file to make sure that there are no design errors.

9.29 Create a Quartus II simulation to verify the functions of the counter in Problem 9.28. The simulation must include the recycle point of the counter and show that the load is really synchronous and that the reset is really asynchronous.

9.30 Refer to the 4-bit counter of Figure 9.46 (p. 543). The block diagram files for the counter are found on the accompanying CD as **4bit_sle.gdf** and **sl_count.gdf** in the folder *drive:*\qdesigns\textbook\ch09. Copy these files to a new folder and modify the synchronous count element **sl_count.gdf** so that it implements an active-HIGH synchronous load and an active-LOW synchronous clear function, as well as the binary count function. Create a symbol for the new element and substitute it in **4bit_sle.gdf** for the existing counter elements **sl_count.** The load function should have priority over count enable, and clear (reset) should have priority over both. Save and compile the new file. *Hints:* (1) The clear function makes $Q = 0$ after a clock pulse. (2) Q follows D.

9.31 Create a Quartus II simulation to verify the functions of the counter in Problem 9.30. The simulation must include the recycle point of the counter and show that the load and clear really are synchronous and that load has priority over count enable and clear has priority over both.

9.32 Derive the Boolean equations for the synchronous DOWN-counter in Figure 9.48.

9.33 Write the Boolean equations for the count logic of the 4-bit bidirectional counter in Figure 9.51. Briefly explain how the logic works.

9.34 Draw a Quartus II Block Diagram File for a bidirectional counter, using T flip-flops. Create a simulation of the counter to verify its function.

9.35 Use Quartus II to create a synchronous bidirectional counter with synchronous load, asynchronous reset, and count enable. The count enable should not affect the operation of the load and reset functions. The functions should have the following priority: (1) clear; (2) load; and (3) count. Create a Quartus II simulation to verify the operation of your design.

9.6 Programming Presettable and Bidirectional Counters for CPLDs

9.36 Write the VHDL code for a counter that uses a behavioral description of the following functions: 12-bit binary UP count, active-LOW asynchronous clear, active-LOW synchronous load, active-LOW count enable, terminal count decoder. The clear function should have the highest priority, followed by load, then count enable. Create a simulation in Quartus II that verifies the functions of this counter.

9.37 Write the VHDL code for a behavioral description of a bidirectional counter with a modulus of 24. The counter should also have an active-LOW synchronous clear function that has priority over the count. Create a Quartus II simulation file to verify the counter operation.

9.38   Write the VHDL code for a 4-bit counter with two decoding outputs called **eq8** and **eq12**. **eq8** goes HIGH when the count equals 8 and **eq12** goes HIGH when the count equals 12 (decimal). The counter should also have an active-LOW asynchronous clear function that has priority over the count. Create a Quartus II simulation file to verify the counter operation.

9.39   **a.** Write a VHDL file for a counter with a generic width (default width of 6) and a parameter that can be used to select an asynchronous set function. The function should asynchronously set the counter output to the value of parameter **aset_value** (default value of 16) when port **aset** = 1. An asynchronous clear function should override all other functions.

   b. Write a set of simulation criteria for the counter that verifies its operation.

   c. Create a Quartus II simulation, based on the simulation criteria you wrote for part **b** of this problem.

9.40   **a.** Write a VHDL file that instantiates the counter designed in Problem 9.39 as a counter with a width of 4 and an asynchronous set value of 7.

   b. Create a Quartus II simulation to verify the operation of the counter in part **a** of this problem.

9.41   **a.** Write a VHDL file for a counter with a generic width (default width of 4) and two parameters that can be used to select an asynchronous set function. If the parameter **aset_type** = "ALL_ONES", set the counter output to all 1's when the port **aset** = 1. If **aset_type** = "PORT_VALUE", set the counter output to the value of parameter **aset_value** (default value of 16) when **aset** = 1. An asynchronous clear function should override all other functions.

   Hints: (1) The parameter **aset_type** must be declared as type STRING. (2) Use an IF statement in the counter description code to select one of the two parameter values. (3) Use the conversion function CONV_STD_LOGIC *(integer_value, bits)* to convert **aset_value** from INTEGER to STD_LOGIC_VECTOR (see Quartus II Help for an example). This conversion function requires the statement USE ieee.std_logic_arith.ALL; to be included in the code.

   b. Write a set of simulation criteria for the counter that verifies its operation.

   c. Create a Quartus II simulation, based on the simulation criteria you wrote for part **b** of this problem.

9.42   **a.** Write a VHDL file that instantiates the counter designed in Problem 9.41 as a counter with a width of 4 and an asynchronous set value of 7.

   b. Create a Quartus II simulation to verify the operation of the counter in part **a** of this problem.

9.43   Modify the VHDL code in Example 9.10 so that the counter synchronously sets to all 1s (= 4095), rather than to 2047. Do not use SVALUE = 4095. Create a simulation in Quartus II that verifies the operation of the counter. State the main difference between the code for Example 9.10 and the solution to this problem.

9.44   Use a counter from the Library of Parameterized Modules to implement the counter described in Problem 9.37. Create a Quartus II simulation file to verify the operation of the counter.

9.45   Write a VHDL file that instantiates an 8-bit LPM count with synchronous load and clear, count enable, and directional control. Also include a terminal count decoder. (The LPM counter has no port for the terminal count function, so it must be done separately.) Create a Quartus II simulation to verify the operation of the counter.

9.46   **a.** Use the Quartus II MegaWizard Plug-In Manager to create a graphical LPM counter with active-HIGH count enable, active-LOW asynchronous clear, active-HIGH synchronous set to 16_{10}, and a modulus of 25_{10}.

   b. Create a Quartus II simulation to verify the operation of the counter.

9.47 **a.** Use the Quartus II Block Editor to manually enter an LPM counter with the following features: 12-bit binary count, active-LOW asynchronous clear, active-LOW asynchronous load, pins for decoded outputs **eq[0]** and **eq[8].**

b. Create a Quartus II simulation that verifies the operation of the counter in part **a** of this problem.

9.7 Shift Registers

9.48 Use the Quartus II Block Editor to draw the circuit of a serial shift register constructed from JK flip-flops. Create a simulation to verify the operation of the shift register.

9.49 Use the Quartus II Block Editor to create the logic diagram of the 4-bit serial shift register based on JK flip-flops that shifts left, rather than right. Create a simulation to verify the operation of the shift register.

9.50 The following bits are applied in sequence to the input of a 6-bit serial right-shift register: 0111111 (0 is applied first). Draw the timing diagram.

9.51 After the data in Problem 9.50 are applied to the 6-bit shift register, the serial input goes to 0 for the next 8 clock pulses and then returns to 1. Write the internal states, Q_5 through Q_0, of the shift register flip-flops after the first 2 clock pulses. Write the states after 6, 8, and 10 clock pulses.

9.52 Complete the timing diagram of Figure 9.115, which is for a serial shift register (right-shift). Assume the shift register is initially cleared. What happens to the state of the circuit if D_7 stays HIGH beyond the end of the diagram and the *CLK* input continues to pulse?

FIGURE 9.115 Problem 9.52: Timing Diagram

9.53 An 8-bit right-shift serial-in serial-out shift register is initially cleared and has the following data clocked into its serial input: 1011001110. Draw a timing diagram of the circuit showing the *CLK, Serial Input,* and *Serial Output.* (Assume that the individual flip-flop outputs are not accessible.)

9.54 Complete the logic circuit shown in Figure 9.116 to make a bidirectional shift register.

FIGURE 9.116 Problem 9.54: Logic Circuit

9.55 Complete the logic circuit shown in Figure 9.117 to make a parallel-in-serial-out shift register.

FIGURE 9.117 Problem 9.55: Logic Circuit

9.8 Programming Shift Registers in VHDL

9.56 Write the VHDL code for an 8-bit serial shift register using a structural design procedure. Use JK flip-flops. (MAX+PLUS II primitive: JKFF.) Create a Quartus II simulation file to verify the operation of your design.

9.57 Repeat Problem 9.56 using a dataflow design procedure.

9.58 Modify the VHDL code for the behaviorally designed shift register **srg4behv.vhd** so that the shift register moves the data left, not right. *Hint:* The statement q (3 downto 0) <= serial_in & q(3 downto 1); is equivalent to the following two statements:

```
q(3) <= serial_in;
q(2 downto 0) <= q(3 downto 1);
```

Create a simulation file to verify the operation of this device.

9.59 Modify the VHDL code for the left-shift register Problem 9.58 to make a shift register of generic width. Use this component in another VHDL file to make a 32-bit shift register that shifts left. Create a simulation file to verify the operation of this design.

9.60 Write the code for a VHDL design entity that implements a 4-bit universal shift register with asynchronous clear. Create a simulation that verifies the design function.

9.61 Use Quartus II to create simulations for the generic-width and the 16-bit universal shift registers in Example 9.18. What is the difference in width between the default value of the generic shift register and the instantiated component in the 16-bit file? Given this difference, why can the generic-width shift register be correctly used as a component in the 16-bit design entity?

9.62 a. Write a VHDL file that uses a GENERIC clause to specify the width and direction of a shift register (default width = 4, default direction = "RIGHT," alternative direction = "LEFT").

   b. Write a set of simulation criteria that test the function of the shift register, using the default direction.

   c. Use the criteria written in part **b** of this question to create a simulation in Quartus II.

9.63 a. Write a VHDL file that instantiates the generic shift register of Problem 9.62 as an 8-bit shift register with a shift direction of "LEFT."

   b. Write a set of simulation criteria for the instantiated shift register.

   c. Use the criteria of part **b** of this problem to create a Quartus II simulation.

9.64 Use an LPM shift register in a VHDL file to instantiate a 48-bit shift register with the following functions: serial input, parallel output, synchronous clear.

9.65 Use an LPM shift register in a VHDL file to instantiate a 10-bit shift register with the following functions: serial input and output whose internal value can be synchronously set to 960. Create a Quartus II simulation to verify the operation of the design.

9.9   Shift Register Counters

9.66 Write the VHDL code for a ring counter of generic width and instantiate it as an 8-bit ring counter. List the sequence of states in a table, assuming the counter is initially cleared, and create a simulation to verify the circuit's operation. Include a clear input (synchronous).

9.67 Construct the count sequence table of a 5-bit Johnson counter, assuming the counter is initially cleared. What changes must be made to the decoder part of the circuit in Figure 9.109 (p. 610) if it is to decode the 5-bit Johnson counter?

9.68 A control sequence has ten steps, each activated by a logic HIGH. Use Quartus II to design a counter and decoder in each of the following configurations to produce the required sequence: binary counter, ring counter, and Johnson counter. You may use a Block Diagram File or VHDL. Create a simulation for each counter and decoder.

9.69 Use the Quartus II Block Editor to design a 4-bit ring counter that can be asynchronously initialized to $Q_3Q_2Q_1Q_0 = 1000$ by using only the clear inputs of its flip-flops. No presets allowed. *Hint:* use a circuit with a "double twist" in the data path.

ANSWERS TO SECTION REVIEW PROBLEMS

9.1

9.1 A mod-24 UP counter goes from 00000 to 10111 (0 to 23). This requires 5 outputs. The counter is a truncated sequence since its modulus is less than $2^5 = 32$.

9.2

9.2 1001, 0000

9.3

9.3 JK flip-flops: $J_3K_3 = X0$, $J_2K_2 = 1X$, $J_1K_1 = X1$, $J_0K_0 = X1$
D flip-flops: $D_3 = 1$, $D_2 = 1$, $D_1 = 0$, $D_0 = 0$

9.4

9.4
```
IF (clock'EVENT AND clock = '0') THEN
        count := count + 1;
END IF;
```

9.5

9.5 The completed timing diagram is shown in Figure 9.118.

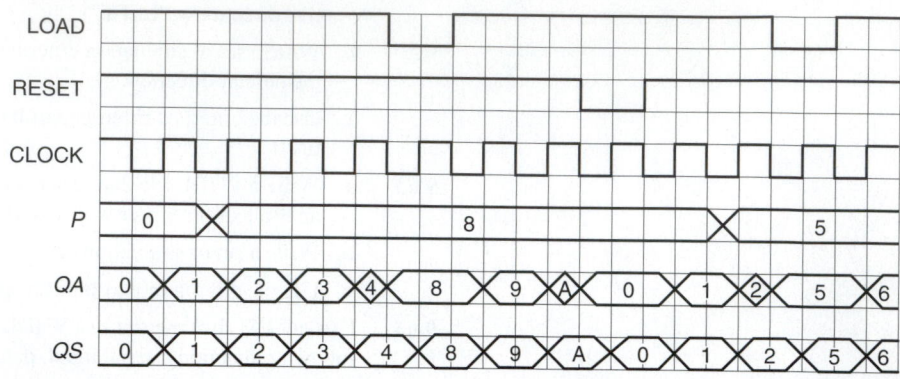

FIGURE 9.118 Answer to Section Review Problem 9.5

9.6

9.6 Asynchronous clear: PROCESS (clock, clear); Synchronous clear: PROCESS (clock)

9.7

9.7 JK flip-flops can be used in the shift register of Figure 9.83. The Q output of any stage connects to the J input of the next stage and the $\overline{Q}$ output of any stage connects to the K input of the next. The **serial_in** input connects directly to the J input of the first flip-flop. **Serial_in** is applied to K of the first flip-flop through an inverter (NOT gate).

9.8

9.8 A shift register output is defined as a port of mode BUFFER because this mode allows a signal to be fed back into the PLD matrix and reused as an input to another part of the circuit.

9.9

9.9 Binary: 5 flip-flops, 24 5-inputs NANDs; Ring: 24 flip-flops, no decoding required; Johnson: 12 flip-flops, 24 2-input NANDs.

KEY TERMS

Bubble

Conditional Transition

Control Input

Enumerated Type

Form A Contact

Form B Contact

Form C Contact

Mealy Machine

Moore Machine

State Machine

State Variables

Unconditional Transition

OUTLINE

10.1   State Machines

10.2   State Machines with No Control Inputs

10.3   State Machines with Control Inputs

10.4   Switch Debouncer for a Normally Open Pushbutton Switch

10.5   Unused States in State Machines

10.6   Traffic Light Controller

CHAPTER OBJECTIVES

Upon successful completion of this chapter you will be able to:

- Describe the components of a state machine.

- Distinguish between Moore and Mealy implementations of state machines.

- Draw the state diagram of a state machine from a verbal description.

- Use the "classical" (state table) method of state machine design to determine the Boolean equations of the state machine.

- Translate the Boolean equations of a state machine into a Block Diagram File in Altera's Quartus II software.

- Write VHDL code to implement state machines.

- Create simulations in Quartus II to verify the function of a state machine design.

- Determine whether the output of a state machine is vulnerable to asynchronous changes of input.

- Design state machine applications, such as a switch debouncer, a single-pulse generator, and a traffic light controller.

CHAPTER 10

State Machine Design

VHDL CONSTRUCTS USED IN THIS CHAPTER

- Enumerated type
- TYPE definition statement
- SIGNAL declaration
- PROCESS statement
- CASE statement
- Selected signal assignment statement

10.1 STATE MACHINES

The synchronous counters and shift registers we examined in Chapter 9 are examples of a larger class of circuits known as **state machines.** As described for synchronous counters in Section 9.2, a state machine consists of a memory section that holds the present state of the machine and a control section that determines the machine's next state. These sections communicate via a series of command and status lines. Depending on the type of machine, the outputs will either be functions of the present state only or of the present and next states.

Figure 10.1 shows the block diagram of a **Moore machine.** The outputs of a Moore machine are determined solely by the present state of the machine's memory section. The output may be directly connected to the Q outputs of the internal flip-flops, or the Q outputs might pass through a decoder circuit. The output of a Moore machine is synchronous to the system clock, since the output can only change when the machine's internal **state variables** change.

The block diagram of a **Mealy machine** is shown in Figure 10.2. The outputs of the Mealy machine are derived from the combinational (control) section of the machine, as well as the sequential (memory) part of the machine. Therefore, the outputs can change asynchronously when the combinational circuit inputs change out

FIGURE 10.1 Moore-Type State Machine

FIGURE 10.2 Mealy-Type State Machine

of phase with the clock. (When we say that the outputs change asynchronously, we generally do not mean a change via a function such as asynchronous reset that directly affects the machine's flip-flops. We simply mean a change that is not synchronized to the system clock.)

■ SECTION 10.1 REVIEW PROBLEM

10.1　What is the main difference between a Moore-type state machine and a Mealy-type state machine?

10.2　STATE MACHINES WITH NO CONTROL INPUTS

■ KEY TERM

Bubble　A circle in a state diagram containing the state name and values of the state variables.

A state machine can be designed using a classical technique, similar to that used to design a synchronous counter. We can also use a VHDL design method. We will design several state machines, using both classical and VHDL techniques.

As an example of these techniques, we will design a state machine whose output depends only on the clock input: a 3-bit counter with a Gray code count sequence. A 3-bit Gray code, shown in Table 10.1, changes only one bit between adjacent codes and is therefore not a binary-weighted sequence. Gray code is described in more detail in Section 7.5 of Chapter 7.

Gray code is often used in situations where it is important to minimize the effect of single-bit errors. For example, suppose the angle of a motor shaft is measured by a detected code on a Gray-coded shaft encoder, shown in Figure 10.3. The encoder indicates a 3-bit number for each of eight angular positions by having three concentric circular segments for each code. A dark band indicates a 1 and a transparent band indicates a 0, with the MSB as the outermost band. The dark or transparent bands are detected by three sensors that detect light shining through a transparent band. (A real shaft encoder has more bits to indicate an angle more precisely. For example, a shaft encoder that measures an angle of one degree would require nine bits, since there are 360 degrees in a circle and $2^8 \leq 360 \leq 2^9$.)

TABLE 10.1　3-Bit Gray Code Sequence

Q_2	Q_1	Q_0
0	0	0
0	0	1
0	1	1
0	1	0
1	1	0
1	1	1
1	0	1
1	0	0

FIGURE 10.3 Gray Code on a Shaft Encoder

For most positions on the encoder, the error of a single bit results in a positional error of only one eighth of the circle. This is not true with binary coding, where single bit errors can give larger positional errors. For example if the positional decoder reads 100 instead of 000, this is a difference of 4 in binary, representing an error of one-half of the circle. The same codes differ by only one position in Gray code.

Classical Design Techniques

We can summarize the classical design technique for a state machine as follows:

1. Define the problem.
2. Draw a state diagram.
3. Make a state table that lists all possible present states and inputs, and the next state and output state for each present state/input combination. *List the present states and inputs in* **binary order.**
4. Use flip-flop excitation tables to determine at what states the flip-flop synchronous inputs must be to make the circuit go from each present state to its next state. *The next state variables are functions of the inputs and present state variables.*
5. Write the output value for each present state/input combination. *The output variables are functions of the inputs and present state variables.*
6. Simplify the Boolean expression for each output and synchronous input.
7. Use the Boolean expressions found in step 6 to draw the required logic circuit.

Let us follow this procedure to design a 3-bit Gray code counter. We will modify the procedure to account for the fact that there are no inputs other than the clock and no outputs that must be designed apart from the counter itself.

1. *Define the problem.* Design a counter whose outputs progress in the sequence defined in Table 10.1.
2. *Draw a state diagram.* The state diagram is shown in Figure 10.4. In addition to the values of state variables shown in each circle (or **bubble**), we also indicate a state name, such as s0, s1, s2, and so on. This name is independent of the value of state variables. We use numbered states (s0, s1, . . .) for convenience, but we could use any names we wanted to.

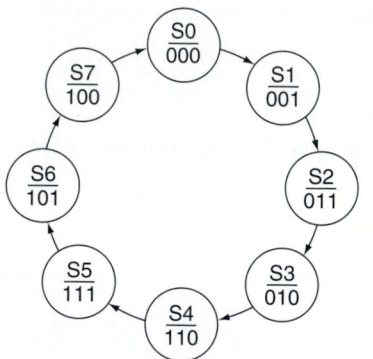

FIGURE 10.4 State Diagram for a 3-Bit Gray Code Counter

TABLE 10.2 State Table for a 3-Bit Gray Code Counter

Present State			Next State			Synchronous Inputs		
Q_2	Q_1	Q_0	Q_2	Q_1	Q_0	D_2	D_1	D_0
0	0	0	0	0	1	0	0	1
0	0	1	0	1	1	0	1	1
0	1	0	1	1	0	1	1	0
0	1	1	0	1	0	0	1	0
1	0	0	0	0	0	0	0	0
1	0	1	1	0	0	1	0	0
1	1	0	1	1	1	1	1	1
1	1	1	1	0	1	1	0	1

3. *Make a state table.* The state table, based on D flip-flops, is shown in Table 10.2. *Since there are eight unique states in the state diagram, we require three state variables* ($2^3 = 8$), *and hence three flip-flops.* Note that the present states are in binary-weighted order, even though the count does not progress in this order. In such a case, it is essential to have an accurate state diagram, from which we derive each next state. For example, if the present state is 010, the next state is not 011, as we would expect, but 110, which we derive by examining the state diagram.

Why list the present states in binary order, rather than the same order as the output sequence? By doing so, we can easily simplify the equations for the *D* inputs of the flip-flops by using a series of Karnaugh maps. This is still possible, but harder to do, if we list the present states in order of the output sequence.

4. *Use flip-flop excitation tables to determine at what states the flip-flop synchronous inputs must be to make the circuit go from each present state to its next state.* This is not necessary if we use *D* flip-flops, since *Q* follows *D*. The *D* inputs are the same as the next state outputs. For JK or T flip-flops, we would follow the same procedure as for the design of synchronous counters outlined in Chapter 9.

5. *Simplify the Boolean expression for each synchronous input.* Figure 10.5 shows three Karnaugh maps, one for each *D* input of the circuit. The K-maps yield three Boolean equations:

$$D_2 = Q_1\overline{Q}_0 + Q_2Q_0$$
$$D_1 = Q_1\overline{Q}_0 + \overline{Q}_2Q_0$$
$$D_0 = \overline{Q}_2\,\overline{Q}_1 + Q_2Q_1$$

6. *Draw the logic circuit for the state machine.* Figure 10.6 shows the circuit for a 3-bit Gray code counter, drawn as a Block Diagram File in Quartus II. A simulation for this circuit is shown in Figure 10.7, with the outputs shown as individual waveforms and as a group with a binary value.

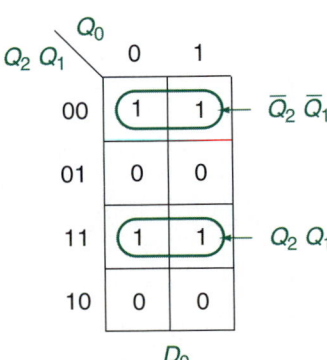

FIGURE 10.5 Karnaugh Maps for 3-Bit Gray Code Counter

FIGURE 10.6 Logic Diagram of a 3-Bit Gray Code Counter

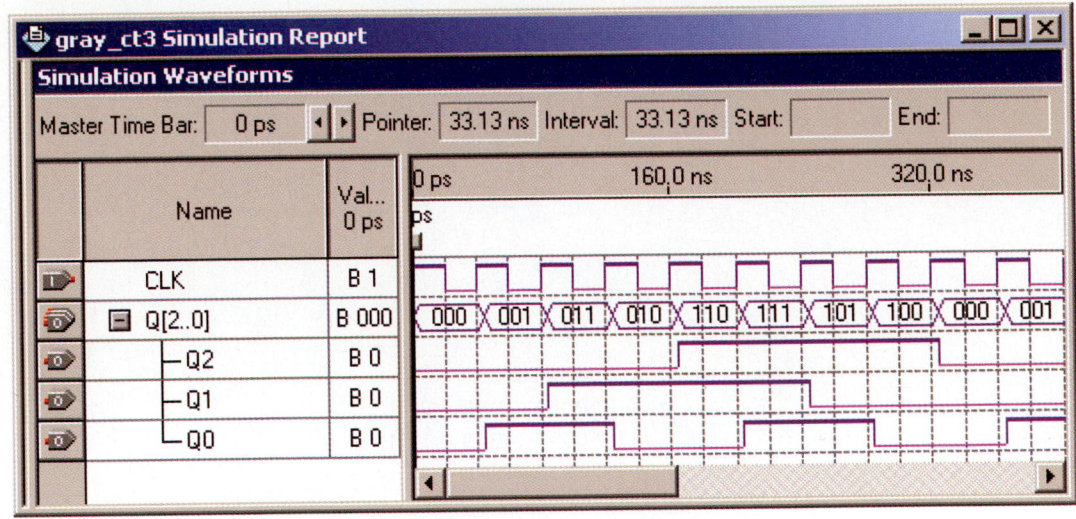

FIGURE 10.7 Simulation of a 3-Bit Gray Code Counter (from Block Diagram File)

VHDL Design of State Machines

> **■ KEY TERM**
>
> **Enumerated Type** A user-defined type in VHDL in which all possible values of a named identifier are listed in a type definition statement.

State machines can be defined in VHDL within a CASE statement. The following VHDL code illustrates the principle, using the 3-bit Gray code counter as an example.

gray_ct1.vhd

```
-- gray_ct1.vhd
-- 3-bit Gray code counter
-- (state machine with decoded outputs)

LIBRARY ieee;
USE ieee.std_logic_1164.ALL;

ENTITY gray_ct1 IS
    PORT(
        clk : IN  STD_LOGIC;
        q   : OUT STD_LOGIC_VECTOR(2 downto 0));
END gray_ct1;

ARCHITECTURE a OF gray_ct1 IS
    TYPE STATE_TYPE IS (s0, s1, s2, s3, s4, s5, s6, s7);
    SIGNAL state: STATE_TYPE;
BEGIN
    PROCESS (clk)
    BEGIN
        IF clk'EVENT AND clk = '1' THEN
            CASE state IS
                WHEN s0 =>
                    state <= s1;
                WHEN s1 =>
                    state <= s2;
                WHEN s2 =>
                    state <= s3;
                WHEN s3 =>
                    state <= s4;
```

```
                        WHEN s4 =>
                            state <= s5;
                        WHEN s5 =>
                            state <= s6;
                        WHEN s6 =>
                            state <= s7;
                        WHEN s7 =>
                            state <= s0;
                END CASE;
            END IF;
        END PROCESS;

        WITH state SELECT
            q <= "000" WHEN s0,
                 "001" WHEN s1,
                 "011" WHEN s2,
                 "010" WHEN s3,
                 "110" WHEN s4,
                 "111" WHEN s5,
                 "101" WHEN s6,
                 "100" WHEN s7;
        END a;
```

Recall that the format of a CASE statement is:

```
CASE __expression IS
    WHEN__constant_value =>
        __statement;
        __statement;
    WHEN__constant_value =>
        __statement;
        __statement;
    WHEN OTHERS =>
        __statement;
        __statement;
END CASE;
```

The keyword **expression** in the CASE statement refers to a signal called **state** that we define to represent the state variables within the machine. For each possible value of **state,** we make an assignment indicating the next state of the machine. For example, the clause WHEN s0 => (state <= s1;) indicates a transition from state s0 to state s1. The actual output values of the counter are assigned in a selected signal assignment statement after the PROCESS statement.

Notice that the signal **state** can have one of eight different values, from s0 to s7. Until now, we have seen signals with values such as '1' (BIT or STD_LOGIC types), "011" (BIT_VECTOR or STD_LOGIC_VECTOR types), or 7 (INTEGER types). The signal **state** is of type STATE_TYPE, which is a user-defined **enumerated type.** An enumerated type is simply a list of all values a signal, variable, or port of that type is allowed to have.

For example, we could define a type called DIRECTION with four values, with the statement:

```
TYPE DIRECTION IS (up, down, left, right);
```

We could then define a signal called **position** of type DIRECTION:

```
SIGNAL position: DIRECTION:
```

An IF statement or other construct could then assign one of the four defined values of type DIRECTION to the signal called **position:**

```
IF (x='0' and y='0') THEN
    position <= down;
ELSIF (x='0' and y='1') THEN
    position <= left;
ELSIF (x='1' and y='0') THEN
    position <= up;
ELSE
    position <= right;
END IF;
```

Thus the named identifier **position** of type DIRECTION can take on only the four values specified in the enumerated type definition.

An alternative way to encode the 3-bit counter is to include output assignments within the body of the CASE statement. Each case then has more than one statement, as indicated in the following VHDL code.

gray_ct2.vhd

```
-- gray_ct2.vhd
-- 3-bit Gray code counter
-- (outputs defined within states)

LIBRARY ieee;
USE ieee.std_logic_1164.ALL;

ENTITY gray_ct2 IS
    PORT(
        clk : IN  STD_LOGIC;
        q   : OUT STD_LOGIC_VECTOR(2 downto 0));
END gray_ct2;

ARCHITECTURE a OF gray_ct2 IS
    TYPE STATE_TYPE IS (s0, s1, s2, s3, s4, s5, s6, s7);
    SIGNAL state: STATE_TYPE;
BEGIN
    PROCESS (clk)
    BEGIN
        IF clk'EVENT AND clk = '1' THEN
            CASE state IS
                WHEN s0 =>
                    state <= s1;
                    q <= "001";
                WHEN s1 =>
                    state <= s2;
                    q <= "011";
                WHEN s2 =>
                    state <= s3;
                    q <= "010";
                WHEN s3 =>
                    state <= s4;
                    q <= "110";
                WHEN s4 =>
                    state <= s5;
                    q <= "111";
                WHEN s5 =>
                    state <= s6;
                    q <= "101";
                WHEN s6 =>
                    state <= s7;
                    q <= "100";
                WHEN s7 =>
                    state <= s0;
```

```
                                q <= "000";
                        END CASE;
                    END IF;
                END PROCESS;
            END a;
```

This VHDL code is identical to that of the previous example, except for the way the outputs are assigned.

■ SECTION 10.2 REVIEW PROBLEM

10.2   Write the Boolean equations for the *J* and *K* inputs of the flip-flops in a 3-bit Gray code counter based on JK flip-flops.

10.3 STATE MACHINES WITH CONTROL INPUTS

➡ *LAB SOURCE*
Lab #29
State Machines

> ### ■ KEY TERMS
>
> **Control Input**   A state machine input that directs the machine from state to state.
>
> **Conditional Transition**   A transition between states of a state machine that occurs only under specific conditions of one or more control inputs.
>
> **Unconditional Transition**   A transition between states of a state machine that occurs regardless of the status of any control inputs.

As an extension of the techniques used in the previous section, we will examine the design of state machines that use **control inputs,** as well as the clock, to direct their operation. Outputs of these state machines will not necessarily be the same as the states of the machine's flip-flops. As a result, this type of state machine requires a more detailed state diagram notation, such as that shown in Figure 10.8.

The state machine represented by the diagram in Figure 10.8 has two states, and thus requires only one state variable. Each state is represented by a bubble (circle) containing the state name and the value of the state variable. For example, the bubble containing the notation $\frac{start}{0}$ indicates that the state called **start** corresponds to a state variable with a value of 0. Each state must have a unique value for the state variable(s).

Transitions between states are marked with a combination of input and output values corresponding to the transition. The inputs and outputs are labeled **in1, in2, . . . , in*x*/out1, out2, . . . , out*x*.** The inputs and outputs are sometimes simply

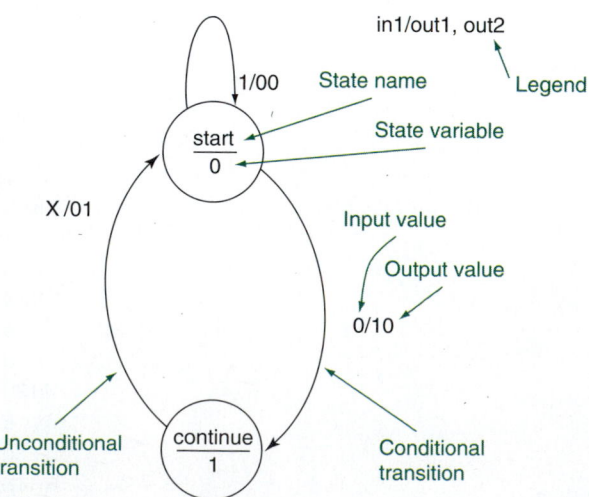

FIGURE 10.8   State Diagram Notation

FIGURE 10.9 Ideal Operation of State Machine in Figure 10.8

indicated by the value of each variable for each transition. In this case, a legend indicates which variable corresponds to which position in the label.

For example, the legend in the state diagram of Figure 10.8 indicates that the inputs and outputs are labeled in the order **in1/out1, out2.** Thus if the machine is in the **start** state and the input **in1** goes to 0, there is a transition to the state **continue.** During this transition, **out1** goes to 1 and **out2** goes to 0. This is indicated by the notation 0/10 beside the transitional arrow. This is called a **conditional transition** because the transition depends on the state of **in1.** The other possibility from the **start** state is a no-change transition, with both outputs at 0, if **in1** = 1. This is shown as 1/00.

If the machine is in the state named **continue,** the notation X/01 indicates that the machine makes a transition back to the **start** state, regardless of the value of **in1,** and that **out1** = 0 and **out2** = 1 upon this transition. Since the transition always happens, it is called an **unconditional transition.**

What does this state machine do? We can determine its function by analyzing the state diagram:

1. There are two states, called **start** and **continue.** The machine begins in the **start** state and waits for a LOW input on **in1.** As long as **in1** is HIGH, the machine waits and the outputs **out1** and **out2** are both LOW.
2. When **in1** goes LOW, the machine makes a transition to **continue** in one clock pulse. Output **out1** goes HIGH.
3. On the next clock pulse, the machine goes back to **start.** The output **out2** goes HIGH and **out1** goes back LOW.
4. If **in1** is HIGH, the machine waits for a new LOW on **in1.** Both outputs are LOW again. If **in1** is LOW, the cycle repeats.

In summary, the machine waits for a LOW input on **in1,** then generates a pulse of one clock cycle duration on **out1,** then on **out2.** A timing diagram describing this operation is shown in Figure 10.9.

Classical Design of State Machines with Control Inputs

We can use the classical design technique of the previous section to design a circuit that implements the state diagram of Figure 10.8.

1. *Define the problem.* Implement a digital circuit that generates a pulse on each of two outputs, as previously described. For this implementation, let us use JK flip-flops for the state logic. If we so choose, we could also use D or T flip-flops.
2. *Draw a state diagram.* The state diagram is shown in Figure 10.8.
3. *Make a state table.* The state table is shown in Table 10.3. The combination of present state and input are listed in binary order, thus making Table 10.3 into a truth table for the next state and output functions. Since there are two states, we require one state variable, Q. The next state of Q, a function of the present state and the input **in1,** is determined by examining the state diagram. (Thus, if you are in state 0, the next state is 1 if **in1** = 0 and 0 if **in1** = 1. If you are in state 1, the next state is always 0.)

TABLE 10.3 State Table for State Diagram in Figure 10.8

Present State	Input	Next State	Sync. Inputs	Outputs	
Q	in1	Q	JK	out1	out2
0	0	1	1X	1	0
0	1	0	0X	0	0
1	0	0	X1	0	1
1	1	0	X1	0	1

TABLE 10.4 JK Flip-Flop
Excitation Table

Transition	JK
0→0	0X
0→1	1X
1→0	X1
1→1	X0

4. *Use flip-flop excitation tables to determine at what states the flip-flop synchronous inputs must be to make the circuit go from each present state to its next state.* Table 10.4 shows the flip-flop excitation table for a JK flip-flop. The synchronous inputs are derived from the present-to-next state transitions in Table 10.4 and entered into Table 10.3. (Refer to the synchronous counter design process in Chapter 9 for more detail about using flip-flop excitation tables.)

5. *Write the output values for each present state/input combination.* These can be determined from the state diagram and are entered in the last two columns of Table 10.3.

6. *Simplify the Boolean expression for each output and synchronous input.* The following equations represent the next state and output logic of the state machine:

$$J = \overline{Q} \cdot \overline{in1} + Q \cdot \overline{in1} = \overline{in1}$$
$$K = 1$$
$$out1 = \overline{Q} \cdot \overline{in1}$$
$$out2 = Q \cdot \overline{in1} + Q \cdot in1 = Q$$

7. *Use the Boolean expressions found in step 6 to draw the required logic circuit.*

Figure 10.10 shows the circuit of the state machine drawn as a Quartus II Block Diagram File. Since **out1** is a function of the control section and the memory section of the machine, we can categorize the circuit as a Mealy machine. (All counter circuits that we have previously examined have been Moore machines since their outputs are derived solely from the flip-flop outputs of the circuit.)

Since the circuit is a Mealy machine, it is vulnerable to asynchronous changes of output due to asynchronous input changes. This is shown in the simulation waveforms of Figure 10.11.

Ideally, **out1** should not change until the first positive clock edge after **in1** goes LOW. However, since **out1** is derived from a combinational output, it will change as soon as **in1** goes LOW, after allowing for a short propagation delay. Also, since **out2** is derived directly from a flip-flop and **out1** is derived from the same flip-flop via a gate, **out1** stays HIGH for a short time after **out2** goes HIGH. (The extra time represents the propagation delay of the gate.)

If output synchronization is a problem (and it may not be), it can be fixed by adding a synchronizing D flip-flop to each output, as shown in Figure 10.12.

The state variable is stored as the state of the JK flip-flop. This state is clocked through a D flip-flop to generate **out2** and combined with **in1** to generate **out1** via another flip-flop. The simulation for this circuit, shown in Figure 10.13, indicates that the two outputs are synchronous with the clock, but delayed by one clock cycle after the state change.

Quartus II:
state_x2a.bdf
state_x2a.vwf
MAX+PLUS II:
state_x2a.gdf
state_x2a.scf

Quartus II:
state_x3a.bdf
state_x3a.vwf
MAX+PLUS II:
state_x3a.gdf
state_x3a.scf

FIGURE 10.10 Implementation of State Machine of Figure 10.8

FIGURE 10.11 Simulation of State Machine Circuit of Figure 10.10

FIGURE 10.12 State Machine with Synchronous Outputs

FIGURE 10.13 Simulation of the State Machine in Figure 10.12

VHDL Implementation of State Machines with Control Inputs

The VHDL code for a state machine with one or more control inputs is similar to that for a machine with no control inputs. The machine states are still defined using a CASE statement, but a case representing a conditional transition will contain an IF statement.

The VHDL code for the state machine just implemented is as follows.

```
-- state_x1.vhd
-- state machine example 1
-- Two states, one input, two outputs
-- Generates a pulse on one output, then the next
-- after receiving a LOW on the input

LIBRARY ieee;
USE ieee.std_logic_1164.ALL;

ENTITY state_x1 IS
    PORT(
        clk, in1  : IN  STD_LOGIC;
        out1, out2 : OUT STD_LOGIC);
END state_x1;

ARCHITECTURE a OF state_x1 IS
    TYPE PULSER IS (start, continue);          Enumerated type lists all
    SIGNAL sequence: PULSER;                   states.
BEGIN                         State variable is called "sequence."
    PROCESS (clk)
    BEGIN
        IF clk'EVENT AND clk = '1' THEN
            CASE sequence IS
              WHEN start =>
                    IF in1 = '1' THEN
                        sequence <= start; -- no change if in1 = 1
                        out1 <= '0';
                        out2 <= '0';
                    ELSE
                        sequence <= continue; -- proceed if in1 = 0
                        out1 <= '1';          -- pulse on out1
                        out2 <= '0';
                    END IF;
              WHEN continue =>
                    sequence <= start;
                    out1 <= '0';
                    out2 <= '1';              -- pulse on out2
            END CASE;
        END IF;
    END PROCESS;
END a;
```

Quartus II:
state_x1.vhd
state_x1.vwf
MAX+PLUS II:
state_x1.vhd
state_x1.scf

Conditional transition

Unconditional transition

The transition from **start** is conditional, so the case for **start** contains an IF statement that defines the possible state transitions and their associated output states. The transition from **continue** is unconditional, so no IF statement is needed in the corresponding case.

Figure 10.14 shows the simulation for the VHDL design entity, **state_x1.vhd.**

The design of the state machine is such that if the input **in1** is held LOW beyond the end of one pulse cycle, the cycle will repeat, as shown in the simulation of Figure 10.15.

FIGURE 10.14 Simulation of the State Machine in VHDL Entity **state_x1**

FIGURE 10.15 Simulation of VHDL State Machine Showing a Repeated Cycle

 EXAMPLE 10.1

A state machine called a single-pulse generator operates as follows:

1. The circuit has two states: **seek** and **find,** an input called **sync** and an output called **pulse.**
2. The state machine resets to the state **seek.** If **sync** = 1, the machine remains in **seek** and the output, **pulse,** remains LOW.
3. When **sync** = 0, the machine makes a transition to **find.** In this transition, **pulse** goes HIGH.
4. When the machine is in state **find** and **sync** = 0, the machine remains in **find** and **pulse** goes LOW.
5. When the machine is in **find** and **sync** = 1, the machine goes back to **seek** and **pulse** remains LOW.

Use classical state machine design techniques to design the circuit for the single-pulse generator, using D flip-flops for the state logic. Use the Quartus II Block Editor to draw the state machine circuit. Create a simulation to verify the design operation. Briefly describe what this state machine does.

(continued)

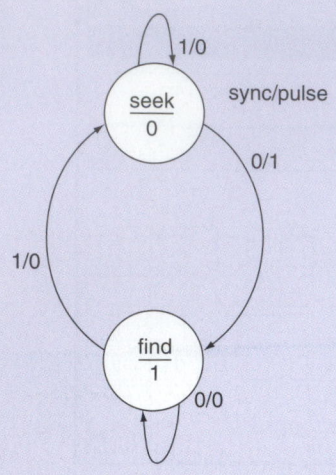

FIGURE 10.16 Example 10.1: State Diagram for a Single-Pulse Generator

Quartus II:
pulse1.bdf
pulse1.vwf
MAX+PLUS II:
pulse1.gdf
pulse1.scf

■ Solution

Figure 10.16 shows the state diagram derived from the description of the state machine. The state table is shown in Table 10.5. Since Q follows D, the D input is the same as the next state of Q.

TABLE 10.5 State Table for Single-Pulse Generator

Present State	Input	Next State	Sync. Input	Output
Q	$sync$	Q	D	$pulse$
0	0	1	1	1
0	1	0	0	0
1	0	1	1	0
1	1	0	0	0

The next-state and output equations are:

$$D = \overline{Q} \cdot \overline{sync} + Q \cdot \overline{sync} = \overline{sync}$$
$$pulse = \overline{Q} \cdot \overline{sync}$$

Figure 10.17 shows the state machine circuit derived from these Boolean equations. The simulation for this circuit is shown in Figure 10.18. The simulation shows that the circuit generates one pulse when the input **sync** goes LOW, regardless of the length of time that **sync** is LOW. The circuit could be used in conjunction with a debounced pushbutton to produce exactly one pulse, regardless of how long the pushbutton was held down. Figure 10.19 shows such a circuit.

FIGURE 10.17 Example 10.1: Single-Pulse Generator

FIGURE 10.18 Example 10.1: Simulation of a Single-Pulse Generator (from BDF)

FIGURE 10.19 Example 10.1: Single-Pulse Generator Used with a Debounced Pushbutton

The state machine of Example 10.1 is vulnerable to asynchronous input changes. How do we know this from the circuit schematic and from the simulation waveform? Modify the circuit to eliminate the asynchronous behavior and show the effect of the change on a simulation of the design. How does this change improve the design?

■ Solution

Quartus II:
pulse1a.bdf
pulse1a.vwf
MAX+PLUS II:
pulse1a.gdf
pulse1a.scf

The output, **pulse,** in the state machine of Figure 10.17 is derived from the state flip-flop and the combinational logic of the circuit. The output can be affected by a change that is purely combinational, thus making the output asynchronous. This is demonstrated on the first pulse of the simulation in Figure 10.18, where **pulse** momentarily goes HIGH between clock edges. Since no clock edge was present when either the input, **sync,** changed or when **pulse** changed, the output pulse must be due entirely to changes in the combinational part of the circuit.

The circuit output can be synchronized to the clock by adding an output flip-flop, as shown in Figure 10.20. A simulation of this circuit is shown in Figure 10.21. With the synchronized output, the output pulse is always the same width: one clock period. This gives a more predictable operation of the circuit.

FIGURE 10.20 Example 10.2: Single-Pulse Generator with Synchronous Output

FIGURE 10.21 Example 10.2: Simulation of a Single-Pulse Generator with Synchronous Output (from BDF)

EXAMPLE 10.3

Write the VHDL code for a design entity that implements the single-pulse generator, as described in Example 10.1. Write a set of simulation criteria that test the correctness of the design. Use the criteria to create a simulation in Quartus II.

■ **Solution**

The required VHDL code is given here in the design entity **sngl_pls.**

Quartus II:
sngl_pls.vhd
sngl_pls.vwf
MAX+PLUS II:
sngl_pls.vhd
sngl_pls.scf

```
LIBRARY ieee;
USE ieee.std_logic_1164.ALL;

ENTITY sngl_pls IS
    PORT(
        clk, sync : IN STD_LOGIC;
        pulse     : OUT STD_LOGIC);
END sngl_pls;

ARCHITECTURE pulser OF sngl_pls IS
    TYPE PULSE_STATE IS (seek, find);
    SIGNAL status: PULSE_STATE;
BEGIN
    PROCESS (clk, sync)
    BEGIN
        IF (clk'EVENT and clk = '1') THEN
            CASE status IS
                WHEN seek =>   IF (sync = '1') THEN
                                   status <= seek;
                                   pulse <= '0';
                               ELSE
                                   status <= find;
                                   pulse <= '1';
                               END IF;
                WHEN find =>   IF (sync = '1') THEN
                                   status <= seek;
                                   pulse <= '0';
                               ELSE
                                   status <= find;
                                   pulse <= '0';
                               END IF;
            END CASE;
        END IF;
    END PROCESS;
END pulser;
```

SIMULATION CRITERIA

■ When **sync** is held LOW for a single positive edge of **clk,** the **pulse** output should produce one pulse, exactly one clock cycle in length.

■ When **sync** is held LOW for more than one clock cycle, the **pulse** output should produce one pulse, exactly one clock cycle in length.

■ When **sync** is held LOW for a period that does not include a positive edge of **clk,** the input at **sync** should be ignored.

The simulation of the VHDL design entity **sngl_pls** is shown in Figure 10.22. The first pusle on **sync** is ignored, since it falls between positive edges of the clock waveform. The other two pulses on **sync** each generate a pulse on the **pulse** output of one clock cycle duration, regardless of the length of the pulse on **sync.**

FIGURE 10.22 Example 10.3: Simulation of a Single-Pulse Generator (from VHDL)

■ **SECTION 10.3 REVIEW PROBLEM**

10.3   Briefly explain why the single-pulse circuit in Figure 10.20 has a flip-flop on its output.

10.4 SWITCH DEBOUNCER FOR A NORMALLY OPEN PUSHBUTTON SWITCH

> ■ **KEY TERMS**
>
> **Form A Contact**   A normally open contact on a switch or relay.
>
> **Form B Contact**   A normally closed contact on a switch or relay.
>
> **Form C Contact**   A pair of contacts, one normally open and one normally closed, that operate with a single action of a switch or relay.

A useful interface function is implemented by a digital circuit that removes the mechanical bounce from a pushbutton switch. The easiest way to debounce a pushbutton switch is with a NAND latch, as shown in Figure 10.23.

The latch eliminates switch bounce by setting or resetting on the first bounce of a switch contact and ignoring further bounces. The limitation of this circuit is that the input switch must have **Form C contacts.** That is, the switch has normally open, normally closed, and common contacts. This is so that the switch resets the latch when pressed (i.e., when the normally open contact closes) and sets the latch when released (normally closed contact recloses). Each switch position activates an opposite latch function.

If the only available switch has a single set of contacts, such as the normally open (**Form A**) pushbuttons on the Altera UP-1 or UP-2 Education Board, a different

FIGURE 10.23   NAND Latch as a Switch Debouncer

debouncer circuit must be used. We will look at two solutions using VHDL: one based on an existing device (the Motorola MC14490 Contact Bounce Eliminator) and another that implements a state machine solution to the contact bounce problem.

Switch Debouncer Based on a 4-Bit Shift Register

The circuit in Figure 10.24 is based on the same principle as the Motorola MC14490 Contact Bounce Eliminator, adapted for use in an Altera CPLD, such as the EPM7128S or the EPF10K20 or EPF10K70 on the Altera UP-1 or UP-2 Education Board.

The heart of the debouncer circuit in Figure 10.24 is a single-bit comparator (an Exclusive NOR gate) and a 4-bit serial shift register, with active-HIGH synchronous load. The XNOR gate compares the shift register serial input and output. When the shift register input and output are *different,* the XNOR output is LOW, and the input data are serially shifted through the register. When input and output of the shift register are *the same,* the XNOR output is HIGH, and the binary value at the serial output is parallel-loaded back into all bits of the shift register.

Figure 10.25 shows the timing of the debouncer circuit with switch bounces on both make and break phases of the switch contact. The line labeled **lpm_shiftreg** refers to the shift register flip-flop outputs. Pushbutton input is **pb_in,** debounced output is **pb_out,** and **clock** is the UP-1 or UP-2 system clock, divided by 2^{16}.

Assume that the shift register is initially filled with 0s. The pushbutton rest state is HIGH. As shown in Figure 10.24, the pushbutton input value is inverted and applied to the shift register input. Therefore, before the switch is pressed, both input and output of the shift register are LOW. Since they are the same, the XNOR output is HIGH, which keeps the shift register in *LOAD* mode, and the LOW at **pb_out** is reloaded to the register on every positive clock edge.

FIGURE 10.24 Switch Debouncer Based on a 4-Bit Shift Register

FIGURE 10.25 Simulation of the Shift Register-Based Debouncer

When the switch is pressed, **pb_in** will bounce, with decreasing lengths of time, as shown in the simulation detail of Figure 10.26. On the LOW part of the bounce, the shift register input goes HIGH and the serial input and output are now different. The XNOR function converts this to a LOW at the **load** input of the shift register, which causes the HIGH at the serial input to shift into the register. The register fills with 1s until **pb_in** bounces HIGH again, about three clock cycles later. The serial input and output of the shift register are now the same again, both at logic 0. The XNOR function converts this condition to a HIGH at the shift register **load** input and loads the 0 at the serial output into all bits of the shift register parallel input.

Note that the shift register is instantiated as an LPM component, which has a default direction of "LEFT". Thus the simulation shows the shift register contents as 0000, 0001, 0011, 0111, 1111. The data still move from **shiftin** to **shiftout,** as shown in Figure 10.24.

When **pb_in** is LOW, the shift register serially shifts data until four 1s in a row are shifted in, making the serial output go HIGH. The serial input and output are again the same, but now at a logic 1 level, and the new logic level is parallel-loaded into the shift register.

A similar process occurs when the waveform goes back to the HIGH state. When the input goes HIGH, a LOW is shifted into the shift register. If the input bounces back LOW, the shift register is parallel-loaded with HIGHs and the process starts over. When **pb_in** is stable at a HIGH level, a LOW is shifted through the register, resulting in the binary sequence 1111, 1110, 1100, 1000, 0000, as shown in the simulation detail of Figure 10.27.

To produce an output change, the shift register input and output must remain different for at least four clock pulses. This implies that the input is stable for that period of time. If the input and output are the same, this could mean one of two things. Either the input is stable and the shift register flip-flops should be kept at a

FIGURE 10.26   Simulation Detail of a Shift Register-Based Debouncer (LOW-to-HIGH Output)

FIGURE 10.27   Simulation of a Shift Register-Based Debouncer (HIGH-to-LOW Output)

constant state or the input has bounced back to its previous level and the shift register should be reinitialized. In either case, the output value should be parallel-loaded back into the shift register. Serial shifting should only occur if there has been an input change.

The debouncer in Figure 10.24 is effective for removing bounce that lasts for no more than four clock periods. Since switch bounce is typically about 10 ms in duration, the clock should have a period of about 2.5 ms. At 25.175 MHz (a clock period of about 40 ns), the Altera UP-1 or UP-2 system clock is much too fast.

If we divide the oscillator frequency by 65536 (= 2^{16}) using a 16-bit counter, we obtain a clock waveform for the debouncer with a period of 2.6 ms. Four clock periods (10.2 ms) are sufficient to take care of switch bounce.

We can use VHDL to synthesize the switch debouncer by instantiating a counter and shift register from the Altera Library of Parameterized Modules and connecting them together with internal signals. The VHDL code is as follows.

Quartus II:
debouncer.vhd
debounce_sim.vwf
MAX+PLUS II:
debounce.vhd

```vhdl
-- debouncer.vhd
-- Debounces a single normally-open (Form A) contact.
-- Similar to a Motorola MC14490 Contact Bounce Eliminator.
-- Uses a 16-bit LPM counter to divide 25.175 MHZ clock
-- to internal clock with a period of ~2.6 ms.
-- 4-bit LPM shift register is clocked every 2.6 ms.
-- If input and output are different for 10.4 ms,
-- output changes. Otherwise, serial output is reloaded
-- into shift register and output does not change.
-- i.e., change on input must be stable for 4 clock cycles
-- before output changes.
LIBRARY ieee;
USE ieee.std_logic_1164.ALL;
LIBRARY lpm;
USE lpm.lpm_components.ALL;

ENTITY debouncer IS
PORT(
    pb_in, clock    : IN        STD_LOGIC;
    pb_out          : OUT       STD_LOGIC);
END debouncer;

ARCHITECTURE debounce OF debouncer IS
    SIGNAL ctr_q :STD_LOGIC_VECTOR(15 downto 0);
    SIGNAL srg_parallel_in : STD_LOGIC_VECTOR (3 downto 0);
    SIGNAL srg_serial_in : STD_LOGIC;
    SIGNAL srg_serial_out : STD_LOGIC;
    SIGNAL srg_load : STD_LOGIC;
BEGIN
    clock_divider: lpm_counter
        GENERIC MAP(LPM_WIDTH => 16)
        PORT MAP(clock => clock, q => ctr_q);

    -- Shift register clocked by MSB of counter (divide by 2^16)
    delay_element : lpm_shiftreg
        GENERIC MAP(LPM_WIDTH => 4)
        PORT MAP(clock   => ctr_q(15),
                load      => srg_load,
                data      => srg_parallel_in,
                shiftin   => srg_serial_in,
                shiftout  => srg_serial_out);
```

```
-- Load serial output into all shift register flip-flops
-- if (not pb_in) is the same as serial_out. Otherwise,
-- shift (not pb_in) through register.
-- Invert pb_in to make output generate a high-going pulse.
srg_load <= (not pb_in) xnor srg_serial_out;
srg_parallel_in(3) <= srg_serial_out;
srg_parallel_in(2) <= srg_serial_out;
srg_parallel_in(1) <= srg_serial_out;
srg_parallel_in(0) <= srg_serial_out;
srg_serial_in <= not pb_in;

pb_out <= srg_serial_out;
END debounce;
```

Quartus II:
two_digit_counter.bdf
count_8.vhd
hex7seg.vhd
MAX+PLUS II:
2digit.gdf
count_8.vhd
sev_segv.vhd

Figure 10.28 shows a fairly easy way to test the switch debouncer. The debouncer output is used to clock an 8-bit counter whose outputs are decoded by two seven-segment decoders. (The decoders are VHDL files developed in a similar way to the seven-segment decoders in Chapter 6.)

Pin numbers, shown in Figure 10.29, are given for the EPM7128S CPLD on the Altera UP-1 or UP-2 circuit board. Since the clock and seven-segment displays are hardwired on the Altera board, the only external connections required for the circuit are wires for the two pushbutton inputs, **reset** and **pb_in.**

FIGURE 10.28 Test Circuit for a Switch Debouncer

FIGURE 10.29   Pin Assignments for Debouncer Test Circuit

If the debouncer is working properly, the seven-segment display should advance by one each time **pb_in** is pressed. If the debouncer is not working, the display will change by an unpredictable number with each switch press.

Behaviorally Designed Switch Debouncer

We can also design a switch debouncer by using a behavioral state machine description in VHDL. In order to do so, we need to define the operation of the circuit with a state diagram, as in Figure 10.30.

Transitions between states are determined by comparing **pb_in** and **pb_out**. If they are the same (00 or 11), the machine advances to the next state; if they are different (01 or 10), the machine reverts to the initial state, s0. At any point in the state diagram (including state s3, the last state), the machine will reset if **pb_in** and **pb_out** are different, indicating a bounce on the input.

If **pb_in** and **pb_out** are the same for four clock pulses, the input is deemed to be stable. Only at this point will the output change to its opposite state.

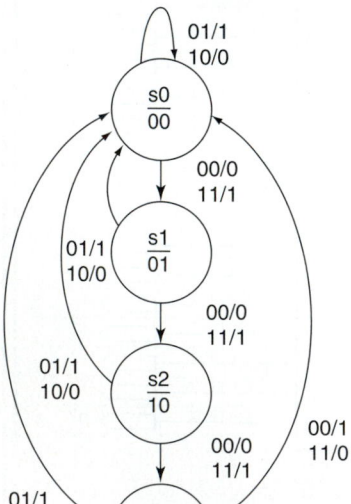

FIGURE 10.30   State Diagram for a Behaviorally Designed Switch Debouncer

> **NOTE . . .**
>
> In the shift register-based debouncer, the circuit advances to the next state if the shift register input and output are different and resets if they are the same. This might appear to be opposite to our behavioral description, but it is not if you look carefully. The shift register debouncer circuit inverts **pb_in** before applying the signal to the serial input of the shift register. Therefore, viewed from the circuit input and output terminals, rather than at the shift register input and output, the description is the same in both cases.

The VHDL code corresponding to the behavioral description of the switch debouncer is given next. The only output change is specified on the transition from state s3 to s0 when **pb_in = pb_out.** Since no change is allowed at any other time, no other output state needs to be specified.

```vhdl
-- dbc_behv.vhd
-- Behavioral definition of a switch debouncer
LIBRARY ieee;
USE ieee.std_logic_1164.ALL;

ENTITY dbc_behv IS
    PORT(
        clk, pb_in : IN STD_LOGIC;
        pb_out     : BUFFER STD_LOGIC);
END dbc_behv;

ARCHITECTURE debounce of dbc_behv IS
    TYPE sequence IS (s0, s1, s2, s3);
    SIGNAL state: sequence;
BEGIN
    PROCESS (clk, pb_in)
    BEGIN
        IF (clk'EVENT and clk='1') THEN
            CASE state IS
                WHEN s0=> IF (pb_in = pb_out) THEN
                                state  <= s1;
                            ELSE
                                state  <= s0;
                            END IF;
                WHEN s1=> IF (pb_in = pb_out) THEN
                                state  <= s2;
                            ELSE
                                state  <= s0;
                            END IF;
                WHEN S2=> IF (pb_in = pb_out) THEN
                                state <= s3;
                            ELSE
                                state <= s0;
                            END IF;
                WHEN s3=> IF (pb_in = pb_out) THEN
                                state <= s0;
                                pb_out <= not pb_out;
                            ELSE
                                state <= s0;
                            END IF;
                WHEN others => state <= s0;
            END CASE;
        END IF;
    END PROCESS;
END debounce;
```

Quartus II:
dbc_behv.vhd
dbc_behv.vwf
MAX+PLUS II:
dbc_behv.vhd
dbc_behv.scf

Note that the behaviorally designed switch debouncer does not have a built-in clock divider. If we were to use the circuit on the Altera UP-1 or UP-2 board, we would need to include a divide-by-2^{16} counter to the circuit, as shown in Figure 10.31.

■ **SECTION 10.4 REVIEW PROBLEM**

10.4 What is the fastest acceptable clock rate for the shift register portion of the debouncer in Figure 10.24 if the pushbutton switch bounces for 15ms?

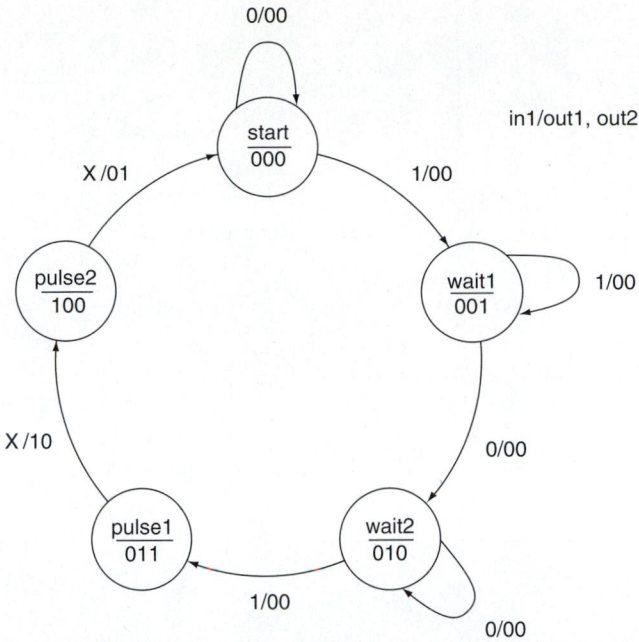

FIGURE 10.32 State Diagram for a Two-Pulse Generator

10.5 UNUSED STATES IN STATE MACHINES

In our study of counter circuits in Chapter 9, we found that when a counter modulus is not equal to a power of two there are unused states in the counter's sequence. For example, a mod-10 counter has six unused states, as the counter requires four bits to express ten states and the maximum number of 4-bit states is sixteen. The unused states (1010, 1011, 1100, 1101, 1110, and 1111) have to be accounted for in the design of a mod-10 counter.

The same is true of state machines whose number of states does not equal a power of two. For instance, a machine with five states requires three state variables. There are up to eight states available in a machine with three state variables, leaving three unused states. Figure 10.32 shows the state diagram of such a machine.

Unused states can be dealt with in two ways: they can be treated as don't care states, or they can be assigned specific destinations in the state diagram. In the latter case, the safest destination is the first state, in this case the state called **start.**

EXAMPLE 10.4

Redraw the state diagram of Figure 10.32 to include the unused states of the machine's state variables. Set the unused states to have a destination state of **start.** Briefly describe the intended operation of the state machine.

■ Solution

Figure 10.33 shows the revised state diagram.

The machine begins in state **start** and waits for a HIGH on **in1.** The machine then makes a transition to **wait1** and stays there until **in1** goes LOW again. The

(continued)

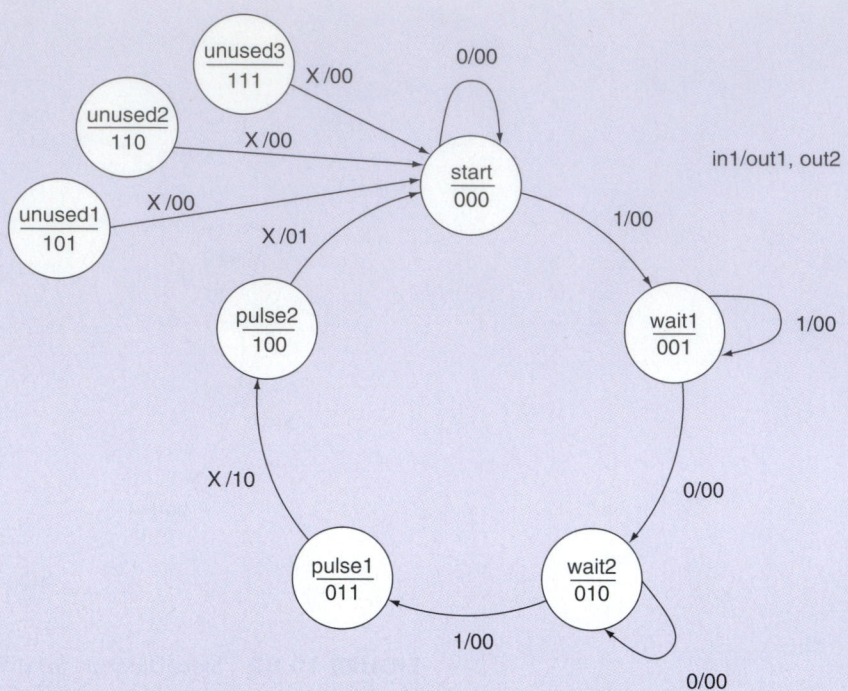

FIGURE 10.33 Example 10.4: State Diagram for Two-Pulse Generator Showing Unused States

machine goes to **wait2** and stays there until **in1** goes HIGH and then makes an unconditional transition to **pulse1** on the next clock pulse. Until this point, there is no change in either output.

The machine makes an unconditional transition to **pulse2** and makes **out1** go HIGH. The next transition, also unconditional, is to **start,** when **out1** goes LOW and **out2** goes HIGH. If **in1** is LOW, the machine stays in **start.** Otherwise, the cycle continues as outlined. In either case, **out2** goes LOW again.

Thus the machine waits for a HIGH-LOW-HIGH input sequence and generates a pulse sequence on two outputs.

EXAMPLE 10.5

Use classical state machine design techniques to implement the state machine described in the modified state diagram of Figure 10.33. Draw the state machine as a Block Diagram File in Quartus II and create a simulation to verify its function.

■ **Solution**

Table 10.6 shows the state table of the state machine represented by Figure 10.33.

TABLE 10.6 State Table for State Machine of Figure 10.33

Present State			Input	Next State			Outputs	
Q_2	Q_1	Q_0	$in1$	Q_2	Q_1	Q_0	$out1$	$out2$
0	0	0	0	0	0	0	0	0
0	0	0	1	0	0	1	0	0
0	0	1	0	0	1	0	0	0
0	0	1	1	0	0	1	0	0
0	1	0	0	0	1	0	0	0
0	1	0	1	0	1	1	0	0
0	1	1	0	1	0	0	1	0
0	1	1	1	1	0	0	1	0
1	0	0	0	0	0	0	0	1
1	0	0	1	0	0	0	0	1
1	0	1	0	0	0	0	0	0
1	0	1	1	0	0	0	0	0
1	1	0	0	0	0	0	0	0
1	1	0	1	0	0	0	0	0
1	1	1	0	0	0	0	0	0
1	1	1	1	0	0	0	0	0

Figure 10.34 shows the Karnaugh maps used to simplify the next-state equations for the state variable flip-flops. The output equations can be simplified by inspection.

The next-state and output equations for the state machine are:

$$D_2 = \overline{Q}_2 Q_1 Q_0$$
$$D_1 = \overline{Q}_2 Q_1 \overline{Q}_0 + \overline{Q}_2 \overline{Q}_1 Q_0 \overline{in1}$$
$$D_0 = \overline{Q}_2 \overline{Q}_0 in1 + \overline{Q}_2 \overline{Q}_1 in1$$
$$out1 = \overline{Q}_2 Q_1 Q_0$$
$$out2 = Q_2 \overline{Q}_1 \overline{Q}_0$$

Figure 10.35 shows the Block Diagram File for the state machine. Figure 10.36 shows the Quartus II simulation waveforms.

FIGURE 10.34 Example 10.5: K-Maps for Two-Pulse Generator

(continued)

FIGURE 10.35   Example 10.5: Two-Pulse Generator

two_pulse_bdf Simulation Report

Simulation Waveforms

| Master Time Bar: | 0 ps | ◄ ► Pointer: | 0 ps | Interval: | 0 ps | Start: | | End: | |

	Name	Value at 0 ps
⭢	in1	B 0
⭢	clk	B 0
⭤	out1	B 0
⭤	out2	B 0
⭤	⊞ state	H 0

FIGURE 10.36 Simulation for a Two-Pulse Generator (BDF)

EXAMPLE 10.6

Write the VHDL code required to implement the two-pulse generator described in Examples 10.4 and 10.5. Create a Quartus II simulation to verify the operation of the design. Based on your examination of the simulations for the VHDL design and the BDF design of the previous example, how do the two designs differ in their operation? What is the reason for the difference?

■ **Solution**

The VHDL code for the state machine in design entity **two_pulse_vhd.vhd** follows. The unused states are accounted for in the **others** clause.

```vhdl
-- two_pulse_vhd.vhd
LIBRARY ieee;
USE ieee.std_logic_1164.ALL;

ENTITY two_pulse_vhd IS
    PORT(
        clk, in1 : IN      STD_LOGIC;
        output   : OUT     STD_LOGIC_VECTOR (1 to 2));
END two_pulse_vhd;

ARCHITECTURE a OF two_pulse_vhd IS
    TYPE SEQUENCE IS (start, wait1, wait2, pulse1, pulse2);
    SIGNAL pulse_state : SEQUENCE;
BEGIN
    PROCESS(clk)
    BEGIN
        IF (clk'EVENT and clk = '1') THEN
            CASE pulse_state IS
            WHEN start =>
                IF in1 = '0' THEN
                    pulse_state  <= start;
                    output    <= "00";
                ELSIF in1 = '1' THEN
                    pulse_state <= wait1;
                    output    <= "00";
                END IF;
```

(continued)

Quartus II:
two_pulse_vhd.vhd
two_pulse_vhd.vwf
MAX+PLUS II:
two_pulse.vhd
two_pulse.scf

```
        WHEN wait1 =>
            IF in1 = '0' THEN
                pulse_state    <= wait2;
                output    <= "00";
            ELSIF in1 = '1' THEN
                pulse_state <= wait1;
                output    <= "00";
            END IF;
        WHEN wait2 =>
            IF in1 = '0' THEN
                pulse_state    <= wait2;
                output    <= "00";
            ELSIF in1 = '1' THEN
                pulse_state <= pulse1;
                output  <= "00";
            END IF;
        WHEN pulse1 =>
            pulse_state    <= pulse2;
            output    <= "10";
        WHEN pulse2 =>
            pulse_state    <= start;
            output    <= "01";
        WHEN others =>
            pulse_state    <= start;
            output    <= "00";
    END CASE;
  END IF;
 END PROCESS;
END a;
```

Figure 10.37 shows the Quartus II simulation of the state machine.

If you closely examine the simulation waveforms in Figures 10.36 and 10.37, you will note that the pulse outputs in Figure 10.37 (VHDL design) occur one clock cycle later than they do in Figure 10.36 (graphical design). This is because the VHDL compiler has synthesized each output with a D flip-flop, as we did for the single-pulse circuit in Figure 10.20, in order to ensure synchronous output operation. (We can verify this by examining the **Fitter Equations** section of the project Compilation report.) Since the outputs are both derived entirely from flip-flop outputs, this synthesis step is not strictly necessary to ensure that the outputs are synchronous with the clock.

FIGURE 10.37   Example 10.6: Simulation of a Two-Pulse Generator (VHDL)

■ **SECTION 10.5 REVIEW PROBLEM**

10.5 Is the state machine designed in Example 10.5 a Moore machine or a Mealy machine? Why?

10.6 TRAFFIC LIGHT CONTROLLER

➡ *LAB SOURCE*
Lab #30
Traffic Light Controller
(Lab Project 3)

A simple traffic light controller can be implemented by a state machine with a state diagram such as the one shown in Figure 10.38.

The control scheme assumes control over a north-south road and an east-west road. The north-south lights are controlled by outputs called **nsr, nsy,** and **nsg** (north-south red, yellow, green). The east-west road is controlled by similar outputs called **ewr, ewy,** and **ewg.** A LOW controller output turns on a light. Thus an output 011110 corresponds to the north-south red and east-west green lights.

An input called *TIMER* controls the length of the two green-light cycles. When *TIMER* = 1, a transition from s0 to s1 or from s2 to s3 is possible (s0 represents the EW green; s2 the NS green). This transition accompanies a change from green to yellow on the active road. The light on the other road stays red. An unconditional transition follows, changing the yellow light to red on one road and the red light to green on the other.

The cycle can be set to any length by changing the signal on the *TIMER* input. (The yellow light will always be on for one clock pulse in this design.) For ease of observation, we will use a cycle of ten clock pulses. For either direction, the cycle consists of 4 clocks GREEN, 1 clock YELLOW, and 5 clocks RED. This cycle can be generated by the MSB of a mod-5 counter, as shown in Figure 10.39. If we model the traffic controller using the Altera UP-1 or UP-2 board, we require a clock divider to slow down the 25.175 MHz clock to a rate of about 0.75 Hz, making it easy to observe the changes of lights. These blocks can all be instantiated in VHDL, which will be left as part of an exercise in the lab manual accompanying this book.

Figure 10.40 shows the simulation of the mod-5 counter that generates the *TIMER* control signal. The MSB goes HIGH for one clock period, then LOW for four. When applied to the *TIMER* input of the output controller, this signal directs the controller from state to state.

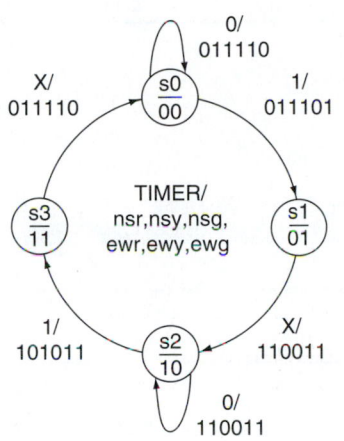

FIGURE 10.38 State Diagram of a Traffic Light Controller

*Cycle: Red for 5 clocks
Green for 4 clocks
Yellow for 1 clock*

FIGURE 10.39 Traffic Control Demonstration Circuit for the Altera UP-1 or UP-2 Board

FIGURE 10.40   Simulation of a Mod-5 Counter

FIGURE 10.41   Simulation of a Traffic Light Decoder

Figure 10.41 shows a simulation of the mod-5 counter and output controller. The north-south lights are red for five clock pulses (shown by 011 in the **north_south** waveform). At the same time, the east-west lights are green for four clock pulses (**east_west** = 110), followed by yellow for one clock pulse (**east_west** = 101). The cycle continues with an east-west red and north-south green and yellow.

According to the state diagram, the yellow light should happen on the transition where *TIMER* = 1. This corresponds to the point on the simulation waveforms where **count** = 4. However, the yellow light does not come on until **count** = 0. This is because the Quartus II VHDL compiler synthesizes the controller outputs with synchronous outputs (flip-flops). As a result, the output states are delayed by one clock cycle. Since the relative lengths of the cycle proportions are preserved, this does not affect the operation of the controller.

SUMMARY

1. A state machine is a synchronous sequential circuit with a memory section (flip-flops) to hold the present state of the machine and a control section (gates) to determine the machine's next state.

2. The number of flip-flops in a state machine's memory section is the same as the number of state variables.

3. Two main types of state machine are the Moore machine and the Mealy machine.

4. The outputs of a Moore machine are entirely dependent on the states of the machine's flip-flops. Output changes will always be synchronous with the system clock.

5. The outputs of a Mealy machine depend on the states of the machine's flip-flops and the gates in the control section. A Mealy machine's outputs can change asynchronously, relative to the system clock.

6. A state machine can be designed in a classical fashion using the same method as in designing a synchronous counter, as follows:

   a. Define the problem and draw a state diagram.

   b. Construct a table of present and next states.

   c. Use flip-flop excitation tables to determine the flip-flop inputs for each state transition.

   d. Use Boolean algebra or K-maps to find the simplest Boolean expression for flip-flop inputs (D, T, or JK) in terms of outputs (Q).

   e. Draw the logic diagram of the state machine.

7. The state names in a state machine can be named numerically (s0, s1, s2, . . .) or literally (start, idle, read, write), depending on the machine function. State names are independent of the values of the state variables.

8. A state machine can be defined in VHDL by using a CASE statement within a PROCESS to define the progression of states. The output values can be defined by a separate decoder construct or they can be assigned within each case of the CASE statement.

9. The possible values of the state variables of a machine are defined within an enumerated type definition. An enumerated type is a list of possible values that a port, variable, or signal of that type is allowed to have.

10. Notation for a state diagram includes a series of bubbles (circles) containing state names and values of state variables in the form $\frac{\text{state\_name}}{\text{state\_variable(s)}}$.

11. The inputs and outputs of a state machine are labeled **in1, in2, . . . , in$x$/out1, out2, . . . , out$x$.**

12. Transitions between states can be conditional or unconditional. A conditional transition happens only under certain conditions of a control input and is labeled with the relevant input condition. An unconditional transition happens under all conditions of input and is labeled with an X for each input variable.

13. Conditional transitions in a VHDL state machine are described by an IF statement within a particular case of the CASE statement that describes the machine.

14. Mealy machine outputs are susceptible to asynchronous output changes if a combinational input changes out of synchronization with the clock. This can be remedied by clocking each output through a separate synchronizing flip-flop.

15. A maximum of 2^n states can be assigned to a state machine that has n state variables. If the number of states is less than 2^n, the unused states must be accounted for. Either they can be treated as don't care states, or they can be assigned a specific destination state, usually the reset state.

16. In a VHDL implementation of a state machine, any unused states can be covered with an **others** clause in the CASE statement that defines the machine.

GLOSSARY

Bubble   A circle in a state diagram containing the state name and values of the state variables.

Conditional Transition   A transition between states of a state machine that occurs only under specific conditions of one or more control inputs.

Control Input   A state machine input that directs the operation of the machine from state to state.

Enumerated Type   A user-defined type in VHDL in which all possible values of a named identifier are listed in a type definition statement.

Form A Contact   A normally open contact on a switch or relay.

Form B Contact   A normally closed contact on a switch or relay.

Form C Contact   A pair of contacts, one normally open and one normally closed, that operate with a single action of a switch or relay.

Mealy Machine   A state machine whose output is determined by both the sequential logic and the combinational logic of the machine.

Moore Machine   A state machine whose output is determined only by the sequential logic of the machine.

State Machine   A synchronous sequential circuit, consisting of a sequential logic section and a combinational logic section, whose outputs and internal flip-flops progress through a predictable sequence of states in response to a clock and other input signals.

State Variables   The variables held in the flip-flops of a state machine that determine its present state.

Unconditional Transition   A transition between states of a state machine that occurs regardless of the status of any control inputs.

PROBLEMS

Problem numbers set in color indicate more difficult problems.

10.1   State Machines

10.1   Is the state machine in Figure 10.42 a Moore machine or a Mealy machine? Explain your answer.

FIGURE 10.42   Problem 10.1: State Machine Circuit

10.2 Is the state machine in Figure 10.43 a Moore machine or a Mealy machine? Explain your answer.

FIGURE 10.43 Problem 10.2: State Machine Circuit

TABLE 10.7 4-Bit Gray Code Sequence

Q_3	Q_2	Q_1	Q_0
0	0	0	0
0	0	0	1
0	0	1	1
0	0	1	0
0	1	1	0
0	1	1	1
0	1	0	1
0	1	0	0
1	1	0	0
1	1	0	1
1	1	1	1
1	1	1	0
1	0	1	0
1	0	1	1
1	0	0	1
1	0	0	0

10.2 State Machines with No Control Inputs

10.3 A 4-bit Gray code sequence is shown in Table 10.7. Use classical design methods to design a counter with this sequence, using D flip-flops. Draw the resulting circuit diagram in a Quartus II Block Diagram File. Create a simulation to verify the circuit operation.

10.4 Use classical state machine design techniques to design a counter whose output sequence is shown in Table 10.8. (This is a divide-by-twelve counter in which the MSB output has a duty cycle of 50%.) Draw the state diagram, derive synchronous equations of the flip-flops, and draw the circuit implementation in Quartus II and create a simulation to verify the circuit's function.

10.5 Write the VHDL code required to implement a 4-bit Gray code counter. Create a simulation in Quartus II to verify the operation of the circuit.

10.6 Write the VHDL code required to implement a counter with the sequence shown in Table 10.8. Create a simulation in Quartus II to verify the operation of the circuit.

10.3 State Machines with Control Inputs

10.7 Use classical state machine design techniques to find the Boolean next state and output equations for the state machine represented by the state diagram in Figure 10.44. Draw the state machine circuit as a Block Diagram File in Quartus II. Create a simulation file to verify the operation of the circuit. Briefly explain the intended function of the state machine.

TABLE 10.8 Counter Sequence for Problem 10.4

Q_3	Q_2	Q_1	Q_0
0	0	0	0
0	0	0	1
0	0	1	0
0	0	1	1
0	1	0	0
0	1	0	1
1	0	0	0
1	0	0	1
1	0	1	0
1	0	1	1
1	1	0	0
1	1	0	1

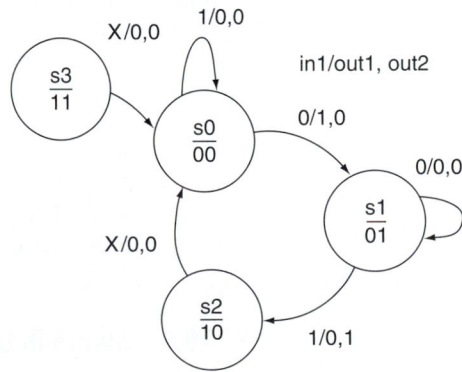

FIGURE 10.44 Problem 10.7: State Diagram

10.8   Referring to the simulation for the state machine in Problem 10.7, briefly explain why it is susceptible to asynchronous input changes. Modify the state machine circuit to eliminate the asynchronous behavior of the outputs. Create a Quartus II simulation to verify the function of the modified state machine.

10.9   Write the VHDL code required to implement the state machine in Problem 10.7. Create a simulation to verify the operation of the state machine.

10.10   A state machine is used to control an analog-to-digital converter, as shown in the block diagram of Figure 10.45.

FIGURE 10.45   Problem 10.10: Analog-to-Digital Converter and Controller

The controller has four states, defined by state variables Q_1 and Q_0 as follows: **idle** (00), **start** (01), **waiting** (11), and **read** (10). There are two outputs: **sc** (Start Conversion; active-HIGH) and **oe** (Output Enable; active-LOW). There are four inputs: clock, **go** (active-LOW) **eoc** (End of Conversion), and asynchronous reset (active-LOW). The machine operates as follows:

a.   In the **idle** state, the outputs are: sc = 0, oe = 1. The machine defaults to the **idle** state when the machine is reset.

b.   Upon detecting a 0 at the **go** input, the machine makes a transition to the **start** state. In this transition, sc = 1, oe = 1.

c.   The machine makes an unconditional transition to the **waiting** state; sc = 0, oe = 1. It remains in this state, with no output change, until input eoc = 1.

d.   When eoc = 1, the machine goes to the **read** state; sc = 0, oe = 0.

e.   The machine makes an unconditional transition to the **idle** state; sc = 0, oe = 1.

Use classical state machine design techniques to design the controller. Draw the required circuit in Quartus II and create a simulation to verify its operation. Is this machine vulnerable to asynchronous input change?

10.11   Use VHDL to implement the controller circuit of Problem 10.10. Create a simulation to verify its operation.

10.12   Write a VHDL file for a state machine that selects a 3-bit binary or Gray code count, depending on the state of an input called **gray.** If **gray = 1,** count in Gray code. Otherwise count in binary. Create a simulation file that verifies the operation of the circuit, clearly showing the full Gray code count, binary count, and reset function. Note that a PROCESS statement can only have one statement that checks the status of the clock (e.g., IF(clock'EVENT and clock = '1')THEN). If two or more such statements are included, the VHDL compiler will generate an error message.

10.4   Switch Debouncer for a Normally Open Pushbutton Switch

10.13   Why is it not possible to debounce the pushbuttons on the Altera UP-1 or UP-2 board using a NAND latch?

10.14   Refer to the switch debouncer circuit in Figure 10.24 (p. 646). For how many clock periods must the input of the debouncer remain stable before the output can change?

10.15   What is the maximum switch bounce time that can be removed by the circuit of Figure 10.24 if the clock at the shift register is running at a rate of 480 Hz?

10.16 Briefly explain how the Exclusive NOR gate in the debounce circuit of Figure 10.24 determines if switch bounce has occurred.

10.17 Refer to the section on the behaviorally designed switch debouncer in Section 10.4. For how many clock periods must the input of the debouncer remain stable before the output can change? What is the maximum switch bounce time that can be removed by this VHDL switch debouncer if the state machine clock is running at a rate of 480 Hz?

10.5 Unused States in State Machines

10.18 Refer to the state diagram in Figure 10.46.

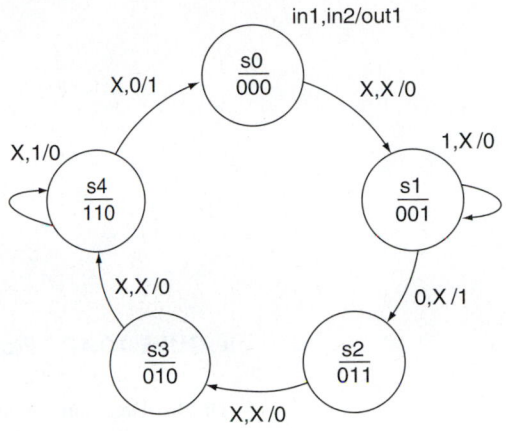

FIGURE 10.46 Problem 10.18: State Diagram

    a. How many state variables are required to implement this state machine? Why?

    b. How many unused states are there for this state machine? List the unused states.

    c. Complete the partial timing diagram shown in Figure 10.47 to illustrate one complete cycle of the state machine represented by the state diagram of Figure 10.46.

FIGURE 10.47 Problem 10.18: Partial Timing Diagram

10.19 Write the VHDL code required to implement the state machine described by the state diagram of Figure 10.46. Create a simulation file to verify the operation of the circuit.

10.20 Use classical state machine design techniques to design a state machine described by the state diagram of Figure 10.48. Briefly describe the intended operation of the circuit. Create a Quartus II simulation to verify the operation of the state machine design. Unused states may be treated as don't care states, but unspecified outputs should always be assigned to 0.

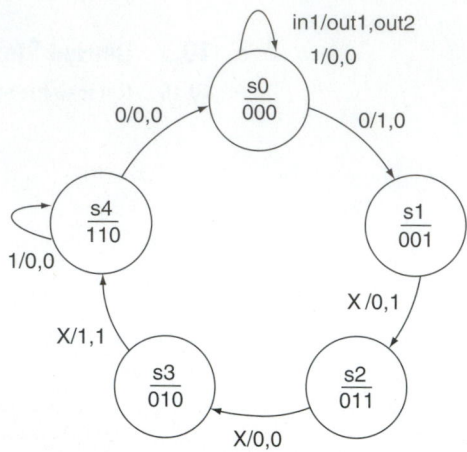

FIGURE 10.48 Problem 10.20: State Diagram

10.21 Determine the next state for each of the unused states of the state machine designed in Problem 10.20. Use this analysis to redraw the state diagram of Figure 10.48 so that it properly includes the unused states. (There is more than one right answer, depending on the result of the Boolean simplification process used in Problem 10.20 to simplify the equation for D_1.)

10.22 Write the VHDL code for the state machine described in Problem 10.20. Create a Quartus II simulation to verify the function of the state machine.

10.23 A state machine is used to control an analog-to-digital converter, as shown in the block diagram of Figure 10.45. (The following description is a modified version of the controller described in Problem 10.10.)

   Five states are used: **idle, start, waiting1, waiting2,** and **read.** There are two outputs: **sc** (Start Conversion; active-HIGH) and **oe** (Output Enable; active-HIGH). There are four inputs: clock, reset, **go,** and **eoc** (End of Conversion). The machine operates as follows:

   a. In the **idle** state, the outputs are: sc = 0, oe = 0. The machine defaults to the **idle** state when asynchronously reset and remains there until go = 0.

   b. When go = 0, the machine makes a transition to the **start** state. In this transition, sc = 1, oe = 0.

   c. The machine makes an unconditional transition to the **waiting1** state; sc = 0, oe = 0. It remains in this state, with no output change, until input eoc = 0.

   d. When eoc = 0, the machine goes to the **waiting2** state; sc = 0, oe = 0. It remains in this state, with no output change, until input eoc = 1.

   e. The machine makes a transition to the **read** state when eoc = 1; sc = 0, oe = 1.

   f. The machine makes an unconditional transition to the **idle** state; sc = 0, oe = 0.

   After reviewing the block diagram and the states just listed,

   a. Draw the state diagram of the controller.

   b. How many state variables are required for the controller described in this question?

10.24 Write the VHDL code for the state machine described in Problem 10.23. Create a simulation file to verify the function of the design.

ANSWERS TO SECTION REVIEW PROBLEMS

10.1

10.1 A Moore state machine has outputs that depend only on the states of the flip-flops in the machine. A Mealy machine's outputs depend on the states of its flip-flops as well as the gates of the machine's control section. This can result in asynchronous output changes in the Mealy machine outputs.

10.2

10.2

$$J_2 = Q_1\overline{Q}_0$$
$$K_2 = \overline{Q}_1 Q_0$$
$$J_1 = \overline{Q}_2 Q_0$$
$$K_1 = Q_2 Q_0$$
$$J_0 = \overline{Q}_2\overline{Q}_1 + Q_2 Q_1 = \overline{Q_2 \oplus Q_1}$$
$$K_0 = \overline{Q}_2 Q_1 + Q_2\overline{Q}_1 = Q_2 \oplus Q_1$$

10.3

10.3 The output flip-flop synchronizes the output to the system clock, yielding the following advantages: (1) the output is always a known width of one clock cycle; and (2) the output is not vulnerable to change due to asynchronous changes of input.

10.4

10.4 $T_c = 3.75$ ms; $f_c = 267$ Hz

10.5

10.5 More machine. The outputs are derived entirely from the output states of the state machine and are not vulnerable to asynchronous changes of input.

KEY TERMS

CMOS
C_{PD}
Cutoff Mode
Driving Gate
ECL
Enhancement-Mode MOSFET
Fanout
High-Speed (Silicon-Gate) CMOS
Hold Time (t_h)
I_{CC}
I_{CCH}
I_{CCL}
I_{DD}
I_{IH}
I_{IL}
I_{OH}
I_{OL}
I_T
Load Gate
MOSFET
n-Channel Enhancement-Mode
   MOSFET
Noise
Noise Margin
n-Type Inversion Layer
Ohmic Region
Open-Collector Output
p-Channel Enhancement-Mode
   MOSFET
Phase Splitter
Power Dissipation
Propagation Delay
Pulse Width (t_w)
Recovery Time (t_{rec})
Saturation Mode
Schottky Barrier Diode

Schottky Transistor
Schottky TTL
Setup Time (t_{su})
Sinking
Sourcing
Speed-Power Product
Storage Time
Substrate
Threshold Voltage $V_{GS(Th)}$
Totem Pole Output
t_{pHL}
t_{pLH}
Tristate Output
TTL
TTL Compatible
V_{CC}
V_{IH}
V_{IL}
V_{OH}
V_{OL}
Wired-AND

OUTLINE

11.1   Electrical Characteristics of Logic Gates

11.2   Propagation Delay

11.3   Flip-Flop Timing Parameters

11.4   Fanout

11.5   Power Dissipation

11.6   Noise Margin

11.7   Interfacing TTL and CMOS Gates

11.8   Internal Circuitry of TTL Gates

11.9   Internal Circuitry of CMOS Gates

11.10   TTL and CMOS Variations

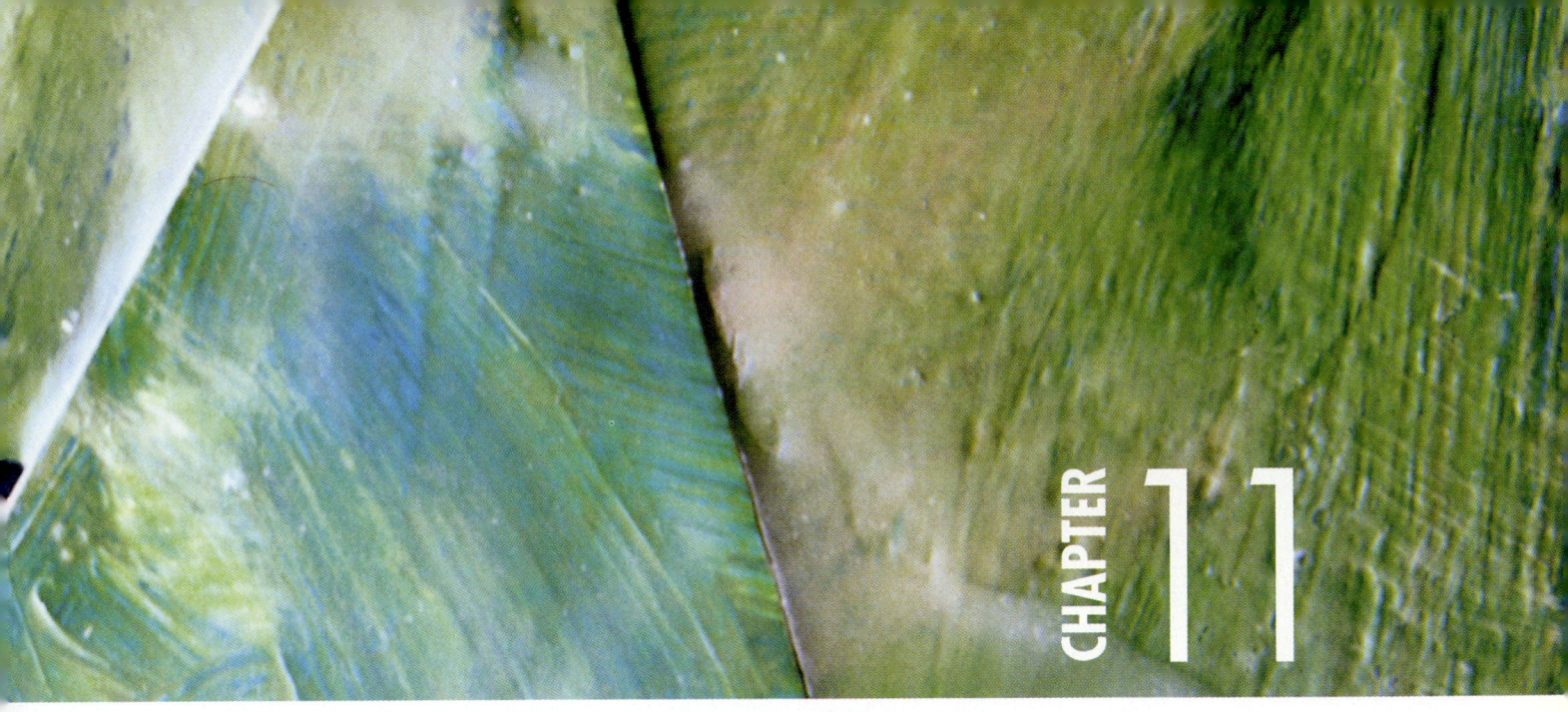

CHAPTER 11

Logic Gate Circuitry

CHAPTER OBJECTIVES

Upon successful completion of this chapter, you will be able to:

- Name the various logic families most commonly in use today and state advantages and disadvantages of each.
- Define propagation delay.
- Calculate propagation delay of simple circuits, using datasheets.
- Define flip-flop timing parameters such as setup time, hold time, pulse width, recovery time, and propagation delay.
- Determine the values of flip-flop timing parameters from a device datasheet.
- Define fanout and calculate its value, using datasheets.
- Calculate power dissipation of TTL and CMOS circuits.
- Calculate noise margin of a logic gate from datasheets.
- Draw circuits that will interface various CMOS and TTL gates.
- Explain how a bipolar junction transistor can be used as a logic inverter.
- Describe the function of a TTL input transistor in all possible input states: HIGH, LOW, and open-circuit.
- Explain the operation of a totem pole output.
- Illustrate how a totem pole output generates power line noise and describe how to remedy this problem.
- Illustrate why totem pole outputs cannot be tied together.
- Explain the difference between open-collector and totem pole outputs of a TTL gate.
- Illustrate the operation of TTL open-collector inverter, NAND, and NOR gates.
- Write the Boolean expression of a wired-AND circuit.
- Design a circuit that uses an open-collector gate to drive a high-current load.
- Calculate the value of a pull-up resistor at the output of an open-collector gate.
- Explain the operation of a tristate gate and name several of its advantages.
- Design a circuit using a tristate bus driver to direct the flow of data from one device to another.

- Describe the basic structure of a MOSFET and state its bias voltage
- Draw the circuit of an CMOS inverter and show how it works.
- Draw the circuits of CMOS NAND and NOR gates and explain the operation of each.
- Design a circuit using a CMOS transmission gate to enable and inhibit digital and analog signals.
- Interpret TTL datasheets to distinguish between the various TTL families.
- Describe the use of the Schottky barrier diode in TTL gates.
- Calculate speed-power products from datasheets.

Our study of logic gates and flip-flops in previous chapters has concentrated on digital *logic* and has largely ignored digital *electronics*. Digital logic devices are electronic circuits with their own characteristic voltages and currents. No serious study of digital circuitry is complete without some examination of this topic.

It is particularly important to understand the inputs and outputs of logic devices as electronic circuits. Knowing the input and output voltages and currents of these circuits is essential, since gate loading, power dissipation, noise voltages, and interfacing between logic families depend on them. The switching speed of device outputs is also fundamental and may be a consideration when choosing the logic family for a circuit design.

Input and output voltages of logic devices are specified in manufacturers' datasheets, which allows us to take a "black box" approach initially.

Later in the chapter, we will examine some basic digital circuits at a transistor level, since digital logic is based on transistor switching. Two major types of transistors, the bipolar junction transistor and the metal-oxide-semiconductor field effect transistor (MOSFET), form the basis of the major logic families in use today. Transistor-transistor logic (TTL) is based on the bipolar transistor. Complementary MOS (CMOS) is based on the MOSFET.

We will briefly study the operating characteristics of both bipolar transistors and MOSFETs and then see how these devices give rise to the electrical characteristics of simple logic gates.

11.1 ELECTRICAL CHARACTERISTICS OF LOGIC GATES

▰ KEY TERMS

CMOS Complementary metal-oxide semiconductor. A logic family based on metal-oxide-semiconductor field effect transistors (MOSFETs).

TTL Transistor-transistor logic. A logic family based on bipolar transistors.

ECL Emitter coupled logic. A high-speed logic family based on bipolar transistors.

When we examine the electrical characteristics of logic circuits, we see them as practical rather than ideal devices. We look at properties such as switching speed, power dissipation, noise immunity, and current-driving capability. There are several commonly available logic families in use today, each having a unique set of electrical characteristics that differentiates it from all the others. Each logic family gives superior performance in one or more of its electrical properties.

CMOS consumes very little power, has excellent noise immunity, and can be used with a wide range of power supply voltages.

TTL has a larger current-driving capability than CMOS. Its power consumption is higher than that of CMOS, and its power supply requirements are more rigid.

ECL (emitter-coupled logic, another bipolar family) is fast, making it the choice for high-speed applications. It is inferior to CMOS and TTL in terms of noise immunity and power consumption.

TTL and CMOS gates come in a wide range of subfamilies. Table 11.1 lists some of the TTL and CMOS variations of the quadruple 2-input NAND gate. All gates listed have the same logic function but different electrical characteristics. Other gates would be similarly designated, with the last two or three digits indicating the gate function (e.g., a quadruple 2-input NOR gate would be designated 74LS02, 74ALS02, 74F02, etc.).

TABLE 11.1 Part Numbers for a Quad 2-Input NAND Gate in Different Logic Families

	Part Number	Logic Family
TTL	74LS00	Low-power Schottky TTL
	74ALS00	Advanced low power Schottky TTL
	74F00	Fast TTL
CMOS	74HC00	High-speed CMOS
	74HCT00	High-speed CMOS (TTL-compatible inputs)
	74LVX00	Low-voltage CMOS

We will examine four electrical characteristics of TTL and CMOS circuits: propagation delay, fanout, noise margin, and power dissipation. The first of these has to do with speed of output response to a change of input. The last three have to do with input and output voltages and currents. All four properties can be read directly from specifications given in a manufacturer's datasheet or derived from these specifications.

Figures 11.1 and Figure 11.2 show how the input and output voltages and currents are defined in a 74XX00 NAND gate. This designation can be generalized to any logic gate input or output.

The voltages and currents are designated with two subscripts, one that designates an input or output and another that indicates the logic level. For example, V_{OL} is the voltage at the gate output when the output is in the logic LOW state. I_{IL} is the input current when the input is in the LOW state.

FIGURE 11.1 Input/Output Voltage Parameters

FIGURE 11.2 Input/Output Current Parameters

SN54ALS00A, SN54AS00, SN74ALS00A, SN74AS00
QUADRUPLE 2-INPUT POSITIVE-NAND GATES

SDAS187A – APRIL 1982 – REVISED DECEMBER 1994

logic diagram (positive logic)

Pin numbers shown are for the D, J, and N packages.

absolute maximum ratings over operating free-air temperature range (unless otherwise noted)[†]

Supply voltage, V_{CC} . . . . 7 V
Input voltage, V_I . . . . 7 V
Operating free-air temperature range, T_A: SN54ALS00A . . . . −55°C to 125°C
                                                                      SN74ALS00A . . . . 0°C to 70°C
Storage temperature range . . . . −65°C to 150°C

[†] Stresses beyond those listed under "absolute maximum ratings" may cause permanent damage to the device. These are stress ratings only, and functional operation of the device at these or any other conditions beyond those indicated under "recommended operating conditions" is not implied. Exposure to absolute-maximum-rated conditions for extended periods may affect device reliability.

recommended operating conditions

		SN54ALS00A			SN74ALS00A			UNIT
		MIN	NOM	MAX	MIN	NOM	MAX	
V_{CC}	Supply voltage	4.5	5	5.5	4.5	5	5.5	V
V_{IH}	High-level input voltage	2			2			V
V_{IL}	Low-level input voltage			0.8[‡]			0.8	V
				0.7[§]				
I_{OH}	High-level output current			−0.4			−0.4	mA
I_{OL}	Low-level output current			4			8	mA
T_A	Operating free-air temperature	−55		125	0		70	°C

[‡] Applies over temperature range −55°C to 70°C
[§] Applies over temperature range 70°C to 125°C

FIGURE 11.3 74ALS00 Data (1 of 2). (Courtesy of Texas Instruments)

These voltages and currents are specified in manufacturers' published datasheets, which are usually available in print form in a data book or in an electronic format, such as Portable Document Format (**pdf**) on a CD or Internet site.

Figure 11.3 shows an extract from a datasheet for a 74ALS00 NAND gate, which also shows parameter values for a 54ALS00, a 74AS00, and a 54AS00 device. A 54-series device is manufactured to military specifications, which require a high range of environmental operating conditions. A 74-series device is suitable for general or commercial use. We will limit ourselves to the 74-series devices.

The voltage and current parameters indicated in Figure 11.1 and Figure 11.2 are all shown in the 74ALS00 datasheet. Some parameters are shown as typical values, as well as maximum or minimum. Typical values should be considered "information only" as device manufacturers do not guarantee these values. An exception to this would be the supply voltage, V_{CC}, whose typical value is simply indicated as the average of maximum and minimum values.

Note that I_{IH} and I_{IL} are shown in Figure 11.2 as flowing in opposite directions, as are I_{OH} and I_{OL}. The arrows show the *actual* directions of the currents. However,

SN54ALS00A, SN54AS00, SN74ALS00A, SN74AS00
QUADRUPLE 2-INPUT POSITIVE-NAND GATES

SDAS187A – APRIL 1982 – REVISED DECEMBER 1994

electrical characteristics over recommended operating free-air temperature range (unless otherwise noted)

PARAMETER	TEST CONDITIONS		SN54ALS00A MIN	SN54ALS00A TYP†	SN54ALS00A MAX	SN74ALS00A MIN	SN74ALS00A TYP†	SN74ALS00A MAX	UNIT
V_{IK}	V_{CC} = 4.5 V,	I_I = −18 mA			−1.2			−1.5	V
V_{OH}	V_{CC} = 4.5 V to 5.5 V,	I_{OH} = −0.4 mA	V_{CC} −2			V_{CC} −2			V
V_{OL}	V_{CC} = 4.5 V	I_{OL} = 4 mA		0.25	0.4		0.25	0.4	V
		I_{OL} = 8 mA					0.35	0.5	
I_I	V_{CC} = 5.5 V,	V_I = 7 V			0.1			0.1	mA
I_{IH}	V_{CC} = 5.5 V,	V_I = 2.7 V			20			20	µA
I_{IL}	V_{CC} = 5.5 V,	V_I = 0.4 V			−0.1			−0.1	mA
I_O‡	V_{CC} = 5.5 V,	V_O = 2.25 V	−20		−112	−30		−112	mA
I_{CCH}	V_{CC} = 5.5 V,	V_I = 0		0.5	0.85		0.5	0.85	mA
I_{CCL}	V_{CC} = 5.5 V,	V_I = 4.5 V		1.5	3		1.5	3	mA

† All typical values are at V_{CC} = 5 V, T_A = 25°C.
‡ The output conditions have been chosen to produce a current that closely approximates one half of the true short-circuit output current, I_{OS}.

switching characteristics (see Figure 1)

PARAMETER	FROM (INPUT)	TO (OUTPUT)	V_{CC} = 4.5 V to 5.5 V, C_L = 50 pF, R_L = 500 Ω, T_A = MIN to MAX§				UNIT
			SN54ALS00A MIN	SN54ALS00A MAX	SN74ALS00A MIN	SN74ALS00A MAX	
t_{PLH}	A or B	Y	3	15	3	11	ns
t_{PHL}			2	9	2	8	

§ For conditions shown as MIN or MAX, use the appropriate value specified under recommended operating conditions.

FIGURE 11.3 74ALS00 Data (2 of 2). (Courtesy of Texas Instruments)

the *defined* direction of a current is always into its corresponding terminal. The actual and defined directions are the same for I_{IH} and I_{OL}, but opposite the I_{IL} and I_{OH}. Therefore, on a datasheet, a current entering a gate terminal (I_{IH} or I_{OL}) is indicated as positive and a current leaving a gate terminal (I_{IL} or I_{OH}) is shown as negative. The reason for these current directions will become apparent when we examine the internal circuits of the gates later in the chapter.

EXAMPLE 11.1

What is the maximum value of V_{OL} for a 74ALS00 NAND gate when the output current is at its maximum value?

■ Solution

When the output is in the LOW state, the output current is given by I_{OL}, which has a maximum value of 8 mA. The output voltage, V_{OL}, is specified for a value of 4 mA and for 8 mA. Since the output condition is specified for maximum I_{OL} (8 mA), then V_{OL} = 0.5 V.

The 74XX00 NAND gate data is sufficient to represent any logic functions having "normal" output current within its particular logic family. This data can be used for most gate or flip-flop circuits within the family. Some specialized devices with higher-current outputs (e.g., 74XX244 octal tristate buffers) have a different set of electrical characteristics within their family.

In the following sections of the chapter, we will use a NAND gate from each of three device families (74ALS00, 74HC00A, and 74HCT00A) for illustrating the general principles of the various electrical characteristics. Devices from other families will also be used in examples and problems. Datasheets for the various devices are included in Appendix C on the CD that accompanies this book.

■ SECTION 11.1 REVIEW PROBLEM

11.1   What are the maximum values of voltage and current we can expect at the output of a 74ALS00 NAND gate when both inputs are LOW? Assume $V_{CC} = 5.0$ V.

11.2   PROPAGATION DELAY

■ KEY TERMS

t_{pHL} Propagation delay when the device output is changing from HIGH to LOW.

t_{pLH} Propagation delay when the device output is changing from LOW to HIGH.

Propagation Delay   The time required for the output of a digital circuit to change states after a change at one or more of its inputs.

Propagation delay occurs because the output of a logic gate or flip-flop cannot respond instantaneously to changes at its input. There is a short delay, on the order of several nanoseconds, between input change and output response. This is largely due to the charging and discharging of capacitances inherent in the switching transistors of the gate or flip-flop.

Figure 11.4 shows propagation delay in two gates: a 74XX00 NAND gate and a 74XX08 AND gate. Each gate has an identical input waveform, a LOW-HIGH-LOW pulse. After each input transition, the output changes after a short delay, t_p.

Two delays are shown for each gate: t_{pLH} and t_{pHL}. The *LH* and *HL* subscripts show the direction of change at the gate *output; LH* indicates that the output goes from LOW to HIGH, and *HL* shows the output changing from HIGH to LOW.

Propagation delay is the time between input and output voltages passing through a standard reference value. The reference voltage for standard TTL is 1.5 V. LSTTL and CMOS have different reference voltages, as follows.

NOTE . . .

Propagation Delay for Various Logic Families:
LSTTL: Time from 1.3 V at input to 1.3 V at output.
Other TTL: Time from 1.5 V at input to 1.5 V at output.
CMOS: Time from 50% of maximum input to 50% of maximum output.

FIGURE 11.4   Propagation Delay in NAND and AND Gates

EXAMPLE 11.2

Use the datasheet in Figure 11.3, as well as those in Appendix C, to find the maximum propagation delays for each of the following gates: 74ALS00 (quadruple 2-input NAND), 74ALS02 (quadruple 2-input NOR), 74ALS08 (quadruple 2-input AND), and 74ALS32 (quadruple 2-input OR).

■ **Solution**

TABLE 11.2 Propagation Delays of 74ALS Gates

	74ALS00	74ALS02	74ALS08	74ALS32
t_{pLH}	11 ns	12 ns	14 ns	14 ns
t_{pHL}	8 ns	10 ns	10 ns	12 ns

Table 11.2 shows the variation of propagation delay among logic gates of the same family (74ALS TTL). Since each logic function has a different circuit, its propagation delay will differ from those of gates with different functions.

 EXAMPLE 11.3

Use datasheets to find the maximum propagation delays for each of the following logic gates: 74F00, 74AS00, 74ALS00, 74HC00, and 74HCT00.

■ **Solution**

TABLE 11.3 Propagation Delays of 74XX00 Gates

	74F00*	74AS00	74ALS00	74HC00**	74HCT00***
t_{pLH}	6 ns	4.5 ns	11 ns	15 ns	22 ns
t_{pHL}	5.3 ns	4 ns	8 ns	15 ns	22 ns

*Temperature range (74F00): 0° C to 70° C.
**V_{CC} = 4.5 V, temperature range (74HC00): –55° C to 25° C
***V_{CC} = 5.5 V, temperature (74HCT00): 25° C.

As indicated by the notes for Table 11.3, propagation delay (and other parameters) vary with certain operating conditions, such as ambient temperature and power supply voltage. Always make sure that the operating conditions are correctly specified when looking up a datasheet parameter.

All gates in Example 11.3 have the same logic function (2-input NAND), but different propagation delay times. We might ask, "Why not always use the advanced Schottky TTL gate (74AS00), since it is the fastest?" The main reason is that it has the highest power dissipation of the gates shown. We wouldn't know this without looking up other specs on the datasheet. (We will learn how to do this later in the chapter.) Thus, it is important to make design decisions based on complete information, not just one parameter.

Propagation Delay in Logic Circuits

A circuit consisting of two or more gates or flip-flops has a propagation delay that is the sum of delays *in the input-to-output path*. Delays in gates that do not affect the circuit are disregarded. Figure 11.5 shows how propagation delay works in a simple logic circuit consisting of a 74HC08 AND gate and a 74HC32 OR gate. Changes

FIGURE 11.5   Propagation Delays in a Logic Gate Circuit

at inputs A and B must propagate through both gates to affect the output. The total delay in such a case is the sum of t_{p1} and t_{p2}. A change at input C must pass only through gate 2. The circuit delay resulting from this change is only t_{p2}.

The timing diagram in Figure 11.5 shows the changes at inputs A, B, and C and the resulting transitions at all gate outputs.

Assume $V_{CC} = 4.5$ V and temperature range is –55° C to 25° C.

1. When A goes LOW, AB, the output of gate 1, also goes LOW after a maximum delay of $t_{pHL} = 15$ ns. This makes Y go LOW after a further delay of up to $t_{pHL} = 15$ ns. Total delay: $t_p = t_{pHL1} + t_{pHL2} = 15$ ns + 15 ns = 30 ns, max.
2. The HIGH-to-LOW transition at input B has no effect, since there is no difference between $0 \cdot 1$ and $0 \cdot 0$. AB is already LOW.
3. The LOW-to-HIGH transition at input C makes Y go HIGH after a maximum delay of $t_{pLH2} = 15$ ns.

■ SECTION 11.2 REVIEW PROBLEM

11.2   Assume that the gates in Figure 11.5 are replaced by a 74ALS08 AND gate and a 74ALS32 OR gate. Repeat the calculations of for the propagation delays if the waveforms of Figure 11.5 are applied to the circuit. The datasheets for the 74ALS08 and 74ALS32 are found in Appendix C.

11.3   FLIP-FLOP TIMING PARAMETERS

■ KEY TERMS

Setup Time (t_{su})   The time required for the synchronous inputs of a flip-flop to be stable before a *CLK* pulse is applied.

Hold Time (t_h)   The time that the synchronous inputs of a flip-flop must remain stable after the active *CLK* transition is finished.

a. Setup, hold, and CLK pulse width

b. $\overline{CLR}$ pulse width, propagation delay, and recovery time

c. Propagation delay from CLK

FIGURE 11.6 Timing Parameters of a D Flip-Flop

Pulse Width (t_w)   Minimum time required for an active-level pulse applied to a *CLK*, $\overline{CLR}$ or $\overline{PRE}$ input, as measured from the midpoint of the leading edge of the pulse to the midpoint of the trailing edge.

Recovery Time (t_{rec})   Minimum time from the midpoint of the trailing edge of a $\overline{CLR}$ or $\overline{PRE}$ pulse to the midpoint of an active *CLK* edge.

Flip-flops are electrical devices with inherent internal switching delays. As such, they have specific requirements for the timing of the input and output waveforms in order for them to operate reliably. Figure 11.6 shows some of the basic timing requirements of a D flip-flop.

Figure 11.6a illustrates the definitions of **setup time (t_{su}), hold time (t_h),** and **pulse width (t_w).** The notation used for the D waveform indicates that the D input could be at either logic level and makes a transition to the opposite level at some point. The setup time is measured from the midpoint of the D transition to the midpoint of the active *CLK* edge. The logic level on the D input must be steady for at least this time for the flip-flop to operate correctly.

Similarly, the hold time is measured from the midpoint of the *CLK* transition to the midpoint of the next D transition. The D level must be held steady for at least this time to ensure dependable operation.

The pulse width, t_w, shows how long the *CLK* needs to be held LOW after an active *CLK* edge. Although the LOW level does not itself latch data into the flip-flop, internal logic levels must reach a steady state before the device can accept a new clock pulse. This minimum pulse width allows the necessary time for these internal transitions.

Figure 11.6b shows the pulse width required at the $\overline{CLR}$ input, the propagation delay from $\overline{CLR}$ to Q and $\overline{Q}$, and the **recovery time** that must be allowed from the end of a $\overline{CLR}$ pulse to the beginning of a CLK pulse. These times also apply to a pulse on the $\overline{PRE}$ input of a flip-flop.

Propagation delay is the result of internal electrical delays, primarily the charging and discharging of internal capacitances of the gate transistor junctions. The practical result of this is that a pulse at the $\overline{CLR}$ input makes Q go LOW, but not immediately; there is a delay of several nanoseconds between input pulse and output response.

The recovery time, t_{rec}, allows the internal logic levels of the flip-flop to reach a steady state after a $\overline{CLR}$ pulse. When the internal levels are stable, the device is ready to accept an active CLK edge. Some datasheets treat this parameter as a species of setup time; it is shown as setup time after the $\overline{CLR}$ is inactive. Same thing.

Finally, Figure 11.6c shows the propagation delay from CLK to Q. This is the time from the midpoint of an active CLK edge to the midpoint of a transition at Q caused by that CLK edge. The parameters are defined, as before, by the direction of the output transition.

The timing restrictions of a flip-flop imply that there is a maximum CLK frequency beyond which the device will not operate reliably.

Table 11.4 summarizes the timing parameters of a number of D flip-flops from different logic families.

TABLE 11.4 Timing Parameters of D Flip-Flops in Several Logic Families

Symbol	Parameter	74LS74A	74F74	74AS74	74HC74	74HCT74	74AC74
t_{su}	Setup time	20 ns	3.0 ns	4.5 ns	16 ns	15 ns	1.0 ns
t_h	Hold time	0 ns	1.0 ns	0 ns	3.0 ns	3 ns	1.5 ns
t_w	$\overline{CLR}$ pulse width	25 ns	4.0 ns	4 ns	12 ns	15 ns	2.5 ns
	CLK pulse width	25 ns	5.0 ns	4 ns	12 ns	15 ns	2.5 ns
t_{rec}	Recovery time	25 ns	2.0 ns	5.5 ns	8.0 ns	6 ns	2.0 ns
t_{pLH}	Propagation delay from CLK (L → H)	25 ns	7.8 ns	7.5 ns	20 ns	24 ns	3.5 ns
t_{pHL}	Propagation delay from CLK (H → L)	40 ns	9.2 ns	10.5 ns	20 ns	24 ns	2.5 ns
t_{pLH}	Propagation delay from $\overline{CLR}$ (L → H)	25 ns	7.1 ns	8 ns	21 ns	24 ns	3.5 ns
t_{pHL}	Propagation delay from $\overline{CLR}$ (H → L)	40 ns	10.5 ns	9 ns	21 ns	24 ns	3.0 ns
f_{max}	Maximum frequency	25 MHz	100 MHz	105 MHz	30 MHz	30 MHz	140 MHz

EXAMPLE 11.4

The timing diagrams in Figure 11.7 represent some of the timing parameters of a D flip-flop. From these diagrams, determine the setup and hold times and the propagation delays from CLK and $\overline{CLR}$ to Q and $\overline{Q}$.

■ **Solution**

The values are as follows:

Setup time = 15 ns
Hold time = 5 ns
Propagation delays (from CLK): 25 ns (t_{pLH} and t_{pHL})
(from $\overline{CLR}$): 20 ns (t_{pLH} and t_{pHL})

FIGURE 11.7 Example 11.4: Timing Parameters

■ **SECTION 11.3 REVIEW PROBLEM**

11.3   An active edge on the clock input of a flip-flop makes Q go from HIGH to LOW. Name the timing parameter that measures the delay between the input and output change. Write the symbol for the parameter.

11.4   FANOUT

■ **KEY TERMS**

Fanout   The number of load gates that a logic gate output is capable of driving without possible logic errors.

I_{OH} Current measured at a device output when the output is HIGH.

I_{OL} Current measured at a device output when the output is LOW.

I_{IH} Current measured at a device input when the input is HIGH.

I_{IL} Current measured at a device input when the input is LOW.

Driving Gate   A gate whose output supplies current to the inputs of other gates.

Load Gate   A gate whose input current is supplied by the output of another gate.

Sourcing   A terminal on a gate or flip-flop is sourcing current when the current flows out of the terminal.

Sinking   A terminal on a gate or flip-flop is sinking current when the current flows into the terminal.

We have assumed that logic gates are able to drive any number of other logic gates. Since gates are electrical devices with finite current-driving capabilities, this is obviously not the case. The number of gates ("loads") a logic gate can drive is referred to as its **fanout**.

> **NOTE . . .**
>
> Fanout is simply an application of Kirchhoff's current law: The algebraic sum of currents at a node must be zero. Thus, the fanout of a logic gate is limited by:
> a. The maximum current its output can supply safely in a given logic state (I_{OH} or I_{OL}), and
> b. The current requirements of the load to which it is connected (I_{IH} or I_{IL}).

Figure 11.8 shows the fanout of an AND gate when its output is in the HIGH and LOW states. The AND gate, or **driving gate,** supplies current to the inputs of the other four gates, which are called the **load gates.**

Each load gate requires a fixed amount of input current, depending on which state it is in. The sum of these input currents equals the current supplied by the driving gate. The fanout is determined by the amount of current the driving gate can supply without degrading its output voltage.

The input and output currents of a gate are established by its internal circuitry. These values are usually the same for two gates in the same family, since the input and output circuitry of a gate is common to all members of the family. Exceptions may occur when the output of a particular gate, such as the 74XX244 octal tri-state buffer, has additional output buffering or an input of a gate such as a 74ALS86 Exclusive OR is equivalent to more than one input load.

a. **HIGH state**　　　　　b. **LOW state**

FIGURE 11.8 Driving Gates and Load Gates

EXAMPLE 11.5

a. Low output on driving gate

b. High output on driving gate

FIGURE 11.9 Example 11.5: Output Current Due to One Load Gate

The gates in Figure 11.9a and b are 74ALS00 NAND gates. Determine the output current of the driving gate in each figure.

■ Solution

From the 74ALS00 data sheet, $I_{IL} = -0.1$ mA and $I_{IH} = 20$ µA. (In some datasheets there are two values given for I_{IH}. Choose the one for the condition $V_{IN} = 2.7$ V, which is the minimum output voltage of a driving gate in the HIGH state (V_{OH}). The other value is not appropriate since a gate will never have a 7 V output, as specified in the condition, if its supply voltage is 5 V. This specification is shown as I_I in our datasheet.)

Since the driving gate is driving one load, its output current is the same as the input current of the load gate. Therefore, the driving gate output currents are given by $I_{OL} = 0.1$ mA (positive, since it is entering the driving gate output) and $I_{OH} = -20$ µA (negative, since it is leaving the driving gate output).

EXAMPLE 11.6

Determine the output current of the driving gate in part a and part b of Figures 11.10 if the gates are all 74ALS00 NAND gates.

■ Solution

Since there are two identical load gates in the circuits of Figure 11.10, the driving gate output current will be twice the load gate input current.

$$I_{OL} = 2 \times 0.1 \text{ mA} = 0.2 \text{ mA.}$$
$$I_{OH} = 2 \times (-20 \text{ μA}) = -40 \text{ μA.}$$

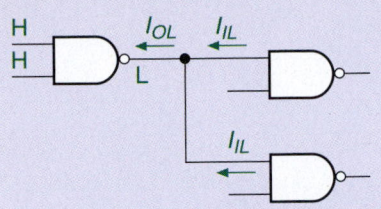

a. Low output on driving gate

b. High output on driving gate

FIGURE 11.10 Example 11.6: Output Current Due to Two Load Gates

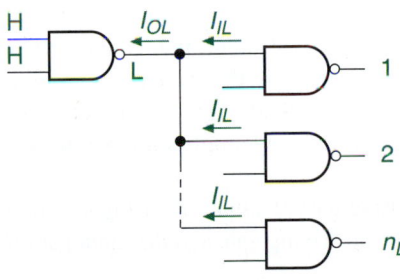

a. Low output on driving gate

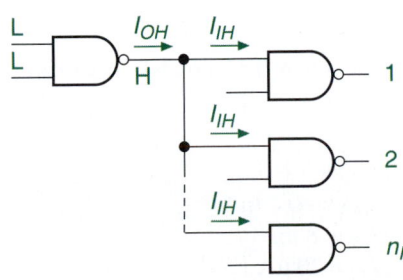

b. High output on driving gate

FIGURE 11.11 Output Current to Fanout Calculation

Figure 11.11 shows the extension of the circuits in Figures 11.9 and 11.10, where the number of load gates is the maximum that can be driven by the driving gate. This is the condition used to calculate fanout.

If the load gates each represent the same load, then by Kirchhoff's current law (KCL):

$$I_{OL} = I_{IL1} + I_{IL2} + \ldots I_{ILnL} = n_L I_{IL}$$
$$\text{and} \quad I_{OH} = I_{IH1} + I_{IH2} + \ldots + I_{IHnH} = n_H I_{IH}$$

The fanout of the driving gate in the LOW and HIGH states can be calculated as:

$$n_L = \frac{I_{OL}}{I_{IL}}$$

$$\text{and} \quad n_H = \frac{I_{OH}}{I_{IH}}$$

Since the defined and actual directions are the same, current entering a gate (I_{IH}, I_{OL}) is denoted as positive. Since the defined and actual current directions are opposite, current leaving the gate (I_{IL}, I_{OH}) is denoted as negative. When current is leaving a gate, we say the gate is **sourcing** current. When current is entering a gate, we say the gate is **sinking** current.

Note that the output of a gate does not always source current, nor does an input always sink current. The current direction changes for the HIGH and LOW states at the same terminal. The reason for this will become apparent when we study the circuitry of logic gate inputs and outputs.

EXAMPLE 11.7

How many 74ALS00 inputs can a 74ALS00 NAND gate drive? (that is, what is the fanout of a 74ALS00 NAND gate?)

■ **Solution**

We must consider the following cases:

a. When the output of the driving gate is LOW
b. When the output of the driving gate is HIGH

Output LOW:

$$I_{OL} = 8 \text{ mA (sinking)}$$
$$I_{IL} = -0.1 \text{ mA (sourcing)}$$
$$n_L = 8 \text{ mA}/0.1 \text{ mA} = 80$$

Output HIGH:

$$I_{OH} = -0.4 \text{ mA (sourcing)}$$
$$I_{IH} = 20 \text{ μA (sinking)}$$
$$n_H = 0.4 \text{ mA}/20 \text{ μA} = 20$$

Fanout is 20, the lower of the two values.

We disregard the negative sign in our calculations, since the input current of the load gate and output current of the driving gate are actually in the same direction. For example, even though I_{OH} is leaving the driving gate (negative), I_{IH} is entering the load gates (positive). These currents flow in the same direction. If we include the minus sign in our calculation, we get a negative value of fanout, which is meaningless.

If the values of HIGH- and LOW-state fanout are different, the smallest value must be used. For example, if a gate can drive four loads in the HIGH state or eight in the LOW state, the fanout of the driving gate is four loads. If we attempt to drive eight loads, we can't guarantee enough driving current to supply all loads in both states.

If a gate from one logic family is used to drive gates from another logic family, we must use the output parameters (I_{OL}, I_{OH}) for the driving gate and the input parameters (I_{IL}, I_{IH}) for the load gates.

EXAMPLE 11.8

Calculate the maximum number of Advanced Schottky TTL loads (74ASXX series) that a 74ALS86 XOR gate can drive.

■ **Solution**

Driving gate:	74ALS86	$I_{OH} = -0.4$ mA,
		$I_{OL} = 8$ mA
Load gates:	74ASXX	$I_{IH} = 20$ μA,
		$I_{IL} = -0.5$ mA

Output LOW:

$$I_{OL} = 8 \text{ mA (sinking)}$$
$$I_{IL} = -0.5 \text{ mA (sourcing)}$$
$$n_L = 8 \text{ mA}/0.5 \text{ mA} = 16$$

Output HIGH:

$$I_{OH} = -0.4 \text{ mA (sourcing)}$$
$$I_{IH} = 20 \text{ μA (sinking)}$$
$$n_H = 0.4 \text{ mA}/20 \text{ μA} = 20$$

Since $n_L < n_H$, fanout = $n_L = 16$.

What happens if we load a gate output beyond its rated layout? Adding more load gates will do this by increasing the value of I_{OL} or I_{ON} beyond its maximum rating. If enough load is added, the output of the driving gate might be destroyed by the heat generated by the excess current. More likely, the performance of the driving gate will be degraded.

Figure 11.12 shows the relationship between output voltage and current for a 74LS00 NAND gate and a 74LS240 tristate buffer gate. Figures 11.12a and b show that the output voltage (LOW state) increases with increasing sink current. Figure 11.12c and d indicate a decrease in HIGH state output voltage with an increase of source current.

a. Output LOW characteristic

b. Output LOW characteristic

c. Output HIGH characteristic

d. Output HIGH characteristic

FIGURE 11.12 Output Characteristics of 74LS00 and 74LS240 Gates. (Courtesy of ON Semiconductor)

In other words, a greater load in either state takes the output voltage further away from its nominal value. This has an effect on other performance factors, such as noise margin, which we will examine in a later section of the chapter.

> **NOTE . . .**
>
> The output voltage of a logic gate is defined in a datasheet for a particular value of output current.

We will examine the fanout of CMOS devices in a later section on interfacing between CMOS and TTL.

■ SECTION 11.4 REVIEW PROBLEM

11.4   The input and output currents I_{OH}, I_{OL}, I_{IH}, and I_{IL}, of a TTL device may be classified as source currents or sink currents. List each input or output current as a source or sink current.

11.5   POWER DISSIPATION

> ### ■ KEY TERMS
>
> **Power Dissipation**   The electrical energy used by a logic circuit in a specified period of time. Abbreviation: P_D
>
> V_{CC} TTL or high-speed CMOS supply voltage.
>
> I_{CC} Total TTL or high-speed CMOS supply current.
>
> I_T When referring to CMOS supply current, the sum of static and dynamic supply currents.
>
> I_{CCL} TTL supply current with all outputs LOW.
>
> I_{CCH} TTL supply current with all outputs HIGH.
>
> C_{PD} Internal capacitance of a high-speed CMOS device used to calculate its power dissipation.

Electronic logic gates require a certain amount of electrical energy to operate. The measure of the energy used over time is called **power dissipation.** Each of the different families of logic has a characteristic range of values for the power it consumes.

For TTL and CMOS, the power dissipation is calculated as follows:

TTL: $P_D = V_{CC} I_{CC}$

High-speed CMOS: $P_D = V_{CC} I_T$ (I_T = quiescent + dynamic supply current)

Figure 11.13 shows the supply voltage and current in a 74XX00 NAND gate.

The main difference between the TTL and CMOS families is the calculation of supply current.

The supply current in a TTL device is different when its outputs are HIGH than when they are LOW. Thus, supply current, I_{CC}, and therefore power dissipation, depends on the states of the device outputs. If the outputs are switching, I_{CC} is proportional to output duty cycle.

In a CMOS device, very little power is consumed when the device outputs are static. Much more current is drawn from the supply when the outputs switch from one state to another. Thus, the power dissipation of a device depends on the switching frequency of its outputs.

FIGURE 11.13   Power Supply Voltage and Current in a 74XX00 NAND Gate

Power Dissipation in TTL Devices

Two values are given for supply current in a TTL datasheet. I_{CCL} is the current drawn from the power supply when all gate outputs are LOW. I_{CCH} is the current drawn from the supply when all outputs are HIGH. If the gate outputs are not all at the same level, the supply current is the sum of currents given by:

$$I_{CC} = \frac{n_H}{n} I_{CCH} + \frac{n_L}{n} I_{CCL}$$

where

n is the total number of gates in the package
n_H is the number of gates whose output is HIGH
n_L is the number of gates whose output is LOW

The power dissipation of a TTL chip also depends on the duty cycle of the gate outputs. That is, it depends on the fraction of time that the chip's outputs are HIGH.

If we assume that, on average, the outputs of a chip are switching with a duty cycle of 50%, the supply current can be calculated as follows:

$$I_{CC} = (I_{CCH} + I_{CCL})/2$$

If the output duty cycle is other than 50%, the supply current is given by:

$$I_{CC} = DC\,I_{CCH} + (1 - DC)\,I_{CCL}$$

where DC = duty cycle.

EXAMPLE 11.9

Figure 11.14 shows a circuit constructed from the gates in a 74XX00 quadruple 2-input NAND gate package. Use the datasheet shown in Figure 11.3 to determine the maximum power dissipation of the circuit if the input is $DCBA = 1001$ and the gates are 74ALS00 NANDs. Refer to the data sheets in Appendix C and repeat the calculation for 74LS00 and 74AS00 gates.

FIGURE 11.14 Example 11.9: Power Dissipation of 74XX00 NAND

■ Solution

Gate 1: $\overline{AB} = 1$
Gate 2: $\overline{CD} = 1$
Gate 3: $AB + CD = 0$
Gate 4: $\overline{AB + CD} = 1$

Since three outputs are HIGH and one is LOW, the supply current is given by:

$$I_{CC} = \frac{n_H}{n} I_{CCH} + \frac{n_L}{n} I_{CCL}$$

$$= \frac{3}{4} I_{CCH} + \frac{1}{4} I_{CCL}$$

(continued)

Maximum supply current for each device is:

$$74LS00: I_{CC} = 0.75(1.6 \text{ mA}) + 0.25(4.4 \text{ mA}) = 2.3 \text{ mA}$$
$$74ALS00: I_{CC} = 0.75(0.85 \text{ mA}) + 0.25(3 \text{ mA}) = 1.3875 \text{ mA}$$
$$74AS00: I_{CC} = 0.75(3.2 \text{ mA}) + 0.25(17.4 \text{ mA}) = 6.75 \text{ mA}$$

Maximum power dissipation for each device is:

$$74LS00: P_D = V_{CC} I_{CC} = (5 \text{ V})(2.3 \text{ mA}) = 11.5 \text{ mW}$$
$$74ALS00: P_D = V_{CC} I_{CC} = (5 \text{ V})(1.3875 \text{ mA}) = 6.94 \text{ mW}$$
$$74AS00: P_D = V_{CC} I_{CC} = (5 \text{ V})(6.75 \text{ mA}) = 33.75 \text{ mW}$$

$(1 \text{ mW} = 1 \text{ milliwatt} = 10^{-3} \text{ W.})$

EXAMPLE 11.10

Find the maximum power dissipation of the circuit in Figure 11.14 if the gates are 74ALS00 and the gate outputs are switching with an average duty cycle of 30%.

■ **Solution**

$$I_{CC} = 0.3 \, I_{CCH} + 0.7 \, I_{CCL}$$
$$I_{CC} = 0.3 \, (0.85 \text{ mA}) + 0.7 \, (3 \text{ mA})$$
$$= 2.355 \text{ mA}$$

$$P_D = V_{CC} I_{CC} = (5 \text{ V}) \, (2.355 \text{ mA}) = 11.775 \text{ mW}$$

Power Dissipation in High-Speed CMOS Devices

CMOS gates draw the most power when their outputs are switching from one logic state to the other. When the outputs are static (not switching), the large internal impedances of the gate limit the supply current. A change of state requires the charging and discharging of internal gate capacitances, resulting in a greater demand on the power supply current. Thus, the faster a CMOS gate switches, the more current, and hence more power, it requires.

CMOS supply current has two components: a quiescent current that flows when the gate is in a steady state and a dynamic component that depends on frequency. For relatively high frequencies (about 1 MHz and up), the quiescent component is small compared to the dynamic component and can be neglected.

The quiescent current is usually specified for an entire chip package, regardless of the number of gates. It is given by $I_{CC} V_{CC}$. For a 74HC00A NAND gate, $I_{CC} = 1 \text{ μA}$ at room temperature for a supply voltage of $V_{CC} = 6.0$ V. The dynamic component calculation accounts for internal and load capacitance and is given, *per gate*, by:

$$(C_L + C_{PD}) \, V_{CC}^2 f$$

where C_L is the gate load capacitance
  C_{PD} is the gate internal capacitance
  V_{CC} is the supply voltage
  f is the switching frequency of the gate output

EXAMPLE 11.11

The circuit in Figure 11.14 is constructed from 74HC00A high-speed CMOS NAND gates. Calculate the power dissipation of the circuit:

a. When the gate inputs are steady at the state $DCBA = 1010$
b. When the outputs are switching at an average frequency of 10 kHz
c. When the outputs are switching at an average frequency of 1 MHz

Supply voltage is 5 V. Temperature range is 25° C to −55° C.

■ **Solution**

Refer to the 74HC00A datasheet in Appendix C.

a. $P_D = V_{CC} I_{CC} = (5\text{ V})(2\ \mu\text{A}) = 10\ \mu\text{W}$. This is the quiescent power dissipation of the circuit.

b. The 74HC00A datasheet indicates that each gate has a maximum input capacitance, C_{in} of 10 pF. Assume that this value represents the load capacitance of gates 1, 2, and 3 of the circuit in Figure 11.14. Further assume that gate 4 has a load capacitance of 0. The total power dissipation of the circuit is given by:

$$P_D = 3(20\text{ pF} + 10\text{ pF})(5\text{ V})^2\ (0.01\text{ MHz})$$
$$+ (20\text{ pF})(5\text{ V})^2(0.01\text{ MHz}) + 10\ \mu\text{W}$$
$$= 3(7.5\ \mu\text{W}) + 5\ \mu\text{W} + 10\ \mu\text{W}$$
$$= 37.5\ \mu\text{W}$$

c. For $f = 1$ MHz, total power dissipation is given by:

$$P_D = 3(20\text{ pF} + 10\text{ pF})(5\text{ V})^2(1\text{ MHz})$$
$$+ (20\text{ pF})(5\text{ V})^2\ (1\text{ MHz}) + 10\ \mu\text{W}$$
$$= 3(750\ \mu\text{W}) + 500\ \mu\text{W} + 10\ \mu\text{W}$$
$$= 2760\ \mu\text{W} = 2.76\text{ mW}$$

 EXAMPLE 11.12

The circuit in Figure 11.14 is constructed using a 74ALS00 quad 2-in NAND gate and again with a 74HC00 quad 2-in NAND. Both circuits have identical waveforms applied to their inputs that make all gate outputs switch with a duty cycle of 50%. Calculate the frequency at which the power dissipation of the 74HC00 circuit exceeds that of the 74ALS00 circuit.

Assume $V_{CC} = 5$ V and temperature = 25° C for both circuits.

■ **Solution**

The power dissipation of the ALS TTL circuit is:

$$P_D = V_{CC} I_{CC} = (V_{CC})\ (I_{CCH} + I_{CCL})/2 = (5\text{V})\ (0.85\text{ mA} + 3\text{ mA})/2$$
$$= (5\text{ V})\ (1.925\text{ mA}) = 9.625\text{ mW}$$

Neglect the quiescent current of the high-speed CMOS circuit.

Per gate: $P_D = (C_L + C_{PD})V_{CC}^2\, f$
$C_{PD} = 20$ pF per gate
$C_L = 10$ pF for 3 gates and 0 pF for 1 gate

Total: $P_D = (3(10\text{pF} + 20\text{ pF}) + 20\text{ pF})(5\text{ V})^2\, f$
$$= (3(30\text{ pF}) + 20\text{ pF})\ (25\text{ V}^2)\, f$$
$$= (90\text{ pF} + 20\text{ pF})\ (25\text{ V}^2)\, f = (110\text{ pF})\ (25\text{ V}^2)\, f$$

For $P_D = 9.625$ mW:

$$f = \frac{9.625\text{mW}}{(110\text{pF})(25\text{V}^2)} = 3.5\text{MHz}$$

The power dissipation of the 74HC00 circuit exceeds that of the 74ALS00 circuit at 3.5 MHz.

> **NOTE . . .**
>
> The power saving in a high-speed CMOS circuit generally results from the fact that most device outputs are not switching at any given time. The power dissipation of a TTL circuit is independent of frequency and therefore draws some power at all times. This is not the case for CMOS, which draws the majority of its power when switching.

■ **SECTION 11.5 REVIEW PROBLEM**

11.5   Why does CMOS power dissipation increase with frequency?

11.6   NOISE MARGIN

> ■ **KEY TERMS**
>
> **Noise**   Unwanted electrical signal, often resulting from electromagnetic radiation.
>
> **Noise Margin**   A measure of the ability of a logic circuit to tolerate noise.
>
> V_{IH} Voltage level required to make the input of a logic circuit HIGH.
>
> V_{IL} Voltage level required to make the input of a logic circuit LOW.
>
> V_{OH} Voltage measured at a device output when the output is HIGH.
>
> V_{OL} Voltage measured at a device output when the output is LOW.
>
> **TTL Compatible**   Able to be driven directly by a TTL output. Usually implies voltage compatibility with TTL.

Electrical circuits are susceptible to **noise,** or unwanted electrical signals. Such signals are often induced by electromagnetic fields of motors, fluorescent lighting, high-frequency electronic circuits, and cosmic rays. They can cause erroneous operation of a digital circuit. Since it is impossible to eliminate all noise from a circuit, it is desirable to build a certain amount of tolerance, or **noise margin,** into digital devices used in the circuit.

In all circuits studied so far, we have assumed that logic HIGH is +5 volts and logic LOW is 0 volts in devices with a 5-volt supply. In practice, there is a certain amount of tolerance on both the logic HIGH and LOW voltages; for TTL devices, a HIGH at a device input is anything above about +2 volts, and a LOW is any voltage below about +0.8 volts. Due to internal voltage drops, the HIGH output of a TTL gate is typically about +3.5 volts.

Figure 11.15 shows one inverter driving another. In Figure 11.15a, the output of the first inverter and the input of the second have the same logic threshold. That is, the input of the second gate recognizes any voltage above 2.7 volts as HIGH ($V_{IH} = 2.7$ V) and any voltage below 0.5 volts as LOW ($V_{IL} = 0.5$ V). The output of the first inverter produces at least 2.7 volts when HIGH ($V_{OH} = 2.7$ V) and no more than 0.5 volts as LOW ($V_{OL} = 0.5$ V).

If there is noise on the line connecting the two gates, it will likely cause the voltage of the second gate input to penetrate into the forbidden region between logic HIGH and LOW levels. This is shown on the graph of the waveform in Figure 11.15a. When the voltage enters the forbidden region, the gate will not operate reliably. Its output may switch states when it is not supposed to.

Figure 11.15b shows the same circuit with different logic thresholds at input and output. The output of the first inverter is guaranteed to be *at least 2.7 volts* when HIGH ($V_{OH} = 2.7$ V) and *no more than 0.5 volts* when LOW ($V_{OL} = 0.5$ V). The second gate recognizes any input voltage *greater than 2 volts* as a HIGH ($V_{IH} = 2$ V) and any input voltage *less than 0.8 volts* ($V_{IL} = 0.8$ V) as a LOW.

a. Zero noise margin

b. Nonzero noise margin

FIGURE 11.15 Noise Margins

The difference between logic thresholds allows for a small noise voltage, equal to or less than the difference, to be superimposed on the desired signal. It will not cause the input voltage of the second inverter to penetrate the forbidden region. This ensures reliable operation even in the presence of some noise.

For the 74ALS04 inverter, the HIGH-state and LOW-state noise margins, V_{NH} and V_{NL}, are:

$$V_{NH} = V_{OH} - V_{IH} = 3.0 \text{ V} - 2.0 \text{ V} = 1.0 \text{ V}$$
$$V_{NL} = V_{IL} - V_{OL} = 0.8 \text{ V} - 0.5 \text{ V} = 0.3 \text{ V}$$

A device with these values of V_{IH} and V_{IL} is deemed to be **TTL compatible.**

EXAMPLE 11.13

Use the 74HC00A datasheet in Appendix C to calculate the noise margins for this gate. Assume $V_{CC} = 4.5$ V, ambient temperature (T_A) is 25° C, and the driving gate is fully loaded $(I_{OUT} = \pm 4$ mA).

■ **Solution**

$$V_{NH} = V_{OH} - V_{IH} = 3.98 \text{ V} - 3.15 \text{ V} = 0.83 \text{ V}$$
$$V_{NL} = V_{IL} - V_{OL} = 1.35 \text{ V} - 0.26 \text{ V} = 1.09 \text{ V}$$

■ **SECTION 11.6 REVIEW PROBLEM**

11.6 Calculate the noise margins of a 74HCT00A NAND gate from the datasheet in Appendix C. $V_{CC} = 4.5$ V, $T_A = 25°$ C, $I_{OUT} = \pm 4$ mA

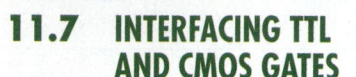

11.7 INTERFACING TTL AND CMOS GATES

Interfacing different logic families is just an extension of the fanout and noise margin problems; you have to know what the load gates of a circuit require and what the driving gates can supply. In practice, this means you must know the specified values of input and output voltages and currents for the gates in question. Table 11.5, which is derived from the manufacturers' datasheets included in Appendix C, gives an overview of input and output parameters for a variety of TTL and CMOS families. Ambient temperature is assumed to be 25° C.

TABLE 11.5 TTL and CMOS Input and Output Parameters

	TTL				High-Speed CMOS				Low-Voltage CMOS	
	74LS	**74F**	**74AS**	**74ALS**	**74HC**	**74HCT**	**74VHC**	**74VHCT**	**74LVX**	**74LCX**
V_{CC} (V)	5.0	5.0	5.0	5.0	4.5	4.5	4.5	4.5	3.0	3.0
V_{OH} (V)	2.7	2.7	3.0	3.0	3.98	3.98	3.94	3.94	2.58	2.2
V_{OL} (V)	0.5	0.5	0.5	0.5	0.26	0.26	0.36	0.36	0.36	0.55
V_{IH} (V)	2.0	2.0	2.0	2.0	3.15	2.0	3.15	2.0	2.0	2.0
V_{IL} (V)	0.8	0.8	0.8	0.8	1.35	0.8	1.35	0.8	0.8	0.8
I_{OH} (mA)	−0.4	−1.0	−2.0	−0.4	−4.0	−4.0	−8.0	−8.0	−4.0	−24.0
I_{OL} (mA)	8.0	20.0	20.0	8.0	4.0	4.0	8.0	8.0	4.0	24.0
I_{IH} (mA)	0.02	0.02	0.02	0.02	0.0001	0.0001	0.0001	0.0001	0.0001	0.0001
I_{IL} (mA)	−0.4	−0.6	−0.5	−0.1	0.0001	0.0001	0.0001	0.0001	0.0001	0.0001

Table 11.5 is useful for comparison of logic families, but it is not a substitute for reading datasheets, as it gives parameters only under a restricted set of conditions. We can, however, make some observations based on the data in Table 11.5.

1. Input currents in a CMOS gate are very low, due to its high input impedance. As a result fanout is generally not a problem with CMOS loads. Interface problems to CMOS loads have to do with input voltage, not current.
2. CMOS devices, such as 74HCT, that have the same values of V_{IH} and V_{IL} as the TTL families in Table 11.5, are considered to be TTL compatible, since they can be driven directly by TTL drivers.

3. LSTTL is usually regarded as the benchmark for measuring TTL loading of a CMOS circuit. For example, a datasheet will claim that a device can drive ten LSTTL loads. This claim depends on the values of I_{OH} and I_{OL} for the driving gate, which are not listed directly in CMOS datasheets, except as absolute maximum ratings. The values in Table 11.5 are the values of current for which the output voltages, V_{OH} and V_{OL}, are defined. (Recall from the section on fanout in this chapter that increasing output current causes output voltages to migrate away from their nominal values, thus reducing device noise margins.)

Let us examine four interfacing problems: high-speed CMOS driving 74LS, 74LS driving 74HC, 74LS driving 74HCT, and 74LS driving low-voltage CMOS.

High-Speed CMOS Driving 74LS

To design an interface between any two logic families, we must examine the output voltages and currents of the driving gate and the input voltages and currents of the load gates.

Assume that a 74HC00 NAND gate drives one or more 74LS00 NAND gates. From the 74HC00 datasheet, we determine that $V_{OH} = 3.98$ V and $V_{OL} = 0.26$ V for $V_{CC} = 4.5$ V. The 74LS00 requires at least 2.0 V at its input in the HIGH state and no more than 0.8 V in the LOW state. The 74HC00 therefore satisfies the input voltage requirement of the 74LS00.

For the defined output voltages, the 74HC00 gate can source or sink 4 mA. The fanout for the circuit is therefore calculated as follows:

$$n_H = \frac{I_{OH}}{I_{IH}} = \frac{4\text{mA}}{20\mu\text{A}} = 200$$

$$n_L = \frac{I_{OL}}{I_{IL}} = \frac{4mA}{0.4\text{A}} = 10$$

$$n = 10$$

Therefore a 74HC00 NAND can drive a 74LS00 directly, with a fanout of 10.

74LS Driving 74HC

As mentioned earlier, CMOS has a very small input current and therefore does not present a fanout problem to a 74LS driving gate. However, we must also examine the interface for voltage compatibility.

From the datasheets, we see that a 74LS00 gate is guaranteed to provide at least 2.7 V in the HIGH state and no more than 0.5 V in the LOW state. A 74HC00 gate will recognize anything less than 1.35 V as a logic LOW and anything more than 3.15 V as a logic HIGH. The 74LS00 meets the LOW-state criterion, but it cannot guarantee sufficient output voltage in the HIGH state.

In order to properly drive a 74HC input with a 74LS output, we must provide a pull-up resistor to ensure sufficient HIGH-state voltage at the 74HC input. The circuit is illustrated in Figure 11.16. The pull-up resistor should be between 1 kΩ and 10kΩ.

74LS Driving 74HCT

74HCT inputs are designed to be compatible with TTL outputs. As with 74HC devices, input currents are sufficiently low that fanout is not a problem with the 74LS-to-74HCT interface. 74HCT input voltages are the same as those for

FIGURE 11.16 LSTTL Driving 74HC CMOS

FIGURE 11.17 74LS-to-74HC Interface Using a 74HC4050 Buffer

TTL ($V_{IH} = 2.0$ V and $V_{IL} = 0.8$ V). Therefore, 74HCT inputs can be driven directly by LSTTL outputs.

74LS Driving Low-Voltage CMOS

CMOS families with supply voltages less than 5 V are rapidly becoming popular in new applications. Two of the reasons for their increasing prominence are reduced power dissipation (inversely proportional to the *square* of the supply voltage) and smaller feature size (i.e., size of the internal transistors) that allows more efficient packaging and faster operation. Low-voltage logic is particularly popular for battery-powered applications such as laptop computing or cell phones. Low voltage families typically operate at $V_{CC} = 3.3$ V or 2.5 V. Newer devices are available for $V_{CC} = 1.8$ V or 1.65 V.

Low-voltage CMOS families such as 74LVX or 74LCX can interface directly with TTL outputs if they are operated with a 3.0 V to 3.3 V power supply. These families are not really suitable for driving 5-volt TTL, as their noise margins are too small when they use a 3.0 V supply voltage.

If we wish to use a 74LS device to drive a 74HC device operating at a power supply voltage of less than 4.5 V, we can use a 74HC4049 or 74HC4050 buffer to translate the TTL logic level down to an appropriate value. The 74HC4049 is a package of six inverting buffers. The 74HC4050 has six noninverting buffers. These buffers can tolerate up to 15 V on their inputs. Their output voltages are determined by the value of their supply voltage.

Figure 11.17 shows an LSTTL-to-74HC interface circuit with a 74HC4050 buffer. Note that the interface buffer has the same power supply voltage as the load gate. Both sides of the interface are referenced to the same ground.

■ SECTION 11.7 REVIEW PROBLEM

11.7 A 74LS00 driving gate is to be interfaced to a 74HC00 load using a 74HC4050 noninverting buffer. The 74HC00 has a power supply voltage of 2.5 V. What supply voltage should the 74HC4050 buffer have? Why?

11.8 INTERNAL CIRCUITRY OF TTL GATES

■ **KEY TERMS**

Cutoff Mode   The operating mode of a bipolar transistor when there is no collector current flowing and the path from collector to emitter is effectively an open circuit. In a digital application, a transistor in cutoff mode is considered OFF.

Saturation Mode   The operating mode of a bipolar transistor when an increase in base current will not cause a further increase in the collector current and the path from collector to emitter is very nearly (but not quite) a short circuit. This is the ON state of a transistor in a digital circuit.

TTL has been around for a long time. The first transistor-transistor logic ICs were developed by Texas Instruments around 1965. Since then, there have been many improvements in the speed and power consumption of these devices, but the basic logic principles remain largely unchanged. Even though they are seldom used in modern designs, it makes sense to examine the internal circuitry of standard TTL gates such as the 7400 NAND, 7402 NOR, and 7404 inverter because the internal logic concepts are similar to the more advanced types of TTL.

The most important parts of the circuit, as far as a designer or technician is concerned, are the input and output circuits, because they are the only parts of the chip to which we have access. It is to these points that we interface other circuits and where we make diagnostic measurements. A basic understanding of the inputs and outputs of logic gate circuitry is helpful when we design or troubleshoot a digital circuit.

Bipolar Transistors as Logic Devices

The basic element of a TTL device is the bipolar junction transistor, illustrated in Figure 11.18. This is not the place to give a detailed analysis of the operation of a bipolar transistor, but a simplified summary of operating modes will be useful.

The bipolar transistor is a current amplifier having three terminals called the collector, emitter, and base. Current flowing into the base controls the amount of current flowing from the collector to the emitter. If base current is below a certain threshold, the transistor is in **cutoff mode** and no current flows in the collector. In this state, the base-emitter voltage is less than 0.6 V and the collector-emitter path acts like an open circuit. We can treat the path as an open switch, as shown in the lefthand diagram in Figure 11.19.

FIGURE 11.18   Currents and Voltages in an NPN Bipolar Transistor

FIGURE 11.19   NPN Bipolar Transistor as a Switch

If the base current increases, the transistor enters the "active region," where the collector current is proportional to the base current by a current gain factor, β. This is the linear, or amplification, region of operation, used by analog amplifiers.

If the base current increases still further, collector current reaches a maximum value and will no longer increase with base current. This is called the **saturation mode** of the transistor. The saturated value of collector current, I_{CS}, is determined by (1) the resistance in the collector-emitter current path, (2) the voltage drop across the collector and emitter, V_{CE}, and (3) the collector supply voltage, V_{CC}. Base-emitter voltage is about 0.7 V and will not increase significantly with increasing base current. The voltage between collector and emitter is in the range from 0.2 V to 0.7 V. In this mode, we can treat the transistor as a closed switch, as shown in the righthand diagram of Figure 11.19.

Table 11.6 summarizes the voltages and currents in the cutoff, active, and saturation regions.

TABLE 11.6   Bipolar Transistor Characteristics

	Cutoff	**Active**	**Saturation**
I_C	0	$= \beta I_B$	$< \beta I_B$
V_{CE}	Open cct.	>0.8 V	0.2 V–0.7
V_{BE}	<0.6 V	0.6 V–0.7 V	≈0.7 V

EXAMPLE 11.14

Figure 11.20 shows an NPN bipolar transistor connected in a common-emitter configuration. With the right choice of input voltages, this circuit acts as a digital inverter.

FIGURE 11.20   Example 11.14: Transistor as Inverter

Analyze the circuit to show that it acts as an inverter if a logic HIGH is defined as ≥3 V and a logic LOW is defined as ≤0.5 V. Assume that $\beta = 100$, and assume that $V_{BE} = 0.7$ V and $V_{CE} = 0.2$ V in saturation.

■ Solution

We will analyze the circuit with two input voltages: 3 V (logic HIGH) and 0.5 V (logic LOW). These two conditions are shown in Figure 11.21.

HIGH input. We must prove that $V_I = 3$ V is sufficient to saturate the transistor. Let us assume that this is true and find out if calculations confirm our assumption.

a. HIGH input (saturation)   **b. LOW input (cutoff)**

FIGURE 11.21   Example 11.14: Voltage and Current Analysis of Inverter

Figure 11.21a shows the circuit with $V_I = 3$ V. By Kirchhoff's voltage law (KVL):

$$V_I = I_B R_B + V_{BE}, \text{ or}$$
$$I_B = (V_I - V_{BE})/R_B$$

If we assume that I_B is sufficient to saturate the transistor, then:

$$I_B = (3 \text{ V} - 0.7 \text{ V})/22 \text{ k}\Omega$$
$$= 105 \text{ }\mu\text{A}$$
$$\beta I_B = (100)(105 \text{ }\mu\text{A}) = 10.5 \text{ mA}$$

Collector current won't increase beyond its saturated value, even if base current increases. Therefore, if the transistor is saturated, βI_B will be larger than the current actually flowing in the collector-emitter path.

In saturation, the collector current can be calculated by KVL:

$$V_{CC} = I_C R_C + V_{CE}, \text{ or}$$
$$I_C = (V_{CC} - V_{CE})/R_C$$
$$I_C = (5 \text{ V} - 0.2 \text{ V})/470 \text{ }\Omega$$
$$= 10.2 \text{ mA}$$

Since $\beta I_B > I_C$, the transistor is saturated. Thus, an input voltage of 3 V will produce sufficient base current to saturate the transistor. The output is given by $V_O = V_{CE} = 0.2$ V, which is within the defined range of a logic LOW.

LOW input. Figure 11.21b shows the circuit with $V_I = 0.5$ V. By KVL:

$$V_I = I_B R_B + V_{BE}$$
$$V_{BE} = 0.5 \text{ V} - I_B R_B$$

Since V_{BE} must be < 0.6 V, the transistor is in cutoff mode. Thus, in the collector circuit:

$$V_{CC} = I_C R_C + V_{CE}$$
$$5 \text{ V} = (0)(470 \text{ }\Omega) + V_{CE}$$
$$V_O = V_{CE} = 5 \text{ V (logic HIGH)}$$

Table 11.7 summarizes the operation of the circuit as an inverter.

TABLE 11.7   Input and Output of Single-Transistor Inverter

Input		Output	
V_I	Logic Level	V_o	Logic Level
0.5V	LOW	5 V	HIGH
3 V	HIGH	0.2V	LOW

TTL Open-Collector Inverter and NAND Gate

> ■ **KEY TERMS**
>
> **Open-Collector Output**   A TTL output where the collector of the LOW-state output transistor is brought out directly to the output pin. Since there is no built-in HIGH-state output circuitry, two or more open-collector outputs can be connected without damage.

> **NOTE . . .**
>
> The TTL gates (7405, 7401, 7404, 7400, and 7402) used in the following sections to illustrate TTL circuit principles are no longer in general use. They are from the original ("standard") TTL family, which has been superceded by faster and more efficient devices. However, the standard TTL devices are easier to understand than devices from the newer TTL subfamilies, since their circuit structure is simpler. The operating principles are similar in both the standard and newer families, so we will use the standard devices to illustrate the general principles of TTL operation.

Figure 11.22 shows the circuit of the simplest TTL gate: a 7405 inverter with **open-collector outputs.** This circuit performs the same function as the single-transistor inverter we examined in Example 11.14. These circuits differ most obviously in their input circuitry. The inverter circuit in Example 11.14 has a resistor as its input; the 7405 inverter has a transistor, Q_1, as its input. The input transistor allows faster switching of input states. This configuration is common to all standard TTL gates and will be examined in detail later in this section.

The logic function of the 7405 is performed by transistors Q_2 and Q_3. Output transistor Q_3 is switched ON and OFF by current flowing in the collector-emitter path of Q_2. When Q_3 is ON, Y is LOW.

However, when Q_3 is OFF, Y is floating. There is a high impedance between Y and ground, so the output is not LOW. But there is no connection to V_{CC} to make the output HIGH. In this condition, Y is neither HIGH nor LOW.

To enable the output to produce a HIGH state, we need to add an external pull-up resistor. The value of this resistor depends on the current sinking capability of Q_3, specified in the data sheet as I_{OL}. We will do such calculations in a later example.

TTL Inputs

Transistor Q_1 and diode D_1 make up the input circuit of the TTL inverter of Figure 11.22. The diode protects the input against small negative voltages. If the input goes more negative than about –0.7 V, the diode will conduct, effectively short-circuiting the input to ground minus one diode drop. This clamps the input to –0.7 V. D_1 has no logic function.

Q_1 can be treated as two back-to-back diodes, as shown in Figure 11.23. Figure 11.24 shows how the input responds to logic HIGH and LOW voltages.

LOW Input   When a TTL input is made LOW, the base-emitter junction of Q_1 acts as a forward-biased diode, creating a current path from V_{CC} to ground via the input pin. This current makes up the majority of current I_{IL}, which has a maximum value of 1.6 mA in standard TTL (0.4mA in LSTTL).

At the moment the input is made LOW, the transistor action of Q_1 transports charge away from the base of Q_2, pulling it LOW and keeping it in cutoff mode. This current dies out when the base charge of Q_2 has been depleted, shortly after the

a. Circuit

b. Logic symbol

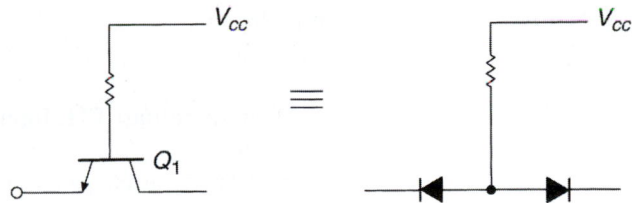

c. 7405 with HIGH-state pull-up

FIGURE 11.22 Open-Collector Inverter (7405)

FIGURE 11.23 Diode Equivalent of TTL Input Transistor

LOW is applied to the input pin. The diode formed by the base-collector junction of Q_1 does not carry sufficient current to turn on Q_2, since the base-emitter path is of much lower impedance.

HIGH Input   A HIGH at a TTL input reverse-biases the base-emitter junction of Q_1. Only a small leakage current, I_{IH}, flows. The maximum value of I_{IH} is 40 µA for standard TTL (20 µA for LSTTL).

Since the low-impedance current path to the input pin has not been established, current flows to the base of Q_2 via the forward-biased base-collector junction of Q_1. This current is sufficient to saturate Q_2.

FIGURE 11.24   HIGH and LOW Inputs at a TTL Gate

a. LOW   **b. HIGH**   **c. OPEN**

FIGURE 11.25   LOW, HIGH, and Open TTL Inputs

Open (Floating) TTL Input   An open-circuit TTL input acts as a logic HIGH, as illustrated by Figure 11.25. A TTL input relies on a logic LOW to establish a low-impedance current path from V_{CC} to the input pin. If the input is open, this LOW is not present and current flows in the base-collector junction of the transistor, by default. This is the same current that flows under the HIGH-input condition.

This HIGH is not stable; it can be converted to logic LOW by induced noise at the input pin. To avoid this uncertainty, an unused input should always be wired to a logic HIGH or LOW state.

TTL Open-Collector Inverter

Figure 11.26 shows the operation of the 7405 open-collector inverter.

LOW Input   As previously described, a LOW input establishes a low-impedance path to ground, which draws current through the base-emitter junction of Q_1. This action also prevents base current from flowing in transistor Q_2, causing it to be in cutoff mode and making $I_{C2} = 0$.

Since I_{B3} is derived from I_{C2}, $I_{B3} = 0$ and Q_3 is in cutoff, making a high-imped-ance path between the collector and emitter of Q_3. As was the case with the single-transistor inverter in Example 11.14, when $I_{C3} = 0$, then $V_O = V_{CE} = V_{CC}$. (Since no

FIGURE 11.26 7405 Operation

a. Circuit b. Input equivalent circuit

FIGURE 11.27 TTL NAND with Open-Collector Output

current flows through the pull-up resistor, the voltage must be the same at both ends.) *Output Y is HIGH.*

HIGH Input When input A is HIGH, the base-emitter junction of Q_1 does not have sufficient voltage across it to be forward-biased. Current flows through the base-collector junction of Q_1, saturating Q_2.

Since Q_2 is ON, current flows to the Q_2 emitter and splits through the 1-kΩ resistor and the base of Q_3. The output transistor, Q_3, turns ON, establishing a low-impedance current path from output Y to ground. Current is limited by the external pull-up resistor, which must be chosen to keep I_{OL} at or under its rated value of 16 mA. V_{CE3} is about 0.2 V to 0.4 V. *Output Y is LOW.*

TTL Open-Collector NAND

Figure 11.27 shows one gate of a 7401 quadruple 2-input NAND gate with open-collector outputs. The circuit is the same as that of the 7405 inverter, except that the

a. **Both inputs LOW**
   (LOW equivalent)

b. **One input LOW**
   (LOW equivalent)

c. **Both inputs HIGH**
   (HIGH equivalent)

FIGURE 11.28   Input Response of Multiple-Emitter Transistor

input transistor has a second emitter. Multiple-emitter transistors of this type are common in TTL circuits and can be modeled by the diode equivalent in Figure 11.27b. Figure 11.28 shows the response of the multiple-emitter input transistor to various combinations of logic levels.

If both inputs are LOW, the NAND acts exactly the same as the 7405 inverter with a LOW input. (A low-impedance path is created through a base-emitter junction.) *Output Y is HIGH,* provided an external pull-up resistor is connected to output. A partial truth table for this condition is:

A	*B*	*Y*
0	0	1

If one input is LOW, the input acts the same as the inverter with a LOW input. The low-impedance current path through the one grounded emitter prevents sufficient base-collector current from flowing to forward-bias that junction. *Output Y is HIGH* if a pull-up resistor is connected to the output. A partial truth table is as follows:

A	*B*	*Y*
0	1	1
1	0	1

If both inputs are HIGH, the NAND circuit acts like the 7405 when its input is HIGH. (There is no base-emitter current path. A collector-emitter path is established by default.) *Output Y is LOW.* This condition can be represented by:

A	*B*	*Y*
1	1	0

Combining all these conditions, we get the standard NAND truth table:

A	*B*	*Y*
0	0	1
0	1	1
1	0	1
1	1	0

The figure at top shows two NAND gate symbols with truth tables.

A	B	Y
0	0	1
0	1	1
1	0	1

a. At least one input LOW makes output HIGH

A	B	Y
1	1	0

b. Both inputs HIGH make output LOW

FIGURE 11.29 DeMorgan Equivalent Forms of a NAND Gate

> **NOTE...**
>
> If one or more emitters of a TTL multiple-emitter input transistor is LOW, the input is a LOW equivalent. All emitters must be HIGH to make the transistor input a HIGH equivalent.

These statements lead to the familiar NAND-gate descriptive sentences, illustrated by the gate symbols in Figure 11.29.

a. At least one input LOW makes the output HIGH.
b. Both inputs HIGH make the output LOW.

■ SECTION 11.8A REVIEW PROBLEM

11.8 What are the two main functions of the pull-up resistor on the output of an open-collector gate?

Open-Collector Applications

> **■ KEY TERM**
>
> **Wired-AND** A connection where open-collector outputs of logic gates are wired together. The logical effect is the ANDing of connected functions.

A more common TTL output than the open collector is the totem pole output, which we will study later in this chapter. The totem pole output has its own internal pull-up circuit for HIGH outputs.

Gates with totem pole outputs cannot be used in all digital circuits. For example, open-collector gates are required when several outputs must be tied together, a connection called **wired-AND.** Totem pole outputs would be damaged by such a connection, since there is the possibility of conflict between an output HIGH and LOW state.

Open-collector outputs can also be used for applications requiring high current drive and for interfacing to circuits having supply voltages other than TTL levels.

A special symbol defined by IEEE/ANSI Standard 91–1984, an underlined square diamond, is shown in Figure 11.30. This symbol is added to a logic gate symbol to indicate that it has an open-collector output. Other symbols, such as a star (*), a dot (•), or the initials OC are also used.

Wired-AND

> **NOTE...**
>
> A wired-AND connection combines the *outputs* of the connected gates in an AND function.

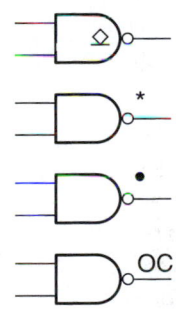

FIGURE 11.30 Open-Collector Symbols Shown for a NAND Gate (e.g., 7401)

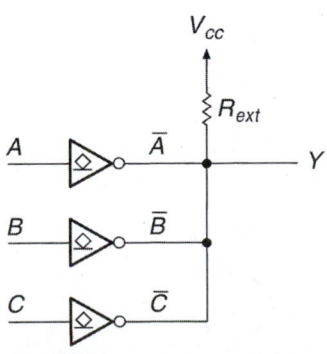

FIGURE 11.31 Three Inverters in a Wired-AND Connection

Figure 11.31 shows three open-collector inverters connected in a wired-AND configuration. The output transistors of the inverters are shown in Figure 11.32, with

a. One inverter with LOW output   b. All inverter outputs LOW

c. All inverter outputs HIGH

FIGURE 11.32 Output Transistors of Open-Collector Inverters in a Wired-AND Connection

a. Pull-up resistor and open-collector gates   **b. Wired-AND symbol**

FIGURE 11.33 NAND Gates in Wired-AND Connection

different possible ON and OFF states. The only way that output Y can remain HIGH is if all the transistors are in their OFF states, as in Figure 11.32c. This can happen only if the outputs of the inverters are all HIGH. This is the same as saying the outputs are ANDed together at Y.

The Boolean expression for Y is:

$$Y = \overline{A} \cdot \overline{B} \cdot \overline{C}$$
$$= \overline{A + B + C}$$

By DeMorgan's theorem, the wired-AND connection of inverter outputs is equivalent to a NOR function. Because of this DeMorgan equivalence, the connection is sometimes called "wired-OR."

Figure 11.33 shows three NAND gates in a wired-AND connection. Since the output functions are ANDed, the Boolean expression for Y is:

$$Y = \overline{AB} \cdot \overline{CD} \cdot \overline{EF}$$
$$= \overline{AB + CD + EF}$$

The resulting function is called AND-OR-INVERT. Normally this requires at least two types of logic gate—AND and NOR. The wired-AND configuration can synthesize any size of AND-OR-INVERT network using only NAND gates.

The wired-AND function is sometimes shown as an AND symbol around a soldered connection, as shown in Figure 11.33b.

High-Current Driver

Standard TTL outputs have higher current ratings in the LOW state than in the HIGH state. Thus, open-collector outputs are useful for driving loads that need more current than a standard TTL output can provide in the HIGH state. There are special TTL gates with higher ratings of I_{OL} to allow even larger loads to be driven. Typical loads would be LEDs, incandescent lamps, and relay coils, all of which require currents in the tens of milliamperes.

EXAMPLE 11.15

FIGURE 11.34 Example 11.15: 74LS06 High-Current Driver

A 74LS06 hex buffer/driver contains six inverting buffers whose outputs are open-collector, rated for $I_{OL\text{max}} = 40$ mA and $V_{OH\text{max}} = 30$ V. That is, even though there is no internal circuit to provide a logic HIGH at the output, the output transistor can withstand a voltage of up to 30 V without damage.

Figure 11.34 shows a 74LS06 buffer driving an incandescent lamp rated at 24 V, with a resistance of 690 Ω. Calculate the current that flows when the lamp is illuminated. What logic level at A turns the lamp on? Could the lamp be driven by a 74LS05 inverter? Why or why not?

(continued)

■ Solution

From the 74LS06 datasheet in Appendix C, we see that $V_{OL} = 0.4$ V for $I_{OL} =$ 16 mA and $V_{OL} = 0.7$ V for $I_{OL} = 40$ mA. Assume the latter value.

$$\text{By KVL: } 24 \text{ V} - (I_{OL})(690 \ \Omega) - V_{OL} = 0$$
$$\text{Thus, } I_{OL} = (24 \text{ V} - 0.7 \text{ V})/690 \ \Omega = 33.8 \text{ mA}$$

Since the buffer is inverting, and current flows when the output of the 74LS06 sinks current to ground (LOW), the lamp is on when A is HIGH.

A 74LS05 open-collector inverter would not be a suitable driver for the circuit for two reasons: its output is only designed to withstand 5.5 V and it can only sink a maximum of 8 mA.

Value of External Pull-up Resistor

The value of the pull-up resistor required by an open-collector circuit is calculated using manufacturer's specifications and the basic principles of circuit theory: Kirchhoff's voltage and current laws (KVL and KCL) and Ohm's law.

Figure 11.35 shows the circuit model for calculating the value of R_{ext}. It accounts for the current requirements of the loads, the LOW-state output voltage, and current-sinking capacity of the open-collector gate.

> **NOTE . . .**
>
> The main rule in resistor selection is to keep the sum of currents into the open-collector output to less than the maximum rated value of I_{OL}.
>
> $$I_{OL} = I_R + nI_{IL}$$
> $$I_R = (V_{CC} - V_{OL})/R_{ext}$$

FIGURE 11.35   Circuit Model for Pull-Up Resistor Calculation

EXAMPLE 11.16

Calculate the minimum value of the pull-up resistor for a 74ALS05 inverter if the circuit drives ten 74ALS00 NAND gate inputs.

■ **Solution**

From 74ALS00 specs: $I_{IL} = 0.1$ mA
For 10 gates: $nI_{IL} = 10I_{IL} = 1$ mA
From 74ALS05 specs: $I_{OL} = 8$ mA

$$I_R = I_{OL} - nI_{IL}$$
$$= 8 \text{ mA} - 1 \text{ mA}$$
$$= 7 \text{ mA}$$

For $I_{OL} = 8$ mA, $V_{OL} = 0.5$ V

$$R_{ext} = (V_{CC} - V_{OL})/I_R$$
$$= (5 \text{ V} - 0.5 \text{ V})/7 \text{ mA}$$
$$= 4.5 \text{ V}/7 \text{ mA} = 643\Omega$$

Use a 680Ω standard value resistor.

■ **SECTION 11.8B REVIEW PROBLEM**

11.9   Calculate the minimum value of pull-up resistor required for a 74ALS05 inverter if it drives one input of a 74ALS00 NAND gate. What is the minimum standard value of this resistor?

Totem Pole Outputs

■ **KEY TERMS**

Totem Pole Output   A type of TTL output with a HIGH and a LOW output transistor, only one of which is active at any time.

Phase Splitter   A transistor in a TTL circuit that ensures that the LOW- and HIGH-state output transistors of a totem pole output are always in opposite phase (i.e., one ON, one OFF).

Figure 11.36 shows one gate of a 7400 quadruple 2-input NAND with **totem pole outputs.** The circuit is the same as that for a 7401 open-collector NAND except for a transistor, resistor, and diode, which make up the HIGH-state output circuitry of the NAND gate.

The totem pole output, shown in Figure 11.36b, has separate transistors to switch the output to the HIGH state (Q_4) and the LOW state (Q_3). These transistors are switched by Q_2, the **phase splitter.** Only one of them is ON at a time; the currents I_{OH} and I_{OL} never flow simultaneously.

The portion of the circuit consisting of Q_4, D_3, and the 130-Ω resistor replaces the external pull-up resistor required by the open-collector TTL output. Since the HIGH state is switched by its own transistor, we say that the circuit has an active pull-up.

The main advantage of the totem pole output over the open collector is that it can change states faster. The external pull-up resistance needed in an open-collector circuit slows down the output switching by contributing to the RC time constant of the output. The HIGH-state transistor circuit, with its relatively low output impedance, reduces this time constant and thus improves switching speed.

a. 7400 NAND gate

b. Current flow in a totem pole output

FIGURE 11.36 NAND Gate with Totem Pole Output

Figure 11.37 shows the operation of the 7400 NAND gate for HIGH and LOW input conditions.

HIGH Input   When both inputs are HIGH, there is no low-impedance base-emitter current path in Q_1. The base-collector junction of Q_1 acts as a forward-biased diode. Base current flows in Q_2, saturating the transistor. Sufficient current flows to Q_3 to saturate it. Y is connected to ground, via the collector-emitter path of Q_3. *The output is LOW.*

LOW Input   Figure 11.37b shows input B of a 7400 NAND gate pulled LOW. The circuit operates the same way if A or both A and B are LOW.

In this condition, a low-impedance path to ground is established through one or both of the base-emitter junctions of Q_1. This pulls the base of Q_2 LOW, causing it to be in cutoff mode. No current flows through the collector-emitter path of Q_2, so no base current flows in Q_3; it is also cut off.

Current flows through the 1.6-kΩ resistor to the base of Q_4, turning it ON. This connects the output, via Q_4, D_3, and the 130-Ω resistor, to V_{CC}. *The output is HIGH.*

Q_4 will not turn ON when Q_3 is ON. We can find out why by calculating $V_{BE4} +$ V_{D3}. For Q_4 to conduct, two *pn* junctions (D_3 and the base-emitter junction of Q_4) must be forward-biased. Thus, ($V_{BE4} + V_{D3}$) must be greater than 0.6 V + 0.6 V = 1.2 V.

By kVL:

$$V_{B4} = V_{BE4} + V_{D3} + V_{CE3}$$

Thus:

$$V_{BE4} + V_{D3} = V_{B4} - V_{CE3}$$

We can calculate V_{B4} by adding up voltage drops, as follows:

$$V_{B4} = V_{CE2} + V_{BE3} = 0.2 \text{ V} + 0.7 \text{ V} = 0.9 \text{ V}$$

a. Both inputs HIGH

b. One input LOW

FIGURE 11.37 NAND Gate Operation

Q_3 is saturated, thus:

$$V_{CE3} = 0.2 \text{ V}$$

The difference between these voltages is:

$$V_{B4} - V_{CE3} = 0.9 \text{ V} - 0.2 \text{ V} = 0.7 \text{ V}$$

This is insufficient to forward-bias BE_4 and D_3. Q_4 stays OFF.

Switching Noise

■ KEY TERM

Storage Time   Time required to transport stored charge away from the base region of a bipolar transistor before it can turn off.

A totem pole output is an inherently noisy circuit. Noise is generated on the supply voltage line when the output switches from LOW to HIGH.

When the output is in a steady HIGH or LOW state, Q_3 and Q_4 are always in opposite phase. The design of the totem pole output is such that when Q_3 is ON, it is saturated, but when Q_4 is ON, it operates in the transistor's active, or linear, region.

A saturated transistor takes longer to shut off than an unsaturated one due to **storage time,** the time required to transport stored charge away from the base region of the transistor. Thus, Q_3 takes longer to turn off than Q_4.

When a totem pole output is LOW, Q_3 is ON and Q_4 is OFF. When the output changes state, Q_4 turns ON before Q_3 can turn OFF, due to the storage time of Q_3. For a few nanoseconds, both transistors are ON. This condition momentarily shorts V_{CC} to ground, causing a surge of supply current, as shown in Figure 11.38.

The inductance of the power line produces a corresponding spike proportional to the instantaneous rate of change of the supply current ($v = L \, di/dt$, where L is the power line inductance and di/dt is the instantaneous rate of change of supply current).

These spikes on the supply voltage line can cause real problems, especially in synchronous circuits. They often cause erroneous switching that is nearly impossible to troubleshoot. The best cure for such problems is prevention.

Figure 11.39 shows the addition of a decoupling capacitor to a totem pole output to eliminate switching spikes. A low-inductance capacitor of about 0.1 µF is placed between the V_{CC} and ground pins of the chip to be decoupled. This capacitor offsets the power line inductance and acts as a low-impedance path to ground for high-frequency noise (i.e., spikes). Since a capacitor is an open circuit for low frequencies, the normal DC supply voltage is not shorted out.

NOTE . . .

It is important that the capacitor be placed *physically close* to the decoupled chip. Inductance of the power line accumulates with distance, and if the capacitor is far away from the chip (say, at the end of the circuit board), the decoupling effect of the capacitor is lost.

It is not necessary to decouple every chip on a circuit board for designs operating at relatively low frequencies (≤1 MHz). In such cases, one capacitor for every two ICs is enough. The capacitor should be connected between V_{CC} and ground *of the same chip,* as shown in Figure 11.40.

FIGURE 11.38   Spikes on Power Line During LOW-to-HIGH Transition of Totem Pole Output

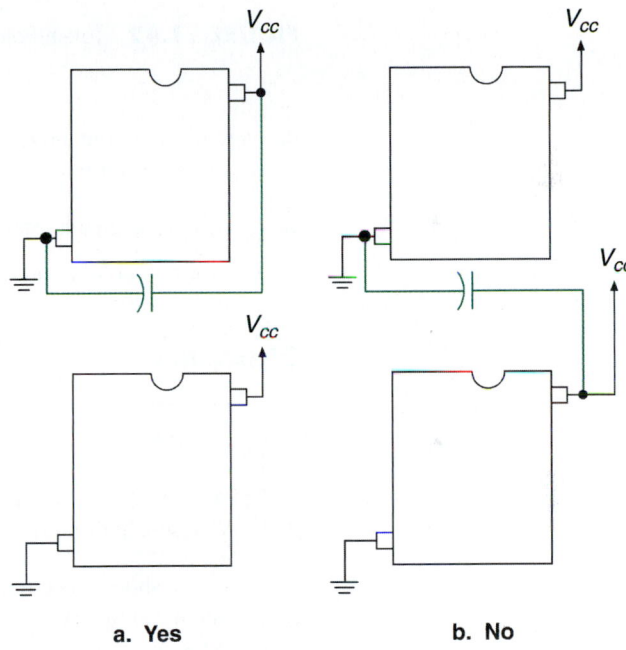

FIGURE 11.39 Decoupling the Power Supply

a. Yes　　　　　**b. No**

FIGURE 11.40 Placement of Decoupling Capacitor (Low-Frequency Designs)

For high-frequency designs, use one capacitor per IC, as shown in Figure 11.41. Connect directly to power and ground traces on a printed circuit board, *as close as possible* to the chip being decoupled.

Connection of Totem Pole Outputs

Totem pole outputs must never be connected together. As shown in Figure 11.42, the problem occurs when two connected outputs are in opposite states.

The active pull-up consisting of Q_4, D_3, and the 130-Ω resistor is designed to supply current to about 10 TTL inputs, each having a large input impedance. It will not withstand the current that flows when the output is forced to ground through the LOW output transistor of another gate.

Under this condition about 30 to 55 mA will flow through Q_{4A} and Q_{3B}. This exceeds the ratings of the outputs in both the HIGH and LOW state and will cause

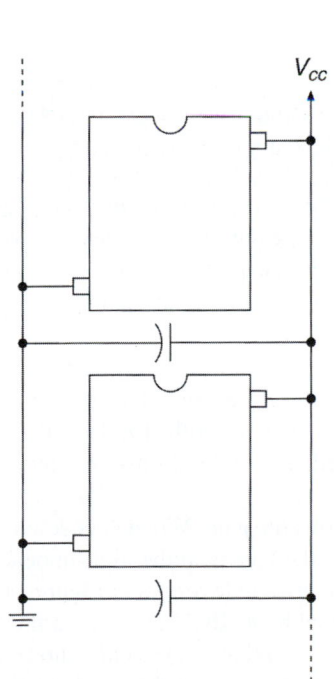

FIGURE 11.41 Placement of Decoupling Capacitors (High-Frequency Designs)

FIGURE 11.42 Totem Poles Connected Together

damage to the outputs over time. The outputs will probably withstand this sort of abuse for several minutes, but eventually will be damaged.

■ **SECTION 11.8C REVIEW PROBLEM**

11.10 A totem pole output is likely to be damaged when shorted to ground. Why?

Tristate Gates

> ■ **KEY TERM**
>
> **Tristate Output**   An output having three possible states: logic HIGH and LOW, and a high-impedance state, in which the output acts as an open circuit.

Figure 11.43 shows the circuits of two TTL inverters with **tristate outputs.** In addition to the usual binary states of HIGH and LOW, the output of the tristate inverter can also be in a "high-impedance" (Hi-Z) state. This state occurs when both Q_3 and Q_4 are OFF. The electrical effect is to produce an open circuit at the output, which is neither HIGH nor LOW.

The output of a tristate gate combines advantages of a totem pole output and an open-collector output. Like the totem pole output, it has an active pull-up with lower output impedance and faster switching than an open collector. Like the open collector, we can connect several outputs together, provided only one output is active at a time.

Input G, the "gating" or "enable" input, controls the gate. When G is active, the gate acts as an ordinary inverter. When inactive, the gate is in the high-impedance state. Table 11.8 summarizes the operation of the tristate inverters in Figure 11.43.

The tristate inverter in Figure 11.43a is enabled by a HIGH at the G input. The circuit is the same as a 7400 NAND gate with two exceptions: (1) an extra diode goes from the base of Q_4 to G, and (2) G connects directly to one of the emitters of Q_1.

When $G = 0$, Q_1 acts as though there was a LOW at a NAND gate input. In a 7400 NAND circuit, this causes Q_2 and Q_3 to be in cutoff mode.

TABLE 11.8   Truth Tables of Tristate Inverters

G	A	Y	$\bar{G}$	A	Y
0	0	Hi-Z	0	0	1
0	1	Hi-Z	0	1	0
1	0	1	1	0	Hi-Z
1	1	0	1	1	Hi-Z

a. Active-HIGH enable

b. Active-LOW enable

FIGURE 11.43 Tristate Inverters

Due to the opposite states of the emitter and collector in Q_2, Q_4 would normally be ON. Instead, the LOW at G pulls the base of Q_4 LOW through the extra diode. Thus, both Q_3 and Q_4 are OFF.

When $G = 1$, the G emitter of Q_1 acts like a HIGH NAND input. By the enable/inhibit rules of a NAND gate, $Y = \overline{A}$. The additional diode prevents the HIGH at G from activating Q_4.

The circuit in Figure 11.43b works the same way, except for the opposite sense of the activating input. This opposite active level is achieved by using an open-collector inverter, consisting of Q_5, Q_6, and Q_7, at input $\overline{G}$.

11.11 Why is the diode from the base of Q_4 necessary in the tristate inverters in Figure 11.43?

Other Basic TTL Gates

Other TTL gates are similar to the NAND and inverter gates we have already examined. A significant variation is the OR/NOR circuit, which has a different input configuration than the AND/NAND/inverter type gates.

7402 NOR Gate

Figure 11.44 shows one gate of a 7402 quadruple 2-input NOR gate package. The difference between this gate and the 7400 NAND gate is the structure of the inputs. The NOR gate does not use the multiple-emitter transistor, but rather an individual transistor (Q_1 or Q_2) for each input. There are two phase splitters (Q_3 and Q_4), which are paralleled, emitter-to-emitter and collector-to-collector.

If either Q_3 or Q_4 is enabled by a HIGH at its corresponding input, it will turn on Q_5, making the output LOW.

If both gate inputs are LOW, both Q_3 and Q_4 are in cutoff mode, and so is Q_5. The output is HIGH through Q_6.

Table 11.9 shows the truth table and the states of the transistors for this gate. It is not strictly correct to refer to Q_1 and Q_2 as being ON or OFF, since there is current flowing in these transistors regardless of whether the inputs are HIGH or LOW. Let us define the ON state of an input transistor as the condition where the base-emitter junction is conducting (LOW input). If the base-collector junction conducts, we will consider the transistor OFF (HIGH input).

FIGURE 11.44 *7402 NOR Gate Circuit*

TABLE 11.9 7402 NOR Function and Truth Table

A	B	Q_1	Q_2	Q_3	Q_4	Q_5	Q_6	Y
0	0	ON	ON	OFF	OFF	OFF	ON	1
0	1	ON	OFF	OFF	ON	ON	OFF	0
1	0	OFF	ON	ON	OFF	ON	OFF	0
1	1	OFF	OFF	ON	ON	ON	OFF	0

7408 AND Gate and 7432 OR Gate

It may not be obvious why we would choose to study NAND and NOR gates before AND and OR. After all, AND and OR are the more basic logic functions.

Electrically, it works the other way around. The simplest TTL circuit is the NAND/inverter, followed by the NOR. AND and OR gates are more complex since they are based on the NAND and NOR and require an extra inverter stage.

Figure 11.45 shows the circuit of a 7408 AND gate, and Figure 11.46 shows a 7432 TTL OR gate circuit. Each of these gates is like its NAND/NOR counterpart, except for an additional inverter, implemented by Q_3 in the AND gate and Q_5 in the OR gate.

Table 11.10 and Table 11.11 show the transistor function and truth table for each gate. In keeping with the convention established for the NOR function table, an input transistor with a conducting base-emitter junction is considered ON.

▪ SECTION 11.8E REVIEW PROBLEM

11.12 Why are noninverting gates more complex than inverting gates?

FIGURE 11.45 7408 AND Gate

FIGURE 11.46   7432 OR Gate

TABLE 11.10   7408 AND Function and Truth Table

A	B	Q_1	Q_2	Q_3	Q_4	Q_5	Q_6	Y
0	0	ON	OFF	OFF	ON	ON	OFF	0
0	1	ON	OFF	OFF	ON	ON	OFF	0
1	0	ON	OFF	OFF	ON	ON	OFF	0
1	1	OFF	ON	ON	OFF	OFF	ON	1

TABLE 11.11   7432 OR Function and Truth Table

A	B	Q_1	Q_2	Q_3	Q_4	Q_5	Q_6	Q_7	Q_8	Y
0	0	ON	ON	OFF	OFF	OFF	ON	ON	OFF	0
0	1	ON	OFF	OFF	ON	ON	OFF	OFF	ON	1
1	0	OFF	ON	ON	OFF	ON	OFF	OFF	ON	1
1	1	OFF	OFF	ON	ON	ON	OFF	OFF	ON	1

11.9   INTERNAL CIRCUITRY OF CMOS GATES

■ **KEY TERMS**

CMOS   A logic family based on the switching of *n*- and *p*-channel ("complementary") enhancement-mode MOSFETs.

MOSFET   Metal-oxide-semiconductor field effect transistor. A MOSFET has three terminals—gate, source, and drain—which are analogous to the base, emitter, and collector of a bipolar junction transistor.

> **Enhancement-Mode MOSFET**   A MOSFET that creates a conduction path (a channel) between its drain and source terminals when the voltage between gate and source exceeds a specified threshold level.
>
> *n*-**Channel Enhancement-Mode MOSFET**   A MOSFET built on a *p*-type substrate with *n*-type drain and source regions. An *n*-type channel is created in the *p*-substrate during conduction.
>
> *p*-**Channel Enhancement-Mode MOSFET**   A MOSFET built on an *n*-type substrate with *p*-type drain and source regions. During conduction, a *p*-type channel is created in the *n*-substrate.
>
> **Substrate**   The foundation of *n*- or *p*-type silicon on which an integrated circuit is built.

All of the logic circuits we have examined so far have been based on the switching of bipolar junction transistors. Another major logic family, **CMOS,** is based on the switching of metal-oxide-semiconductor field effect transistors, or **MOSFETS.**

There are two major types of MOSFETs, called depletion-mode and **enhancement-mode MOSFETs.** We will concentrate on the enhancement-mode devices, as they are the type used in the manufacture of digital ICs. Details of the differences between depletion- and enhancement-mode transistors can be found in any good textbook on electronic devices.

MOSFETs can be categorized in another way: as *n*-**channel** and *p*-**channel** devices, much as bipolar transistors are classified as NPN or PNP.

CMOS logic is constructed from both *n*- and *p*-channel MOSFETs. CMOS ("Complementary MOS") refers to the opposite, or complementary, operation of *n*- and *p*-channel transistors.

MOSFET Structure

Figure 11.47 shows the structure and symbol of an *n*-channel enhancement-mode MOSFET in an integrated circuit. The device is built on a **substrate** of *p*-type silicon, which has a deficiency of electrons in its structure. The drain and source regions are "wells" of *n*-type silicon, which has an excess of electrons. The drain and source are roughly equivalent to the emitter and collector of a bipolar transistor.

The substrate is shown as a terminal with an arrow. The arrow points in for an *n*-channel device and out for a *p*-channel device. In nearly all cases, the substrate is shorted to the source terminal. (Some exceptions to this general rule will be examined when we look at circuits of CMOS gates.)

a. Structure   **b. Symbol**

FIGURE 11.47   *n*-Channel MOSFET

The gate terminal is similar to the base of a bipolar transistor in that it controls the flow of current between the drain and source. The difference is that a MOSFET uses gate *voltage* to control drain current, whereas a bipolar transistor uses base *current* to control collector current.

The gate consists of an insulating layer of silicon dioxide (SiO_2) and a layer of metal over the substrate between the drain and source. This gate structure is what gives the MOSFET its name (*metal-oxide-semiconductor* field effect transistor).

> **NOTE . . .**
>
> The oxide layer of the gate structure is subject to damage if excessive voltage (greater than about 100 V) is applied. This especially includes static electricity, or electrostatic discharge (ESD). There are standard precautions for working with MOS devices that should be followed carefully.
>
> Most important are ensuring that MOS devices are stored in antistatic or conducting material, that work surfaces are not likely to generate static, that unused inputs are not left open or floating, that you avoid touching the pins of a MOS device, and that if you must handle a MOS IC, you discharge any static on your person *before* touching it.
>
> A conductive wrist strap with a high series resistance to ground (about 1 MΩ) is often worn to reduce static. The high resistance protects the operator from shock injury in the event of a short circuit.
>
> A list of handling precautions is included in Appendix D.

■ SECTION 11.9A REVIEW PROBLEM

11.13 Why are MOSFET circuits particularly susceptible to static damage?

Bias Requirement for MOS Transistors

> ### ■ KEY TERMS
>
> **Ohmic Region**   The MOSFET equivalent of saturation. When a MOSFET is biased ON, it acts like a relatively low resistance, or "ohmically."
>
> **Threshold Voltage** $V_{GS(Th)}$ The minimum voltage between gate and source of a MOSFET for the formation of the conducting inversion layer (channel).
>
> ***n*-Type Inversion Layer**   The conducting layer formed between drain and source when an enhancement-mode *n*-channel MOSFET is biased ON. Also referred to as the channel.

When we studied the operation of TTL gate circuits, we discovered that, for the most part, the bipolar transistors in the gates operated either in the saturation or the cutoff regions. In MOS-type gates, we make use of two similar operating regions in the constituent MOSFETs:

1. The cutoff region is the same as that for a bipolar transistor. Under this condition, there is a very high impedance between the drain and source terminals of the MOSFET.
2. The **ohmic region** is analogous to the saturation region of a bipolar transistor. In this state, there is a relatively low resistance between the MOSFET's drain and source.

The MOSFET switches between cutoff and ohmic regions when the voltage between gate and source, V_{GS}, is less than or greater than a value called the **threshold voltage.** abbreviation for this voltage is $V_{GS(Th)}$; its value is between 1 and 5 volts, typically 1.5 V.

a. Bias voltages   **b. Equivalent circuit**

FIGURE 11.48   *n*-Channel MOSFET in Cutoff Region

$V_{GS} < V_{GS(Th)}$
No conduction between
drain and source

$V_{GS} > V_{GS(Th)}$
Conduction between
drain and source

FIGURE 11.49   Channel Formation in an *n*-Channel MOSFET

Figure 11.48 shows an *n*-channel MOSFET operating in the cutoff region. The gate-source voltage, V_{GS} is less than $V_{GS(Th)}$. There is no conduction between the drain and source. The resistance, $R_{DS(OFF)}$, between drain and source is very large, typically in the thousands of megohms.

When the value of V_{GS} increases and exceeds the threshold voltage, the MOSFET enters the ohmic region. A conduction channel, called the ***n*-type inversion layer,** is created in the *p*-substrate of the transistor, as shown in Figure 11.49. This layer is like an artificially created region of *n*-type silicon, which allows conduction between the drain and source, provided there is sufficient potential difference between them.

Figure 11.50 shows a MOSFET operating in the ohmic region. $R_{DS(ON)}$, the equivalent resistance of a MOSFET in the ohmic region, is typically around 500 Ω to 2 kΩ. The drain-source current, I_{DS}, is determined by Ohm's law: $I_{DS} = V_{CC}/R_{DS(ON)}$.

The operation of a *p*-channel MOSFET is similar, but with polarities reversed. If $V_{GS(Th)}$ is + 1.5 V for an *n*-channel device, an equivalent *p*-channel MOSFET has a threshold voltage of –1.5 V. $V_{GS} > +1.5$ V turns ON an *n*-channel transistor; $V_{GS} < -1.5$ V turns ON a *p*-channel device.

Figure 11.51 summarizes the bias requirements for *n*- and *p*-channel enhancement-mode MOSFETs.

a. Bias voltages   b. Equivalent circuit

FIGURE 11.50   n-Channel MOSFET in Ohmic Region

FIGURE 11.51   Bias Requirements of n- and p-Channel MOSFETs

CMOS Inverter

Figure 11.52 shows the circuit of a CMOS inverter, which consists of one n-channel and one p-channel MOSFET.

Recall the bias conditions of the two transistors:

n-channel: threshold voltage, $V_{GS(Th)} \approx +1.5$ V
ON when $V_{GS} > V_{GS(Th)}$ (e.g., $V_{GS} = V_{CC}$)
OFF when $V_{GS} < V_{GS(Th)}$ (e.g., $V_{GS} = 0$ V)
p-channel: threshold voltage, $V_{GS(Th)} \approx -1.5$ V
ON when $V_{GS} < V_{GS(Th)}$ (e.g., $V_{GS} = -V_{CC}$)
OFF when $V_{GS} > V_{GS(Th)}$ (e.g., $V_{GS} = 0$ V)

The operation of the CMOS inverter, and any other CMOS gate, depends on arranging the bias conditions of each complementary pair of transistors so that they are always in opposite states. Whenever Q_1 is ON, Q_2 is OFF, and vice versa. Figure 11.53 shows how this is accomplished.

FIGURE 11.52 CMOS Inverter

a. **LOW input**   b. **HIGH input**

FIGURE 11.53 Operation of CMOS Inverter

Assume that a LOW input is at ground potential and that a HIGH input is equal to V_{CC}.

When *input A is LOW*, the gate voltage of Q_2 is the same as its source voltage; $V_{GS2} = 0$ and Q_2 is OFF. This places a high-impedance path between output Y and ground. At the same time, the gate voltage of Q_1 is 0 V and its source voltage is V_{CC}; $V_{GS1} = V_{G1} - V_{S1} = 0 - V_{CC} = -V_{CC}$. (The p-channel transistor, Q_1, is drawn "upside down" to make the complementary pair symmetrical.) Q_1 is ON, forming a low-impedance path from V_{CC} to the output Y. *Output Y is HIGH.*

When *input A is HIGH,* the gate-source voltage of the n-channel transistor is V_{CC}, causing Q_2 to turn ON. The gate of Q_1 is also at V_{CC}. Since the source of the p-channel transistor is at V_{CC}, $V_{GS1} = V_{G1} - V_{S1} = V_{CC} - V_{CC} = 0$ V; Q_1 is OFF. This combination creates a high impedance between V_{CC} and output Y and a low-impedance path from output Y to ground, as shown in Figure 11.53b. *Output Y is LOW.*

CMOS NAND/NOR Gates

CMOS NAND and NOR gates are constructed from complementary pairs of MOS-FETs. Each MOSFET pair has an n-channel transistor that is turned ON by a HIGH input and a p-channel transistor that is turned ON by a LOW input. The n-channel devices switch the output to ground; the p-channel ones switch the output to V_{CC}. NAND and NOR functions are generated by arranging the MOSFET drain-source paths in series (AND) and parallel (OR) configurations.

In Figure 11.54, we see the DeMorgan equivalent forms of a NAND gate. Each form illustrates an aspect of NAND operation that can be described with a brief sentence and implemented by a MOSFET circuit. The combination of forms describes the complete operation of the device.

Figure 11.54a states that both NAND inputs must be HIGH to make the output LOW. A logic HIGH activates an n-channel MOSFET. The gate output is switched to ground by an n-channel MOSFET. Thus, the drain-source paths of two n-channel transistors must be connected in series to make the output LOW under the stated conditions.

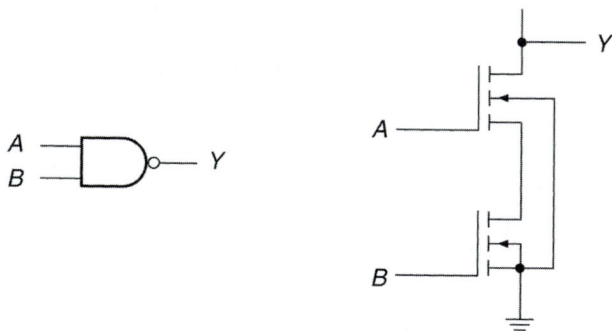

a. Both inputs HIGH make output LOW
(***n*-channels in series to ground)**

b. Either input LOW makes output HIGH
(***p*-channels in parallel to V_{CC})**

FIGURE 11.54   NAND Functions of MOSFETs

FIGURE 11.55 CMOS NAND Gate

Figure 11.54b shows that the NAND output is HIGH if either input is LOW. A p-channel transistor will turn ON when its gate is LOW and will switch a HIGH to the output. A parallel combination of p-channel MOSFETs will satisfy these conditions.

The stated conditions are combined in the CMOS NAND circuit shown in Figure 11.55. Transistors Q_1 and Q_4 form a complementary pair, as do Q_2 and Q_3. When A and B are both HIGH, Q_1 and Q_2 are both OFF, cutting off the connection between V_{CC} and output Y. Q_3 and Q_4 are both ON, supplying a low-impedance path from output Y to ground, making the output LOW. This is shown in the partial truth table in Table 11.12.

TABLE 11.12 Partial CMOS NAND Function and Truth Table

A	B	Q_1	Q_2	Q_3	Q_4	Y
1	1	OFF	OFF	ON	ON	0

When A is LOW and B is HIGH, Q_1 is ON. This creates a path from V_{CC} to output Y. At the same time, Q_4 is OFF. This cuts the Y-to-ground path through the n-channel MOSFETs: the series path from output to ground is broken. One parallel path from V_{CC} to output has been established. Output Y is HIGH.

The remaining input combinations also make the output HIGH, as shown in Table 11.13. They do so by breaking the n-channel path from output to ground and enabling one or both p-channel paths from V_{CC} to output.

TABLE 11.13 Partial CMOS NAND Function and Truth Table

A	B	Q_1	Q_2	Q_3	Q_4	Y
0	0	ON	ON	OFF	OFF	1
0	1	ON	OFF	ON	OFF	1
1	0	OFF	ON	OFF	ON	1

Each MOSFET in a logic circuit must have its own independent substrate bias. This ensures that the transistor will operate as expected when a logic HIGH or LOW is applied to its gate.

Normally, the substrate of a MOSFET is shorted to its source terminal. If a MOSFET source terminal is isolated from V_{CC} or ground, the substrate must be biased separately. For example, in the NAND gate in Figure 11.55 the substrate of Q_3 connects directly to ground.

NOR gates are similar to NANDs in construction. Figure 11.56 shows the DeMorgan equivalent forms of the NOR function and the MOSFET implementations of each aspect of the gate operation.

When either input is HIGH, the output is LOW. This function is implemented by two parallel n-channel MOSFETs. Both inputs must be LOW to make the output HIGH, which implies a series connection of two p-channel transistors. The complete NOR gate circuit is shown in Figure 11.57. (Note that the substrate of Q_2 is connected directly to V_{CC} to ensure that it has its own bias voltage.)

As was the case with the NAND circuit, transistors Q_1 and Q_4 form a complementary MOSFET pair. Transistors Q_2 and Q_3 form the second pair.

When both inputs are LOW, both p-channel transistors are ON. This creates a low-impedance path from V_{CC} to output Y. The n-channel transistors, Q_3 and Q_4, are both OFF. This isolates the output from ground. Output Y is HIGH. Table 11.14 shows the MOSFET states under this condition.

a. Either input HIGH makes output LOW
(*n*-channels in parallel to ground)

b. Both inputs LOW make output HIGH
(*p*-channels in series to *V*_{cc})

FIGURE 11.56 NOR Functions of MOSFETs

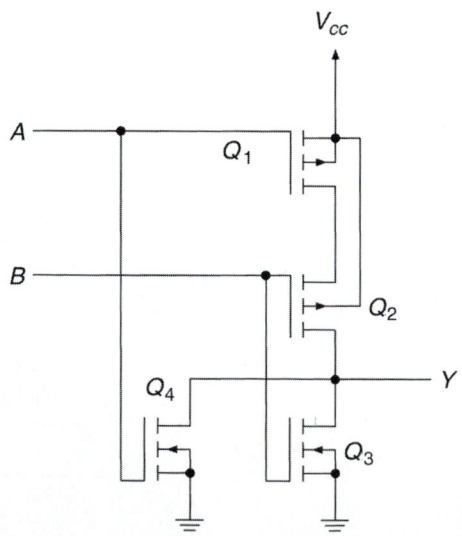

FIGURE 11.57 CMOS NOR Gate

TABLE 11.14 Partial CMOS NOR Function and Truth Table

A	B	Q_1	Q_2	Q_3	Q_4	Y
0	0	ON	ON	OFF	OFF	1

If either input is HIGH, one or both of the p-channel transistors will turn OFF. This action breaks the path from V_{CC} to output Y. The complementary n-channel transistor will turn ON. This creates a low-impedance path from output Y to ground. Output Y is LOW. Table 11.15 summarizes the possible input conditions and MOSFET states when the NOR output is LOW.

TABLE 11.15 Partial CMOS NOR Function and Truth Table

A	B	Q_1	Q_2	Q_3	Q_4	Y
0	1	ON	OFF	ON	OFF	0
1	0	OFF	ON	OFF	ON	0
1	1	OFF	OFF	ON	ON	0

CMOS AND and OR Gates

Figures 11.58 and 11.59 show the circuits of CMOS AND and OR gates. The AND gate is the same as the NAND circuit, except for the output inverter section constructed from Q_5 and Q_6. The OR gate is the same as the NOR with an output inverter section.

▮ SECTION 11.9B REVIEW PROBLEM

11.14 Why is the source of a p-channel MOSFET connected to V_{CC} in a CMOS gate?

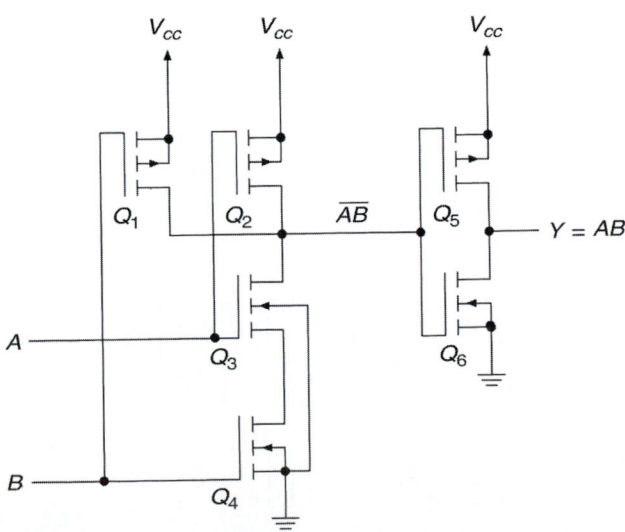

FIGURE 11.58 CMOS AND Gate

FIGURE 11.59 CMOS OR Gate

CMOS Transmission Gate

Figure 11.60 shows the circuit of a CMOS transmission gate. A CMOS transmission gate, or analog switch, conducts in both directions. This makes it possible to enable or inhibit time-varying analog signals having both positive and negative values. Conduction takes place between the input and output terminals through MOSFETs Q_1 and Q_2. Positive current (left to right in the diagram) flows through Q_2, and negative current (right to left) flows through Q_1. Two inverters, consisting of the Q_3/Q_4 and Q_5/Q_6 pairs of MOSFETs, control the ON/OFF state of the circuit.

When *CONTROL* = 1, the inverters bias both Q_1 and Q_2 ON, allowing them to conduct. When *CONTROL* = 0, the circuit inhibits conduction between input and output.

The substrate terminal of Q_1 is connected, not to the source terminal of that transistor, but directly to V_{CC} thus providing the correct bias to Q_1 in the ON state.

A particular device with this function is the 74HC4066 quad analog switch, whose circuit symbol is shown in Figure 11.61. When the *CONTROL* input is HIGH, analog and digital signals can pass between the bidirectional input terminals.

FIGURE 11.60 CMOS Transmission Gate

In/Out ——————⧓—————— Out/In

CONTROL ——————

FIGURE 11.61   One of Four Analog Switches from 74HC4066

EXAMPLE 11.17

Figure 11.62 shows a circuit where the analog switches in a 74HC4066 package are used to control the selection and muting of two pairs of speakers in a stereophonic audio system. Briefly explain the circuit operation.

FIGURE 11.62   Example 11.17: 74HC4066 Analog Switches as Audio Selectors

(continued)

▪ Solution

The audio signal to each speaker is passed or blocked by a CMOS transmission gate. The speakers are paired into *A* and *B* groups. Each pair has a left and a right channel speaker. The same logic gate controls both speakers of each group.

The Select *A* switch enables the *A* speakers when it is open (logic HIGH). The Select *B* switch enables the *B* speakers when it is open. The Mute Toggle flip-flop mutes (disables) both sets of speakers when *Q* is LOW. This action inhibits both AND gates, making all transmission gate *CONTROL* inputs LOW. The mute function toggles ON and OFF with each push of the Mute ON/OFF switch.

11.10 TTL AND CMOS VARIATIONS

Standard (7400-series) TTL and CMOS represented the two main standards of logic design for many years, and their influence is still visible in other, more advanced types of logic. The changes that have been made in newer logic families are not fundamental changes in the working concepts, but improvements to the specifications, particularly switching speed and power dissipation.

TTL Logic Families

> ▪ **KEY TERMS**
>
> **Schottky Barrier Diode**   A specialized diode with a forward voltage drop of about +0.4 V.
>
> **Schottky Transistor**   A bipolar transistor with a Schottky diode across its base-collector junction, which prevents the transistor from going into deep saturation.
>
> **Schottky TTL**   A series of unsaturated TTL logic families based on Schottky transistors. Schottky TTL switches faster than standard TTL due to decreased storage time in its transistors.
>
> **Speed-Power Product**   A measure of a logic circuit's efficiency, calculated by multiplying its propagation delay by its power dissipation. Unit: picojoule (pJ)

Probably the most important development in TTL technology was the introduction, in the early 1970s, of the **Schottky barrier diode** into circuit designs. This made possible the first family of nonsaturated bipolar logic, with its resultant improvement in switching speed.

Figure 11.63 shows a bipolar transistor with a Schottky diode connected across its base and collector and the equivalent circuit symbol of this combination. We call this configuration a **Schottky transistor** and logic devices using such transistors **Schottky TTL.**

Normally the base-collector junction of a saturated bipolar transistor has a drop of about 0.5 volts, as shown in Figure 11.64. The Schottky diode clamps this junction voltage to about 0.4 volts. This keeps the transistor out of deep saturation in its ON state. The base region of the Schottky-clamped transistor holds less charge than does a standard bipolar transistor. Its storage time, the time required to dissipate base charge upon turn-off, is substantially reduced. The transistor can switch faster with the Schottky diode than without.

a. Schottky diode-clamped base-collector junction

b. Equivalent circuit symbol

FIGURE 11.63 Schottky Transistor

a. Saturated bipolar transistor

b. Base-collector voltage in Schottky transistor

FIGURE 11.64 ON-State Operating Voltages of Bipolar Transistors

Figure 11.65 shows the circuits of the 74S00 Schottky and 74LS00 low-power Schottky NAND gates. Compare these circuits to each other and to the 7400 standard TTL NAND gate in Figure 11.36.

In the 74S00 circuit, Q_1 acts as the input and Q_2 as the phase splitter, as in the 7400 gate. The HIGH output circuit consists of Q_3 and Q_4 connected as a modified Darlington pair. When Q_2 is OFF (at least one input is LOW), enough base current flows in Q_3 to turn it on. Collector-emitter current in Q_3 turns on Q_4, making the output HIGH.

When Q_2 is ON (both inputs are HIGH), the base of Q_3 is pulled LOW, turning it OFF. Sufficient current flows in the base of Q_5 to turn it ON. The resultant current through Q_5 will turn on Q_6, making the output LOW. A similar analysis can be made for the 74LS00 gate.

One difference between the 74S00 and 74LS00 circuits is the size of the resistors; the LS device has larger resistors. Less current flows in the gate circuit. This reduces power dissipation of the chip. The larger resistor values also slow down the switching times of the various transistors by increasing the *RC* time constants of the circuit elements.

Speed-Power Product

One measure of logic circuit efficiency is its **speed-power product,** calculated by multiplying switching speed and power dissipation, usually expressed in picojoules (pJ). (The joule is the SI unit of energy. Power is the rate of energy used per unit time.) A major goal of logic device design is the reduction of a device's speed-power product.

a. 74S00 Schottky NAND gate

b. 74LS00 low-power Schottky NAND gate

FIGURE 11.65   Schottky TTL Circuits

Table 11.16 shows the propagation delay, supply current, and speed-power product for a NAND gate in six TTL families: standard TTL (7400), Schottky (74S00), low-power Schottky (74LS00), fast TTL (74F00), advanced Schottky (74AS00), and advanced low-power Schottky (74ALS00).

TABLE 11.16 TTL Speed and Power Specifications

	7400	**74LS00**	**74S00**	**74F00**	**74ALS00**	**74AS00**
t_{pLH} (max)	22 ns	15 ns	4.5 ns	6 ns	11 ns	4.5 ns
t_{pHL} (max)	15 ns	15 ns	5 ns	5.3 ns	8 ns	4 ns
$I_{CCH}/4$ (max)	2 mA	0.4 mA	4 mA	0.7 mA	0.21 mA	0.8 mA
$I_{CCL}/4$ (max)	5.5 mA	1.1 mA	9 mA	2.6 mA	0.75 mA	4.35 mA
Speed-power product (per gate)	605 pJ	82.5 pJ	225 pJ	78.0 pJ	41.25 pJ	97.9 pJ

The speed-power product shown is the worst-case value. This is calculated by multiplying the largest value of $I_{CC}/4$ by the slowest switching speed by 5 volts for each family. We use $I_{CC}/4$ because I_{CC} is specified per chip (four gates).

A faster switching speed results in an overall increase in speed-power product, other factors being equal. For example, the AS series (the faster advanced Schottky family) has a higher speed-power product than the ALS series.

The smaller resistors used to speed up output switching imply a proportional drop in propagation delay (higher speed) but an increased supply current. Power dissipation increases in proportion to the square of the supply current, thus overwhelming the effect of the increased switching speed.

CMOS Logic Families

> **■ KEY TERM**
>
> V_{DD} Metal-gate CMOS Supply Volage.

The CMOS gates we have looked at in this chapter are simpler than most gates actually in use. There are two main families of CMOS devices: metal-gate CMOS, and silicon-gate, or high-speed, CMOS.

Metal-Gate CMOS

There are two main variations on this type of circuit, designated B-series and UB-series CMOS. Most CMOS gates are B-series; UB-series is available in a limited number of inverting-type gates, such as inverters and 2-, 3-, and 4-input NAND and NOR gates. Figure 11.66 shows the difference in the two configurations.

Figure 11.66b shows one gate from a 4011UB quadruple 2-input NAND package. Its circuit is the same as the NAND configuration examined in Section 11.9. Power supply voltages in metal-gate CMOS are designated V_{DD} (power) and V_{SS} (ground). High-speed, or silicon-gate, CMOS uses the same power supply designations as TTL: V_{CC} and ground.

The B-series configuration of this circuit has two additional inverter outputs in cascade with the NAND logic. (The same gate becomes an AND when we add a third output inverter.) The inverter configuration is actually an amplifier; extra inverter stages provide additional gain and increase noise margin by allowing the circuit to accept smaller input signals.

CMOS gates are sometimes used in analog applications, such as oscillators. The UB-series gates, with their lower gain, are more desirable for such applications. Due to its low switching speed, metal-gate CMOS is rarely used in new designs.

a. B-series AND/NAND gate (4081B/4011B)

b. UB-series NAND gate (4011UB)

FIGURE 11.66   Metal-Gate CMOS Circuits

High-Speed CMOS

> ### ■ KEY TERM
>
> **High-Speed (Silicon-Gate) CMOS**   A CMOS logic family with a smaller device structure and thus higher speed than standard (metal-gate) CMOS.

Metal-gate CMOS has been considered a nearly ideal family for logic designs, with its high noise immunity, low power consumption, and flexible power supply requirements. Unfortunately, its propagation delay times, typically 10 to 20 times greater than those of equivalent TTL devices, are just not fast enough for use in modern microprocessor-based systems.

High-speed CMOS was developed to address the problem of switching speed, while striving to keep the other advantages of CMOS. This is achieved by using MOSFETs with a polysilicon material for the gate, rather than metal, as in standard CMOS. Because of advantages gained in this manufacturing process, each transistor is physically smaller and has a lower gate capacitance than metal-gate MOSFETs. Both these factors contribute to a lower propagation delay for the logic gate circuit.

Several subfamilies of high-speed CMOS are available for various logic and linear applications, designated by the labels 74HC, 74HC4NNN, 74HCT, and 74HCU.

The 74HC series duplicates equivalent LSTTL functions in packages having identical pinouts to LSTTL. The 74HC4NNN replaces CMOS functions pin for pin. Both these series have CMOS-equivalent input and output levels, within the power supply limits (2.0 V to 6.0 V) of high-speed CMOS.

The 74HCT devices are designed to be directly compatible with LSTTL devices, and thus have LSTTL-equivalent inputs and CMOS-equivalent outputs.

74HCU devices have no output buffers, like the 4000 UB-series standard CMOS devices. The 74HCU devices are used, as are the 4000UB devices, for linear applications such as oscillators and multivibrators.

Table 11.17 shows the relative performance of the various CMOS families. As in TTL, the 2-input NAND gate is used as the standard, except for the HCU family, where this gate is not available. The quiescent speed-power product of all CMOS

TABLE 11.17 CMOS Speed and Power Specifications

	Metal-Gate CMOS		High-Speed CMOS			Advanced High-Speed CMOS		Low-Voltage CMOS	
	4011B	**4011UB**	**74HC00A**	**74HCT00A**	**74HCU04**	**74VHC00**	**74VHCT00**	**74LVX00**	**74LCX00**
t_{pLH}, t_{pHL}	250 ns	180 ns	15 ns	19 ns	14 ns	5.5 ns	6.9 ns	6.2 ns	5.2 ns
I_{DD} or I_{CC}	0.25 µA	0.25 µA	0.25 µA	0.25 µA	0.17 µA	0.5 µA	0.5 µA	0.5 µA	0.25 µA
V_{DD} or V_{CC}	5.0 V	5.0 V	4.5 V	4.5 V	4.5 V	4.5 V	4.5 V	3.3 V	3.3 V
P_D (1 MHz)	1.5 mW	1.5 mW	446 µW	304 µW	303 µW	385 µW	385 µW	208 µW	272 µW
Speed-power product (quiescent)	0.31 pJ	0.23 pJ	0.017 pJ	0.021 pJ	0.011 pJ	0.012 pJ	0.015 pJ	0.010 pJ	0.043 pJ
Speed-power product (1 MHz)	375 pJ	270 pJ	6.68 pJ	5.77 pJ	4.25 pJ	2.12 pJ	2.65 pJ	1.29 pJ	1.42 pJ

families is much smaller than that of any TTL family. The high-speed CMOS families have propagation delays comparable to those of LSTTL.

The power dissipation of a CMOS device increases directly with frequency. The speed-power product also goes up with higher frequencies.

Table 11.16 shows CMOS speed-power product for a switching speed of 1 MHz. At these speeds, B-series CMOS has no advantage over the common TTL families in terms of its efficiency. It still has the edge on TTL with respect to noise immunity and power supply flexibility.

■ **SECTION 11.10 REVIEW PROBLEM**

11.15 Assuming that power dissipation of a 74HC00A NAND gate is directly proportional to its switching frequency, what is the speed-power product of the gate at 2 MHz, 5 MHz, and 10 MHz?

SUMMARY

1. TTL (transistor-transistor logic) and CMOS (complementary metal-oxide semiconductor) are two major logic families in use today. TTL is constructed from bipolar junction transistors. CMOS is made from metal-oxide-semiconductor field effect transistors (MOSFETs).

2. The main CMOS advantages include low power consumption, high noise immunity, and a flexibility in choosing a power supply voltage.

3. The main advantages of TTL include relatively high switching speed and an ability to drive loads with relatively high current requirements.

4. TTL and high-speed CMOS logic families are alphabetically designated by a part number having the form 74XXNN, where XX is the family and NN is a numeric logic function designator. (For example, 74HC00 and 74LS00 have the same logic function, but are from different logic families.) Devices from earlier CMOS families are designated by a part number of the form 4NNNB or 4NNNUB.

5. Devices of the same logic family generally have the same electrical characteristics.

6. Data such as input/output voltages and currents are specified in manufacturers' datasheets. Only the maximum or minimum values of these parameters should be used as design information. "Typical" values should be regarded as "information only."

7. The time required for a logic circuit output to change as a result of an input change is called propagation delay.

8. Propagation delay is specified as t_{pLH} when an output changes from LOW to HIGH and t_{pHL} when the output goes from HIGH to LOW.

9. Propagation delay in a circuit is the sum of all delays in the slowest input-to-output path. Gates whose outputs do not change are ignored in the calculation.

10. Several important timing parameters for a flip-flop include: setup and hold time, propagation delay, minimum pulse width, and recovery time.

11. Setup time is the time before a clock edge that a synchronous input must be held steady. Hold time is the time after an applied clock edge that an input level must be held constant.

12. Propagation delay in a flip-flop is the time for an input change, such as on CLK or $\overline{CLR}$, to have an effect on an output, such as Q. Propagation time is always indicated with respect to the change in output level: t_{pLH} for a LOW-to-HIGH output transition and t_{pHL}, for a HIGH-to-LOW output change.

13. Minimum pulse width, t_w, indicates how long a CLK or $\overline{CLR}$ input must be held after an active edge or level is applied before returning to the original level.

14. Recovery time is the minimum time required from the end of an active level on one input (such as $\overline{CLR}$) to an active CLK edge.

15. Fanout is the maximum number of device inputs that can be driven by the output of a logic device.

16. The actual value of output current in a driving gate is the sum of all load currents, which are the input currents of the load gates. For n loads,

$$I_{OL} = I_{IL1} + I_{IL2} + \ldots + I_{ILn_L} = n_L \, I_{IL}$$
$$\text{and } I_{OH} = I_{IH1} + I_{IH2} + \ldots + I_{IHn_H} = n_H I_{IH}$$

17. The fanout of the driving gate in the LOW and HIGH states can be calculated as:

$$n_L = \frac{I_{OL}}{I_{IL}}$$

$$\text{and } n_H = \frac{I_{OH}}{I_{IH}}$$

18. If the fanout is unequal for LOW and HIGH states, the smaller value must be used.

19. If the fanout of a gate is exceeded, the output voltage of the driving gate will drop if the output is HIGH and rise if the output is LOW. This move away from the nominal value degrades the general performance of the driving gate.

20. Power supply current (I_{CC}), and therefore power dissipation (P_D), of a TTL device depends on the number of outputs in the device that are HIGH or LOW. $P_D = V_{CC} I_{CC} = V_{CC}\left(\dfrac{n_H}{n} I_{CCH} + \dfrac{n_L}{n} I_{CCL}\right)$ for a device with n outputs, n_H of which are HIGH and n_L of which are LOW.

21. CMOS devices draw most current from the power supply when their outputs are switching and very little when they are static. Power dissipation of a high-speed CMOS device with n outputs has a static and a dynamic component, given by:

$$P_D = (C_L + C_{PD})V_{CC}^2 f + \frac{V_{CC} I_{CC}}{n}$$

At high frequencies (≥ 1 MHz), the quiescent current can be neglected.

22. Noise margin is a measure of the noise voltage that can be tolerated by a logic device input. In the HIGH state, it is given by $V_{NH} = V_{OH} - V_{IH}$. In the LOW state, it is given by $V_{NL} = V_{IL} - V_{OL}$. CMOS devices generally have higher noise margins than TTL.

23. When interfacing two devices from different logic families, the driving gate must satisfy the voltage and current requirements of the load gates.

24. Input current in a CMOS gate is very low, due to its high input impedance. Thus, fanout is generally not a problem with CMOS loads.

25. CMOS devices that have the same values of V_{IH} and V_{IL} as TTL are considered to be TTL compatible, since they can be driven directly by TTL drivers.

26. A 74HC or 74HCT device can drive 10 LSTTL loads directly. To calculate fanout, we use the output currents for which the driving gate output voltages are defined.

27. A 74LS device can drive one or more 74HC devices, provided each 74HC input has a pull-up resistor (about 1 kΩ to 10 kΩ) to supply sufficient voltage in the HIGH state.

28. A 74LS device can drive one or more 74HCT inputs directly.

29. Low-voltage CMOS (e.g., 74LVX or 74LCX) can be driven directly by a TTL device if the CMOS device is operated with a 3.3 V power supply. Noise margins are too small for a low-voltage CMOS driver to drive TTL loads.

30. 74HC or 74HCT gates can be operated at a low value of V_{CC} (e.g., 3 volts) and interfaced to a higher-voltage driver by an inverting or noninverting buffer, such as the 74HC4049 or 74HC4050. The interface buffer can tolerate relatively high input voltages (up to 15 V) and, if it shares the same supply voltage as the load gate, can provide correct input voltages to the load.

31. A bipolar transistor with a grounded emitter acts as an inverter or a digital switch. A HIGH at the base causes the transistor to conduct, pulling the collector to near-ground potential. If there is a pull-up resistor on the collector, there will be a HIGH state at the collector when the base is LOW.

32. The simplest TTL input is a transistor with its base connected to V_{CC} through a resistor. It can be treated as two diodes, back-to-back.

33. A TTL LOW input forward-biases the base-emitter junction of the input transistor, supplying a path to ground for input current.

34. A TTL HIGH input reverse-biases the base-emitter junction of the input transistor and forward-biases its base-collector junction. Input current in the HIGH state is restricted to reverse leakage current through the base-emitter junction.

35. An open TTL input is equivalent to a HIGH, as it provides no path to ground.

36. Some types of TTL gates, such as NAND, have multiple-emitter input transistors. Any one input LOW acts as a LOW for the whole circuit.

37. Other TTL gates, such as NOR, have separate transistors for each input. Any HIGH input acts as a HIGH for the whole circuit.

38. An open-collector output has one output transistor that switches on a path to ground (logic LOW) when it is turned on. There is no separate internal circuit for a HIGH output. This must be provided by an external pull-up resistor.

39. Open-collector outputs can be used to parallel outputs (wired-AND), drive high-current loads, or interface to a circuit with a different power supply voltage than the driving gate.

40. A totem pole output has a transistor that switches on for a LOW output and another that switches on for a HIGH output. These output transistors are always in opposite states, except briefly during times when the output is changing states.

41. Totem pole outputs generate noise spikes on the power line of a circuit when they switch between logic states. These spikes can be amplified by inductance of the power line. Decoupling capacitors placed close to each device help minimize this problem.

42. TTL outputs should never be connected together, as they can be damaged when the outputs are in opposite states. (Too much output current flows.) The output logic level under such conditions is not certain.

43. Gates with tristate outputs can generate logic LOW, logic HIGH, or high-impedance states. A high-impedance state is like an open circuit or electrical disconnection of the gate output from the circuit. In this state, both HIGH- and LOW-state output transistors are off.

44. The operation of a tristate output is controlled by the state of a control input. In one control state, the output is either HIGH or LOW. In the opposite control state, the output is in the high-impedance state.

45. CMOS (complementary MOS) devices are based on *n*-channel and *p*-channel MOS-FETs (metal-oxide-semiconductor field effect transistors).

46. A MOSFET consists of a silicon substrate of a particular type of silicon (e.g., *p*-type), embedded with wells of the opposite type (e.g., *n*-type) that form the drain and source regions of the MOSFET. A gate electrode can bias the substrate to create a conduction channel between drain and source.

47. An n-channel enhancement mode MOSFET is biased on when its gate voltage exceeds its source voltage by a given amount called the threshold voltage.

48. A p-channel enhancement mode MOSFET is biased on when its gate voltage is less than its source voltage by a given amount called the threshold voltage.

49. An n-channel and p-channel MOSFET can be connected in such a way that one of the pair of MOSFETs is always on and one is always off. This connection is called a complementary pair and forms the basis for CMOS logic.

50. Logic functions, such as NAND and NOR, can be implemented with a complementary pair of MOSFETs for each input, with the MOSFETs in series or parallel to V_{CC} or ground, as required.

51. Many TTL families have been designed to incorporate Schottky barrier diodes, which limit the saturation of their transistors, allowing faster internal and output switching speeds.

52. Metal-gate CMOS has been superceded by high-speed (silicon-gate) CMOS, which has a smaller MOSFET size, resulting in faster switching and lower gate capacitance.

53. Speed-power product is a measure of the energy used by a gate. More advanced logic families have smaller values of speed-power product.

GLOSSARY

CMOS   Complementary metal-oxide semiconductor. A logic family based on the switching of n- and p-channel metal-oxide-semiconductor field effect transistors (MOSFETs).

C_{PD} Internal capacitance of a high-speed CMOS device used to calculate its power dissipation.

Cutoff Mode   The operating mode of a transistor when there is no collector or drain current flowing and the path from collector to emitter or drain to source is effectively an open circuit.

Driving Gate   A gate whose output supplies current to the inputs of other gates.

ECL   Emitter coupled logic. A high-speed logic family based on bipolar transistors.

Enhancement-Mode MOSFET   A MOSFET which creates a conduction path (a channel) between its drain and source terminals when the voltage between gate and source exceeds a specified threshold level.

Fanout   The number of gate inputs that a gate output is capable of driving without possible logic errors.

High-Speed (Silicon-Gate) CMOS   A CMOS logic family with a smaller device structure and thus higher speed than standard (metal-gate) CMOS.

Hold Time (t_h)   The time that the synchronous inputs of a flip-flop must remain stable after the active *CLK* transition is finished.

I_{CC} Total supply current in a TTL or high-speed CMOS device.

I_{CCH} TTL supply current with all outputs HIGH.

I_{CCL} TTL supply current with all outputs LOW.

I_{DD} CMOS supply current under static (nonswitching) conditions.

I_{IH} Current measured at a device input when the input is HIGH.

I_{IL} Current measured at a device input when the input is LOW.

I_{OH} Current measured at a device output when the output is HIGH.

I_{OL} Current measured at a device output when the output is LOW.

I_T When referring to CMOS supply current, the sum of static and dynamic supply currents.

Load Gate   A gate whose input current is supplied by the output of another gate.

MOSFET   Metal-oxide-semiconductor field effect transistor. A MOSFET has three terminals—gate, source, and drain—which are analogous to the base, emitter, and collector of a bipolar junction transistor.

n-Channel Enhancement-Mode MOSFET   A MOSFET built on a p-type substrate with n-type drain and source regions. An n-type channel is created in the p-substrate during conduction.

Noise   Unwanted electrical signal, often resulting from electromagnetic radiation.

Noise Margin   A measure of the ability of a logic circuit to tolerate noise.

n-Type Inversion Layer   The conducting layer formed between drain and source when an enhancement-mode n-channel MOSFET is biased ON. Also referred to as the channel.

Ohmic Region   The MOSFET equivalent of saturation. When a MOSFET is biased ON, it acts like a relatively low resistance, or "ohmically."

Open-Collector Output   A TTL output where the collector of the LOW-state output transistor is brought out directly to the output pin. Since there is no built-in HIGH-state output circuitry, two or more open collector outputs can be connected without possible damage.

p-Channel Enhancement-Mode MOSFET   A MOSFET built on an n-type substrate with p-type drain and source regions. During conduction, a p-type channel is created in the n-substrate.

Phase Splitter   A transistor in a TTL circuit which ensures that the LOW- and HIGH-state output transistors of a totem pole output are always in opposite phase (i.e., one ON, one OFF).

Power Dissipation   The electrical energy used by a logic circuit in a specified period of time. Abbreviation: P_D.

Propagation Delay   time required for the output of a digital circuit to change states after a change at one or more of its inputs.

Pulse Width (t_w)   Minimum time required for an active-level pulse applied to a CLK, $\overline{CLR}$, or $\overline{PRE}$ input, as measured from the midpoint of the leading edge of the pulse to the midpoint of the trailing edge.

Recovery Time (t_{rec})   Minimum time from the midpoint of the trailing edge of a $\overline{CLR}$ or $\overline{PRE}$ pulse to the midpoint of an active CLK edge.

Saturation Mode   The operating mode of a bipolar transistor when an increase in base current will not cause a further increase in the collector current and the path from collector to emitter is very nearly (but not quite) a short circuit. This is the ON state of a transistor in a digital circuit.

Schottky Barrier Diode   A specialized diode with a forward drop of about +0.4 V.

Schottky Transistor   A bipolar transistor with a Schottky diode across its base-collector junction, which prevents the transistor from going into deep saturation.

Schottky TTL   A series of unsaturated TTL logic families based on Schottky transistors. Schottky TTL switches faster than standard TTL due to decreased storage time in its transistors.

Setup Time (t_{su})   The time required for the synchronous inputs of a flip-flop to be stable before a CLK pulse is applied.

Sinking   A terminal on a gate or flip-flop is sinking current when the current flows into the terminal.

Sourcing   A terminal on a gate or flip-flop is sourcing current when the current flows out of the terminal.

Speed-Power Product   A measure of a logic circuit's efficiency, calculated by multiplying its propagation delay by its power dissipation. Unit: picojoule (pJ)

Storage Time   Time required to transport stored charge away from the base region of a bipolar transistor before it can turn off.

Substrate   The foundation of n- or p-type silicon on which an integrated circuit is built.

Threshold Voltage, $V_{GS(Th)}$ The minimum voltage between gate and source of a MOSFET for the formation of the conducting inversion layer (channel).

Totem Pole Output   A type of TTL output with a HIGH and a LOW output transistor, only one of which is active at any time.

t_{pHL}   Propagation delay when the device output is changing from HIGH to LOW.

t_{pLH}   Propagation delay when the device output is changing from LOW to HIGH.

Tristate Output   An output having three possible states: logic HIGH, logic LOW, and a high-impedance state, in which the output acts as an open circuit.

TTL   Transistor-transistor logic. A logic family based on bipolar transistors.

TTL Compatible   Able to be driven directly by a TTL output. Usually implies voltage compatibility with TTL.

V_{CC} Supply voltage for TTL and high-speed CMOS devices.

V_{DD} Metal-gate CMOS supply voltage.

V_{IH} Voltage level required to make the input of a logic circuit HIGH.

V_{IL} Voltage level required to make the input of a logic circuit LOW.

V_{OH} Voltage measured at a device output when the output is HIGH.

V_{OL} Voltage measured at a device output when the output is LOW.

Wired-AND   A connection where open-collector outputs of logic gates are wired together. The logical effect is the ANDing of connected functions.

PROBLEMS

Problem numbers set in color indicate more difficult problems.

11.1 Electrical Characteristics of Logic Gates

11.1   Briefly list the advantages and disadvantages of TTL, CMOS, and ECL logic gates.

11.2 Propagation Delay

11.2   Explain how propagation delay is measured in TTL devices and CMOS devices. How do these measurements differ?

11.3   Figure 11.67 shows the input and output waveforms of a logic gate. Use the graph to calculate t_{pHL} and t_{pLH}. The gate is high-speed CMOS.

FIGURE 11.67   Problem 11.3: Waveforms

11.4   The inputs of the logic circuit in Figure 11.68 are in state 1 in the following table. The inputs change to state 2, then to state 3.

74LS02

74LS00

$Y = A + B + \overline{C}$

FIGURE 11.68   Problems 11.4 and 11.5: Logic Circuit

	A	B	C
State 1	1	0	1
State 2	0	0	1
State 3	0	0	0

a. Draw a timing diagram that uses these changes of input state to illustrate the effect of propagation delay in the circuit.

b. Calculate the maximum time it takes for the output to change when the inputs change from state 1 to state 2.

c. Calculate the maximum time it takes for the output to change when the inputs change from state 2 to state 3.

11.5 Repeat Problem 11.4, parts b and c, for a 74HC00 NAND and a 74HC02 NOR gate.

11.3 Flip-Flop Timing Parameters

11.6 Use a TTL or high-speed CMOS datasheet, as appropriate, to look up the setup and hold times of the following devices:

    a. 74LS74A

    b. 74AC74

    c. 74HCT74

    d. 74HC74

    e. 74ALS74

11.7 Draw a timing diagram showing the setup and hold times for a 74LS74A flip-flop.

11.8 Draw timing diagrams (to *scale*) showing setup and hold times, minimum *CLK* and $\overline{CLR}$ pulse widths, recovery time, and propagation delay times from *CLK* and $\overline{CLR}$ for both 74AS74 and 74HC74 flip-flops.

11.9 Write names and values of the JK flip-flop timing parameters illustrated in Figure 11.69.

FIGURE 11.69 Problem 11.9: Timing Parameters

11.10 Repeat Problem 11.9 for the timing diagram in Figure 11.70.

FIGURE 11.70 Problem 11.10: Timing Diagram

11.4 Fanout

11.11 Calculate the maximum number of advanced low-power Schottky TTL loads (74ALSNN series) that a 74AS86 XOR gate can drive.

11.12 What is the maximum number of 74AS32 OR gates that a 74ALS00 NAND gate can drive?

11.13   What is the maximum number of 74ALS00 NAND gates that a 74AS32 OR gate can drive?

11.14   An ALSTTL gate is driving seven ALSTTL gate inputs, each equivalent to the load presented by a 74ALS00 NAND input. Calculate the source and sink currents required from the driving gate.

11.5   Power Dissipation

11.15   The circuit in Figure 11.71 is constructed from the gates of a 74ALS08 AND device. Calculate the power dissipation of the circuit for the following input logic levels:

	A	B	C	D	E
a.	0	0	0	0	0
b.	1	1	0	1	1
c.	1	1	1	1	0
d.	1	1	1	1	1

FIGURE 11.71   Problems 11.15 to 11.17: Logic Circuit

11.16   The gate outputs in Figure 11.71 are switching at an average frequency of 100 kHz, with an average duty cycle of 60%. Calculate the power dissipation if the gates are all 74AS08 AND gates.

11.17   The gates in Figure 11.71 are 74HC08A high-speed CMOS gates.

a. Calculate the power dissipation of the circuit if the input state is $ABCDE = 010101$. ($V_{CC} = 4.5$ V, $T_A = 25°$ C)

b. Calculate the circuit power dissipation if the outputs are switching at a frequency of 10 kHz, 50% duty cycle.

c. Repeat part b for a frequency of 2 MHz.

11.18   The circuit in Figure 11.72 consists of two 74ALS00 NAND gates (gates 4 and 5) and three 74ALS02 NOR gates (gates 1, 2, and 3). When this circuit is actually built, there will be two unused NAND gates and one unused NOR gate in the device packages.

FIGURE 11.72   Problem 11.18: Logic Circuit

Calculate the maximum total power dissipation of the circuit when its input state is $ABCDE = 01100$. Include all unused gates. (Connect unused gate inputs so that they will dissipate the least amount of power.)

11.19   a. Calculate the no-load power dissipation of a single gate at 1 MHz for a 74HC00A quad 2-input NAND gate ($V_{CC} = 5$ V). (Neglect quiescent current.)

b. Calculate the percent change in power dissipation if the gate in part a of this question is operated with a new value of $V_{CC} = 3.3$ V. ($f = 1$ MHz)

11.6 Noise Margin

11.20 Calculate the maximum noise margins, in both HIGH and LOW states, of:

a. A 74S00 NAND gate

b. A 74LS00 NAND gate

c. A 74AS00 NAND gate

d. A 74ALS00 NAND gate

e. A 74HC00 NAND gate ($V_{CC} = 5$ V)

f. A 74HCT00 NAND gate ($V_{CC} = 5$ V)

11.7 Interfacing TTL and CMOS Gates

11.21 Why can an LSTTL gate drive a 74HCT gate directly, but not a 74HC? Show calculations.

11.22 Draw a circuit that allows an LSTTL gate to drive a 74HC gate. Explain briefly how it works.

11.23 How many LSTTL loads (e.g., 74LS00) can a 74HC00A NAND gate drive? Use datasheet parameters to support your answer. Assume $V_{CC} = 4.5$ V. Show all calculations.

11.8 Internal Circuitry of TTL Gates

11.24 In what logic state is an open TTL input? Why?

11.25 Briefly describe the operation of the TTL open-collector inverter shown in Figure 11.22. What is the purpose of the diode?

11.26 Briefly explain the operation of a multiple-emitter input transistor used in a TTL NAND gate. Describe how the transistor responds to various combinations of HIGH and LOW inputs.

11.27 Draw a wired-AND circuit consisting of three open-collector NAND gates and an output pull-up resistor. The gate inputs are as follows:

Gate 1: Inputs A, B

Gate 2: Inputs C, D

Gate 3: Inputs E, F

Write the Boolean function of the circuit output.

11.28 Calculate the minimum value of the pull-up resistor if the circuit drawn in Problem 11.27 is to drive a logic gate having input current $I_{IL} = 0.8$ mA and the NAND gates can sink 12 mA in the LOW output state. (Assume that $V_{OL} = 0.4$ V.)

11.29 Draw a circuit consisting only of open-collector gates whose Boolean expression is the product-of-sums expression

$$(A + B)(C + D)(E + F)(G + H).$$

11.30 Is an open-collector TTL output likely to be damaged if shorted to ground? Why or why not?

11.31 Is an open-collector TTL output likely to be damaged if shorted to V_{CC}? Why or why not?

11.32 Draw the totem pole output of a standard TTL gate.

11.33 Refer to the TTL NAND gate in Figure 11.36.

a. Why are Q_3 and Q_4 never on at the same time (ideally)?

b. How does switching noise originate in a totem pole output? How can the problem be controlled?

11.34 Explain briefly why two totem pole outputs should not be connected together.

11.35 Two LED driver circuits are shown in Figure 11.73. For each circuit, calculate the current flowing when the LED is ON. Calculate the ratio between the LED ON current and I_{OL} or I_{OH} of the inverter, whichever is appropriate for each circuit. State which is the best connection for LED driving and explain why.

FIGURE 11.73 Problem 11.35: LED Drivers

FIGURE 11.74 Problem 11.36: Lamp Driver

11.36 Calculate the current flowing when the lamp in Figure 11.74 is illuminated. Choose one of the following devices as a suitable driver: 74ALS04, 74ALS05, 74LS06. Explain your choice. (Datasheets for these devices are found in Appendix C.)

11.9 Internal Circuitry of CMOS Gates

11.37 State several precautions that should be taken to prevent electrostatic damage to MOSFET circuits.

11.38 **a.** Draw the circuit symbols for an *n*-channel and a *p*-channel enhancement-mode MOSFET.

   b. Describe the required bias conditions for each type of MOSFET in the cutoff and ohmic regions.

   c. State the approximate channel resistance for a MOSFET in the cutoff and ohmic regions.

11.39 Draw the circuit diagram of a CMOS AND gate. Derive the truth table of the gate by analyzing the operation of all the transistors under all possible input conditions.

11.40 Repeat Problem 11.39 for a CMOS OR gate.

11.41 Figure 11.75 shows a circuit that can switch two analog signals to an automotive speedometer/tachometer. Each sensor produces an analog voltage proportional to its measured quantity. Briefly explain how these analog signals are switched to the display output circuitry.

FIGURE 11.75 Problem 11.41: Speedometer/Tachometer Switching Circuit

11.10 TTL and CMOS Variations

11.42 Briefly explain how a Schottky barrier diode can improve the performance of a transistor in a TTL circuit.

11.43 Is the speed-power product of a TTL gate affected by the switching frequency of its output? Explain.

11.44 Use data sheets to calculate the speed-power products of the following gates:

   a. 74LS00

   b. 74S00

   c. 74ALS00

   d. 74AS00

   e. 74HC00A (quiescent and 10 MHz)

   f. 74HCT00A (quiescent and 10 MHz)

   g. 74F00

11.45 Briefly explain the differences among the following high-speed CMOS logic families: 74HC, 74HC4NNN, 74HCT, and 74HCU.

11.46 Assume that the power dissipation of a metal-gate or high-speed CMOS gate increases in proportion to the switching frequency of its output. Calculate the speed-power product of the following gates at 2 MHz, 5 MHz, and 10 MHz:

   a. 4011B

   b. 74HCT04

   c. 74HCU04

ANSWERS TO SECTION REVIEW PROBLEMS

11.1

11.1 $V_{OH} = 3.0$ V min. $I_{OH} = -0.4$ mA (The negative sign indicates that the current is leaving the gate. See Figure 11.2.)

11.2

11.2 $t_{pHL1} + t_{pHL2} = 10$ ns + 12 ns = 22 ns; $t_{pLH2} = 14$ ns

11.3

11.3   The parameter is called propagation delay. For the specified output transition, the symbol is t_{pHL}.

11.4

11.4   Source currents: I_{OH}, I_{IL}; sink currents: I_{OL}, I_{IH}

11.5

11.5   CMOS draws very little current when its outputs are not switching. Since the majority of current is drawn when the outputs switch, the more often the outputs switch, the more current is drawn from the supply. This is the same as saying that power dissipation increases with frequency.

11.6

11.6 $V_{NH} = 1.98$ V, $V_{NL} = 0.54$ V

11.7

11.7   2.5 V. The interface buffer and load should have the same supply voltage so that the output voltage of the buffer and input voltage of the load are compatible.

11.8A

11.8   **a.** Provision of logic HIGH when output transistor is OFF

b. Limitation of I_{OL} when output transistor is ON

11.8B

11.9 $R_{ext} = 570$ Ω. Minimum standard value: 680 Ω

11.8C

11.10   When the output is HIGH, current flows to ground through a low-impedance path, causing I_{OH} to exceed its rating.

11.8D

11.11   The diode allows the base of Q_4 to be pulled LOW through G, but will not allow a HIGH at G to turn it on. This keeps both output transistors OFF in the high-impedance state and allows them to be in opposite states when the output is enabled.

11.8E

11.12   Noninverting gates are actually double-inverting gates. They require an extra transistor stage to cancel the inversion introduced by NAND or NOR transistor logic.

11.9A

11.13 The thin oxide layer in the gate region can be damaged by overvoltage, such as that caused by electrostatic discharge. If the oxide layer is damaged, it may no longer insulate the gate terminal from the MOSFET substrate, which causes the transistor to malfunction.

11.9B

11.14 It allows complementary operation with an *n*-channel MOSFET. Specifically, a gate voltage of 0 V turns OFF an *n*-channel device having a grounded source. The same voltage turns ON the *p*-channel device whose source is tied to V_{CC}. It does so by making the *p*-channel gate-source voltage more negative than the required threshold.

11.10

11.15 13.36 pJ, 33.4 pJ, and 66.8 pJ.

KEY TERMS

Aliasing
Analog
Analog-to-Digital Converter
Anti-Aliasing Filter
Bipolar ADC
Bipolar DAC
Continuous
Data Acquisition Network
Digital
Digital-to-Analog Converter
Discrete
Dual Slope ADC
Flash Converter
Full Scale (FS)
Integrator
Multiplying DAC
Nyquist Sampling Theorem
Offset Binary
Priority Encoder
Quantization
Quantization Error
Resolution
Sample
Sample and Hold Circuit
Sampling Frequency
Sigma-Delta ADC
Successive Approximation Register
Transfer Characteristic
Unipolar ADC
Unipolar DAC

OUTLINE

12.1 Analog and Digital Signals
12.2 Digital-to-Analog Conversion
12.3 Analog-to-Digital Conversion
12.4 Data Acquisition

CHAPTER OBJECTIVES

Upon successful completion of this chapter, you will be able to:

- Define the terms "analog" and "digital" and give examples of each.

- Explain the sampling of an analog signal and the effects of sampling frequency and quantization on the quality of the converted digital signal.

- Describe the coding of data in analog-to-digital and digital-to-analog converters for unipolar, offset binary, and 2's complement coding.

- Perform calculations to convert input voltages to unipolar, offset binary, or 2's complement codes in an analog-to-digital converter.

- Perform calculations to determine output voltages of a digital-to-analog converter, given an input code in unipolar, offset binary, or 2's complement format.

- Draw the block diagram of a generic digital-to-analog converter (DAC) and circuits of a weighted resistor DAC and an R-2R ladder DAC.

- Calculate analog output voltages of a DAC, given a reference voltage and a digital input code.

- Configure an MC1408 integrated circuit DAC for unipolar and bipolar output, and calculate output voltage from known component values, reference voltage, and digital inputs.

- Describe important performance specifications of a digital-to-analog converter.

- Draw the circuit for a flash analog-to-digital converter (ADC) and briefly explain its operation.

- Define "quantization error" and describe its effect on the output of an ADC.

CHAPTER 12

Interfacing Analog and Digital Circuits

- Explain the basis of the successive approximation ADC, draw its block diagram, and briefly describe its operation.
- Describe the operation of an integrator with constant input voltage.
- Draw the block diagram of a dual slope (integrating) ADC and briefly explain its operation.
- Draw the block diagram of a sigma-delta analog-to-digital converter, briefly explain its operation, and perform calculations to determine its output data stream.
- Explain the necessity of a sample and hold circuit in an ADC and its operation.
- State the Nyquist sampling theorem and do simple calculations of maximum analog frequencies that can be accurately sampled by an ADC system.
- Describe the phenomenon of aliasing and explain how it arises and how it can be remedied.
- Interface an ADC0808 analog-to-digital converter to a CPLD-based state machine.
- Design a 4-channel data acquisition system, including an ADC0808 analog-to-digital converter and a CPLD-based state machine.

Electronic circuits and signals can be divided into two main categories: analog and digital. Analog signals can vary continuously throughout a defined range. Digital signals take on specific values only, each usually described by a binary number.

Many phenomena in the world around us are analog in nature. Sound, light, heat, position, velocity, acceleration, time, weight, and volume are all analog quantities. Each of these can be represented by a voltage or current in an electronic circuit. This voltage or current is a copy, or analog, of the sound, velocity, or whatever.

We can also represent these physical properties digitally, that is, as a series of numbers, each describing an aspect of the property, such as its magnitude at a particular time. To translate between the physical world and a digital circuit, we must be able to convert analog signals to digital and vice versa.

We will begin by examining some of the factors involved in the conversion between analog and digital signals, including sampling rate, resolution, range, and quantization.

We will then examine circuits for converting digital signals to analog, since these have a fairly standard form. Analog-to-digital conversion has no standard method. We will study several of the most popular: simultaneous (flash) conversion, successive approximation, dual slope (integrating) and sigma-delta conversion. ■

12.1   ANALOG AND DIGITAL SIGNALS

■ KEY TERMS

Analog   A way of representing some physical quantity, such as temperature or velocity, by a proportional continuous voltage or current. An analog voltage or current can have any value within a defined range.

Continuous   Smoothly connected. An unbroken series of consecutive values with no instantaneous changes.

Digital   A way of representing a physical quantity by a series of binary numbers. A digital representation can have only specific discrete values.

Discrete   Separated into distinct segments or pieces. A series of discontinuous values.

Analog-to-Digital Converter   A circuit that converts an analog signal at its input to a digital code. (Also called an A-to-D converter, A/D converter, or ADC.)

Digital-to-Analog Converter   A circuit that converts a digital code at its input to an analog voltage or current. (Also called a D-to-A converter, D/A converter, or DAC.)

Electronic circuits are tools to measure and change our environment. Measurement instruments tell us about the physical properties of objects around us. They answer questions such as "How hot is this water?", "How fast is this car going?", and "How many electrons are flowing past this point per second?" These data can correspond to voltages and currents in electronic instruments.

If the internal voltage of an instrument is directly proportional to the quantity being measured, with no breaks in the proportional function, we say that it is an **analog** voltage. Like the property being measured, the voltage can vary continuously throughout a defined range.

For example, sound waves are **continuous** movements in the air. We can plot these movements mathematically as a sum of sine waves of various frequencies. The patterns of magnetic domains on an audio tape are analogous to the sound waves that produce them and electromagnetically represent the same mathematical functions. When the tape is played, the playback head produces a voltage that is also proportional to the original sound waves. This analog audio voltage can be any value between the maximum and minimum voltages of the audio system amplifier.

If an instrument represents a measured quantity as a series of binary numbers, the representation is **digital.** Since the binary numbers in a circuit necessarily have a fixed number of bits, the instrument can represent the measured quantities only as having specific **discrete** values.

A compact disc stores a record of sound waves as a series of binary numbers. Each number represents the amplitude of the sound at a particular time. These numbers are decoded and translated into analog sound waves upon playback. The values of the stored numbers (the encoded sound information) are limited by the number of bits in each stored digital "word."

The main advantage of a digital representation is that it is not subject to the same distortions as an analog signal. Nonideal properties of analog circuits, such as stray inductance and capacitance, amplification limits, and unwanted phase shifts, all degrade an analog signal. Storage techniques, such as magnetic tape, can also introduce distortion due to the nonlinearity of the recording medium.

Digital signals, on the other hand, do not depend on the shape of a waveform to preserve the encoded information. All that is required is to maintain the integrity of the logic HIGHs and LOWs of the digital signal. Digital information can be easily moved around in a circuit and stored in a latch or on some magnetic or optical medium. When the information is required in analog form, the analog quantity is reproduced as a new copy every time it is needed. Each copy is as good as any previous one. Distortions are not introduced between copy generations, as is the case with analog copying techniques, unless the constituent bits themselves are changed.

Digital circuits give us a good way of measuring and evaluating the physical world, with many advantages over analog methods. However, most properties of the physical world are analog. How do we bridge the gap?

We can make these translations with two classes of circuits. An **analog-to-digital converter** accepts an analog voltage or current at its input and produces a corresponding digital code. A **digital-to-analog converter** generates a unique analog voltage or current for every combination of bits at its inputs.

Sampling an Analog Voltage

> ■ **KEY TERMS**
>
> **Sample**   An instantaneous measurement of an analog voltage, taken at regular intervals.
>
> **Sampling Frequency**   The number of samples taken per unit time of an analog signal.
>
> **Quantization**   The number of bits used to represent an analog voltage as a digital number.
>
> **Resolution**   The difference in analog voltage corresponding to two adjacent digital codes. Analog step size.

Before we examine actual D/A and A/D converter circuits, we need to look at some of theoretical issues behind the conversion process. We will look at the concept of **sampling** an analog signal and discover how the **sampling frequency** affects the accuracy of the digital representation. We will also examine **quantization,** or the number of bits in the digital representation of the analog sample, and its effect on the quality of a digital signal.

Figure 12.1 shows a circuit that converts an analog signal (a sine pulse) to a series of 4-bit digital codes, then back to an analog output. The analog input and output voltages are shown on the two graphs.

There are two main reasons why the output is not a very good copy of the input. First, the number of bits in the digital representation is too low. Second, the input signal is not sampled frequently enough. To help us understand the effect of each of these factors, let us examine the conversion process in more detail.

The analog input signal varies between 0 and 8 volts. This is evenly divided into 16 ranges, each corresponding to a 4-bit digital code (0000 to 1111). We say that the

FIGURE 12.1   Analog Input and Output Signals

signal is quantized into 4 bits. The **resolution,** or analog step size, for a 4-bit quantization is 8 V/16 steps = 0.5 V/step. Table 12.1 shows the codes for each analog range.

TABLE 12.1   4-Bit Digital Codes for 0 to 8 V Analog Range

Analog Voltage	Digital Code
0.00–0.25	0000
0.25–0.75	0001
0.75–1.25	0010
1.25–1.75	0011
1.75–2.25	0100
2.25–2.75	0101
2.75–3.25	0110
3.25–3.75	0111
3.75–4.25	1000
4.25–4.75	1001
4.75–5.25	1010
5.25–5.75	1011
5.75–6.25	1100
6.25–6.75	1101
6.75–7.25	1110
7.25–8.00	1111

The analog input is sampled and converted at the beginning of each time division on the graph. The 4-bit digital code does not change until the next conversion, 1 ms later. This is the same as saying that the system has a sampling frequency of 1 kHz ($f = 1/T = 1/(1 \text{ ms}) = 1 \text{ kHz}$).

Table 12.2 shows the digital codes for samples taken from $t = 0$ to $t = 18$ ms. The analog voltages in Table 12.2 are calculated by the formula

$$V_{analog} = 8 \text{ V} \sin (t \times (10°/\text{ms}))$$

For example at $t = 2$ ms, $V_{analog} = 8 \text{ V} \sin (2 \text{ ms} \times (10°/\text{ms})) = 8 \text{ V} \sin (20°) = 2.736 \text{ V}$.

The calculated analog values are compared to the voltage ranges in Table 12.1 and assigned the appropriate code. The value 2.736 V is between 2.25 V and 2.75 V and therefore is assigned the 4-bit value of 0101.

The digital-to-analog converter in Figure 12.1 continuously converts the digital codes to their analog equivalents. Each code produces an analog voltage whose value is the midpoint of the range corresponding to that code.

For this particular analog waveform, the A/D converter introduces the greatest inaccuracy at the peak of the waveform, where the magnitude of the input voltage changes the least per unit time. There is not sufficient difference between the values of successive analog samples to map them into unique codes. As a result, the output waveform flattens out at the top.

This is the consequence of using a 4-bit quantization, which allows only 16 different analog ranges in the signal. By using more bits, we could divide the analog signal into a greater number of smaller ranges, allowing more accurate conversion of a signal having small changes in amplitude. For example, an 8-bit code would give us 256 steps (a resolution of 8 V/256 = 31.25 mV). This would yield the code assignments shown in Table 12.3. Note that for an 8-bit code, there is a unique value for every sampled voltage.

TABLE 12.2 4-Bit Codes for a Sampled Analog Signal

Time (ms)	Analog Amplitude (volts)	Digital Code
0	0.000	0000
1	1.389	0011
2	2.736	0101
3	4.000	1000
4	5.142	1010
5	6.128	1100
6	6.928	1110
7	7.518	1111
8	7.878	1111
9	8.000	1111
10	7.878	1111
11	7.518	1111
12	6.928	1110
13	6.128	1100
14	5.142	1010
15	4.000	1000
16	2.736	0101
17	1.389	0011
18	0.000	0000

TABLE 12.3 8-Bit Codes for a Sampled Analog Signal

Time (ms)	Analog Amplitude (volts)	Digital Code
0	0.000	00000000
1	1.389	00101100
2	2.736	01010111
3	4.000	10000000
4	5.142	10100100
5	6.128	11000100
6	6.928	11011101
7	7.518	11110000
8	7.878	11111100
9	8.000	11111111
10	7.878	11111100
11	7.518	11110000
12	6.928	11011101
13	6.128	11000100
14	5.142	10100100
15	4.000	10000000
16	2.736	01010111
17	1.389	00101100
18	0.000	00000000

Figure 12.2 shows how different levels of quantization affect the accuracy of a digital representation of an analog signal. The analog input is a sine wave, converted to digital codes and back to analog, as in Figure 12.1. The graphs show the analog input and three analog outputs, each of which has been sampled 28 times per cycle, but with different quantizations. The corresponding digital codes range from a maximum negative value of n 0s to a maximum positive value of n 1s for an n-bit quantization (e.g., for a 4-bit quantization, maximum negative = 0000, maximum positive = 1111). This is a bipolar coding called offset binary.

a. Analog signal   b. Infinite quantization

c. 4-bit quantization   d. 3-bit quantization

FIGURE 12.2   Effect of Quantization

The first output signal has an infinite number of bits in its quantization. Even the smallest analog change between samples has a unique code. This ideal case is not attainable, since a digital circuit always has a finite number of bits. We can see from the codes in Table 12.3 that an 8-bit quantization is sufficient to give unique codes for this waveform. An infinite quantization implies that the resolution is small enough that each sampled voltage can be represented, not only by a unique code, but as its exact value rather than as a range of values.

The 4-bit and 3-bit quantizations in the next two graphs (Figures 12.2c and d) show progressively worse representation of the original signal, especially at the peaks. The change in analog voltage is too small for each sample to have a unique code at these low quantizations.

Figure 12.3 shows how the digital representation of a signal can be improved by increasing its sampling frequency. It shows an analog signal and three analog wave-forms resulting from an analog-digital-analog conversion. All waveforms have infinite quantization, but different numbers of samples in the analog-to-digital conversion. As the number of samples decreases, the output waveform becomes a poorer copy of the input.

It is worth noting that any of the sampled waveforms in Figure 12.2 or Figure 12.3 contain sufficient information to reproduce the original analog sine wave of Figure 12.2a or Figure 12.3a. The original waveform can be recovered by running any of the digi-tized waveforms through a suitable low-pass filter, called a reconstruction filter.

In general, the sampling frequency affects the horizontal resolution of the digi-tized waveform and the quantization affects the vertical resolution.

■ SECTION 12.1A REVIEW PROBLEM

12.1   An analog signal has a range of 0 to 24 mV. The range is divided into 32 equal steps for conversion to a series of digital codes. How many bits are in the resultant digital codes? What is the resolution of the A/D converter?

a. Analog signal   b. 72 samples

c. 18 samples   d. 8 samples

FIGURE 12.3   Effect of Sampling Frequency

Data Coding for Analog-to-Digital and Digital-to-Analog Converters

■ KEY TERMS

Transfer Characteristic A graph or mathematical function that shows the relationship between the input and output of an electronic circuit.

Unipolar ADC An analog-to-digital converter whose input voltages are positive only.

Full Scale (FS) The maximum range of analog reference voltage or current of an analog-to-digital or digital-to-analog converter.

Bipolar ADC An analog-to-digital converter whose input voltages have a positive and negative range.

Offset Binary A binary code in which the lower half of the range represents negative numbers and the upper half represents positive numbers.

Unipolar DAC A digital-to-analog converter whose output voltages are positive only.

Bipolar DAC A digital-to-analog converter whose output voltages have a positive and negative range.

Before we look at actual circuits of A/D and D/A converters, let us examine several ways in which analog voltages and digital codes can relate to one another in an A/D or D/A converter.

Digital output code

$$code = \frac{V_a}{FS} \cdot 2^n \qquad V_a = \text{analog voltage}$$

$$FS = \text{Full scale}$$

$$= \frac{V_a}{8V} \cdot 8 \qquad n = \text{number of output bits}$$

FIGURE 12.4   Transfer Characteristic of a Unipolar ADC

Unipolar Analog-to-Digital Converter (ADC)

Figure 12.4 shows a graph that represents the **transfer characteristic** of a 3-bit **unipolar ADC**. This ADC accepts positive input voltages from 0 to 8 volts and generates a 3-bit binary code for any given input voltage.

We begin at 0 volts on the horizontal axis, which gives an output code of 000. As we increase the voltage, the code remains constant at 000 until 0.5 volts, where the output changes to 001. The code remains the same for any voltage up to 1.5 volts, where the code changes again to 010. This process continues until 6.5 volts, where the output changes to the highest possible 3-bit code, 111.

Although the resolution of this ADC is 1V/step, the first step occurs at 0.5 volts, one half-step from 0 volts. Why? If we examine one of the steps in the transfer characteristic, such as the voltage for output 110, we see that this step ranges from 5.5 volts to 6.5 volts, with the center of the step at 6.0 volts.

The output code can be calculated as follows:

$$code = \frac{V_a}{FS} \times 2^n$$

where V_a = analog input voltage of the sample to be converted,

FS = **Full scale** range of input voltage, and

n = number of bits in the output code.

If we calculate the code for an input of 6.0 volts, we find that the code will be $(6\,V/8\,V) \times 8 = 6_{10} = 110_2$. The half-step offset allows the nominal voltage for a particular code to be in the center of the voltage range for that code. Thus, the voltage range for any output code except the first and last is $V_{nominal} \pm \frac{1}{2}$ LSB, where the value of one least significant bit (LSB) is the resolution of the converter. For example, the input voltage for output code 100 is 4 V $\pm \frac{1}{2}$ LSB, or 4.0 V $\pm$ 0.5 V.

The half-step offset is made up at the top of the range so that the input range for output 111 is 7 V $- \frac{1}{2}$ LSB, +1 LSB (6.5 V to 8.0 V), a total of $1\frac{1}{2}$ steps.

Table 12.4 shows a summary of the output coding of a unipolar ADC. Notice that the highest nominal voltage is 7.0 V, even though there are eight steps at 1 V/step. This occurs because we require a code for 0 V, leaving only seven more codes. This is true for any size of converter. The nominal voltage of the highest step of the converter will be 1 LSB less than the full scale voltage.

TABLE 12.4 Output Codes for a 3-Bit Unipolar ADC (FS = 8 V)

Nominal Voltage of Input Step (volts)	Range (volts)	Output Code
0.0	0.0 – 0.5	000
1.0	0.5 – 1.5	001
2.0	1.5 – 2.5	010
3.0	2.5 – 3.5	011
4.0	3.5 – 4.5	100
5.0	4.5 – 5.5	101
6.0	5.5 – 6.5	110
7.0	6.5 – 8.0	111

EXAMPLE 12.1

a. Determine the range of voltages that will generate a code of 00000000 for an 8-bit unipolar ADC with a full scale value of 8 volts.
b. Determine the range of voltages that will generate a code of 11111111 for the same ADC.
c. Calculate the output code when the input voltage is at the halfway point of its range.

■ Solution

To calculate these answers, we first need to know the resolution of the ADC.

$$\text{Resolution} = FS/2^n = 8 \text{ V}/2^8 = 8 \text{ V}/256 = 31.25 \text{ mV} = 1 \text{ LSB}$$
$$\tfrac{1}{2} \text{ LSB} = 31.25 \text{ mV}/2 = 15.625 \text{ mV}$$

a. Range for 00000000 is: 0 V to $(0 + \frac{1}{2}$ LSB$) = 0$V to 15.625 mV
b. Nominal voltage for 11111111 is $FS - 1$ LSB $= 8$ V $- 0.03125$ V $= 7.96875$ V.

This can also be calculated as $(255/256) \times 8$ V $= 7.96875$ V.

Range is $V_{\text{nominal}} - \frac{1}{2}$ LSB, +1 LSB: $(7.96875$ V $- \frac{1}{2}(0.03125$ V$))$ to 8.0 V
$$= 7.953125\text{V to } 8.0 \text{ V}$$

c. Halfway point = 4 V. Code = $(4 \text{ V}/8 \text{ V}) \times 256 = 128_{10} = 10000000$.

Note that in a unipolar ADC the code for the input of $FS/2$ is always 100 . . ., regardless of the number of bits. The code for $FS/4$ is always 010. . . . The code for $3FS/4$ is always 110. . . .

EXAMPLE 12.2

An ADC has a full scale voltage of 8 V. Write the codes for 0 V, 2 V, 4 V, and 6 V for the cases where the code is 4 bits, 6 bits, and 8 bits. Also write the voltage for the highest code for each case.

■ **Solution**

$$FS = 8 \text{ V}$$
$$FS/4 = 2 \text{ V: code} = 010 \ldots$$
$$FS/2 = 4 \text{ V: code} = 100 \ldots$$
$$3FS/4 = 6 \text{ V: code} = 110 \ldots$$

Nominal voltages for highest codes: (highest code$/2^n$) $\times FS = \left(\dfrac{2^n - 1}{2^n} \right) \times FS$

$$\text{4-bit: } (15/16) \times 8 \text{ V} = 7.5 \text{ V}$$
$$\text{6-bit: } (63/64) \times 8 \text{ V} = 7.875 \text{ V}$$
$$\text{8-bit: } (255/256) \times 8 \text{ V} = 7.96875 \text{ V}$$

(Note that the nominal voltage for the highest code gets closer to full scale with more bits, but it never reaches full scale. The maximum nominal voltage is always FS - 1 LSB.)

Input Voltage	4-bit code	6-bit code	8-bit code
0 V	0000	000000	00000000
2 V	0100	010000	01000000
4 V	1000	100000	10000000
6 V	1100	110000	11000000
8 V – 1 LSB	1111 (7.5 V)	111111(7.875 V)	11111111 (7.96875 V)

Bipolar ADC (Offset Binary Coding)

If we want to represent positive and negative input voltages in an analog-to-digital converter, one way we can do so is to offset the transfer characteristic to include negative input values in the range. Figure 12.5 shows the transfer characteristic of a **bipolar ADC** with **offset binary** coding.

The graph of Figure 12.5 is similar to that of Figure 12.4, except that the inputs range from –4 volts to +4 volts, rather than 0 to +8 volts. The range of values is the same in both cases, but the second graph has an offset of 4 volts. We calculate the output code for the center voltage of a step as follows:

$$\text{code} = \left(\frac{V_a}{FS} \times 2^n \right) + \text{offset}$$

$$= \left(\frac{V_a}{FS} \times 2^n \right) + \frac{2^n}{2} = \left(\frac{V_a}{4\text{V} - (-4\text{V})} \times 8 \right) + 4$$

For example, the code for –2 volts is $((-2 \text{ V}/8 \text{ V}) \times 8) + 4 = 2_{10} = 010_2$.

The output code in this type of converter is an unsigned binary number, where the numbers from 0 to just below the halfway point represent negative voltages, the code at the halfway point (100) is for 0 volts, and the numbers above the halfway point are for positive voltages. The output codes are summarized in Table 12.5.

Digital output
code

Code is an unsigned binary value.

FIGURE 12.5 Transfer Characteristic of a Bipolar ADC with Offset Binary Coding

TABLE 12.5 Output Codes for a 3-Bit Bipolar ADC with Offset Binary Coding ($FS = \pm 4$ V)

Nominal Voltage of Input Step (volts)	Range (volts)	Output Code
–4.0	–4.0 to –3.5	000
–3.0	–3.5 to –2.5	001
–2.0	–2.5 to –1.5	010
–1.0	–1.5 to –0.5	011
0.0	–0.5 to +0.5	100
+1.0	+0.5 to +1.5	101
+2.0	+1.5 to +2.5	110
+3.0	+2.5 to +4.0	111

EXAMPLE 12.3

a. Determine the range of voltages that will generate a code of 00000000 for an 8-bit bipolar ADC (offset binary coding) with a full scale range of –4 V to +4 V.

b. Determine the range of voltages that will generate a code of 11111111 for the same ADC.

c. Calculate the output code when the input voltage is at the halfway point of its range.

(continued)

■ **Solution**

$$\text{Resolution} = (4\text{ V} - (-4\text{ V}))/256 = 31.25\text{ mV} = 1\text{ LSB}$$
$$\tfrac{1}{2}\text{ LSB} = 31.25\text{ mV}/2 = 15.625\text{ mV}$$

a. For 00000000, range is -4 V to $(-4\text{ V} + \tfrac{1}{2}\text{ LSB}) = -4$ V to -3.984375 V
b. For 11111111, range is $V_{nominal} - \tfrac{1}{2}\text{ LSB} + 1\text{ LSB}$

$$V_{nominal} = V_{max} - 1\text{ LSB} = +4\text{ V} - 0.03125\text{ V} = +3.96875\text{ V}$$
$$\text{Range: } (+3.96875\text{ V} - \tfrac{1}{2}\text{ LSB}) \text{ to } (+3.96875\text{ V} + 1\text{ LSB})$$
$$= +3.953125\text{ V to } +4.0\text{ V}$$

c. Halfway point in range is 0 V.

$$\text{Code} = ((V_a/FS) \times 2^n) + (2^n/2) = ((0\text{ V}/8\text{ V}) \times 256) + 128 = 128_{10} = 10000000_2$$

TABLE 12.6 Decimal Values of 3-Bit 2's Complement Numbers

Decimal	2's Complement (3-bit)
+3	011
+2	010
+1	001
0	000
−1	111
−2	110
−3	101
−4	100

Bipolar ADC (2's Complement Coding)

Another way to represent positive and negative input voltages in an ADC is to use a signed binary output, such as a 2's complement number. In this case the most significant bit of the output code is a sign bit, with the sign bit 0 for positive and 1 for negative. Table 12.6 shows the range of 3-bit 2's complement numbers.

Figure 12.6 shows the transfer characteristic of a bipolar ADC with 2's complement coding. We can find the code for the center voltage of a step by:

$$\text{code} = \frac{V_a}{FS} \times 2^n$$

where code is a 2's complement signed value.

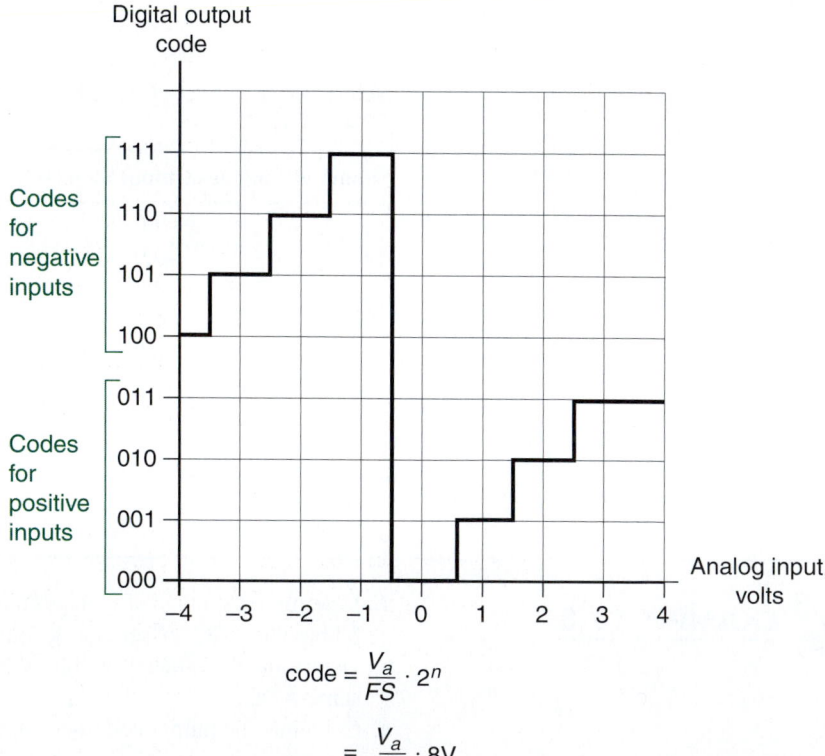

Code is a signed 2's complement number.

FIGURE 12.6 Transfer Characteristic of a Bipolar ADC with 2's Complement Coding

For example, for $V_a = -2$ volts, code = $(-2\ V/(+4\ V - (-4\ V)) \times 8 = -2_{10} = 110_2$. Table 12.7 summarizes the output codes for a 2's complement-coded ADC.

TABLE 12.7 Output Codes for a 3-Bit Bipolar ADC with 2's Complement Coding ($FS = \pm4$ V)

Nominal Voltage of Input Step (volts)	Range (volts)	Output Code
−4.0	−4.0 to −3.5	100
−3.0	−3.5 to −2.5	101
−2.0	−2.5 to −1.5	110
−1.0	−1.5 to −0.5	111
0.0	−0.5 to +0.5	000
+1.0	+0.5 to +1.5	001
+2.0	+1.5 to +2.5	010
+3.0	+2.5 to 4.0	011

EXAMPLE 12.4

a. Determine the range of voltages that will generate a code of 00000000 for an 8-bit bipolar ADC (2's complement coding) with a full scale range of −4 V to +4 V.
b. Determine the range of voltages that will generate a code of 11111111 for the same ADC.
c. Calculate the output code when the input voltage is at the halfway point of its range.

■ **Solution**

$$\text{Resolution} = (4\ V - (-4\ V))/256 = 31.25\ mV = 1\ LSB$$
$$\tfrac{1}{2}\ LSB = 31.25\ mV/2 = 15.625\ mV$$

a. For 00000000, range is $0\ V \pm \tfrac{1}{2}\ LSB = -15.625\ mV$ to $+15.625\ mV$
b. The decimal equivalent of the 2's complement number 11111111 is −1. The nominal voltage of this range is 1 LSB less than 0 V or simply −1 LSB.

$$V_{nominal} = -31.25mV$$
$$\text{Range: } V_{nominal} \pm \tfrac{1}{2}\ LSB$$
$$= (-31.25\ mV - 15.625\ mV) \text{ to } (-31.25\ mV + 15.625\ mV)$$
$$= -46.875\ mV \text{ to } -15.625\ mV$$

c. Halfway point in the range is 0 V. Range is same as for part **a** of this example.

Unipolar DAC

Unlike an analog-to-digital converter, a digital-to-analog converter (DAC) does not have a range of voltages corresponding to a single digital code. Rather, for the DAC one input code corresponds to one output voltage, as shown in the transfer characteristic of a 3-bit **unipolar DAC** in Figure 12.7. The DAC has a full scale of 8 volts.

If the DAC is working correctly, we can draw a straight-line approximation through the data points of the transfer characteristic to show the general trend of the output voltages. However, the DAC can only have 2^n discrete voltage values.

We calculate the output voltage as follows:

$$V_a = \left(\frac{\text{code}}{2^n}\right) \times FS \text{ for an } n\text{-bit code.}$$

For example, for a code of 101, $V_a = (5/8) \times (8\ V) = 5\ V$.

Analog output
volts

$$V_a = \frac{\text{code}}{2^n} \cdot FS$$

$$= \frac{\text{code}}{8} \cdot 8V$$

FIGURE 12.7 Transfer Characteristic of a Unipolar DAC

EXAMPLE 12.5

a. Calculate the output voltage of an 8-bit unipolar DAC for code = 00000000. $FS = 8$ V.
b. Calculate the output voltage of the same DAC for code = 11111111.
c. Write the code for $FS/2$ and state the value of the output voltage.

■ **Solution**

a. $V_a = (0/256) \times 8$ V $= 0$ V
b. $11111111_2 = 255_{10}$

$$V_a = (255/256) \times 8 \text{ V} = 7.96875 \text{ V}$$

c. $10000000_2 = 128_{10}$

$$V_a = (128/256) \times 8 \text{ V} = 4 \text{ V}$$

Bipolar DAC (Offset Binary Coding)

Figure 12.8 shows the transfer characteristic of a 3-bit **bipolar DAC** with a full scale range of –4 V to +4 V. The extremes of this range represent a difference of 8 V. The transfer characteristic is the same as that of the unipolar DAC in Figure 12.7, but with an offset of –4 V. Thus, we can calculate the output voltage by:

$$V_a = \left(\left(\frac{\text{code}}{2^n} \right) \times FS \right) - \left(\frac{FS}{2} \right)$$

For example, for a code of 010, $V_a = ((2/8) \times 8$ V$) - (8$ V$/2) = 2$ V $- 4$ V $= -2$ V.

$$V_a = \frac{\text{code}}{2^n} \cdot FS - \frac{FS}{2}$$

$$= \left(\frac{\text{code}}{8} \cdot 8V\right) - 4V$$

Code is an unsigned binary number.

FIGURE 12.8 Transfer Characteristic of a Bipolar DAC with Offset Binary Coding

 EXAMPLE 12.6

a. Calculate the output voltage of an 8-bit bipolar DAC (offset binary) for code = 00000000. $FS = 8$ V.

b. Calculate the output voltage of the same DAC for code = 11111111.

c. Write the code for $FS/2$ and state the value of the output voltage.

■ Solution

a. $V_a = ((0/256) \times 8\text{ V}) - 4\text{ V} = -4\text{ V}$

b. $11111111_2 = 255_{10}$

$$V_a = ((255/256) \times 8\text{ V}) - 4\text{ V} = 3.96875\text{ V}$$

c. $10000000_2 = 128_{10}$

$$V_a = ((128/256) \times 8\text{ V}) - 4\text{ V} = 0\text{ V}$$

Bipolar DAC (2's Complement Coding)

If a DAC is designed to accept digital codes in 2's complement format, it will have a transfer characteristic similar to that in Figure 12.9.

The output voltage is calculated by:

$$V_a = \left(\frac{\text{code}}{2^n}\right) \times FS$$

where code is a 2's complement signed number.

For example, for code = 110, which is –2 in decimal, $V_a = (-2/8) \times 8\text{ V} = -2\text{ V}$.

$$V_a = \frac{\text{code}}{2^n} \cdot FS$$

$$= \frac{\text{code}}{8} \cdot 8V$$

Code is a signed 2's complement number.

FIGURE 12.9 Transfer Characteristic of a Bipolar DAC with 2's Complement Coding

EXAMPLE 12.7

a. Calculate the output voltage of an 8-bit bipolar DAC (2's complement) for code = 00000000. $FS = 8$ V.
b. Calculate the output voltage of the same DAC for code = 11111111.
c. Write the code for $FS/2$ and state the value of the output voltage.

■ **Solution**

a. $V_a = ((0/256) \times 8\text{ V}) = 0$ V
b. $11111111_2 = -1_{10}$

$$V_a = ((-1/256) \times 8\text{ V}) = -31.25\text{ mV}$$

c. Code for $FS/2 = 00000000$

$$V_a = 0\text{ V}$$

Table 12.8 shows a summary of the various types of coding for an 8-bit DAC.

TABLE 12.8 Summary of DAC Coding

	Input Code		
Output Value	**Unipolar DAC**	**Bipolar DAC (Offset Binary)**	**Bipolar DAC (2's Complement)**
Maximum Positive	11111111	11111111	01111111
+1 LSB	00000001	10000001	00000001
0	00000000	10000000	00000000
−1 LSB	N/A	01111111	11111111
Maximum Negative	N/A	00000000	10000000

■ SECTION 12.1B REVIEW PROBLEM

12.2 A 12-bit DAC has a full scale range of 10 volts. Calculate the analog output voltage for a hexadecimal code of C35 if the DAC is unipolar, if it is bipolar (offset binary coding), and if it is bipolar (2's complement coding).

12.2 DIGITAL-TO-ANALOG CONVERSION

Figure 12.10 shows the block diagram of a generalized digital-to-analog converter. Each digital input switches a proportionally weighted current on or off, with the current for the MSB being the largest. The second MSB produces a current half as large. The current generated by the third MSB is one quarter of the MSB current, and so on.

These currents all sum at the operational amplifier's (op amp's) inverting input. The total analog current for an n-bit circuit is given by:

$$I_a = \frac{b_{n-1}2^{n-1} + \cdots + b_2 2^2 + b_1 2^1 + b_0 2^0}{2^n} I_{\text{ref}}$$

The bit values b_0, b_1, . . . b_n can be only 0 or 1. The function of each bit is to include or exclude a term from the general expression.

The op amp acts as a current-to-voltage converter. The analysis, illustrated in Figure 12.10b, is the same as for an inverting op amp circuit with a constant input current.

The input impedance of the op amp is the impedance between its inverting (−) and noninverting (+) terminals. This value is very large, on the order of 2 MΩ. If this is large compared to other circuit resistances, we can neglect the op amp input current, I_{in}.

This implies that the voltage drop across the input terminals is very small; the inverting and noninverting terminals are at approximately the same voltage. Since the noninverting input is grounded, we can say that the inverting input is "virtually grounded."

Current I_F flows in the feedback loop, through resistor R_F. Since $I_a - I_{in} - I_F = 0$ and $I_{in} \approx 0$, then $I_F \approx I_a$. By Ohm's law, the voltage across R_F is given by $V_F = I_a R_F$. The feedback resistor is connected to the output at one end and to virtual ground at the other. Since the op amp output voltage is measured with respect to ground, the two voltages are effectively in parallel. Thus, the output voltage is the same as the

Digital inputs

a. **Generic DAC circuit**

b. **Op amp analysis**

FIGURE 12.10   Analysis of a Generalized Digital-to-Analog Converter

voltage across the feedback resistor, with a polarity opposite to V_F, calculated previously.

$$V_a = -V_F = -I_a R_F$$

$$= \frac{-(b_{n-1}2^{n-1} + \cdots + b_2 2^2 + b_1 2^1 + b_0 2^0)}{2^n} I_{ref} R_F$$

The range of analog output voltage is set by choosing the appropriate value of R_F.

EXAMPLE 12.8

Write the expression for analog current, I_a, of a 4-bit D/A converter. Calculate values of I_a for input codes $b_3 b_2 b_1 b_0$ = 0000, 0001, 1000, 1010, and 1111, if I_{ref} = 1 mA.

■ **Solution**

The analog current of a 4-bit converter is:

$$I_a = \frac{b_3 2^3 + b_2 2^2 + b_1 2^1 + b_0 2^0}{2^4} \cdot I_{ref}$$

$$= \frac{8b_3 + 4b_2 + 2b_1 + b_0}{16} (1 \text{ mA})$$

$$b_3 b_2 b_1 b_0 = 0000, \; I_a = \frac{(0 + 0 + 0 + 0)(1 \text{ mA})}{16} = 0$$

$$b_3 b_2 b_1 b_0 = 0001, \; I_a = \frac{(0 + 0 + 0 + 1)(1 \text{ mA})}{16} = \frac{1 \text{ mA}}{16} = 62.5 \text{ μA}$$

$$b_3b_2b_1b_0 = 1000, \quad I_a = \frac{(8+0+0+0)(1\text{ mA})}{16} = \frac{8}{16}(1\text{ mA}) = 0.5\text{ mA}$$

$$b_3b_2b_1b_0 = 1010, \quad I_a = \frac{(8+0+2+0)(1\text{ mA})}{16} = \frac{10}{16}(1\text{ mA}) = 0.625\text{ mA}$$

$$b_3b_2b_1b_0 = 1111, \quad I_a = \frac{(8+4+2+1)(1\text{ mA})}{16} = \frac{15}{16}(1\text{ mA}) = 0.9375\text{ mA}$$

Example 12.8 suggests an easy way to calculate D/A analog current. I_a is a fraction of the reference current I_{ref}. The denominator of the fraction is 2^n for an n-bit converter. The numerator is the decimal equivalent of the binary input. For example, for input $b_3b_2b_1b_0 = 0111$, $I_a = (7/16)(I_{ref})$.

Note that when $b_3b_2b_1b_0 = 1111$, the analog current is not the full value of I_{ref}, but 15/16 of it. This is one least significant bit less than full scale.

This is true for any D/A converter, regardless of the number of bits. The maximum analog current for a 5-bit converter is 31/32 of full scale. In an 8-bit converter, I_a cannot exceed 255/256 of full scale. This is because the analog value 0 has its own code. An n-bit converter has 2^n input codes, ranging from 0 to $2^n - 1$.

As we saw in the previous section of this chapter, the difference between the full scale *(FS)* of a digital-to-analog converter and its maximum output is the resolution of the converter. Since the resolution is the smallest change in output, equivalent to a change in the least significant bit, we can define the maximum output as $FS - 1$ LSB. (As an example, in the case of an 8-bit converter $FS - 1\text{ LSB} = (255/256) \cdot I_{ref}$.)

■ SECTION 12.2A REVIEW PROBLEM

12.3 Calculate the range of analog voltage of a 4-bit D/A converter having values of $I_{ref} = 1$ mA and $R_F = 10$ kΩ. Repeat the calculation for an 8-bit D/A converter.

Weighted Resistor D/A Converter

Figure 12.11 shows the circuit of a 4-bit weighted resistor D/A converter. The heart of this circuit is a parallel network of binary-weighted resistors. The MSB has a resistor value of R. Successive branches have resistor values that double with each bit: $2R$, $4R$, and $8R$. The branch currents decrease by halves with each descending bit value.

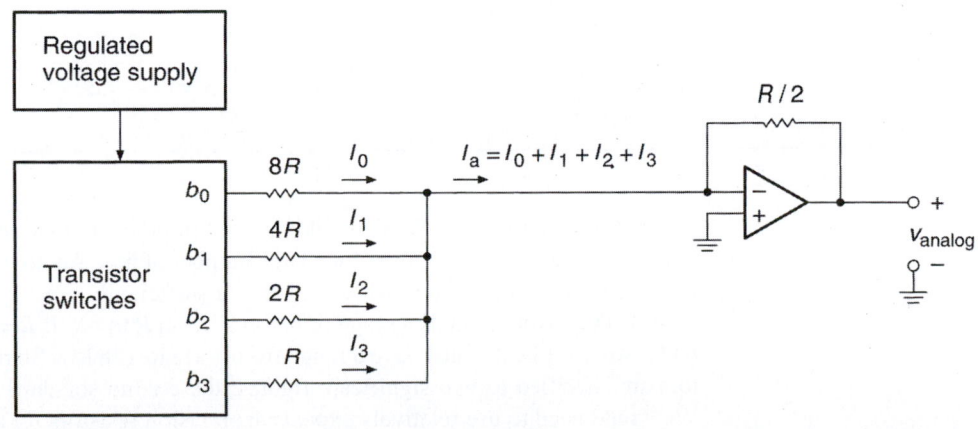

FIGURE 12.11 Weighted Resistor D-to-A Converter

The bit inputs, b_3, b_2, b_1, and b_0, are either 0 V or V_{ref}. When the corresponding bits are HIGH, the branch currents are:

$$I_3 = V_{ref}/R$$
$$I_2 = V_{ref}/2R$$
$$I_1 = V_{ref}/4R$$
$$I_0 = V_{ref}/8R$$

The sum of branch currents gives us the analog current I_a.

$$I_a = \frac{b_3 \ V_{ref}}{R} + \frac{b_2 \ V_{ref}}{2R} + \frac{b_1 \ V_{ref}}{4R} + \frac{b_0 \ V_{ref}}{8R}$$

$$= \left[\frac{b_3}{1} + \frac{b_2}{2} + \frac{b_1}{4} + \frac{b_0}{8} \right] \frac{V_{ref}}{R}$$

We can calculate the analog voltage by Ohm's law:

$$V_a = -I_a R_F = -I_a (R/2)$$

$$= -\left[\frac{b_3}{1} + \frac{b_2}{2} + \frac{b_1}{4} + \frac{b_0}{8} \right] \frac{V_{ref}}{R} \frac{R}{2}$$

$$= -\left[\frac{b_3}{1} + \frac{b_2}{2} + \frac{b_1}{4} + \frac{b_0}{8} \right] \frac{V_{ref}}{2}$$

$$= -\left[\frac{b_3}{2} + \frac{b_2}{4} + \frac{b_1}{8} + \frac{b_0}{16} \right] V_{ref} = -\frac{code}{16} \cdot V_{ref}$$

The choice of $R_F = R/2$ makes the analog output a binary fraction of V_{ref}.

EXAMPLE 12.9

Calculate the analog voltage of a weighted resistor D/A converter when the binary inputs have the following values: $b_3b_2b_1b_0$ = 0000, 1000, 1111. V_{ref} = 5 V.

■ **Solution**

$$b_3b_2b_1b_0 = 0000$$

$$V_a = -\left[\frac{0}{2} + \frac{0}{4} + \frac{0}{8} + \frac{0}{16} \right] V_{ref} = -\frac{0}{16} \cdot V_{ref} = 0$$

$$b_3b_2b_1b_0 = 1000$$

$$V_a = -\left[\frac{1}{2} + \frac{0}{4} + \frac{0}{8} + \frac{0}{16} \right] V_{ref} = -\frac{8}{16} \cdot V_{ref} = -\frac{1}{2}(5 \text{ V}) = -2.5 \text{ V}$$

$$b_3b_2b_1b_0 = 1111$$

$$V_a = -\left[\frac{1}{2} + \frac{1}{4} + \frac{1}{8} + \frac{1}{16} \right] V_{ref} = -\frac{15}{16}(5 \text{ V}) = -4.69 \text{ V}$$

The weighted resistor DAC is seldom used in practice. One reason is the wide range of resistor values required for a large number of bits. Another reason is the difficulty in obtaining resistors whose values are sufficiently precise.

A 4-bit converter needs a range of resistors from R to $8R$. If R = 1 kΩ, then $8R$ = 8 kΩ. An 8-bit DAC must have a range from 1 kΩ to 128 kΩ. Standard value resistors are specified to two significant figures; there is no standard 128-kΩ resistor. We would need to use relatively expensive precision resistors for any value having more than two significant figures.

Another DAC circuit, the R-2R ladder, is more commonly used. It requires only two values of resistance for any number of bits.

■ SECTION 12.2B REVIEW PROBLEM

12.4 The resistor for the MSB of a 12-bit weighted resistor D/A converter is 1 kΩ. What is the resistor value for the LSB?

R-2R Ladder D/A Converter

Figure 12.12 shows the circuit of an R-2R ladder D/A converter. Like the weighted resistor DAC, this circuit produces an analog current that is the sum of binary-weighted currents. An operational amplifier converts the current to a proportional voltage.

The circuit requires an operational amplifier with a high slew rate such as an MC34071 or TL071. Slew rate is the rate at which the output changes after a step change at the input. If a standard op amp (e.g., 741C) is used, the circuit will not accurately reproduce changes introduced by large changes in the digital input.

The method of generating the analog current for an R-2R ladder DAC is a little less obvious than for the weighted resistor DAC. As the name implies, the resistor network is a ladder that has two values of resistance, one of which is twice the other. This circuit is expandable to any number of bits simply by adding one resistor of each value for each bit.

The analog output is a function of the digital input and the value of the op amp feedback resistor. If logic HIGH = V_{ref}, logic LOW = 0 V, and $R_F = R$, the analog output is given by:

$$V_a = -\left[\frac{b_3}{2} + \frac{b_2}{4} + \frac{b_1}{8} + \frac{b_0}{16}\right]V_{ref}$$

One way to analyze this circuit is to replace the R-2R ladder with its Thévenin equivalent circuit and treat the circuit as an inverting amplifier. Figure 12.13 shows the equivalent circuit for the input code $b_3b_2b_1b_0 = 1000$.

Figure 12.14a shows the equivalent circuit of the R-2R ladder when $b_3b_2b_1b_0 = 1000$. All LOW bits are grounded, and the HIGH bit connects to V_{ref}. We can reduce the network to two resistors by using series and parallel combinations.

The two resistors at the far left of the ladder are in parallel: $2R \parallel 2R = R$. This equivalent resistance is in series with another: $R + R = 2R$. The new resistance is in parallel with yet another: $2R \parallel 2R = R$. We continue this process until we get the simplified circuit shown in Figure 12.14b.

FIGURE 12.12 R-2R Ladder DAC

a. $b_3b_2b_1b_0 = 1000$

b. Equivalent circuit

FIGURE 12.13 Equivalent Circuit for $b_3b_2b_1b_0 = 1000$

a. Equivalent circuit

b. Simplified equivalent circuit   **c. Thévenin equivalent**

FIGURE 12.14 R-2R Circuit Analysis for $b_3b_2b_1b_0 = 1000$

Next, we find the Thévenin equivalent of the simplified circuit. To find E_{Th}, calculate the terminal voltage of the circuit, using voltage division.

$$E_{Th} = \frac{2R}{2R + 2R} V_{ref} = V_{ref}/2$$

R_{Th} is the resistance of the circuit, as measured from the terminals, with the voltage source short-circuited. Its value is that of the two resistors in parallel: $R_{Th} = 2R \parallel 2R = R$.

NOTE . . .

The value of the Thévenin resistance of the R-2R ladder will always be R, regardless of the digital input code. This is because we short-circuit any voltage sources when we make this calculation, which grounds the corresponding bit resistors. The other resistors are already grounded by logic LOWs. We reduce the circuit to a single resistor, R, by parallel and series combinations of R and $2R$. Figure 12.15 shows the equivalent circuit.

$$R_{Th} = (((((2R \parallel 2R) + R) \parallel 2R) + R) \parallel 2R) + R) \parallel 2R = R$$

FIGURE 12.15 Equivalent Circuit for Calculating R_{Th}

On the other hand, the value of E_{Th} will be different for each different binary input. It will be the sum of binary fractions of the full-scale output voltage, as previously calculated for the generic DAC.

Similar analysis of the R-2R ladder shows that when $b_3 b_2 b_1 b_0 = 0100$, $V_a = -V_{ref}/4$, when $b_3 b_2 b_1 b_0 = 0010$, $V_a = -V_{ref}/8$, and when $b_3 b_2 b_1 b_0 = 0001$, $V_a = -V_{ref}/16$.

If two or more bits in the R-2R ladder are active, each bit acts as a separate voltage source. Analysis becomes much more complicated if we try to solve the network as we did for one active bit.

There is no need to go through a tedious circuit analysis to find the corresponding analog voltage. We can simplify the process greatly by applying the Superposition theorem. This theorem states that the effect of two or more sources in a network can be determined by calculating the effect of each source separately and adding the results.

The Superposition theorem suggests a generalized equivalent circuit of the R-2R ladder DAC. This is shown in Figure 12.16. A Thévenin equivalent source and

FIGURE 12.16 Equivalent Circuit of R-2R DAC

resistance corresponds to each bit. The source and resistance are switched in and out of the circuit, depending on whether or not the corresponding bit is active.

This model is easily expanded. The source for the most significant bit always has the value $V_{ref}/2$. Each source is half the value of the preceding bit. Thus, for a 5-bit circuit, the source for the least significant bit has a value of $V_{ref}/32$. An 8-bit circuit has an LSB equivalent source of $V_{ref}/256$.

EXAMPLE 12.10

A 4-bit DAC based on an R-2R ladder has a reference voltage of 10 volts. Calculate the analog output voltage, V_a, for the following input codes:

a. 0000
b. 1000
c. 0100
d. 1100

■ **Solution**

a. $V_a = -(0/16)\ V_{ref} = 0\ \text{V}$
b. $V_a = -(8/16)\ V_{ref} = -(1/2)\ V_{ref} = -5\ \text{V}$
c. $V_a = -(4/16)\ V_{ref} = -(1/4)\ V_{ref} = -2.5\ \text{V}$
d. $V_a = -(12/16)\ V_{ref} = -(3/4)\ V_{ref} = -7.5\ \text{V}$

EXAMPLE 12.11

Calculate the output voltage of an 8-bit DAC based on an R-2R ladder for the following input codes. What general conclusion can be drawn about each code when compared to the solutions in Example 12.10?

a. 00000000
b. 10000000
c. 01000000
d. 11000000

■ **Solution**

a. $V_a = -(0/256)\ V_{ref} = 0\ \text{V}$
b. $V_a = -(128/256)\ V_{ref} = -(1/2)\ V_{ref} = -5\ \text{V}$
c. $V_a = -(64/256)\ V_{ref} = -(1/4)\ V_{ref} = -2.5\ \text{V}$
d. $V_a = -(192/256)\ V_{ref} = -(3/4)\ V_{ref} = -7.5\ \text{V}$

In general, a DAC input code consisting of 1 followed by all 0s generates an output value of $\frac{1}{2}$ full scale. A code of 01 followed by all 0s yields an output of $\frac{1}{4}$ full scale. An output of 11 followed by all 0s generates an output of $\frac{3}{4}$ full scale.

■ **SECTION 12.2C REVIEW PROBLEM**

12.5   Calculate V_a for an 8-bit R-2R ladder DAC when the input code is 10100001. Assume that V_{ref} is 10 V.

MC1408 Integrated Circuit D/A Converter

■ **KEY TERM**

Multiplying DAC   A DAC whose output changes linearly with a change in DAC reference voltage.

FIGURE 12.17 MC1408 DAC

A common and inexpensive DAC is the MC1408 8-bit multiplying digital-to-analog converter. This device also goes by the designation DAC0808. A logic symbol for this DAC is shown in Figure 12.17.

The DAC output current, I_o, flows into pin 4. I_o is a binary fraction of the current flowing into pin 14, as specified by the states of the digital inputs. Other inputs select the range of output voltage and allow for phase compensation.

Figure 12.18 shows the MC1408 in a simple D/A configuration. R_{14} and R_{15} are approximately equal. Pin 14 is approximately at ground potential. This implies:

1. That the DAC reference current can be calculated using only V_{ref} (+) and R_{14} ($I_{\text{ref}} = V_{\text{ref}}$ (+)/R_{14})
2. That R_{15} is not strictly necessary in the circuit. (It is used primarily to stabilize the circuit against temperature drift.)

The reference voltage *must* be set up so that current flows into pin 14 and out of pin 15. Thus, V_{ref} (+) must be positive with respect to V_{ref} (−). (Although V_{ref} (−) is usually grounded, it is permissible to ground pin 14 if pin 15 is at a negative voltage.)

I_o is given by:

$$I_o = \left[\frac{b_7}{2} + \frac{b_6}{4} + \frac{b_5}{8} + \frac{b_4}{16} + \frac{b_3}{32} + \frac{b_2}{64} + \frac{b_1}{128} + \frac{b_0}{256} \right] \frac{V_{\text{ref}}(+)}{R_{14}} = \frac{\text{code}}{256} \cdot \frac{V_{\text{ref}}(+)}{R_{14}}$$

Since the output is proportional to V_{ref} (+), we refer to the MC1408 as a **multiplying DAC.**

I_o should not exceed 2 mA. We calculate the output voltage by Ohm's law: $V_o = -I_o R_L$. The output voltage is negative because current flows from ground into pin 4.

FIGURE 12.18 MC1408 Configured for Unbuffered Analog Output

The open pin on the Range input allows the output voltage dropped across R_L to range from +0.4 V to –5.0 V without damaging the output circuit of the DAC. If the Range input is grounded, the output can range from +0.4 to –0.55 V. The lower voltage range allows the output to switch about four times faster than it can in the higher range.

EXAMPLE 12.12

The DAC circuit in Figure 12.18 has the following component values: $R_{14} = R_{15}$ = 5.6 kΩ; R_L = 3.3 kΩ. V_{ref} (+) is +8 V, and V_{ref} (–) is grounded.

Calculate the value of V_o for each of the following input codes: $b_7b_6b_5b_4b_3b_2b_1b_0 =$ 00000000, 00000001, 10000000, 10100000, 11111111.

What is the resolution of this DAC?

■ **Solution**

First, calculate the value of I_{ref}.

$$I_{ref} = V_{ref} (+)/R_{14}$$
$$= +8 \text{ V}/5.6 \text{ k}\Omega = 1.43 \text{ mA}$$

Calculate the output current by using the binary fraction for each code. Multiply $-I_o$ by R_L to get the output voltage.

$b_7b_6b_5b_4b_3b_2b_1b_0 = 00000000$
$I_o = 0, V_o = 0$

$b_7b_6b_5b_4b_3b_2b_1b_0 = 00000001$
$I_o = (1/256) (1.43 \text{ mA}) = 5.58 \text{ μA}$
$V_o = -(5.58 \text{ μA})(3.3 \text{ k}\Omega) = -18.4 \text{ mV}$

$b_7b_6b_5b_4b_3b_2b_1b_0 = 10000000$
$I_o = (1/2) (1.43 \text{ mA}) = 714 \text{ μA}$
$V_o = -(714 \text{ μA})(3.3 \text{ k}\Omega) = -2.36 \text{ V}$

$b_7b_6b_5b_4b_3b_2b_1b_0 = 10100000$
$I_o = -(1/2 + 1/8)(1.43 \text{ mA}) = (5/8)(1.43 \text{ mA}) = 893 \text{ μA}$
$V_o = -(893 \text{ μA})(3.3 \text{ k}\Omega) = -2.95 \text{ V}$

$b_7b_6b_5b_4b_3b_2b_1b_0 = 11111111$
$I_o = (255/256) (1.43 \text{ mA}) = 1.42 \text{ mA}$
$V_o = -(1.42 \text{ mA})(3.3 \text{ k}\Omega) = -4.70 \text{ V}$

Resolution is the same as the output resulting from the LSB: 18.4 mV/step

■ **SECTION 12.2D REVIEW PROBLEM**

12.6   The output voltage range of an MC1408 DAC can be limited by grounding the Range pin. Why would we choose to do this?

Op Amp Buffering of MC1408

The MC1408 DAC will not drive much of a load on its own, particularly when the Range input is grounded. We can use an operational amplifier to increase the output voltage and current. This allows us to select the lower voltage range for faster switching while retaining the ability to drive a reasonable load. The output voltage is limited only by the op amp supply voltages. We can use an MC34071 or TL071 or other high slew rate op amp for fast switching.

Figure 12.19 shows such a circuit. The 0.1-μF capacitor decouples the +5-V supply. (The manufacturer actually recommends that the +5-V logic supply not be used as a reference voltage. It doesn't matter for a demonstration circuit, but may introduce noise that is unacceptable in a commercial design.) The 75-pF capacitor is for phase compensation.

FIGURE 12.19   DAC with Op Amp Buffering

V_a is positive because the voltage drop across R_F is positive with respect to the virtual ground at the op amp (–) input. This feedback voltage is in parallel with (i.e., the same as) the output voltage, since both are measured from output to ground.

We can develop the formula for the analog voltage, V_a, in three stages:

1. Calculate the reference current:

$$I_{\text{ref}} = V_{\text{ref}}(+)/R_{14}$$

2. Determine the binary-weighted fraction of reference current to get DAC output current:

$$I_o = \left[\frac{b_7}{2} + \frac{b_6}{4} + \frac{b_5}{8} + \frac{b_4}{16} + \frac{b_3}{32} + \frac{b_2}{64} + \frac{b_1}{128} + \frac{b_0}{256} \right] I_{\text{ref}}$$

$$= \left[\frac{b_7}{2} + \frac{b_6}{4} + \frac{b_5}{8} + \frac{b_4}{16} + \frac{b_3}{32} + \frac{b_2}{64} + \frac{b_1}{128} + \frac{b_0}{256} \right] \frac{V_{\text{ref}}}{R_{14}}$$

$$= \left(\frac{\text{digital code}}{256} \right) \left(\frac{V_{\text{ref}}}{R_{14}} \right)$$

3. Use Ohm's law to calculate the op amp output voltage:

$$V_a = I_o R_F = \left(\frac{\text{digital code}}{256} \right) \left(\frac{R_F}{R_{14}} \right) V_{\text{ref}}$$

The resistor values in these formulae are the total resistances for the corresponding part of the circuit. That is, $R_{14} = R_{14A} + R_{14B}$ and $R_F = R_{FA} + R_{FB}$. These both consist of a fixed and a variable resistor, which has two advantages: (a) The reference current and output voltage can be independently adjusted within a specified range by the variable resistors. (b) The resistances defining the reference and feedback currents cannot go below a specified minimum value, determined by the fixed resistance, ensuring that excessive current does not flow into the reference input or the DAC output terminal.

V_a can, in theory, be any positive value less than the op amp positive supply (+ 12 V in this case). Any attempt to exceed this voltage makes the op amp saturate. The actual maximum value, if not the same as the op amp's saturation voltage, depends on the values of R_F and R_{14}.

EXAMPLE 12.13

Describe a step-by-step procedure that calibrates the DAC circuit in Figure 12.19 so that it has a reference current of 1 mA and a full-scale analog output voltage of 10 V, using only a series of measurements of the analog output voltage. When the procedure is complete, what are the resistance values in the circuit? What is the range of the DAC?

■ Solution

Since the maximum output of the DAC is 1 LSB less than full scale, we must indirectly measure the full scale value. We can do so by setting the digital input code to 10000000, which exactly represents the half-scale value of output current, and making appropriate adjustments.

Set the variable feedback resistor to zero so that the output voltage is due only to the fixed feedback resistor and the feedback current. Measure the output voltage of the circuit and adjust R_{14B} so that $V_a = 2.35$ volts. Ohm's law tells us that this sets the feedback current to $I_F = 2.35$ V/4.7 kΩ = 0.5 mA. Since the digital code is set for half scale, $I_{ref} = 2\ I_F = 1$ mA. (This calculation assumes that R_{FA} is exactly 4.7 kΩ, which is probably not the case. The calculation is accurate to within the tolerance of R_{FA}, probably ±5%.)

Adjust R_{FB} so that the half-scale output voltage is 5.00 V.

After adjustment, $R_{14} = 2.7$ kΩ + 2.3 kΩ = 5 kΩ and $R_F = 4.7$ kΩ + 4.3 kΩ = 10 kΩ. In both cases the variable resistors were selected so that their final values are about halfway through their respective ranges.

The range of the DAC is 0 V to 9.961 V.

$$(FS - 1\ \text{LSB} = 10\ \text{V} - (10\ \text{V}/256) = 9.961\ \text{V})$$

EXAMPLE 12.14

Figure 12.20 shows the circuit of an analog ramp (sawtooth) generator built from an MC1408 DAC, an op amp, and an 8-bit synchronous counter. (A ramp generator has numerous analog applications, such as sweep generation in an oscilloscope and frequency sweep in a spectrum analyzer.)

Briefly explain the operation of the circuit and sketch the output waveform. Calculate the step size between analog outputs resulting from adjacent codes. Assume that the DAC is set for +6-V output when the input code is 10000000. Calculate the output sawtooth frequency when the clock is running at 1 MHz.

■ Solution

The 8-bit counter cycles from 00000000 to 11111111 and repeats continuously. This is a total of 256 states.

The DAC output is 0 V for an input code of 00000000 and (12 V – 1 LSB) for a code of 11111111. We know this because a code of 10000000 always gives an output voltage of half the full-scale value (6 V = 12 V/2), and the maximum code gives an output that is one step less than the full-scale voltage. The step size is 12 V/256 steps = 46.9 mV/step. The DAC output advances in linear steps from 0 to (12 V – 1 LSB) in 256 clock cycles.

FIGURE 12.20 Example 12.14: DAC Ramp Generator

Figure 12.21 shows the analog output plotted against the number of input clock cycles. The ramp looks smooth at the scale shown. A section enlarged 32 times shows the analog steps resulting from eight clock pulses.

One complete cycle of the sawtooth waveform requires 256 clock pulses. Thus, if $f_{CLK} = 1$ MHz, $f_o = 1$ MHz/256 = 3.9 kHz.

(Note that if we do not use a high slew rate op amp, the sawtooth waveform will not have vertical sides.)

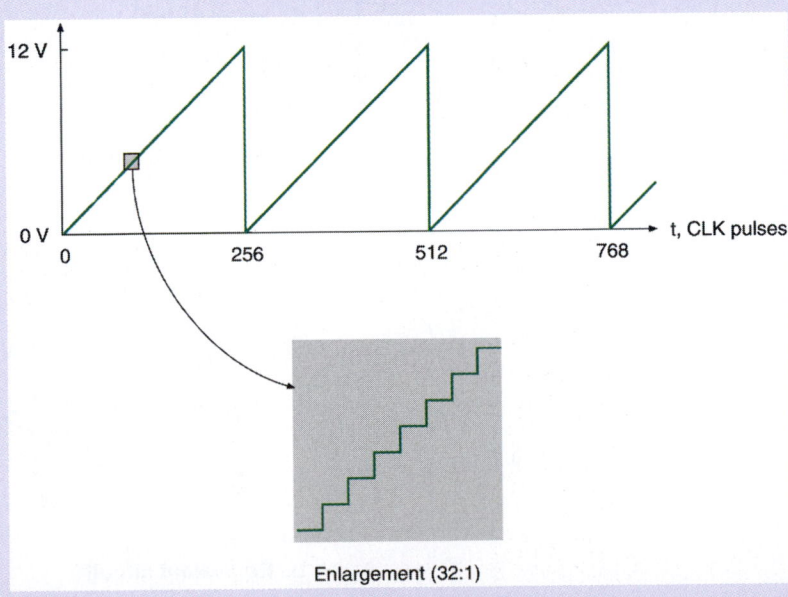

Enlargement (32:1)

FIGURE 12.21 Example 12.14: Sawtooth Waveform Output of Circuit in Figure 12.19

Bipolar Operation of MC1408

Many analog signals are bipolar, that is, they have both positive and negative values. We can configure the MC1408 to produce a bipolar output voltage with offset binary coding. Such a circuit is shown in Figure 12.22.

We can model the bipolar DAC as shown in Figure 12.22b. The amplitude of the constant-current sink, I_o, is set by V_{ref} (+), R_{14}, and the binary value of the digital inputs. I_s is determined by Ohm's law: $I_s = V_{ref} (+)/R_4$.

a. Bipolar DAC circuit

b. Equivalent circuit

FIGURE 12.22 MC1408 as a Bipolar D/A Converter

The output voltage is set by the value of I_F:

$$V_a = I_F R_F = I_F (R_{FA} + R_{FB})$$

By Kirchhoff's current law:

$$I_s + I_F - I_o = 0$$

or

$$I_F = I_o - I_s$$

Thus, output voltage is given by:

$$V_a = (I_o - I_s)R_F = I_o R_F - I_s R_F$$

$$= \left[\frac{b_7}{2} + \frac{b_6}{4} + \frac{b_5}{8} + \frac{b_4}{16} + \frac{b_3}{32} + \frac{b_2}{64} + \frac{b_1}{128} + \frac{b_0}{256} \right] \frac{R_F}{R_{14}} V_{\text{ref}} - \frac{R_F}{R_4} V_{\text{ref}}$$

$$= \left(\frac{\text{digital code}}{256} \right) \left(\frac{R_F}{R_{14}} \right) V_{\text{ref}} - \frac{R_F}{R_4} V_{\text{ref}}$$

How do we understand the circuit operation from this mathematical analysis?

The current sink, I_o, is a variable element. The voltage source, V_{ref} (+), remains constant. To satisfy Kirchhoff's current law, the feedback current, I_F, must vary to the same degree as I_o. Depending on the value of I_o with respect to I_s, I_F can be positive or negative.

We can get some intuitive understanding of the circuit operation by examining several cases of the equation V_a.

Case 1: $I_o = 0$. This corresponds to the digital input $b_7b_6b_5b_4b_3b_2b_1b_0 = 00000000$. The output voltage is:

$$V_a = (I_o - I_s)R_F = -I_s R_F = -\frac{R_F}{R_4} V_{\text{ref}}$$

This is the maximum negative output voltage.

Case 2: $0 < I_o < I_s$. The term $(I_o - I_s)$ is negative, so output voltage is also a negative value.

Case 3: $I_o = I_s$. The output is given by:

$$V_a = (I_o - I_s)R_F = 0$$

The digital code for this case could be any value, depending on the setting of R_{14}. To set the zero-crossing to half-scale, set the digital input to 10000000 and adjust R_{14} for 0 V.

Case 4: $I_o > I_s$. Since the term $(I_o - I_s)$ is positive, output voltage is positive. The largest value of I_o (and thus the maximum positive output voltage) corresponds to the input code $b_7b_6b_5b_4b_3b_2b_1b_0 = 11111111$.

The magnitude of the maximum positive output voltage of this particular circuit is 1 LSB less than the magnitude of the maximum negative voltage. Specifically, $V_a = (127/256)(R_F/R_{14})(V_{\text{ref}})$ if $R_4 = 2R_{14}$.

To summarize:

Input Code	Output Voltage
00000000	Maximum negative*
10000000	0 V**
11111111	Maximum positive

*As adjusted by R_{FB}
**As adjusted by R_{14B}

Negative Range:

00000000 to 01111111   (128 codes)

Positive Range:

10000001 to 11111111   (127 codes)

Zero:

00000000   (1 code)

$\overline{}$256 codes

EXAMPLE 12.15

Calculate the values to which R_{14} and R_F must be set to make the output of the bipolar DAC in Figure 12.22 range from -12 V to $(+12$ V $- 1$ LSB). Describe the procedure you would use to set the circuit output as specified.

Confirm that the calculated resistor settings generate the correct values of maximum and minimum output.

■ **Solution**

Set R_{14} so that the DAC circuit has an output of 0 V when input code is $b_7b_6b_5b_4b_3b_2b_1b_0 = 10000000$. We can calculate the value of R_{14} as follows:

$$\frac{R_F}{2R_{14}}V_{\text{ref}} - \frac{R_F}{R_4}V_{\text{ref}} = 0$$

The first term is set by the value of the input code. Solving for R_{14}, we get:

$$\left[\frac{1}{2R_{14}} - \frac{1}{R_4}\right]R_F V_{\text{ref}} = 0$$

$$\frac{1}{2R_{14}} - \frac{1}{R_4} = 0$$

$$\frac{1}{2R_{14}} = \frac{1}{R_4}$$

$$2R_{14} = R_4$$

$$R_{14} = R_4/2 = 10\ \text{k}\Omega/2 = 5\ \text{k}\Omega$$

To set the maximum negative value, set the input code to 00000000 and adjust R_{FB} for -12 V. $R_{FB} = R_F - R_{FA}$. Solve the following equation for R_F:

$$-\frac{R_F}{R_4}V_{\text{ref}} = -12\ \text{V}$$

$$-\frac{R_F}{10\ \text{k}\Omega}(5\ \text{V}) = -12\ \text{V}$$

$$R_F = (12\ \text{V})(10\ \text{k}\Omega)/5\ \text{V} - 24\ \text{k}\Omega$$

$$R_{FB} = 24\ \text{k}\Omega - 18\ \text{k}\Omega = 6\ \text{k}\Omega$$

Settings

$$R_{14} = R_4/2 = 5\ \text{k}\Omega \text{ for zero-crossing at half-scale.}$$
$$R_F = 24\ \text{k}\Omega \text{ for output of } \pm 12\ \text{V.}$$

Check Output Range

For $b_7b_6b_5b_4b_3b_2b_1b_0 = 00000000$:

$$V_a = \left[\frac{0}{R_{14}} - \frac{1}{R_4}\right]R_F V_{\text{ref}} = -\frac{(24\ \text{k}\Omega)(5\ \text{V})}{10\ \text{k}\Omega} = -12\ \text{V}$$

For $b_7b_6b_5b_4b_3b_2b_1b_0 = 11111111$:

$$V_a = \left[\frac{255}{256\,R_{14}} - \frac{1}{R_4} \right] R_F V_{ref}$$

$$= \left[\frac{255}{(256)(5\text{ k}\Omega)} - \frac{1}{10\text{ k}\Omega} \right] (24\text{ k}\Omega)(5\text{ V}) = 11.906\text{ V}$$

(Note: 12 V – 1 LSB = 12 V – (24 V/256) = 12 V – 94 mV = 11.906 V.)

■ SECTION 12.2E REVIEW PROBLEM

12.7 Why is the actual maximum value of an 8-bit DAC less than its reference (i.e., its apparent maximum) voltage?

DAC Performance Specifications

A number of factors affect the performance of a digital-to-analog converter. The major factors are briefly described in the following paragraphs.

Monotonicity. The output of a DAC is monotonic if the magnitude of the output voltage increases every time the input code increases. Figure 12.23 shows the output of a DAC that increases monotonically and the output of a DAC that does not.

We show the output response of a DAC as a series of data points joined by a straight-line approximation. One input code produces one voltage, so there is no value that corresponds to anything in between codes, but the straight-line approximation allows us to see a trend over the whole range of input codes.

a. Ideal DAC response (monotonically increasing)

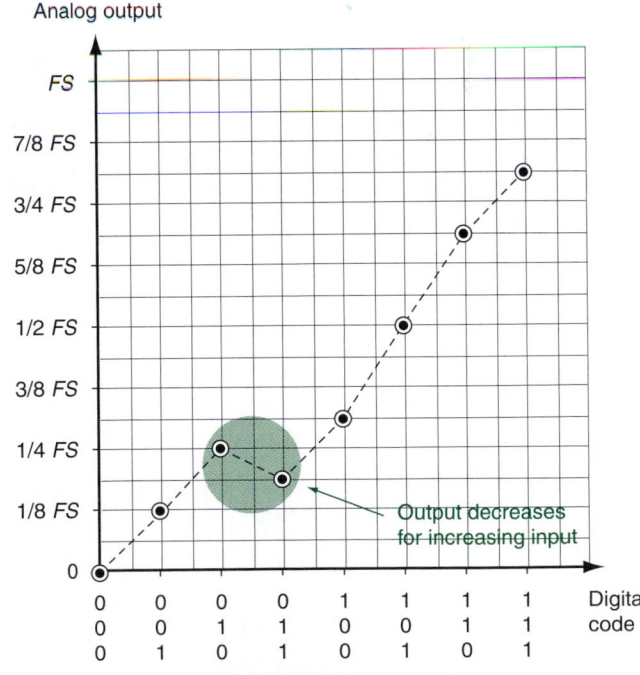

b. Nonmonotonically increasing

FIGURE 12.23   DAC Monotonicity

Absolute accuracy. This is a measure of DAC output voltage with respect to its expected value.

Relative accuracy. Relative accuracy is a more frequently used measurement than absolute accuracy. It measures the deviation of the actual from the ideal output voltage as a fraction of the full-scale voltage. The MC1408 DAC has a relative accuracy of $\pm\frac{1}{2}$ LSB = $\pm 0.195\%$ of full scale.

Settling time. The time required for the outputs to switch and settle to within $\pm\frac{1}{2}$ LSB when the input code switches from all 0s to all 1s. The MC1408 has a settling time of 300 ns for 8-bit accuracy, limiting its output switching frequency to 1/300 ns = 3.33 MHz. Depending on the value of R_4, the output resistor, the settling time of the MC1408 may increase to as much as 1.2 μs when the Range input is open.

Gain error. Gain error primarily affects the high end of the output voltage range. If the gain of a DAC is too high, the output saturates before reaching the maximum output code. Figure 12.24 shows the effect of gain error in a 3-bit DAC. In the high gain response, the last two input codes (110 and 111) produce the same output voltage.

Linearity error. Linearity error is present when the analog output does not follow a straight-line increase with increasing digital input codes. Figure 12.25 shows this error. A linearity error of more than $\pm\frac{1}{2}$ LSB can result in a nonmonotonic output. For example, in Figure 12.23b, the transition from 010 to 011 should result in an output change of +1 LSB. Instead, it results in a change of $-\frac{1}{2}$ LSB. This is an error of $-1\frac{1}{2}$ LSB, resulting in a nonmonotonic output.

In Figure 12.25, the code for 011 has a linearity error of $+\frac{1}{2}$ LSB and the adjacent code (100) has a linearity error of $-\frac{1}{2}$ LSB, yielding a flat output for the two codes. This makes it impossible to distinguish the value of input code for that analog output value.

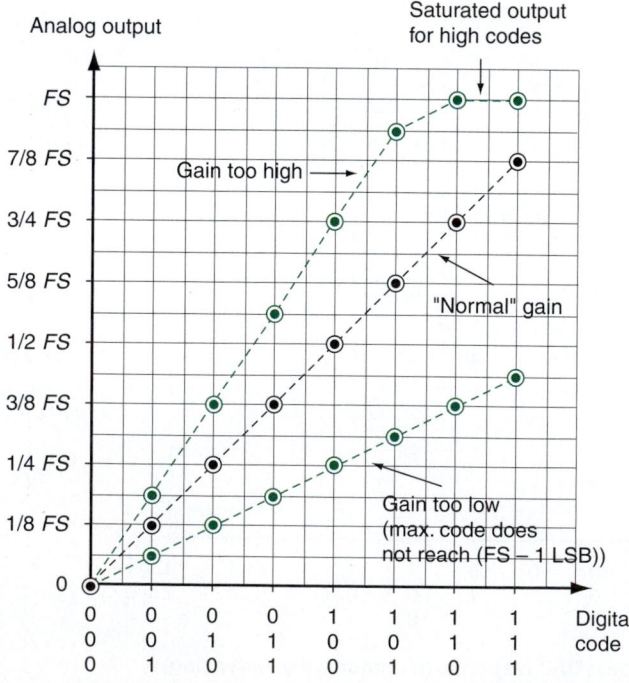

FIGURE 12.24 DAC Gain Errors

FIGURE 12.25 DAC Linearity Error

Differential nonlinearity. This specification measures the difference between actual and expected step size of a DAC when the input code is changed by 1 LSB. An actual step that is smaller than the expected step can result in a nonmonotonic output.

Offset error. This error occurs when the analog output of a positive-value DAC is not 0 V when the input code is all 0s. Figure 12.26 shows the effect of offset error.

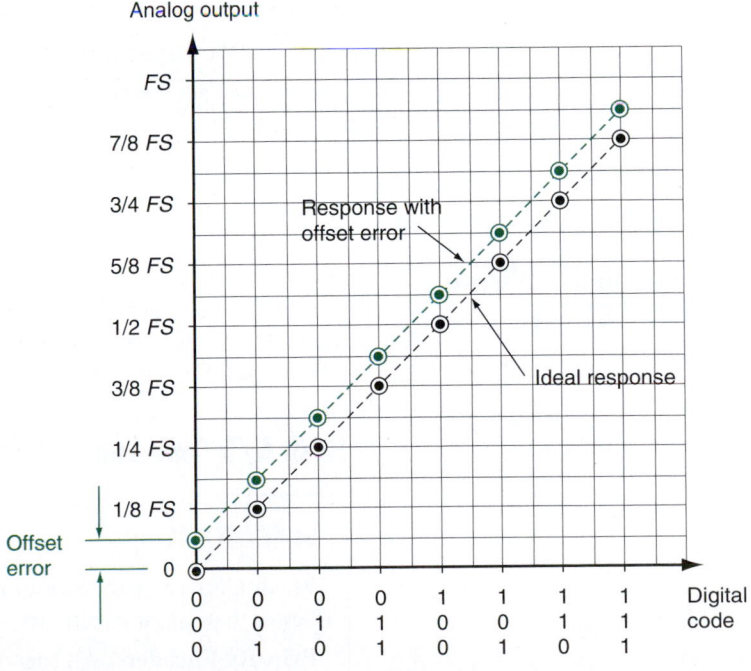

FIGURE 12.26 DAC Offset Error

EXAMPLE 12.16

An 8-bit DAC has an output range of 0 to (+8 volts – 1 LSB). The hexadecimal value of the input is symbolized by x.

a. What is the value of 1 LSB?
b. Assuming an ideal DAC, what would the output be for a binary input $x =$ C0H?
c. If the DAC has an input of $x =$ 00H and the output voltage is 0.008 V, calculate the offset error *(OE)* of the DAC in LSB and as a percentage of the full scale *(FS)*. Assume no other errors.
d. If the DAC has an input of $x =$ FFH and the output voltage is 7.98 V, calculate the gain error *(GE)* of the DAC in LSB and as a percentage of the full scale *(FS)*. Assume no other errors.
e. If the DAC output is 4 V for an input $x =$ 80H and the output is 0.015 V for an input of $x =$ 00H, calculate the linearity error *(LE)* and offset error *(OE)* of the DAC in LSB and as a percentage of the full scale *(FS)*.

■ **Solution**

a. 1 LSB = $FS/2^n$ = 8 V/2^8 = 8 V/256 = 31.25 mV
b. C0H = 192_{10}
   V_a = (code/256) 8 V = (192/256) 8 V = 6 V
   alternatively: C0H = 11000000, which corresponds to $\frac{3}{4}$ FS = 6 V
c. When $x =$ 00H, V_a should be 0 V. Therefore, OE = 8 mV.
   OE[LSB] = 8 mV/31.25 mV = 0.256 LSB
   OE[%FS] = (8 mV/8 V) × 100% FS = 0.1% FS
d. When $x =$ FFH, V_a should be (255/256) 8 V = 7.96875 V. GE = 7.98 V – 7.96875 V = 11.25 mV.
   GE[LSB] = 11.25 mV/31.25 mV = 0.36 LSB
   GE[%FS] = (11.25 mV/8 V) × 100% FS = 0.14% FS
e. Without accounting for other possible errors, the output value for an input of 80H appears to be correct. However, we find an offset error of 0.015 V that must be subtracted out of all measured values in the DAC output.

   Adjusted value at 80H = 4 V – 0.015 V = 3.985 V. This error is exactly balanced by the offset error, so both have the same value.

   LE[LSB] = OE[LSB] = 15 mV/31.25 mV = 0.48 LSB
   LE[%FS] = OE[%FS] = (15 mV/8 V) × 100% FS = 0.188% FS

12.3   ANALOG-TO-DIGITAL CONVERSION

We saw in an earlier section of this chapter that all digital-to-analog converters can be described by a generic form. This is not true of analog-to-digital converters. There are many circuits for converting analog signals to digital codes, each with its own advantages. We will look at several of the most popular.

Flash A/D Converter

■ **KEY TERMS**

Flash Converter (or **Simultaneous Converter**)   An analog-to-digital converter that uses comparators and a priority encoder to produce a digital code.

Priority Encoder   An encoder that will produce a binary output corresponding to the subscript of the highest-priority active input. This is usually defined as the input with the largest subscript.

FIGURE 12.27   Flash Converter (ADC)

Figure 12.27 shows the circuit for a 3-bit **flash analog-to-digital converter.** The circuit consists of a resistive voltage divider, seven analog comparators, a **priority encoder,** and an output latch array.

The voltage divider has a total resistance of $8R$. The resistors are selected to produce seven equally spaced reference voltages ($V_{ref}/16$, $3V_{ref}/16$, $5V_{ref}/16$, . . . $15V_{ref}/16$; each is separated by $V_{ref}/8$). Each reference voltage is fed to the inverting input of a comparator.

A comparator output goes HIGH if the voltage at its noninverting (+) input is higher than the voltage at its inverting (–) input. If the (–) input voltage is greater than the (+) input voltage, the comparator output is LOW.

The analog voltage, V_a, is applied to the noninverting inputs of all comparators simultaneously. Thus, if the analog voltage exceeds the reference voltage of a particular comparator, that comparator switches its output to the HIGH state.

For most analog input values, more than one comparator will have a HIGH output. For example, the reference voltage of comparator 3 is ($5V_{ref}/16$). Comparator 4 has a reference voltage of ($7V_{ref}/16$). If the analog voltage is in the range ($5V_{ref}/16$) $\leq V_a < (7V_{ref}/16)$, comparators 3, 2, and 1 all have HIGH outputs and comparators 4, 5, 6, and 7 all have LOW outputs.

The priority encoder recognizes that input D_3 is the highest-priority active input and produces the digital code 011 at its outputs. The output latches store this value when the *CLK* input is pulsed.

We can regularly sample an analog signal by applying a pulse waveform to the *CLK* input of the latch circuit. The sampling frequency is the same as the clock frequency.

The D_0 input of the priority encoder is grounded, rather than connected to a comparator output. No comparator is needed for this input; if $V_a < (V_{ref}/16)$, all comparator outputs are LOW and the resulting digital code is 000.

Figure 12.28 shows the transfer characteristic of the flash ADC with a reference voltage of 8 V. The digital steps are centered on the analog voltages that are whole-number fractions ($\frac{1}{8}$, $\frac{1}{4}$, $\frac{3}{8}$, ... $\frac{7}{8}$) of the reference voltage. The transitions are

FIGURE 12.28   Transfer Characteristic of Flash ADC

midway these points. This is why the resistor for the least significant bit is $R/2$, rather than R.

The general form of this circuit has $2^n - 1$ comparators for an n-bit output. For example, an 8-bit flash converter has $2^8 - 1 = 255$ comparators. For any large number of bits, the circuit becomes overly complex.

The main advantage of this circuit is its speed. Since the analog input is compared to the threshold values of all possible input codes at one time, conversion occurs in one clock cycle, limited only by the propagation delays of the components.

■ **SECTION 12.3A REVIEW PROBLEM**

12.8   The flash ADC circuit in Figure 12.27 has a V_{ref} of 12 volts and an input voltage of 7 volts.

    a. Which comparators are on and which are off?
    b. What is the output code of the ADC?

Successive Approximation A/D Converter

■ **KEY TERMS**

Successive Approximation Register   A state machine used to generate a sequence of closer and closer binary approximations to an analog signal.

Quantization Error   Inaccuracy introduced into a digital signal by the inability of a fixed number of bits to represent the exact value of an analog signal.

Probably the most widely used type of analog-to-digital converter is the successive approximation ADC. The idea behind this type of converter is a technique a computer programmer would call "binary search."

The analog voltage to be converted is a number within a defined range. The search technique works by narrowing down progressively smaller binary fractions of the known range of numbers.

Suppose we know that the analog voltage is a number between 0 and 255, inclusive. We can find the binary value of any randomly chosen number in this range in no more than eight guesses, or approximations, since $2^8 = 256$. Each approximation adds one more bit to the estimated digital value.

The first approximation determines which half of the range the number is in. The second test finds which quarter of the range, the third test which eighth, the fourth test which sixteenth, and so on until we run out of bits.

EXAMPLE 12.17

Use a binary search technique to find the value of a number in the range 0 to 255. (The number is 44.)

■ **Solution**

See Figure 12.29.

1. The number must be in the upper or lower half of the range. Cut the range in half:
   0–127, 128–255.
   Is $x \geq 128$? No. $0 \leq x < 128$.
2. Cut the remaining range in half: 0–63, 64–128.
   Is $x \geq 64$? No. $0 \leq x < 64$.
3. Cut the remaining range in half: 0–31, 32–64.
   Is $x \geq 32$? Yes. $32 \leq x < 64$.
4. Cut the remaining range in half: 32–47, 48–64.
   Is $x \geq 48$? No. $32 \leq x < 48$.
5. Cut the remaining range in half: 32–39, 40–48.
   Is $x \geq 40$? Yes. $40 \leq x < 48$.
6. Cut the remaining range in half: 40–43, 44–48.
   Is $x \geq 44$? Yes. $44 \leq x < 48$.
7. Cut the remaining range in half: 44–45, 46–48.
   Is $x \geq 46$? No. $44 \leq x < 46$.
8. Cut the remaining range in half: 44–45, 46.
   Is $x \geq 45$? No. $x = 44$.

Figure 12.29 shows the binary search process in two graphical forms. Figure 12.29a shows the ranges of values that are narrowed down in the search procedure. The last four ranges, beginning with 32 to 40, are rescaled so that they are more easily visible. Figure 12.29b shows the search procedure as a graphed line that comes closer to the final value as the approximations become more accurate.

(continued)

a. Number ranges

b. Graphical representation

FIGURE 12.29 Example 12.17: Binary Search Technique

FIGURE 12.30 Successive Approximation ADC

The test criteria for each step in Example 12.17 are phrased so that the answer is always yes or no. (For example, $x \geq 64$? can only be answered yes or no.) Assume that a 1 means yes and a 0 means no. The tests in Example 12.17 give the following sequence of results: 00101100. The decimal equivalent of this binary number is 44, our original value.

A successive approximation ADC such as the one shown in Figure 12.30 applies a similar technique. The circuit has three main components: an analog comparator, a digital-to-analog converter, and a state machine called a **successive approximation register** (SAR). The SAR is an 8-bit register whose bits can be set and cleared individually, to a specific control sequence and the logic value at the output of the analog comparator.

When a pulse activates the Start Conversion (SC) input, bit Q_7 of the SAR is set. This makes the SAR output 10000000. The DAC converts the SAR output to an analog equivalent. When only the MSB is set, this is one half the reference voltage of the DAC.

The DAC output voltage is compared to an analog input voltage. (In effect, the SAR asks, "Is this approximation greater or less than the actual analog voltage?")

If $V_{analog} > V_{DAC}$, the comparator output is HIGH and the MSB remains set. Otherwise, the comparator output is LOW and the MSB is cleared. The process is repeated for all bits.

After all bits have been set or cleared, the End of Conversion (EOC) output changes state. This can be used to load the final digital value into an 8-bit latch.

EXAMPLE 12.18

An 8-bit successive approximation ADC has an analog input voltage of 9.5 V. Describe the steps the circuit performs to generate an 8-bit digital equivalent value if the DAC in the circuit has a reference voltage of 12 V.

▪ **Solution**

Figure 12.31 shows the steps the converter performs to generate the 8-bit digital equivalent of 9.5 V. The conversion process is also summarized in Table 12.9.

(continued)

FIGURE 12.31 Example 12.18: Successive Approximation A/D Conversion

TABLE 12.9 8-Bit Successive Approximation Conversion

Bit	New Digital Value	Analog Equivalent	$V_{analog} \geq V_{DAC}$?	Comparator Output	Accumulated Digital Value
Q_7	10000000	6 V	Yes	1	10000000
Q_6	11000000	9 V	Yes	1	11000000
Q_5	11100000	10.5 V	No	0	11000000
Q_4	11010000	9.75 V	No	0	11000000
Q_3	11001000	9.375 V	Yes	1	11001000
Q_2	11001100	9.5625 V	No	0	11001000
Q_1	11001010	9.46875 V	Yes	1	11001010
Q_0	11001011	9.515625 V	No	0	11001010

The following steps occur for each bit:

1. The bit is set.
2. The digital output is converted to an analog voltage and compared to the actual analog input.
3. If the analog voltage is greater than the DAC output voltage, the bit remains set. Otherwise it is cleared.

Figure 12.32 shows the output of the SAR during the conversion process. The values are shown in decimal.

FIGURE 12.32   Example 12.18: SAR Output Values for 8-Bit Successive Approximation ADC

There is no exact 8-bit binary value for the analog voltage specified in Example 12.18 (9.5 V). The final answer is the binary equivalent of 9.46875 V, which is within 31.25 mV, out of 12 V, which is pretty close but not exact. This difference is called **quantization error.** The maximum value of quantization error is $\pm\frac{1}{2}$ LSB for any ADC, except on the lowest step, where the error is $+\frac{1}{2}$, -0 LSB, and on the highest step, where the error is $+1$, $-\frac{1}{2}$ LSB.

As more bits are added to the accumulated digital value, the analog equivalent of the approximation acquires more decimal places of accuracy. Note that once the analog value extends beyond the decimal point, the last decimal digit is always 5.

An advantage of a successive approximation ADC is that the conversion time is always the same, regardless of the analog input voltage. This is not true with all types of analog-to-digital converters. The constant conversion time allows the output to be synchronized so that it can be read at known intervals.

The conversion time can be as few as $(n + 1)$ clock pulses for an n-bit device, if a bit is set by a clock edge and cleared asynchronously or by the opposite clock edge. Some SARs require four or more clock pulses per bit.

■ SECTION 12.3B REVIEW PROBLEM

12.9 State the purpose of the comparator in a successive approximation ADC.

Dual Slope A/D Converter

> ### ■ KEY TERMS
>
> **Dual Slope ADC** Also called an integrating ADC. An analog-to-digital converter based on an integrator. The name derives from the fact that during the conversion process the integrator output changes linearly over time, with two different slopes.
>
> **Integrator** A circuit whose output is the accumulated sum of all previous input values. The integrator's output changes linearly with time when the input voltage is constant.

A **dual slope analog-to-digital converter** is based on an **integrator** circuit, such as the one shown in Figure 12.33. The circuit output is proportional to the integral of the input voltage as a function of time. Integration with respect to time is the summing of instantaneous values of a function over a specified period of time. In other words, the output of an integrator is the accumulated total of all previous values of input voltage.

We can analyze the circuit without calculus under special conditions, such as when the input voltage is constant. An integrator is similar to an inverting amplifier and can be analyzed using similar techniques. Since the input impedance of the op amp is large, there is very little current flowing into its (−) terminal. Ohm's law thus implies that there is very little voltage difference between the (+) and (−) terminals. Since they are at almost the same potential and the (+) terminal is grounded, we can say that the (−) terminal is "virtually grounded."

If the input voltage is constant, a DC current, I, flows in R. Since R is connected to the positive terminal of the input voltage source at one end and virtual ground at the other, the entire source voltage drops across the resistor. By Ohm's law,

$$I = V_{in}/R$$

a. General circuit   **b. Constant input voltage**

FIGURE 12.33 Integrator

Since the op amp input impedance is large, most current flows into the capacitor, causing it to charge over time. The current direction defines a polarity for V_c, the capacitor voltage.

The op amp output voltage is measured with respect to ground. The capacitor is connected from the op amp output to virtual ground. Therefore, the output voltage, V_o, is dropped across the capacitor. Notice that the polarities defined for V_o and V_c are opposite:

$$V_o = -V_c$$

The capacitor voltage is determined by the stored charge, Q, and the value of capacitance, C:

$$V_c = Q/C$$

The current I is the amount of charge flowing past a given point in a fixed time:

$$I = Q/t$$

Thus,

$$V_c = It/C$$

and

$$V_o = -It/C$$

Substitute the expression for I into this equation to get

$$V_o = -(t/RC)V_{in}$$

The output of an integrator with a constant input changes linearly with time, with a slope equal to $-\dfrac{V_{in}}{RC}$.

This equation describes the *change* in output voltage due to a constant input. When the input goes to 0 V, the capacitor holds its charge (ideally forever; in practice until it leaks away through circuit impedances) and maintains the output voltage at its final value. If a new input voltage is applied, we can use the integrator equation to calculate the change in output, which must then be added to the previous value.

EXAMPLE 12.19

The integrator circuit of Figure 12.33 has the following component values:

$$C = 0.025 \ \mu F, \ R = 10 \ k\Omega$$

Sketch the graph of the output voltage if the waveform shown in the graph of Figure 12.34a is applied to the integrator input. The integrator output is originally at 0 V.

■ Solution

We must examine the graph in two sections:

1. From 0 to 3 ms
2. From 3 to 9 ms

A different constant input voltage is applied for each section of the input graph. From this, we can determine the shape of the output graph, shown in Figure 12.34b.

(continued)

a. Input voltage

b. Output voltage

FIGURE 12.34 Example 12.19: Integrator Operation

0 to 3 ms:

The output at 3 ms is given by:

$$\text{slope} = -\frac{V_{in}}{RC} = -\frac{1V}{(0.025\mu F)(10\ k\Omega)}$$

$$= -4\ \text{V/ms}$$

$$v_o(3\ \text{ms}) = v_o(0) + (t/RC)V_{in}$$

$$= 0V + [(3\ \text{ms})(-4\ \text{V/ms})]$$

$$= -12\ \text{V}$$

The output changes at a rate of –4 V/ms for 3 ms.

3 to 9 ms:

The output at 9 ms is given by:

$$\text{slope} = -\frac{V_{in}}{RC} = -\frac{(-0.5\ \text{V})}{(0.025\mu F)(10\ k\Omega)}$$

$$= +2\ \text{V/ms}$$

$$v_o\,(9\text{ ms}) = v_o(3\text{ ms}) - (t/RC)V_{in}$$

$$= -12\text{ V} + [(6\text{ ms})(+2\text{ V/ms})]$$

$$= -12\text{ V} + (+12\text{ V})$$

$$= 0\text{ V}$$

The output changes at a rate of +2 V/ms for 6 ms. This cancels the effect of the original input.

Figure 12.35 shows the block diagram of an 8-bit dual slope analog-to-digital converter. Integrator output voltages for several input values are shown in Figure 12.36. Assume that the integrator has the same R and C values as in Example 12.19.

1. Before conversion starts, an auto-zero circuit sets the comparator output to 0 V by applying a compensating voltage to the comparator.
2. The input analog voltage causes the integrator output to increase in magnitude, as shown in the left half of Figure 12.36. As soon as this integrator voltage is nonzero, the comparator enables a counter via the control logic.
3. When the counter overflows (i.e., recycles to 00000000), the integrator input is switched from the analog input to $-V_{ref}$.

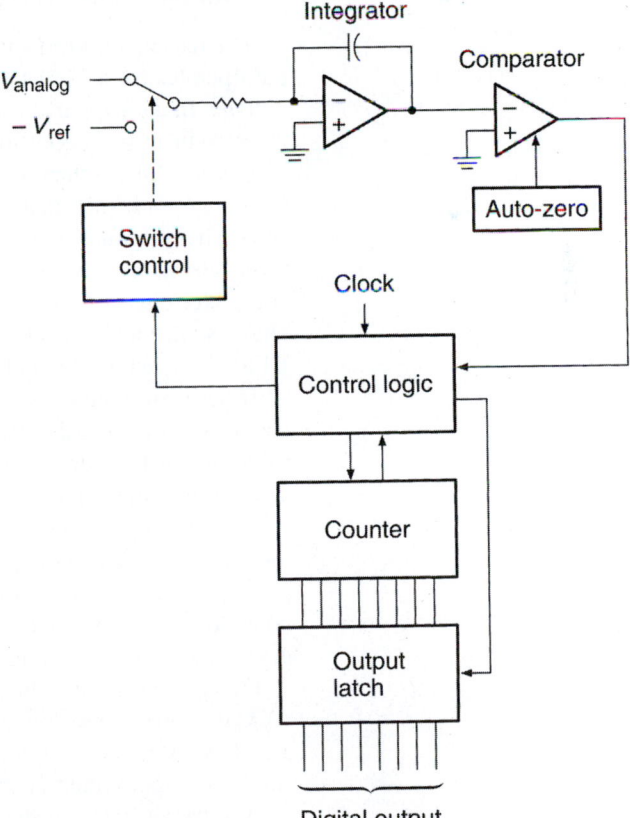

FIGURE 12.35   Dual Slope ADC

FIGURE 12.36 Integrator Outputs for Various Input Voltages

4. The reference voltage causes the integrator output to move toward 0 V at a known rate, as shown in the right half of Figure 12.36. During this rezeroing time, the counter continues to clock. When the integrator output voltage reaches 0 V, the comparator disables the counter. The digital equivalent of the analog voltage is now contained in the counter.

The reason this works is that in the initial integrating phase, the integrator output operates for a *known time,* producing a final output proportional to the input voltage. In the second phase, the output moves toward zero at a *known rate,* reaching zero in a time proportional to the final voltage of the first phase.

For example, assume that the components of the integrator and the clock rate of the counter are such that a 1-V input corresponds to the full-scale digital output *(FS).* The integrator output reaches a value of –12 V in 3 ms. The time required to rezero the integrator is the same as the initial integrating phase, 3 ms. The counter completes one cycle in the integrating phase and another cycle in the rezeroing phase, so that its final value is 00000000. (Note that this is the result obtained when 1 LSB is added to 11111111.)

If the input voltage is 0.25 volts, the integrator output is –3 V after 3 ms (one counter cycle). Since the integrator always rezeros at the same rate (4 V/ms), the rezeroing time is 0.75 ms, or one fourth of a counter cycle (since 12 V/4 = 3 V). The counter has time to reach state 01000000 or $\frac{1}{4}$ *FS.*

If we attempt to measure a voltage beyond that corresponding to full scale, the integrator output cannot rezero within the second counter cycle. Usually, an output pin on the ADC activates to show this condition. Some digital multimeters that use dual slope ADCs show an overvoltage or out-of-range condition by blanking the display, except for a leading digit 1.

One advantage of a dual slope ADC is its accuracy. One particular dual slope ADC is accurate to within ±0.05% ± 1 count. This accuracy is balanced against a relatively slow conversion time, in the milliseconds, compared to microseconds for a successive approximation ADC and nanoseconds for a flash converter.

Another advantage is the ability of the integrator to reject noise. If we assume that noise voltage is random, then it will be positive about half the time and negative about half the time. Over time it should average out to zero.

As was alluded to above, a common application of this device is as a voltmeter circuit, where speed is less important than accuracy.

■ **SECTION 12.3C REVIEW PROBLEMS**

12.10 Suppose that the dual slope ADC just described (same component values) has an input voltage of 0.375 V ($\frac{3}{8}$ full scale).

    a. What is the slope of the integrator voltage during the integrating phase?

    b. What is its slope during the rezeroing phase?

    c. How much time elapses during the rezeroing phase?

    d. What digital code is contained in the output latch after the conversion is complete?

Sigma-Delta ADC

> ■ **KEY TERM**
>
> **Sigma-Delta ADC**   An analog-to-digital converter that uses an integrator, a comparator, and a DAC to generate a serial stream of bits based on the sum of voltage changes within the circuit during the conversion process. Also called delta-sigma ADC.

The general idea behind the sigma-delta analog-to-digital converter can be understood by looking at the meaning of the symbols that make up its name. Sigma and delta are Greek letters with conventional meanings as mathematical symbols. Sigma (Σ) is the symbol used for summation. Delta(Δ) is the symbol used to signify a small change in value. Combined, we can understand sigma-delta to be the "sum of changes." Let us examine what we mean by this.

Figure 12.37 shows the circuit of a sigma-delta ADC. The sigma-delta ADC, also called a delta-sigma ADC by some manufacturers, is unusual in that its output is a serial stream of bits rather than a multibit parallel value. It is this property that allows the ADC to generate a highly accurate digital output, with outputs of up to 24 bits, a level of precision that is impossible to attain with converters that use parallel-output methods.

The circuit begins by integrating an input value and then sending a 0 or 1 to the circuit output depending on the value of the integrator output relative to ground. This output bit is converted to one of two values ($-V_{ref}$ or $+V_{ref}$) by a 1-bit DAC. The DAC output is subtracted from the input voltage at a summing junction and this sum is inverted and added to the previous output value of the integrator. In other words the integrator (Σ) sums the changes introduced by the DAC (Δ) in the circuit's feedback loop. The process continues for a defined number of iterations, each of which represents a new sample of the input voltage. Each iteration generates a bit in the serial output stream. The number of 1s in the bit stream is proportional to the fraction of full scale that represents the analog input voltage, V_a.

Figure 12.38 shows the steps involved within the ADC during the first two iterations of the conversion. Initially, all circuit voltages are set to zero. The value of

FIGURE 12.37 Sigma-Delta ADC

V_{ref} for the DAC is ±5 volts. An analog input of +2.5 V is sampled and applied in Figure 12.38a. This voltage is inverted by the integrator and summed to the previous value of 0 V, giving an output of –2.5 V. (Clearly, for this to work, the integrator must have the sample voltage applied for a known time.) Since –2.5 V is less than the ground voltage at the comparator's (+) input, the comparator sends a 1 to the circuit output.

In Figure 12.38b, the 1 at the bit stream output is fed back to the DAC, which generates an output of +5 V. In Figure 12.38c, the DAC output voltage is subtracted from the input sample, giving a total of –2.5 V. This sum is applied to the integrator in

a. +2.5 volts applied to initialized ADC

b. Present output bit converted by 1-bit DAC

c. Feedback from DAC subtracted from input at summing junction

d. Summed voltage is inverted and added to previous value of integrator

FIGURE 12.38 Example Iteration for Sigma-Delta ADC

Figure 12.38d, where it is inverted and added to the previous integrator total. This now gives an integrator output of (+2.5 V + (−2.5 V) = 0 V). The comparator uses this to send a 0 to the output.

Figure 12.39 shows the states of the circuit after each of four iterations with the same input voltage as in Figure 12.38 (+2.5 V). Figure 12.39a shows the same state of the circuit as Figure 12.38a. Figure 12.39b shows the combined result of the steps in Figures 12.38b to 12.38d. In each of Figures 12.39b to 12.39d, the circuit voltages are consistent if you start at the DAC input and follow the voltages around the

a. +2.5 volts applied to initialized ADC

b. End of Step 2

c. End of Step 3

d. End of Step 4

FIGURE 12.39 Four Calculation Steps in a Sigma-Delta ADC (V_a = +2.5V) loop, stopping at the comparator output. The resulting bit stream output is then applied to the DAC input of the next figure.

For example, in Figure 12.39c, we start at the DAC input, which has a value of 0 from the bit stream output in Figure 12.39b. This generates a value of –5.0 V at the DAC output, which is subtracted from the sampled analog input to give +2.5 V – (–5.0 V) = +7.5 V. This voltage is inverted and added to the previous integrator output to give 0 V + (–7.5 V) = –7.5 V. Since –7.5 V is less than the voltage of the comparator (+) input, the comparator output is 1.

Over a series of four interations, the output bit stream is 1011. For a full scale range of ±5 V, +2.5 V = $\frac{3}{4}$ FS. In the bit stream, 3 out of 4 bits are HIGH. Since the process is repetitive, we can easily represent the voltages at the summing junction, integrator, comparator, output bit stream, and DAC using a spreadsheet program, such as Microsoft Excel®, Corel Quattro Pro®, or Lotus 1-2-3®. We can then use the spreadsheet to graph several of the parameters to give us a visual idea of what's going on in the circuit.

Figure 12.40 shows part of a spreadsheet created using Microsoft Excel. The sigma-delta function is iterated 64 times in the spreadsheet, representing 64 samples of the analog input, with the comparator and DAC outputs represented by bar graphs. Looking at the graphs and the data in the comparator column, we can see that the bit stream 1011 repeats itself indefinitely.

Figure 12.41 shows how the circuit generates a bit stream for the analog input, $V_a = -2.5$ V. For a value of $V_{ref} = \pm 5$ V, this represents an input of $\frac{1}{4}$ FS. Thus, we should expect a bit stream where 1 out of 4 bits is HIGH. For the resulting bit stream of 0010, this is indeed the case. Figure 12.42 shows this calculation over 64 iterations on an Excel spreadsheet.

FIGURE 12.40 Partial Spreadsheet and Graphed Output for Sigma-Delta ADC ($V_a = +2.5V$; $\frac{3}{4}$ FS)

FIGURE 12.41 Four Calculation Steps in a Sigma-Delta ADC ($V_a = -2.5V$)

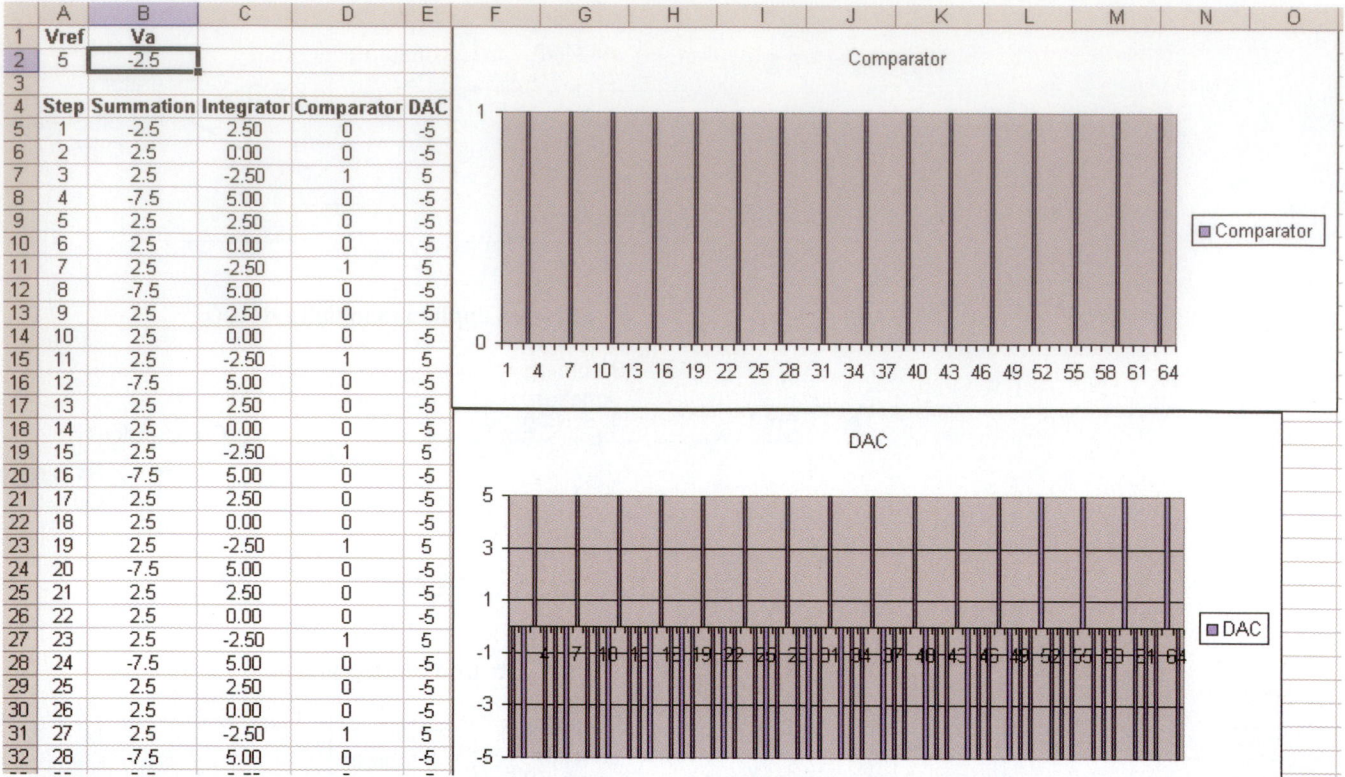

Step	Summation	Integrator	Comparator	DAC
Vref	Va			
5	-2.5			
1	-2.5	2.50	0	-5
2	2.5	0.00	0	-5
3	2.5	-2.50	1	5
4	-7.5	5.00	0	-5
5	2.5	2.50	0	-5
6	2.5	0.00	0	-5
7	2.5	-2.50	1	5
8	-7.5	5.00	0	-5
9	2.5	2.50	0	-5
10	2.5	0.00	0	-5
11	2.5	-2.50	1	5
12	-7.5	5.00	0	-5
13	2.5	2.50	0	-5
14	2.5	0.00	0	-5
15	2.5	-2.50	1	5
16	-7.5	5.00	0	-5
17	2.5	2.50	0	-5
18	2.5	0.00	0	-5
19	2.5	-2.50	1	5
20	-7.5	5.00	0	-5
21	2.5	2.50	0	-5
22	2.5	0.00	0	-5
23	2.5	-2.50	1	5
24	-7.5	5.00	0	-5
25	2.5	2.50	0	-5
26	2.5	0.00	0	-5
27	2.5	-2.50	1	5
28	-7.5	5.00	0	-5

FIGURE 12.42   Partial Spreadsheet and Graphed Output for Sigma-Delta ADC ($V_a = -2.5V$; $\frac{1}{4}$ *FS*)

EXAMPLE 12.20

A voltage of –4 V is applied to a sigma-delta ADC with a reference voltage of ±5 V. What fraction of full scale does this input represent? What pattern should we expect to see in the output bit stream? How does a value of 64 iterations affect the accuracy of the result for this input value? How would this accuracy change for 1024 iterations?

■ Solution

An input of –4 V represents a value of 1/10 *FS*, which we can calculate as follows:

$$Fraction\ of\ Full\ Scale = \frac{V_a - V_{\min}}{V_{\max} - V_{\min}} = \frac{(-4V) - (-5V)}{(+5V) - (-5V)} = \frac{1}{10}$$

We should expect to see a bit stream where 1 out of 10 bits is HIGH.

Figure 12.43 shows an Excel spreadsheet with an input voltage of –4 volts. This shows a repeating bit pattern of 0000010000 at the comparator output. (You can tell where the pattern begins to repeat by examining the integrator output. The iteration where the integrator is the same as the initial value marks the beginning of a new cycle of the pattern. In Figure 12.43, this is in cell C15. The values of the integrator and comparator for one cycle of the pattern are highlighted by the box around cells C4 through D14.)

Over 64 iterations, the spreadsheet shows 6 HIGH outputs, which gives an output of 6/64 *FS* = 0.09375 *FS*. This gives an error of:

$$\frac{0.09375 - 0.1}{0.1} \times 100\% = -6.25\%$$

For 1024 iterations, the pattern repeats 102 times plus 4 iterations. The 1 normally appears on the sixth iteration of the pattern, so we wouldn't see it in the last

FIGURE 12.43 Example 12.20: Partial Spreadsheet and Graphed Output for Sigma-Delta ADC ($V_a = -4.0V$; $\frac{1}{10}$ FS)

pattern, which is only partial. The output over 1024 iterations is 102/1024 *FS* = 0.099609375 *FS*. This gives an error of:

$$\frac{0.09960375 - 0.1}{0.1} \times 100\% = -0.390625\%$$

This implies that more iterations have the effect of reducing the quantization error of the bit stream.

The output of a sigma-delta ADC is usually modified to change the serial bit stream to a parallel *n*-bit output. A simple way to do this is to use the bit stream to clock a binary counter for a known number of iterations of the ADC, as shown in Figure 12.44a. The total number of iterations can be counted with a separate counter, which is not shown. For an *n*-bit parallel output, the ADC comparator must output a bit stream of 2^n bits. Thus a 16-bit output requires 65,536 sample-and-iterate cycles. A 20-bit ADC requires 1,048,576 samples.

A more common, but more complicated, technique is to use a digital filter, such as a decimation filter, which converts a low-resolution, high-sample rate input, such as from the sigma-delta ADC, to a high-resolution, low-sample rate output. Figure 12.44b shows this configuration. This technique results in a better-quality output than the binary counter method. The operation of digital filters is beyond the scope of this textbook.

Sigma-delta ADCs allow very high-accuracy conversions, since every iteration relies on the output of a very low-resolution (1-bit) DAC, which is less subject to error than a high-resolution DAC.

As we saw in earlier sections of the chapter, a parallel-output DAC has a smaller and smaller output resolution with a greater number of input bits. At some point, the

a. Bit stream clocks *n*-bit binary counter

b. Bit stream is digitally filtered

FIGURE 12.44   Creating an *n*-Bit Output from a Sigma-Delta ADC

LSB voltage of such a DAC is indistinguishable from the noise voltages present in the circuit. For example, the resolution of a 16-bit DAC for V_{ref} of ±5 V is 152.6 μV. For a 20-bit DAC with the same range, the resolution is 9.54 μV. These are very small voltages, relative to full scale, and they may well be less than the noise floor of the circuit.

A 1-bit DAC is not subject to this problem. It's output is always ±FS/2, which is very much larger than the noise floor of the circuit (120 dB greater for a ±5-V reference and a 10-μV noise floor). Thus, a more-accurate sigma-delta ADC does not rely on more outputs with LSB voltages decreasing with each added bit. Rather, its accuracy increases with an increased number of iterations used to convert a voltage to a bit stream. Simply put, more iterations, more accuracy.

The limiting factor for such a device is the speed at which an analog input voltage is changing. If the output cannot be made to hold still for the number of iterations required to convert the input, inaccuracies will be introduced. Thus, sigma-delta ADCs are most suited for applications where high accuracy is required, but where the input is changing relatively slowly, such as digital audio.

■ SECTION 12.3D REVIEW PROBLEM

12.11   A sigma-delta ADC has a reference voltage of V_{ref} = ±5 V. For an input of +2 V, what fraction of full scale is the input? What bit values would you expect to see in the output bit stream of the ADC?

Sample and Hold Circuit

■ KEY TERM

Sample and Hold Circuit   A circuit that samples an analog signal at periodic intervals and holds the sampled value long enough for an ADC to convert it to a digital code.

FIGURE 12.45 Sample and Hold Circuit

FIGURE 12.46 Sample and Hold Output

For the sake of analysis, we have been assuming that the analog input voltage of any analog-to-digital converter is constant. This is an actual requirement. Most of these circuits will not produce a correct digital code if the analog voltage at the input changes during conversion time.

Unfortunately, most analog signals are not constant. Usually, we want to sample these signals at periodic intervals and generate a series of digital codes that tells us something about the way the input signal is changing over time. A circuit called a **sample and hold circuit** must be used to bridge the gap between a changing analog signal and a requirement for a constant ADC input voltage.

Figure 12.45 shows a basic sample and hold circuit. The voltage followers act as buffers with high input and low output impedances. The transmission gate is enabled during the sampling period, during which it charges the hold capacitor to the current value of the analog signal. During the hold period, the capacitor retains its charge, thus preserving the sampled analog voltage. The high input impedance of the second voltage follower prevents the capacitor from discharging significantly during the hold period.

Figure 12.46 shows how a sample and hold circuit produces a steady series of constant analog voltages for an ADC input. Since these sampled values have yet to be converted to digital codes, they can take on any value within the analog range; they are not yet limited by the number of bits in the quantization.

Ideally, a sample and hold circuit should charge quickly in sample mode and discharge slowly in hold mode. These characteristics are facilitated by the low output impedance and high input impedance of the voltage follower circuits.

Figure 12.47 shows the equivalent circuits of the sample and hold modes of the circuit in Figure 12.45. In sample mode, the capacitor charges through the output impedance, Z_o, of the first voltage follower. Since this is a very small value (about

a. Sampling phase   **b. Hold phase**

FIGURE 12.47   Equivalent Circuits for Sample-and-Hold Circuit

$75 \times 10^{-5}\,\Omega$), the capacitor will charge quickly. In the hold mode, the capacitor discharges slowly through the very high input impedance, Z_i, of the second voltage follower (about $2 \times 10^{11}\,\Omega$).

> **NOTE . . .**
>
> The input and output impedances of the voltage follower are significantly different from the open-loop op amp values. This is because, in the voltage follower configuration, the output impedance is divided by the open loop gain (about $75\,\Omega/100{,}000$) and the input impedance is multiplied by the open loop gain (about $2\,M\Omega \times 100{,}000$).

A variation of the sample and hold circuit is the track and hold circuit. The difference is not so much in the circuit as in the way it is operated. A sample and hold circuit is restricted by the charging speed of its hold capacitor. If there is a large change in signal level between samples, the hold capacitor may not be able to keep up with the change. A track and hold circuit samples the analog signal continuously, minimizing charging delays of the hold capacitor. When the analog signal needs to be converted, the track and hold circuit reverts to hold mode by closing the analog transmission gate. Many high-speed ADCs have a track and hold circuit as an integral part of the device.

Sampling Frequency and Aliasing

> **■ KEY TERMS**
>
> **Nyquist Sampling Theorem**   A theorem from information theory stating that in order to preserve all information in a signal, it must be sampled at a rate of at least twice the highest-frequency component of the signal. ($f_s \geq 2f_{max}$)
>
> **Aliasing**   A phenomenon that produces an unwanted low-frequency component in a sampled analog signal due to a sampling frequency that is too slow relative to the sampled analog signal.
>
> **Anti-Aliasing Filter**   A low-pass filter with a corner frequency of twice the sampling frequency, used to prevent aliasing in an ADC.

In the first section of this chapter, we saw that the sampling frequency of an ADC has a great effect on the quality of the digital representation of an analog signal. We may ask, what is the minimum value of the sampling frequency for any particular analog signal and what happens if this criterion is not met?

A theorem in information theory, called the **Nyquist sampling theorem,** states that a periodic signal must be sampled at least twice a cycle to preserve all its information. In practice, this means that the sampling frequency of a particular system must be twice the maximum frequency of any signal to be sampled by the system.

(These frequencies might also include harmonics of a signal that add to the basic signal to give it its characteristic shape.) This can be expressed mathematically as $f_s \geq 2f_{max}$ for a sampling frequency f_s and a maximum-frequency component of f_{max}.

For example, the sampling frequency for compact disc audio is 44.1 kHz, which allows signals of up to 22.05 kHz to be sampled accurately. This fits in nicely with the statistical range of human hearing: 20 Hz–20 kHz. (People who have listened to any amount of rock music in their youth can probably only get up to 12 kHz.) Telephone-quality signals are sampled at 8 kHz, yielding a maximum frequency of 4 kHz, which is a bit more than the classical telephone-line bandwidth of 300 Hz–3300 Hz.

A sampling frequency of an ADC system that does not meet the criterion required by the Nyquist sampling theorem results in **aliasing,** a phenomenon that generates a false low-frequency component of the digital sample.

To get an idea of how aliasing works, let us examine a sine wave with a period of 12 μs (f = 83.3 kHz), shown in Figure 12.48. If we sampled the signal every 1 μs, we would capture the values listed in Table 12.10.

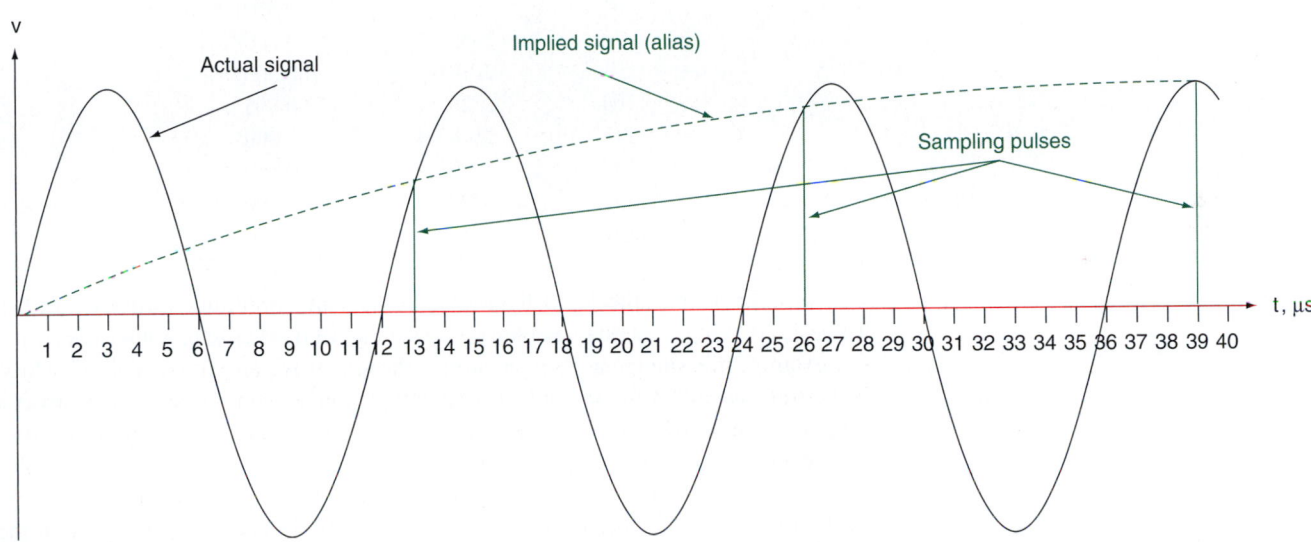

FIGURE 12.48 Effect of Sampling Too Slowly

TABLE 12.10 Sampled Values of an 83.3 kHz Sine Wave (1 μs Sampling)

Time (μs)	Degrees	Fraction of Peak
0	0°	0.000
1	30°	0.500
2	60°	0.866
3	90°	1.000
4	120°	0.866
5	150°	0.500
6	180°	0.000
7	210°	−0.500
8	240°	−0.866
9	270°	−1.000
10	300°	−0.866
11	330°	−0.500
12	360°	0.000

The points in Table 12.10 can be used to accurately reconstruct the original sine wave. (The reconstructed output would need to be filtered to eliminate introduced high-frequency components, but the fundamental frequency would be correct.)

Suppose now that we sample the same sine wave at less than twice a cycle. Table 12.11 shows the samples captured by a series of sampling pulses that are spaced by 13 μs. The first four samples in the table are shown by vertical lines in Figure 12.48.

TABLE 12.11   Sampled Values of an 83.3 kHz Sine Wave (13 μs Sampling)

Time (μs)	Degrees	Fraction of Peak
0	0°	0.000
13	390°	0.500
26	780°	0.866
39	1170°	1.000
52	1560°	0.866
65	1950°	0.500
78	2340°	0.000
91	2730°	−0.500
104	3120°	−0.866
117	3510°	−1.000
130	3900°	−0.866
143	4290°	−0.500
156	4680°	0.000

The samples in Table 12.11 have exactly the same amplitude as those taken in Table 12.10. However, the samples are spaced at 13 μs intervals, rather than 1 μs. For example, the sample at 13 μs measures the sine wave amplitude at 390°, which is the same as 30° of the *second* cycle. If these samples were used to reconstruct a sine wave, it would have a period of 156 μs, rather than 12 μs. This false low-frequency component is shown by the broken line connecting the first four samples in Figure 12.48, which represent measurements, not in a single cycle, but in four cycles of the original analog signal. Figure 12.49 shows one complete cycle of the alias frequency as a broken line and 13 cycles of the sampled signal as a solid line.

FIGURE 12.49 Aliasing

FIGURE 12.50 Anti-Aliasing Filtering

Aliasing can be prevented by filtering the analog input to an ADC with an **anti-aliasing filter,** as shown in Figure 12.50.

The anti-aliasing filter is a low-pass filter with the corner frequency set to $2f_{max}$. Frequencies less than $2f_{max}$ are allowed to pass to the analog input of the ADC. Frequencies greater than $2f_{max}$ are attenuated. In this way, the ADC never converts any signal with a frequency greater than $2f_{max}$ and thus an alias frequency cannot develop.

■ SECTION 12.3E REVIEW PROBLEM

12.12 An analog signal consists of the following combined frequency components: 1 kHz, 2 kHz, 4 kHz, and 16 kHz. State the minimum frequency required to sample the signal without loss of information.

12.4 DATA ACQUISITION

➡ **LAB SOURCE**
Lab #33
Analog-to-Digital
Conversion

■ KEY TERM

Data Acquisition Network   A circuit that gathers and digitizes data from several analog sources.

CPLD Interface for an ADC

Figure 12.51 shows the symbol for an ADC0808 analog-to-digital converter. This successive approximation ADC can form the basis of a **data acquisition network,** a system that can convert analog information from up to eight channels and store the converted values in a series of output latches.

The ADC0808 has a built-in 8-channel analog multiplexer with inputs *IN0* through *IN7*, which are selected by the states of three address inputs, *ADD C, ADD B,* and *ADD A,* where *ADD C* is the most significant bit. Before an analog input can be converted, its address must be stored in an internal address latch by a high-going pulse on *ALE* (Address Latch Enable).

The conversion process starts with a high-going pulse on the *START* input. (*START* and *ALE* can be tied together.) End-of-conversion is indicated by the *EOC* output. The conversion process is driven by the *CLOCK* input. After conversion is complete, the digital output can be read by making *OE* (Output Enable) HIGH. When *OE* is not active, the digital outputs are in the high-impedance state.

Figure 12.52 shows a timing diagram relating the various control signals of the ADC0808. Figure 12.53 is an excerpt from the ADC0808 datasheet that shows relevant timing information. The ADC is reset on the rising edge of *START*. After

FIGURE 12.51   ADC0808 Analog-to-Digital Converter

FIGURE 12.52   Timing Diagram for an ADC0808

START/ALE goes LOW, the ADC makes *EOC* go LOW within 8 clock cycles + 2 μs. *EOC* stays LOW until conversion is complete. The simplest way to operate the ADC on a stand-alone basis is to tie the *EOC* line to the *START/ALE* line so that the ADC starts as soon as *EOC* goes HIGH and the ADC continuously updates the value of the digital output.

We can design a state machine that controls the ADC and stores output values in an octal latch automatically. Figure 12.54 shows such a circuit. The controller is a state machine that will accept a LOW pulse from a debounced pushbutton switch labeled **go,** perform one analog-to-digital conversion from one of eight analog channels and store the resulting 8-bit digital value in an octal latch. The analog channel is manually selected by DIP switches at the address select inputs. The entire state machine and latch portion of the circuit is contained in one CPLD, such as the Altera EPM7128SLC84-7 on the Altera UP-1 or UP-2 circuit board.

Figure 12.55 shows the state diagram of the controller, with two synchronous inputs called **go** and **eoc.** The asynchronous reset, which sets the machine to the **idle** state, is not shown on the state diagram. Outputs are **sc** (start conversion), **oe** (output enable), and **en** (latch enable). The states are as follows:

- **idle**—Wait for **go** = 0 (switch pressed). All outputs are LOW.
- **start**—Wait for **go** = 1 (switch released). Transition to **wait1** makes **sc** = 1 (*START/ALE* pulse). Other outputs are LOW.
- **wait1**—Wait for **eoc** = 0. (Wait for conversion to start. *EOC* is LOW during, but not before, conversion. Do not test for **eoc** = 1 until *after* conversion.) All outputs LOW.
- **wait2**—Wait for **eoc** = 1. (Conversion complete.) When complete, transition to **read.** At that time, **oe** = 1, **en** = 1.

Electrical Characteristics (Continued)

Digital Levels and DC Specifications: ADC0808CCN, ADC0808CCV, ADC0809CCN and ADC0809CCV, $4.75 \leq V_{CC} \leq 5.25V$, $-40\,C \leq T_A \leq +85°\,C$ unless otherwise noted

Symbol	Parameter	Conditions	Min	Typ	Max	Units
DATA OUTPUTS AND EOC (INTERRUPT)						
$V_{OUT(1)}$	Logical ™"1" Output Voltage	$V_{CC} = 4.75V$				
		$I_{OUT} = -360\ \mu A$		2.4		V(min)
		$I_{OUT} = -10\ \mu A$		4.5		V(min)
$V_{OUT(0)}$	Logical ™"0" Output Voltage	$I_O = 1.6$ mA			0.45	V
$V_{OUT(0)}$	Logical ™"0" Output Voltage EOC	$I_O = 1.2$ mA			0.45	V
I_{OUT}	TRI-STATE Output Current	$V_O = 5V$			3	μA
		$V_O = 0$	-3			μA

Electrical Characteristics

Timing Specifications $V_{CC} = V_{REF(+)} = 5V$, $V_{REF(,)} = GND$, $t_r = t_f = 20$ ns and $T_A = 25°C$ unless otherwise noted.

Symbol	Parameter	Conditions		Min	Typ	Max	Units
t_{WS}	Minimum Start Pulse Width	(Figure 5)			100	200	ns
t_{WALE}	Minimum ALE Pulse Width	(Figure 5)			100	200	ns
t_s	Minimum Address Set-Up Time	(Figure 5)	2		25	50	ns
t_H	Minimum Address Hold Time	(Figure 5)	2		25	50	ns
t_D	Analog MUX Delay Time From ALE	$R_S = 0\Omega$ (Figure 5)			1	2.5	μs
t_{H1}, t_{H0}	OE Control to Q Logic State	$C_L = 50$ pF, $R_L = 10k$ (Figure 8)			125	250	ns
t_{1H}, t_{0H}	OE Control to Hi-Z	$C_L = 10$ pF, $R_L = 10k$ (Figure 8)			125	250	ns
t_c	Conversion Time	$f_c = 640$ kHz, (Figure 5) (Note 7)		90	100	116	μs
f_c	Clock Frequency			10	640	1280	kHz
t_{EOC}	EOC Delay Time	(Figure 5)		0		$8 + 2\mu S$	Clock Periods
C_{IN}	Input Capacitance	At Control Inputs			10	15	pF
C_{OUT}	TRI-STATE Output Capacitance	At TRI-STATE Outputs			10	15	pF

Note 1: Absolute Maximum Ratings indicate limits beyond which damage to the device may occur. DC and AC electrical specifications do not apply when operating the device beyond its specified operating conditions.

Note 2: All voltages are measured with respect to GND, unless othewise specified.

Note 3: A zener diode exists, internally, from V_{CC} to GND and has a typical breakdown voltage of 7 V_{DO}.

Note 4: Two on-chip diodes are tied to each analog input which will forward conduct for analog input voltages one diode drop below ground or one diode drop greater than the V_{CC}n supply. The spec allows 100 mV forward bias of either diode. This means that as long as the analog V_{IN} does not exceed the supply voltage by more than 100 mV, the output code will be correct. To achieve an absolute $0V_{DC}$ to $5V_{DC}$ input voltage range will therefore require a minimum supply voltage of 4.900 V_{DC} over temperature variations, initial tolerance and loading.

Note 5: Total unadjusted error includes offset, full-scale, linearity, and multiplexer errors. See Figure 3. None of these A/Ds requires a zero or full-scale adjust. How-ever, if an all zero code is desired for an analog input other than 0.0V, or if a narrow full-scale span exists (for example: 0.5V to 4.5V full-scale) the reference voltages can be adjusted to achieve this. See Figure 13.

Note 6: Comparator input current is a bias current into or out of the chopper stabilized comparator. The bias current varies directly with clock frequency and has little temperature dependence (Figure 6). See paragraph 4.0.

Note 7: The outputs of the data register are updated one clock cycle before the rising edge of EOC.

Note 8: Human body model, 100 pF discharged through a 1.5 kΩ resistor.

FIGURE 12.53 Extract from ADC0808 Datasheet (Reprinted with permission of National Semiconductor)

FIGURE 12.54 ADC Interface with One Output Channel and Manual Input Channel Selection

- **read**—Enable ADC output (**oe** = 1) and make latch transparent (**en** = 1).
- **store**—Keep ADC output enabled, put latch in store mode (**en** = 0). ADC digital output is now stored in the output latch.

Figure 12.56 shows a simulation of the controller.

Both the latch and controller can be implemented as VHDL design entities and instantiated as components in the top level of a VHDL hierarchy. This and later VHDL examples will be saved as exercises for the lab manual accompanying this book. The VHDL files are available to instructors in the e-Resource for this book.

We should note that if the controller/latch circuit is to be implemented on the Altera UP-1/UP-2 board or the RSR PLDT-2, we must also include a clock divider circuit. The Altera UP-1/UP-2 board has an on-board oscillator that runs at 25.175 MHz. The RSR PLDT-2 has a 4 MHz oscillator. The clock rate, as defined by the ADC0808 datasheet, must be in the range 10 kHz $\leq f_e \leq$ 1280 kHz.

A 5-bit counter can serve as a divide-by-32 circuit (2^5 = 32), as shown in Figure 12.57. If the UP-1/UP-2 oscillator is applied to the counter clock, the Q_4 output frequency is given by 25.175 MHz/32 = 786.7 kHz, which is within the required

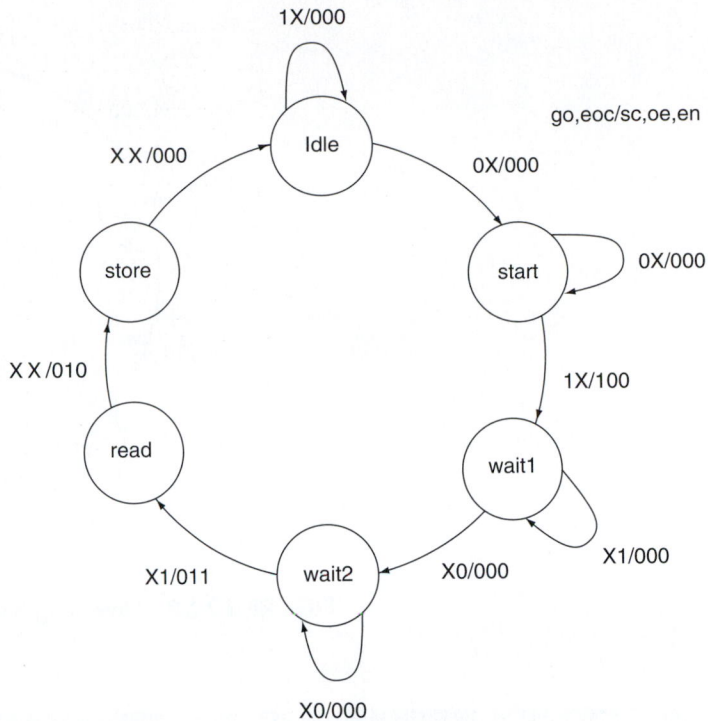

FIGURE 12.55   State Diagram for an ADC Controller

FIGURE 12.56   Simulation of State Machine ADC Controller

FIGURE 12.57   5-Bit Counter as Divide-by-32 Clock Divider

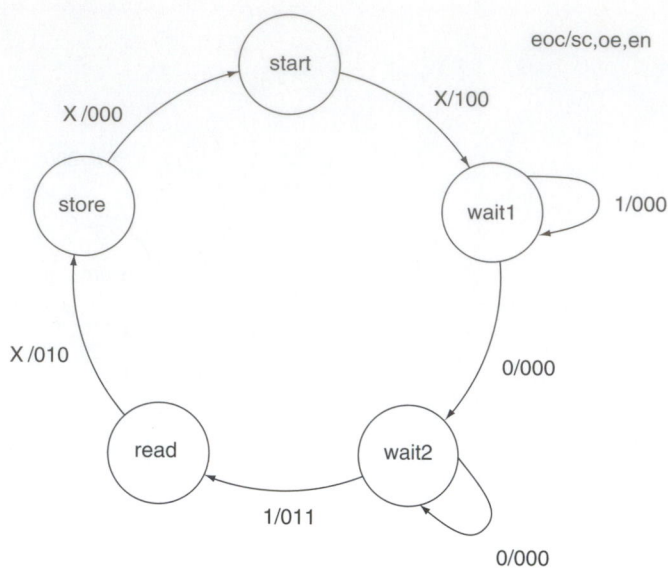

FIGURE 12.58   State Diagram for Continuous-Convert ADC Controller

FIGURE 12.59   Simulation of a Continuous-Convert ADC Controller

range for the ADC clock. Q_4 should then be used to clock the state machine, as well as any other synchronous circuitry used in conjunction with the ADC0808.

Similarly, a divide-by-4 counter can be used to divide the RSR PLDT-2 clock down to 1 MHz, which is within the range of the ADC0808 clock specification.

We can make a few minor changes to the ADC interface in Figure 12.54 to make it run as a continuous-conversion circuit. First, we eliminate the **go** input and associated pushbutton. Second, we change the state diagram to eliminate the **idle** state, as shown in Figure 12.58. With no pushbutton to press and then release, we eliminate two wait transitions (previously associated with the **idle** and **start** states) from the state diagram. Otherwise, the circuit and controller remain the same. A simulation of the modified controller is shown in Figure 12.59.

EXAMPLE 12.21

From the ADC0808 datasheet extract in Figure 12.53, determine the number of clock cycles required for the conversion of an analog signal.

■ **Solution**

For a clock frequency of 640 kHz, typical conversion time is given as 100 μs.

$$\frac{640 \times 10^3 \text{ clock cycles}}{\text{second}} \times (100 \times 10^{-6} \text{ seconds}) = 64 \text{ clock cycles}$$

EXAMPLE 12.22

Calculate the highest-frequency analog input that can be accurately converted by an ADC0808 controlled by a state machine represented by the state diagram of Figure 12.58 if the system clock frequency is 787 kHz.

■ **Solution**

One conversion cycle, T_s, requires 64 clock cycles for the ADC and an overhead of 13 clocks +2 μs for the state machine for a total of 77 clock cycles +2 μs. (**Start** to **wait1** requires one clock cycle. An additional 8 cycles +2 μs are needed before **eoc** goes LOW. According to Note 7 in Figure 12.53, the ADC conversion is complete one clock cycle before *EOC* goes HIGH. From this point back to **start** is 4 clocks.)

$$T_s = \frac{1}{787 \times 10^3 \text{ clock cycles/second}} \times 77 \text{ clock cycles} + 2\mu s$$

$$= 99.8 \times 10^{-6} \text{ seconds}$$

$$f_s = \frac{1}{T_s} = \frac{1}{99.8 \times 10^{-6} \text{ s}} = 10.02 \text{ kHz}$$

According to the Nyquist sampling theorem, the maximum-frequency component of the sampled analog signal is $f_{\max} = f_s/2 = 10.02 \text{ kHz}/2 = 5.01 \text{ kHz}$. This is of the same order of magnitude as a telephone-quality audio signal.

CPLD-Based Data Acquisition Network

Figure 12.60 shows a data acquisition system that continuously converts and stores data from four analog channels. All the circuitry within the broken line is contained within a single CPLD, such as the Altera EPM7128SLC84. The operation is similar to the system in Figure 12.54, except that with multiple latches in the circuit, a counter and decoder are required to keep track of the selected channel.

The controller, whose state diagram is shown in Figure 12.61, generates the same control signals for the ADC as the system in Figure 12.54. When the conversion is complete and the controller detects a HIGH on its **eoc** input, it reads the ADC output and transfers the contents to the selected 8-bit latch. The latch is selected, via the decoder, by the value of the counter (e.g., $Q_1Q_0 = 11$ selects analog input channel 3, decoder output Y_3, and latch 3). The selected latch input is enabled (i.e., made transparent) by the controller during the transition from **wait2** to **read.** At all other times all decoder outputs are LOW, disabling all latches, thus placing them in store mode. After the ADC data have been stored, the controller sets **cnt_en** HIGH, which allows the counter to be incremented on the next clock pulse. The next channel is now ready for a convert-and-store cycle. After all channels have been sampled, converted, and stored, the cycle begins again at channel 0 and continues indefinitely.

FIGURE 12.60   4-Channel Data Acquisition System

Figure 12.62 shows a simulation of the controller, counter, and decoder for the data acquisition system of Figure 12.61. During the read-and-store part of the cycle, only one of the latch enables, **y[0]** to **y[3]**, is active when **oe** is active. The number of the active latch enable is the same as the counter value on the second last waveform. The last line in the simulation (**adc_auto:controller|outputs[3]**) is the internal **cnt_en** line from the controller to the counter. The counter is incremented on the first positive edge of the clock after this line goes HIGH. This point is indicated by the cursor line on the transition from channel 0 to channel 1.

The circuit in Figure 12.60 could be expanded to convert all eight analog channels from the ADC, but the chosen CPLD (EPM7128SLC84) does not have enough I/O pins. Eight 8-bit latch outputs require 64 pins; the CPLD only has 60 user I/Os. An 8-channel system could be implemented if it used eight external latches, such as

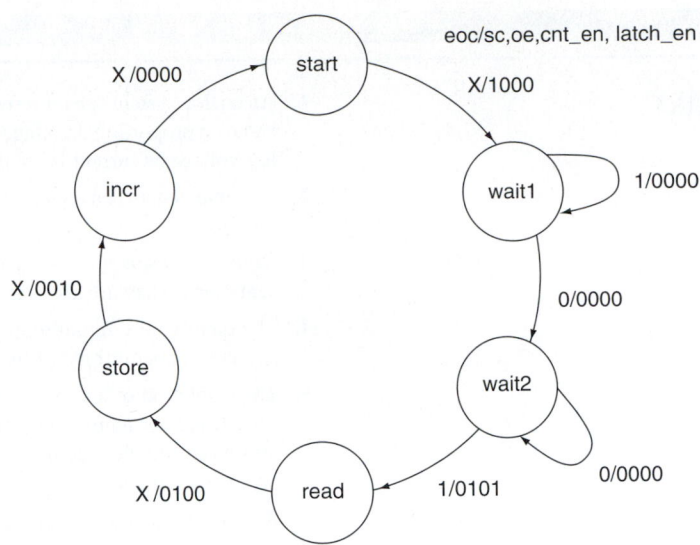

FIGURE 12.61 State Diagram for 4-Channel Data Acquisition System

FIGURE 12.62 Simulation of a 4-Channel Data Acquisition System

eight 74HC373 octal latches, or internal latches on a different CPLD. Note that the CPLD has enough logic cells to implement the system, just not enough I/O pins. The identical device in a different package (EPM7128SQC100; 100-pin quad flat-pack) can accommodate the entire system.

For an 8-channel system, the counter would need to be expanded to 3 bits and the decoder to a 3-line-to-8-line device.

■ **SECTION 12.4 REVIEW PROBLEM**

12.13 Calculate the highest-frequency component of an analog signal that can be accurately converted by the 4-channel data acquisition system in Figure 12.60. Assume the system clock is running at 787 kHz.

SUMMARY

1. An analog system can represent a physical property (e.g., temperature, pressure, or velocity) by a proportional voltage or current. The mathematical function describing the analog voltage or current is continuous throughout a defined range.

2. A digital system can represent a physical property by a series of binary numbers of a fixed bit size.

3. Digital representations of data are not subject to the same distortions as analog representations. They are also easier to store and reproduce than analog.

4. The quality of a digital representation depends on the sampling frequency and quantization (number of bits) of the system that converts an analog input to a digital output.

5. The resolution of a system is a function of the number of bits in its digital representation. A greater number of bits implies that the sampled analog input can be broken up into more, smaller segments, allowing each segment to more closely approximate the original input value.

6. A unipolar analog-to-digital converter (ADC) has a transfer characteristic in which the output code increases by steps when the analog input voltage increases from 0 to full scale.

7. The steps in an ADC transfer characteristic have a width of 1 LSB, where 1 LSB equals the resolution of the converter.

8. The first step in an ADC transfer characteristic is offset by $\frac{1}{2}$ LSB. This has the effect of centering the steps of the characteristic on the nominal voltage for each output code ($V_{\text{nominal}} \pm \frac{1}{2}$ LSB). The offset is made up at the top of the range, where the last step is $1\frac{1}{2}$ LSB wide.

9. The code for a unipolar ADC can be calculated by: $\text{code} = \dfrac{V_a}{FS} \times 2^n$

   where V_a = analog input voltage of the sample to be converted,
   FS = full scale range of input voltage, and
   n = number of bits in the output code.

10. A bipolar ADC can accommodate positive and negative values of input voltage. Outputs can be coded as offset binary or 2's complement.

11. Offset binary codes are unsigned binary numbers where the lower half of the codes represent negative inputs and the upper half represent positive inputs. The output code of an n-bit offset binary bipolar ADC can be calculated by:

$$\text{code} = \left(\frac{V_a}{FS} \times 2^n \right) + \text{offset}$$

$$= \left(\frac{V_a}{FS} \times 2^n \right) + \frac{2^n}{2}$$

12. 2's complement codes are signed binary values, where the most significant bit indicates the sign. The MSB is 0 for positive and 1 for negative. The output code for this type of device can be calculated by:

$$\text{code} = \frac{V_a}{FS} \times 2^n$$

13. The transfer characteristic for a digital-to-analog converter (DAC) consists of 2^n data points for an n-bit input. The points can be joined to see a straight-line approximation of the DAC function.

14. A unipolar DAC generates positive voltages from input codes. The analog output voltage, V_a, is calculated by: $V_a = \left(\dfrac{\text{code}}{2^n} \right) \times FS$ for an n-bit code.

15. A bipolar DAC generates positive and negative voltages from an input code. The inputs can be coded in offset binary or 2's complement.

16. An offset binary bipolar DAC uses unsigned binary numbers to code an output voltage. The lower half of the codes are for negative values. The upper half represent positive values. The output can be calculated by: $V_a = \left(\left(\dfrac{code}{2^n}\right) \times FS\right) - \left(\dfrac{FS}{2}\right)$

17. A 2's complement bipolar DAC uses signed binary numbers to code an output voltage. The output can be calculated by: $V_a = \left(\dfrac{code}{2^n}\right) \times FS$ where *code* is a 2's complement signed number.

18. A digital-to-analog converter (DAC) uses electronic switches to sum binary-weighted currents to a total analog output current. Analog current can be calculated by:

$$I_a = \dfrac{b_{n-1}2^{n-1} + b_{n-2}2^{n-2} + \cdots b_2 2^2 + b_1 2^1 + b_0 2^0}{2^n} I_{ref}$$

or, more simply:

$$I_a = \dfrac{\text{digital code}}{2^n} I_{ref}$$

for an *n*-bit DAC, where $b_{n-1}b_{n-2} \ldots b_2 b_1 b_0$ is the digital input code, I_a is the analog output current, and I_{ref} is the DAC reference (full scale) current.

19. The maximum output of a DAC is full scale *(FS)* minus the value represented by a change in the least significant bit of the input *(FS − 1 LSB)*. For example, for a 4-bit converter (1 LSB = 1/16 *FS*), the maximum output is *(FS − 1/16 FS) = 15/16 FS*. For an 8-bit converter (1 LSB = 1/256 *FS*), the maximum output is *(FS − 1/256 FS) = 255/256 FS*.

20. A weighted-resistor DAC derives its binary-weighted currents from binary-weighted resistors connected to the reference voltage supply.

21. An R-2R ladder DAC derives its binary weighted currents from a resistor ladder network that consists of resistors of two values only, one of which is twice the other. The R-2R ladder is more common than the weighted resistor DAC.

22. A DAC input code consisting of a 1 followed by all 0s represents an output of $\frac{1}{2}$ *FS*, regardless of the number of bits in the DAC input. A code of 01 followed by all 0s represents an output of $\frac{1}{4}$ *FS*. A code of 11 followed by all 0s is $\frac{3}{4}$ *FS*.

23. The MC1408 DAC is an example of a monolithic (single-chip) DAC. Output current at pin 4 is a binary-weighted fraction of the reference current at pin 14:

$$I_o = -\left(\dfrac{\text{digital code}}{256}\right)\left(\dfrac{V_{ref}}{R_{14}}\right)$$

24. If the output of an MC1408 DAC is buffered by an op amp with a feedback resistance of R_F, the output voltage is given by:

$$V_a = I_o R_F = \left(\dfrac{\text{digital code}}{256}\right)\left(\dfrac{R_F}{R_{14}}\right)V_{ref}$$

25. An 8-bit DAC can be used as a ramp generator by connecting an 8-bit binary counter to the digital inputs.

26. An MC1408 DAC can be configured for bipolar output with offset binary coding by connecting a pull-up resistor (R_4) from the output (pin 4) to the reference voltage supply. Output is given by:

$$V_a = I_o R_F - I_s R_F = \left(\dfrac{\text{digital code}}{256}\right)\left(\dfrac{R_F}{R_{14}}\right)V_{ref} - \dfrac{R_F}{R_4}V_{ref}$$

27. A DAC is monotonic if every increase in binary input results in an increase in analog output.

28. DAC errors include: offset error (nonzero output for zero input code), gain error (output falling above or below *FS − 1 LSB* for maximum input code due to an incorrect slope), linearity error (deviation from straight-line approximation between codes), and differential nonlinearity (deviation of step sizes from ideal of one step per LSB).

29. DAC linearity error of greater than $\pm\frac{1}{2}$ LSB can result in a nonmonotonic output.

30. Several popular types of analog-to-digital converters (ADC) are flash or simultaneous, successive approximation, dual slope or integrating, and sigma-delta.

31. A flash ADC consists of a voltage divider with the same number of steps as output codes, a set of comparators (one for every output code), and a priority encoder. All comparators whose reference input is less than the analog input will fire, the priority encoder will detect the highest-value active comparator, and generate the corresponding output code. A flash ADC is fast, but requires 2^n comparators for an n-bit output code.

32. A successive approximation ADC consists of a state machine called a successive approximation register (SAR) whose bits can be set and cleared individually in a specific sequence, a digital-to-analog converter, and an analog comparator.

33. A successive approximation ADC sets each bit of the SAR in turn as an approximation of the required digital code. For each bit, the approximation is converted back to analog form and compared with the incoming analog value. If the converted value is less than the actual analog value, the bit remains set and the next bit is tried. If the converted value is greater than the actual analog input, the bit is cleared and the next bit is tried.

34. A dual slope ADC consists of an integrator, comparator, counter, and control logic. The integrator output changes with a slope of $-V_{in}/RC$ for a constant input. This ADC allows the integrator to charge for the time required for the counter to complete one full cycle (known time). At that time, the integrator input is switched to a reference voltage of opposite polarity. The reference voltage discharges the integrator at a known rate. The time required to do this is stored in the counter and represents the fraction of full scale analog voltage applied to the converter.

35. A sigma-delta ADC consists of an integrator, comparator, 1-bit DAC in a feedback loop, and a summing junction. The integrator adds the values at the summing junction to the previous integrator value. The comparator sends a single bit to the output based on the integrator value. The DAC converts the comparator output to an analog voltage ($\pm V_{ref}$), which is subtracted from the current input sample, thus compensating for changes in the integrator output.

36. The sigma-delta ADC sends out a stream of bits to represent the analog input value. The bits are generated by sampling the input once for each output bit.

37. The number of 1s at the output of a sigma-delta ADC is proportional to the fraction of full scale represented by the analog input.

38. The bit stream of the sigma-delta ADC can be converted to an n-bit output by a binary counter or a digital filter.

39. Precision in a sigma-delta ADC is increased by increasing the number of samples or iterations of calculation. The precision is limited by the speed of change of the analog input. This type of ADC is best suited for applications where high accuracy is required, but the input voltage changes relatively slowly (e.g., digital audio).

40. A sample and hold circuit may be required to hold the input value of an ADC constant for the conversion time of the ADC. It samples an analog signal at periodic intervals and holds the sampled value in a capacitor until the next sample is taken. A track and hold circuit performs a similar function, but allows the capacitor to charge and discharge along with the changing analog signal, holding its value only during the conversion time of the ADC.

41. In order to preserve the information in an analog signal, it must be sampled at a frequency of at least twice the maximum-frequency component of the signal ($f_s \geq 2f_{max}$). This criterion is called the Nyquist sampling theorem.

42. If the Nyquist sampling theorem is violated, an alias frequency, or false low-frequency component, will be added to the digital representation of the analog signal.

43. Alias frequencies can be eliminated with an anti-aliasing filter, a low-pass filter used to pass only frequencies less than $2f_{max}$ to the input of an ADC. This input frequency range automatically satisfies the Nyquist criterion at the ADC input.

44. An ADC0808 successive approximation ADC contains an 8-channel analog MUX and can be used as the basis for an 8-channel data acquisition system.

45. The conversion sequence for the ADC0808 is as follows:

    a. an analog input channel is selected by setting the appropriate address on lines *ADD C, ADD B,* and *ADD A.*

    b. *ALE* and *START* are pulsed HIGH.

    c. *EOC* (end-of-conversion) goes LOW no later than 8 clock cycles +2 μs after *START.*

    d. *EOC* goes HIGH when conversion is complete.

    e. *OE* (output enable) is set HIGH to read converted output.

This sequence can be controlled by a CPLD-based state machine.

46. A data acquisition system based on an ADC0808 requires an octal latch for each analog channel, a state-machine controller, and a counter/decoder circuit to select the active analog channel and latch.

GLOSSARY

Aliasing   A phenomenon that produces an unwanted low-frequency component in a sampled analog signal due to a sampling frequency that is too slow relative to the sampled analog signal.

Analog   A way of representing some physical quantity, such as temperature or velocity, by a proportional continuous voltage or current. An analog voltage or current can have any value within a defined range.

Analog-to-Digital Converter   A circuit that converts an analog signal at its input to a digital code. (Also called an A-to-D converter, A/D converter, or ADC.)

Anti-Aliasing Filter   A low-pass filter with a corner frequency of twice the maximum frequency of a sampled signal, used to prevent aliasing in an ADC.

Bipolar ADC   An analog-to-digital converter whose input voltages have a positive and negative range.

Bipolar DAC   A digital-to-analog converter whose output voltages have a positive and negative range.

Continuous   Smoothly connected. An unbroken series of consecutive values with no instantaneous changes.

Data Acquisition Network   A circuit that gathers and digitizes data from several analog sources.

Digital   A way of representing a physical quantity by a series of binary numbers. A digital representation can have only specific discrete values.

Digital-to-Analog Converter   A circuit that converts a digital code at its input to an analog voltage or current. (Also called a D-to-A converter, D/A converter, or DAC.)

Discrete   Separated into distinct segments or pieces. A series of discontinuous values.

Dual Slope ADC   Also called an integrating ADC. An analog-to-digital converter based on an integrator. The name derives from the fact that during the conversion process the integrator output changes linearly over time, with two different slopes.

Flash Converter (or Simultaneous Converter)   An analog-to-digital converter that uses comparators and a priority encoder to produce a digital code.

Full Scale (FS)   The maximum range of analog reference voltage or current of an analog-to-digital or digital-to-analog converter.

Integrator   A circuit whose output is the accumulated sum of all previous input values. The integrator's output changes linearly with time when the input voltage is constant.

Multiplying DAC   A DAC whose output changes linearly with a change in DAC reference voltage.

Nyquist Sampling Theorem   A theorem from information theory stating that in order to preserve all information in a signal, it must be sampled at a rate of at least twice the highest-frequency component of the signal. ($f_s \geq 2f_{max}$)

Offset Binary   A binary code in which the lower half of the range represents negative numbers and the upper half represents positive numbers.

Priority Encoder   An encoder that will produce a binary output corresponding to the subscript of the highest-priority active input. This is usually defined as the input with the largest subscript.

Quantization   The number of bits used to represent an analog voltage as a digital number.

Quantization Error   Inaccuracy introduced into a digital signal by the inability of a fixed number of bits to represent the exact value of an analog signal.

Resolution   The difference in analog voltage corresponding to two adjacent digital codes. Analog step size.

Sample   An instantaneous measurement of an analog voltage, taken at regular intervals.

Sample and Hold Circuit   A circuit that samples an analog signal at periodic intervals and holds the sampled value long enough for an ADC to convert it to a digital code.

Sampling Frequency   The number of samples taken per unit time of an analog signal.

Sigma-Delta ADC   An analog-to-digital converter that uses an integrator, a comparator, and a DAC to generate a serial stream of bits based on the sum of voltage changes within the circuit during the conversion process. Also called delta-sigma ADC.

Successive Approximation Register   A state machine used to generate a sequence of closer and closer binary approximations to an analog signal.

Transfer Characteristic   A graph or mathematical function that shows the relationship between the input and output of an electronic circuit.

Unipolar ADC   An analog-to-digital converter whose input voltages are positive only.

Unipolar DAC   A digital-to-analog converter whose output voltages are positive only.

PROBLEMS

Problem numbers set in color indicate more difficult problems.

12.1   Analog and Digital Signals

12.1   An analog signal with a range of 0 to 12 V is converted to a series of 3-bit digital codes. Make a table similar to Table 12.1 showing the analog range for each digital code.

12.2   Sketch the positive half of a sine wave with a peak voltage of 12 V. Assume that this signal will be quantized according to the table constructed in Problem 12.1. Write the digital codes for the points 0, $T/8$, $T/4$, $3T/8$, . . . , T where T is the period of the half sine wave.

12.3   Repeat Problems 12.1 and 12.2 for a 4-bit quantization.

12.4   Write the 3-bit and 4-bit digital codes for the points 0, $T/16$, $T/8$, $3T/16$, . . ., T for the half sine wave described in Problem 12.2.

12.5   An analog-to-digital converter divides the range of an analog signal into 64 equal parts. The analog input has a range of 0 to 500 mV. How many bits are there in the digital codes? What is the resolution of the A/D converter?

12.6   Repeat Problem 12.5 if the analog range is divided into 256 equal parts.

12.7   The analog range of a signal is divided into m equal parts, yielding a digital quantization of n bits. If the range is divided into $2m$ parts, how many bits are in the equivalent digital codes? (That is, how many extra bits do we get for each doubling of the number of codes?)

12.8   Sketch the transfer characteristic of a 4-bit unipolar ADC with a full scale range of 0 to +12V.

12.9   Sketch the transfer characteristic of a 4-bit bipolar ADC (offset binary coding) with a full scale range of –6 V to +6 V.

12.10   Sketch the transfer characteristic of a 4-bit bipolar ADC (2's complement coding) with a full scale range of –6 V to +6 V.

12.11   A 4-bit analog-to-digital converter has a full scale of 0 V to +12 V.

   a.   Calculate the output code for an input voltage of 9 V.

   b.   Determine the range of input voltages that will generate the same code as in part **a** of this problem.

   c.   Determine the output code from the ADC if the input is 10 V.

12.12 An 8-bit ADC has a full scale range of 0 to 10 V.

    a. Calculate the output code for an input voltage of 2.5 V.

    b. Determine the range of input voltages that will generate the same code as in part **a** of this problem.

    c. Determine the output code from the ADC if the input is 2.56 V.

12.13 A 4-bit analog-to-digital converter has a full scale range of –6 V to +6 V and uses offset binary coding.

    a. Calculate the output code for an input voltage of +3 V.

    b. Determine the range of input voltages that will generate the same code as in part **a** of this problem.

    c. Determine the output code from the ADC if the input is +4 V.

12.14 An 8-bit ADC has a full scale range of –5 to +5 V and uses offset binary coding.

    a. Calculate the output code for an input voltage of –2.5 V.

    b. Determine the range of input voltages that will generate the same code as in part **a** of this problem.

    c. Determine the output code from the ADC if the input is –2.56 V.

12.15 A 4-bit analog-to-digital converter has a full scale range of –6 V to +6 V and uses 2's complement coding.

    a. Calculate the output code for an input voltage of +3 V.

    b. Determine the range of input voltages that will generate the same code as in part **a** of this problem.

    c. Determine the output code from the ADC if the input is +4 V.

12.16 An 8-bit ADC has a full scale range of –5 to +5 V and uses 2's complement coding.

    a. Calculate the output code for an input voltage of –2.5 V.

    b. Determine the range of input voltages that will generate the same code as in part **a** of this problem.

    c. Determine the output code from the ADC if the input is –2.56 V.

12.17 Sketch the transfer characteristic of a 4-bit unipolar DAC with a full scale range of 0 to +12V.

12.18 Sketch the transfer characteristic of a 4-bit bipolar DAC (offset binary coding) with a full scale range of –6 V to +6 V.

12.19 Sketch the transfer characteristic of a 4-bit bipolar DAC (2's complement coding) with a full scale range of –6 V to +6 V.

12.20 An 8-bit unipolar DAC has a full scale range of 0 to 10 V. Calculate the output voltages for input codes 01010101 and 10101010.

12.21 An 8-bit DAC has a full scale range of –5 V to +5 V. Calculate the output voltages for input codes 01010101 and 10101010 for a bipolar DAC with offset binary coding and a bipolar DAC with 2's complement coding.

12.2 Digital-to-Analog Conversion

12.22   **a.** Calculate the analog output voltage, V_a, for a 4-bit DAC when the input code is 1010.

    b. Calculate V_a for an 8-bit DAC when the input code is 10100000.

    c. Compare the results of parts **a** and **b**. What can you conclude from this comparison?

12.23   **a.** Calculate the analog output voltage, V_a, for a 4-bit DAC when the input code is 1100.

    b. Calculate V_a for an 8-bit DAC when the input code is 11001000.

    c. Compare the results of parts **a** and **b**. What can you conclude from this comparison? How does this differ from the comparison made in Problem 12.22?

12.24 Refer to the generalized D/A converter in Figure 12.10. For $I_{ref} = 500 \ \mu A$ and $R_F = 22 \ k\Omega$, calculate the range of analog output voltage, V_a, if the DAC is a 4-bit circuit. Repeat the calculation for an 8-bit DAC.

12.25   The resistor for the MSB of a 16-bit weighted resistor D/A converter is 1 kΩ. List the resistor values for all bits. What component problem do we encounter when we try to build this circuit?

12.26   Draw the circuit for an 8-bit R-2R ladder DAC.

12.27   Calculate the value of V_a of an R-2R ladder DAC when digital inputs are as follows. $V_{ref} = 12$ V.

   DCBA

   a. 1111

   b. 1011

   c. 0110

   d. 0011

12.28   An MC1408 DAC is configured as shown in Figure 12.18. $R_{14} = R_{15} = 6.8$ kΩ, $V_{ref}(+) = +12$ V, $V_{ref}(-) =$ ground, and $R_L = 2.2$ kΩ. Calculate the output voltage, V_a, for the following digital input codes: 00000000, 00000001, 10000000, 10101010, 11100010, 11111111.

12.29   Calculate the resolution of the DAC in Problem 12.28.

12.30   Refer to the op amp-buffered DAC in Figure 12.19. Assume that the resistor values are changed as follows: $R_{14A} = 270$ Ω, $R_{14B} = 2$ kΩ (max), $R_{FA} = 1.2$ kΩ, $R_{FB} = 5$ kΩ (max). Describe a step-by-step procedure that calibrates the DAC so that it has a reference current of 4 mA and a full scale analog output voltage of 12 volts, using only a series of measurements of the analog output voltage. When the procedure is complete, what are the resistance values in the circuit? What is the range of the DAC?

12.31   The resistor networks shown in the DAC circuit of Figure 12.19 allow us to set our input reference current and output gain to values within a specified range. Using the values shown in Figure 12.19, fill in Table 12.12 for the cases when V_a is at minimum and maximum, and when the potentiometers are at their midpoint values. Assume that the DAC input is set to 1111 1111. Show all calculations.

TABLE 12.12   DAC Output Range

	R_{14} (Ω)	R_F(Ω)	I_{ref} (mA)	I_o(mA)	V_a(V)
Minimum V_a					
Maximum V_a					
Pots at midpoint					

12.32   The waveform in Figure 12.63 is observed at the output of the DAC ramp generator of Figure 12.20. (Compare this to the proper waveform, found in Figure 12.21.) What is likely to be the problem with the circuit? Can it be easily fixed? How?

FIGURE 12.63   Problem 12.32: Waveform

12.33 The waveform in Figure 12.64 is observed at the output of the DAC ramp generator in Figure 12.20. What is likely to be the problem with the circuit?

FIGURE 12.64 Problem 12.33: Waveform

12.34 Refer to the bipolar DAC circuit in Figure 12.22. Describe how you would adjust the output for a range of –10 V to (+10 V – 1 LSB). Include values of variable components. Calculate the resolution of this circuit.

12.35 A 3-bit DAC has a reference voltage of 12 V and a transfer characteristic summarized in Table 12.13. Plot the data on a graph similar to those in Figures 12.24 through 12.26. From the data in Table 12.13, determine the offset error, gain error, and linearity error of the DAC, both in % of full scale and as a fraction of an LSB.

TABLE 12.13 DAC Transfer Characteristic for Problem 12.35

Digital Code	Analog Output (volts)
000	0.5
001	2.0
010	3.5
011	5.0
100	6.5
101	8.0
110	9.5
111	11.0

12.36 A 3-bit DAC has a reference voltage of 8 V and a transfer characteristic summarized in Table 12.14. Plot the data on a graph. From the data in Table 12.14, determine the offset error, gain error, linearity error, and differential nonlinearity of the DAC, both in % of full scale and as a fraction of an LSB.

TABLE 12.14 DAC Transfer Characteristic for Problem 12.36

Digital Code	Analog Output (volts)
000	0.000
001	1.036
010	2.071
011	3.107
100	4.143
101	5.179
110	6.214
111	7.250

12.37   A 3-bit DAC has a reference voltage of 4 V and a transfer characteristic summarized in Table 12.15. Plot the data on a graph. From the data in the Table 12.15, determine the offset error, gain error, and linearity error of the DAC, both in % of full scale and as a fraction of an LSB.

TABLE 12.15 DAC Transfer Characteristic for Problem 12.37

Digital Code	Analog Output (volts)
000	0.000
001	0.500
010	1.025
011	1.525
100	1.985
101	2.675
110	3.000
111	3.500

12.3 Analog-to-Digital Conversion

12.38   How many comparators are needed to construct an 8-bit flash converter? Sketch the circuit of this converter. (It is only necessary to show a few of the comparators and indicate how many there are.)

12.39   Briefly explain the operation of a flash ADC. What is the purpose of the priority encoder? Explain how the latch can be used to synchronize the output to a particular sampling frequency.

12.40   Why do we choose a value of $R/2$ for the LSB resistor of a flash ADC?

12.41   An 8-bit successive approximation ADC has a reference voltage of +16 V. Describe the conversion sequence for the case where the analog input is 4.75 V. Summarize the steps in Table 12.16. (Refer to Example 12.18.)

TABLE 12.16 Table for Problem 12.41

Bit	New Digital Value	Analog Equivalent	$v_{analog} \geq v_{DAC}$?	Comparator Output	Accumulated Digital Value
Q_7					
Q_6					
Q_5					
Q_4					
Q_3					
Q_2					
Q_1					
Q_0					

12.42 What is displayed on the seven-segment display in Figure 12.65 when $v_{analog} = 5.25$ V? Assume that the reference voltage is 12 V and that the display can show hex digits.

FIGURE 12.65 Problem 12.42: Successive Approximation ADC and Seven-Segment Display

12.43 Describe the operation of each part of the successive approximation ADC shown in Figure 12.65 when the analog input changes from 5.25 V to 8.0 V. What is the new number displayed on the seven-segment display?

12.44 **a.** An 8-bit successive approximation ADC has a reference voltage of 12 V. Calculate the resolution of this ADC.

b. The analog input voltage to the ADC in part **a** is 8 V. Can this input voltage be represented exactly? What digital code represents the closest value to 8 V? What exact analog value does this represent? Calculate the percent error of this conversion.

12.45 What is the maximum quantization error of an ADC, relative to a fraction of 1 LSB?

12.46 An 8-bit dual slope analog-to-digital converter has a reference voltage of 16 V. The integrator component values are: R = 80 kΩ, C = 0.1 μF. The analog input voltage is 14 V.

Calculate the slope of the integrator voltage during:

a. the integrating phase, and

b. the rezeroing phase.

c. How much time elapses during the rezeroing phase?
(Assume that [1] the integrating and rezeroing time are equal if the integrator output is at full scale, and [2] the reference voltage will rezero the integrator from full scale in exactly one counter cycle.)

d. Sketch the integrator output waveform.

e. What digital code is contained in the output latch after the conversion is complete?

12.47 Repeat Problem 12.46 if the analog input voltage is 3 V.

12.48 Repeat Problem 12.46 if the analog input voltage is 18 V.

12.49 List and give a brief description of each component of a sigma-delta ADC.

12.50   If you are proficient with spreadsheet software, create the spreadsheet calculator used in Figures 12.40, 12.42, and 12.43 to calculate the output of a sigma-delta ADC with 64 iterations. Use the spreadsheet to determine the bit pattern for the following analog input values:

    a.   +3 V;

    b.   –3 V;

    c.   –3.75 V
       (Assume $V_{ref} = \pm 5$ V)

    Also add in a bar graph for the integrator output. Try a few values of input voltage to see how the integrator responds (e.g., try +4 V and –4 V).

    Hint: In Excel, the Comparator and DAC columns can be calculated with the IF function, which has the form = **IF(logical_test, value_if_true, value_if_false).** For example, = IF(G7>0, 2, –4) returns the value of 2 if the contents of cell G7 is greater than 0 and the value of –4 if it is not.

12.51   A sigma-delta ADC has a reference voltage of ±8 V.

    a.   For a bit stream output of 01001010, determine the input voltage and calculate the percentage of full scale that this represents.

    b.   For a bit stream output of 01010, determine the input voltage and calculate the percentage of full scale that this represents.

    c.   For the bit stream in part **b** of this question, determine the quantization error for a value of 128 iterations of the ADC.

    d.   For the bit stream in part **b** of this question, determine the quantization error for a value of 2048 iterations of the ADC.

12.52   A sigma-delta ADC has a reference voltage of ±8 V. Assume that the output of the comparator of this ADC drives a binary counter that stores the final parallel output value of the circuit.

    a.   For a bit stream output of 1011011101, determine the input voltage and calculate the percentage of full scale that this represents.

    b.   Determine the minimum number of iterations required for 16-bit accuracy of the ADC. Calculate the quantization error of this ADC for the bit stream given in part **a** of this problem.

    c   Determine the minimum number of iterations required for 20-bit accuracy of the ADC. Calculate the quantization error of this ADC for the bit stream given in part **a** of this problem.

12.53   **a.**   State how to change the design of a parallel-output ADC to increase its precision.

    b.   State how to change the design of a sigma-delta ADC to increase its precision.

12.54   **a.**   State the factor that limits the precision of a parallel-output ADC.

    b.   State the factor that limits the precision of a sigma-delta ADC.

12.55   Make a sketch of a basic sample and hold circuit and briefly explain its operation.

12.56   Explain why a sample and hold circuit may be needed at the input of an analog-to-digital converter.

12.57   What is the highest-frequency component of an analog signal that can be accurately represented digitally if it is sampled at a rate of 100 kHz?

12.58   Calculate the minimum sampling frequency required to preserve all information when sampling a sine wave with a frequency of 130 kHz.

12.59   Suppose a sine wave with a period of 4.8 μs is sampled every 5.2 μs. What alias frequency will result? (*Hint:* see Figure 12.49.)

12.60   Calculate the corner frequency of an anti-aliasing filter for an ADC with a sampling frequency of 8 kHz. What type of filter (low-pass, high-pass, bandpass, etc.) is required?

12.4 Data Acquisition

12.61 Refer to the data acquisition system in Figure 12.54. Write a VHDL file to implement the continuous-convert version of the ADC controller, as represented in the state diagram of Figure 12.58. Create a simulation in Quartus II to verify the operation of the controller.

12.62 Use the state machine controller from Problem 12.61 and an octal latch as components in a VHDL hierarchy that represents the ADC interface of Figure 12.54. Create a simulation in Quartus II to verify the operation of the design.

12.63 The data acquisition system in Figure 12.54 is designed with the controller from Problem 12.61. (The controller state diagram is shown in Figure 12.62.) Assume the controller and latch are interfaced with a different ADC that has a conversion time of 16 μs, which is equivalent to 64 clock cycles. Calculate the highest-frequency component that can be accurately converted with this system for a clock rate of 787 kHz.

12.64 Repeat Problem 12.63 for a 4-channel data acquisition system, assuming the same conversion rate for the ADC and the controller state diagram of Figure 12.61.

ANSWERS TO SECTION REVIEW PROBLEMS

12.1A

12.1 5 bits ($2^5 = 32$). Resolution = 24 mV/32 steps = 0.75 mV/step.

12.1B

12.2 C35H = $110000110101_2 = 3125_{10}$ (unsigned)
C35H = $110000110101_2 = -971_{10}$ (2's complement)
unipolar: $V_a = (3125/4096) \times 10$ V = 7.63 V
bipolar (offset binary): $V_a = ((3125/4096) \times 10V) - 5$ V = 2.63 V
bipolar (2's complement) = $(-971/4096) \times 10$ V = -2.37 V

12.2A

12.3 4-bit: I_a = 0 to (15/16)(1 mA) = 0 to 0.9375 mA; $V_a = -I_a R_F$ = 0 to -9.375 V
8-bit: I_a = 0 to (255/256)(1 mA) = 0 to 0.9961 mA; V_a = 0 to -9.961 V

12.2B

12.4 2.048 MΩ.

12.2C

12.5 $V_a = -((10 \text{ V}/2) + (10 \text{ V}/8) + (10 \text{ V}/256)) = -6.29$ or $V_a = -(161/256) \cdot 10$ V = -6.29 V

12.2D

12.6 The maximum switching speed is higher if we choose the lower range of output voltage.

12.2E

12.7 The output 0 V requires its own code. This leaves 255, not 256, codes for the remaining output values. The maximum value of a positive-only output is 255/256 of the reference voltage. A bipolar DAC ranges from $-FS/2$ to $(+FS/2 - 1 \text{ LSB})$.

12.3A

12.8   **a.** $(7 \text{ V}/12 \text{ V}) \times 16 = 9.33$. Input voltage is between 9/16 and 11/16 of V_{ref}.
Comparator 4 has a reference voltage of $9/16 \times 12 \text{ V} = 6.75 \text{ V}$.
Comparator 5 has a reference voltage of $11/16 \times 12 \text{ V} = 8.25 \text{ V}$.
Comparators 1, 2, 3, and 4 are ON. Comparators 5, 6, and 7 are OFF.

b. Code = 100

12.3B

12.9   The comparator determines if the analog input voltage is greater than the analog
equivalent of the present approximation to the final output code.

12.3C

12.10   **a.** −1.5 V/ms

b. +4 V/ms

c. 1.125 ms

d. 01100000

12.3D

12.11   0.7 FS; 7 out of 10 bits in the output bit stream should be HIGH.

12.3E

12.12   8 kHz

12.4

12.13   1.26 kHz

KEY TERMS

Address
Address Decoder
Address Multiplexing
Address Space
b
B
Bit-Organized
Boot Block
Bottom Boot Block
Bus Contention
Byte
$\overline{CAS}$ (Column Address Strobe)
Data
DDR (Double Data Rate)
Dual In-Line Memory Module (DIMM)
Dynamic RAM
EEPROM (or E²PROM)
EPROM
FAMOS FET
FIFO (First-In First-Out)
Firmware
Flash Memory
G
Hardware
K
LIFO (Last-In First-Out)
M
Mask-Programmed ROM
Memory
Memory Map
Memory Module
Queue
RAM Cell
Rambus
Rambus Channel

Random Access Memory (RAM)
RAS (Row Address Strobe)
Read
Read-Only Memory (ROM)
Refresh Cycle
SDRAM (Synchronous DRAM)
Sector
Sequential Memory
Single In-Line Memory Module (SIMM)
Software
Stack
Static RAM
Top Boot Block
Volatile
Word
Word Length
Word-Organized
Write

OUTLINE

13.1   Basic Memory Concepts
13.2   Random Access Read/Write Memory (RAM)
13.3   Read-Only Memory (ROM)
13.4   Sequential Memory: FIFO and LIFO
13.5   Dynamic RAM Modules
13.6   Memory Systems

CHAPTER 13

Memory Devices and Systems

CHAPTER OBJECTIVES

Upon successful completion of this chapter, you will be able to:

- Describe basic memory concepts of address and data.
- Understand how latches and flip-flops act as simple memory devices and sketch simple memory systems based on these devices.
- Distinguish between random access read/write memory (RAM) and read-only memory (ROM).
- Describe the uses of tristate logic in data bussing.
- Sketch the circuits of static and dynamic RAM cells.
- Sketch a block diagram of a static or dynamic RAM chip.
- Describe various types of ROM cells and arrays: mask-programmed, UV erasable, and electrically erasable.
- Use various types of ROM in simple applications, such as digital function generation.
- Describe the basic configuration of flash memory.
- Describe the basic configuration and operation of two types of sequential memory: first-in-first-out (FIFO) and last-in-first-out (LIFO).
- Describe how dynamic RAM is configured into high-capacity memory modules.
- Sketch a basic memory system, consisting of several memory devices, an address and a data bus, and address decoding circuitry.
- Represent the location of various memory device addresses on a system memory map.
- Recognize and eliminate conditions leading to bus contention in a memory system.
- Expand memory capacity by parallel bussing and CPLD-based decoding.

In recent years, memory has become one of the most important topics in digital electronics. This is tied closely to the increasing prominence of cheap and readily available microprocessor chips. The simplest memory is a device we are already familiar with: the D flip-flop. This device stores a single bit of information as long as necessary. This simple concept is at the heart of all memory devices.

The other basic concept of memory is the organization of stored data. Bits are stored in locations specified by an "address," a unique number which tells a digital system how to find data that have been previously stored. (By analogy, think of your street address: a unique way to find you and anyone you live with.)

Some memory can be written to and read from in random order; this is called random access read/write memory (RAM). Other memory can be read only: read-only memory (ROM). Yet another type of memory, sequential memory, can be read or written only in a specific sequence. There are several variations on all these basic classes.

Memory devices are usually part of a larger system, including a microprocessor, peripheral devices, and a system of tristate busses. If dynamic RAM is used in such a system, it is often in a memory module of some type. The capacity of a single memory chip is usually less than the memory capacity of the microprocessor system in which it is used. In order to use the full system capacity, it is necessary to use a method of memory address decoding to select a particular RAM device for a specified portion of system memory. ■

13.1   BASIC MEMORY CONCEPTS

■ KEY TERMS

Memory   A device for storing digital data in such a way that they can be recalled for later use in a digital system.

Data   Binary digits (0s and 1s) that contain some kind of information. The digital contents of a memory device.

Byte   A group of 8 bits.

Address   A number, represented by the binary states of a group of inputs or outputs, uniquely defining the location of data stored in a memory device.

Read   Retrieve data from a memory device.

Write   Store data in a memory device.

Address and Data

Quartus II:
octal_latch.bdf
1×8mem.bdf
MAX+PLUS II:
octal_latch.gdf
1×8mem.gdf

A **memory** is a digital device or circuit that can store one or more bits of **data.** The simplest memory device, a D-type latch, shown in Figure 13.1, can store 1 bit. A 0 or 1 is stored in the latch and remains there until changed.

A simple extension of the single D-type latch is an array of latches, shown in Figure 13.2, that can store 8 bits (1 **byte**) of data. Figure 13.3 shows this octal latch used as a component in a Quartus II Block Diagram File and configured as an 8-bit memory.

When the *WRITEn* line goes LOW, then HIGH, data at the *DATA_IN* are stored in the eight latches. Data are available at the *DATA_OUT* pins when *READ* is HIGH. Note that although the *READ* and *WRITEn* inputs are separate in this design, their functions would often be implemented as opposite logic levels of the same pin.

FIGURE 13.1   D-Type Latch

FIGURE 13.2   Octal Latch

octal_latch

READ ⊐ INPUT → OE

DATA_IN0 ⊐ INPUT → D0   Q0 → OUTPUT ⊳ DATA_OUT0

DATA_IN1 ⊐ INPUT → D1   Q1 → OUTPUT ⊳ DATA_OUT1

DATA_IN2 ⊐ INPUT → D2   Q2 → OUTPUT ⊳ DATA_OUT2

DATA_IN3 ⊐ INPUT → D3   Q3 → OUTPUT ⊳ DATA_OUT3

DATA_IN4 ⊐ INPUT → D4   Q4 → OUTPUT ⊳ DATA_OUT4

DATA_IN5 ⊐ INPUT → D5   Q5 → OUTPUT ⊳ DATA_OUT5

DATA_IN6 ⊐ INPUT → D6   Q6 → OUTPUT ⊳ DATA_OUT6

DATA_IN7 ⊐ INPUT → D7   Q7 → OUTPUT ⊳ DATA_OUT7

WRITEn ⊐ INPUT → Gn

FIGURE 13.3 Octal Latch as 8-Bit Memory

EXAMPLE 13.1

Write a set of simulation criteria for the octal latch of Figure 13.3, when it is used as an 8-bit memory.

■ Solution

SIMULATION CRITERIA

- Test if data can be written into memory, then read back. Do this by applying data to the input and pulsing the *WRITEn* input LOW, while keeping *READ* LOW, then making *READ* HIGH to read back the newly stored value.
- Test if new storage results in new data being read back.
- Test all stored bits HIGH and LOW in a series of combinations.
- When nothing is being being read, the output should revert to the high-impedance state.

Quartus II:
1×8mem.vwf
MAX+PLUS II:
1×8mem.scf

Figure 13.4 shows a simulation of the 8-bit memory. The LOW pulses on *WRITEn* write the data, shown as two hexadecimal digits on the *DATA_IN* line, into the latches. To read the values stored in the eight latches, we set *READ* HIGH. In between read states, all *DATA_OUT* lines are in the high-impedance state, indicated by the notation ZZ.

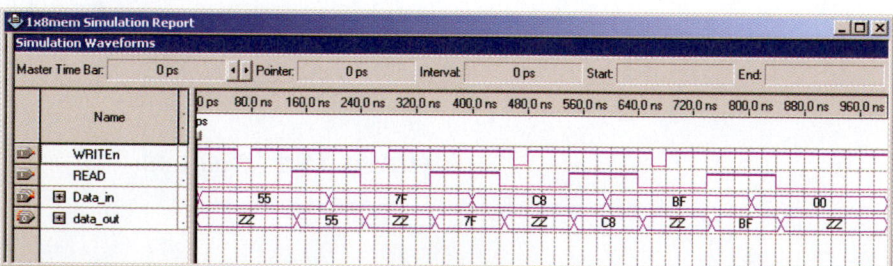

FIGURE 13.4 Simulation of 8-Bit Memory

FIGURE 13.5 4 × 8-Bit Memory from Octal Latches

Quartus II:
4×8reg.bdf
dcdr2to4.vhd
oct4to1.vhd

Figure 13.5 shows an expanded version of the octal latch memory circuit. Four octal latches are configured to make a 4 × 8-bit memory that can store and recall four separate 8-bit words. The four octal latches are instantiated as megafunctions based on the LPM_LATCH component from the Altera Library of Parameterized Modules (LPM). The 8-bit tristate output is a megafunction instance of the LPM_BUSTRI component. The remaining components of Figure 13.5 are behaviorally-designed VHDL components.

The 8-bit input data are applied to the inputs of all four octal latches simultaneously. Data are written to a particular latch when a 2-bit **address** and a LOW on *WRITEn* cause an output of a 2-line-to-4-line decoder to enable the selected latch. For example, when *ADDR[1..0]* = 01 AND *WRITEn* = 0, decoder output y[1] goes HIGH, activating the *ENABLE* input on latch 1. The values at *DATA_IN[7..0]* are transferred to latch 1 and stored there when *WRITEn* goes HIGH.

The latch outputs are applied to the data inputs of an octal 4-to-1 multiplexer. Recall that this circuit will direct one of four 8-bit inputs to an 8-bit output. The selected set of inputs correspond to the binary value at the MUX select inputs, which is the same as the address applied to the decoder in the write phase. The MUX output is directed to the *DATA_OUT* lines by an octal tristate bus driver, which is enabled by the *READ* line. To read the contents of latch 1, we set the address to 01, as before, and make the *READ* line HIGH. If *READ* is LOW, the *DATA_OUT* lines are in the high-impedance state.

Figure 13.6 shows a simulation of the 4 × 8-bit memory. The address inputs change in a continuous binary sequence. For each address, a write pulse loads 8-bit data into the selected latch. After all four latches have been loaded, the latches are read in a rotating sequence. To **read** any new data from the memory, we would first have to **write** the new data into one or more of the latch locations.

4×8reg.vwf

FIGURE 13.6 Simulation of a 4 × 8 Memory

RAM and ROM

> ■ **KEY TERMS**
>
> **Random Access Memory (RAM)** A type of memory device where data can be accessed in any order, that is, randomly. The term usually refers to random access read/write memory.
>
> **Read-Only Memory (ROM)** A type of memory where data are permanently stored and can only be read, not written.

➡ *LAB SOURCE*
Lab #34
RAM and ROM
in Altera CPLDs

The memory circuit in Figure 13.5 is one type of **random access memory,** or **RAM.** Data can be stored in or retrieved from any address at any time. The data can be accessed randomly, without the need to follow a sequence of addresses, as would be necessary in a sequential storage device such as magnetic tape.

RAM has come to mean random access read/write memory, memory that can have its data changed by a write operation, as well as have its data read. The data in another type of memory, called **read-only memory,** or **ROM,** can also be accessed randomly, although it cannot be changed, or at least not changed as easily as RAM; there is no write function; hence the name "read only." Even though both types of memory are random access, we generally do not include ROM in this category.

Memory Capacity

> ■ **KEY TERMS**
>
> **b** Bit.
>
> **B** Byte.
>
> **K** 1024 (= 2^{10}). Analogous to the metric prefix "k" (kilo-).
>
> **M** 1,048,576 (= 2^{20}). Analogous to the metric prefix "M" (mega-).
>
> **G** 1,073,741,824 (= 2^{30}). Analogous to the metric prefix "G" (giga-).

The capacity of a memory device is specified by the address and data sizes. The circuit shown in Figure 13.5 has a capacity of 4 × 8 bits ("four-by-eight"). This tells us that the memory can store 32 bits (32**b**), organized in groups of 8 bits at 4 different locations. This memory can also be described as having a capacity of 4 bytes (4**B**), since there are 8 bits per byte.

For large memories, with capacities of thousands or millions of bits, we use the shorthand designations **K** or **M** as prefixes for large binary numbers. The prefix K is analogous to, but not the same as, the metric prefix k (kilo). The metric kilo (lowercase k) indicates a multiplier of $10^3 = 1000$; the binary prefix K (uppercase) indicates a multiplier of $2^{10} = 1024$. Thus, one kilobit (Kb) is 1024 bits.

Similarly, the binary prefix M is analogous to the metric prefix M (mega). Both, unfortunately, are represented by uppercase M. The metric prefix represents a multiplier of $10^6 = 1,000,000$; the binary prefix M represents a value of $2^{20} = 1,048,576$. One megabit (Mb) is 1,048,576 bits. The next extension of this system is the multiplier G ($= 2^{30}$), which is analogous to the metric prefix **G** (giga; 10^9).

There is a move afoot to untangle all the inconsistencies in this notation and develop separate units for binary and metric applications, but to date, such new notation is not widely used.

EXAMPLE 13.2

A small microcontroller system (i.e., a stand-alone microcomputer system designed for a particular control application) has a memory with a capacity of 64 Kb, organized as $8K \times 8$. What is the total memory capacity of the system in bits? What is the memory capacity in bytes?

■ **Solution**

The total number of bits in the system memory is:

$$8K \times 8 = 8 \times 8 \times 1K = 64 \text{ Kb} = 64 \times 1024 \text{ bits} = 65,536 \text{ bits}$$

The number of bytes in system memory is:

$$\frac{64 \text{ Kb}}{8b/B} = 8 \text{ KB}$$

Usually, the range of numbers spanning 1K is expressed as the 1024 numbers from 0_{10} to 1023_{10} (0000000000_2 to 1111111111_2). This is the full range of numbers that can be expressed by 10 bits. In hexadecimal, the range of numbers spanning 1K is from 000H to 3FFH. The range of numbers in 1M is given as the full hexadecimal range of 20-bit numbers: 00000H to FFFFFH.

The range of numbers spanning 8K can be written in 13 bits ($8 \times 1K = 2^3 \times 2^{10} = 2^{13}$). The binary addresses in an $8K \times 8$ memory range from 0000000000000 to 1111111111111, or 0000 to 1FFF in hexadecimal. Thus, a memory device that is organized as $8K \times 8$ has 13 address lines and 8 data lines.

Figure 13.7 shows the address and data lines of an $8K \times 8$ memory and a map of its contents. The addresses progress in binary order, but the contents of any location

	D_8							D_1	Binary	Hexadecimal
	1	0	1	1	0	1	0	1	0 0000 0000 0000	0000
	0	0	0	1	1	0	1	1	0 0000 0000 0001	0001
	1	1	0	1	0	0	1	1	0 0000 0000 0010	0002
	0	0	0	0	0	1	1	1	0 0000 0000 0011	0003
	0	1	1	1	0	1	1	1	0 0000 0000 0100	0004
	1	0	0	0	1	0	1	0	0 0000 0000 0101	0005
	0	1	0	1	1	1	1	1	0 0000 0000 0110	0006

Addresses

									Binary	Hexadecimal
	1	0	1	0	1	0	1	0	1 1111 1111 1101	1FFD
	0	0	0	1	1	1	1	1	1 1111 1111 1110	1FFE
	1	1	0	0	1	0	1	1	1 1111 1111 1111	1FFF

a. Address and data lines   **b. Contents (data) and location (address)**

FIGURE 13.7 Address and Data in an $8K \times 8$ Memory

are the last data stored there. Since there is no way to predict what those data are, they are essentially random. For example, in Figure 13.7, the byte at address 0000000000100_2 (0004H) is 01110111_2 (77H). (One can readily see the advantage of using hexadecimal notation.)

EXAMPLE 13.3

How many address lines are needed to access all addressable locations in a memory that is organized as $64K \times 4$? How many data lines are required?

■ Solution

Addressable locations: $2^n = 64K$

$$64K = 64 \times 1K = 2^6 \times 2^{10} = 2^{16}$$
$$n = 16 \text{ address lines}$$

Data lines: There are 4 data bits for each addressable location. Thus, the memory requires 4 data lines.

Control Signals

Two memory devices are shown in Figure 13.8. The device in Figure 13.8a is a $1K \times 4$ random access read/write memory (RAM). Figure 13.8b shows $8K \times 8$ erasable programmable read-only memory (EPROM). The address lines are designated by A and the data lines by DQ. The dual notation DQ indicates that these lines are used for both input (D) and output (Q) data, using the conventional designations of D-type latches. (Note that the data inputs on the ROM can only be used under special conditions. They are used to load permanent or semipermanent data into the device, a process known as programming or burning.) The input and output data are prevented from interfering with one another by a pair of opposite-direction tristate

a. 1K x 4 RAM   b. 1K x 8 ROM (EPROM)

FIGURE 13.8   Address, Data, and Control Signals

buffers on each input/output pin. One buffer goes to a memory cell input; the other comes from the memory cell output. The tristate outputs on the devices in Figure 13.8 allow the outputs to be electrically isolated from a system data bus that would connect several such devices to a microprocessor.

In addition to the address and data lines, most memory devices, including those in Figure 13.8, have one or more of the following control signal inputs. (Different manufacturers use different notation, so several alternate designations for each function are listed.)

$\bar{E}$ **(or $\overline{CE}$ or $\overline{CS}$)**. $\overline{\text{Enable}}$ (or $\overline{\text{Chip Enable}}$ or $\overline{\text{Chip Select}}$). The memory is enabled when this line is pulled LOW. If this line is HIGH, the memory cannot be written to or read from.

$\overline{W}$ **(or $\overline{WE}$ or $R/\overline{W}$)**. $\overline{\text{Write}}$ (or $\overline{\text{Write Enable}}$ or Read/$\overline{\text{Write}}$). This input is used to select the read or write function when data input and output are on the same lines. When HIGH, this line selects the read (output) function if the chip is selected. When LOW, the write (input) function is selected.

$\bar{G}$ **(or $\overline{OE}$)**. $\overline{\text{Gate}}$ (or $\overline{\text{Output Enable}}$). Some memory chips have a separate control to enable their tristate output buffers. When this line is LOW, the output buffers are enabled and the memory can be read. If this line is HIGH, the output buffers are in the high-state. The chip select performs this function in devices without output enable pins.

The electrical functions of these control signals are illustrated in Figure 13.9.

FIGURE 13.9 Memory Control Signals

13.2 RANDOM ACCESS READ/WRITE MEMORY (RAM)

■ KEY TERMS

Volatile   A memory is volatile if its stored data are lost when electrical power is lost.

Static RAM   A random access memory that can retain data indefinitely as long as electrical power is available to the chip.

Dynamic RAM   A random access memory that cannot retain data for more than a few milliseconds without being "refreshed."

RAM Cell   The smallest storage unit of a RAM, capable of storing 1 bit.

Random access read/write memory (RAM) is used for temporary storage of large blocks of data. An important characteristic of RAM is that it is **volatile.** It can retain its stored data only as long as power is applied to the memory. When power is lost, so are the data. There are two main RAM configurations: static (SRAM) and dynamic (DRAM).

Static RAM (SRAM) consists of arrays of memory cells that are essentially flip-flops. Data can be stored in a static RAM cell and left there indefinitely, as long as power is available to the RAM.

A **dynamic RAM** cell stores a bit as the charged or discharged state of a small capacitor. Since the capacitor can hold its charge for only a few milliseconds, the charge must be restored ("refreshed") regularly. This makes a dynamic RAM (DRAM) system more complicated than SRAM, as it introduces a requirement for memory refresh circuitry.

DRAMs have the advantage of large memory capacity over SRAMs. At the time of this writing, the largest SRAMs have a capacity of about 32 Mb, whereas the largest DRAMs have a capacity of 512 Mb. DRAM modules, that is, groups of DRAM chips on a small circuit board, have capacities of up to 2 GB. These figures are constantly increasing and are never up to date for very long. (The most famous estimate of the growth rate of semiconductor memory capacity, Moore's law, estimates that it doubles every 18 months. My casual observation is that this is accurate to within an order of magnitude.)

Static RAM Cells

The typical static **RAM cell** consists of at least two transistors that are cross-coupled in a flip-flop arrangement. Other parts of the cell include pull-up circuitry that can be active (transistor switches) or passive (resistors) and some decoding/switching logic. Figure 13.10 shows an SRAM cell in three technologies: bipolar, NMOS, and CMOS.

Each of these cells can store 1 bit of data, a 0 or a 1, as the state of one of the transistors in the cell. The data are available in true or complement form, as the *BIT* and $\overline{BIT}$ outputs of the flip-flop.

All types of SRAM cells operate in more or less the same way. We will analyze the operation of the NMOS cell (Figure 13.10b) and then compare it to the other types.

Transistors Q_1 and Q_2 are permanently biased ON, making them into pull-up resistors. Channel width and length are chosen to give a resistance of about 1 kΩ. These NMOS load transistors are considered passive pull-ups, as they do not switch on and off.

A bit is stored as V_{DS3}, the drain voltage of Q_3 with respect to its source. If this voltage is HIGH, the gate of Q_4 is HIGH with respect to its source and Q_4 is biased ON. This completes a conduction path from the drain of Q_4 to its source, making V_{DS4}

a. **Bipolar SRAM cell**

b. **NMOS SRAM cell**

c. **CMOS SRAM cell**

FIGURE 13.10 SRAM Cells

logic LOW. This LOW is fed back to the gate of Q_3, turning it OFF. There is no conduction path between the drain and source of Q_3, so $V_{DS3} = V_{DD}$ or logic HIGH. The cell is storing a 1.

This bit can be read by making the *ROW SELECT* line HIGH. This turns Q_5 and Q_6 ON, which puts the data onto the *BIT* and $\overline{BIT}$ lines where it can be read by other circuitry inside the RAM chip.

To change the cell contents to a 0, we make the *BIT* line LOW and the *ROW SELECT* line HIGH. The *ROW SELECT* line gives access to the cell by turning on Q_5 and Q_6, completing the conduction path between the *BIT* lines and the flip-flop inputs. The LOW on the *BIT* line pulls the gate of Q_4 LOW, turning it OFF. This breaks the conduction path from Q_4 drain to source and makes $V_{DS4} = V_{DD}$, a logic HIGH. This HIGH is applied to the gate of Q_3, turning it ON. A conduction path is established between Q_3 drain and source, pulling the drain of Q_3 LOW. The cell now stores a logic 0.

> **NOTE . . .**
>
> The contents of an SRAM cell must be changed by introducing a LOW on the BIT or the $\overline{BIT}$ line. The data cannot be changed by pulling an input HIGH without pulling the opposite input LOW. If a MOSFET gate is at the LOW state, a HIGH applied to that gate will be pulled down by the LOW level already existing there and will not cause the cell to change state.

The CMOS cell (Figure 13.10c) functions in the same way, except for the actions of Q_1 and Q_2. Q_1 and Q_3 are a complementary pair, as are transistors Q_2 and Q_4. For each of these pairs, when the p-channel transistor is ON, the n-channel is OFF, and vice versa. This arrangement is more energy efficient than the NMOS cell, since there is not the constant current drain associated with the load transistors. Power is consumed primarily during switching between states.

The main design goal of new memory technology is to increase speed and capacity while reducing power consumption and chip area. The NMOS cell has the advantage of being constructed from only one type of component. This makes it possible to manufacture more cells in the same chip area than can be done in either the CMOS or bipolar technologies. NMOS chips, however, are slower than bipolar. New advances in high-speed CMOS technologies have made possible CMOS memories that are as dense or denser than NMOS and faster. Because of this, NMOS will probably decline in importance over time.

Bipolar SRAMs can be either TTL, as shown in Figure 13.10a, or ECL, which is not shown. Of the two bipolar technologies, ECL is the faster. Historically, all bipolar SRAMs have had the advantage of speed over NMOS and CMOS chips. New CMOS devices however, have exceeded the speeds of TTL.

The bipolar SRAM cell is the least suitable for high-density memory. Both bipolar transistors and resistors are large components compared to a MOSFET. Thus, the bipolar cell is inherently larger than the CMOS or NMOS cell. Bipolar memories historically have been used when a small amount of high-speed memory is required.

The operation of the bipolar SRAM cell is similar to that of the MOSFET cells. In the quiescent state, the ROW $SELECT$ line is LOW. In either the Read or the Write mode, the ROW $SELECT$ line is HIGH. To change the data in the cell, pull one of the emitters LOW. When the emitter of Q_1 goes LOW, the cell contents become 0. When the emitter of Q_2 is pulled LOW, the cell contents are 1.

Static RAM Cell Arrays

> **■ KEY TERMS**
>
> **Word**   Data accessed at one addressable location.
>
> **Word-Organized**   A memory is word-organized if one address accesses one word of data.
>
> **Word Length**   Number of bits in a word.

Static RAM cell arrays are arranged in a square or rectangular format, accessible by groups in rows and columns. Each column corresponds to a complementary pair of BIT lines and each row to a ROW $SELECT$ line, as shown in Figure 13.11.

The column lines have MOSFETs configured as pull-up resistors at one end and a circuit called a sense amplifier at the other. The sense amp is a large RAM cell that amplifies the charge of an active storage cell on the same BIT line. Having a larger RAM cell as a sense amp allows the storage cells to be smaller, since each individual cell need not carry the charge required for a logic level output.

Figure 13.12 shows the block diagram of a 4-megabit (Mb) SRAM array, including blocks for address decoding and output circuitry. The RAM cells are arrayed in

FIGURE 13.11 SRAM Cell Array

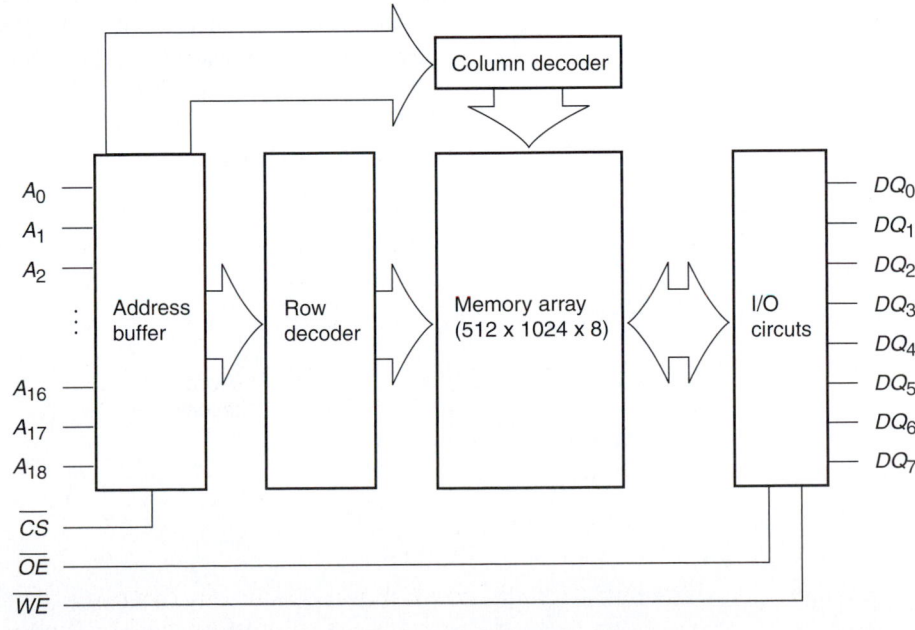

FIGURE 13.12 Block Diagram of a 4-Mb (512 KB) SRAM

a pattern of 512 rows and 8192 columns for efficient packaging. When a particular address is applied to address lines $A_{18} \ldots A_0$, the row and column decoders select an SRAM cell in the memory array for a read or write by activating the associated sense amps for the column and the row select line for the cell.

The columns are further subdivided into groups of eight, so that one column address selects eight bits (one byte) for a read or write operation. Thus, there are 512 separate row addresses (9 bits) and 1024 separate column addresses (10 bits) for every unique group of 8 data bits, requiring a total of 19 address lines and 8 data lines. The capacity of the SRAM can be written as $512 \times 1024 \times 8$.

Since one address reads or writes 8 cells (an 8-bit **word**), we say that the SRAM in Figure 13.12 is **word-organized** and that the **word length** of the SRAM is 8 bits. Other popular word lengths for various memory arrays are 4, 16, 32, and 64 bits.

■ SECTION 13.2A REVIEW PROBLEM

13.1   If an SRAM array is organized as $512 \times 512 \times 16$, how many address and data lines are required? How does the bit capacity of this SRAM compare to that of Figure 13.12?

Dynamic RAM Cells

> ### ■ KEY TERM
>
> **Refresh Cycle**   The process that periodically recharges the storage capacitors in a dynamic RAM.

A dynamic RAM (DRAM) cell consists of a capacitor and a pass transistor, as shown in Figure 13.13. A bit is stored in the cell as the charged or discharged state of the capacitor. The bit location is read from or written to by activating the cell MOSFET via the Word Select line, thus connecting the capacitor to the *BIT* line.

The major disadvantage of dynamic RAM is that the capacitor will eventually discharge by internal leakage current and must be recharged periodically to maintain integrity of the stored data. The recharging of the DRAM cell capacitors, known as refreshing the memory, must be done every 8 to 64 ms, depending on the device.

The **refresh cycle** adds an extra level of complication to the DRAM hardware and also to the timing of the read and write cycles, since the memory might have to be refreshed between read and write tasks. DRAM timing cycles are much more complicated than the equivalent SRAM cycles.

This inconvenience is offset by the high bit densities of DRAM, which are possible due to the simplicity of the DRAM cell. Up to 512 megabits of data can be stored on a single chip.

FIGURE 13.13   Dynamic RAM Cell

DRAM Cell Arrays

■ **KEY TERMS**

Bit-Organized A memory is bit-organized if one address accesses one bit of data.

Address Multiplexing A technique of addressing storage cells in a dynamic RAM that sequentially uses the same inputs for the row address and column address of the cell.

$\overline{RAS}$ **(Row Address Strobe)** A signal used to latch the row address into the decoding circuitry of a dynamic RAM with multiplexed addressing.

$\overline{CAS}$ **(Column Address Strobe)** A signal used to latch the column address into the decoding circuitry of a dynamic RAM with multiplexed addressing.

Dynamic RAM is sometimes **bit-organized** rather than word-organized. That is, one address will access one bit rather than one word of data. A bit-organized DRAM with a large capacity requires more address lines than a static RAM (e.g., 4 Mb × 1 DRAM requires 22 address lines ($2^{22} = 4,194,304 = 4M$) and 1 data line to access all cells).

In order to save pins on the IC package, a system of **address multiplexing** is used to specify the address of each cell. Each cell has a row address and a column address, which use the same input pins. Two negative-edge signals called **row address strobe** ($\overline{RAS}$) and **column address strobe** ($\overline{CAS}$) latch the row and column addresses into the DRAM's decoding circuitry. Figure 13.14 shows a simplified block diagram of the row and column addressing circuitry of a 1 M × 1 dynamic RAM.

Figure 13.15 shows the relative timing of the address inputs of a dynamic RAM. The first part of the address is applied to the address pins and latched into the row address buffers when $\overline{RAS}$ goes LOW. The second part of the address is then applied to the address pins and latched into the column address buffers by the $\overline{CAS}$ signal. This allows a 20-bit address to be implemented with 12 pins: 10 address and 2 control lines. Adding another address line effectively adds 2 bits to the address, allowing access to 4 times the number of cells.

The memory cell array in Figure 13.14 is rectangular, not square. One of the Row Address lines is connected internally to the Column Address decoder, resulting in a 512-row-by-2048-column memory array.

FIGURE 13.14 Row and Column Decoding in a 1M × 1 Dynamic RAM

FIGURE 13.15 DRAM Address Latch Signals

One advantage to the rectangular format shown is that it cuts the memory refresh time in half, since all the cells are refreshed by accessing the rows in sequence. Fewer rows means a faster refresh cycle. All cells in a row are also refreshed by normal read and write operations.

■ SECTION 13.2B REVIEW PROBLEMS

13.2 How many address and data lines are required for the following sizes of dynamic RAM, assuming that each memory cell array is organized in a square format, with common Row and Column Address pins?

   a. $1M \times 1$

   b. $1M \times 4$

   c. $4M \times 1$

13.3 READ-ONLY MEMORY (ROM)

> ### ■ KEY TERMS
>
> **Software** Programming instructions required to make hardware perform specified tasks.
>
> **Hardware** The electronic circuit of a digital or computer system.
>
> **Firmware** Software instructions permanently stored in ROM.

The main advantage of read-only memory (ROM) over random access read/write memory (RAM) is that ROM is nonvolatile. It will retain data even when electrical power is lost to the ROM chip. The disadvantage of this is that stored data are difficult or impossible to change.

ROM is used for storing data required for tasks that never or rarely change, such as **software** instructions for a bootstrap loader in a personal computer or microcontroller (the **hardware**). (The bootstrap loader—a term derived from the whimsical idea of pulling oneself up by one's bootstraps, that is, starting from nothing—is the software that gives the personal computer its minimum startup information. Generally, it contains the instructions needed to read a magnetic disk containing further operating instructions. This task is always the same for any given machine and is needed every time the machine is turned on, thus making it the ideal candidate for ROM storage.)

Software instructions stored in ROM are called **firmware.**

Mask-Programmed ROM

> ### ■ KEY TERM
>
> **Mask-Programmed ROM** A type of read only memory (ROM) where the stored data are permanently encoded into the memory device during the manufacturing process.

The most permanent form of read-only memory is the **mask-programmed ROM**, where the stored data are manufactured into the memory chip. Due to the inflexibility of this type of ROM and the relatively high cost of development, it is used only for well-developed high-volume applications. However, even though development cost of a mask-programmed ROM is high, volume production is cheaper than for some other types of ROM.

Examples of applications suitable to mask-programmed ROM include:

■ Bootstrap loaders and BIOS (basic input/output system) for PCs
■ Character generators (decoders that convert ASCII codes into alphanumeric characters on a CRT display)
■ Function lookup tables (tables corresponding to binary values of trigonometric, exponential, or other functions)
■ Special software instructions that must be permanently stored and never changed (firmware)

In recent years, mask-programmed ROM has been superceded by flash memory for many of these applications.

Figure 13.16 shows a ROM based on a matrix of MOSFETs. Each cell is manufactured with a MOSFET and its gate and source connections. LOWs are programmed by making a connection between the drain of the cell's MOSFET and the corresponding Bit line. When the appropriate Row Select goes HIGH, the MOSFET turns ON, providing a path to ground from the selected Bit line. Cells

FIGURE 13.16 Mask-Programmed ROM

programmed HIGH have no connection between the MOSFET drain and the Bit line, which thus cannot be pulled LOW when the cell is selected.

These connections can be made by a custom overlay of connections (a mask) on top of the standard-cell layer. The standard-cell-plus-custom-overlay format is cheaper to manufacture than custom cells for each bit, even if many of the MOSFETs are never used.

EPROM

> **■ KEY TERMS**
>
> **EPROM**   Erasable programmable read-only memory. A type of ROM that can be programmed ("burned") by the user and erased later, if necessary, by exposing the chip to ultraviolet radiation.
>
> **FAMOS FET**   Floating-gate avalanche MOSFET. A MOSFET with a second, "floating" gate in which charge can be trapped to change the MOSFET's gate-source threshold voltage.

Mask-programmed ROM is useful because of its nonvolatility, but it is hard to program and impossible to erase. **Erasable programmable read-only memory (EPROM)** combines the nonvolatility of ROM with the ability to change the internal data if necessary.

This erasability is particularly useful in the development of a ROM-based system. Anyone who has built a complex circuit or written a computer program knows that there is no such thing as getting it right the first time. Modifications can be made easily and cheaply to data stored in an EPROM. Later, when the design is complete, a mask ROM version can be prepared for mass production. Alternatively, if the design will be produced in small numbers, the ROM data can be stored in EPROMs, saving the cost of preparing a mask-programmed ROM.

The basis of the EPROM memory cell is the **FAMOS FET,** whose circuit symbol shown in Figure 13.17. FAMOS stands for floating-gate avalanche metal-oxide-semiconductor. ("Avalanche" refers to electron behavior in a semiconductor under certain bias conditions.) This is a MOSFET with a second, or floating, gate that is insulated from the first by a thin oxide layer.

The floating gate has no electrical contact with either the first gate or the source and drain terminals. As is the case in a standard MOSFET, conduction between drain and source terminals is effected by the voltage of the gate terminal with respect to the source. If this voltage is above a certain threshold level, the transistor will turn ON, allowing current to flow between drain and source.

In the unprogrammed state the FAMOS transistor's threshold voltage is low enough for the transistor to be turned ON by a 5-V read signal on the Row Select line. During the programming operation, a relatively high voltage pulse (about 12 V to 25 V, depending on the device) on the Row Select line drives high-energy electrons into the floating gate and traps them there. This raises the threshold voltage of the programmed cell to a level where the cell won't turn ON when selected by a 5-V read.

The EPROM cells are configured so that an unprogrammed location contains a logic HIGH and the programming signal forces it LOW.

To erase an EPROM, the die (i.e., the silicon chip itself) must be exposed for about 20 to 45 minutes to high-intensity ultraviolet light of a specified wavelength (2537 angstroms) at a distance of 2.5 cm (1 inch). The high-energy photons that make up the UV radiation release the electrons trapped in the floating gate and restore the cell threshold voltages to their unprogrammed levels.

FIGURE 13.17   FAMOS FET

EPROMS are manufactured with a quartz window over the die to allow the UV radiation in. Since both sunlight and fluorescent light contain UV light of the right wavelength to erase the EPROM over time (several days to several years, depending on the intensity of the source), the quartz window should be covered by an opaque label after the EPROM has been programmed.

EPROM Application: Digital Function Generator

An EPROM can be used as the central component of a digital function generator. Other components in the system include a clock generator, a counter, a digital-to-analog converter, and an output op amp buffer. The portion of the circuit including the last three of these components is shown in Figure 13.18.

The generator can produce the usual analog waveforms—sine, square, triangle, sawtooth—and any other waveforms that you wish to store in the EPROM. A single cycle of each waveform is stored as 256 consecutive 8-bit numbers. For example, the data for one cycle of the sine waveform are stored at addresses 0000H to FFFFH, as shown in hex form in Table 13.1. (The data are coded in offset binary format. That is, FF is maximum positive, 80 is zero, and 00 is maximum negative.) The square wave data are stored at addresses 0100 to 01FF, also shown in Table 13.1. The data for other functions, stored in subsequent 256-byte blocks, are not shown. A full list of the function data and an ANSI C program to generate an EPROM record file (Intel format) are included in Appendix E found on the CD that accompanies this book.

FIGURE 13.18　Digital Function Generator

TABLE 13.1 EPROM Sine and Square Wave Data

SINE

Base Address	Byte Addresses															
	0	1	2	3	4	5	6	7	8	9	A	B	C	D	E	F
0000	80	83	86	89	8C	8F	92	95	98	9C	9F	A2	A5	A8	AB	AE
0010	B0	B3	B6	B9	BC	BF	C1	C4	C7	C9	CC	CE	D1	D3	D5	D8
0020	DA	DC	DE	E0	E2	E4	E6	E8	EA	EC	ED	EF	F0	F2	F3	F5
0030	F6	F7	F8	F9	FA	FB	FC	FC	FD	FE	FE	FF	FF	FF	FF	FF
0040	FF	FF	FF	FF	FF	FF	FE	FE	FD	FC	FC	FB	FA	F9	F8	F7
0050	F6	F5	F3	F2	F0	EF	ED	EC	EA	E8	E6	E4	E2	E0	DE	DC
0060	DA	D8	D5	D3	D1	CE	CC	C9	C7	C4	C1	BF	BC	B9	B6	B3
0070	B0	AE	AB	A8	A5	A2	9F	9C	98	95	92	8F	8C	89	86	83
0080	7F	7C	79	76	73	70	6D	6A	67	63	60	5D	5A	57	54	51
0090	4F	4C	49	46	43	40	3E	3B	38	36	33	31	2E	2C	2A	27
00A0	25	23	21	1F	1D	1B	19	17	15	13	12	10	0F	0D	0C	0A
00B0	09	08	07	06	05	04	03	03	02	01	01	00	00	00	00	00
00C0	00	00	00	00	00	00	01	01	02	03	03	04	05	06	07	08
00D0	09	0A	0C	0D	0F	10	12	13	15	17	19	1B	1D	1F	21	23
00E0	25	27	2A	2C	2E	31	33	36	38	3B	3E	40	43	46	49	4C
00F0	4F	51	54	57	5A	5D	60	63	67	6A	6D	70	73	76	79	7C

SQUARE

Base Address	Byte Addresses															
	0	1	2	3	4	5	6	7	8	9	A	B	C	D	E	F
0100	FF	FF	FF	FF	FF	FF	FF	FF	FF	FF	FF	FF	FF	FF	FF	FF
0110	FF	FF	FF	FF	FF	FF	FF	FF	FF	FF	FF	FF	FF	FF	FF	FF
0120	FF	FF	FF	FF	FF	FF	FF	FF	FF	FF	FF	FF	FF	FF	FF	FF
0130	FF	FF	FF	FF	FF	FF	FF	FF	FF	FF	FF	FF	FF	FF	FF	FF
0140	FF	FF	FF	FF	FF	FF	FF	FF	FF	FF	FF	FF	FF	FF	FF	FF
0150	FF	FF	FF	FF	FF	FF	FF	FF	FF	FF	FF	FF	FF	FF	FF	FF
0160	FF	FF	FF	FF	FF	FF	FF	FF	FF	FF	FF	FF	FF	FF	FF	FF
0170	FF	FF	FF	FF	FF	FF	FF	FF	FF	FF	FF	FF	FF	FF	FF	FF
0180	00	00	00	00	00	00	00	00	00	00	00	00	00	00	00	00
0190	00	00	00	00	00	00	00	00	00	00	00	00	00	00	00	00
01A0	00	00	00	00	00	00	00	00	00	00	00	00	00	00	00	00
01B0	00	00	00	00	00	00	00	00	00	00	00	00	00	00	00	00
01C0	00	00	00	00	00	00	00	00	00	00	00	00	00	00	00	00
01D0	00	00	00	00	00	00	00	00	00	00	00	00	00	00	00	00
01E0	00	00	00	00	00	00	00	00	00	00	00	00	00	00	00	00
01F0	00	00	00	00	00	00	00	00	00	00	00	00	00	00	00	00

The counter and EPROM can also be implemented in a CPLD. Alternatively, a VHDL-designed state machine can replace the counter and EPROM, except for the sine function. These configurations are designed and built as exercises in Lab 32 in the lab manual that accompanies this book.

The most significant bits of the EPROM address select the waveform function by selecting a block of 256 address. The 8 least significant bits of the EPROM address are connected to an 8-bit (mod-256) counter, which continuously cycles through the 256 selected addresses. A 27C64 EPROM (8K × 8) has 13 address lines. After the eight lower lines are accounted for, the remaining five lines can be used to select up to 32 digital functions. With the two binary Function Select switches, we can potentially select 4 functions.

For example, to select the Sine function, inputs A_9 and A_8, which comprise the most significant digit of the EPROM address, are set to 00. Thus, the 8-bit counter cycles through addresses 0000–00FF, the location of the sine data. The Square Wave function is selected by setting A_9 and A_8 to 01, thus selecting the address block 0100–01FF. For a third function, $A_9 = 1$ and $A_8 = 0$, which selects the address block 0200–02FF. Other functions can be similarly selected.

The data at each address are sent to the D/A converter (MC1408), which, in combination with the op amp, is configured to produce a bipolar (both positive and negative) output. (We use a high slew rate op amp, such as an MC34071 or TL071, so that the generated square waves will have vertical sides.) The circuit generates a continuous waveform by retracing the data points in one 256-byte section of the EPROM over and over.

The DAC/op amp combination produces a maximum negative voltage for a hex input of 00, a 0 V output for an input of 80, and a maximum positive voltage for an input of FF. (You might wish to refer to the section Bipolar Operation of MC1408 in Chapter 12 for details of the DAC operation.)

You can see from the Sine function data in Table 13.1 that 8 bits are not sufficient to represent each of the 256 steps of a digital sine function as a unique number. The peaks of the waveform are changing too slowly to be represented accurately by an 8-bit quantization, and as a result, the top of the sine wave is flat for several clock pulses. (Mathematically, a sine function is tangential to a horizontal line at its peak. However, since tangential means touching at one point, the flat top is a distortion.) A unique number for each of 256 steps of a sine function needs at least 13 bits,[1] but this requires additional bits on the D/A converter input, and therefore a different DAC and an expanded memory word length. Alternatively, unique codes for each step can be generated by using fewer than 256 steps.

The output frequency of the function generator is 1/256 of the clock rate. Given that the settling time of the MC1408 DAC is about 300ns, the maximum clock rate of the circuit is 1/300 ns = 3.33 MHz. At this rate, the output frequency is 3.33 MHz/256 = 13 kHz.

EEPROM

■ KEY TERM

EEPROM (or E²PROM)   Electrically erasable programmable read-only memory. A type of read only memory that can be field-programmed and selectively erased while still in a circuit.

As was discussed in the previous section, EPROMs have the useful property of being erasable. The problem is that they must be removed from the circuit for erasure, and bits cannot be selectively erased; the whole memory cell array is erased as a unit.

Electrically erasable programmable read-only memory (**EEPROM** or **E²PROM**) provides the advantages of EPROM along with the additional benefit of allowing erasure of selected bits while the chip is in the circuit; it combines the read/write properties of RAM with the nonvolatility of ROM. EEPROM is useful for storage of data that need to be changed occasionally, but that must be retained when

---

[1]Bits required:

- 360°/256 steps = 1.40625°/step.
- Sine function changes most slowly at peak, so calculate $A \sin (90° \pm 1.40625°)$ to find smallest amplitude change.
- The smallest power-of-2 amplitude, A, for which $A \sin (90°) - A \sin (90° \pm 1.40625°) \geq 1$ is 4096.
- The amplitude range $-4096 \leq A \leq 4095$ can be represented by a 13-bit number.

00000H	S0 (Boot block)	16 KB
04000H	S1	8 KB
06000H	S2	8 KB
08000H		
	S3	32 KB
10000H		
	S4	64 KB
20000H		
	S5	64 KB
30000H		
	S6	64 KB
40000H		
	S7	64 KB
50000H		
	S8	64 KB
60000H		
	S9	64 KB
70000H		
	S10	64 KB
7FFFFH		

FIGURE 13.20 Sectors in a 512K × 8b Flash Memory (Bottom Boot Block)

power is lost to the EEPROM chip. One example is the memory circuit in an electronically tuned car radio that stores the channel numbers of local stations.

Like the UV-erasable EPROM, the memory cell of the EEPROM is based on the FAMOS transistor. Unlike the EPROM, the FAMOS FET is coupled with a standard MOSFET, as shown in Figure 13.19.

The FAMOS FET is programmed in the same way as UV-erasable EPROM: a programming voltage pulse (V_{PP}) drives high-energy electrons into the floating gate of the FAMOS transistor, where they remain trapped and change the threshold voltage of the transistor. The cell is read by keeping the programming line at 5 V and making the cell's Row Select line HIGH. The FAMOS transistor will or will not turn on, depending on its programmed state.

The FAMOS transistors used in EPROM and EEPROM differ in one important respect. The EEPROM transistor is manufactured with a very thin oxide layer between the drain and the upper (nonfloating) gate. This construction allows trapped electrons in the floating gate to be forced out electrically, thus erasing the cell contents.

Given the obvious advantages of EEPROM, why doesn't it replace all other types of memory? There are several reasons:

1. EEPROM has a much slower access time than RAM and is thus not good for high-speed applications.
2. The currently available EEPROMs have significantly smaller bit capacities than commercially available RAM (especially dynamic RAM) and EPROM.
3. EEPROM has a fixed number of write/erase cycles, typically 100,000. After that, new data cannot be programmed into the device.

FIGURE 13.19 EEPROM cell

Flash Memory

A popular variation on EEPROM is **flash memory.** This type of nonvolatile memory generally has a larger byte capacity than EEPROM devices (up to 32 Mb) and thus can be used to store fairly large amounts of firmware, such as the BIOS (basic input/output system) of a PC.

A flash memory is divided into **sectors,** groups of bytes that are programmed and erased at one time. One sector is designated as the **boot block,** which is either the sector with the highest (**top boot block**) or lowest (**bottom boot block**) address. The primary firmware is usually stored in the boot block, with the idea that the system using the flash memory is configured to look there first for firmware instructions. The boot block can also be protected from unauthorized erasure or modification (e.g., by a virus), thus adding a security feature to the device.

Figure 13.20 shows the arrangement of sectors of a 512K × 8-bit (4 Mb) flash memory with a bottom boot block architecture. The range of addresses are shown alongside the blocks. For example, sector S0 (the boot block) has a 16 KB address range of 00000H to 03FFFH. Sector S1 has an 8 KB address range from 04000H to 05FFFH. The first 64 KB of the memory are divided into one 16 KB, two 8 KB, and one 32 KB sectors. The remainder of the memory is divided into equal 64 KB sectors. Note that even though the boot block is drawn at the top of Figure 13.20, it is a bottom boot block because it is the sector with the lowest address.

A flash memory with a top boot block would have the same proportions given over to its sectors, but mirror-image to the diagram in Figure 13.20. That is, S10 (boot block) would be a 16 KB sector from 7C000H to 7FFFFH. The other sectors would be identical to the bottom boot block architecture, but in reverse order.

As with other EEPROM devices, a flash memory can be erased and reprogrammed while installed in a circuit. The memory cells in a flash device have a limited number of program/erase cycles, like other EEPROMs. The sector architecture of the flash memory makes it faster to erase and program than other EEPROM-based memories which must erase or program bytes one at a time. This same characteristic makes it unsuitable for use as system RAM, which must be able to program single bytes.

■ SECTION 13.3 REVIEW PROBLEM

13.3 A flash memory has a capacity of 8 Mb, organized as 1M × 8-bit. List the address range for the 32 KB boot block sector of the memory if the device has a bottom boot block architecture and if it has a top boot block architecture.

13.4 SEQUENTIAL MEMORY: FIFO AND LIFO

■ KEY TERMS

Sequential Memory   Memory in which the stored data cannot be read or written in random order, but must be addressed in a specific sequence.

Queue   A FIFO memory.

Stack   A LIFO memory.

FIFO (First-In First-Out)   A sequential memory in which the stored data can be read only in the order in which it was written.

LIFO (Last-In First-Out)   A sequential memory in which the last data written are the first data read.

The RAM and ROM devices we have examined up until now have all been random access devices. That is, any data could be read from or written to any sequence of addresses in any order. There is another class of memory in which the data must be accessed in a particular order. Such devices are called **sequential memory.**

a. Queue (FIFO)

b. Stack (LIFO)

FIGURE 13.21 Sequential Memory

There are two main ways of organizing a sequential memory—as a **queue** or as a **stack.** Figure 13.21 shows the arrangement of data in each of these types of memory.

A queue is a **first-in first-out (FIFO)** memory, meaning that the data can be read only in the same order they are written, much as railway cars always come out of a tunnel in the same order they go in.

One common use for FIFO memory is to connect two devices that have different data rates. For instance, a computer can send data to a printer much faster than the printer can use it. To keep the computer from either waiting for the printer to print everything or periodically interrupting the computer's operation to continue the print task, data can be sent in a burst to a FIFO, where the printer can read them as needed. The only proviso is that there must be some logic signal to the computer telling it when the queue is full and not to send more data and another signal to the printer letting it know that there are some data to read from the queue.

The **last-in-first-out (LIFO),** or stack, memory configuration, also shown in Figure 13.21, is not available as a special chip, but rather is a way of organizing RAM in a memory system.

The term "stack" is analogous to the idea of a spring-loaded stack of plates in a cafeteria line. When you put a bunch of plates on the stack, they settle into the recessed storage area. When a plate is removed, the stack springs back slightly and brings the second plate to the top level. (The other plates, of course, all move up a notch.) The top plate is the only one available for removal from the stack, and plates are always removed in reverse order from that in which they were loaded.

Figure 13.21b shows how data are transferred to and from a LIFO memory. A block of addresses in a RAM is designated as a stack, and one or two bytes of data in the RAM store a number called the stack pointer, which is the current address of the top of the stack. (The size of the stack pointer depends on the width of the system address bus.)

In Figure 13.21, the value of the stack pointer changes with every change of data in the stack, pointing to the last-in data in every case. When data are removed from the stack, the stack pointer is used to locate the data that must be read first. After the read, the stack pointer is modified to point to the next-out data. Some stack configurations have the stack pointer pointing to the next empty location on the stack.

A common application for LIFO memory is in a computer system. If a program is interrupted during its execution by a demand from the program or some piece of hardware that needs attention, the status of various registers within the computer are stored on a stack and the computer can pay attention to the new demand, which will certainly change its operating state. After the interrupting task is finished, the original operating state of the computer can be taken from the top of the stack and reloaded into the appropriate registers, and the program can resume where it left off.

■ SECTION 13.4 REVIEW PROBLEM

13.4   State the main difference between a stack and a queue.

13.5   DYNAMIC RAM MODULES

> **■ KEY TERMS**
>
> **Memory Modules**   A small circuit board containing several dynamic RAM chips.
>
> **Single In-Line Memory Module (SIMM)**   A memory module with DRAMs and connector pins on one side of the board only.
>
> **Dual In-Line Memory Module (DIMM)**   A memory module with DRAMs and connector pins on both sides of the board.

Dynamic RAM chips are often combined on a small circuit board to make a **memory module.** This is because the data bus widths of systems requiring the DRAMs are not always the same as the DRAMs themselves. For example, Figure 13.22 shows how four $64M \times 8$ DRAMs are combined to make a $64M \times 32$ memory module. The block diagram of the module is shown in Figure 13.22, and the mechanical outline is shown in Figure 13.23. The data input/output lines are separate from one another so that there are 32 data I/Os *(DQ).* The address lines *(ADDR*[12..0]) for the module are parallel on all chips. With address multiplexing, this 13-bit address bus yields a 26-bit address, giving a 64M address range. Chip selects *(CS)* for all devices are connected together so that selecting the module selects all chips on the module.

This particular memory module is configured as a **single in-line memory module (SIMM),** which has the DRAM chips and pin connections on one side of the board only. A **dual in-line memory module (DIMM)** has the DRAMs mounted on both sides of the circuit board and pin connections on both sides of the board as well.

Figure 13.24 shows the layout of a 168-pin DRAM. There are 84 pins on each side of the board.

Recent DRAM Types

> **■ KEY TERMS**
>
> **SDRAM (Synchronous DRAM)**   Dynamic RAM whose data are synchronously transferred to and from the data bus.
>
> **DDR (Double Data Rate)**   SDRAM that uses both edges of the clock to transfer data.

FIGURE 13.22 SIMM Block Diagram

FIGURE 13.23 SIMM Layout

> **Rambus**   A proprietary DRAM technology that uses a special controller and bus architecture to effect high-speed transfer of data.
>
> **Rambus Channel**   A bus with special physical and electrical characteristics that connect several DRAMs in a Rambus system.

A limitation of standard dynamic RAM is the speed of data transfer between the RAM and the system using it. Historically, data transfer in DRAM has been asyn-

FIGURE 13.24 168-Pin DIMM

chronous; after various control inputs ($\overline{WE}$, $\overline{RAS}$, $\overline{CAS}$) have been applied, data are available on the bus as soon as they can get there. This asynchronous behavior has made data transfer to and from these DRAMs difficult to interface with other system components, such as a microprocessor, often requiring wait states that add up over time, slowing down general system performance.

One solution has been the development of **SDRAM** (synchronous DRAM), which uses a clock signal to synchronously transfer data to the system bus. The synchronous nature of this type of DRAM makes the availability of data more predictable, thus speeding the transfer of data to and from other system components.

A further development along this line is **DDR (Double Data Rate)** SDRAM, which uses both the rising and falling edges of the clock signal to transfer data. This allows a doubling of the data I/O rate, while using the same clock frequency as for the original SDRAM system.

A completely different approach to high-speed, high-capacity memory has been the development of a proprietary system called **Rambus.** This architecture relies on a special bus layout, with parallel bus lines and termination resistors at one end, that connect to a number of DRAM chips in a bus called **Rambus channel.** Address pins are multiplexed, as with other DRAM systems, but the controls for row and column select are carried on separate control busses to reduce data transfer time.

■ **SECTION 13.5 REVIEW PROBLEM**

13.5 A SIMM has a capacity of 16M × 32. How many 16M × 8 DRAMs are required to make this SIMM? How many address lines does the SIMM require? How should the DRAMs be connected?

13.6 MEMORY SYSTEMS

■ **KEY TERMS**

Bus Contention The condition that results when two or more devices try to send data to a bus at the same time. Bus contention can damage the output buffers of the devices involved.

Address Decoder A circuit enabling a particular memory device to be selected by the address bus of a larger memory system.

Memory Map A diagram showing the total address space of a memory system and the placement of various memory devices within that space.

Address Space A block of addresses in a memory system.

In the section on memory modules, we saw how multiple memory devices can be combined to make a system that has the same number of addressable locations as the

a. Single 32 K × 8 SRAM

**b. Two 32 K × 8 SRAMS connected to
make 64 K × 8 SRAM system**

FIGURE 13.25   Expanding Memory Space

individual devices making up the system, but with a wider data bus. We can also create memory systems where the data I/O width of the system is the same as the individual chips, but where the system has more addressable locations than any chip within the system.

In such a system, the data I/O and control lines from the individual memory chips connected in parallel, as are the lower bits of an address bus connecting the chips. However, it is important that only one memory device be enabled at any given time, in order to avoid **bus contention,** the condition that results when more than one output attempts to drive a common bus line. To avoid bus contention, one or more additional address lines must be decoded by an **address decoder** that allows only one chip to be selected at a time.

Figure 13.25 shows two 32K × 8 SRAMs connected to make a 64K × 8 memory system. A single 32K × 8 SRAM, as shown in Figure 13.25a, requires 15 address lines, 8 data lines, a write enable ($\overline{WE}$), and chip select ($\overline{CS}$) line. To make a 64K × 8 SRAM system, all of these lines are connected in parallel, except the ($\overline{CS}$) lines. In order to enable only one at a time, we use one more address line, A_{15}, and enable the top SRAM when $A_{15} = 0$ and the bottom SRAM when $A_{15} = 1$.

The address range of one 32K × 8 SRAM is given by the range of states of the address lines $A[14..0]$:

Lowest single-chip address: 000 0000 0000 0000 = 0000H
Highest single-chip address: 111 1111 1111 1111 = 7FFFH

The address range of the whole system must also account for the A_{15} bit:

Lowest system address: 0000 0000 0000 0000 = 0000H
Highest system address: 1111 1111 1111 1111 = FFFFH

FIGURE 13.26   Memory Map

Within the context of the system, each individual SRAM chip has a range of addresses, depending on the state of A_{15}. Assume $SRAM_0$ is selected when $A_{15} = 0$ and $SRAM_1$ is selected when $A_{15} = 1$.

Lowest $SRAM_0$ address:	0000 0000 0000 0000 = 0000H
Highest $SRAM_0$ address:	0111 1111 1111 1111 = 7FFFH
Lowest $SRAM_1$ address:	1000 0000 0000 0000 = 8000H
Highest $SRAM_1$ address:	1111 1111 1111 1111 = FFFFH

Figure 13.26 shows a **memory map** of the 64K × 8 SRAM system, indicating the range of addresses for each device in the system. The total range of addresses in the system is called the **address space.**

EXAMPLE 13.4

0000H	SRAM₁
4000H	
8000H	SRAM₂
C000H	
FFFFH	

FIGURE 13.27 Example 13.4: Memory Map Showing Noncontiguous Decoded Blocks

Figure 13.27 shows a memory map for a system with an address space of 64K (16 address lines). Two 16K × 8 blocks of SRAM are located at start addresses of 0000H and 8000H, respectively. Sketch a memory system that implements the memory map of Figure 13.27.

■ Solution

A 16K address block requires 14 address lines, since

$$16K = 16 \times 1024 = 2^4 \times 2^{10} = 2^{14}$$

The entire 64K address space requires 16 address lines, since

$$64K = 64 \times 1024 = 2^6 \times 2^{10} = 2^{16}$$

The highest address in a block is the start address plus the block size.

	16K block size:	11 1111 1111 1111 = 3FFFH
$SRAM_0$:	Lowest address:	0000 0000 0000 0000 = 0000H
	Highest address:	0011 1111 1111 1111 = 3FFFH
$SRAM_2$:	Lowest Address:	1000 0000 0000 0000 = 8000H
	Highest Address:	1011 1111 1111 1111 = BFFFH

$A_{15}A_{14} = 00$ for the entire range of the $SRAM_0$ block. $A_{15}A_{14} = 10$ for the entire $SRAM_2$ range. These can be decoded by the gates shown in Figure 13.28.

FIGURE 13.28 Example 13.4: 32 K × 8 SRAM with Noncontinguous Blocks

FIGURE 13.29   64K Memory System

FIGURE 13.30   Memory Map
for Figure 13.29

Address Decoding with *n*-Line-to-*m*-Line Decoders

Figure 13.29 shows a 64K memory system with four 16K chips: one EPROM at 0000H and three SRAMs at 4000H, 8000H, and C000H, respectively. In this circuit, the address decoding is done by a 2-line-to-4-line decoder, which can be an off-the-shelf MSI decoder, such as a 74HC139 decoder or a PLD-based design.

Table 13.2 shows the address ranges decoded by each decoder output. The first two address bits are the same throughout any given address range. Figure 13.30 shows the memory map for the system.

TABLE 13.2 Address Decoding for Figure 13.29

A_{15}	A_{14}	Active Decoder Output	Device	Address Range
0	0	Y_0	EPROM	0000 0000 0000 0000 = 0000H
				0011 1111 1111 1111 = 3FFFH
0	1	Y_1	SRAM$_1$	0100 0000 0000 0000 = 4000H
				0111 1111 1111 1111 = 7FFFH
1	0	Y_2	SRAM$_2$	1000 0000 0000 0000 = 8000H
				1011 1111 1111 1111 = BFFFH
1	1	Y_3	SRAM$_3$	1100 0000 0000 0000 = C000H
				1111 1111 1111 1111 = FFFFH

■ **SECTION 13.6 REVIEW PROBLEM**

13.6 Calculate the number of 128K memory blocks that will fit into a 1M address space. Write the start addresses for the blocks.

SUMMARY

1. A memory is a device that can accept data and store them for later recall.

2. Data are located in a memory by an address, a binary number at a set of address inputs that uniquely locates the block of data.

3. The operation that stores data in a memory is called the write function. The operation that recalls the stored data is the read function. These functions are controlled by functions such as write enable ($\overline{\text{WE}}$), chip select ($\overline{\text{CS}}$), and output enable ($\overline{\text{OE}}$).

4. RAM is random access memory. RAM can be written to and read from in any order of addresses. RAM is volatile. That is, it loses its data when power is removed from the device.

5. ROM is read-only memory. Original ROM devices could not be written to at all, except at the time of manufacture. Modern variations can also be written to, but not as easily as RAM. ROM is nonvolatile; it retains its data when power is removed from the device.

6. Memory capacity is given as $m \times n$ for m addressable locations and an n-bit data bus. For example, a 64K $\times$ 8 memory has 65,536 addressable locations, each with 8-bit data.

7. Large blocks of memory are designated with the binary prefixes K (2^{10} = 1024), M (2^{20} = 1,048,576), and G (2^{30} = 1,073,741,824).

8. RAM can be divided into two major classes: static RAM (SRAM) and dynamic RAM (DRAM). SRAM retains its data as long as power is applied to the device. DRAM requires its data to be refreshed periodically.

9. Typically DRAM capacity is larger than SRAM because DRAM cells are smaller than SRAM cells. An SRAM cell is essentially a flip-flop consisting of several transistors. A DRAM cell has only one transistor and a capacitor.

10. RAM cells are arranged in rectangular arrays for efficient packaging. Internal circuitry locates each cell at the intersection of a row and column within the array.

11. For packaging efficiency, DRAM addresses are often multiplexed so that the device receives half its address as a row address, latched in to the device by a $\overline{\text{RAS}}$ (row address strobe) signal and the second half as a column address, latched in by a $\overline{\text{CAS}}$ (column address strobe) signal.

12. Read-only memory (ROM) is used where it is important to retain data after power is removed.

13. Mask-programmed ROM is programmed at the time of manufacture. Programming is done by making a custom overlay of connections onto a standard cell array. Data cannot be changed. This is suitable for mature designs in high-volume production.

14. Erasable programmable read-only memory (EPROM) can be programmed by the user and erased by exposure to ultraviolet light of a specified frequency and intensity. An EPROM must be removed from its circuit for erasing and reprogramming.

15. Electrically erasable read-only memory (EEPROM or E²PROM) can be programmed and erased in-circuit. It is nonvolatile, but unsuitable for use as system RAM due to its long programming/erase times and finite number of program/erase cycles.

16. Flash memory is a type of EEPROM that is organized into sectors that are erased all at once. This is faster than other EEPROM, which must be erased byte-by-byte.

17. Flash memory is often configured with one sector as a boot block, where primary firmware is stored. A bottom boot block architecture has the boot block at the lowest chip address. A top boot block architecture has the boot block at the highest chip address.

18. Sequential memory must have its data accessed in sequence. Two major classes are first-in first-out (FIFO) and last-in first-out (LIFO). FIFO is also called a queue and LIFO is called a stack.

19. Dynamic RAM chips are often configured as memory modules, small circuit boards with multiple DRAMs. The modules usually have the same number of address locations as the individual chips on the module, but a wider data bus.

20. Memory systems can be configured to have the same data width as individual memory devices comprising the system, but with more addressable locations than any chip in the system. The additional addresses require additional system address lines, which are decoded to enable one chip at a time within the system.

GLOSSARY

Address   A number, represented by the binary states of a group of inputs or outputs, uniquely defining the location of data stored in a memory device.

Address Decoder   A circuit enabling a particular memory device to be selected by the address bus of a larger memory system.

Address Multiplexing   A technique of addressing storage cells in a dynamic RAM which sequentially uses the same inputs for row address and column address of the cell.

Address Space   A block of addresses in a memory system.

b   Bit.

B   Byte.

Bit-Organized   A memory is bit-organized if one address accesses one bit of data.

Boot Block   A sector in a flash memory reserved for primary firmware.

Bottom Boot Block   A boot block sector in flash memory placed at the lowest address in the memory.

Bus Contention   The condition that results when two or more devices try to send data to a bus at the same time. Bus contention can damage the output buffers of the devices involved.

Byte   A group of 8 bits.

$\overline{\text{CAS}}$ (Column Address Strobe)   A signal used to latch the column address into the decoding circuitry of a dynamic RAM with multiplexed addressing.

Data   Binary digits (0s and 1s) which contain some kind of information. In the context of memory, the digital contents of a memory device.

DDR (Double Data Rate)   SDRAM that uses both edges of the clock to transfer data.

Dual In-Line Memory Module (DIMM)   A memory module with DRAMs and connector pins on both sides of the board.

Dynamic RAM   A random access memory which cannot retain data for more than a few (e.g., 64) milliseconds without being "refreshed."

EEPROM (or E² PROM)   Electrically erasable programmable read-only memory. A type of read-only memory that can be field-programmed and selectively erased while still in a circuit.

EPROM   Erasable programmable read-only memory. A type of ROM that can be programmed ("burned") by the user and erased later, if necessary, by exposing the chip to ultraviolet radiation.

FAMOS FET Floating-gate avalanche. MOSFET. A MOSFET with a second, "floating" gate in which charge can be trapped to change the MOSFET's gate-source threshold voltage. A FAMOS transistor is the memory element in an EPROM cell.

FIFO (First-In First-Out) A sequential memory in which the stored data can only be read in the order in which it was written.

Firmware Software instructions permanently stored in ROM.

Flash Memory A nonvolatile type of memory, similar to EEPROM, that can be programmed and erased in sectors, rather than byte-at-a-time.

G 1,073,741,824 (= 2^{30}). Analogous to the metric prefix "G" (giga).

Hardware The electronic circuit of a digital or computer system.

K 1024 (=2^{10}) Analogous to the metric prefix "k" (kilo).

LIFO (Last-In First-Out) A sequential memory in which the last data written is the first data read.

M 1,048,576 (=2^{20}) Analogous to the metric prefix "M" (mega).

Mask-Programmed ROM A type of read-only memory (ROM) where the stored data are permanently encoded into the memory device during the manufacturing process.

Memory A device for storing digital data in such a way that it can be recalled for later use in a digital system.

Memory Map A diagram showing the total address space of a memory system and the placement of various memory devices within that space.

Memory Module A small circuit board containing several dynamic RAM chips.

PROM Programmable read-only memory. A type of ROM whose data need not be manufactured into the chip, but can be programmed by the user.

Queue A FIFO memory.

RAM Cell The smallest storage unit of a RAM, capable of storing one bit.

Rambus A proprietary DRAM technology that uses a special controller and bus architecture to effect high-speed transfer of data.

Rambus channel A bus with special physical and electrical characteristics that connect several DRAMs in a Rambus system.

Random Access Memory (RAM) A type of memory device where data at any address can be accessed in any order, that is, randomly. The term usually refers to random access read/write memory.

RAS (Row Address Strobe) A signal used to latch the row address into the decoding circuitry of a dynamic RAM with multiplexed addressing.

Read Retrieve data from a memory device.

Read-Only Memory (ROM) A type of memory where data is permanently stored and can only be read, not written.

Refresh Cycle The process which periodically recharges the storage capacitors in a dynamic RAM.

SDRAM (Synchronous DRAM) Dynamic RAM whose data are synchronously transferred to and from the data bus.

Sector A segment of flash memory that forms the smallest amount that can be erased and reprogrammed at one time.

Sequential Memory Memory in which the stored data cannot be read or written in random order, but must be addressed in a specific sequence.

Single In-Line Memory Module (SIMM) A memory module with DRAMs and connector pins on one side of the board only.

Software Programming instructions required to make hardware perform specified tasks.

Stack A LIFO memory.

Static RAM A random access memory which can retain data indefinitely as long as electrical power is available to the chip.

Top Boot Block A boot block sector in a flash memory placed at the highest address in the memory.

Volatile   A memory is volatile if its stored data is lost when electrical power is lost.

Word   Data accessed at one addressable location.

Word Length   Number of bits in a word.

Word-Organized   A memory is word-organized if one address accesses one word of data.

Write   Store data in a memory device.

PROBLEMS

Problem numbers set in color indicate more difficult problems.

13.1   Basic Memory Concepts

13.1   How many address lines are necessary to make an 8×8 memory similar to the 4×8 memory in Figure 13.5? How many address lines are necessary to make a 16×8 memory?

13.2   Briefly explain the difference between RAM and ROM.

13.3   Calculate the number of address lines and data lines needed to access all stored data in each of the following sizes of memory:

  a. $64K \times 8$

  b. $128K \times 16$

  c. $128K \times 32$

  d. $256K \times 16$

Calculate the total bit capacity of each memory.

13.4   Explain the difference between the chip enable $(\overline{E})$ and the output enable $(\overline{G})$ control functions in a RAM.

13.5   Refer to Figure 13.9. Briefly explain the operation of the $\overline{W}$, $\overline{E}$, and $\overline{G}$ functions of the RAM shown.

13.2   Random Access Read/Write Memory (RAM)

13.6   Draw the circuit for an NMOS static RAM cell. Label one output BIT and the other $\overline{BIT}$.

13.7   Refer to the NMOS static RAM cell drawn in Problem 13.6. Assume that $BIT = 1$. Describe the operation required to change BIT to 0.

13.8   Describe the main difference between a CMOS and an NMOS static RAM cell.

13.9   Explain how a particular RAM cell is selected from a group of many cells.

13.10   How many address lines are required to access all elements in a $1M \times 1$ dynamic RAM with address multiplexing?

13.11   What is the capacity of an address-multiplexed DRAM with one more address line than the DRAM referred to in Problem 13.10? With two more address lines?

13.12   How many address lines are required to access all elements in a $256M \times 16$ DRAM with address multiplexing?

13.3   Read-Only Memory (ROM)

13.13   Briefly list some of the differences between mask-programmed ROM, UV-erasable EPROM, EEPROM, and flash memory.

13.14   Briefly describe the programming and erasing process of a UV-EPROM.

13.15   Briefly explain the difference between flash memory and other EEPROM. What is the advantage of each configuration?

13.16   A flash memory has a capacity of 8 Mb, organized as $512K \times 16$-bit. List the address range for the 16 KB boot block sector of the memory if the device has a bottom boot block architecture and if it has a top boot block architecture.

13.17   Briefly state why EEPROM is not suitable for use as system RAM.

13.18   Briefly state why flash memory is unsuitable for use as system RAM.

13.4   Sequential Memory

13.19   State one possible application for a FIFO and for a LIFO memory.

13.5 Dynamic RAM Modules

13.20 A SIMM has a capacity of 32M × 64. How many 32M × 8 DRAMs are required to make this SIMM? How many address lines does the SIMM require? How should the DRAMs be connected?

13.6 Memory Systems

13.21 A microcontroller system with a 16-bit address bus is connected to a 4K × 8 RAM chip and an 8K × 8 RAM chip. The 8K address begins at 6000H. The 4K address block starts at 2000H.

Calculate the end address for each block and show address blocks for both memory chips on a 64K memory map.

13.22 Draw the memory system of Problem 13.21.

13.23 A microcontroller system with a 16-bit address bus has the following memory assignments:

Memory	Size	Start Address
RAM_0	16K	4000H
RAM_1	8K	8000H
RAM_2	8K	A000H

Show the blocks on a 64 Kmemory map.

13.24 Draw the memory system described in Problem 13.23.

13.25 The memory map of a microcontroller system with a 16-bit address bus is shown in Figure 13.31. Make a table of start and end addresses for each of the blocks shown. Indicate the size of each block.

13.26 Sketch the memory system described in Problem 13.25.

13.27 How many 16M × 32 DIMMs are required to make a 256M × 32 memory system? Make a table showing the start and end addresses of each block.

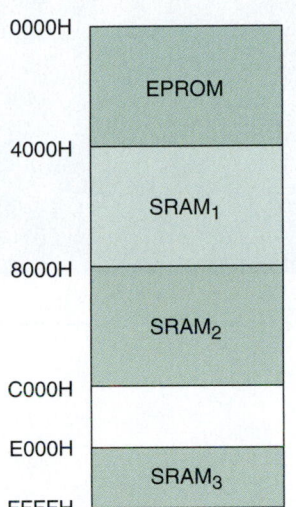

0000H	
	EPROM
4000H	
	$SRAM_1$
8000H	
	$SRAM_2$
C000H	
E000H	
	$SRAM_3$
FFFFH	

FIGURE 13.31 Problem 13.25: Memory Map

ANSWERS TO SECTION REVIEW PROBLEMS

13.2A

13.1 18 address lines, 16 data lines; capacity = 4Mb, same as Figure 3.12

13.2B

13.2
  a. 10 address, 1 data
  b. 10 address, 4 data
  c. 11 address, 1 data

13.3

13.3 Bottom boot block: 00000H to 07FFFH; top boot block: F8000H to FFFFFH.

13.4

13.4 A stack is a last-in first-out (LIFO) memory and a queue is a first-in first-out (FIFO) memory.

13.5

13.5 Four DRAMs. 12 address lines. Address and control lines are in parallel with all DRAMs. Data I/O lines are separate.

13.6

13.6 Eight blocks. Start addresses: 00000H, 20000H, 40000H, 60000H, 80000H, A0000H, C0000H, E0000H.

KEY TERMS

Accumulator
Address Bus
Arithmetic Logic Unit (ALU)
Block
Bus Contention
Conduit Line
Control Bus
Controller
CPU (Central Processing Unit)
Data Bus
Embedded Controller
Execute
Fetch
Flow Control
Instruction Decoder
Instruction Register (IR)
Mapper
Memory Address Register (MAR)
Memory Data Register (MDR)
Microcomputer
Microprocessor
Op Code (Operation Code)
Output Register
Peripheral Devices (or Peripherals)
Program Counter (PC)
Program Instructions
Register
RISC (Reduced Instruction Set
    Computer)
Tristate Outputs

OUTLINE

14.1 Basic Structure of a Microcomputer
14.2 Register Level Structure of a Microcomputer System
14.3 Tristate Busses in Altera CPLDs
14.4 Quartus II Implementation of the RISC8v1 MCU
14.5 Creating New Instructions (RISC8v2)
14.6 Branch Instructions (RISC8v3)
14.7 RISC8 Test Circuit
14.8 Possible Enhancements to RISC8v3 MCU

CHAPTER OBJECTIVES

After successful completion of this chapter, you will be able to:

- Draw the block diagram of a simplified microcomputer system, showing blocks for the various components, interconnected by address, data, and control busses.
- Describe bus contention and what can be done to remedy this problem.
- Describe synchronous and asynchronous register data transfers and draw timing diagrams to represent them.
- Describe the functions of the registers in a model microcomputer system.
- Indicate the sequence of control signals required to fetch and execute an instruction in a model microcomputer system and represent the transfers on a timing diagram.
- Describe the functions of simple RISC instructions, such as LOAD, ADD, OUTPUT, and HALT.
- List the various functions of an 8-function ALU and analyze a VHDL file that implements these functions.
- Draw logic diagrams showing different methods of creating a tristate bus in an Altera CPLD. Write VHDL files to implement a tristate bus.
- Write VHDL files to implement the function of registers in a RISC CPU.

CHAPTER 14

Introduction to Microprocessors

- Use the Quartus II Block Editor to create block diagrams with the blocks connected by conduit lines.
- Analyze VHDL files that implement registers and multiplexers for the RISC8v1.
- Create Memory Initialization Files to store program instructions and operands in the system ROM for a RISC microcomputer.
- Analyze the structure of a VHDL file that defines the function of the RISC8v1 controller.
- Modify the VHDL code for the controller to add new instructions to the RISC8v1 MCU.
- Describe the operation of a branch instruction and the hardware modifications required to implement it.

VHDL CONSTRUCTS USED IN THIS CHAPTER

- PROCESS statement
- CASE statement
- Setting all elements of a vector to '0' or 'Z' using "others"
- Synchronous register clocking
- IN, OUT, and BUFFER modes
- std_logic_1164, std_logic_unsigned, and std_logic_arith packages
- State machine
- TYPE declaration
- Selected signal assignment statement
- Concurrent signal assignment statement
- Concatenation

14.1   BASIC STRUCTURE OF A MICROCOMPUTER

■ KEY TERMS

Microcomputer   A self-contained computer system that consists of a microprocessor, memory components, such as RAM or ROM, and peripheral devices. Also called a microcomputer unit (MCU or μC).

Microprocessor   The component of a computer that generates control signals for the other components of the computer and performs arithmetic and logic functions. Also called a microprocessor unit, MPU, or μP.

Program Instructions   A set of binary or hexadecimal codes that are interpreted by the CPU of a computer system to perform various functions.

Peripheral Devices (or **Peripherals**)   Devices connected to the address, data, and control busses of a computer system that act as interface circuits between the computer and the external world (e.g., switches, keyboards, LEDs, numeric displays, video monitors, printers, disk drives, etc.).

CPU (Central Processing Unit)   Another name for microprocessor.

Address Bus   A set of parallel conductors that select the required address within the computer system memory to access data or program instructions.

Data Bus   A set of parallel conductors that transfer data within and outside the computer.

Control Bus   A set of conductors that control the flow of data among different modules of a computer system.

In the modern world, **microcomputers** and **microprocessors** have become so familiar and pervasive that often we don't even realize that they are embedded within many devices we encounter daily, such as microwave ovens, VCRs, automobiles, fax machines, and so forth.

What do we mean by a microcomputer and by a microprocessor? The distinction is that a microcomputer is a complete computer system, whereas a microprocessor is a component in that system. In addition to the microprocessor, a microcomputer also has memory, containing **program instructions** and data, as well as **peripheral devices,** which communicate with the world external to the microcomputer system.

The functions of the microprocessor, or **CPU (Central Processing Unit),** are to coordinate the functioning of the various components of the system by generating and responding to control signals and to perform various arithmetic and logic functions. The program instructions stored in memory are decoded by the CPU, which then acts to perform those instructions.

Figure 14.1 shows a simplified block diagram of a microcomputer system. The CPU, memory, and peripheral components are interconnected by a series of busses, or parallel groups of connectors, that carry address, data, and control signals among the components.

The **address bus** is used to specify which part of memory or which peripheral should receive data from or send data to the CPU. The **data bus** sends or receives the actual data. The **control bus** is not a parallel set of lines going to all components, but rather a set of control lines going from the CPU to various components, not all of which have the same control signals. The purpose of these control lines is to direct the flow of data among the modules of the microcomputer unit (MCU) by enabling and disabling the various data paths that interconnect them.

The address bus is shown as unidirectional, since these signals are always sent from the CPU to the memory and peripherals. The data bus is shown as bidirectional since data can flow in both directions between CPU and RAM and peripherals. (Data only flows *from* ROM, as it is a read-only device.) The control bus shows

FIGURE 14.1 Simplified Microcomputer System

signals going to and from the CPU, although in most cases, signal flow is from the CPU.

Tristate Bussing

> ■ **KEY TERMS**
>
> **Bus Contention** Conflict that arises when two or more devices attempt to write data onto a bus at the same time.
>
> **Tristate Outputs** Device outputs that can be in three possible states: logic HIGH, logic LOW, and high-impedance. The high-impedance state acts like an open circuit.
>
> **Register** A multibit latch or flip-flop that acts as a temporary data storage location within a computer.

If multiple devices are connected to a bus, there must be some system to prevent more than one device from writing to the bus at the same time. This is necessary to avoid conflicting signals, or **bus contention,** which prevents the bus from having a clear HIGH or LOW state on each line and can also cause damage to the outputs of the driving devices.

Bus contention is avoided by designing the driver circuits so that they have **tristate outputs,** as shown in Figure 14.2. Each driver is controlled by an enable signal. The CPU generates control signals that ensure that only one output is enabled at any given time.

Figure 14.2 shows a system that has two source **registers,** or storage locations, Source 1 and Source 2, with tristate drivers, and two destination registers, Destination 1 and Destination 2. The four registers are interconnected by a data bus. Either source can send data to either destination, depending on the status of the four control lines and the clock.

The source registers write their data to the bus when their respective control lines, *SRC1_OE* or *SRC2_OE,* are *HIGH.* (*OE* stands for Output Enable). The destination

FIGURE 14.2 Connections on an Asynchronous Tristate Data Bus

FIGURE 14.3 Source-Destination Timing on a Synchronous Tristate Bus

registers accept data when their control lines (*DEST1_LD* or *DEST2_LD*) are HIGH and a positive clock edge is applied. (*LD* stands for LoaD).

Assume that all registers are 8 bits wide and that Source 1 contains hexadecimal data 55 and Source 2 contains AA. Figure 14.3 shows the timing diagram of a data transfer from Source 1 to Destination 2, followed by a transfer from Source 2 to Destination 1. The clock waveform is slightly offset from the grid in the diagram to indicate propagation delays within the system. The portions of the data bus waveform shown as ZZ indicate a high-impedance state on the bus. The portions of the destination register waveforms with XX indicate an unknown state, prior to latching new data. (Since the bus notation follows hexadecimal convention, ZZ or XX indicates the states of greater than 4 bits and up to 8, i.e., two hex digits.)

The control waveforms in Figure 14.3 are assumed to be under control of an unseen CPU, which makes them active on the positive edge of the clock. When *SRC1_OE* goes HIGH, the Source 1 data (55H) appears on the data bus immediately. *DEST2_LD* goes HIGH at the same time, but the data is not latched into Destination 2 until the next clock edge, at the end of the period where *DEST2_LD* is HIGH. A similar transfer occurs from Source 2 to Destination 1.

FIGURE 14.4 Connections on a Synchronous Tristate Data Bus

FIGURE 14.5 Source-Destination Timing on a Synchronous Tristate Bus

The bus system in Figure 14.3 is asynchronous. That is, the source data is present on the bus as soon as the *SRC1_OE* line is HIGH. This can make the data transfer susceptible to errors due to propagation delay and asynchronous changes of data in the circuit. A more stable arrangement is shown in Figure 14.4.

The two flip-flops for each source in Figure 14.4 make the source outputs synchronous with the system clock. For example, the current Source 1 data is clocked into a D flip-flop with every clock pulse, making it stable for the remainder of the clock period.

To get the data to the bus still requires a HIGH signal at the tristate enable input. If *SRC1_OE* is HIGH, that HIGH state is clocked into a flip-flop that enables the tristate buffer for the Source 1 data. Thus, both the data itself and the enable signal are held stable for an entire clock cycle, protecting them both from asynchronous changes at the register inputs.

The price one pays for the synchronous transfer is a delay of data to the bus by one clock cycle. Figure 14.5 shows the same data transfers as Figure 14.3, but with a synchronous bus system. *SRC1_OE* goes HIGH, but the Source 1 data does not appear on the data bus until the end of that clock cycle. Only then can the *DEST2_LD* signal go HIGH to transfer data to Destination 2. The transfer occurs at the end of the cycle when the Source 1 data (55H) is on the bus. A similar transfer occurs from Source 2 to Destination 1.

In the case of the asynchronous transfer, data moved from source to destination in one clock cycle. In the synchronous bus, it takes two.

■ SECTION 14.1 REVIEW PROBLEM

14.1   State the difference between an asynchronous and a synchronous bus transfer. State an advantage of each.

14.2 REGISTER LEVEL STRUCTURE OF A MICROCOMPUTER SYSTEM

■ KEY TERMS

Embedded Controller   A self-contained microcomputer system (often a single chip) in which program instructions are permanently stored in ROM, making the system a stand-alone control device.

RISC (reduced instruction set computer)   A computer that is capable of performing relatively few instructions, with the goal of making a simpler and faster computer architecture.

Controller   A state machine within the CPU that generates control signals required to transfer data among the various CPU registers.

Fetch   Retrieve a program instruction from memory.

Execute   Decode a fetched instruction and perform required CPU operations.

Accumulator   A CPU register that holds the accumulated result of arithmetic and logic operations performed by the ALU.

Memory Data Register (MDR)   A CPU register that holds data transferred from memory as a second operand in an arithmetic or logical operation.

Arithmetic Logic Unit (ALU)   A combinational circuit within the CPU that performs arithmetic and logical operations, such as add, subtract, AND, OR and XOR.

Output Register   A peripheral register to which data is transferred from the accumulator. This can also be called a parallel output port.

Op Code (Operation Code)   A binary or hexadecimal code that represents an instruction for a CPU to execute.

Instruction Decoder   A combinational circuit within the CPU that interprets the binary value of the op code in the instruction register, then directs the controller to generate the required control signals to execute the instruction.

Program Counter (PC)   A counter within a CPU that keeps track of the address of the next program instruction to be fetched from memory.

Memory Address Register (MAR)   A CPU register that holds the memory address required for the current instruction or data.

Instruction Register (IR)   A CPU register that holds the current instruction op code being executed by the CPU.

Microcomputer systems are often configured so that a single chip contains at least a partial system implementation, with limited amounts of RAM, ROM, and some on-board peripherals. Such a system usually contains its program within nonvolatile

FIGURE 14.6 RISC8v1 Block Diagram

on-board memory, such as ROM, EPROM, EEPROM, or flash. It is designed to begin operating its program as soon as it is powered up.

Since the software is permanently "embedded" within the system memory, this is called an **embedded controller.** This is probably the most common computer configuration we encounter in our daily lives, although we are usually unaware of it. For example, when we punch in a time value to heat up our coffee in the microwave, we don't have to think about booting up the computer that controls the microwave. It just takes our instructions and acts on them, since the computer begins operating as soon as it is powered up and requires no intervention to run its embedded program.

Figure 14.6 shows the block diagram of a self-contained computer system, called the **RISC8v1 MCU. RISC** stands for **reduced instruction set computer.** This computer has an 8-bit data bus and a 4-bit address bus and can be programmed into a CPLD, such as the FLEX 10K device on the Altera UP-1 or UP-2 board. (The RISC8v1 MCU won't fit into the MAX 7000S chip. If you do not have the Altera UP-1 or UP-2 board available, you can still simulate the system in Quartus II.)

Each block in the system has a function that can be programmed in VHDL. We will show the VHDL implementation of the system later in the chapter, but for the moment, we will examine the system at a general register level.

The various blocks of the systems are controlled by Output Enable (*OE*) and Load (*LD*) signals, generated by a state machine called a **controller.** Program instructions for arithmetic, logic, and data transfer operations are stored in ROM. In order to perform these functions, each instruction must be **fetched** from ROM in the correct sequence, then **executed.**

The version 1 model of our RISC machine has an extremely reduced instruction set—just four instructions—listed in Table 14.1.

TABLE 14.1   Instruction Set for RISC8v1 Microcomputer

Load	Transfers data from ROM to the **accumulator** (the register where the accumulated total of arithmetic and logic operations is stored).
Add	Transfers data from ROM to the **memory data register (MDR),** uses the **arithmetic logic unit (ALU)** to add the data in the MDR to that in the accumulator, then transfers the total sum to the accumulator.
Output	Transfers data from the accumulator to the **output register.**
Halt	Stops the program. The computer can only start again with a reset signal.

To add two numbers, display the result, and stop, we must perform the following steps:

- Load accumulator with an operand
- Add a second operand to the accumulator value
- Output the result
- Halt

TABLE 14.2   Hexadecimal Values of RISC8v1 Op Codes

Add	1
Load	8
Output	9
Halt	F

Each of these steps breaks down into a sequence of data transfers between registers, including a fetch cycle (the same for every instruction) and an execute cycle (dependent on the individual instruction).

The four instructions of the RISC8v1 computer each correspond to a 4-bit code called an operation code or simply **op code.** These op codes are determined by the design of the **instruction decoder** in the CPU. They are listed in Table 14.2, in hexadecimal form.

The first two instructions in the list require an operand. We cannot just say "Load"; we must also specify what to load. The way the RISC8v1 computer is designed, the Load and Add instructions also specify a ROM address where the required data is found. (Technically, this is known as "direct addressing.")

Thus an op code and operand of 8C means "load the accumulator with the contents of ROM address C (hexadecimal)." Similarly, the op code and operand 1D means "add the contents of ROM address D to the contents of the accumulator." The other two instructions require no operand. However, due to the architecture of the computer, a second hex digit is automatically included in the operand field, even though it is ignored. We can simply write this as a 0.

These op codes can be stored in ROM as an embedded program. Since our computer is only a small demonstration model, the ROM is very small—only 16 bytes, the maximum available to a 4-bit address bus. The program would be stored in ROM as shown in Table 14.3.

TABLE 14.3   Simple Program for the RISC8v1 Microcomputer

Address	Data	Comment
0	8C	Load contents of C
1	1D	Add contents of D
2	90	Send accumulator contents to output register
3	F0	Halt
4 – B	Blank (00)	
C	55	Data for Load instruction
D	64	Data for Add instruction
E – F	Blank (00)	

Fetch Cycle

Figures 14.7 through 14.11 show the block diagram of the RISC machine through the five clock cycles that make up the fetch cycle of the machine.

Every instruction is fetched the same way. The shaded registers and the bold highlighting of the relevant control lines in Figures 14.7 through 14.11 show the data transfers in the fetch cycle. The five steps are as follows:

Fetch 1 (Figure 14.7)—Transfer the contents of the **program counter (PC)** to the 4-bit address bus. This value is the address of the current instruction to be executed. Active control line: **pc_oe.**

Fetch 2 (Figure 14.8)—Transfer the PC address from the address bus to the **memory address register (MAR).** Increment the program counter so that it is ready to point to the next instruction. Active control lines: **mar_ld, pc_inc.**

Fetch 3 (Figure 14.9)—The value in the MAR points to a ROM address containing an instruction to be executed, consisting of a 4-bit op code and a 4-bit operand address. This is now transferred from ROM to the 8-bit data bus. Active control line: **rom_oe.**

Fetch 4 (Figure 14.10)—The op code/address pair is transferred to the **instruction register (IR)** from the 8-bit data bus.

Fetch 5 (Figure 14.11)—This is a "do nothing" state, required in the RISC8v1 CPU for data from the ROM to stabilize on the data bus. This state is also used to decode the IR contents and direct the CPU to begin executing the selected instruction.

FIGURE 14.7 Fetch 1—PC Contents to Address Bus

FIGURE 14.8   Fetch 2—MAR Loads PC Value from Address Bus; PC Increments

FIGURE 14.9   Fetch 3—Data from ROM to Data Bus

Fetch 4

FIGURE 14.10 Fetch 4—IR Loads Instruction/Address Pair from Data Bus

Fetch 5

FIGURE 14.11 Fetch 5—Wait State

Fetch Cycle (for LOAD instruction)

FIGURE 14.12 Timing Diagram for Fetch Cycle

The sequence of events for the fetch cycle for the first (Load) instruction is shown in Figure 14.12 as a timing diagram. The program counter (PC) contents of 0 is transferred to the address bus, then to the memory address register (MAR). The PC increments to 1, so as to be ready to point to the next instruction on the next fetch cycle. The contents of ROM address 0 (8C) is transferred to the data bus, then to the instruction register (IR). This sequence fetches the instruction 8C, which means, "load the accumulator with the data at ROM address C."

Execute Cycle (Load Instruction)

Figures 14.13 to 14.16 show the register transfers for the execute cycle of the Load instruction. The Load instruction consists of the following actions.

Load 1 (Figure 14.13)—The instruction/operand address pair fetched from ROM is split into the op code (4 MSBs) and operand address (4 LSBs). The op code goes by a direct connection to the instruction decoder in the controller state machine, which determines the op code value and generates the correct control signals for the remainder of the cycle. The operand address is transferred to the address bus. Active control line: **ir_oe.**

Load 2 (Figure 14.14)—The MAR loads the contents of the address bus, thus latching the ROM address of the operand for the Load instruction. Active control line: **mar_ld.**

Load 3 (Figure 14.15)—Data transfers from the ROM to data bus. Active control line: **rom_oe.**

Load 4 (Figure 14.16)—Data transfers from the data bus to the accumulator. Active control line: **acc_ld.**

Figure 14.17 shows the timing of the Load instruction execute cycle. Operand address C appears on the address bus and is latched into the MAR. Operand 55 is transferred, via the data bus, to the accumulator from ROM address C.

FIGURE 14.13   Load 1—Operand Address Transfers from IR to Address Bus

FIGURE 14.14   Load 2—Operand Address Latched into MAR

FIGURE 14.15   Load 3—Operand Transferred from ROM to Data Bus

FIGURE 14.16   Load 4—Operand Latched into Accumulator

Execute Cycle (Load)

FIGURE 14.17 Timing Diagram of Load Instruction Execute Cycle

Arithmetic Logic Unit (ALU)

The RISC8v1 CPU has an arithmetic logic unit (ALU) capable of performing four 8-bit arithmetic functions and four 8-bit bitwise logic functions. The ALU functions are selected by the states of three select inputs, S[2..0]. The functions can be summarized as listed in Table 14.4.

TABLE 14.4 Function Codes for RISC8v1 ALU

S[2..0]	Function	Operation
000	Increment	Acc + 1
001	Add	Acc + MDR
010	Subtract	Acc – MDR
011	Decrement	Acc – 1
100	Complement	NOT Acc
101	AND	Acc AND MDR
110	OR	Acc OR MDR
111	XOR	Acc XOR MDR

The ALU is designed to operate either on the accumulator alone (increment, decrement, and complement) or on the accumulator and memory data register (all other instructions).

In the RISC8v1, only the Add instruction is used; the remaining functions are reserved for later versions of the CPU. The execute cycle of the Add instruction must set the ALU inputs to S[2..0] = 001.

The ALU functions can be specified by the following series of VHDL statements.

```
-- Arithmetic Logic Unit
-- Capable of implementing 4 arithmetic and 4 logic functions
-- Input: from accumulator and memory data register
-- Output: tristate data bus via Data_MUX

LIBRARY ieee;
```

```
USE ieee.std_logic_1164.ALL;
USE ieee.std_logic_signed.ALL;
USE ieee.std_logic_arith.ALL;

ENTITY alu IS
PORT(
    operand_a   : IN  STD_LOGIC_VECTOR(7 downto 0);
    s           : IN  STD_LOGIC_VECTOR(2 downto 0);
    memory_data : IN  STD_LOGIC_VECTOR(7 downto 0);
    alu_data    : OUT STD_LOGIC_VECTOR(7 downto 0));
END alu;

ARCHITECTURE a OF alu IS
BEGIN
    PROCESS (operand_a, memory_data, s)
    BEGIN
        CASE s IS
            WHEN "000" =>
                alu_data <= operand_a + 1;            -- Increment A
            WHEN "001" =>
                alu_data <= operand_a + memory_data;  -- Add
            WHEN "010" =>
                alu_data <= operand_a - memory_data;  -- Subtract
            WHEN "011" =>
                alu_data <= operand_a - 1;            -- Decrement A
            WHEN "100" =>
                alu_data <= not operand_a;            -- Complement A
            WHEN "101" =>
                alu_data <= operand_a and memory_data; -- AND
            WHEN "110" =>
                alu_data <= operand_a or memory_data;  -- OR
            WHEN "111" =>
                alu_data <= operand_a xor memory_data; -- XOR
            WHEN others =>
                alu_data <= (others => '0');
        END CASE;
    END PROCESS;
END a;
```

Alu.vhd

Execute Cycle (Add Instruction)

The first three steps of the execute cycle for the Add instruction, shown in Figures 14.18 to 14.20, are the same as for the Load instruction, except that the S inputs to the ALU are set to 001 for the Add function. In these steps, the instruction/address pair is transferred to the instruction register, the operand address is transferred to the memory address register, and the operand is transferred from the ROM to the CPU data bus.

The remaining steps are shown in Figures 14.21 to 14.23.

Add 4 (Figure 14.21)—Transfer data from data bus to the memory data register (MDR). Active control lines: **mdr_ld, s[2..0]=001.**

Add 5 (Figure 14.22)—The ALU adds the accumulator contents to the MDR contents. The result transfers to the data bus. Active control lines: **alu_oe, s[2..0]=001.**

Add 6 (Figure 14.23)—The accumulator transfers the final result from the data bus to the accumulator. Active control lines: **acc_ld, s[2..0]=001.**

Figure 14.24 and Figure 14.25 show the timing diagram for the fetch and execute cycles of the Add instruction. In Figure 14.24, the program counter contents (1) are

FIGURE 14.18 Add 1—Operand Address Transfers from IR to Address Bus

FIGURE 14.19 Add 2—Operand Address Latched into MAR

FIGURE 14.20   Add 3—Operand Transferred from ROM to Data Bus

FIGURE 14.21   Add 4—Data Transferred from Data Bus to MDR

Add 5

FIGURE 14.22 Add 5—Contents of ALU Transferred to Data Bus

Add 6

FIGURE 14.23 Add 6—Final Result Transferred from Data Bus to Accumulator

Fetch Cycle (for ADD instruction)

FIGURE 14.24   Timing Diagram for Add Instruction Fetch Cycle

Execute Cycle (Add)

FIGURE 14.25   Timing Diagram for Add Instruction Execute Cycle transferred to the address bus and then to the memory address register. The PC is incremented to 2. Data at ROM address 1 (1D) is transferred to the instruction register.

In Figure 14.25, the ROM address of the operand (D) is placed on the address bus from the instruction register and latched into the memory address register. The data at address D (64) is transferred to the MDR via the data bus. The sum of (55 + 64 = B9) is transferred via the data bus to the accumulator.

Other Instructions

Output—The contents of the accumulator are transferred to the output register by making **acc_oe** = 1 and **out_ld** = 1.

Halt—All processing stops after the fetch cycle. This is accomplished by making the halt state in the CPU controller always transition to itself.

■ SECTION 14.2 REVIEW PROBLEM

14.2   Which register is data transferred to in the Load instruction of the RISC8v1 CPU?

14.3 TRISTATE BUSSES IN ALTERA CPLDs

Figure 14.26 shows the simplified configuration of a CPLD output with a tristate driver, such as a MAX 7000S macrocell or a FLEX 10K logic element. Since the tristate driver is connected only to an I/O pin, internal tristate bussing is not available in the CPLD. (The feedback line cannot be used for this purpose.)

Two possibilities remain for tristate bussing: connect I/O pins external to the CPLD, as shown in Figure 14.27, or multiplex internal logic to a single tristate pin, as shown in Figure 14.28.

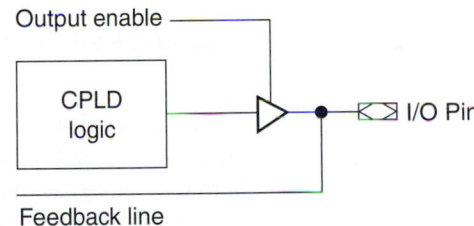

FIGURE 14.26 Simplified CPLD Tristate Output

FIGURE 14.27 Tristate Bussing Using External Connections

FIGURE 14.28 Tristate Bussing Using an Internal Multiplexer

FIGURE 14.29 Tristate Bus with Two Sources and Two Destinations

If we are interested in keeping our entire design internal to the CPLD, then we must use the multiplexing scheme in Figure 14.28. In this scheme, data is transferred to the tristate bus, represented by the I/O pin, from a source (CPLD Logic 1 or 2) when the source is selected by the MUX and the Output Enable is active.

To use the MUX connection with the two-source, two-destination model we have discussed earlier, we require the I/O pin to be bidirectional, using a Quartus II graphical component called BIDIR. This configuration is shown in Figure 14.29. Connections are made automatically in Quartus II to assign source and destination logic to various macrocells or logic elements, as required.

Tristate Address Bussing in the RISC8v1 CPU

Figure 14.30 shows a Quartus II Block Diagram File that implements the tristate address bus of the RISC8v1 CPU. The addressing system consists of the program counter, instruction register, memory address register, and the synchronous tristate address MUX. The VHDL files of these components are listed in a later part of this section. A similar configuration exists for the data bus structure.

The simulation of the tristate bus system is shown in Figure 14.31. Note that when there is a bidirectional pin in the Block Diagram File, Quartus II creates an input and output node for that pin in the simulation. The address bus input is shown as **addr_bus[3..0]** in the Vector Waveform File. In order for the simulation to avoid bus contention, this node must be set to the value "Z" (high-impedance) when the Vector Waveform File is created. The output value of the bus is written to a node

addr_bus.bdf
addr_bus.vwf

FIGURE 14.30 Address Bus Block Diagram

called **addr_bus~result[3..0]** when the simulation is run. This node is not part of the original Vector Waveform File, but is automatically generated as part of the simulation result waveforms.

Figure 14.31 shows the transfer of a 4-bit address (0) first from the program counter to the memory address register (MAR). The program counter is updated at this point to a value of 1. The 8-bit value of the data bus (8C) is transferred to the instruction register, after which the lower 4-bits of this value (C) are transferred from the instruction register to the MAR.

In a simulation of the next instruction fetch, address 1 is placed on the address bus from the program counter and transferred to the MAR. The data bus value (1D) is latched into the address register, then split into upper and lower digits. The upper digit (1) appears in the line labeled **Instruction** and the lower digit (D) goes out the address bus and to the MAR.

All data transfers are synchronous; an address appears on the bus one clock cycle after its output enable line is HIGH. This is accomplished by asynchronously transferring an address digit to the address multiplexer, then synchronously clocking the selected value to the address bus in the MUX. Transfers into the IR and MAR are clocked synchronously from the address bus.

FIGURE 14.31 Simulation of Address Bus Components

The inputs to the address bus MUX are labeled **pc_addr[3..0]** and **ir_address[3..0]** in the simulation. One of these is selected and sent to the address bus, **addr_bus~result[3..0],** when the appropriate output enable line is active. The bus carries the value of the program counter one clock cycle after **pc_oe** is HIGH and **ir_oe** is LOW. One clock cycle after **ir_oe** is HIGH and **pc_oe** is LOW, the address bus carries the value of the lower four bits of the instruction register. In all other cases, the address bus is in the high-impedance state.

Block Diagram Format

> ### ■ KEY TERMS
>
> **Block**   A design unit in a Quartus Block Diagram File that represents a sub-unit, such as a lower-level Block Diagram File or a VHDL file, and indicates its inputs and outputs by a table embedded in the block.
>
> **Conduit Line**   A line in a Quartus II Block Diagram File that connects signals going between blocks by matching names on the connected block I/O lists.
>
> **Mapper**   An interface point between a block and a node, bus, or conduit line that maps the signal names between the block and the node, bus, or conduit.

The form of the RISC block diagram in Figure 14.30 may be unfamiliar. It is a form of Block Diagram File in Quartus II that uses **blocks** rather than graphic symbols as we have been using.

In previous Quartus II diagrams, we have used graphic symbols that have specific positions for input and output ports that are connected by lines representing the various nodes and busses in the design. When we use a block format, each block has its input and output ports listed on a table embedded in the block. The blocks can be connected by node and bus lines, as before, but we also have a new type of line, called a **conduit line,** available to us.

A conduit line, represented by a "hollow" line on the diagram, can be used to convey multiple signals between blocks. The signal names do not have to be the same in the different blocks, but the design is simpler if they are. If the names are the same in every block connected by a conduit, the compiler automatically maps the connections by simply examining the I/O tables in the blocks and connecting ports that have the same names.

A line drawn *from* a block is a conduit line by default. The point of connection between the conduit and the block is an arrow, double-ended by default, called a **mapper.** If the signal names are not the same in the different blocks, the mapper must be labeled to map the names in the various locations. Since our signal names are all the same, we don't need to do this.

A very nice feature of the conduit line is the way it can clean up a diagram by reducing the number of interconnecting lines. For example, in Figure 14.30, all of the control lines are tied to a single conduit line. The conduit line is then connected to all four blocks in the diagram. The compiler sorts out which control line is required in which block, without us having to specifically say so. We have, in effect, created a control bus. The diagram also includes a bus for the **clock** and **reset** lines and another for the address bus.

The next several figures show how to create a block in a Block Diagram File. Figure 14.32 shows the Block Editor toolbar with the Block Tool highlighted. We can use this tool to create a block, as shown in Figure 14.33. Select the Block Tool and use it to make a window on the Block Editor workspace. The size of the block is not important at this time.

Click on the toolbar button for the Selection Tool, as shown in Figure 14.34. Highlight the block by clicking on it. Right-click to get the menu in Figure 14.35 and select **Block Properties.**

In the **General** tab of the **Block Properties** dialog box (Figure 14.36), type the block name (e.g., **program_counter**). The program counter is defined by a VHDL design entity. The block name must be the same as the VHDL file name for the program counter and must be included in the project for the top-level design. (More about this later.)

Block tool

FIGURE 14.32 Block Tool

FIGURE 14.33 Entering a Block

Selection
tool

FIGURE 14.34   Selection Tool

✂	Cut	Ctrl+X
📋	Copy	Ctrl+C
📋	Paste	Ctrl+V
✕	Delete	Del

AutoFit

🖉 Assignment Editor   Ctrl+Shift+A

📇 Locate in Timing Closure Floorplan
📇 Locate in Last Compilation Floorplan
📇 Locate in Chip Editor

Create Design File from Selected Block...
Update Design File from Selected Block...
Open Design File

📄 Update Symbol or Block...

Zoom   ▶

Mapper Properties

📇 Block Properties

FIGURE 14.35   Pop-Up Menu (Block Properties)

Block Properties

General | I/Os | Parameters | Format

Name: program_counter

Instance name: inst1

OK   Cancel

FIGURE 14.36   Naming a Block

FIGURE 14.37 Entering Port Names

In the **I/Os** tab of the **Block Properties** box (Figure 14.37), enter all the input and output names. For single nodes, enter the name and whether the port is an input or an output. For busses, be sure to include bus sizes (e.g., **pc_addr[3..0]**).

Once all the ports have been entered, click **OK** to close the box. Highlight the block and right-click. Select **Autofit,** as shown in Figure 14.38, to automatically size the box to accommodate the I/O list. The finished block is shown in Figure 14.39.

All blocks included in a Block Diagram File must be included in the current project. To add these files or examine the files in a project, select **Add/Remove Files in Project . . .** from the **Project** menu or select **Files and Directories, Add/Remove** from the **Settings** menu. Both selections open the same box, shown in Figure 14.40. If necessary, add a design file for each block.

VHDL Code for Address Bus Components

The VHDL code for the instruction register, as follows, is typical of a CPU block that has a synchronous parallel load function. Data is latched from the data bus into the block on a positive clock edge when the load input, **ir_ld,** is HIGH. An internal signal holds the latched data, which is split into an instruction portion and an operand address portion. The latter is presented to the address multiplexer and sent to the address bus, when the address MUX is enabled.

instruction_register.vhd

```
-- Instruction register
-- 8-bit flip-flop with input and output enables
-- Input: op code (4-bit) and
--        address of operand (4-bit) from system ROM
-- Output: opcode to Instruction Decoder module in controller,
--        address to tristate address bus via Address_MUX
-- Register has active-LOW asynchronous reset
```

FIGURE 14.38 Using Autofit to Size a Block

FIGURE 14.39 Finished Block

FIGURE 14.40 Files Selected as Part of Project

```
LIBRARY ieee;
USE ieee.std_logic_1164.ALL;

ENTITY instruction_register IS
PORT(
    data_bus            : IN  STD_LOGIC_VECTOR(7 downto 0);
    clock, reset, ir_ld : IN  STD_LOGIC;
    ir_addr             : OUT STD_LOGIC_VECTOR(3 downto 0);
    instruction         : OUT STD_LOGIC_VECTOR(3 downto 0));
END instruction_register;

ARCHITECTURE instr OF instruction_register IS
    -- Internal flip-flop values
    SIGNAL q_int : STD_LOGIC_VECTOR(7 downto 0);
BEGIN
    PROCESS(clock, reset)
    BEGIN
        IF (reset = '0') THEN
            -- Clear all register bits
            q_int <= (others => '0');
```

```
        ELSIF (clock'EVENT and clock = '1') THEN
            -- Load register on positive clock edge, if input is enabled
            IF (ir_ld = '1') THEN
                q_int <= data_bus;
            END IF;
        END IF;
    END PROCESS;
    -- Split data into opcode and operand address
    instruction    <= q_int(7 downto 4);
    ir_addr        <= q_int(3 downto 0);
END instr ;
```

The program counter (PC) is the other CPU block whose output is connected to the address bus via the tristate multiplexer. This block, whose VHDL code follows, is a simple 4-bit binary counter with a count enable input. In normal operation, the PC increments once every fetch/execute cycle. Its output connects directly to an input of the address bus MUX.

program_counter.vhd

```
-- Program Counter
-- 4-bit counter with active-LOW asynchronous reset
-- Holds address of next instruction to be fetched
-- Increments when PC_INC is HIGH
-- Output is multiplexed onto tristate address bus by Address_MUX

LIBRARY ieee;
USE ieee.std_logic_1164.ALL;
USE ieee.std_logic_unsigned.ALL;

ENTITY program_counter IS
    PORT(
        clock, reset, pc_inc : IN STD_LOGIC;
        pc_addr : BUFFER STD_LOGIC_VECTOR(3 downto 0));
END program_counter;

ARCHITECTURE pc OF program_counter IS
BEGIN
    PROCESS(clock, reset)
    BEGIN
        IF(reset = '0')THEN
            pc_addr <= (others => '0');
        ELSIF(clock'EVENT and clock = '1')THEN
            IF(pc_inc = '1')THEN
                pc_addr <= pc_addr + 1;
            END IF;
        END IF;
    END PROCESS;
END pc;
```

The VHDL code for the memory address register follows. Data is latched from the address bus into the register on a positive clock edge, when enabled. The register outputs are directly connected to the ROM address lines.

memory_address_register.vhd

```
-- Memory address register
-- 4-bit flip-flop
-- Output is always applied to address inputs of ROM
-- Input: 4-bit address bus
-- Register has active-LOW asynchronous reset

LIBRARY ieee;
USE ieee.std_logic_1164.ALL;
```

```
ENTITY memory_address_register IS
    PORT(
        clock, reset, mar_ld : IN   STD_LOGIC;
        addr_bus              : IN   STD_LOGIC_VECTOR(3 downto 0);
        rom_addr              : OUT STD_LOGIC_VECTOR(3 downto 0));
END memory_address_register;

ARCHITECTURE mar OF memory_address_register IS
BEGIN
    PROCESS(reset, clock)
    BEGIN
        IF(reset = '0')THEN
            rom_addr <= (others => '0');
        ELSIF(clock'EVENT and clock = '1')THEN
            IF(mar_ld = '1')THEN
                rom_addr <= addr_bus;
            END IF;
        END IF;
    END PROCESS;
END mar;
```

The VHDL code for a tristate multiplexer follows. In this design, there is a separate select input for each logic signal to be multiplexed. That is, there is a select input for the instruction register (**ir_oe**) and another for the program counter (**pc_oe**). These are combined into an internal signal called **controls,** which is used in a CASE statement to choose which logic input is to be switched to the MUX output. Data is applied asynchronously to the MUX inputs and latched synchronously to the address bus.

The high-impedance state of the MUX output is selected by default, if neither signal is active. This is indicated in the default clause as (others => 'Z'), which is a way of setting all elements of a bus to the high-impedance state without specifying the bus width.

```
-- Address multiplexer with synchronous tristate outputs
-- Places instruction register or program counter contents
-- on address bus when selected. Otherwise output is high-impedance.
LIBRARY ieee;
USE ieee.std_logic_1164.ALL;

ENTITY address_mux IS
    PORT(
        clock, reset, ir_oe, pc_oe : IN STD_LOGIC;
        ir_addr, pc_addr  : IN  STD_LOGIC_VECTOR(3 downto 0);
        addr_bus          : OUT STD_LOGIC_VECTOR(3 downto 0));
END address_mux;
```

address_mux.vhd

```
ARCHITECTURE mux OF address_mux IS
    SIGNAL controls : STD_LOGIC_VECTOR(1 downto 0);
BEGIN
    -- Concatenate output enable lines to use in
    -- CASE statement
    controls <= pc_oe & ir_oe;

    PROCESS(clock, reset)
    BEGIN
        IF(reset = '0')THEN
            addr_bus <= (others => 'Z');
        ELSIF(clock'EVENT and clock = '1')THEN
            CASE controls IS
                WHEN "01" =>
```

```
                          addr_bus <= ir_addr;
                  WHEN "10" =>
                          addr_bus <= pc_addr;
                  WHEN others =>
                          addr_bus <= (others => 'Z');
              END CASE;
          END IF;
      END PROCESS;
  END mux;
```

■ SECTION 14.3 REVIEW PROBLEM

14.3 Write a default clause in a CASE statement that can be used to set a signal called **data_bus** to the high-impedance state.

14.4 QUARTUS II IMPLEMENTATION OF THE RISC8v1 MCU

risc8vl.bdf
risc8vl.vwf

Figure 14.41 shows a Quartus II Block Diagram File for the RISC8v1 MCU. Compare this to the block diagram of the MCU in Figure 14.6. In addition to the various registers, the Block Diagram File has two bus multiplexers: the address bus MUX described in the previous section and a similar MUX for the data bus.

We have already examined several of the registers. The remaining ones are described in the following paragraphs.

ROM

Figure 14.42 shows the block for the system ROM. The design file for the block is a lower-level block diagram file with an LPM megafunction component called LPM_ROM. This file also contains input and output pins with the same names as in the top-level block diagram.

To create the ROM component, create a new block diagram file in Quartus II and save it as **rom.bdf**. Add the component LPM_ROM to the file, as shown in Figure 14.43. Figures 14.44 to 14.48 show the MegaWizard Plug-In Manager screens used to create the ROM.

The MegaWizard Plug-In Manager is used to select VHDL output (Figure 14.44), FLEX10K device family, 4-bit address, and 8-bit output bus width (Figure 14.45), unregistered address and output ports (Figure 14.46), and to set a parameter called LPM_FILE to the name of the Memory Initialization File for the ROM (Figure 14.47). The screen in Figure 14.48 lists the files created for the ROM component. The finished component along with address and data pins is shown in Figure 14.49.

The system ROM is used to store programs that are run on the RISC8v1 MCU. A simple program to load a number, add another, and send the result to the output is listed in Table 14.5. (This is the same program as was listed in Table 14.3, earlier in the chapter.)

TABLE 14.5 Simple Program for the RISC8v1 MCU

Address	Data	Comment
0	8C	Load contents of C
1	1D	Add contents of D
2	90	Send accumulator contents to output register
3	F0	Halt
4 – B	Blank (00)	
C	55	Data for Load instruction
D	64	Data for Add instruction
E – F	Blank (00)	

instruction_register

I/O	Type
data_bus[7..0]	INPUT
clock	INPUT
reset	INPUT
ir_ld	INPUT
ir_addr[3..0]	OUTPUT
instruction[3..0]	OUTPUT

inst2

data_mux

I/O	Type
rom_data[7..0]	INPUT
acc_data[7..0]	INPUT
alu_data[7..0]	INPUT
rom_oe	INPUT
acc_oe	INPUT
alu_oe	INPUT
clock	INPUT
reset	INPUT
data_bus[7..0]	OUTPUT

inst5

BIDIR   data_bus[7..0]

controller_v1

I/O	Type
clock	INPUT
reset	INPUT
instruction[3..0]	INPUT
fetch	OUTPUT
pc_inc	OUTPUT
pc_oe	OUTPUT
ir_ld	OUTPUT
ir_oe	OUTPUT
mar_ld	OUTPUT
rom_oe	OUTPUT
acc_ld	OUTPUT
acc_oe	OUTPUT
alu_oe	OUTPUT
mdr_ld	OUTPUT
or_ld	OUTPUT
s[2..0]	OUTPUT

inst9

OUTPUT   fetch

address_mux

I/O	Type
pc_addr[3..0]	INPUT
pc_oe	INPUT
ir_oe	INPUT
ir_addr[3..0]	INPUT
clock	INPUT
reset	INPUT
addr_bus[3..0]	OUTPUT

inst4

BIDIR   addr_bus[3..0]

program_counter

I/O	Type
clock	INPUT
reset	INPUT
pc_inc	INPUT
pc_addr[3..0]	OUTPUT

inst1

accumulator

I/O	Type
clock	INPUT
reset	INPUT
acc_ld	INPUT
data_bus[7..0]	INPUT
acc_data[7..0]	OUTPUT
operand_a[7..0]	OUTPUT

inst6

memory_address_register

I/O	Type
clock	INPUT
reset	INPUT
mar_ld	INPUT
addr_bus[3..0]	INPUT
rom_addr[3..0]	OUTPUT

inst3

alu

I/O	Type
operand_a[7..0]	INPUT
memory_data[7..0]	INPUT
s[2..0]	INPUT
alu_data[7..0]	OUTPUT

inst7

rom

I/O	Type
rom_addr[3..0]	INPUT
rom_data[7..0]	OUTPUT

inst

memory_data_register

I/O	Type
clock	INPUT
reset	INPUT
data_bus[7..0]	INPUT
mdr_ld	INPUT
memory_data[7..0]	OUTPUT

inst8

output_register

I/O	Type
clock	INPUT
reset	INPUT
data_bus[7..0]	INPUT
or_ld	INPUT
output[7..0]	OUTPUT

inst12

OUTPUT   output[7..0]

reset   INPUT

clock   INPUT

FIGURE 14.41 Quartus Block Diagram of RISC8v1

rom

I/O	Type
rom_addr[3..0]	INPUT
rom_data[7..0]	OUTPUT

inst

FIGURE 14.42 ROM Block

FIGURE 14.43 Select LPM-ROM Component

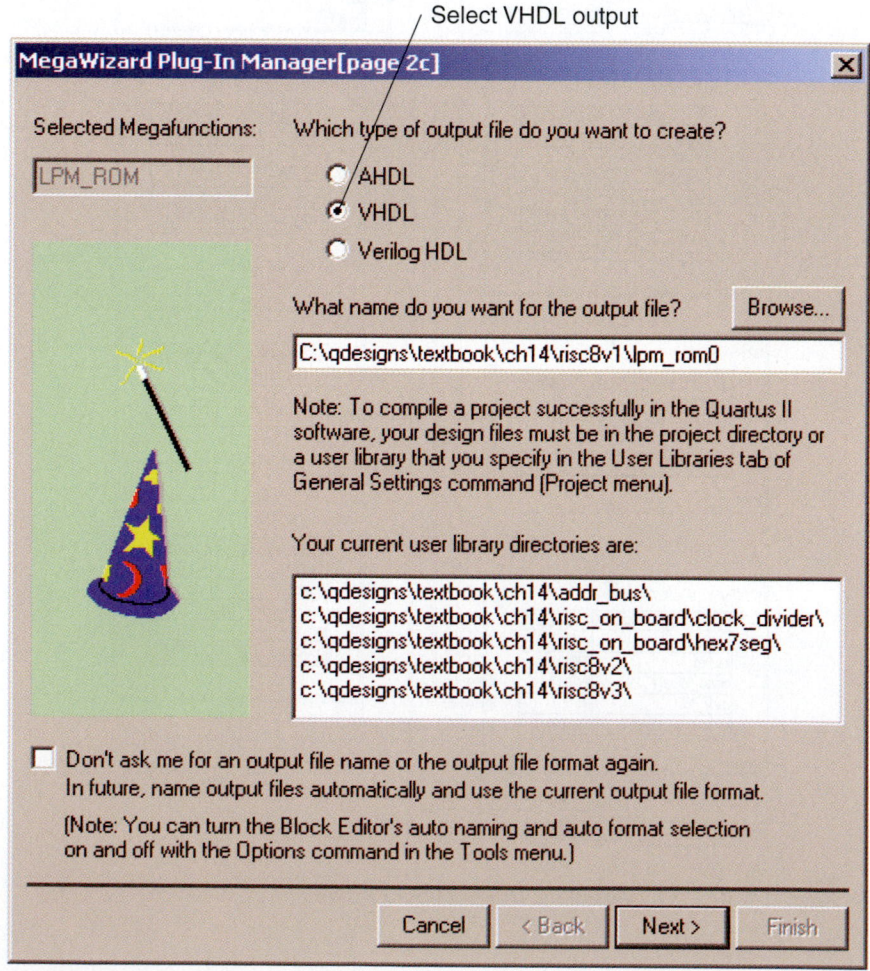

FIGURE 14.44 Selecting VHDL Output for ROM Megafunction

FIGURE 14.45 Selecting Device Family and Bus Width for ROM Megafunction

FIGURE 14.46 Selecting Unregistered Ports for ROM Megafunction

Filename for memory intialization

FIGURE 14.47 Selecting Memory Initialization File for ROM Megafunction

FIGURE 14.48 ROM Megafunction Summary

FIGURE 14.49 ROM Megafunction Symbol

rom.bdf

We can load this program into the ROM by creating a Memory Initialization File (MIF). When the design for RISC8v1 is compiled, the contents of the MIF will be loaded into the ROM component.

To create a Memory Initialization File, start the Quartus II Memory Editor by creating a new file, as shown in Figure 14.50. Select the **Other Files** tab in the **New** dialog box, and then select **Memory Initialization File.** A dialog box, shown in

FIGURE 14.50 New File Dialog (MIF)

FIGURE 14.51 Specifying Memory Dimensions

FIGURE 14.52 Setting Address Radix

Figure 14.51, will appear. Specify the dimensions of the memory as **Number of words: 16** and **Word size: 8.**

The Memory Editor has the form of a spreadsheet that shows the individual memory locations. Right-click on the top row of cells and select the **Address Radix** as **Hexadecimal,** as shown in Figure 14.52. Repeat the procedure to select the **Memory Radix** as **Hexadecimal,** as shown in Figure 14.53. Right-click on one of the lower two rows of cells and select **Cells Per Row/AutoFit,** as shown in Figure 14.54. This should yield a single row of cells, as shown in Figure 14.55. If it does not, adjust the width of the window and repeat the procedure.

Fill in the cells as shown in Figure 14.55 to enter the program in Table 14.5. Save the file as **program.mif.** The MIF takes effect the next time the project is compiled.

To change the program, edit the ROM component to point to a different MIF, or else edit the existing MIF. To point to the new MIF, double-click the ROM block to open the Block Diagram File that contains the LPM_ROM component. Double-click the LPM_ROM symbol to start the MegaWizard Plug-In Manager and follow the prompts until the screen that specifies the MIF name is shown. Make the change, click **Finish,** and accept all the prompts to exit the file.

In some earlier versions of Quartus II, double-clicking does not open the design file of a component. In this case, click the ROM block to highlight it, then right-click and select **Open Design File** from the resulting pop-up menu. This should open the Block Diagram File that contains the LPM_ROM component. Highlight and right-

FIGURE 14.53 Setting Memory (Data) Radix

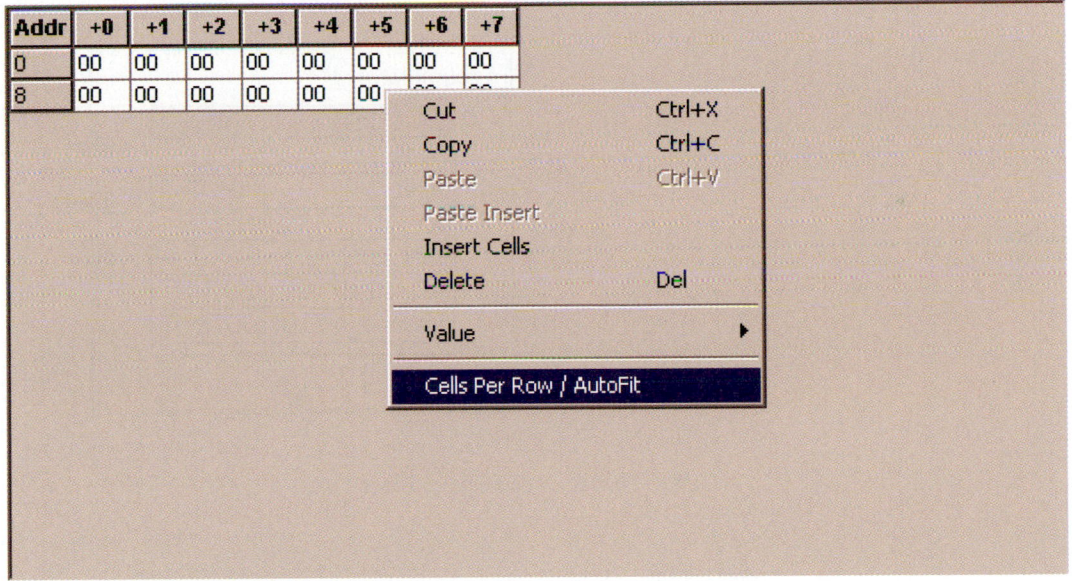

FIGURE 14.54 Setting Row/Column Size Using Autofit

click this component. In the pop-up menu select **MegaWizard Plug-In Manager** and proceed as explained.

Accumulator

Figure 14.56 shows the Quartus II block symbol for the accumulator. This register is an 8-bit flip-flop with input and output enables. Data is synchronously latched into the accumulator from the data bus when **acc_ld** is HIGH. This data is directly applied to one 8-bit input port of the ALU via the port called **operand_a[7..0].** No enable input is required for this connection. Data is also asynchronously sent to the data bus multiplexer when **acc_oe** is HIGH. The VHDL code for the accumulator is as follows.

FIGURE 14.55 Final MIF

FIGURE 14.56 Accumulator Block

```
-- Accumulator
-- 8-bit flip-flop with input and output enables
-- Input: 8-bit operand from data bus
-- Output: 8-bit operand to ALU (direct connection) and
-- 8-bit output to data bus via Data_MUX
-- Register has active-LOW asynchronous reset

LIBRARY ieee;
USE ieee.std_logic_1164.ALL;

ENTITY accumulator IS
    PORT(
        clock, reset, acc_ld : IN  STD_LOGIC;
        data_bus             : IN  STD_LOGIC_VECTOR(7 downto 0);
        acc_data, operand_a  : OUT STD_LOGIC_VECTOR(7 downto 0));
END accumulator;
```

accumulator.vhd

```
ARCHITECTURE acc OF accumulator IS
BEGIN
    PROCESS(clock, reset)
    BEGIN
        IF(reset = '0')THEN
            acc_data  <= (others => '0');
            operand_a <= (others => '0');
        ELSIF(clock'EVENT and clock = '1')THEN
            IF(acc_ld = '1')THEN
                acc_data  <= data_bus;
                operand_a <= data_bus;
            END IF;
        END IF;
    END PROCESS;
END acc;
```

Memory Data Register

The symbol for the memory data register is shown in Figure 14.57. This is similar to the accumulator, but has no output to the system data bus. Data is loaded synchronously when **mdr_ld** is HIGH. The port labeled **memory_data[7..0]** connects directly to an input port of the ALU without any enable. The VHDL code of the MDR is as follows.

```
-- Memory data register
-- 8-bit flip-flop with input
-- Output is always applied to "B" inputs of ALU
-- Input: 8-bit data bus
-- Register has active-LOW asynchronous reset

LIBRARY ieee;
USE ieee.std_logic_1164.ALL;

ENTITY memory_data_register IS
PORT(
    data_bus               : IN  STD_LOGIC_VECTOR(7 downto 0);
    clock, reset, mdr_ld   : IN  STD_LOGIC;
    memory_data            : OUT STD_LOGIC_VECTOR(7 downto 0));
END memory_data_register;
```

FIGURE 14.57 Memory Data Register Block

memory_data_register.vhd

```
ARCHITECTURE mdr OF memory_data_register IS
BEGIN
    PROCESS(clock, reset)
    BEGIN
        IF(reset = '0')THEN
            -- Clear all register bits
            memory_data <= (others => '0');
        ELSIF (clock'EVENT and clock = '1') THEN
            -- Load register on positive clock edge, if input is enabled
            IF(mdr_ld = '1')THEN
                memory_data <= data_bus;
            END IF;
        END IF;
    END PROCESS;
END mdr ;
```

Data Bus MUX

The symbol for the data bus multiplexer is shown in Figure 14.58. This block is necessary to implement a tristate bus in an Altera CPLD. The MUX combines output data from the ROM, ALU, and accumulator and selects one of these lines to go to the 8-bit data bus. When none of these signals is selected, the bus is in the high-impedance state.

Note that one of the conduit lines connected to the MUX is also connected to a bidirectional pin, allowing signals to go into and out of the various blocks on the same bidirectional tristate bus. This line acts as the data bus for the RISC8v1 MCU, carrying both the input and output signals for to and from the MUX. The VHDL code for this block is as follows.

beta_mux.vhd

```
-- Multiplexes data from ROM, Accumulator, or ALU onto tristate data
-- bus. Each data input is separately enabled; only one enable
-- permitted at a time.
LIBRARY ieee;
USE ieee.std_logic_1164.ALL;

ENTITY data_mux IS
    PORT(
        clock, reset            : IN  STD_LOGIC;
        -- ROM instruction/data
```

FIGURE 14.58 Data Bus MUX

```
        rom_data                : IN  STD_LOGIC_VECTOR(7 downto 0);
        -- Accumulator contents
        acc_data                : IN  STD_LOGIC_VECTOR(7 downto 0);
        -- ALU contents
        alu_data                : IN  STD_LOGIC_VECTOR(7 downto 0);
        rom_oe, acc_oe, alu_oe  : IN STD_LOGIC;
        data_bus                : OUT STD_LOGIC_VECTOR(7 downto 0));
END data_mux;

ARCHITECTURE a OF data_mux IS
    SIGNAL controls : STD_LOGIC_VECTOR(2 downto 0);
BEGIN
    controls <= rom_oe & acc_oe & alu_oe;

    PROCESS(clock, reset)
    BEGIN
        IF(reset = '0')THEN
            data_bus <= (others => 'Z');
        ELSIF(clock'EVENT and clock = '1')THEN
            CASE controls IS
                WHEN "100" =>
                    data_bus <= rom_data;
                WHEN "010" =>
                    data_bus <= acc_data;
                WHEN "001" =>
                    data_bus <= alu_data;
                WHEN others =>
                    data_bus <= (others => 'Z');
            END CASE;
        END IF;
    END PROCESS;
END a;
```

Output Register

The output register, shown in Figure 14.59, is an 8-bit flip-flop that is synchronously loaded when **or_ld** is HIGH. This register is an example of a parallel output port. The VHDL code for this register is as follows.

FIGURE 14.59   Output Register Block

```
-- Output register
-- 8-bit flip-flop with input from data bus
-- Input: 8-bit data bus
-- Register has active-LOW asynchronous reset

LIBRARY ieee;
USE ieee.std_logic_1164.ALL;

ENTITY output_register IS
PORT(
    data_bus             : IN  STD_LOGIC_VECTOR(7 downto 0);
    clock, reset, or_ld  : IN  STD_LOGIC;
    output               : OUT STD_LOGIC_VECTOR(7 downto 0));
END output_register;

ARCHITECTURE output OF output_register IS
BEGIN
    PROCESS(clock, reset)
    BEGIN
        IF(reset = '0')THEN
            -- Clear all register bits
            output <= (others => '0');
        ELSIF (clock'EVENT and clock = '1') THEN
            -- Load register on positive clock edge, if input is enabled
            IF(or_ld = '1')THEN
                output <= data_bus;
            END IF;
        END IF;
    END PROCESS;
END output ;
```

output_register.vhd

Controller

The controller for the RISC8v1, shown in Figure 14.60, is by far the most complex block in the CPU. Its function is to interpret the CPU instruction codes and generate the control signals for all registers to correctly execute the program functions.

The controller is implemented as a VHDL-based state machine whose architecture body has four main parts:

1. A section defining the state type and listing all states in the machine.
2. Two CASE statements defining the actual progress of states.
3. A series of concurrent signal assignment statements that map the output ports to a 16-bit internal signal called **control_word.**
4. A selected signal assignment statement that decodes the states of the state machine and generates the required control signals by assigning them to the control word.

The complete VHDL listing for the controller is as follows.

```
-- Controller for 8-bit RISC CPU
-- Generates signals to transfer data between CPU modules
-- and controls arithmetic/logic functions.
-- Version 1 has four instructions, indicated in hexadecimal:
--
-- 1 ADD    (Adds contents of Accumulator and MDR)
-- 8 LOAD   (Loads data from ROM to Accumulator)
-- 9 OUTPUT (Transfers accumulator contents to Output Register)
-- F HALT   (Stops processing)
```

FIGURE 14.60 Controller Block

```
--
-- Opcodes 0 to 7 are reserved for ALU functions, as follows:
--
-- 0 INC    (Increments Accumulator (ACC))
-- 1 ADD    (Adds contents of ACC and MDR)
-- 2 SUB    (Subtracts contents of MDR from ACC)
-- 3 DEC    (Decrements ACC)
-- 4 NOT    (Complements ACC)
-- 5 AND    (ACC AND MDR)
-- 6 OR     (ACC OR MDR)
-- 7 XOR    (ACC XOR MDR)
--
-- Above codes the same for opcodes and ALU function select codes.
--

ENTITY controller_v1 IS
   PORT(
        clock, reset               : IN  BIT;
        instruction                : IN  BIT_VECTOR(3 downto 0);
        fetch, pc_inc, pc_oe       : OUT BIT;
        ir_ld, ir_oe, mar_ld, rom_oe : OUT BIT;
        acc_ld, acc_oe, alu_oe, mdr_ld : OUT BIT;
        or_ld                      : OUT BIT;
        s                          : OUT BIT_VECTOR(2 downto 0));
END controller_v1;
```

controller_v1.vhd

```vhdl
ARCHITECTURE ctrl OF controller_v1 IS
    TYPE state_type IS (start, fetch1, fetch2, fetch3, fetch4,
                        fetch5, load1, load2, load3, load4, add1,
                        add2, add3, add4, add5, add6, output1,
                        output2, halt);
    SIGNAL state : state_type;
    SIGNAL control_word   : BIT_VECTOR(15 downto 0);
BEGIN
    PROCESS(clock, reset)
    BEGIN
        IF(reset = '0')THEN
            state <= start;
        ELSIF(clock'EVENT and clock = '1')THEN
            -- Create state machine for instruction sequences.
            CASE state IS
                WHEN start =>
                    state <= fetch1;
                WHEN fetch1 =>          -- Fetch cycle
                    state <= fetch2;    -- (same for all instructions)
                WHEN fetch2 =>
                    state <= fetch3;
                WHEN fetch3 =>
                    state <= fetch4;
                WHEN fetch4 =>
                    state <= fetch5;
                WHEN fetch5 =>
                    CASE instruction IS  -- Decode instruction
                        WHEN x"1" =>
                            state <= add1;
                        WHEN x"8" =>
                            state <= load1;
                        WHEN x"9" =>
                            state <= output1;
                        WHEN x"F" =>
                            state <= halt;
                        WHEN others =>
                            state <= halt;
                    END CASE;
                WHEN add1 =>              -- ADD
                    state <= add2;
                WHEN add2 =>
                    state <= add3;
                WHEN add3 =>
                    state <= add4;
                WHEN add4 =>
                    state <= add5;
                WHEN add5 =>
                    state <= add6;
                WHEN add6 =>
                    state <= fetch1;

                WHEN load1 =>            -- LOAD
                    state <= load2;
                WHEN load2 =>
                    state <= load3;
                WHEN load3 =>
                    state <= load4;
```

```vhdl
            WHEN load4 =>
                state <= fetch1;

            WHEN output1 =>          -- OUTPUT
                state <= output2;
            WHEN output2 =>
                state <= fetch1;

            WHEN halt =>            -- HALT
                state <= halt;
            WHEN others =>
                state <= halt;
         END CASE;
      END IF;
   END PROCESS;

-- Assign control bus outputs.
fetch  <= control_word(14);
pc_inc <= control_word(13);
pc_oe  <= control_word(12);

ir_ld  <= control_word(11);
ir_oe  <= control_word(10);
mar_ld <= control_word(9);
rom_oe <= control_word(8);

acc_ld <= control_word(7);
acc_oe <= control_word(6);
alu_oe <= control_word(5);
mdr_ld <= control_word(4);

or_ld  <= control_word(3);
s      <= control_word(2 downto 0);

-- Assign output control lines for each control state
-- "Fetch" output goes LOW to turn on LED during fetch cycle
WITH state SELECT
      control_word <=  x"4000" WHEN start,      -- Fetch LED OFF
                       x"1000" WHEN fetch1,     -- pc_oe
                       x"2200" WHEN fetch2,     -- pc_inc, mar_ld
                       x"0100" WHEN fetch3,     -- rom_oe
                       x"0800" WHEN fetch4,     -- ir_ld
                       x"0000" WHEN fetch5,     -- wait state

                       x"4400" WHEN load1,      -- ir_oe
                       x"4200" WHEN load2,      -- mar_ld
                       x"4100" WHEN load3,      -- rom_oe
                       x"4080" WHEN load4,      -- acc_ld

                       x"4401" WHEN add1,       -- ir_oe, s=001
                       x"4201" WHEN add2,       -- mar_ld, s=001
                       x"4101" WHEN add3,       -- rom_oe, s=001
                       x"4011" WHEN add4,       -- mdr_ld, s=001
                       x"4021" WHEN add5,       -- alu_oe, s=001
                       x"4081" WHEN add6,       -- acc_ld, s=001

                       x"4040" WHEN output1,    -- acc_oe
                       x"4008" WHEN output2,    -- or_ld

                       x"4000" WHEN others;     -- Fetch LED stays OFF

END ctrl;
```

State Definitions

The controller states are defined as shown. There is a state for each step of the fetch cycle, which is common to all instructions, and for each step of the execute cycle for each individual instruction.

```
TYPE state_type IS (start, fetch1, fetch2, fetch3, fetch4,
                    fetch5, load1, load2, load3, load4,
                    add1, add2, add3, add4, add5, add6,
                    output1, output2, halt);
SIGNAL state : state_type;
```

State Machine Description Statements

The state machine transitions are defined by the VHDL code that follows. The CASE statement defining the state machine directs the machine unconditionally through the five states that make up the machine's fetch cycle. In state **fetch5,** another CASE statement evaluates the instruction op code and branches to the execute cycle of the selected code. For example, if the op code is for the ADD instruction, the next state is **add1.** The machine then follows through the execute cycle of the ADD instruction until it reaches the last state, **add6,** at which point the machine reverts to the **fetch1** state and retrieves the next instruction.

The exception to this sequence is the **halt** state, which is also the default state, where the next state is always **halt.** The machine can only leave this state by a reset.

The op codes are listed in a hexadecimal format (e.g., x"1" for the ADD instruction).

```
PROCESS(clock, reset)
BEGIN
    IF(reset = '0')THEN
       state <= start;
    ELSIF(clock'EVENT and clock = '1')THEN
    -- Create state machine for instruction sequences.
    CASE state IS
        WHEN start =>
            state <= fetch1;
        WHEN fetch1 =>         -- Fetch cycle
            state <= fetch2;  -- (same for all
instructions)
        WHEN fetch2 =>
            state <= fetch3;
        WHEN fetch3 =>
            state <= fetch4;
        WHEN fetch4 =>
            state <= fetch5;
        WHEN fetch5 =>
            CASE instruction IS   -- Decode instruction
                WHEN x"1" =>
                    state <= add1;
                WHEN x"8" =>
                    state <= load1;
                WHEN x"9" =>
                    state <= output1;
                WHEN x"F" =>
                    state <= halt;
                WHEN others =>
                    state <= halt;
            END CASE;
```

```
        WHEN add1 =>              -- ADD
             state <= add2;
        WHEN add2 =>
             state <= add3;
        WHEN add3 =>
             state <= add4;
        WHEN add4 =>
             state <= add5;
        WHEN add5 =>
             state <= add6;
        WHEN add6 =>
             state <= fetch1;

        WHEN load1 =>             -- LOAD
             state <= load2;
        WHEN load2 =>
             state <= load3;
        WHEN load3 =>
             state <= load4;
        WHEN load4 =>
             state <= fetch1;

        WHEN output1 =>          -- OUTPUT
             state <= output2;
        WHEN output2 =>
             state <= fetch1;

        WHEN halt =>             -- HALT
             state <= halt;

        WHEN others =>
             state <= halt;
        END CASE;
    END IF;
END PROCESS;
```

Mapping of Outputs to Control Word

The following statements map the output ports of the controller to a 16-bit control word, which has previously been defined as a BIT_VECTOR. We must define the signal as 16 bits in order to assign the outputs using four hex digits. A separate signal called **fetch** is assigned to **control_word(14),** whose purpose to turn on an LED during the fetch cycle of an instruction.

```
-- Assign control bus outputs.
fetch  <= control_word(14);
pc_inc <= control_word(13);
pc_oe  <= control_word(12);

ir_ld  <= control_word(11);
ir_oe  <= control_word(10);
mar_ld <= control_word(9);
rom_oe <= control_word(8);

acc_ld <= control_word(7);
acc_oe <= control_word(6);
alu_oe <= control_word(5);
mdr_ld <= control_word(4);

or_ld  <= control_word(3);
s      <= control_word(2 downto 0);
```

Output Decoder

The controller output decoder generates the required control output signals for each state. These are listed as hexadecimal values. The last hex digit in the control word contains the control line for the output register (**or_ld**) and the function select code for the ALU. Since the output register is not used in the ADD instruction, the last digit is 1H (0001_2) for all steps of the instruction.

```
-- Assign output control lines for each control state
-- "Fetch" output goes LOW to turn on LED during fetch cycle
WITH state SELECT
    control_word <= x"4000" WHEN start,      -- Fetch LED OFF

                    x"1000" WHEN fetch1,     -- pc_oe
                    x"2200" WHEN fetch2,     -- pc_inc, mar_ld
                    x"0100" WHEN fetch3,     -- rom_oe
                    x"0800" WHEN fetch4,     -- ir_ld
                    x"0000" WHEN fetch5,     -- wait state

                    x"4400" WHEN load1,      -- ir_oe
                    x"4200" WHEN load2,      -- mar_ld
                    x"4100" WHEN load3,      -- rom_oe
                    x"4080" WHEN load4,      -- acc_ld

                    x"4401" WHEN add1,       -- ir_oe,  s=001
                    x"4201" WHEN add2,       -- mar_ld, s=001
                    x"4101" WHEN add3,       -- rom_oe, s=001
                    x"4011" WHEN add4,       -- mdr_ld, s=001
                    x"4021" WHEN add5,       -- alu_oe, s=001
                    x"4081" WHEN add6,       -- acc_ld, s=001

                    x"4040" WHEN output1,    -- acc_oe
                    x"4008" WHEN output2,    -- or_ld

                    x"4000" WHEN others;     -- Fetch LED stays OFF
```

■ SECTION 14.4 REVIEW PROBLEM

14.4   To which state do most instructions go after they have finished executing? Which instruction is the exception?

14.5   CREATING NEW INSTRUCTIONS (RISC8v2)

The ALU for the RISC8v1 CPU has the capability of implementing a total of eight instructions without modification. At present, the only one implemented is the ADD instruction, with an ALU select input value of **s[2..0]** = 001.

A new version of the CPU can implement as many as sixteen different op codes, assuming the structure of a 4-bit op code and optional 4-bit operand. This gives us sufficient room to implement all eight ALU functions and have eight codes remaining for data transfer and flow control functions. A sample list of possible instructions for version 2 of the RISC machine is given in Table 14.6.

TABLE 14.6 Selected Possible Op Codes for the RISC8v2 MCU

Op Code	ALU S[2..0]	Function	Operation
0	000	Increment	Acc ← Acc + 1
1	001	Add	Acc ← Acc + MDR
2	010	Subtract	Acc ← Acc − MDR
3	011	Decrement	Acc ← Acc − 1
4	100	Complement	Acc ← NOT Acc
5	101	AND	Acc ← Acc AND MDR
6	110	OR	Acc ← Acc OR MDR
7	111	XOR	Acc ← Acc XOR MDR
8	N/A	Load	Acc ← [ROM]
9	N/A	Output	Output Register ← Acc
F	N/A	Halt	Stop processing

Of the eight ALU functions, five (Add, Subtract, AND, OR, XOR) require two operands and three (Increment, Decrement, Complement) operate only on the accumulator contents. This implies that the states for these two groups of instructions will be a bit different from each other. Within each group, the operations will be quite similar, differing only by the ALU selection code generated by the controller for each instruction.

Let us add two more instructions, one from each group, to show the general procedure. For our examples, we will use the Increment and Subtract functions. Neither instruction requires additional ports to be added to the controller block. However, the VHDL code of the controller must be modified to add the new instructions.

Each instruction requires the following modifications:

1. Add the states for the new instruction to the TYPE declaration statement.
2. Add the op code for the instruction and the first state of the instruction to the CASE statement within the **fetch5** clause of the higher-level CASE statement.
3. Add the state transitions to the higher-level CASE statement.
4. Add the output states to the output decoder.

The Increment function has two states, listed in Table 14.7.

TABLE 14.7 Execute States for Increment Instruction

State Name	Action	s[2..0]	Active Control Lines
incr1	Transfer ALU to data bus	000	alu_oe
incr2	Transfer data from bus to Acc	000	acc_ld

The Subtract function has five states, shown in Table 14.8.

TABLE 14.8 Execute States for Subtract Instruction

State Name	Action	s[2..0]	Active Control Lines
sub1	Transfer address of operand from IR to address bus	010	ir_oe
sub2	Transfer address from bus to MAR	010	mar_ld
sub3	Transfer operand from ROM to data bus	010	rom_oe
sub4	Transfer data from bus to MDR	010	mdr_ld
sub5	Transfer ALU to data bus	010	alu_oe
sub6	Transfer data from bus to Acc	010	acc_ld

The TYPE declaration is modified to hold these states as follows:

```
TYPE state_type IS (reset, fetch1, fetch2, fetch3, fetch4,
                    fetch5, load1, load2, load3, load4,
                    incr1, incr2,
                    add1, add2, add3, add4, add5, add6,
                    sub1, sub2, sub3, sub4, sub5, sub6);
```

We modify the CASE statements to account for the new instructions, first by decoding the op codes:

```
WHEN fetch5 =>
    CASE instruction IS--Decode instruction
        WHEN x"0" =>
            state <= incr1;
        WHEN x"1" =>
            state <= add1;
        WHEN x"2" =>
            state <= sub1;
        WHEN x"8" =>
            state <= load1;
        WHEN x"9" =>
            state <= output1;
        WHEN x"F" =>
            state <= halt;
        WHEN others =>
            state <= halt;
        END CASE;
```

Next, we add in the remaining state transitions for the new instructions by adding WHEN clauses to the main CASE statement:

```
CASE state IS
    WHEN reset =>
        state <= fetch1;
    -- Cases for fetch, load, add, output, halt

    WHEN incr1 =>          -- INCREMENT
        state <= incr2;
    WHEN incr2 =>
        state <= fetch1;

    WHEN sub1 =>           -- SUBTRACT
        state <= sub2;
    WHEN sub2 =>
        state <= sub3;
    WHEN sub3 =>
        state <= sub4;
    WHEN sub4 =>
        state <= sub5;
    WHEN sub5 =>
        state <= sub6;
    WHEN sub6 =>
        state <= fetch1;
END CASE;
```

Finally, we add in the additional states for the output decoder:

```
WITH state SELECT
        control_word <=   x"4000" WHEN start,      -- Fetch LED OFF

                          x"1000" WHEN fetch1,     -- pc_oe
                          x"2200" WHEN fetch2,     -- pc_inc, mar_ld
                          x"0100" WHEN fetch3,     -- rom_oe
                          x"0800" WHEN fetch4,     -- ir_ld
                          x"0000" WHEN fetch5,     -- wait state

                          x"4400" WHEN load1,      -- ir_oe
                          x"4200" WHEN load2,      -- mar_ld
                          x"4100" WHEN load3,      -- rom_oe
                          x"4080" WHEN load4,      -- acc_ld

                          x"4020" WHEN incr1,      -- alu_oe, s=000
                          x"4080" WHEN incr2,      -- acc_ld, s=000

                          x"4401" WHEN add1,       -- ir_oe, s=001
                          x"4201" WHEN add2,       -- mar_ld, s=001
                          x"4101" WHEN add3,       -- rom_oe, s=001
                          x"4011" WHEN add4,       -- mdr_ld, s=001
                          x"4021" WHEN add5,       -- alu_oe, s=001
                          x"4081" WHEN add6,       -- acc_ld, s=001

                          x"4402" WHEN sub1,       -- ir_oe,  s=010
                          x"4202" WHEN sub2,       -- mar_ld, s=010
                          x"4102" WHEN sub3,       -- rom_oe, s=010
                          x"4012" WHEN sub4,       -- mdr_ld, s=010
                          x"4022" WHEN sub5,       -- alu_oe, s=010
                          x"4082" WHEN sub6,       -- acc_ld, s=010

                          x"4040" WHEN output1,    -- acc_oe
                          x"4008" WHEN output2,    -- or_ld

                          x"4000" WHEN others;     -- Fetch LED stays OFF
```

Additional instructions, such as AND, OR, and XOR are implemented in the same manner as the Subtract instruction, except with the correct ALU function code. Instructions that do not require a second operand, such as Decrement and Complement, are implemented like the Increment function, but with the appropriate ALU code.

◼ SECTION 14.5 REVIEW PROBLEM

14.5 Which registers are used in the instructions that only require one operand, such as Increment, Decrement, and Complement? What distinguishes these instructions from one another?

14.6 BRANCH INSTRUCTIONS (RISC8v3)

> ### ◼ KEY TERM
>
> **Flow Control** Control of the sequence in which program instructions are executed. Examples in VHDL include IF, CASE and WITH . . . SELECT statements. Examples in C include **if** and **while** statements.

A branch instruction is a basic tool of **flow control** in an MCU instruction set. A branch allows the stored program to evaluate data, make a decision, then execute an instruction that is not in a straight-line relation to the previous instruction.

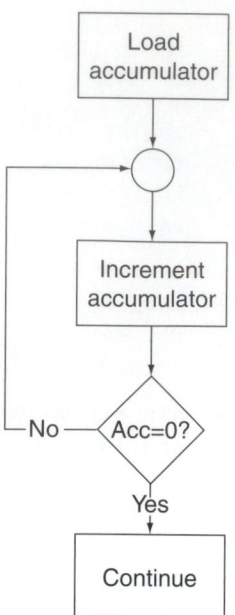

FIGURE 14.61 Flowchart Showing "Branch If Accumulator Not Zero (BNZ)"

Figure 14.61 shows the basic flow of a simple branch instruction that goes to a specified ROM address when the accumulator contents do not equal zero.

The Branch if Not Zero (BNZ) instruction works by loading the program counter with an operand representing the address of the next instruction to be executed. Our current design only allows for "direct addressing," that is, having the address of the operand as the second digit of the instruction/operand byte. Branch instructions in real CPUs often use "relative addressing," which directs the program flow to an instruction address a certain number of bytes away from the branch instruction, regardless of the instruction's absolute address. This would involve a calculation that uses the current value of the program counter as an operand, while finding a way to preserve the present values of the accumulator and program counter. Storing these register values requires the use of RAM, which is not part of the present design.

For version 3 of our RISC machine, let us assign x"A" as the op code for BNZ. The following program code segment uses the branch instruction to generate a count sequence on the output register of the MCU. (The following code represents a text version of a Memory Initialization File. The numbers on the left column are the ROM address. The next column, after the colons, is ROM data. The text between the % signs are comments. Quartus II allows MIFs to be read in this form, but not saved. Saving requires the use of the Memory Editor.)

```
0   :   8C;       % LDA C            (= F2H) %
1   :   90;       % Output                   %
2   :   00;       % Increment                %
3   :   90;       % Output                   %
4   :   A2;       % Branch to 2 if Acc != 0  %
A   :   F0;       % HALT                     %
C   :   F2;       % DATA Space               %
```

The program loads the hex value F2 from address C in ROM and displays it on the Output Register. The accumulator is incremented and the resultant value sent to output. If the accumulator value is not zero, the program branches back to ROM address 2 and executes the increment and output instructions again. After the accumulator is incremented 14 times, its value will roll over from FF to 00 and the branch test will yield a false condition. At that point, the program flow falls through the branch statement and the CPU executes the next instruction, which is HALT.

In order to implement this function, we need to modify three of the registers for the RISC8v3 CPU:

1. The program counter must be modified to load a parallel input from the address bus.
2. The accumulator requires a **zero_flag** output that is set HIGH when the accumulator contents is zero.
3. The controller must be modified to accept the **zero_flag** input from the accumulator and, based on the status of the flag, activate control lines to load the program counter from the operand output (the lower four bits) of the instruction register.

Program Counter Modifications

Figure 14.62 shows the block symbol for the modified program counter, showing the additional new input, **pc_ld,** for the parallel load function. The VHDL code for the new counter is as follows.

```
-- Program Counter
-- 4-bit counter with active-LOW asynchronous reset
-- Increments when PC_INC is HIGH, loads address when PC_LD is HIGH
-- Outputs count value when PC_OE is HIGH
-- Output is multiplexed onto tristate address bus by Address_MUX
```

FIGURE 14.62 Modified Program Counter

```
LIBRARY ieee;
USE ieee.std_logic_1164.ALL;
USE ieee.std_logic_unsigned.ALL;

ENTITY program_counter IS
    PORT(
        clock, reset, pc_inc, pc_ld : IN STD_LOGIC;
        addr_bus : IN    STD_LOGIC_VECTOR(3 downto 0);
        pc_addr  : BUFFER STD_LOGIC_VECTOR(3 downto 0));
END program_counter;

ARCHITECTURE pc OF program_counter IS
BEGIN
    PROCESS(clock, reset)
    BEGIN
        IF(reset = '0')THEN
            pc_addr <= (others => '0');
        ELSIF(clock'EVENT and clock = '1')THEN
            IF(pc_inc = '1')THEN
                pc_addr <= pc_addr + 1;
            ELSIF(pc_ld = '1')THEN
                pc_addr <= addr_bus;
            END IF;
        END IF;
    END PROCESS;
END pc;
```

Accumulator Modifications

Figure 14.63 shows the symbol for the accumulator with the zero flag output. The VHDL code for the accumulator is as follows.

```
-- Accumulator
-- 8-bit flip-flop with input and output enables
-- Input: 8-bit operand from data bus
-- Output:   8-bit operand to ALU (direct connection),
--           8-bit output to data bus via Data_MUX, and
--           zero_flag (HIGH when accumulator contents are zero).
-- Register has active-LOW asynchronous reset
LIBRARY ieee;
USE ieee.std_logic_1164.ALL;
```

FIGURE 14.63 Modified Accumulator, Showing Zero-Flag Output

```
ENTITY accumulator IS
    PORT(
        clock, reset, acc_ld : IN       STD_LOGIC;
        data_bus             : IN       STD_LOGIC_VECTOR(7 downto 0);
        acc_data             : OUT      STD_LOGIC_VECTOR(7 downto 0);
        operand_a            : BUFFER   STD_LOGIC_VECTOR(7 downto 0);
        zero_flag            : OUT      STD_LOGIC);
END accumulator;

ARCHITECTURE acc OF accumulator IS
BEGIN
    PROCESS(clock, reset)
    BEGIN
        IF(reset = '0')THEN
            acc_data  <= (others => '0');
            operand_a <= (others => '0');
        ELSIF(clock'EVENT and clock = '1')THEN
            IF(acc_ld = '1')THEN
                acc_data  <= data_bus;
                operand_a <= data_bus;
            END IF;
        END IF;
    END PROCESS;

    -- Set zero_flag when accumulator contents equal 0.
    WITH operand_a SELECT
        zero_flag <=  '1' WHEN "00000000",
                      '0' WHEN others;
END acc;
```

Controller Modifications

Figure 14.64 shows the modified symbol for the controller, with an new input to monitor the accumulator zero flag.

Partial VHDL code for the instruction decode CASE is as follows. Note that BNZ is executed, conditional upon the status of the zero flag. (The state **bnz1** must also be added to the TYPE definition of the file.)

FIGURE 14.64   Modified Controller Block Showing Zero-Flag Input

```
WHEN fetch5 =>              -- Decode instruction and execute first cycle
    CASE instruction IS    -- of selected instruction

    --Cases for load, add, and other instructions

        WHEN x"A" =>                    -- BRANCH   (op code A)
            IF (zero_flag = '0') THEN
                state <= bnz1;
            ELSE
                state <= fetch1;
            END IF;
    WHEN x"F" =>
        state <= halt;                 -- HALT     (op code F)
    WHEN others =>
        state <= halt;
END CASE;
```

The BNZ instruction has two steps, one to transfer the operand address to the address bus and one to transfer the address from the bus to the program counter. The statements for this case are:

```
WHEN bnz1 =>
    state <= bnz2;
WHEN bnz2 =>
    state <= fetch1;
```

The control line **pc_ld** must be added to the control word mapping.

```
pc_ld  <= control_word(15);
```

FIGURE 14.65   Simulation of RISC Machine Executing BNZ Instruction

Program counter
to memory address register

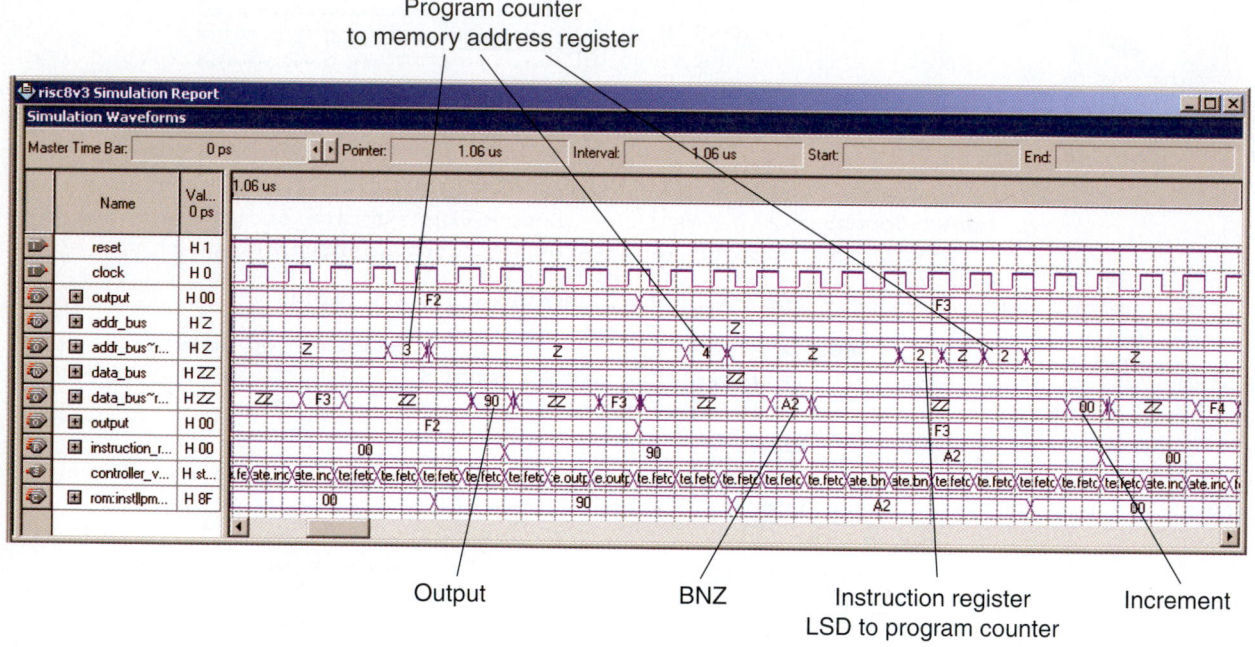

Output   BNZ   Instruction register   Increment
                                 LSD to program counter

FIGURE 14.66   Simulation Detail (BNZ Instruction)

The state decoder must now include the states for the branch instruction, which just consists of the two states that transfer the operand address in the instruction register to the program counter.

```
WITH state SELECT
    control_word <=   x"4000" WHEN start,    -- Fetch LED OFF

                      x"1000" WHEN fetch1,   -- pc_oe

                      -- states for other functions

                      x"4400" WHEN bnz1,     -- ir_oe
                      x"C000" WHEN bnz2,      -- pc_ld

                      x"4000" WHEN others;   -- Fetch LED stays OFF
```

Figures 14.65 and 14.66 show the simulation of the RISC8v3 MCU running the op code sequence that uses the BNZ instruction. Figure 14.65 shows how the value F2 is loaded then incremented until the accumulator rolls over to 00. Figure 14.66 shows a detail of the simulation. Notice that the address bus holds the value 3, just as the CPU fetches the OUTPUT instruction, followed by 4 for the BNZ instruction. When BNZ is executed, the address bus now shows 2, the value which is transferred into the program counter to execute the Increment instruction at ROM address 2.

■ SECTION 14.6 REVIEW PROBLEM

14.6 State how the program counter must be modified to be able to implement a branch instruction. Why is the modification necessary to the branch instruction?

14.7 RISC8 TEST CIRCUIT

In order to test the RISC8v1, v2, or v3 MCU on the Altera UP-1 or UP-2 board, some interface components must be added and pin numbers assigned. Figure 14.67 shows a possible test circuit.

In addition to the RISC8v3 component (created from the Block Diagram File), the circuit also has two hexadecimal-to-seven-segment decoders connected to the output register and a clock divider to slow down the RISC8 operation. The on-board oscillator runs at about 25 MHz. A 20-bit clock divider slows down the system clock to about 25 Hz. For a program with, say, a sequence of Load, Output, Add, Output, Halt, the display changes every $\frac{2}{3}$ to $\frac{3}{4}$ of a second, depending on the number of clock cycles per instruction.

Table 14.9 shows the pin assignments for the RISC8 test circuit. The device must be assigned as EPF10K**20**RC240-4 for the Altera UP-1 board and as

FIGURE 14.67   Test Circuit for RISC8 MCU

TABLE 14.9   Pin Assignments for RISC8 Test Circuit

Function	Pin Number	Function	Pin Number
clock	91	reset (pushbutton FLEX_PB2)	29
a1	6	a2	17
b1	7	b2	18
c1	8	c2	19
d1	9	d2	20
e1	11	e2	21
f1	12	f2	23
g1	13	g2	24
		fetch (dp2 LED)	25
addr_bus[3]	149	data_bus[7]	159
addr_bus[2]	148	data_bus[6]	158
addr_bus[1]	147	data_bus[5]	157
addr_bus[0]	146	data_bus[4]	156
		data_bus[3]	154
		data_bus[2]	153
		data_bus[1]	152
		data_bus[0]	151

EPF10K**70**RC240-4 for the UP-2. Pins must be assigned to the bidirectional address and data busses to prevent them from inadvertently being assigned by the compiler to FLEK 10K pins that are connected to hardwired input devices, such as the on-board DIP switches. The **fetch** output is connected to the decimal point of digit 2 of the seven-segment display. This output goes ON during the fetch cycle of an instruction. The clock is connected to the global clock pin of the FLEX 10K device. The reset input connects to a hardwired pushbutton (FLEX_PB2).

After pins have been assigned, the project for the test circuit must be compiled. Jumpers on the Altera UP-1 or UP-2 board must be set to configure the EPF10K20/70, as shown in Figure 14.68. When this has been done, the board can be configured by the Quartus II Programmer. The RISC8 program should automatically run immediately after configuration and can be run again at any time by pressing the reset pushbutton.

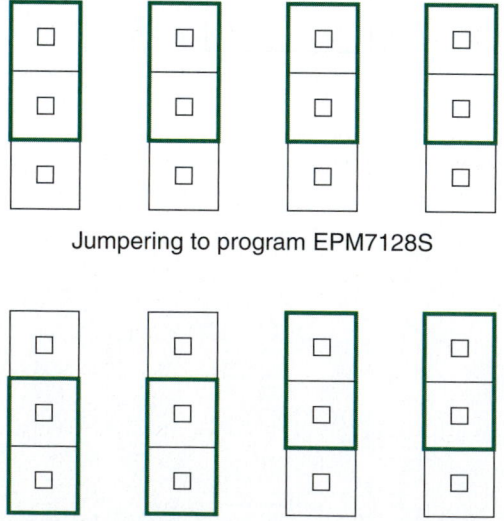

Jumpering to program EPM7128S

Jumpering to configure EPF10K20 or EPF10K70

FIGURE 14.68   Jumpering for Altera UP-1 and UP-2 Boards

14.8 POSSIBLE ENHANCEMENTS TO RISC8v3 MCU

A number of enhancements could be made to the RISC8v3 MCU. In no particular order, here are a couple of suggestions.

- Expand the address bus to 8 bits, allowing for a 256-instruction program space. The present configuration could be maintained by making the data bus 16 bits wide, thus allowing for an 8-bit op code and 8-bit operand. This, in turn, allows for 256 different op codes. Further expansion is possible (e.g., 16-bit address and 32-bit data, allowing $65,536 \times 4$ bytes of program and data storage and 65,536 op codes).

- Add RAM to the system and instructions to load from and store to RAM.

- Add instructions allowing the ALU to operate on data stored in RAM, rather than ROM. This could be done by keeping the current architecture and having a separate set of instructions for RAM and for ROM. Alternatively, the RAM and ROM could be mapped such that, say, the lower half of the memory map is ROM and the upper half is RAM. The instructions would thus be very similar to the present ones and would differentiate between ROM-based and RAM-based instructions by the address of the operand. This would require the addition of an address decoder.

- Allow for multiple-byte op codes, which would enable more op codes and larger addresses to be used. This would require the controller to be able to increment the program counter more than once for a given instruction, depending on the size of instruction.

- Add a status flag register (negative, overflow, carry, etc.) and other branch instructions based on those flags.

- Add two stack instructions: PUSH, which stores the contents of a register in RAM and POP, which restores that value to the right place. This requires a new register called the stack pointer, which starts by pointing to the top of RAM and decrementing each time a value is stacked. The stack pointer is incremented when a register is unstacked.

- A real challenge would be to allow for a hardware interrupt function, where, after receipt of an interrupt request (e.g., notification of incoming data on a port), the CPU would finish executing the current instruction, then stack the program counter, accumulator, and status flags, load the program counter with the address of program code to service the interrupt, execute the code, unstack all registers, and continue the original code.

- Add parallel inputs, as well as outputs.

- Add serial input and output.

SUMMARY

1. A microcomputer is a complete computer system. A microprocessor is a component in that system.

2. A microcomputer system includes a microprocessor or Central Processing Unit (CPU), as well as memory (RAM or ROM), and peripheral devices.

3. The components in a microcomputer system are connected by a series of tristate busses. The address bus carries data that points to a particular memory location or a particular peripheral. The data bus transfers data between peripherals, memory, and the CPU. The control bus carries signals that control the flow of data between modules.

4. Tristate bussing is a powerful and flexible technique for controlling the flow of data within a microcomputer. However, care must be taken that only one source register drives data onto the bus at any given time. Failure to do so results in bus contention, where two or more data sources are connected to the bus at the same time, giving rise to ambiguous bus states and possible damage to the driving outputs.

5. Data can be transferred between registers synchronously or asynchronously. In an asynchronous transfer, data is available on the bus as soon as the source register output is enabled. In a synchronous transfer, data is available one clock cycle after the register output is enabled. In both cases, data is latched into the source register at the end of the clock cycle in which the data is present on the bus. While the synchronous transfer may take longer, it is more stable.

6. RISC stands for *reduced instruction set computer* and is used to describe a computer with a simplified architecture.

7. Program instructions for a microcomputer are stored in the microcomputer's memory and must be fetched and executed in a particular sequence. The fetch cycle and execute cycle each involve the transfer of data between several register pairs.

8. A very simple RISC machine, the RISC8v1, can be implemented with four instructions: Load, Add, Output, and Halt. These instructions can be used to load a number into a register called the accumulator, add another number and save the result in the accumulator, transfer the result to an output register, and then stop processing.

9. The RISC8v1 can be implemented in a CPLD such as the Altera EPF10K20RC240-4 (UP-1 board) or the Altera EPF10K70RC240-4 (UP-2 board).

10. The fetch cycle is the same for all instructions in the RISC8v1. It consists of the following synchronous register transfers:

   a. The program counter (PC) contains the address of the next instruction to be executed. Its contents are transferred to the memory address register (MAR). The PC is incremented so as to be ready to point to the next instruction. Control lines: **pc_oe, mar_ld, pc_inc.**

   b. The address in the MAR is applied directly to the ROM address port. The data at the selected ROM address consists of an instruction and the address of an operand for that instruction (e.g., the data 8C represents the Load instruction [8] and an operand at address C in ROM. Thus, "load data from ROM address C to the accumulator.") The instruction/address byte is transferred to the instruction register (IR). Control lines: **rom_oe, ir_ld.**

11. The execute cycle for the Load instruction consists of the following synchronous register transfers:

   a. The contents of the IR are split into the instruction (upper 4 bits) and operand address (lower 4 bits). The instruction is decoded by the instruction decoder in the CPU controller, which then directs the register transfers for the execution cycle of the instruction. The address is transferred to the MAR, where it points to the address that holds the operand for the instruction. Control lines: **ir_oe, mar_ld**

   b. Data is transferred from the ROM to the accumulator. Control lines: **rom_oe, acc_ld.**

12. The first register transfer in the execute cycle for the Add instruction is the same as for the Load instruction (IR to MAR). The MAR now points to the ROM address for the operand. The remaining transfers are:

   a. Transfer data from ROM to the memory data register (MDR). Control lines: **rom_oe, mdr_ld.**

   b. Data from the accumulator and MDR are added in the arithmetic logic unit (ALU). The result is transferred to the ALU back to the accumulator. Control lines: **alu_oe, acc_ld.**

13. The RISC8 ALU has four arithmetic functions (Increment, Add, Subtract, Decrement) and four logic functions (Complement (NOT), AND, OR, XOR). The functions are selected by a 3-bit select input, **s[2..0].** Only the Add function is used in version 1 of the RISC8.

14. The Output instruction of the RISC8v1 MCU transfers data from the accumulator to the output register. Control lines: **acc_oe, or_ld.**

15. The Halt instruction of the RISC8v1 MCU stops all processing when executed. The machine can only be restarted by an asynchronous reset signal.

16. MAX 7000S and FLEX 10K CPLDs do not allow for internal tristate bussing. Two methods of creating a tristate bus are:

    a. Individual tristate outputs that are connected externally to the CPLD, and

    b. Separate functions that are multiplexed internally to the CPLD, then applied to a tristate output.

17. Tristate busses must always be brought to a bidirectional I/O pin in the CPLD.

18. The CPLD implementation of the RISC8v1 MCU requires a multiplexer for the address bus and another for the data bus.

19. Bus transfers in the RISC8v1 are synchronous. Data is applied asynchronously from a register to either the address bus MUX or data bus MUX. The MUX output is synchronously transferred to the bus via a flip-flop and a tristate output buffer, then synchronously from the bus to the destination register.

20. Block diagrams can be created in Quartus II that allow subdesigns to be represented as blocks with input and output ports listed in an embedded table. Blocks can be interconnected with node, bus, or conduit lines, with interface points to each block, called mappers.

21. Conduit lines are the most flexible type of connection, allowing construction of bus lines that carry multiple signals between blocks. If the signals are named identically in all blocks, the mappers connecting the conduit lines do not need to be labeled. Connections are determined by the software at compile time.

22. Program instructions and operands in the RISC8v1 are stored in an LPM_ROM component, instantiated as a megafunction. The MegaWizard Plug-In Manager can be used to set a parameter called LPM_FILE, which points to a Memory Initialization File (MIF). At compile time, the ROM is loaded with the contents of the MIF.

23. The controller for the RISC8v1 CPU is implemented as a VHDL-based state machine whose architecture body has four main parts:

    a. a section defining the state type and listing all states in the machine;

    b. two CASE statements defining the actual progress of states;

    c. a series of concurrent signal assignment statements that map the output ports to a 16-bit internal signal called **control_word;** and

    d. a selected signal assignment statement that decodes the states of the state machine and generates the required control signals by assigning them to the control word.

24. There are two types of instructions in version 2 of the RISC8 machine (RISC8v2): instructions that require the accumulator contents only (Increment, Decrement, Complement) and those that require a second operand (Add, Subtract, AND, OR, XOR). The former instructions only require a correct function code at the ALU and a transfer of data from the ALU to the accumulator. The latter instructions first require a transfer of the second operand from ROM to the memory data register.

25. New instructions require several modifications to the controller's VHDL code:

    a. new states for each instruction must be added to the TYPE declaration;

    b. state transitions must be added to the CASE statements;

    c. states must be added to the output decoder.

26. A branch instruction can be added to create the RISC8v3 CPU. This instruction allows execution of statements in an order other than a linear sequence.

27. A Branch if Not Zero (BNZ) instruction examines the contents of the accumulator and, if not equal to zero, loads a new address into the program counter. This is the address of the next instruction to be executed.

28. The RISC8 CPU must be modified as follows to implement the BNZ instruction:

    a. The program counter must be modified to load a parallel input from the address bus.

    b. The accumulator requires a **zero_flag** output that is set HIGH when the accumulator contents is zero.

    c. The controller must be modified to accept the **zero_flag** input from the accumulator and, based on the status of the flag, activate control lines to load the program counter from the operand output (the lower four bits) of the instruction register.

GLOSSARY

Accumulator   A CPU register that holds the accumulated result of arithmetic and logic operations performed by the ALU.

Address Bus   A set of parallel conductors that select the required address within the computer system memory to access data or program instructions.

Arithmetic Logic Unit (ALU)   A combinational circuit within the CPU that performs arithmetic and logical operations, such as add, subtract, AND, OR and XOR.

Block   A design unit in a Quartus Block Diagram File that represents a subunit, such as a lower-level Block Diagram File or a VHDL file, and indicates its inputs and outputs by a table embedded in the block.

Bus Contention   Conflict that arises when two or more devices attempt to write data onto a bus at the same time.

Conduit Line   A line in a Quartus II Block Diagram File that connects signals going between blocks by matching names on the connected block I/O lists.

Control Bus   A set of conductors that control the flow of data among different modules of a computer system.

Controller   A state machine within the CPU that generates control signals required to transfer data among the various CPU registers.

CPU (Central Processing Unit)   Another name for microprocessor.

Data Bus   A set of parallel conductors that transfer data within and outside the computer.

Embedded Controller   A self-contained microcomputer system (often a single chip) in which program instructions are permanently stored in ROM, making the system a stand-alone control device.

Execute   Decode a fetched instruction and perform required CPU operations.

Fetch   Retrieve a program instruction from memory.

Flow Control   Control of the sequence in which program instructions are executed. Examples in VHDL include IF, CASE and WITH. . .SELECT statements. Examples in C include **if** and **while** statements.

Instruction Decoder   A combinational circuit within the CPU that interprets the binary value of the op code in the instruction register, then directs the controller to generate the required control signals to execute the instruction.

Instruction Register (IR)   A CPU register that holds the current instruction op code being executed by the CPU.

Mapper   An interface point between a block and a node, bus, or conduit line that maps the signal names between the block and the node, bus, or conduit.

Memory Address Register (MAR)   A CPU register that holds the memory address required for the current instruction or data.

Memory Data Register (MDR)   A CPU register that holds data transferred from memory as a second operand in an arithmetic or logical operation.

Microcomputer   A self-contained computer system that consists of a microprocessor, memory components, such as RAM or ROM, and peripheral devices. Also called a microcomputer unit (MCU or μC).

Microprocessor   The component of a computer that generates control signals for the other components of the computer and performs arithmetic and logic functions. Also called a microprocessor unit, MPU, or μP.

Op Code (Operation Code)   A binary or hexadecimal code that represents an instruction for a CPU to execute.

Output Register   A peripheral register to which data is transferred from the accumulator. This can also be called a parallel output port.

Peripheral Devices (or Peripherals)   Devices connected to the address, data, and control busses of a computer system that act as interface circuits between the computer and the external world (e.g., switches, keyboards, LEDs, numeric displays, video monitors, printers, disk drives, etc.).

Program Counter (PC)   A counter within a CPU that keeps track of the address of the next program instruction to be fetched from memory.

Program Instructions   A set of binary or hexadecimal codes that are interpreted by the CPU of a computer system to perform various functions.

Register   A multibit latch or flip-flop that acts as a temporary data storage location within a computer.

RISC (Reduced Instruction Set Computer)   A computer that is capable of performing relatively few instructions, with the goal of making a simpler and faster computer architecture.

Tristate Outputs   Device outputs that can be in three possible states: logic HIGH, logic LOW, and high-impedance. The high-impedance state acts like an open circuit.

PROBLEMS

Problem numbers set in color indicate more difficult problems.

14.1   Basic Structure of a Microcomputer

14.1   State the difference between a microcomputer and a microprocessor.

14.2   Draw the block diagram of a simple microcomputer system, showing the CPU as a single block (i.e., do not draw all the CPU registers). Label all required busses.

14.3   State the function of each of the following busses:

    a. address bus

    b. data bus

    c. control bus

14.4   What is the name of the condition that arises when two or more devices attempt to write data to a bus simultaneously? What are the possible consequences of this condition?

14.5   How can the condition referred to in Problem 14.4 be avoided?

14.6   A tristate bus system that transfers data from one of two 8-bit sources to one of two 8-bit destinations is shown in Figure 14.2. Is the system synchronous or asynchronous? Complete the timing diagram in Figure 14.69 for the following conditions:

    ■ Source 1 contents = C7H.

    ■ Source 2 contents = 38H.

    ■ Transfer source 1 to destination 1.

    ■ Transfer source 1 to destination 2.

    ■ Transfer source 2 to destination 1.

FIGURE 14.69   Timing Diagram for Problem 14.6

14.7 A tristate bus system that transfers data from one of two 8-bit sources to one of two 8-bit destinations is shown in Figure 14.4. Is the system synchronous or asynchronous? Complete the timing diagram in Figure 14.70 for the conditions listed in Problem 14.6.

FIGURE 14.70 Timing Diagram for Problem 14.7

14.2 Register Level Structure of a Microcomputer System

14.8 Explain what is meant by an "embedded controller."

14.9 A CPU has the following instructions:

Op Code	Function
1	Add ROM contents to accumulator.
2	Subtract ROM contents from accumulator.
8	Load accumulator with ROM contents.
9	Send accumulator contents to output register.
F	Halt processing.

a. Write a sequence of addresses and instructions to perform the following functions:

- Load the hex value 8AH to the accumulator.
- Add 71H.
- Output the result.
- Load 75H to the accumulator.
- Subtract 3DH.
- Output the result.
- Halt.

b. What values appear in the output register of the CPU and in what sequence?

NOTE . . .

Refer to the RISC8v1 block diagram in Figure 14.6 when answering Problems 14.10 to 14.12.

14.10 Fill in the following table to indicate the active registers and control lines of the RISC8v1 for the Fetch cycle of any CPU instruction.

State	Registers	Control Lines
fetch1	PC	pc_oe
fetch2	MAR, PC	pc_inc, mar_ld

14.11 Fill in the following table to indicate the active registers and control lines of the RISC8v1 for the Execute cycle of a Load instruction.

State	Registers	Control Lines

14.12 Fill in the following table to indicate the active registers and control lines of the RISC8v1 for the Execute cycle of an Add instruction.

State	Registers	Control Lines

14.13 Shade in the registers used and highlight the appropriate control lines in Figure 14.71 for the execute cycle of the OUTPUT instruction of the RISC8v1MCU.

FIGURE 14.71 Problem 14.13: RISC8v1 Block Diagram

14.14 Complete the timing diagram in Figure 14.72 for the fetch and execute cycles of the OUTPUT instruction for the RISC8v1 MCU. Assume that the accumulator has the contents E7H.

FIGURE 14.72 Problem 14.14: Timing Diagram for Output Instruction

14.15 State the function of each of the following blocks of the RISC8v1 MCU.

a. Program counter

b. Memory address register

c. Read-only memory

d. Instruction register

e. Instruction decoder/controller

f. Accumulator

g. Memory data register

h. Arithmetic logic unit

i. Output register

14.3 Tristate Busses in Altera CPLDs

14.16 State two methods for making tristate busses using Altera MAX 7000S or FLEX 10K CPLDs.

14.17　Complete the following VHDL code for the address multiplexer of the RISC8v1 CPU. The MUX selects 4-bit data from the PC or IR and sends it to the address bus. When disabled, the MUX output is in the high-impedance state.

```
--Address multiplexer with synchronous tristate outputs
--Places instruction register or program counter contents
--on address bus when selected. Otherwise output is high-impedance.
LIBRARY ieee;
USE ieee.std_logic_1164.ALL;

ENTITY address_mux IS
    PORT(
        clock, reset, ir_oe, pc_oe : IN STD_LOGIC;
        ir_addr, pc_addr : IN STD_LOGIC_VECTOR(3 downto 0);
        addr_bus : OUT STD_LOGIC_VECTOR(3 downto 0));
END address_mux;

ARCHITECTURE mux OF address_mux IS
    SIGNAL controls : STD_LOGIC_VECTOR(1 downto 0);
BEGIN
    -- Concatenate output enable lines to use in
    -- CASE statement
    controls <= _____;

    PROCESS (clock, reset)
    BEGIN
        IF(reset = '0')THEN
            addr_bus <= _____;
        ELSIF(clock'EVENT and clock = '1')THEN
            CASE controls IS
                -- Write CASEs here
            END CASE;
        END IF;
    END PROCESS;
END mux;
```

14.4　Quartus II Implementation of RISC8v1 MCU

14.17　List the four major parts of the ARCHITECTURE body of the VHDL design entity that describe the controller for the RISC8v1 CPU.

14.18　How can we move the controller state machine out of the HALT state?

14.19　How does the RISC8v1 CPU controller decode a hexadecimal instruction?

14.5　Creating New Instructions (RISC8v2)

14.20　List all modifications required to the VHDL code for the RISC8v1 CPU controller to add an instruction for "decrement" that is assigned to an op code of x"3". Make your answer sufficiently detailed so that it includes the actual statements.

14.21　List all modifications required to the VHDL code for the RISC8v1 CPU controller to add an instruction for "XOR" that is assigned to an op code of x"7". Make your answer sufficiently detailed so that it includes the actual statements.

Converting MAX+PLUS II Projects to Quartus II

If you have been using Altera's MAX+PLUS II CPLD design and programming software, either from the first edition of this book, another book, or your own work, you probably have a large number of projects that you really do not want to enter again in Quartus II. Fortunately, the conversion process between MAX+PLUS II and Quartus II is fairly straightforward. Basically, it comes down to the following conversions:

- A Graphic Design File (*.gdf) in MAX+PLUS II must be converted to a Block Diagram File (*.bdf) in Quartus II.
- To convert a Quartus II simulation file (*.scf), you must first create a Table File (*.tbl) in MAX+PLUS II. The Table File must then be opened in Quartus II and saved as a Vector Waveform File (*.vwf).
- VHDL files do not require conversion. However, a VHDL file must be included in a Quartus II project.
- A Quartus II project must be created to hold all design and vector waveform files. Each project should be created in a separate Windows folder.
- Bus notation changes in Quartus II. In a MAX+PLUS II Graphic Design File, a bus element is indicated by a numerical suffix. For example, for a bus called **s[2..0],** the individual elements are labeled as **s2, s1,** and **s0** in a MAX+PLUS II **gdf.** In a Quartus II Block Design File, the elements are indicated by a numeral in square brackets. For a bus called **s[2..0],** the individual elements are called **s[2], s[1],** and **s[0].**
- The elements of a vector in VHDL are labeled the same for both MAX+PLUS II and Quartus II. For a port called `s : IN STD_LOGIC_VECTOR(2 downto 0);`, the individual elements are labeled `s(2)`, `s(1)`, and `s(0)`.

NOTE . . .

In recent versions of Quartus II, there is an option in the **File** menu labeled **Convert MAX+PLUS II Project**... This utility will convert device and pin assignments and design files to Quartus II format, but it does not convert simulations. Follow the steps outlined next to make sure everything works as expected for all aspects of the Quartus II design flow. We will examine the use of this utility in a later part of this appendix.

CONVERTING A MAX+PLUS II GRAPHIC DESIGN FILE

Figure A.1 shows a MAX+PLUS II Graphic Design File (**2to4dcd.gdf**) of a 2-line-to-4-line binary decoder with an active-LOW enable input, **G.** The MAX+PLUS II simulation for the decoder is shown in Figure A.2.

With the simulation file open in MAX+PLUS II, open the **File** menu and select **Create Table File...,** as shown in Figure A.3. In the dialog box, shown in Figure A.4, save the Table File with the same name as the simulation file, but with a *.**tbl**

FIGURE A.1   MAX+PLUS II Graphic Design File for 2-to-4 Decoder

FIGURE A.2   MAX+PLUS II Simulation of 2-to-4 Decoder

extension. (This is the default option. If you choose this option, the file will be saved in the same folder as the original simulation file.) Click **OK.** The box, as shown in Figure A.5, confirms that the Table File has been generated.

Create a new Windows folder for the new Quartus II project and copy the files **2to4dcd.gdf** and **2to4dcd.tbl** from their MAX+PLUS II folder to the new folder. (**Tip:** create a master folder called **qdesigns** to hold all your Quartus II projects in subfolders.)

Open the graphic file in Quartus II, as shown in Figure A.6. From the **File** menu, select **Save As.** The resultant dialog box is shown in Figure A.7. Change the file type by selecting **Block Diagram/Schematic File (*.bdf)** from the drop-down box labeled **Save as type:.** Also make sure that the box labeled **Create new project based on this file** is checked. Click **OK.** The box shown in Figure A.8 will appear. Click **Yes** to confirm the creation of a new project.

FIGURE A.3   Create Table File
(File Menu)

FIGURE A.4   Create Table File Dialog

FIGURE A.5   Table File Confirmation

FIGURE A.6   Open **gdf** in Quartus

FIGURE A.7 Save as BDF and Create New Project

FIGURE A.8 New Project Confirmation

Figures A.9 to A.15 show the screens of the New Project Wizard. For most cases, you can simply click through each screen by accepting the default choices and clicking **Next.** (You can short-circuit this process by clicking **Finish** at any time, but it's usually a good idea to go through the screens to make sure that you have chosen the correct target device.)

Figure A.9 is an introduction screen that has a few basic tips and instructions. Figure A.10 shows some hierarchy information, such as directory and file names. Design and simulation files can be added to the project in the screen shown in Figure A.11. Figure A.12 allows you to specify various optional software plug-ins that can be used with Quartus II. We do not use this feature. The MAX 7000S device family is selected, shown in Figure A.13, and the specific target device (EPM7128SLC84-7) is selected, shown in Figure A.14. The screen shown in Figure A.15 summarizes the project settings.

Once you have gone through all of these steps, you can compile the new project, either by selecting the **Start Compilation** button on the Quartus II toolbar or by selecting this option from the **Processing** menu.

FIGURE A.9 New Project Wizard (Intro)

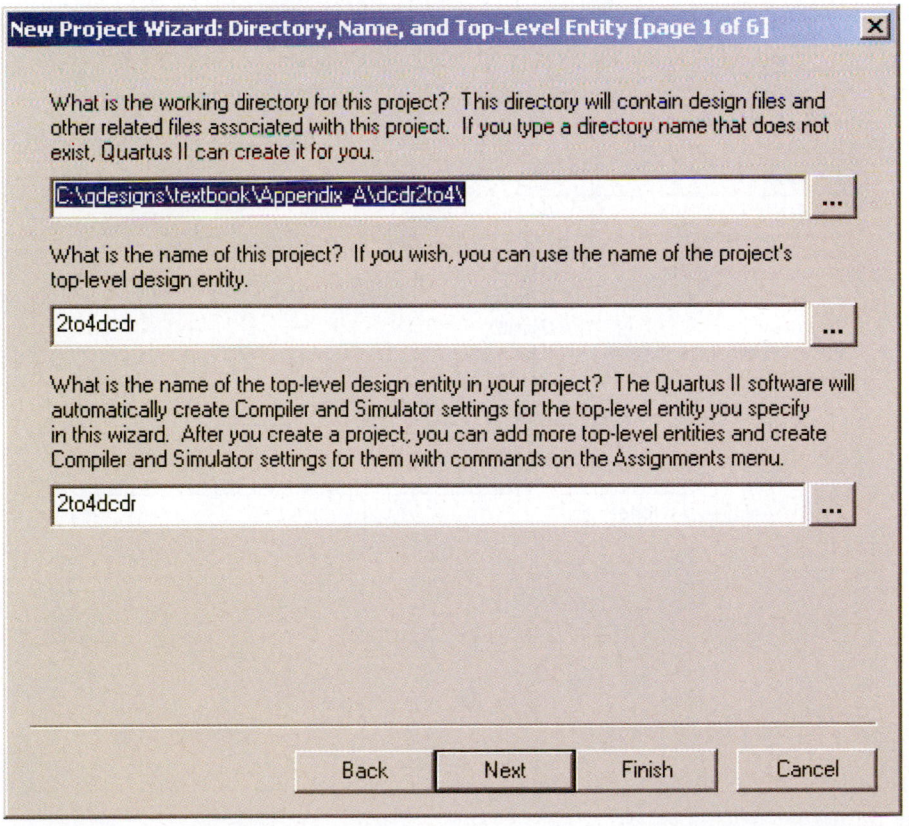

FIGURE A.10 New Project Wizard 1 (Directory)

FIGURE A.11   New Project Wizard 2 (Add Files)

FIGURE A.12   New Project Wizard 3 (EDA Tools)

FIGURE A.13   New Project Wizard 4 (Device Family)

FIGURE A.14   New Project Wizard 5 (Target Device)

FIGURE A.15   New Project Wizard 6 (Summary)

CONVERTING A MAX+PLUS II SIMULATION FILE

Make sure that you have copied the Table File created from the MAX+PLUS II simulation to the working folder for the Quartus II Project. Open the Table File in Quartus II, as shown in Figure A.16. Make sure that the drop-down box labeled **Files of type:** contains the type **Waveform/Vector Files (*.vwf, *.vec, *.tbl.)** The file will appear as in Figure A.17.

Optional step: Highlight the output lines on the Table File, as shown in Figure A.18, by dragging the mouse from Y3 to Y1. On the Waveform Editor Toolbar, click the "0" button to set all output waveforms to 0. (Output waveforms do not need to be initialized, as they are overwritten in the simulation process. However, they are only overwritten in the simulation output file and are not modified in the Quartus II Vector Waveform File.)

Save the Table File as a Vector Waveform File, as shown in Figure A.19. Make sure that the box labeled **Add file to current project** is checked. Run the simulation by clicking the **Start Simulation** toolbar button or by selecting this function from the **Processing** menu. The simulation for the 2-to-4 decoder should appear as shown in Figure A.20.

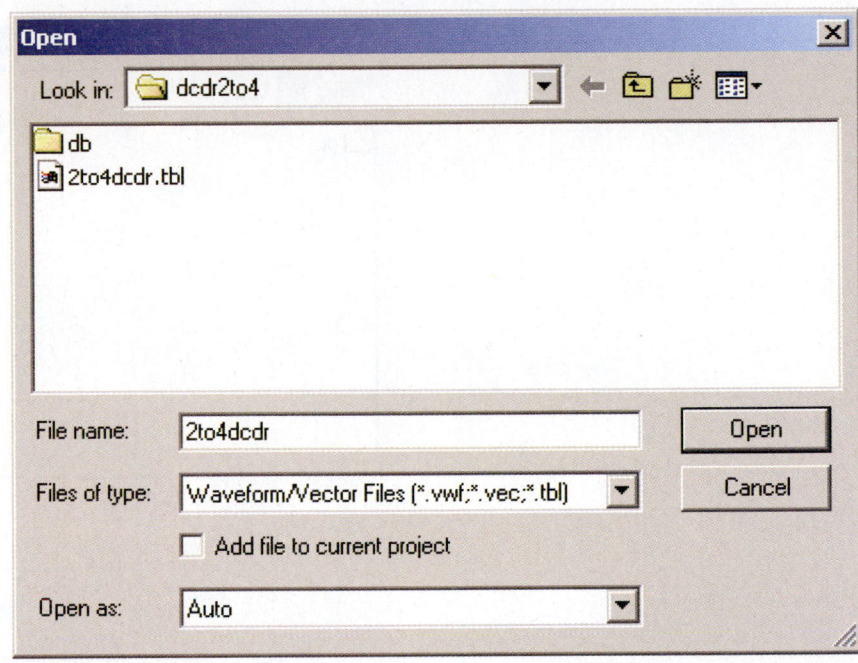

FIGURE A.16 Open Table File

FIGURE A.17 Table File in Quartus II

FIGURE A.18 Highlight Output Lines and Set to Zero (Optional)

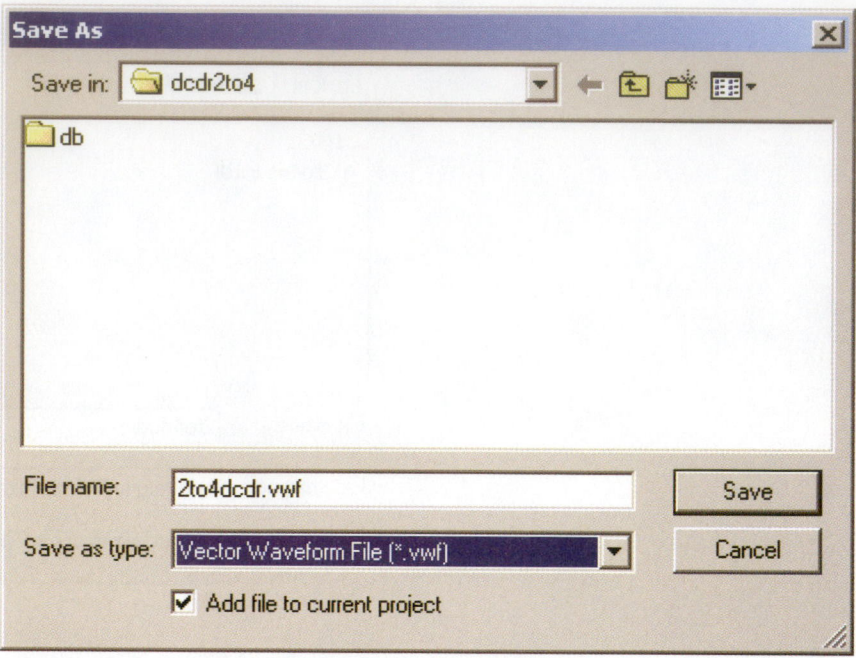

FIGURE A.19   Save as Vector Waveform File

FIGURE A.20   Simulation Result for 2-to-4 Decoder

CONVERTING A VHDL-BASED MAX+PLUS II PROJECT

A VHDL design entry file does not require conversion, but it must be included in a new Quartus II project. Also, vectors are treated differently in VHDL, in a MAX+PLUS II simulation, and in a Quartus II simulation.

Let us create a Quartus II project that includes a VHDL-based 2-to-4 decoder called **decode3.vhd.** First, create a folder for the file with a name such as ***drive:\ qdesigns\decode3\,*** where ***drive*** represents the local or network disk drive where the folder is located. Copy the file **decode3.vhd** to this drive and open in it in Quartus II, as shown in Figure A.21.

From the **File** menu, select **Save As** and save the file with the same name, but make sure that you create a new project based on the file name. Follow the procedure outlined earlier in this appendix to a create a new project based on the VHDL

```
decode3                                                    _ □ ×
  1  LIBRARY ieee;
  2  USE ieee.std_logic_1164.ALL;
  3
  4  ENTITY decode3 IS
  5      PORT(
  6          d    : IN     STD_LOGIC_VECTOR (1 downto 0);
  7          y    : OUT    STD_LOGIC_VECTOR (3 downto 0));
  8  END decode3;
  9
 10  ARCHITECTURE decoder OF decode3 IS
 11  BEGIN
 12      WITH d SELECT
 13          y <=      "0001" WHEN "00",
 14                    "0010" WHEN "01",
 15                    "0100" WHEN "10",
 16                    "1000" WHEN "11",
 17                    "0000" WHEN others;
 18  END decoder;
 19
 20
```

FIGURE A.21   VHDL File

file. Compile the project by clicking the **Start Compilation** toolbar button or by selecting this function from the **Processing** menu.

The previously-created MAX+PLUS II simulation file for this design entity is shown in Figure A.22. Open the file in MAX+PLUS II and create a Table File, as described earlier in this appendix. Copy the Table File to the Quartus II project folder. Open the Table File in Quartus II and save it as a Vector Waveform File, as described in the previous section. Run the simulation by clicking the **Start Simulation** toolbar button or selecting this function from the **Processing** menu. The simulation result should appear as shown in Figure A.23.

Note that in the MAX+PLUS II simulation, individual elements of the output vector, **y,** are labeled with a numerical suffix: **y3, y2, y1,** and **y0.** In the Quartus II

FIGURE A.22   MAX+PLUS II Simulation

FIGURE A.23   Quartus II Simulation

simulation, the output vector elements have their numbers in square brackets: **y[3]**, **y[2]**, **y[1]**, and **y[0]**. This conversion is done automatically when the Table File is saved as a Vector Waveform File.

CONVERTING A HIERARCHICAL DESIGN FROM MAX+PLUS II TO QUARTUS II

Figure A.24 shows a MAX+PLUS II Graphic Design File with two VHDL-based components: an 8-to-1 multiplexer (**mux_8ch.vhd**) and a clock divider (**clockdiv. vhd**). The Graphic Design File is called **mux8test.gdf.**

Create a folder for the new Quartus II project. Copy the three design files (**mux8test.gdf, mux_8ch.vhd,** and **clockdiv.vhd**) to the new folder. Open the top-

FIGURE A.24   Hierarchical MAX+PLUS II Graphic Design File

FIGURE A.25 Hierarchical Quartus II Block Diagram File

level file (**mux8test.gdf**) in Quartus II, as shown in Figure A.25, and save it as a Block Diagram File, making sure that the box labeled **Create new project based on this file** is checked in the **Save As** dialog box.

In screen 2 of 6 in the New Project Wizard, add all three design files to the project, as shown in Figure A.26. You can do this by clicking the button labeled "**. . .**" and browsing for the required files. Alternatively, you can click the **Add All** button. Since this will also add the old **gdf** file, you will have to delete it from the list that appears in the **File name** box. Continue clicking through the remaining screens of the New Project Wizard.

If you need to add a file to a Quartus II project after it has been created, you can do this by using the **Add/Remove** option in the **Settings** dialog box, as shown in Figure A.27. This box can be opened from the Quartus II **Assignments** menu.

After all files have been added to the project, you should be able to open the design files by double-clicking on their corresponding symbols. The VHDL files for the components are shown in Figure A.28 and Figure A.29. In some earlier versions of Quartus II, you may have to open the design files by clicking on a component to highlight it, then right-clicking and selecting **Open Design File** from the pop-up menu that appears.

New Project Wizard: Add Files [page 2 of 6]   ✕

Select the design files and software source files you want to include in your project. Click
Add All to add all design files and software source files in the project directory.
Note: it is optional to add files here unless you have design files not contained in the project
directory, or files in which the file name is not the same as the entity name.

File name: | | ... | Add

File name	Type
clockdiv.vhd	VHDL File
mux_8ch.vhd	VHDL File
mux8test.bdf	Block Dia...

Add All

Remove

Properties

Up

Down

If your project includes libraries of custom functions, specify their
pathnames:

User Library Pathnames...

Back   Next   Finish   Cancel

FIGURE A.26   Adding Files to a Project Using the New Project Wizard

FIGURE A.27 Adding Files to a Project Using the Settings Dialog Box

```
abc clockdiv.vhd                                                    _ □ X

  1  -- clockdiv.vhd
  2  -- Eight-bit binary counter based on a component
  3  --    from the Library of Parameterized Modules (LPM)
  4
  5  LIBRARY ieee;
  6  USE ieee.std_logic_1164.ALL;
  7  LIBRARY lpm;
  8  USE lpm.lpm_components.ALL;
  9
 10  ENTITY clockdiv IS
 11     PORT(
 12        clk                             : IN    STD_LOGIC;
 13        q7, q6, q5, q4, q3, q2, q1, q0  : OUT   STD_LOGIC);
 14  END clockdiv;
 15
 16  ARCHITECTURE count OF clockdiv IS
 17     SIGNAL qd : STD_LOGIC_VECTOR (29 downto 0);
 18  BEGIN
 19  -- Instantiate 8-bit counter
 20     count: lpm_counter
 21        GENERIC MAP (LPM_WIDTH  =>  30)
 22        PORT MAP (  clock       =>  clk,
 23                    q           =>  qd(29 downto 0));
 24
 25     q7  <=  qd(29);
 26     q6  <=  qd(28);
 27     q5  <=  qd(27);
 28     q4  <=  qd(26);
 29     q3  <=  qd(25);
 30     q2  <=  qd(24);
 31     q1  <=  qd(23);
 32     q0  <=  qd(22);
 33  END count;
 34
```

FIGURE A.28 VHDL Design File for **clockdiv** Component

```
abc mux_8ch.vhd                                                    _ □ X
    1  ENTITY mux_8ch IS
    2     PORT(
    3           sel : IN    BIT_VECTOR(2 downto 0);
    4           d   : IN    BIT_VECTOR(7 downto 0);
    5           y   : OUT   BIT);
    6  END mux_8ch;
    7
    8  ARCHITECTURE a OF mux_8ch IS
    9  BEGIN
   10     --  Selected Signal Assignment
   11  MUX8:    WITH sel SELECT
   12         y   <=  d(0) WHEN "000",
   13                 d(1) WHEN "001",
   14                 d(2) WHEN "010",
   15                 d(3) WHEN "011",
   16                 d(4) WHEN "100",
   17                 d(5) WHEN "101",
   18                 d(6) WHEN "110",
   19                 d(7) WHEN "111";
   20  END a;
   21
```

FIGURE A.29 VHDL Design File for **mux_8ch** Component

IMPORTING PIN NUMBERS FROM A MAX+PLUS II PROJECT

A conversion utility called **Convert MAX+PLUS II Project** in the Quartus II **File** menu can be used to import pin numbers from a MAX+PLUS II project. This can be done as the first step in creating a Quartus II project or later, after a project has been created. If you choose to import pins to an existing project, the original project will be overwritten. This has no effect on the existing project design files.

Device and pin assignments in MAX+PLUS II are stored in an Assignment and Configuration File (**acf**). To import these assignments, create a folder for the new Quartus II project, if you have not already done so, and copy the **acf** from the MAX+PLUS II directory to the Quartus II directory.

> **NOTE . . .**
>
> The **acf** must have the same name as the Quartus II project, which should also be the same as the name of the top-level design file in the hierarchy.

From the **File** menu in Quartus II, select the function that says **Convert MAX+PLUS II Project. . .,** as shown in Figure A.30. The dialog box shown in Figure A.31 will open.

Click the ". . ." button to browse for the **acf** file, in this case a file called **ct8bit_a.acf,** as shown in the **Select File** dialog box shown in Figure A.32. When the file has been selected, the **Convert MAX+PLUS II Project** dialog box appears as shown in Figure A.33, with the MAX+PLUS II and Quartus II file names filled in. (Note that it is not possible in this dialog box to select a folder for the new Quartus II project. Therefore the **acf** selected must be in the folder intended for the Quartus II project.) Click **OK** to close the box and select this choice. The Quartus II project now has the same pin assignments as the original MAX+PLUS II project and now can be compiled as usual.

FIGURE A.30   Convert MAX+PLUS II
Project (File Menu)

FIGURE A.31   Convert MAX+PLUS II Dialog

FIGURE A.32   Select File Dialog

FIGURE A.33   Convert MAX+PLUS II Dialog (File Selected)

Answers to Selected Odd-Numbered Problems

Chapter 1

1.1 Analog quantities:

    a. Water temperature at the beach

    b. weight of a bucket of sand

    e. height of a wave

    Digital quantities:

    c. grains of sand in a bucket

    d. waves hitting the beach in one hour

    f. people in a square mile
        Generally, any quantity that can be expressed as "the number of. . ." is digital.

1.3   **a.** 4   **b.** 8   **c.** 25   **d.** 6   **e.** 21

      f. 29   **g.** 59   **h.** 93   **i.** 33   **j.** 185

1.5   101, 110, 111, 1000

1.7   16

1.9   **a.** 0.625   **b.** 0.375   **c.** 0.8125

1.11 $\frac{1}{3}$

1.13   **a.** 0.11   **b.** 0.101   **c.** 0.0011   **d.** 0.10$\overline{1001}$ **e.** 1.11

      f. 11.11$\overline{1100}$ **g.** 1000011.1101011100001. . . (nonrepeating)

1.15   9F7, 9F8, 9F9, 9FA, 9FB, 9FC, 9FD, 9FE, 9FF, A00, A01, A02, A03

1.17   **a** 2C5   **b.** 761   **c.** FFF   **d.** 1000   **e.** 2790

      f. 7D00   **g.** 8000

1.19   **a.** 5E86   **b.** B6A   **c.** C5B   **d.** 6BC4   **e.** 15785

      f. 198B7   **g.** 28000

1.21   **Periodic: b., c., e.** Each of these waveforms repeats itself in a fixed period of time. (Note that waveform **b.** may not immediately appear to be periodic. However, if we count the sequence of short pulse, short space, medium pulse, medium space, short pulse, long space, we will find that each repetition of this sequence takes the same time.)

      Aperiodic: a., d. Neither of these waveforms repeats in a fixed period of time. Waveform **a.** has three equally-spaced pulses of equal width, but this pattern does not repeat in the time shown. Waveform **d.** has pulses of equal duration, spaced at increasing (i.e., unequal) intervals.

1.23   Waveform **a., c., d.,** and **e.** are periodic.

Waveform	Time HIGH	Time LOW	Period	Frequency	Duty Cycle
a.	1 µs	1 µs	2 µs	500 kHz	50%
c.	4 µs	4 µs	8 µs	125 kHz	50%
d.	2 µs	6 µs	8 µs	125 kHz	25%
e.	6 µs	2 µs	8 µs	125 kHz	75%

1.25 From the graph in Figure 1.16 read the times corresponding to the 10%, 50%, and 90% values of the pulse on both leading and trailing edges.

Leading edge:	10%: 5 μs	**Trailing edge:** 90%:	40 μs
	50%: 7.5 μs	50%:	45 μs
	90%: 10 μs	10%:	50 μs

Pulse width: 50% of leading edge to 50% of trailing edge.

$$t_w = 45\ \mu s - 7.5\ \mu s = 37.5\ \mu s$$

Rise time: 10% of rising edge to 90% of rising edge.

$$t_r = 10\ \mu s - 5\ \mu s = 5\ \mu s$$

Fall time: 90% of falling edge to 10% of falling edge.

$$t_f = 50\ \mu s - 40\ \mu s = 10\ \mu s$$

TABLE ANS2.5 4-Input OR
Truth Table

J	K	L	M	N
0	0	0	0	0
0	0	0	1	1*
0	0	1	0	1*
0	0	1	1	1*
0	1	0	0	1*
0	1	0	1	1*
0	1	1	0	1*
0	1	1	1	1*
1	0	0	0	1*
1	0	0	1	1*
1	0	1	0	1*
1	0	1	1	1*
1	1	0	0	1*
1	1	0	1	1*
1	1	1	0	1*
1	1	1	1	1*

TABLE ANS.11 4-Input NOR
Truth Table

A	B	C	D	Y
0	0	0	0	1
0	0	0	1	0
0	0	1	0	0
0	0	1	1	0
0	1	0	0	0
0	1	0	1	0
0	1	1	0	0
0	1	1	1	0
1	0	0	0	0
1	0	0	1	0
1	0	1	0	0
1	0	1	1	0
1	1	0	0	0
1	1	0	1	0
1	1	1	0	0
1	1	1	1	0

Chapter 2

2.5 N is HIGH if J OR K OR L OR M IS HIGH. See Table ANS2.5.

2.11 **a.** Output Y is LOW when A OR B OR C OR D are HIGH. The truth table is shown in Table ANS2.11.

　　　b. $Y = \overline{A + B + C + D}$

2.13 Output Y is LOW if inputs A AND B AND C AND D AND E are all HIGH. The truth table has 32 lines.

2.15 Required gate is a 2-input AND.

2.21 XNOR.

2.23 Output is HIGH if an odd number of inputs is HIGH.

2.25 **a.** and **c.** The attributes of shape, input level, and output level are all different between these two symbols.

2.27 Active LOW. When the switch is pressed, it generates a logic LOW.

2.29 The anode must be at a higher voltage than the cathode by a specified amount.

2.37 A HIGH is required to enable the AND gate. This allows the lamp to flash.

2.39 No. An XOR gate has no inhibit state. The lamp always flashes.

2.41 Transistor-Transistor Logic (TTL) and Complementary Metal-Oxide-Semiconductor (CMOS). Typically, TTL can drive higher-current loads. CMOS has more flexible power supply requirements and uses less power.

2.43 Low-power Schottky TTL: 74LS02; High-speed CMOS: 74HC02. NANDs and NORs are differentiated by the last two digits in their part numbers.

Chapter 3

3.1 **a.** $Y = ABC$

　　　b. $X = PQ + RS$

　　　c. $M = HJKL$

　　　d. $A = W + X + Y + Z$

　　　e. $Y = (A + B)(C + D)$

　　　f. $Y = \overline{(A + B)(C + D)}$

　　　g. $Y = (\bar{A} + \bar{B})(\bar{C} + \bar{D})$

　　　h. $X = \bar{P}\bar{Q} + \bar{R}\bar{S}$

　　　i. $X = \overline{\bar{P}\bar{Q} + \bar{R}\bar{S}}$

3.3 Boolean expressions:

　　　a. $X = \bar{T} + \bar{U} + V + \bar{W}$

　　　e. $Y = AB + AC$

　　　f. $Y = (A + B)(A + C)$

　　　h. $Y = \overline{\bar{A}\,\bar{B}} + \bar{B}\,\bar{C} + AC$

i. $Y = (A + B) + (B + C) + (\overline{A} + \overline{C}) = 1$

j. $Y = (A + B + C + D)AB\overline{C} = AB\overline{C}$

3.5   **a.** Unsimplified expressions:

Figure 3.101a: $Y = \overline{(A + \overline{B} + C)B\overline{C}}$

Figure 3.101b: $Y = \overline{(A(\overline{B + C}))((\overline{A + C})B)((\overline{A + B})C)}$

b. Expressions of redrawn circuits:

Figure 3.101a: $Y = \overline{A} \, B \, \overline{C}(\overline{B} + C)$ or $Y = \overline{(A + \overline{B} + C)} + B\overline{C}$

Figure 3.101b: $Y = A \, \overline{B} \, \overline{C} + \overline{A} \, B \, \overline{C} + \overline{A} \, \overline{B} \, C$

3.7 $Y = \overline{D_3}D_2D_1D_0 + D_3\overline{D_2}D_1D_0 + D_3D_2\overline{D_1}D_0 + D_3D_2D_1\overline{D_0}$ for a circuit that indicates that *exactly* three inputs are HIGH. If *at least* three inputs are HIGH, the equation simplifies to $Y = D_2D_1D_0 + D_3D_1D_0 + D_3D_2D_0 + D_3D_2D_1$.

3.9   **e.** $Y = (\overline{A} + \overline{C}) + \overline{B}\overline{C}$

f. $Y = A\overline{B}C + C$

g. $(\overline{A}BD)(B + \overline{C}) + \overline{A}\overline{C}$

h. $Y = (\overline{A}B)(\overline{A}C)(BC)$

i. $Y = (A + \overline{B}) + (\overline{A} \, C)(BC)$

All of the above equations could be simplified further with Boolean algebra.

3.11   **a.**

T	U	V	W	X
0	0	0	0	1
0	0	0	1	1
0	0	1	0	1
0	0	1	1	1
0	1	0	0	1
0	1	0	1	1
0	1	1	0	1
0	1	1	1	1
1	0	0	0	1
1	0	0	1	1
1	0	1	0	1
1	0	1	1	1
1	1	0	0	1
1	1	0	1	0
1	1	1	0	1
1	1	1	1	1

h.

A	B	C	Y
0	0	0	0
0	0	1	0
0	1	0	1
0	1	1	1
1	0	0	0
1	0	1	0
1	1	0	1
1	1	1	0

i.

A	B	C	Y
0	0	0	1
0	0	1	1
0	1	0	1
0	1	1	1
1	0	0	1
1	0	1	1
1	1	0	1
1	1	1	1

j.

A	B	C	D	Y
0	0	0	0	0
0	0	0	1	0
0	0	1	0	0
0	0	1	1	0
0	1	0	0	0
0	1	0	1	0
0	1	1	0	0
0	1	1	1	0
1	0	0	0	0
1	0	0	1	0
1	0	1	0	0
1	0	1	1	0
1	1	0	0	1
1	1	0	1	1
1	1	1	0	0
1	1	1	1	0

3.13   SOP: $Y = \bar{A}\,\bar{B}\,\bar{C} + \bar{A}\,\bar{B}\,C + \bar{A}\,B\,\bar{C} + \bar{A}\,B\,C$

POS: $Y = (\bar{A} + B + C)(\bar{A} + B + \bar{C})(\bar{A} + \bar{B} + C)(\bar{A} + \bar{B} + \bar{C})$

3.15   SOP: $Y = \bar{A}\bar{B}C + \bar{A}B\bar{C} + A\bar{B}C + AB\bar{C} + ABC$

POS: $Y = (A + B + C)(A + \bar{B} + \bar{C})(\bar{A} + B + C)$

3.17 $Y = (A + B)(\bar{A} + \bar{B})$

3.19 $Y = (A + B + C)\bar{D} = A\bar{D} + B\bar{D} + C\bar{D}$

3.21   **a.** $Y = AB + C;$ **b.** $Y = C;$ **c.** $J = K;$

　　　　　 d. $S = 0;$ **e.** $S = T;$ **f.** $Y = B\,\bar{C}\,\bar{D} + A\,\bar{B}\,F + \bar{C}\,F$

3.23   **a.** $Y = \bar{A} + \bar{B};$

　　　　　 b. $Y = C\,D + \bar{C}\bar{D} + A\,B;$

　　　　　 c. $K = M\bar{N} + ML$

3.25   SOP: $Y = \bar{A}\bar{C} + B\,\bar{C};$ POS: $Y = (\bar{A} + B)\bar{C}$

3.27 $Y = AD + B\bar{C}$

3.29 $Y = \bar{A}D + \bar{C}D + BC\bar{D}$

3.31 $Y = \bar{A}\,\bar{B}\,\bar{C}\,D + A\,\bar{B}\,\bar{C}\,\bar{D} + BC$

3.33 $Y = \bar{A}\,\bar{B}\,\bar{C} + A\,B\,\bar{C} + \bar{A}\,\bar{B}\,D + A\,\bar{D}$

3.35 $Y = \bar{A}\,B + CD$

3.37 $Y = AD + \bar{B}\,C$

3.39 $Y = A\,\bar{B}\,\bar{C}\,\bar{D} + \bar{A}D + CD$

3.41 $Y = A\bar{B}\bar{C}D + \bar{A}B + B\bar{D}$

3.43 $Y = D$

3.45 $Y = A\overline{C} + A\overline{B} + BCD$

3.47 $Y = (\overline{A} + C)(A + \overline{C})(A + B + \overline{D})$

3.49 $E_4 = D_4 + D_3D_1 + D_3D_2$

$E_3 = \overline{D}_3D_2 + \overline{D}_3D_1 + D_3\overline{D}_2\overline{D}_1$

$E_2 = \overline{D}_2\overline{D}_1 + D_2D_1$

$E_1 = \overline{D}_1$

3.51 Unsimplified: $Y = (\overline{\overline{\overline{AB}} + D}) + (\overline{C + \overline{D}}) + (\overline{A + C})$

Simplified: $Y = ABD + \overline{C}D + \overline{A}\overline{C}$

3.53 Unsimplified: $Y = \overline{(A + \overline{B} + C)\overline{B}\overline{C}}$

Simplified: $Y = \overline{A}B\overline{C}(\overline{B} + C)$

3.69 **a.** The truth table is shown in Table ANS3.69.

b. The Boolean equations are:

$Y1 = D1 + D0$

$Y2 = D1$

$Y3 = D1 \cdot D0$

TABLE ANS3.69

D1	D0	Y1	Y2	Y3
0	0	0	0	0
0	1	1	0	0
1	0	1	1	0
1	1	1	1	1

Chapter 4

4.1 Advantages of programmable logic: User is not restricted to standard digital functions from a device manufacturer; only required functions need be implemented; package count can be reduced; design can be reprogrammed or reconfigured without changing the circuit board.

4.3 PAL (Programmable Array Logic); GAL (Generic Array Logic); EPLD (Erasable Programmable Logic Device); FPGA (Field-Programmable Gate Array)

4.5 The AND gates have too many inputs to be shown individually and still maintain a neat and readable appearance in the PLD logic diagram.

4.7 A PAL AND array has programmable junctions so that any possible input or its complement could be an input to the AND gate. These inputs are connected to the product lines of an AND array via physical fuse links or programmable transistors. Connections are made by leaving the fuse links intact and are broken by blowing the fuse links or by programming the transistors ON or OFF.

4.9 **a.** Ten dedicated inputs **b.** 2 dedicated outputs **c.** 6 I/O pins.

4.11 The buffer will be always enabled if all fuses on its product line are blown and always disabled if all fuses on its product line are intact.

4.19 There are many different CPLDs with different input and output pin layouts. In order to accurately create our design in a particular chip, we need to specify which chip is the target of our design.

4.21 A project can be created when a design file is initially saved (provided there is no other active project in the folder) or later by using the **Project** menu or the **Settings** dialog box.

4.29 If the inputs to the CPLD from the external world were changing at a rate comparable to the input-to-output delay, it would be important to test whether our design could respond to the inputs without introducing errors in output due to the CPLD delay times.

4.31 ISP = In-System Programmable. ICR = In-Circuit Reconfigurable. ISP applies to nonvolatile devices (e.g., MAX 7000S family). ICR applies to volatile devices (e.g., FLEX 10K family).

4.33 The programming hardware must be set up when the PC running Quartus II has never been used to program a CPLD.

Chapter 5

5.1 VHDL = VHSIC Hardware Description Language; VHSIC = Very High-Speed Integrated Circuit.

5.3 A port of type BIT can have the values '0' and '1' only.

5.5
```
ENTITY prob_5_5 IS
  PORT(
       a, b, c: IN  BIT;
       y      : OUT BIT);
END prob_5_5;

ARCHITECTURE cct OF prob_5_5 IS
BEGIN
 y <= ((a xor b) and c) or (a and b);
END cct;
```

5.7 All concurrent signals in a VHDL file are evaluated simultaneously. This is required to make a VHDL design behave like combinational hardware, where two or more circuits with the same inputs will change their outputs at the same time if their common inputs change state.

5.9 The mode of a port defines the direction of signal flow at the port. Four modes are: IN (input only), OUT (output only), INOUT (bidirectional), and BUFFER (special case of output with feedback into the PLD matrix).

5.11 **a.** addr : IN BIT_VETOR (7 downto 0);

b. addr : IN BIT_VECTOR (0 to 7);

5.13
```
ENTITY prob5_13 IS
    PORT(
         a, b : IN  BIT_VECTOR(7 downto 0);
         y    : OUT BIT_VECTOR(7 downto 0));
END prob5_13;

ARCHITECTURE xor_gate OF prob5_13 IS
BEGIN
    y <= a xor b;
END xor_gate;
```

5.15
```
x(0) <= 1;
x(1) <= 0;
x(2) <= 0;
x(3) <= 1;
x(4) <= 0;
x(5) <= 1;
```

5.17
```
ENTITY prob5_17 IS
    PORT(
         d : IN  BIT_VECTOR(3 downto 0);
         y : OUT BIT);
END prob5_17;

ARCHITECTURE sig OF prob5_17 IS
BEGIN
    WITH d SELECT
        y  <= '1' WHEN "0111",
              '1' WHEN "1011",
              '1' WHEN "1101",
              '1' WHEN "1110",
              '0' WHEN others;
END sig;
```

5.19   **a.** The truth table for the SOP circuit shown in Figure 5.38 is as follows.

A	B	C	D	Y
0	0	0	0	0
0	0	0	1	1
0	0	1	0	0
0	0	1	1	0
0	1	0	0	0
0	1	0	1	0
0	1	1	0	0
0	1	1	1	1
1	0	0	0	1
1	0	0	1	0
1	0	1	0	0
1	0	1	1	1
1	1	0	0	0
1	1	0	1	1
1	1	1	0	0
1	1	1	1	0

The Boolean equation for this circuit is given by:

$$Y = \overline{A}\,\overline{B}\,\overline{C}D + \overline{A}BCD + A\overline{B}\,\overline{C}\,\overline{D} + A\overline{B}CD + AB\overline{C}D$$

b. This circuit is best suited to implementation with a selected signal assignment statement. The Boolean equation is too complex for a concurrent signal assignment statement and the truth table is fairly easy to implement.

c.
```
LIBRARY ieee;
USE ieee.std_logic_1164.ALL;

ENTITY prob5_19 IS
    PORT(
        d : IN  STD_LOGIC_VECTOR(3 downto 0);
        y : OUT STD_LOGIC);
END prob5_19;

ARCHITECTURE sig OF prob5_19 IS
BEGIN
    WITH d SELECT
        y  <=  '1' WHEN "0001",
               '1' WHEN "0111",
               '1' WHEN "1000",
               '1' WHEN "1011",
               '1' WHEN "1101",
               '0' WHEN others;

END sig;
```

5.21   A circuit, such as a full adder or equality comparator, with one or more fairly simple Boolean equations is most suited to a concurrent signal assignment statement. A circuit, such as the SOP network in Figure 5.38, has a complicated Boolean expression that would required lots of parentheses in a VHDL format. However, it has a straightforward truth table that lends itself to the use of a selected signal assignment statement.

5.23   **a.** `control <= enable & read & write & selector & compare;`

b.
```
address_latch <= status (2);
data_enable   <= status (1);
strobe        <= status (0);
```

5.25
```
--More efficient coding of this circuit will be examined
--in Chapter 6
LIBRARY ieee;
USE ieee.std_logic_1164.ALL;

ENTITY prob5_25 IS
   PORT(
         a, b   : IN  STD_LOGIC_VECTOR (3 downto 0);
         a_eq_b : OUT STD_LOGIC);
END prob5_25;

ARCHITECTURE compare OF prob5_25 IS
      SIGNAL inputs : STD_LOGIC_VECTOR (7 downto 0);
BEGIN
      Inputs (7 downto 4) <= a;
      inputs (3 downto 0) <= b;

      -- Alternate coding;
      -- inputs <= a & b;

      WITH inputs SELECT
            a_eq_b <=  '1' WHEN "00000000",
                       '1' WHEN "00010001",
                       '1' WHEN "00100010",
                       '1' WHEN "00110011",
                       '1' WHEN "01000100",
                       '1' WHEN "01010101",
                       '1' WHEN "01100110",
                       '1' WHEN "01110111",
                       '1' WHEN "10001000",
                       '1' WHEN "10011001",
                       '1' WHEN "10101010",
                       '1' WHEN "10111011",
                       '1' WHEN "11001100",
                       '1' WHEN "11011101",
                       '1' WHEN "11101110",
                       '1' WHEN "11111111",
                       '0' WHEN others;
END compare;
```

Chapter 6

6.1 $1100, 0001, 1111; Y = D_3 D_2 \bar{D}_1 \bar{D}_0; Y = \bar{D}_3 \bar{D}_2 \bar{D}_1 D_0; Y = D_3 D_2 D_1 D_0$

6.5   **a.** 32   **b.** 64   **c.** 256; $m = 2^n$

6.7   **a.**

```
-- dcd6_7a.vhd
-- Solution to Problem 6.7a
-- One output goes LOW for each value of "d". (e.g. When d = 5, y5 = 0)
-- and all other outputs are HIGH.

LIBRARY ieee;
USE ieee.std_logic_11.64.ALL;

ENTITY dcd6_7a IS
   PORT(
        d   : IN  STD_LOGIC_VECTOR (2 downto 0);
        y   : OUT STD_LOGIC_VECTOR (0 to 7));
END dcd6_7a;
```

```
ARCHITECTURE a OF dcd6_7a IS
BEGIN
    WITH d SELECT
    y <=    "01111111" WHEN "000",
            "10111111" WHEN "001",
            "11011111" WHEN "010",
            "11101111" WHEN "011",
            "11110111" WHEN "100",
            "11111011" WHEN "101",
            "11111101" WHEN "110",
            "11111110" WHEN "111",
            "11111111" WHEN others;
END a;
```

b. Simulation Criteria

■ Every binary combination of **d** should make one **y** output go LOW, while all others remain HIGH.

■ The subscript of the LOW output should be the same as the decimal equivalent of the binary value applied to **d.**

6.9

```
-- dcd_138.vhd
-- dcd_138.vhd
-- Solution to Problem 6.9a
-- VHDL implementation of a 74138 decoder
-- Enable inputs: G1 (active-HIGH), G2AN and G2BN (active-LOW)
-- Decode inputs: CBA (C = MSB)
-- Decode outputs: Y[0..7]
-- When all enables are active, one output of 8 is LOW
-- Subscript of active input is same as decimal equivalent of CBA
-- When any enable is inactive, all outputs are HIGH

LIBRARY ieee;
USE ieee.std_logic_1164.ALL;

ENTITY dcd_138 IS
    PORT(
        a, b, c      : IN  STD_LOGIC;
        g1, g2an, g2bn: IN  STD_LOGIC;
        y            : OUT STD_LOGIC_VECTOR(0 to 7));
END dcd_138;

ARCHITECTURE a OF dcd_138 IS
    SIGNAL inputs : STD_LOGIC_VECTOR (5 downto 0);
BEGIN
    inputs <= g1 & g2an & g2bn & c & b & a;
    WITH inputs SELECT
    y <=   "01111111" WHEN "100000",
           "10111111" WHEN "100001",
           "11011111" WHEN "100010"
           "11101111" WHEN "100011",
           "11110111" WHEN "100100",
           "11111011" WHEN "100101",
           "11111101" WHEN "100110",
           "11111110" WHEN "100111",
           "11111111" WHEN others;
END a;
```

Simulation Criteria

■ When **g1** = 1, **g2an** = 0, and **g2bn** = 0, every binary combination of **cba** should make one **y** output go LOW, while all others remain HIGH.

■ The subscript of the LOW output should be the same as the decimal equivalent of the binary value applied to **cba.**

■ If any enable input is not active, all **y** outputs should be HIGH.

6.11 $a = \bar{D}_3 \bar{D}_2 \bar{D}_1 D_0 + \bar{D}_3 D_2 \bar{D}_1 \bar{D}_0 + D_3 \bar{D}_2 D_1 D_0 + D_3 D_2 \bar{D}_1 D_0$

$b = \bar{D}_3 D_2 \bar{D}_1 D_0 + D_3 D_2 D_1 + D_3 D_2 \bar{D}_0 + D_3 D_1 D_0 + D_2 D_1 \bar{D}_0$

$c = \bar{D}_3 \bar{D}_2 D_1 \bar{D}_0 + D_3 D_2 \bar{D}_1 \bar{D}_0 + D_3 D_2 D_1$

$d = \bar{D}_3 \bar{D}_2 \bar{D}_1 D_0 + \bar{D}_3 D_2 \bar{D}_1 \bar{D}_0 + D_3 \bar{D}_2 D_1 \bar{D}_0 + D_2 D_1 D_0$

$e = \bar{D}_3 D_0 + \bar{D}_3 D_2 \bar{D}_1 + \bar{D}_2 \bar{D}_1 D_0$

$f = \bar{D}_3 \bar{D}_2 D_0 + \bar{D}_3 \bar{D}_2 D_1 + \bar{D}_3 D_1 D_0 + D_3 D_2 \bar{D}_1 D_0$

$g = \bar{D}_3 \bar{D}_2 \bar{D}_1 + \bar{D}_3 D_2 D_1 D_0 + D_3 D_2 \bar{D}_1 \bar{D}_0$

6.15   **a.** 1000   **b.** 1001   **c.** 1001

6.19   **a.**
```
-- hi_pri16.vhd
ENTITY hi_pri16 IS
    PORT(
        d : IN  BIT_VECTOR (15 downto 0);
        q : OUT BIT_VECTOR (3 downto 0));
END hi_pri16;

ARCHITECTURE a OF hi_pri16 IS
BEGIN
    -- Conditional Signal Assignment
encoder:
    q  <= "1111" WHEN d(15) = '1' ELSE
          "1110" WHEN d(14) = '1' ELSE
          "1101" WHEN d(13) = '1' ELSE
          "1100" WHEN d(12) = '1' ELSE
          "1011" WHEN d(11) = '1' ELSE
          "1010" WHEN d(10) = '1' ELSE
          "1001" WHEN d(9)  = '1' ELSE
          "1000" WHEN d(8)  = '1' ELSE
          "0111" WHEN d(7)  = '1' ELSE
          "0110" WHEN d(6)  = '1' ELSE
          "0101" WHEN d(5)  = '1' ELSE
          "0100" WHEN d(4)  = '1' ELSE
          "0011" WHEN d(3)  = '1' ELSE
          "0010" WHEN d(2)  = '1' ELSE
          "0001" WHEN d(1)  = '1' ELSE
          "0000";
END a;
```

b. Simulation Criteria

■ With all input bits HIGH, d(15) is highest priority of active inputs. Expected output is 1111_2 (= 15_{10}).

■ With d(15) LOW and all other inputs HIGH, d(14) is now highest priority active input. Expected output is 1110_2 (= 14_{10}).

■ With d(15) and d(14) LOW and remaining bits HIGH, highest priority is d(13). Expected output is 1101_2 (= 13_{10}).

■ As each bit goes LOW in order, output code should count down by one with each step.

6.21 Truth Table for an 8-to-1 MUX

S_2	S_1	S_0	Y
0	0	0	D_0
0	0	1	D_1
0	1	0	D_2
0	1	1	D_3
1	0	0	D_4
1	0	1	D_5
1	1	0	D_6
1	1	1	D_7

Truth Table for a 16-to-1 MUX

S_3	S_2	S_1	S_0	Y
0	0	0	0	D_0
0	0	0	1	D_1
0	0	1	0	D_2
0	0	1	1	D_3
0	1	0	0	D_4
0	1	0	1	D_5
0	1	1	0	D_6
0	1	1	1	D_7
1	0	0	0	D_8
1	0	0	1	D_9
1	0	1	0	D_{10}
1	0	1	1	D_{11}
1	1	0	0	D_{12}
1	1	0	1	D_{13}
1	1	1	0	D_{14}
1	1	1	1	D_{15}

6.23 $Y = \bar{S}_2\bar{S}_1\bar{S}_0 D_0 + \bar{S}_2\bar{S}_1 S_0 D_1 + \bar{S}_2 S_1 \bar{S}_0 D_2 + \bar{S}_2 S_1 S_0 D_3 + S_2 \bar{S}_1 \bar{S}_0 D_4 + S_2 \bar{S}_1 S_0 D_5$
$+ S_2 S_1 \bar{S}_0 D_6 + S_2 S_1 S_0 D_7$

$= \bar{1}\cdot\bar{0}\cdot\bar{1}\cdot D_0 + \bar{1}\cdot\bar{0}\cdot 1 \cdot D_1 + \bar{1}\cdot 0 \cdot \bar{1}\cdot D_2 + \bar{1}\cdot 0 \cdot 1 \cdot D_3 + 1 \cdot \bar{0}\cdot\bar{1}\cdot D_4 + 1 \cdot \bar{0}\cdot 1 \cdot D_5 + 1 \cdot 0 \cdot \bar{1}\, D_6 + 1 \cdot 0 \cdot 1 \cdot D_7$

$= 0\cdot D_0 + 0 \cdot D_1 + 0 \cdot D_2 + 0 \cdot D_3 + 0 \cdot D_4 + 1 \cdot D_5 + 0 \cdot D_6 + 0 \cdot D_7$

$= D_5$

6.25 **a.**

```
-- quad8to1.vhd

-- Eight-channel 4-bit multiplexer
-- One of eight sets four inputs
-- (d03..d00), (d13..d10), (d23..d20), (d33..d30),
-- (d43..d40), (d53..d50), (d63..d60), (d73..d70)
-- is directed to an output (y), based on the status of three
-- select inputs (s2, s1, s0).

ENTITY quad8to1 IS
     PORT(
           s  : IN  INTEGER RANGE 0 to 7;
           d0 : IN  BIT_VECTOR (3 downto 0);
           d1 : IN  BIT_VECTOR (3 downto 0);
           d2 : IN  BIT_VECTOR (3 downto 0);
           d3 : IN  BIT_VECTOR (3 downto 0);
           d4 : IN  BIT_VECTOR (3 downto 0);
           d5 : IN  BIT_VECTOR (3 downto 0);
           d6 : IN  BIT_VECTOR (3 downto 0);
           d7 : IN  BIT_VECTOR (3 downto 0);
           y  : OUT BIT_VECTOR (3 downto 0));
END quad8to1;

ARCHITECTURE mux8 OF quad8to1 IS
BEGIN
     -- Selected Signal Assignment
MUX4: WITH s SELECT
           y <=      d0 WHEN 0,
                     d1 WHEN 1,
```

```
                                        d2  WHEN 2,
                                        d3  WHEN 3,
                                        d4  WHEN 4,
                                        d5  WHEN 5,
                                        d6  WHEN 6,
                                        d7  WHEN 7;
END mux8;
```

Simulation Criteria

■ An increasing binary count applied to the select inputs should send each of the 4-bit data inputs to the output in the sequence d0[3..0] to d7[3..0].

■ Each data input should have a recognizable 4-bit value. Constant hexadecimal values on the data inputs allow us to easily see the data change.

6.27   a.

```
-- oct4to1.vhd
-- Four-channel 8-bit multiplexer
-- One of four sets eight inputs
-- (d07..d00), (d17..d10), (d27..d20), or (d37..d30)
-- is directed to an output (y), based on the status of two
-- select inputs (s1, s0).

ENTITY oct4to1 IS
      PORT(
                s  : IN   INTEGER RANGE 0 to 3;
                d0 : IN   BIT_VECTOR (7 downto 0);
                d1 : IN   BIT_VECTOR (7 downto 0);
                d2 : IN   BIT_VECTOR (7 downto 0);
                d3 : IN   BIT_VECTOR (7 downto 0);
                y  : OUT  BIT_VECTOR (7 downto 0));
END oct4to1;

ARCHITECTURE mux4 OF oct4to1 IS
BEGIN
      -- Selected Signal Assignment
MUX8: WITH s SELECT
                y      <=   d0 WHEN 0,
                           d1 WHEN 1,
                           d2 WHEN 2,
                           d3 WHEN 3;
END mux4;
```

Simulation Criteria

■ An increasing binary count applied to the select inputs should send each of the 8-bit data inputs to the output in the sequence d0[3..0] to d3[3..0].

■ Each data input should have a recognizable 8-bit value. Constant hexadecimal values on the data inputs allow us to easily see the data change.

6.29
```
ENTITY mux_8ch IS
         PORT(
                  sel : IN  BIT_VECTOR (2 downto 0);
                  d   : IN  BIT_VECTOR (7 downto 0);
                  y   : OUT BIT);
         END mux_8ch;
```

```
ARCHITECTURE a OF mux_8ch IS
BEGIN
        -- Selected Signal Assignment
MUX8: WITH sel SELECT
            y       <=      d(0) WHEN "000",
                            d(1) WHEN "001",
                            d(2) WHEN "010",
                            d(3) WHEN "011",
                            d(4) WHEN "100",
                            d(5) WHEN "101",
                            d(6) WHEN "110",
                            d(7) WHEN "111";
END a;
```

An 8-to-1 MUX can be easily extended to a 16-bit device by adding one select input, eight data inputs, and eight lines to the selected signal assignment statement.

6.35

```
-- dmux16.vhd
-- 1-to-16 demultiplexer/decoder
-- Decoder: set d to '0'; outputs are activated by
-- binary combination of s.
-- Demultiplexer: apply data stream to d; data directed to
-- y output with subscript same as value of s.
-- Outputs and d are active-LOW. DMUX data are inverted twice
-- to keep them true.

ENTITY dmux16 IS
    PORT(
        s   : IN  INTEGER Range 0 to 15;
        d   : IN  BIT;
        y   : OUT BIT_VECTOR (0 to 15));
END dmux16;

ARCHITECTURE a OF dmux16 IS
    SIGNAL output : BIT VECTOR (0 to 15);
BEGIN
    PROCESS (d, s)
    BEGIN
        IF (d = '1') THEN
            output <= "1111111111111111";
        ELSE
            CASE s IS
                WHEN 0        => output <= "0111111111111111";
                WHEN 1        => output <= "1011111111111111";
                WHEN 2        => output <= "1101111111111111";
                WHEN 3        => output <= "1110111111111111";
                WHEN 4        => output <= "1111011111111111";
                WHEN 5        => output <= "1111101111111111";
                WHEN 6        => output <= "1111110111111111";
                WHEN 7        => output <= "1111111011111111";
                WHEN 8        => output <= "1111111101111111";
                WHEN 9        => output <= "1111111110111111";
                WHEN 10       => output <= "1111111111011111";
                WHEN 11       => output <= "1111111111101111";
                WHEN 12       => output <= "1111111111110111";
                WHEN 13       => output <= "1111111111111011";
                WHEN 14       => output <= "1111111111111101";
```

```
                                   WHEN 15        => output <= "1111111111111110";
                                   WHEN OTHERS    => output <= "1111111111111111";
                           END CASE;
                    END IF;
                    y  <= output;
              END PROCESS;
        END a;
```

6.37 An analog switch can transmit a range of positive and negative voltages, not just 0V and 5V. Since positive and negative voltages are allowed, signal flow can be in either direction.

6.41 $AEQB = \overline{(\overline{A_5 \oplus B_5})(\overline{A_4 \oplus B_4})(\overline{A_3 \oplus B_3})(\overline{A_2 \oplus B_2})(\overline{A_1 \oplus B_1})(\overline{A_0 \oplus B_0})}$

$AGTB = A_5\overline{B_5} + A_4\overline{B_4}(\overline{A_5 \oplus B_5}) + A_3\overline{B_3}(\overline{A_5 \oplus B_5})(\overline{A_4 \oplus B_4})$
$\qquad + A_2\overline{B_2}(\overline{A_5 \oplus B_5})(\overline{A_4 \oplus B_4})(\overline{A_3 \oplus B_3}) + A_1\overline{B_1}(\overline{A_5 \oplus B_5})(\overline{A_4 \oplus B_4})$
$\qquad (\overline{A_3 \oplus B_3})(\overline{A_2 \oplus B_2}) + A_0\overline{B_0}(\overline{A_5 \oplus B_5})(\overline{A_4 \oplus B_4})(\overline{A_3 \oplus B_3})$
$\qquad (\overline{A_2 \oplus B_2})(\overline{A_1 \oplus B_1})$

$ALTB = \overline{A_5}B_5 + \overline{A_4}B_4(\overline{A_5 \oplus B_5}) + \overline{A_3}B_3(\overline{A_5 \oplus B_5})(\overline{A_4 \oplus B_4})$
$\qquad + \overline{A_2}B_2(\overline{A_5 \oplus B_5})(\overline{A_4 \oplus B_4})(\overline{A_3 \oplus B_3}) + \overline{A_1}B_1(\overline{A_5 \oplus B_5})(\overline{A_4 \oplus B_4})(\overline{A_3 \oplus B_3})$
$\qquad (\overline{A_2 \oplus B_2}) + \overline{A_0}B_0(\overline{A_5 \oplus B_5})(\overline{A_4 \oplus B_4})(\overline{A_3 \oplus B_3})(\overline{A_2 \oplus B_2})(\overline{A_1 \oplus B_1})$

6.43

```
-- cmp4x6.vhd

LIBRARY ieee;
USE ieee.std_logic_1164.ALL;

ENTITY cmp4x6 IS
      PORT(
            a, b  : IN INTEGER RANGE 0 TO 15;
            altb, aleb, aeqb, aneb, ageb, agtb : OUT STD_LOGIC);
END cmp4x6;

ARCHITECTURE a OF cmp4x6 IS
      SIGNAL compare    : STD_LOGIC_VECTOR (5 downto 0);
BEGIN
      PROCESS (a,b)
      BEGIN
            IF a<b THEN
                  compare      <=      "001011";
            ELSIF a=b THEN
                  compare      <=      "100101";
            ELSIF a>b THEN
                  compare      <=      "111000";
            ELSE
                  compare      <=      "111111";
            END IF;
            altb  <=    compare (5); -- a is less than b
            aleb  <=    compare (4); -- a is less than or equal to b
            aeqb  <=    compare (3); -- a equals b
            aneb  <=    compare (2); -- a is not equal to b
            ageb  <=    compare (1); -- a is greater than or equal to b
            agtb  <=    compare (0); -- a is greater than b
      END PROCESS;
END a;
```

6.45   **a.** 1111100; five 1s; $P_E = 1$; $P_O = 0$;

       b. 1010110; four 1s; $P_E = 0$; $P_O = 1$;

       c. 0001101; three 1s; $P_E = 1$; $P_O = 0$

6.47   **a.** *ABCDEFGHP* = 110101100; $P' = 1$; Error in bit *D*.

       b. *ABCDEFGHP* = 110001101; $P' = 1$; Error in parity bit.

       c. *ABCDEFGHP* = 110001100; $P' = 0$; Data received correctly.

       d. *ABCDEFGHP* = 110010100; $P' = 0$; Errors in bits *E* and *F* undetected

6.51   **a.**

```
-- 4-bit parity generate using GENERATE statement
-- nEven_odd = 0 for EVEN parity
-- nEven_odd = 1 for ODD parity
LIBRARY ieee;
USE ieee.std_logic_1164.ALL;

ENTITY parity_gen4_even_odd IS
   PORT(
          d         : IN  STD_LOGIC_VECTOR(0 to 3);
          nEven_odd : IN  STD_LOGIC;
          p_out     : OUT STD_LOGIC);
END parity_gen4_even_odd;

ARCHITECTURE parity OF parity_gen4_even_odd IS
   SIGNAL p      : STD_LOGIC_VECTOR(1 to 4);
   SIGNAL inputs : STD_LOGIC_VECTOR(0 to 4);
BEGIN
   inputs <= d & nEven_odd;

   p(1) <= inputs(0) xor inputs(1);

   -- Combine 4 inputs to generate parity (inputs 0 to 3),
   -- select EVEN or ODD parity bit (input 4)
   parity check:
   FOR 1 IN 2 to 4 GENERATE
      p(i) <= p(i - 1) xor inputs(i);
   END GENERATE;

   p_out <= p(4);
END parity;
```

   b. Simulation Criteria

     ■ Apply an increasing binary count to all **d** inputs and input **nEven_odd.**

     ■ **p_out** output should make total number of 1s even with **nEven_odd** = 0.

     ■ **p_out** output should make total number of 1s odd with **nEven_odd** = 1.

6.53   **a.**

```
-- n-bit parity checker (default = 8-bit) using GENERATE statement
LIBRARY ieee;
USE ieee.std_logic_1164.ALL;

ENTITY parity_chk_n IS
   PORT(
          d         : IN  STD LOGIC VECTOR(0 to 7);
          p_in      : IN  STD_LOGIC;
          nEven_odd : IN  STD_LOGIC;
          p_out     : OUT STD_LOGIC);
END parity_chk_n;
```

```
                        ARCHITECTURE parity OF parity_chk_n IS
                            SIGNAL p      : STD_LOGIC_VECTOR(1 to (d'HIGH + 2));
                            SIGNAL inputs : STD_LOGIC_VECTOR(0 to (d'HIGH + 2));
                        BEGIN
                            inputs <= d & nEven_odd & p_in;

                            p(1) <= inputs (0) xor inputs (1);

                            -- Combine n inputs to generate parity (inputs 0 to n-2),
                            -- select EVEN or ODD parity bit (input n-1),
                            -- then compare with parity bit input (input n) to check for
                            -- correctly received parity bit
                            parity_check:
                            FOR i IN 2 to (d'HIGH + 2) GENERATE
                                p(1) <= p (i-1) xor inputs (1);
                            END GENERATE;

                            p_out <= p (p'HIGH);
                        END parity;
```

b. Simulation Criteria

- Apply an increasing binary count to all **d** inputs, combined with inputs **nEven_odd** and **p_in**.
- **p_out** output should make total number of 1s even with **nEven_odd** = 0.
- **p_out** output should make total number of 1s odd with **nEven_odd** = 1.

Chapter 7

7.1   **a.** 11111   **b.** 100000   **c.** 11110
          d. 101010   **e.** 101100   **f.** 1100100

7.3

	Decimal	True Magnitude	1's Complement	2's Complement
a.	−110	11101110	10010001	10010010
b.	67	01000011	01000011	01000011
c.	−54	10110110	11001001	11001010
d.	−93	11011101	10100010	10100011
e.	0	00000000	00000000	00000000
f.	−1	10000001	11111110	11111111
g.	127	01111111	01111111	01111111
h.	−127	11111111	10000000	10000001

7.5   Largest: $01111111_2 = + 127_{10}$
          smallest: $10000000_2 = -128_{10}$

7.7   Overflow in an 8-bit signed addition results if the sum is outside the range $-128 \leq$ sum $\leq + 127$. The sums in parts **a.** and **f.** do not generate an overflow. The sums in parts **b., c., d.,** and **e.** do.

7.9   **a.** $+ 19 = 010011$ (minimum number of bits)

          b. $+ 19 = 00010011$ (8-bit number)

          c. $+ 19 = 000000010011$ (12-bit number)

7.11 **a.** 3D

          b. 120

          c. B1A

          d. FFF

          e. 2A7F

7.13

Decimal	True Binary	8421 BCD	Excess-3
709	1011000101	0111 0000 1001	1010 0011 1100
1889	11101100001	0001 1000 1000 1001	0100 1011 1011 1100
2395	100101011011	0010 0011 1001 0101	0101 0110 1100 1000
1259	10011101011	0001 0010 0101 1001	0100 0101 1000 1100
3972	111110000100	0011 1001 0111 0010	0110 1100 1010 0101
7730	1111000110010	0111 0111 0011 0000	1010 1010 0110 0011

7.17 The sequence of codes yields the following text:

```
43 41 55 54 49 4F 4E 21 20 45 72 61 73 69 6E 67 20
 C  A  U  T  I  O  N  !  SP E  r  a  s  i  n  g  SP

61 6C 6C 20 64 61 74 61 21 20 41 72 65 20 79 6F 75 20
 a  l  l  SP d  a  t  a  !  SP A  r  e  SP y  o  u  SP

73 75 72 65 3F
 s  u  r  e  ?
```

SP = Space

7.23 A fast carry circuit is "flatter", but "wider" than a ripple carry circuit. There are more gate levels for an input change to propagate through in a ripple carry circuit. The ripple carry is thus slower. The limitation on a fast carry circuit is its width, both in the number of gates and on the number of inputs on the gates. Both factors increase with adder bit size.

7.25 A carry is generated if the MSB of either A or B is HIGH AND the second bit of either A or B is HIGH AND the third bits of both A and B are HIGH.

7.27 To generate all possible combinations of input for an 8-bit adder requires $2^{16} = 65,536$ combinations. (A simulation with one change every 40 ns would have an end time of 2.62144 ms.)

7.31 $SUB = 1$: Input carry is forced HIGH, automatically adding 1 to the output sum; the XOR gates act as inverters, making the inputs to the adder equal to the one's complement of B; the output is $A +$ (one's complement of B) $+ 1 = A - B$

$SUB = 0$: Input carry is forced LOW, adding 0 to the output sum; the XOR gates act as noninverting buffers, making the inputs to the adder equal the true binary value of B; the output is $A + B + 0 = A + B$.

7.35
```
-- addsubv1.vhd
-- 4-bit parallel adder with overflow detection,
-- using a generate statement and components
-- overflow: SOP network

ENTITY addsubv1 IS
    PORT(
        sub    : IN     BIT;
        a, b  : IN     BIT_VECTOR(4 downto 1);
        c4, v : OUT    BIT;
        sum    : BUFFER BIT_VECTOR(4 downto 1));
END addsubv1;

ARCHITECTURE adder OF addsubv1 IS
    -- Component declaration
    COMPONENT full_add
        PORT(
            a, b, c_in : IN  BIT;
            c_out, sum : OUT BIT);
    END COMPONENT;
    -- Define a signal for internal carry bits
    SIGNAL c        : BIT_VECTOR (4 downto 0);
    SIGNAL b_comp   : BIT_VECTOR (4 downto 1);
```

```
        BEGIN
                -- Carry input depends on add or subtract (sub=1 for subtract)
                c(0)   <=    sub;
                adders:
                FOR I IN 1 to 4 GENERATE
                        -- invert b for subtract function (b(i) xor 1)
                        -- do not invert b for add function (b(i) xor 0)
                        b_comp(i) <= b(i) xor sub;
                adder: full_add PORT MAP (a(i), b_comp(i), c(i-1), c(i), sum(i));
                END GENERATE;
                c4      <= c(4);
                v <= (a(4) and b(4) and (not sum(4)))
                        or ((not a(4)) and (not b(4)) and sum(4));
        END adder;

        -- addsubv2.vhd
        -- 4-bit parallel adder with overflow detection,
        -- using a generate statement and components
        -- overflow: xor gate

        ENTITY addsubv2 IS
              PORT(
                        sub             : IN  BIT;
                        a, b            : IN  BIT_VECTOR(4 downto 1);
                        c4, v          : OUT BIT;
                        sum            : OUT BIT_VECTOR(4 downto 1));
        END addsubv2;

        ARCHITECTURE adder OF addsubv2 IS
             -- Component declaration
             COMPONENT full_add
                  PORT(
                          a, b, c_in   : IN  BIT;
                          c_out, sum   : OUT BIT);
             END COMPONENT;
             -- Define a signal for internal carry bits
             SIGNAL c            : BIT_VECTOR (4 downto 0);
             SIGNAL b_comp       : BIT_VECTOR (4 downto 1);
        BEGIN
                -- Carry input depends on add or subtract (sub=1 for subtract)
                c(0) <=    sub;
                adders:
                FOR i IN 1 to 4 GENERATE
                        -- invert b for subtract function (b(i) xor 1)
                        -- do not invert b for add function (b(i) xor 0)
                        b_comp(i) <= b(i) xor sub;
                        adder: full_add PORT MAP (a(i), b_comp(i), c(i-1), c(i), sum(i));
                        END GENERATE;
                        c4     <=     c(4);
                        v <= c(4) xor c(3);
        END adder;
```

7.37   1999; 3 1/2 digits

7.39   1 followed by n 9s.

7.41   See Figure 7.24 in text.

7.43

```
-- add4bcd.vhd
-- 4-bit bcd adder
LIBRARY ieee;
USE ieee.std_logic_1164.ALL;
ENTITY add4bcd IS
    PORT(
        c0                  : IN STD_LOGIC:
        a_bcd, b_bcd        : IN STD_LOGIC_VECTOR(4 down to 1);
        sum_bcd             : OUT STD_LOGIC_VECTOR(5 downto 1));
END add4bcd;

ARCHITECTURE adder OF add4bcd IS
    -- Component declaration
    COMPONENT add4gen
    PORT(
        c0          : IN  STD_LOGIC;
        a, b        : IN  STD_LOGIC_VECTOR(4 downto 1);
        c4          : OUT STD_LOGIC;
        sum         : OUT STD_LOGIC_VECTOR(4 downto 1));
    END COMPONENT;

    COMPONENT bin2bcd
    PORT(
        bin : IN  STD_LOGIC_VECTOR(5 downto 1);
        bcd : OUT STD_LOGIC_VECTOR(5 downto 1);
    END COMPONENT;
    -- Define a signal for internal carry bits
    SIGNAL connect : STD_LOGIC_VECTOR (5 downto 1);
BEGIN
        adder: add4gen PORT MAP (c0, a_bcd, b_bcd, connect(5),
            connect (4 downto 1));
        converter: bin2bcd PORT MAP (connect, sum_bcd);
END adder;
```

7.25 The circuit will be like Figure 7.25 in the text, minus the thousands digit. It will generate a $3\frac{1}{2}$ digit output.

Chapter 8

8.5

$\bar{S}$	$\bar{R}$	
0	0	Latch tries to set and reset at the same time. Forbidden state.
0	1	Set input active. $Q = 1$.
1	0	Reset input active. $Q = 0$.
1	1	Neither set nor reset active. No change.

8.11 **b.** Both set and reset are active at the same time.

  c. i. R is last input active. Latch resets; **ii.** S is last input active. Latch sets; **iii.** S and R go from both active to no change state. The latch cannot predictably resolve this transition. Output unknown.

8.21

```
-- latch8_primitive.vhd
-- 8-bit D-latch with active-HIGH level-sensitive enable

LIBRARY ieee;
USE ieee.std_logic_1164.ALL;

LIBRARY altera;
USE altera.maxplus2.ALL;

ENTITY latch8_primitive IS
    PORT(
        d in   : IN  STD_LOGIC_VECTOR(7 downto 0);
        enable : IN  STD_LOGIC;
        q_out  : OUT STD_LOGIC_VECTOR(7 downto 0));
END latch8_primitive;

ARCHITECTURE a of latch8_primitive IS
BEGIN
    -- Instantiate a latch from MAX+PLUS II (Quartus II) primitive
    latch8:
    FOR i IN 7 downto 0 GENERATE
        latch_primitive: latch
        PORT MAP (d => d_in(i), ena => enable, q => q_out (i));
    END GENERATE;
END a;
```

Simulation Criteria

- If **enable** = 0 and **d_in** changes, **q_out** should not change.
- The next time that **enable** goes HIGH, **q_out** should immediately change to follow **d_in**.
- If **d_in** changes value when **enable** = 1, **q_out** should immediately change to follow **d_in**.

8.31

```
-- dff12lpm.vhd
-- 12-BIT D flip-flop
-- Uses a flip-flop component from the Library of Parameterized Modules (LPM)

LIBRARY ieee;
USE ieee.std_logic_1164.ALL;
LIBRARY lpm;
USE lpm.lpm_components.ALL;
ENTITY dff12lpm IS
    PORT(d_in  : IN  STD_LOGIC_VECTOR(11 downto 0);
         clk   : IN  STD_LOGIC;
         q_out : OUT STD_LOGIC_VECTOR(11 downto 0));
END dff12lpm;

ARCHITECTURE a OF dff12lpm IS
BEGIN
    -- Instantiate flip-flop from an LPM component
    dff12: lpm_ff
        GENERIC MAP (LPM_WIDTH => 12)
        PORT MAP (data  => d_in,
                  clock => clk,
                  q     => q_out);
END a;
```

8.35 The circuit generates the following repeating pattern: 111, 110, 101, 100, 011, 010, 001, 000. This is a 3-bit binary down-count sequence.

8.37 The circuit generates a 4-bit binary sequence from 0000 to 1111, then repeats indefinitely.

8.43 Similarity: an asynchronous circuit and an asynchronous input cause outputs to change out of synchronization with a system clock. Difference: an asynchronous circuit may be clocked, but at different times throughout the circuit; an asynchronous input is independent of the clock function altogether.

8.45

```
-- dff12_lpm
-- 12-BIT D flip-flop with positive edge-triggered clock and
-- active-LOW asynchronous set and clear
-- Uses a flip-flop component from the Library of Parameterized
-- Modules (LPM)

LIBRARY ieee;
USE ieee.std_logic_1164.ALL;
LIBRARY lpm;
USE lpm.lpm_components.ALL;

ENTITY dff12_lpm IS
    PORT(d_in                : IN  STD_LOGIC_VECTOR (11 downto 0);
         clk, clrn, setn    : IN  STD_LOGIC;
         q_out               : OUT STD_LOGIC_VECTOR (11 downto 0));
END dff12_lpm;

ARCHITECTURE a OF dff12_lpm IS
    SIGNAL clr, set: STD_LOGIC;
BEGIN
    -- Instantiate flip-flop from an LPM component
    latch12: lpm_ff
        GENERIC MAP (LPM_WIDTH => 12)
        PORT MAP ( data     => d_in,
                   clock    => clk,
                   aset     => set,
                   aclr     => clr,
                   q        => q_out);

    -- Invert asychronous inputs
    set <= not setn;
    clr <= not clrn;

END a;
```

8.49

```
-- syn4tprm.vhd
-- 4-bit sync counter (TFF primitives)

LIBRARY ieee;
USE ieee.std_logic_1164.ALL;
LIBRARY altera;
USE altera.maxplus2.ALL;

ENTITY syn4tprm IS
        PORT (clock, reset : IN  STD_LOGIC;
              q             : OUT STD_LOGIC_VECTOR(3 downto 0));
END syn4tprm;
```

```
ARCHITECTURE a OF syn4tprm IS
        -- Declare component only with ports actually used
        COMPONENT TFF
            PORT ( t   : IN  STD_LOGIC;
                   clk : IN  STD_LOGIC;
                   clrn: IN  STD_LOGIC;
                   q   : OUT STD_LOGIC);
        END COMPONENT;
        SIGNAL q_int : STD_LOGIC_VECTOR(2 downto 0);
        SIGNAL t_int : STD_LOGIC_VECTOR(3 downto 0);
BEGIN
        -- Instantiate 4 T flip-flops.
        ff0: tff
            PORT MAP (t_int(0), clock, reset, q_int(0));
        ff1: tff
            PORT MAP (t_int(1), clock, reset, q_int(1));
        ff2: tff
            PORT MAP (t_int(2), clock, reset, q_int(2));
        ff3: tff
            PORT MAP (t_int(3), clock, reset, q(3));

-- Connect flip-flops internally
t_int(0)    <=  '1';
t_int(1)    <=  q_int(0);
t_int(2)    <=  q_int(0) and q_int(1);
t_int(3)    <=  q_int(0) and q_int(1) and q_int(2);

q(0)        <= q_int(0);
q(1)        <= q_int(1);
q(2)        <= q_int(2);
END a;
```

8.51 **a.** 4   **b.** 6   **c.** 8

8.53 Registered/active LOW; registered/active HIGH; combinatorial/active LOW; combinatorial/active HIGH

8.55 Yes. Asynchronous reset.

8.57 Global. These functions operate simultaneously on all macrocells.

8.59 **a.** 32   **b.** 64   **c.** 128   **d.** 160

8.61 $n/16$ Logic Array Blocks for n macrocells (e.g. 128/16 = 8 LABs for an EPM7128S).

8.63 Macrocells without pin connections can be used for internal logic.

8.65 A MAX7000S macrocell can be reset from a global clear pin (GCLRn) or locally from a product term.

8.67 Five dedicated product terms; by using terms from shared logic expanders and parallel logic expanders; 5 dedicated, up to 15 from parallel logic expanders; up to 16 from shared logic expanders.

8.69 A sum-of-products network constructs Boolean expressions by switching signals into an OR-gate output via a programmable matrix of AND gates. A look-up table network stores the output values of the network in a small memory whose storage locations are selected by combinations of the input signals.

8.71 A carry chain allows for efficient fast-carry implementation of adders, comparators, and other circuits whose inputs become wider as higher-order bits are added.

8.73 2048

Chapter 9

9.1 A 12-bit counter recycles to 0 after 4096 cars have entered the parking lot. The last car causes all bits to go LOW. The negative edge on the MSB clocks a flip flop whose output enables the LOT FULL sign. Every car out of the gate resets the flip-flop and turns off the sign.

A better circuit would have the exit gate make the counter output decrease by 1 with every vehicle exiting.

9.5 **b.** **i.** 0100

ii. 0110

iii. 0011

9.7

Q_3	Q_2	Q_1	Q_0
0	0	0	0
0	0	0	1
0	0	1	0
0	0	1	1
0	1	0	0
0	1	0	1
0	1	1	0
0	1	1	1
1	0	0	0
1	0	0	1

9.9 Q_0: 24 kHz; Q_1: 12 kHz; Q_2: 6 kHz; Q_3: 3 kHz

9.13 $J_0 = K_0 = 1$
$J_1 = K_1 = Q_0$
$J_2 = K_2 = Q_1 Q_0$
$J_3 = K_3 = Q_2 Q_1 Q_0$
$J_4 = K_4 = Q_3 Q_2 Q_1 Q_0$
$J_5 = K_5 = Q_4 Q_3 Q_2 Q_1 Q_0$
$J_6 = K_6 = Q_5 Q_4 Q_3 Q_2 Q_1 Q_0$
$J_7 = K_7 = Q_6 Q_5 Q_4 Q_3 Q_2 Q_1 Q_0$

9.15 **a.** $J_3 = Q_2 Q_1 Q_0$
$K_3 = Q_1 Q_0$
$J_2 = \bar{Q}_3 Q_1 Q_0$
$K_2 = Q_1 Q_0$
$J_1 = Q_0$
$K_1 = Q_0$
$J_0 = 1$
$K_0 = 1$

b. 1011, 0000, 0001

9.19 The equations for the D flip-flops in the counter are:

$$D_3 = Q_3 \bar{Q}_0 + Q_2 Q_1 Q_0$$
$$D_2 = \bar{Q}_2 Q_1 Q_0 + Q_2 \bar{Q}_1 + Q_2 \bar{Q}_0.$$
$$D_1 = Q_3 \bar{Q}_1 Q_0 + Q_1 \bar{Q}_0$$
$$D_0 = \bar{Q}_0$$

9.21 Boolean equations:

$$D_3 = \bar{Q}_3 Q_2 + Q_3 \bar{Q}_2$$
$$D_2 = Q_1 Q_0$$
$$D_1 = Q_1 Q_0 + Q_1 \bar{Q}_0$$
$$D_0 = \bar{Q}_2 \bar{Q}_0$$

9.27 See Figure 9.35 in the text. Asynchronous load transfers data directly to the flip-flops of a counter as soon as the load input is asserted; it does not wait for a clock edge. Synchronous load waits for an active clock edge to load a value into the counter flip-flops.

9.33 $D_0 = \overline{Q_0}$

$D_1 = (Q_0 DIR + \overline{Q_0}\,\overline{DIR}) \oplus Q_1$

$D_2 = (Q_1 Q_0 DIR + \overline{Q_1}\,\overline{Q_0}\,\overline{DIR}) \oplus Q_2$

$D_3 = (Q_2 Q_1 Q_0 DIR + \overline{Q_2}\,\overline{Q_1}\,\overline{Q_0}\,\overline{DIR}) \oplus Q_3$

The right-hand product term of each equation represents the down-count logic, which is enabled whenever $DIR = 0$. The left-hand product term is the up-count logic, enabled when $DIR = 1$. D_0 is always the opposite of Q_0, regardless of whether the count is up or down.

9.37

```
-- ct_mod24
-- Presettable counter with synchronous clear and load
-- and a modulus of 24

ENTITY ct_mod24 IS
     PORT(
       clk                     :  IN  BIT;
       clear, direction        :  IN  BIT;
       q                       :  OUT INTEGER RANGE 0 TO 23);
END ct_mod24;

ARCHITECTURE a OF ct_mod24 IS
     BEGIN
     PROCESS (clk)
        VARIABLE cnt : INTEGER RANGE 0 TO 23;
          BEGIN
             IF (clk'EVENT AND clk = '1') THEN
                 IF (clear = '0') THEN -- Synchronous clear
                    cnt := 0;
                 ELSIF (direction = '0') THEN
                    IF cnt = 0 THEN
                       cnt := 23;
                     ELSE
                       cnt := cnt - 1;
                     END IF;
                 ELSIF (direction = '1') THEN
                    IF cnt = 23 THEN
                        cnt := 0;
                     ELSE
                        cnt := cnt + 1;
                     END IF;
                  END IF;
               END IF;
               q      <=      cnt;
           END PROCESS;
      END a;
```

9.39   a.

```
-- aset_counter.vhd
-- Counter of generic width, based on a behavioral description
-- Includes an asynchronous set function

LIBRARY ieee;
USE ieee.std_logic_1164.ALL;
USE ieee.std_logic_unsigned.ALL;
USE ieee.std_logic_arith.ALL;
```

```
ENTITY aset_counter IS
    GENERIC (width    : INTEGER := 6;
             async_set : INTEGER := 16);
    PORT(
         clock, reset, aset : IN      STD_LOGIC;
         q: BUFFER STD_LOGIC_VECTOR(width-1 downto 0));
END aset_counter;

ARCHITECTURE counter of aset_counter IS
BEGIN
    PROCESS (clock, reset, aset)
    BEGIN
        IF (reset = '0') THEN
            q <= (others => '0');
        ELSIF (aset = '1') THEN
            -- Convert INTEGER to STD_LOGIC_VECTOR
            q <= CONV_STD_LOGIC_VECTOR(async_set, width);
        ELSIF (clock'EVENT and clock = '1') THEN
            q <= q + 1;
        END IF;
    END PROCESS;
END Counter;
```

Simulation Criteria

- With **reset** = 1 and **aset** = 0 (both disabled), a series of pulses on **clock** should cause **q** to advance by 1 until it reaches 3FH (= 111111_2 = 63_{10}). The counter should roll over to 0 on the next clock pulse.
- When **reset** = 0, the counter should set to 0 without waiting for the next clock pulse.
- When **aset** = 1, the counter should set to the default value without waiting for the next clock pulse.
- If **aset** and **reset** are both active, **reset** should have priority.

9.41   a.

```
-- selectable_aset_counter.vhd
-- Counter of generic width, based on a behavioral description
-- Includes an asynchronous set function

LIBRARY ieee;
USE ieee.std_logic_1164.ALL;
USE ieee.std_logic_unsigned.ALL;
USE ieee.std_logic_arith.ALL;

ENTITY selectable_aset_counter IS
    GENERIC (width     : INTEGER := 6;
             aset_type  : STRING  := "ALL_ONES";
             aset_value : INTEGER := 16);
    PORT(
         clock, reset, aset : IN  STD_LOGIC;
         q : BUFFER STD_LOGIC_VECTOR (width-1 downto 0));
END selectable_aset_counter;

ARCHITECTURE counter of selectable_aset_counter IS
BEGIN
    PROCESS (clock, reset, aset)
    BEGIN
        IF (reset = '0') THEN
            q <= (others => '0');
        ELSIF (aset = '1') THEN
            IF aset_type = "ALL_ONES" THEN
                q <= (others => '1');
```

```
                          ELSIF aset_type = "PORT_VALUE" THEN
                             -- Convert INTEGER to STD LOGIC VECTOR
                             q <= CONV_STD_LOGIC_VECTOR (aset_value, width);
                          END IF;
                       ELSIF (clock'EVENT and clock = '1') THEN
                          q <= q + 1;
                       END IF;
                    END PROCESS;
                 END counter;
```

b. Simulation criteria for this counter are the same as for the counter in Problem 9.39, except that the default value for **aset** is all 1s on the output.

9.43

```
-- sst1_lpm.vhd
-- 12-bit LPM counter with sst1 and aclr (Chapter problem)

LIBRARY ieee;
USE ieee.std_logic_1164.ALL;
LIBRARY lpm;
USE lpm.lpm_components.ALL;

ENTITY sst1_lpm IS
       PORT(
          clk            : IN  STD_LOGIC;
          clear, set     : IN  STD_LOGIC;
          q              : OUT STD_LOGIC_VECTOR (11 downto 0));
END sst1_lpm;

ARCHITECTURE a OF sst1_lpm IS
BEGIN
       counter1: lpm_counter
          GENERIC MAP (LPM_WIDTH => 12)
          PORT MAP ( clock => clk,
                     sset  => set,
                     aclr  => clear,
                     q     => q);
       END a;
```

Since the parameter LPM_SVALUE is not specified, the counter in this problem sets to all 1s (1111 1111 1111 = FFFH), rather than 0111 1111 1111 (= 7FFH).

9.45

```
-- lpm8term
-- 8-bit presettable counter with synchronous clear and load,
-- count enable, a directional control port,
-- and terminal count decoding

LIBRARY ieee;
USE ieee.std_logic_1164.ALL;
LIBRARY lpm;
USE lpm.lpm_components.ALL;

ENTITY lpm8term IS
       PORT(
          clk, count_ena           : IN  STD_LOGIC;
          clear, load, direction   : IN  STD_LOGIC;
          p                        : IN  STD_LOGIC_VECTOR(7 downto 0);
          max_min                  : OUT STD_LOGIC;
          q                        : OUT STD_LOGIC_VECTOR(7 downto 0));
END lpm8term;
```

```
ARCHITECTURE a OF lpm8term IS
        SIGNAL cnt : STD_LOGIC_VECTOR(7 downto 0);
BEGIN
        counter1: lpm_counter
                GENERIC MAP (LPM_WIDTH => 8)
                PORT MAP ( clock  => clk,
                                updown => direction,
                                cnt_en => count_ena,
                                data   => p,
                                sload  => load,
                                sclr   => clear,
                                q      => cnt);
        q <= cnt;
        PROCESS (clk, cnt)
        BEGIN
            -- Terminal count decoder
            IF (cnt = "00000000" and direction = '0') THEN
                    max_min    <= '1';
            ELSIF (cnt = "11111111" and direction = '1') THEN
                    max_min <=  '1';
            ELSE
                    max_min <=  '0';
            END IF;
        END PROCESS;
END a;
```

9.51   001111, 000000, 000000, 110000

9.59

```
-- Left-shift register of generic width
LIBRARY ieee;
USE ieee.std_logic_1164.ALL;

ENTITY slt_bhv IS
    GENERIC (width : POSITIVE);
    PORT(
            serial_in, clk    : IN       STD_LOGIC;
            q                 : BUFFER   STD_LOGIC_VECTOR(width-1 downto 0));
END slt_bhv;

ARCHITECTURE left_shift of slt_bhv IS
BEGIN
    PROCESS (clk)
    BEGIN
        IF (clk'EVENT and clk = '1') THEN
                q(width-1 downto 0) <= q(width-2 downto 0) & serial_in;
        END IF;
    END PROCESS;
END left_shift;
```

```
-- 32-bit left-shift register
LIBRARY ieee;
USE ieee.std_logic_1164.ALL;

ENTITY slt32_bhv IS
    PORT(
            data_in, clock    : IN       STD_LOGIC;
            qo                : BUFFER   STD_LOGIC_VECTOR(31 downto 0));
END slt32_bhv;
```

```
ARCHITECTURE left_shift of slt32_bhv IS
COMPONENT slt_bhv
        GENERIC (width : POSITIVE);
        PORT(
                serial_in, clk   : IN  STD_LOGIC;
                q                : OUT STD_LOGIC_VECTOR(31 downto 0));
END COMPONENT;
BEGIN
        Shift_left_32: slt_bhv
                GENERIC MAP (width=> 32)
                PORT MAP (serial_in => data_in,
                        clk      => clock,
                        q        => qo);
END left_shift;
```

9.61 The generic component has a default width of 8 bits. The instantiated component has an assigned width of 16 bits. The generic map in the instantiated component overrides the default parameter.

9.63 **a.**

```
LIBRARY ieee;
USE ieee.std_logic_1164.ALL;

ENTITY srg_dir_param_instance IS
    PORT(
        data_in, clock: IN    STD_LOGIC;
        qo              : BUFFER STD_LOGIC VECTOR(7 downto 0));
END srg_dir_param_instance;

ARCHITECTURE shift of srg_dir_param_instance IS
    COMPONENT srg_dir_param
        GENERIC (width : POSITIVE;
                 direction : STRING);
        PORT(
            serial_in, clk: IN STD_LOGIC;
            q               : OUT STD_LOGIC_VECTOR(7 downto 0));
    END COMPONENT;
BEGIN
    Shift_dir_8: srg_dir_param
        GENERIC MAP (width=> 8, direction => "LEFT")
        PORT MAP (serial_in => data_in,
                  clk      => clock,
                  q        => qo);
END shift;
```

Simulation Criteria

- Data applied to **data_in** should be clocked into **qo(0).**
- Data should move to the flip-flop of the next highest subscript with each applied clock pulse.
- Data will reach the end of the shift register in 8 clock pulses.

9.65
```
-- srg10lpm.vhd
-- 10-bit serial shift register (shift right)
LIBRARY ieee;
USE ieee.std_logic_1164.ALL;
LIBRARY lpm;
USE lpm.lpm_components.ALL;
```

```
ENTITY srg10lpm IS
      PORT(
               clk         : IN  STD_LOGIC;
               serial_in   : IN  STD_LOGIC;
               sync_set    : IN  STD_LOGIC;
               serial_out  : OUT STD_LOGIC);
END srg10lpm;

ARCHITECTURE lpm_shift of srg10lpm IS
COMPONENT lpm_shiftreg
      GENERIC(LPM_WIDTH: POSITIVE; LPM_SVALUE: STRING);
      PORT(
               clock, shiftin  : IN STD_LOGIC;
               sset            : IN STD_LOGIC;
               shiftout        : OUT STD_LOGIC);
END COMPONENT;
BEGIN
      Shift_10: lpm_shiftreg
               GENERIC MAP (LPM_WIDTH=> 10, LPM_SVALUE => "960")
               PORT MAP (clk, serial_in, sync_set, serial_out);
END lpm_shift;
```

The parameter LPM_SVALUE is set to 960, the decimal equivalent of H"3C0".

9.67

Q_4	Q_3	Q_2	Q_1	Q_0
0	0	0	0	0
1	0	0	0	0
1	1	0	0	0
1	1	1	0	0
1	1	1	1	0
1	1	1	1	1
0	1	1	1	1
0	0	1	1	1
0	0	0	1	1
0	0	0	0	1

All gates used in the decoder of Figure 9.109 remain unchanged except those decoding the MSB/LSB pairs (Q_3Q_0 and $\bar{Q}_3\bar{Q}_0$). Change these to decode Q_4Q_0 and $\bar{Q}_4\bar{Q}_0$. Add two new gates to decode $Q_4\bar{Q}_3$ (2nd state) and $\bar{Q}_4\bar{Q}_3$ (7th state).

Chapter 10

10.1   Mealy machine. The output is fed by combinational, as well as sequential, logic.

10.3 $D_3 = Q_2\bar{Q}_1\bar{Q}_0 + Q_3Q_1 + Q_3Q_0$
$D_2 = \bar{Q}_3Q_1\bar{Q}_0 + Q_2\bar{Q}_1 + Q_2Q_0$
$D_1 = \bar{Q}_3\bar{Q}_2Q_0 + \bar{Q}_3Q_2\bar{Q}_0 + Q_1\bar{Q}_0$
$D_0 = \bar{Q}_3\bar{Q}_2\bar{Q}_1 + \bar{Q}_3Q_2\bar{Q}_1 + Q_3Q_2Q_1 + Q_3\bar{Q}_2Q_1$

10.5
```
-- prob10_5.vhd
-- 4-bit Gray code counter
-- (state machine with decoded outputs)

LIBRARY ieee;
USE ieee.std_logic_1164.ALL;
```

```
ENTITY prob10_5 IS
    PORT(
        clk : IN  STD_LOGIC;
        q   : OUT STD_LOGIC_VECTOR(3 downto 0));
END prob10_5;

ARCHITECTURE a OF prob10_5 IS
    TYPE STATE_TYPE IS (s0, s1, s2, s3, s4, s5, s6, s7, s8,
                        s9, s10, s11, s12, s13, s14, s15):

    SIGNAL state: STATE_TYPE;
BEGIN
    PROCESS (clk)
    BEGIN
        IF clk'EVENT AND clk = '1' THEN
            CASE state IS
                WHEN s0 => state <= s1;
                WHEN s1 => state <= s2;
                WHEN s2 => state <= s3;
                WHEN s3 => state <= s4;
                WHEN s4 => state <= s5;
                WHEN s5 => state <= s6;
                WHEN s6 => state <= s7;
                WHEN s7 => state <= s8;
                WHEN s8 => state <= s9;
                WHEN s9 => state <= s10;
                WHEN s10 => state <= s11;
                WHEN s11 => state <= s12;
                WHEN s12 => state <= s13;
                WHEN s13 => state <= s14;
                WHEN s14 => state <= s15;
                WHEN s15 => state <= s0;
            END CASE;
        END IF;
    END PROCESS;

    WITH state SELECT
            q   <=  "0000" WHEN     s0,
                    "0001" WHEN     s1,
                    "0011" WHEN     s2,
                    "0010" WHEN     s3,
                    "0110" WHEN     s4,
                    "0111" WHEN     s5,
                    "0101" WHEN     s6,
                    "0100" WHEN     s7,
                    "1100" WHEN     s8,
                    "1101" WHEN     s9,
                    "1111" WHEN     s10,
                    "1110" WHEN     s11,
                    "1010" WHEN     s12,
                    "1011" WHEN     s13,
                    "1001" WHEN     s14,
                    "1000" WHEN     s15;
END a;
```

10.7 $D_1 = \bar{Q}_1 Q_0 in1$
$D_0 = \bar{Q}_1 \overline{in1}$
$out1 = \bar{Q}_1 Q_0 \overline{in1}$
$out2 = Q_1 Q_0 in1$

The circuit generates a HIGH pulse on **out1** when **in1** goes LOW and a HIGH pulse on **out2** when the input goes back HIGH.

10.9

```vhdl
-- prob10_9.vhd

LIBRARY ieee;
USE ieee.std_logic_1164.ALL;

ENTITY prob10_9 IS
    PORT(
        clk, in1    : IN  STD_LOGIC;
        out1, out2  : OUT STD_LOGIC);
END prob10_9;

ARCHITECTURE a OF prob10_9 IS
    TYPE PULSER IS (s0, s1, s2, s3);
    SIGNAL sequence: PULSER;
BEGIN
    PROCESS (clk)
    BEGIN
        IF clk'EVENT AND clk = '1' THEN
            CASE sequence IS
                WHEN s0 =>
                    IF in1 = '1' THEN
                        sequence <= s0;    -- no change if in1 = 1
                        out1 <= '0';
                        out2 <= '0';
                    ELSE
                        sequence <= s1;    -- proceed if in1 = 0
                        out1 <= '1';       -- pulse on out1
                        out2 <= '0';
                    END IF;
                WHEN s1 =>
                    IF in1 = '0' THEN
                        sequence <= s1;    -- outputs LOW
                        out1 <= '0';
                        out2 <= '0';
                    ELSE
                        sequence <= s2;
                        out1 <= '0';
                        out2 <= '1';       -- pulse on out2
                    END IF;
                WHEN s2 =>
                        sequence <= s0;
                        out1 <= '0';
                        out2 <= '0';
                WHEN others =>
                        sequence <= s0;
                        out1 <= '0';
                        out2 <= '0';
            END CASE;
        END IF;
    END PROCESS;
END a;
```

10.11

```
LIBRARY ieee;
USE ieee.std_logic_1164.ALL;
ENTITY prob10_11 IS
     PORT(
          clk, go, reset, eoc   : IN STD_LOGIC;
          sc, oe    : OUT STD_LOGIC);
END prob10_11;

ARCHITECTURE a OF prob10_11 IS
     TYPE ADC IS (idle, start, waiting, read);
     SIGNAL state: ADC;
     SIGNAL outputs: STD_LOGIC_VECTOR(1 downto 0);
BEGIN
     sc <= outputs(1);
     oe <= outputs(0);
     PROCESS (clk)
     BEGIN
          IF clk'EVENT AND clk = '1' THEN
               IF reset = '0' THEN
                    state <= idle;
                    outputs <= "01";
               ELSE
                    CASE state IS
                         WHEN idle =>
                              IF go = '0' THEN
                                   state <= idle;
                                   outputs <= "01";
                              ELSIF go = '1' THEN
                                   state <= start;
                                   outputs <= "11";
                              END IF;
                         WHEN start =>
                              state <= waiting;
                              outputs <= "01";
                         WHEN waiting =>
                              IF eoc = '0' THEN
                                   state <= waiting;
                                   outputs <= "01";
                              ELSIF eoc = '1' THEN
                                   state <= read;
                                   outputs <= "00";
                              END IF;
                         WHEN read =>
                              state <= idle;
                              outputs <= "01";
                    END CASE;
               END IF;
          END IF;
     END PROCESS;
END a;
```

10.13 A NAND latch can only debounce a switch with a normally open and a normally closed contact: one to set and the other to reset the latch. The pushbutton on the Altera UP-1 or UP-2 board has only a normally open contact.

10.15 8.33 ms

10.17 Four clock periods. 8.33 ms

10.19
```
-- prob10_19.vhd
LIBRARY ieee;
USE ieee.std_logic_1164.ALL;

ENTITY prob10_19 IS
      PORT(
            clk, in1, in2    : IN  STD_LOGIC;
            out1             : OUT STD_LOGIC);
END prob10_19;

ARCHITECTURE a OF prob10_19 IS
      TYPE STATE_TYPE IS (s0, s1, s2, s3, s4);
      SIGNAL state: STATE_TYPE;
BEGIN
      PROCESS (clk)
      BEGIN
            IF clk'EVENT AND clk = '1' THEN
                  CASE state IS
                        WHEN s0 =>
                              state <= s1;
                              out1 <= '0';
                        WHEN s1 =>
                              IF in1 = '1' THEN
                                    state <= s1;
                                    out1 <= '0';
                              ELSIF in1 = '0' THEN
                                    state <= s2;
                                    out1 <= '1';
                              END IF;
                        WHEN s2 =>
                              state <= s3;
                              out1 <= '0';
                        WHEN s3 =>
                              state <= s4;
                              out1 <= '0';
                        WHEN s4 =>
                              IF in2 = '1' THEN
                                    state <= s4;
                                    out1 <= '0';
                              ELSIF in2 = '0' THEN
                                    state <= s0;
                                    out1 <= '1';
                              END IF;
                  END CASE;
            END IF;
      END PROCESS;
END a;
```

Chapter 11

11.1 **TTL:** advantages—relatively high speed, high current driving capability; disadvantages—high power consumption, rigid power supply requirements. **CMOS:** advantages—low power consumption, high noise immunity, flexible power supply requirements; disadvantages—low output current **ECL:** advantages—high speed; disadvantages—high susceptibility to noise, high power consumption.

11.3 $t_{pHL} = 12$ ns, $t_{pLH} = 10$ ns

11.5   Transition from state 1 to state 2: $t_p = t_{pLH02} + t_{pHL00} = 16$ ns $+ 15$ ns $= 31$ ns. Transition from state 2 to state 3: $t_p = t_{pLH00} = 15$ ns (Assume $V_{CC} = 4.5$ volts; $T = 25°$ C to $-55°$ C)

11.9   Clock pulse width: $t_w = 12$ ns; Setup time: $t_{su} = 10$ ns; Hold time: $t_h = 5$ ns.

11.11

74ALS86 Driver	74ALS00 Load
$I_{OH} = -0.4$ mA	$I_{IH} = 20$ μA
$I_{OL} = 8$ mA	$I_{IL} = -0.1$ mA

$n_H = 0.4$ mA/20 μA $= 20$
$n_L = 8$ mA/0.1 mA $= 80$
$n = n_H = 20$

11.13

74AS32 Driver	74ALS00 Load
$I_{OH} = -2$ mA	$I_{IH} = 20$ μA
$I_{OL} = 20$ mA	$I_{IL} = -0.1$ mA

$n_H = 2$ mA/20 μA $= 100$
$n_L = 20$ mA/0.1 mA $= 200$
$n = n_H = 100$

11.15   **a.** 20 mW   **b.** 18 mW   **c.** 14 mW   **d.** 12 mW

11.17   **a.** 4.5 μW   **b.** 28.3 μW   **c.** 4.46 mW

11.19   **a.** 550 μW   **b.** −56.4%

11.21   The outputs of a 74LS00 gates are guaranteed to produce output voltages of $V_{OH} \geq 2.7$ V and $V_{OL} \leq 0.5$ V. The inputs of a 74HCT series gate are voltage compatible with LSTTL outputs since $V_{IH} \geq 2$ V and $V_{IL} \leq 0.8$ V. This is not the case for 74HC series gates, where $V_{IH} \geq 3.15$V and $V_{IL} \leq 1.35$V. The 74LS gate is not guaranteed to drive the 74HC gate in the HIGH state.

11.23   10 loads, since the 74HC output voltages are defined for an output current of ±4mA.

11.25   HIGH input: The base-emitter junction of transistor Q_1 is reverse-biased. Current flows, by default, through the base-collector junction of Q_1, supplying base current to Q_2, saturating it. This in turn, saturates Q_3, making its collector LOW.

LOW input: Current has a path to ground via the base-emitter junction of Q_1. This transports charge away from the base region of Q_2, making it cutoff. Since Q_2 is cutoff, no base current flows in Q_3. If an external pull-up resistor is connected to the collector of Q_3, the output will be HIGH. Otherwise, it is floating.

11.27   See Figure 11.33 in text. $Y = \overline{AB} \cdot \overline{CD} \cdot \overline{EF} = \overline{AB + CD + EF}$

11.31   Yes. When the output transistor saturates (output LOW), there will be a direct connection from the output to V_{CC}. Since there is relatively little resistance in the current path, the current will likely exceed the rated output current, I_{OL}.

11.33   **a.** Q_3 and Q_4 are never on at the same time because the phase splitter, Q_2, keeps them in opposite states. The voltage in the circuit is divided such that when Q_2 is on, it pulls the base of Q_4 into the cutoff region for that transistor. At the same time, Q_3 is supplied with base current and thus saturated. When Q_2 is off, there is no base current in Q_3, making it cutoff. The base voltage at Q_4 is now such that it is on.

b. Switching noise originates in a totem pole output because the HIGH output transistor, Q_4, can switch on faster than the LOW output transistor, Q_3, can switch off. For a brief time, both transistors are on, causing a supply current spike. This can be counteracted by connecting a small capacitor between the supply voltage, V_{CC}, and ground.

11.35   7.58 mA, 95% of I_{OL}; 2.12 mA, 530% of I_{OH} The first circuit is more suitable, as it can drive a higher current to the LED and still remain within the output specification of the inverter.

11.37 Store MOS devices in antistatic or conducting material. Work only on an antistatic work surface and wear a conductive wrist strap. Connect all unused device inputs to power or ground. Do not touch the pins of the MOS device.

11.39 See Figure 11.58 in text.

A	B	Q_1	Q_2	Q_3	Q_4	Q_5	Q_6	Y
0	0	ON	ON	OFF	OFF	OFF	ON	0
0	1	OFF	ON	OFF	ON	OFF	ON	0
1	0	ON	OFF	ON	OFF	OFF	ON	0
1	1	OFF	OFF	ON	ON	ON	ON	1

11.41 The state of flip-flop output Q selects which signal is switched to the data converter/display driver by enabling one of the CMOS transmission gates. When $Q = 1$, the wheel rotation sensor is selected. $Q = 0$ selects the engine rotation sensor.

11.43 No. TTL power dissipation, and therefore the speed-power product, depends on the logic states of the device outputs, not on frequency.

11.45 74HC: pin replacement for TTL device; CMOS-compatible inputs; TTL-compatible outputs. 74HC4NNN: pin replacement for CMOS device; CMOS-compatible inputs; TTL-compatible outputs. 74HCT: pin replacement for TTL device; TTL-compatible inputs; TTL-compatible outputs. 74HCU: unbuffered CMOS outputs.

Chapter 12

12.1

Analog Voltage	Code
0–0.75	000
0.75–2.25	001
2.25–3.75	010
3.75–5.25	011
5.25–6.75	100
6.75–8.25	101
8.25–9.75	110
9.75–12.00	111

Analog Voltage	Code
0.000–0.375	0000
0.375–1.125	0001
1.125–1.875	0010
1.875–2.625	0011
2.625–3.375	0100
3.375–4.125	0101
4.125–4.875	0110
4.875–5.625	0111
5.625–6.375	1000
6.375–7.125	1001
7.125–7.875	1010
7.875–8.625	1011
8.625–9.375	1100
9.375–10.125	1101
10.125–10.875	1110
10.875–12.000	1111

12.3

Fraction of T	Sine Voltage	Digital Code
0	0 V	0000
T/8	4.59 V	0110
T/4	8.48 V	1011
3T/8	11.09 V	1111
T/2	12.00 V	1111
5T/8	11.09 V	1111
3T/4	8.48 V	1011
7T/8	4.59 V	0110
T	0 V	0000

12.5   Six bits, since $64 = 2^6$. Resolution = 500 mV/64 = 7.8125 mV.

12.7 $n + 1$ (One extra bit for each doubling of the number of codes.)

12.11 $2^4 = 16$

    a. $code = (9 \text{ V}/12 \text{ V}) \times 16 = 12_{10} = 1100_2$

    b. range $= 9 \text{ V} \pm \frac{1}{2} \text{ LSB}$
        1 LSB = 12 V/16 = 0.75 V; $\frac{1}{2}$ LSB = 0.375 V
        Range = 8.625 V to 9.375 V

    c. $code = (10 \text{ V}/12 \text{ V}) \times 16 = 13.33$
        Code must be a whole number.
        $V_a = (code \times 12 \text{ V})/16$
        For code = 13: $V_a = (13 \times 12 \text{ V})/16 = 9.75$ V
            Range of V_a for *code* = 13: 9.75 V $\pm \frac{1}{2}$ LSB = 9.375 V to 10.125 V
            10 V is within this range. $Code = 13_{10} = 1101_2$

12.13   **a.** $code = ((3 \text{ V}/12 \text{ V}) \times 16) + (16/2) = 12_{10} = 1100_2$

    b. range: 3 V $\pm \frac{1}{2}$ LSB
        1 LSB = 12 V/16 = 0.75 V; $\frac{1}{2}$ LSB = 0.375 V
        Range = 2.625 V to 3.375 V

    c. $code = ((4\text{V}/12 \text{ V}) \times 16) + (16/2) = 13.33$
        $V_a = (code - 8) \times 12/16$
        For *code* = 13: $V_a = (13 - 8) \times 12/16 = 3.75$ V
        Range for *code* = 13: 3.75 V $\pm$ 0.375 V = 3.375V to 4.125 V
        4 V is within this range, so $code = 13_{10} = 1101_2$

12.15   **a.** $code = (3 \text{ V}/12 \text{ V}) \times 16 = 4_{10} = 0100_2$

    b. range $= 3 \text{ V} \pm \frac{1}{2}$ LSB
        1 LSB = 12 V/16 = 0.75 V; $\frac{1}{2}$ LSB = 0.375 V
        Range = 2.625 V to 3.375 V

    c. $code = (4 \text{ V}/12 \text{ V}) \times 16 = 5.33_{10}$
        $V_a = (code \times 12 \text{ V}/16) \pm \frac{1}{2}$ LSB
        For *code* = 5: $V_a = +3.75 \text{ V} \pm 0.375 \text{ V} = +3.375$ V to +4.125 V
        V_a is within this range, so $code = 5_{10} = 0101_2$

12.21.   Offset binary:
        $code = 01010101_2 = 85_{10}$: $V_a = ((85/256) \times 10 \text{ V}) - 5 \text{ V} = -1.68$ V
        $code = 10101010_2 = 170_{10}$: $V_a = ((170/256 \times 10 \text{ V}) - 5 \text{ V} = +1.64$ V
    2's complement:
        $code = 01010101_2 = 85_{10}$: $V_a = (85/256) \times 10 \text{ V} = +3.32$ V
        $code = 10101010_2 = -86_{10}$: $V_a = (-86/256) \times 10 \text{ V} = -3.36$ V

12.23   **a.** $V_a = (code/2^4) V_{ref} = (12/16) V_{ref} = 0.75 V_{ref}$;

    b. $V_a = (code/2^8) V_{ref} = (200/256) V_{ref} = 0.78125 V_{ref}$;

    c. A 4-bit and 8-bit quantization of the same analog voltage are the same in the first four bits. The additional bit in the lower 4 bits adds an extra voltage to the analog output.

12.25   From most to least significant bits: 1 kΩ, 2 kΩ, 4 kΩ, 8 kΩ, 16 kΩ, 32 kΩ, 64 kΩ, 128 kΩ, 256 kΩ, 512 kΩ, 1024 kΩ, 2048 kΩ, 4096 kΩ, 8192 kΩ, 16,384 kΩ, 32,768 kΩ. All resistors greater than 64 kΩ are specified to three or more significant figures. These values, which are necessary to maintain conversion accuracy, are not available as standard commercial components.

12.27   **a.** $V_a = (15/16) \times 12 \text{ V} = 11.25$ V

    b. $V_a = (11/16) \times 12 \text{ V} = 8.25$ V

    c. $V_a = (6/16) \times 12 \text{ V} = 4.5$ V

    d. $V_a = (3/16) \times 12 \text{ V} = 2.25$ V

12.29   Resolution = $(1/256)(2.2 \text{ kΩ} \times 12 \text{ V})/6.8 \text{ kΩ} = 15.16$ mV

12.33   There are only 16 steps in the waveform and they reach to 15/16 of the reference value. Therefore, the four least significant bits are stuck at logic LOW.

12.35   Offset error *(OE)* = 0.5 V; *OE* = 0.333 LSB; *OE* = 4.167% *FS*. Gain error = 0; Linearity error = 0

12.37 Linearity error *(LE)* = 0.175 V; *LE* = 0.35 LSB; *LE* = 4.375%. Gain error = 0; Offset error = 0

12.39 The priority encoder converts the highest active comparator voltage to a digital code. The enable input of the latch can be pulsed with a waveform having the same frequency as the sampling frequency.

12.41

Bit	New Digital Value	Analog Equivalent from DAC	$V_{analog} \geq V_{DAC}$	Comparator Output	Accumulated Digital Value
Q_7	10000000	8 V	No	0	00000000
Q_6	01000000	4 V	Yes	1	01000000
Q_5	01100000	6 V	No	0	01000000
Q_4	01010000	5 V	No	0	01000000
Q_3	01001000	4.5 V	Yes	1	01001000
Q_2	01001100	4.75 V	Yes	1	01001100
Q_1	01001110	4.875 V	No	0	01001100
Q_0	01001101	4.8125 V	No	0	01001100

12.43 (8 V/12 V) × 16 = 10.667. Since the SAR method of A/D conversion truncates a result, the new code value will be 1010. The new hex digit is A.

12.45 $\pm\frac{1}{2}$LSB

12.47   **a.** Integrating phase: The slope for a Full Scale input is given by:

$$-(v_{in})/RC = -16 \text{ V}/(80 \text{ k}\Omega)(0.1 \text{ }\mu\text{F}) = -2 \text{ V/ms}.$$

Since the slope is proportional to the input voltage, the slope for a 3 V input is:

$$(3/16) \times (-2 \text{ V/ms}) = -0.375 \text{ V/ms}$$

   b. Rezeroing phase: At +2 V/ms, the integrator would take 8 seconds to rezero from Full Scale. This is always the slope when the circuit rezeros.

   c. It would take (3/16) × (8 s) = 1.5 s to rezero for an input of 3 V.

   d. The integrator waveform is similar to that for the input of 1/4 full scale shown in Figure 12.36 in the text.

   e. *Code* = (3/16) × 256 = 48_{10} = 00110000_2.

12.49 Integrator—Sums the voltage at the summing junction to the previous integrator output value.

Comparator—Sets output bit HIGH or LOW, depending on integrator output voltage.

DAC—Converts output bit to a voltage of $\pm V_{ref}$.

Summing junction—Subtracts the DAC output from the analog input sample.

12.51   **a.** For an 8-bit pattern, there are 3 HIGHs. Input is 3/8 *FS*.

$$V_a = ((3/8) \times 16 \text{ V}) - 8 \text{ V} = -2 \text{ V}.$$

   b. For a 5-bit pattern, there are 2 HIGHs. Input is 2/5 *FS*.

$$V_a = ((0.4) \times 16 \text{ V}) - 8 \text{V} = -1.6 \text{ V}.$$

   c. 128 iterations = 25 patterns + 3 iterations = 51 HIGH outputs. 51/128 = 0.3984375
Nominal value = 2/5 = 0.4
%Error = ((0.3984375 − 0.4)/0.4) × 100% = 0.39062%

   d. 2048 iterations = 409 patterns + 3 iterations = 819 HIGH outputs.
819/2048 = 0.399902
%Error = ((0.399902 − 0.4)/0.4) × 100% = 0.0244%

12.53   **a.** Increase the number of output bits.

   b. Increase the number of iteration cycles.

12.55 See Figure 12.45 in text.

12.57 200 kHz

12.59 The sampling frequency is 13/12 times the period of the sampled analog waveform. This is 1-1/12 periods, or 30 degrees greater than the sampled waveform. Thus one full cycle of the alias frequency is 5.2 μs × 12 = 62.4 μs. The alias frequency is approximately 16 kHz.

12.61

```
-- adc_cont.vhd
-- State machine interface to ADC0808
-- Continuous conversion, single latch,
-- analog channel selected externally

LIBRARY ieee;
USE ieee.std_logic_1164.ALL;

ENTITY adc_cont IS
PORT(
        clock, reset, eoc : IN STD_LOGIC;
        sc, oe, en : OUT STD_LOGIC);
END adc_cont;

ARCHITECTURE adc OF adc_cont IS
        TYPE state_type IS (start, wait1, wait2, read, store);
        SIGNAL state: state_type;
        SIGNAL outputs: STD_LOGIC_VECTOR (1 to 3);
BEGIN
        PROCESS (clock, reset)
        BEGIN
                IF (reset = '0') THEN
                        state <= start;
                        outputs <= "000";
                ELSIF (clock'EVENT and clock = '1') THEN
                        CASE state IS
                                WHEN start =>
                                        state <= wait1;
                                        outputs <= "100";
                                WHEN wait1 =>
                                        IF (eoc = '1') THEN
                                                state <= wait1;
                                                outputs <= "000";
                                        ELSIF (eoc = '0') THEN
                                                state <= wait2;
                                                outputs <= "000";
                                        END IF;
                                WHEN wait2 =>
                                        IF (eoc = '0') THEN
                                                state <= wait2;
                                                outputs <= "000";
                                        ELSIF (eoc = '1') THEN
                                                state <= read;
                                                outputs <= "011";
                                        END IF;
                                WHEN read =>
                                        state <= store;
                                        outputs <= "010";
                                WHEN store =>
                                        state <= start;
                                        outputs <= "000";
```

```
                        END CASE;
                END IF;
                sc              <= outputs(1);
                oe              <= outputs(2);
                en              <= outputs(3);
        END PROCESS;
END adc;
```

12.63 23.5 kHz

Chapter 13

13.1 The number of address lines is n for 2^n memory locations. Thus, an 8×8 memory requires 3 address lines ($2^3 = 8$). A 16×8 memory requires 4 address lines ($2^4 = 16$).

13.3 **a.** $64K = 2^6 \times 2^{10} = 2^{16}$; 16 address lines, 8 data lines

 b. $128K = 2^7 \times 2^{10} = 2^{17}$; 17 address lines, 16 data lines

 c. $128K = 2^7 \times 2^{10} = 2^{17}$; 17 address lines, 32 data lines

 d. $256K = 2^8 \times 2^{10} = 2^{18}$; 18 address lines, 16 data lines

13.5 The inputs $\overline{W}$ (Write), $\overline{G}$ (Gate), and $\overline{E}$ (Enable) control the flow of data into or out of the RAM shown by enabling or disabling the two tristate buffers on each pin. There is an output (read) buffer and an input (write) buffer for each pin.

 The read buffers are enabled when $\overline{W} = 1$, $\overline{E} = 0$, and $\overline{G} = 0$. The write buffers are enabled when $\overline{W} = 0$ and $\overline{E} = 0$. $\overline{G}$ is not required for the write buffer. Thus $\overline{W}$ controls the direction of the data (read or write), $\overline{E}$ enables the tristate buffers in either direction, and $\overline{G}$ enables the output buffers only.

13.7 To change the cell contents to a 0, we make the *BIT* line LOW and the ROW SELECT line HIGH. The ROW SELECT line gives access to the cell by turning on Q_5 and Q_6, completing the conduction path between the *BIT* lines and the flip flop inputs. The LOW on the *BIT* line pulls the gate of Q_4 LOW, turning it OFF. This breaks the conduction path from Q_4 drain to source and makes $V_{DS4} = V_{DD}$, a logic HIGH. This HIGH is applied to the gate of Q_3, turning it ON. A conduction path is established between Q_3 drain and source, pulling the drain of Q_3 LOW. The cell now stores a logic 0.

13.9 A selected RAM cell is at the junction of an active ROW line and an active COLUMN line in a rectangular matrix of cells.

13.11 The DRAM in Problem 13.10 has 10 multiplexed ROW/COLUMN address lines. Adding one more line makes 11 lines, each of which are used for a ROW address and also a column address. This make a total of 22 lines, giving an address capacity of $2^{22} = 4M$ locations. Adding another address line gives 12 multiplexed lines, each used for ROW and COLUMN, giving a total of $2^{24} = 16M$ locations.

13.13 The primary difference between the different types of ROM is how easy each type is to program and erase.

 Mask-programmed ROM has the data manufactured into the device, making it difficult to program and impossible to erase. It is relatively cheap to mass-produce and is useful for storing unchanging data that must always be retained, including after power failure. An example is the "boot ROM" in a personal computer that contains data for minimal start-up instructions.

 UV-erasable EPROM is fairly expensive because of the specialized packaging it requires. It is user-programmable and can be easily erased by exposure to ultraviolet light when removed from the circuit. It is useful for unfinished designs, since stored data can be changed as development of a product proceeds.

 EEPROM can be used for applications which require data to be stored after power is removed from a device, but which require periodic in-circuit changes of data. One example might be an EEPROM which stores the numbers of several local channels in a digitally-programmed car radio.

13.15 Unlike EEPROM, flash memory is organized into sectors that can be erased all at one time. One sector, called the boot block, can be protected against unauthorized erasure or modification, thus adding a level of security to the memory.

13.17 EEPROM has slower access time and smaller bit capacity than RAM. It also has a finite number of program/erase cycles.

13.19 FIFO: buffer for serial data transmission; LIFO: memory stack in a microcomputer

13.21 $4K = 2^{12}$. Range = 0000 0000 0000 to 1111 1111 1111 (000H to FFFH); End address = Start + Maximum = 2000H + FFFH = 2FFFH.

$8K = 2^{13}$. Range = 0 0000 0000 0000 to 1 1111 1111 1111 (0000H to 1FFFH);

End = Start + Maximum = 6000H + 1FFFH = 7FFFH

13.25

Device	Start Address	End Address	Size
EPROM	0000H	3FFFH	16K
SRAM$_1$	4000H	7FFFH	16K
SRAM$_2$	8000H	BFFFH	16K
SRAM$_3$	E000H	FFFFH	8K

13.27 16 DIMMs

Device	Start Address	End Address
0	0000000H	0FFFFFFH
1	1000000H	1FFFFFFH
2	2000000H	2FFFFFFH
3	3000000H	3FFFFFFH
4	4000000H	4FFFFFFH
5	5000000H	5FFFFFFH
6	6000000H	6FFFFFFH
7	7000000H	7FFFFFFH
8	8000000H	8FFFFFFH
9	9000000H	9FFFFFFH
10	A000000H	AFFFFFFH
11	B000000H	BFFFFFFH
12	C000000H	CFFFFFFH
13	D000000H	DFFFFFFH
14	E000000H	EFFFFFFH
15	F000000H	FFFFFFFH

Chapter 14

14.1 A microcomputer is a system that includes a microprocessor as a component, as well as peripheral devices and memory.

14.3 **a.** The address bus transfers information regarding the location of instructions or data in the system memory.

b. The data bus transfers program instructions or operands between registers.

c. The control bus activates input and output enable lines to direct the flow of addresses or data between registers.

14.5 Bus contention can be avoided by the use of tristate buffering. Any signal driving a bus must be controlled so that only one signal is active on the bus at any time. All other drivers must be in the high-impedance state, effectively disconnecting them from the bus.

14.7 The system is synchronous, since both the output enable line and data are stored in flip-flops, synchronized to the system clock.

14.9 **a.** 0 : 8C

1 : 2D

2 : 90

3 : 8E

4 : 3F

5 : 90

6 : F0

C : 8A
D : 71
E : 75
F : 3D

b. FB, B2

14.11

State	Registers	Control Lines
load1	IR	ir_oe
load2	MAR	mar_ld
load3	ROM	rom_oe
load4	Accumulator	acc_ld

14.15 State the function of each of the following blocks of the RISC8v1 MCU.

a. PC: holds address of next instruction to be executed;

b. MAR: holds address of present instruction or address of present operand;

c. ROM: contains program instructions and operand data;

d. IR: Splits instruction into op code and operand address;

e. Instruction decoder: interprets 4 most significant bits of op code and sets control lines to perform required data transfers and arithmetic/logic operations;

f. Accumulator: holds accumulated total of arithmetic and logic operations;

g. MDR: holds second operand in a two-operand function;

h. ALU: combines contents of accumulator and memory data register in one of several selectable arithmetic or logic operations;

i. Output register: receives contents of accumulator in an Output instruction.

14.17

```
-- Address multiplexer with synchronous tristate outputs
-- Places instruction register or program counter contents
-- on address bus when selected. Otherwise output is high impedance.
LIBRARY ieee;
USE ieee.std_logic_1164.ALL;

ENTITY address_mux IS
    PORT(
        clock, reset, ir_oe, pc_oe : IN  STD_LOGIC;
        ir_addr, pc_addr  : IN  STD_LOGIC_VECTOR(3 downto 0);
        addr_bus          : OUT STD_LOGIC_VECTOR(3 downto 0));
END address_mux;

ARCHITECTURE mux OF address_mux IS
    SIGNAL controls : STD_LOGIC_VECTOR(1 downto 0);
BEGIN
    -- Concatenate output enable lines to use in
    -- CASE statement
    controls <= ir_oe & pc_oe;

    PROCESS (clock, reset)
    BEGIN
        IF (reset = '0') THEN
            addr_bus <= (others => 'Z');
        ELSIF (clock'EVENT and clock = '1') THEN
            CASE controls IS
                WHEN "10" =>
                    addr_bus <= ir_addr;
                WHEN "01" =>
                    addr_bus <= pc_addr;
```

```
                              WHEN others =>
                                  addr_bus <= (others => 'Z');
                          END CASE;
                      END IF;
                  END PROCESS;
              END mux;
```

14.19 A nested CASE statement in the state machine code reads the contents of the instruction register and selects the appropriate path in the controller's state diagram.

14.23 **a.** 0 : 8D
1 : 5E
2 : 6F
3 : 30
4 : 90
5 : F0

D : 75
E : 0F
F : A7

b. A6

14.25 0 : 8F
1 : 90
2 : 30
3 : 90
4 : A2
5 : F0

F : 7F

Index

Absolute accuracy, 778
Accumulator, 870, 872, 904
  modifications, 919–20
Active levels, 37–38
  HIGH, 37
  LOW, 37
ADC. *See* Analog-to-digital conversion
ADC808 analog-to-digital converter, 805–11
Add instruction (execute cycle), 880–85
Addend, 349
Addition
  hexadecimal, 360–61
  signed binary arithmetic, 353
  unsigned binary arithmetic, 349–50
  See also Binary adders and subtractors
Address, 830, 833
Address bus, 866
Address bus components, VHDL code for,
  891–96
Address decoding, 268, 273–74, 855–56
  with n-line-to-m line decoders, 858–59
Address multiplexing, 943
Address space, 855, 857
Adjacent cells, 104–5
Aliasing, 802–4
All-NAND forms, 122–27
All-NOR forms, 123–27
Alphanumeric code, ASCII, 365–64
Altera Corporation, 166
  See also Quartus II
Altera FLEX 10K, 166
Altera FLEX 10K block diagram, 479
Altera FLEX 10K CPLD, 473–79
Altera FLEX 10K Device, 182
Altera Library of Parameterized Modules, 252
Altera MAX 7000S, 166, 182
Altera MAX 7000S CPLD, 471–73
*Altera University Program Design Laboratory
  Package User Guide,* 211
Altera UP-1 or UP-2 circuit board,
  programming on, 209–11
Altera UP-2 Board, 179–80
ALU. *See* Arithmetic logic unit
American National Standards Institute
  (ANSI), 32
American Standard Code for Information
  Interchange. *See* ASCII
Amplitude, 20, 22
Analog, vs. digital electronics, 4–5
Analog signals
  analog-to-digital converter, 746–47
  defined, 746
  digital-to-analog converter, 746–47
  examples of, 745
  sampling an analog voltage, 747–51

Analog sound wave reproduction, 4–5
Analog-to-digital converter (ADC), 780–805
  bipolar (2's complement coding), 756–57
  bipolar (offset binary coding), 751, 754–56
  CPLD interface for, 805–11
  dual slope ADC, 788–93
  flash ADC, 780–82
  full scale, 751–52
  sample and hold circuit, 800–802
  sampling frequency and aliasing, 802–5
  sigma-delta ADC, 793–800
  successive approximation ADC, 782–88
  unipolar, 751–54
AND function, 34–35
  See also Logic functions and gates; Logical
    product
AND gates, 34–35, 49–51
  CMOS, 723–24
ANSI. *See* American National Standards
  Institute
Anti-aliasing filter, 802, 805
Aperiodic waveforms, 3, 19–20
Architecture body, 234–35
Architecture cell, 467–68
Arithmetic logic unit (ALU), 870, 872, 879–80
Arrays, 244, 246
ASCII (American Standard Code for
  Information Interchange) characters,
  244, 246, 365–67
Associative properties, 87
Asynchronous inputs (preset and clear), 457–60
Asynchronous load, 538–40
Asynchronous vs. synchronous circuits,
  455–57
Attribute, 523
Augend, 349

b (bit), 834
B (byte), 830, 834
Ball grid array (BGA), 56, 58
BCD. *See* Binary coded decimal
Bidirectional counter, 534–35, 544–46
Bidirectional shift register, 575, 583–85
Binary adders and subtractors, 367–88
  half and full adders, 367–71
  Boolean algebra design method, 368–71
  Karnaugh map design method, 368
  overflow detection, 386–88
  parallel binary adder/subtractor, 371–73
  cascade, 371
  fast carry, 373–74
  ripple carry, 371–72
  using VHDL components to implement a
    parallel adder, 373–86
  2's complement subtractor, 380–81

GENERATE statements, 378–80
  parallel binary adder/subtractor, 381–86
Binary coded decimal (BCD), 267
  8421 code, 363
  adders, 388–93
  excess-3 code, 363–64
  Gray code, 364–65
  priority encoder, 296–97
Binary decoder, 279–81
Binary inputs, 7–10
Binary numbers, 3, 6–14
  binary inputs, 7–10
  decimal-to-binary conversion, 10–12
  defined, 6
  fractional, 12–14
  positional notation, 6–7
Binary point, 12–14
Binary sequence, constructing, 9–10
Binary-coded decimal (BCD), 283–84
BIOS (basic input/output system), 845
Bipolar ADC (2's complement coding), 756–57
Bipolar ADC (offset binary coding), 754–56
Bipolar DAC (2's complement coding),
  759–61
Bipolar DAC (offset binary coding), 758–59
Bipolar SRAM cell, 839
Bipolar transistors as logic devices, 693–95
BIT (VHDL type), 246–48
Bit multiplexing, 310–11
Bit-organized memory, 843
Bits, 6
  least significant bit (LSB), 7–10
  most significant bit (MSB), 7–10
BIT_VECTOR, 246–48
Block, defined, 888
Block Diagram File, 182–83
  creating, 185–91
Block diagrams, 888–91
Boldface, 271
Boolean algebra, 32
  binary adders, 368–71
  rules, 99
Boolean algebra theorems, 86–99
  commutative, associative, and distributive
    properties, 87–89
  multivariable theorems, 94–98
    DeMorgan's, 94
  rules summarized, 99
  single-variable theorems, 89–93
    double inversion, 93
    x AND/OR/XOR 0/1, 91
    x AND/OR/XOR $x/\bar{x}$, 91–93
Boolean expressions, 31–33, 72–80
  from logic diagrams, 72–75
  See also Logic functions and gates

Boolean functions. *See* Boolean expressions
Boolean variable, 32
Boot block, 850–51
Bootstrap loaders, 845
Borrowing rules, 350
Bottom boot block, 850–51
Breadboard, 56, 58
Bubble-to-bubble convention, 72–73
Bubbles, 33–34, 629–30
Buffers, 33–34
Buried logic, 471
Bus, defined, 52–53
Bus contention, 855–56, 867
Bus form, 80, 82
Byte (B), 830, 834
Byte (or word) multiplexing, 310–11
ByteBlaster, 209–10
ByteBlaster MV, 209–10

Carry, 349
Carry bit, 349
Carry chain, 474–77
Carry generation, 393–95
Carry output, 389–90
Cascade, 371
Cascade chain, 473, 475–76
Case shift, 365
CASE statement, 287
   within a PROCESS, 301–2
Cells (in K-map), 104, 171, 173
   grouping along outside edges, 107–8
   product line first cell number, 171, 173
Central processing unit. *See* CPU
Character generators, 845
Characters, 244, 246
Checkers. *See* Parity generators and checkers
Checksum, 171, 175
Chips, 55–56
Clear, 457–60, 534–35
CLOCK, 447–48
Clocks, 305
CMOS. *See* Complementary metal-oxide
   semiconductor; Logic gate circuitry
CMOS analog switch, 317
Coincidence gate, 38, 41
Column address strobe, 843
Combinational logic circuit, 72
Combinational logic functions, 266–345
   decoders, 268–91
      multiple-output, 270–74
      seven-segment, 281–91
         decoder, 283–86
         display, 282–83
         ripple blanking, 287–91
      simulation criteria for *n*-line-to-*m* line,
         275–79
      single-gate, 268–70
      VHDL binary, 279–81
   demultiplexers, 313–18
      CMOS analog multiplexer/demultiplexer,
         317–21
      demultiplexng a TDM signal, 314–17
      VHDL magnitude comparators, 322–25
   encoders, 291–97
      BCD priority encoder, 296–97
      priority encoder, 291–94
      VHDL priority encoder, 294–96

multiplexers, 297–313
   multiplexing of time-varying signals,
      299–300
   VHDL implementation of, 300–302
   multiplexers applications, 302–13
      multiple-channel data selection, 303–5
      single-channel data selection, 302–3
      time division multiplexing, 310–13
      time-dependent, 305–10
   parity generators and checkers, 325–34
      parity circuits in VHDL using the
         GENERATE statement, 330–34
Combinatorial logic circuit, 72
Command lines, 508
Comments, 245
Common anode display, 281, 283
Common cathode display, 281, 283
Commutative properties, 87
Compilation, in Quartus II, 195–96
Compile, 166, 168
Complement form, 49, 51
Complementary metal-oxide-semiconductor
      (CMOS), 55–56, 670–71
   74LS driving low-voltage CMOS, 692
   CMOS AND and OR gates, 723–24
   CMOS inverter, 719–20
   CMOS NAND/NOR gates, 720–23
   CMOS SRAM cell, 839
   high-speed, 730–31
   high-speed CMOS driving 74LS, 691
   interfacing TTL with, 690–92
   logic families, 729–31
   metal-gate CMOS, 729–30
   MOSFET structure, 714–18
   power dissipation in high-speed, 686–88
   transmission gate, 724–25
   See also Logic gate circuitry
Complex programmable logic devices (CPLDs),
      166–67
   CPLD interface for an ADC, 805–11
   CPLD-based data acquisition network, 811–13
   FLEX 10K CPLD, 473–79
   MAX 7000S CPLD, 471–73
   programming on the Altera UP-1 or UP-2
      circuit board, 209–11
   programming binary counters for, 523–34
   programming presettable and bidirectional
      counters for, 552–75
   transferring a design to a target CPLD, 205–15
   tristate busses in Altera CPLDs, 885–96
   using Quartus II, 178–82
Component, 374–75
Component declaration statement, 374, 376
Component instantiation statement, 374, 376
Concatenate, 254, 256
Concurrent signal assignment statement,
      234, 237
Concurrent signal assignment, 300
Concurrent statement, 234, 237
Conditional signal assignment statement, 279
Conditional transition, 636–37
Conduit line, 888–89
Continuous, 4, 746
Control bus, 866
Control inputs
   state machines with, 636
   state machines with no, 629–36

Control section, 508
Control signals, 836–37
Controller (in CPU), 870–71, 908–14
   modifications, 920–22
Conversions between hexadecimal and
      binary, 17
Count enable, 534–35, 542–44
Count sequence, 503–4
Counters, 305, 502–75
   bits and maximum modulus, 504–5
   count-sequence table and timing diagram,
      505–8
   defined, 502–3
   programming binary counters for complex
      programmable logic devices (CPLDs),
      523–34
      behavioral description of, 523–26
      Library of Parameterized Modules in
         VHDL, 526–27
      Quartus II Block Editor, 527–34
      VHDL counters, 523–27
   programming presettable and bidirectional
      counters for complex programmable
      logic devices (CPLDs), 552–75
      behavioral design, 552–58
      LPM counters, 562–67
      Quartus II block diagram files, 567–75
      VHDL generic counters, 559–62
   shift register, 601–11
      Johnson counters, 601, 604–11
      ring counters, 601–4
   state diagram, 504
   synchronous counters, 508–14
      bidirectional counter, 534–35, 544–46
      clear, 534–35
      count enable, 534–35, 544–46
      flip-flop excitation tables, 514–15
      mod-12 counter, 515–22
      modulus of, 511
      output decoding, 534–35, 546–51
      parallel loading, 534–42
      presettable counter, 534–35
      ripple carry out or ripple clock out,
         534–35
      terminal count, 534–35
CPLDs. *See* Complex programmable logic
      devices
CPU (central processing unit), 866
Cutoff mode, 693

D latches, 429, 431–32, 830–31
   in Quartus II, 432–35
Data, defined, 830
Data acquisition, 805–13
   CPLD interface for an ADC, 805–11
   CPLD-based acquisition network, 811–13
Data book, 56, 59
Data bus, 866
Data bus MUX, 906–7
Data inputs, 297–98
Datasheet, 56, 59
DDR. *See* Double data rate
Decimal-to-binary conversion, 10–12
Decimal-to-hexadecimal conversion, 16–17
Decoders, 268–91
   instruction, 870, 872
   multiple-output, 270–74

output, 913–14
seven-segment, 281–91
simulation criteria for *n*-line-to-*m* line, 275–79
single-gate, 268–70
VHDL binary, 279–81
See also Combinational logic functions
Decoding
  Johnson counter modulus and, 609–11
  the output of a counter, 534–35, 546–51
  ring counter modulus and, 602–4
DeMorgan equivalent forms, 42–45
DeMorgan equivalent gates simplification, 120–22
DeMorgan's theorems, 94
  and gate equivalence, 42–45
Demultiplexers, 313–18
  CMOS analog multiplexer/demultiplexer, 317–21
  demultiplexing a TDM signal, 314–17
  VHDL magnitude comparators, 322–25
  See also Combinational logic functions
Derived logic functions, 38–42
  exclusive OR and exclusive NOR functions, 41
  expanding NAND and NOR gates, 40
  NAND and NOR functions, 38–41
  NAND and NOR gates as inverters, 40–41
  See also Logic functions and gates
Design entry, 178–79, 238–39
Design flow, 182–84
DeVry eSOC Board, 180
Difference, 350
Differential nonlinearity, 779
Digital, 4
  vs. analog electronics, 4–5
Digital arithmetic, 348–65
  addition, 349–50
  borrowing rules, 350
  signed binary numbers, 351–53
  subtraction, 350–51
  unsigned binary numbers, 348–51
  See also Binary adders and subtractors; Binary coded decimal (BCD) adders; Hexadecimal arithmetic; Signed binary arithmetic
Digital electronics, defined, 3–4
Digital function generator, 847–49
Digital logic levels, 5–6
Digital signals, 48–49, 745
  defined, 746
Digital sound reproduction, 4–5
Digital waveforms, 18–23
  aperiodic, 19–20
  periodic, 18–19
  pulse, 20–23
Digital-to-analog conversion (DAC), 746–47, 761–80
  bipolar (2's complement coding), 759–61
  bipolar (offset binary coding), 751, 758–59
  MC1408 integrated circuit D/A converter, 768–77
    bipolar operation of, 774–77
    op amp buffering of, 770–73
  performance specifications, 777–80
  R-2R ladder D/A converter, 765–68
  unipolar, 751, 757–58
  weighted resistor D/A converter, 763–65

DIMM. *See* Dual in-line memory module
Discrete, 4, 746
Distinctive-shape symbols, 33–34
Distributive properties, 87–89
Don't care states, 115–18
Double data rate, 853, 855
Double inversion, 93
Double-rail inputs, 75
Double-subscript notation, 297, 299
DOWN counter, 503–4
Download, 179, 181
DRAM. *See* Dynamic random access memory
Driving gate, 679–80
Dual in-line memory module (DIMM), 853–55
Dual in-line package (DIP), 55, 57
Dual slope A/D converter, 788–93
Duty cycle (DC), 18–19
Dynamic random access memory (DRAM), 838, 842
  cell arrays, 843–44
  modules, 853–55

EABs. *See* Embedded Array Blocks
ECL. *See* Emitter coupled logic
Edge, 20–23, 447–48
Edge detector, 448–49
Edge of a K-map, 107–8
Edge of a positive pulse, 305–6
Edge-sensitive, 447–48
Edge-triggered, 447–48
Edge-triggered D flip-flops, 447–52
  See also Sequential logic
Edge-triggered JK flip-flops, 452–60
  See also Sequential logic
Edge-triggered T flip-flops, 460–62
  See also Sequential logic
EEPROM. *See* Electrically erasable programmable read-only memory 8421 code, 363
Electrically erasable programmable read-only memory (EEPROM), 849–50
Embedded Array Blocks (EABs), 474, 478
Embedded controller, 870–71
EMP7128SLC84 device, 179
Enable, 48–55
Encoders, 291–97
  BCD priority encoder, 296–97
  priority encoder, 291–94
  VHDL priority encoder, 294–96
Enhancement-mode MOSFET, 715
Entity declaration, 234–35
Enumerated type, 631, 634
EPROM. *See* Erasable programmable read-only memory
Equivalent circuits, 766–67
Erasable programmable read-only memory (EPROM), 166, 846
  application: digital function generator, 847–49
Even parity, 325–30
Excess-3 code, 363–64
Excitation tables, 514
  flip-flop, 514–15
Exclusive NOR gates, 38, 41, 52
Exclusive OR (XOR) gates, 38, 41, 52
Execute, 870–71
Execute cycle (add instruction), 880–85

Execute cycle (load instruction), 876–79
Expanding NAND and NOR gates, 40

Fall time, 20, 22
Falling edge, 20–23
FAMOS FET. *See* Floating-gate avalanche MOSFET
Fanout, 679–84
Fast carry (look-ahead carry), 373–74
Fetch, 870–71
Fetch cycle, 873–76
FIFO (first-in-first-out), 851–53
Firmware, defined, 844
First-in-first out. *See* FIFO
Fitting, 179
Fixed-function logic, 166–67
Flash A/D converter, 780–82
Flash memory, 850–51
Flip-flops
  defined, 448
  edge-triggered D flip-flops, 447–52, 830
  edge-triggered JK, 452–60
  edge-triggered T, 460–62
  excitation tables, 514–15
  in programmable logic devices (PLDs) (registered outputs), 462–66
  timing parameters, 676–79
  See also Sequential logic
Floating-gate avalanche MOSFET (FAMOS FET), 846
Flow control, 917
Form A contact, 645
Form B contact, 645
Form C contact, 645
Fractional binary numbers, 12–14
Fractional-decimal-to-fractional binary conversion, 13–14
Frequency (f), 18–19
FS. *See* Full scale (FS) current
Full adder, 367–68
Full scale (FS) current, 751–52
Function lookup tables, 845

G, 834–35
Gain error, 778
GAL22VI0, 166, 467–70
GAL. *See* Generic array logic
Gated D latch, 429, 431–32
  in Quartus II, 432–35
Gated latches, 429–47
  gated SR latch, 429–31
  See also Sequential logic
Gates. *See* Logic functions and gates; Sequential logic
GENERATE statements, 330–34, 378–80, 436–37
Generic array logic (GAL), 467–70
  See also Sequential logic
Generic map, 435–36
Generic width, shift registers, 594–97
Global clock, 467–68
Graphical user interface, 184–85
Gray code, 364–65, 629–33

Half adder, 367
Halt instruction, 885

Hardware, defined, 844
Hardware Description Language (HDL), 182–83
HDL. *See* Hardware Description Language
Hexadecimal arithmetic, 360–62
　addition, 360–61
　subtraction, 361–62
Hexadecimal numbers, 3, 14–17
　conversions between hexadecimal and binary, 17
　counting, 15
　decimal-to-hexadecimal conversion, 16–17
　defined, 14
　hexadecimal-to-decimal conversion, 15–16
Hexadecimal-to-decimal conversion, 15–16
Hierarchical design, 185–86
　using Quartus II to create, 216–22
Hierarchy, 185–86, 373, 375
High-current driver, 703–4
High-impedance state, 52–53
High-speed CMOS, 730–31
High-speed CMOS driving 74LS, 691
Hold time, 676–77
HVW Technologies Intro-FPGA Board, 181

I/O Control Block, 471
ICR. *See* In-Circuit Reconfigurability
IEEE. *See* Institute of Electrical and Electronic Engineers
IEEE Std. 1164 Multi-Value Logic, 244, 251
IEEE/ANSI Standard 91–1984, 32–34
IF statement, 287, 523
In phase, 49
In-Circuit Reconfigurability (ICR), 209–10
In-system programmability (ISP), 209–10
Index variable, 330–31
Inhibit, 49–55
Input line, 169
Input line number, 171, 173
Inputs, 280, 286
　asynchronous, 457–60
　data, 297–98
　select, 297–98
　synchronous, 457
Instance, 191
Instantiate, 374, 377
Institute of Electrical and Electronic Engineers (IEEE), 32, 234
Instruction decoder, 870, 872
Instruction register, 870, 873
INTEGER, 253–54
Integrated circuit (IC) logic gates, 55–60
Inverters, 33–34
　NAND and NOR gates as, 40–41
ISP. *See* In-System Programmability

JEDEC. *See* Joint Electron Device Engineering Council
Johnson counters, 601, 604–11
　modulus and decoding, 609–11
Joint Electron Device Engineering Council (JEDEC), 171, 173
JEDEC file, 171, 173
Joint Test Action Group (JTAG), 209–10
　JTAG Chain, 209, 211
　port, 467
　Test Clock (TCK), 209, 211

Test Data In (TDI), 209–10
Test Data Out (TDO), 209–10
Test Mode Select (TMS), 209–10
JTAG. *See* Joint Test Action Group

K, 834–35
Karnaugh map simplification, 368
　conditions for maximum simplication, 111–12
　counters, 517, 521
　don't care states, 115–18
　grouping cells along outside edges, 107–8
　loading a K-map from a truth table, 108–10
　overlapping groups, 110–11
　product-of-sums (POS), 118–20
　sum-of-products (SOP), 104–18
　two-variable map, 105–6
　using K-maps for partially simplified circuits, 112–15

LAB. *See* Logic Array Block
Large scale integration (LSI), 55–56
Last-in-first-out. *See* LIFO
Latches, 411–47
　gated latches, 429–47
　NAND/NOR latches, 414–28
　in Quartus II, 420–26
　as switch debouncer, 426–28
　See also Sequential logic
LE. *See* Logic element
Leading edge, 20–23
Least significant bit (LSB), 7–10
LED indicators, 47–48
Left shift, 575–76
Level-sensitive, 447–48
Levels of gating, 75–76
Library, 244, 251
Library of Parameterized Modules (LPM), 435–37
　with Active-Low ports, 446–47
　counters in, 523, 526–34, 562–67
　manual entry of, 443–46
　shift registers, 597–601
LIFO (last-in-first-out), 851–53
Light switch/dimmer analogy, 4
Light-emitting diode. *See* LED
Linearity error, 778
Load gate, 679–80
Load instruction (execute cycle), 876–79
Logic Aarray Blocks (LABs), 471–72
Logic circuits
　general approach to designing, 133–39
　pulsed operations of, 131–33
Logic diagrams, 72–80
　from Boolean expressions, 75–78
Logic element (LE), 473, 475–76
Logic functions
　defined, 31–33
　See also Logic functions and gates
Logic functions and gates, 30–68
　active levels, 37–38
　AND function, 34–35
　DeMorgan's theorems and gate equivalence, 42–45
　derived logic functions, 38–42
　　exclusive NOR, 41
　　exclusive OR, 41
　　NAND, 38–41

　　NOR, 38–41
　enable and inhibit properties of logic gates, 48–55
　　AND or OR gates, 49–51
　　exclusive OR and NOR gates, 52
　　NAND and NOR gates, 51
　　tristate buffers, 52–55
　integrated circuit logic gates, 55–60
　LED indicators, 47–48
　logic gate defined, 32–33
　logic switches, 45–47
　NOT function, 33–34
　OR function, 35–36
　See also Boolean expressions
Logic gate circuitry, 668–743
　complementary metal-oxide semiconductor (CMOS), 670–71, 714–26
　　bias requirement for MOS transistors, 716–18
　　CMOS AND and OR gates, 723–24
　　CMOS inverter, 719–20
　　CMOS NAND/NOR gates, 720–23
　　CMOS transmission gate, 724–26
　　MOSFET structure, 715–16
　electrical characteristics of, 670–74
　emitter coupler logic (ECL), 670–71
　fanout, 679–84
　flip-flop timing parameters, 676–79
　interfacing TTL and CMOS gates, 690–92
　　74LS driving 74HC, 691
　　74LS driving 74HCT, 691–92
　　74LS driving low-voltage CMOS, 692
　　high-speed CMOS driving 74LS, 691
　noise margin, 688–90
　power dissipation, 684–88
　　in high speed CMOS devices, 686–88
　　in TTL devices, 685–86
　propagation delay, 674–76
　transistor-transistor logic (TTL), 670–71
　TTL and CMOS variations, 726–31
　　CMOS logic families, 729–31
　　TTL logic families, 726–29
　of TTL gates, 693–714
　　7402 NOR gate, 712–13
　　7408 AND gate and 7432 OR gate, 713–14
　　bipolar transistors, 693–95
　　open-collector applications, 701–5
　　totem pole outputs, 705–10
　　tristate gates, 710–12
　　TTL open-collector inverter and NAND gate, 696–701
Logic gate networks, 72
Logic gates
　defined, 31–33
　enable and inhibit properties of, 48–55
　integrated circuit (IC), 55–60
　See also Logic functions and gates
Logic HIGH (or logic 1), 5–6
Logic levels, defined, 5–6
Logic LOW (or logic 0), 5–6
Logic switches, 45–44
Logical product, 34
　See also AND function
Logical sum, 35
　See also OR function
Look-ahead carry. *See* Fast carry
Look-up tables (LUTs), 166, 473–75

LPM. *See* Library of Parameterized Modules
LPM latch, 437–38
LUTs. *See* Look-up tables

M, 834–36
Magnitude bits, 351
Magnitude comparators, 318–25
   VHDL, 322–25
Mapper, 888–89
Mapping of outputs to control word, 913
MAR. *See* Memory address register
Mask-programmed ROM, 844–46
Master reset, 459–60
Maximum modulus, 504–5
Maximum POS simplication, 100–104
Maximum SOP simplication, 100–104
Maxterm, 80, 83
MC1408 integrated circuit D/A converter,
   768–77
   See also Digital-to-analog conversion (DAC)
MDR. *See* Memory data register
Mealy machine, 628–29
Media, digital ways of storing, 5
Medium scale integration (MSI), 55–56
MegaWizard Plug-In Manager, 438–42,
   527–33, 567–70
Memory address register (MAR), 870, 873
Memory capacity, 834–36
Memory data register (MDR), 870, 872, 905
Memory devices and systems, 828–63
   address, 830, 833
   basic concepts, 830–37
   byte defined, 830
   control signals, 836–37
   data defined, 830
   dynamic RAM modules, 853–55
   dynamic random access memory (DRAM),
     842–44
   electrically erasable programmable read-only
     memory (EEPROM), 849–50
   erasable programmable read-only memory
     (EPROM), 846–6
   flash memory, 850–51
   memory capacity, 834–36
   memory defined, 830
   memory systems, 855–59
   random access memory (RAM), 834
   read, 830, 833
   read-only memory (ROM), 834, 844–51
   sequential memory: FIFO and LIFO, 851–53
   static random access memory (SRAM),
     838–42
   write, 830, 833
Memory map, 855, 857
Memory modules, 853–55
Memory section, 508
Metal-gate CMOS, 729–30
Metal-oxide-semiconductor field effect
   transistor (MOSFET), 714–18
Microcomputers, 866–70
Microprocessors, 864–935
   branch instructions (RISC8v3), 917–22
     accumulator modifications, 919–20
     controller modifications, 920–22
     program counter modifications, 918–19
   creating new instructions (RISC8v2), 914–17
   microcomputer stucture, 866–70

possible enhancements to RISC8v3 MCU,
   924–25
Quartus II implementation of the RISC8v1
   MCU, 896–914
   controller, 908–14
   data bus MUX, 906–7
   mapping of outputs to control word, 913
   memory data register, 905
   output decoder, 913–14
   output register, 907–8
   ROM, 896–903
register level structure of, 870–85
   arithmetic logic unit (ALU), 879–80
   execute cycle (add instruction), 880–85
   execute cycle (load instruction), 876–79
   fetch cycle, 873–76
RISC8 test circuit, 922–24
tristate busses in Altera CPLDs, 885–96
   block diagram format, 888–91
   in the RISC8v1 CPU, 886–88
   VHDL code for address bus components,
     891–96
tristate bussing in microcomputers, 867–70
Minterm, 80–81
Minuend, 350
Mod-12 counter, 515–22
Modes, 234, 236, 245–46
Modulo arithmetic, 503–4
Modulo-*n* (mod-*n*) counter, 503
Modulus, 503–4
   Johnson counter and decoding, 609–11
   maximum, 504–5
   ring counter and decoding, 602–4
   of a synchronous counter, 511
Monotonicity, 777
Moore machine, 628
MOS transistors, 716–18
MOSFET. *See* Metal-oxide-semiconductor field
   effect transistor; Logic gate circuitry
Most significant bit (MSB), 7–10
Multi-bit signals, 255–57
   combining with single-bit, 258–59
Multi-channel data selection, 303–5
Multibit latches in VHDL, 435–38
Multiple-digit BCD adders, 393
Multiple-input NAND and NOR gates, 39
Multiple-output decoders, 270–74
Multiplexers, 171, 175, 297–313
   applications, 302–13
     multi-channel data selection, 303–5
     single-channel data selection, 302–3
     time division, 310–13
     time-dependent, 305–10
   multiplexing of time-varying signals, 299–300
   VHDL implementation of multiplexers,
     300–302
     CASE statement within a PROCESS,
       301–2
     concurrent signal assignment, 300
     selected signal assignment statement, 301
   See also Combinational logic functions
Multiplying DAC, 768–67

n-channel enhancement-mode MOSFET, 715
n-line-to-*m* line decoders, 275–79, 858–59
n-type inversion layer, 716–17
Named identifier, 244

NAND gates, 38–41, 51, 671–74
   TTL open-collector, 699–701
   universal property of, 122–27
NAND/NOR gates, CMOS, 720–23
NAND/NOR latches, 414–28
   See also Sequential logic
NATURAL, 253
Negative logic, 5–6
Negative sum or difference, 355
Negative sums, overflow in, 358
Next state, 508
9's complement, 363
Noise margin, 688–90
NOR gates, 38–41, 51
   universal property of, 122–27
NOR/NAND latches, 414–28
NOT function, 33–34
   See also Logic functions and gates;
      Truth tables
Numeric codes, 362–65
   binary-coded decimal (BCDs), 362–63
   binary-coded decimal (BCDs)
     8421 code, 363
      excess-3 code, 363–64
     Gray code, 364–65
Nyquist sampling theorem, 802

Octal latch, 830–32
Octal tristate buffers, 54–55
Octets, 104, 107
Odd parity, 325–30
Offset binary coding
   ADC, 754–56
   DAC, 758–59
Offset error, 779
Ohmic region, 716
OLMC. *See* Output logic macrocell
One-time programmable (OTP), 467
1's complement, 351–52, 354
Op code, 870, 872
Open-collector inverter, 696–705
Operand, 349
OR function, 35–36
   See also Logical sum
OR gates, 49–51
   CMOS, 723–24
Order of precedence, 72, 74
OTP. *See* One-time programmable
Out of phase, 49, 51
Output, 885
Output decoding, 534–35, 546–51, 913–14
Output logic macrocell (OLMC), 467–68
Output register, 870, 872, 907–8
Overflow, sign bit, 357–60
Overflow detection, 386–88

p-channel enhancement-mode MOSFET, 715
Package, 244, 251
Pairs, 104, 106
PAL. *See* Programmable array logic
Parallel binary adder/subtractor, 371–73
   using VHDL, 373–86
Parallel loading, 534–42
Parallel logic expanders, 471–72
Parallel transfer, 575
Parallel-load shift register, 575, 585–89
Parameter, 435–36

Parity, 325
Parity bit, 325
Parity generators and checkers, 325–34
  parity circuits in VHDL using the
    GENERATE statement, 330–34
  See also Combinational logic functions
Period (T), 18–19
Periodic waveforms, 3
Peripheral devices, 866
Phase splitter, 705
PIA. *See* Programmable Interconnect Array
Pin name assignment, 193–94, 205–6, 243
Plastic leaded chip carrier (PLCC), 56, 58
PLDs. *See* Programmable logic devices
Port, 234, 236, 374, 377
Port map, 435–36
Portable document format (PDF), 56, 59
Positional notation, 6–7
POSITIVE, 253
Positive logic, 5–6
Postitive edge, 305–6
Power dissipation, 684–88
Practical circuit implementation in SSI logic,
    128–31
Present state, 508
Preset, 457–60
Presettable counter, 534–35
Primitives, 191–92
Printed circuit board (PCB), 55, 57
Priority encoder, 291–94, 780–81
PROCESS, 287, 289
  CASE statement within a, 301–2
Product line, 169
Product line first cell number, 171, 173
Product term, 80–81
Product-of-sums (POS), 80, 83
  Karnaugh map method, 104–18
Product-of-sums (POS) simplification, 100–104
Program counter, 870, 873
  modifications, 918–19
Program instructions, 866
Programmable array logic (PAL), 169–70
  fuse matrix and combinational outputs,
    171–75
  outputs with programmable polarity, 175–78
Programmable Interconnect Array (PIA),
    471–72
Programmable logic devices (PLDs), 165–69
  defined, 166–69
  flip-flops in, 462–66
Programmable sum-of-products arrays, 169–70
Projects, 182
  See also Quartus II
Propagation delay, 674–76
Pull-up resistor, 45–46, 704–5
Pulse waveforms, 3, 20–23, 48–49
  See also Digital signals
Pulse width, 20, 22, 677
Pulsed operation, of logic circuits, 131–33
Pulses, 20–23

qdesigns, 185
Quad flat pack (QFP), 56, 58
Quads, 104, 107
Qualifying symbol, 33–34
Quantization, 747, 750
Quantization error, 782, 787

Quartus II, 166, 168
  assigning pin names, 193–94
  Assignment Editor, 206–9
  carry generation in, 393–95
  compilation, 195–96
  connecting components, 193
  counters, 527–34, 567–75
  creating Block Diagram File (new project),
    185–91
  design flow, 182–84
  entering components, 191–92
  graphical user interface, 184–85
  implementation of the RISC8v1 MCU,
    896–914
  implementing D latches in, 432–35
  making a VHDL file in, 237–43
  MegaWizard Plug-In Manager, 438–43
  multibit latchs in Block Editor, 438–47
  NAND latch synthesis, 420–26
  programming complex programmable logic
    devices (CPLDs), 178–82
  Quartus II Programmer, 211–15
  simulation, 196–204
  transferring a design to a target complex
    programmable logic device (CPLD),
    205–15
  using Block Editor to create a hierarchical
    design, 216–22
    symbol files and user libaries, 216–18
  working with previously defined projects,
    194–95
Quartus II Installation & Licensing for PCs
    (Altera), 184
Queue (FIFO memory), 851–52

R-2R ladder D/A converter, 765–68
Radix point, 13–14
RAM. *See* Random access memory
Rambus, 854–55
Rambus channel, 854–55
Random access memory (RAM), 834, 838–44
  dynamic RAM (DRAM) cell arrays, 843–44
  dynamic RAM (DRAM) cells, 842–44
  static RAM (SRAM) cell arrays, 840–42
  static RAM (SRAM) cells, 838–40
Range (VHDL), 244, 246
Range of signed numbers, 355, 357
RCO. *See* Ripple carry out or ripple clock out
Read, 830, 833
Read-only memory (ROM), 834, 844–51
  electrically erasable programmable read-only
    memory (EEPROM), 849–50
  erasable progammable read-only memory
    (EPROM), 846–49
  flash memory, 850–51
  mask-programmed ROM, 844–46
  Quartus II implementation of the RISC8v1
    MCU, 896–903
Recovery time, 677
Rectangular-outline symbols, 33–34
Recycle, 503
Refresh cycle, 842
Register level structure of a microcomputer
    system, 870–85
  See also Microprocessors
Registered outputs, 462–66
  See also Flip-flops

Registers, 462, 867
Relative accuracy (DAC), 778
Repeated division by 16, 17
Repeated division by 2, 12
Reserved keywords, 244–45
RESET, 411–12
Reset, master reset, 459–60
Resolution, 747–48
Response waveforms, 275
Right shift, 575–76
Ring counters, 601–4
  modulus and decoding, 602–4
Ripple blanking, 287–91
Ripple blanking input, 287
Ripple blanking output, 287
Ripple carry, 371–72
Ripple carry out or ripple clock out (RCO),
    534–35
  terminal count and, 548–50
RISC (reduced instruction set computer),
    870–71
RISC8 test circuit, 922–24
RISC8v1 MCU, Quartus II implementation of,
    896–914
RISC8v1 tristate address bussing in, 886–88
RISC8v2, creating new instructions in,
    914–17
RISC8v3, branch instructions, 917–22
RISC8v3 MCU, possible enhancements to,
    924–25
Rise time, 20–23
Rising edge, 20–23
ROM. *See* Read-only memory
Rotation, 575
Row address strobe, 843
RSR Electronics PLDT-2 Board, 181

Sample, 747
Sample and hold circuit, 800–802
Sampling frequency, 747
Sampling frequency and aliasing, 802–5
Saturation mode, 693–94
Schematic entry, 166, 168
Schottky barrier diode, 726
Schottky transistor, 726
Schottky TTL, 726, 728
SDRAM. *See* Synchronous DRAM
Sector, 850–51
Select inputs, 297–98
Selected signal assignment statement, 244,
    248–50, 279, 301
Self-complementing code, 363
Sensitivity list, 287, 289
Sequential logic, 72, 408–98
  FLEX 10K CPLD, 473–79
  flip-flops, 447–66
    edge-triggered D, 447–52
    edge-triggered JK, 452–60
      asynchronous inputs (preset and clear),
        457–60
      synchronous vs. asynchronous circuits,
        455–57
    edge-triggered T (toggle), 460–62
    in programmable logic devices (PLDs)
      (registered outputs), 462–66
    vs. latches, 448–52
  gated latches, 429–47

generic array logic (GAL), 467–70
    GAL22V10, 467–70
latches, 411–47
    gated SR latch, 429–31
    implementing gated D latches in Quartus II, 432–35
    multibit latches in VHDL, 435–38
        4 LATCH primitives and a GENERATE statement, 436–37
        behavioral description, 436
        LPM latch, 437–38
    multibit in Quartus II Block Editor, 438–47
        LPM components with Active-LOW ports, 446–47
        manual entry of an LPM component, 443–46
        MegaWizard Plug-In Manager, 438–43
NAND/NOR latches, 414–28
    functions, 414
    operation, 415–28
    RESET, 411–12
    SET, 411–12
    transparent latch (gated D latch), 431–32
MAX 7000S CPLD, 471–73
Sequential memory: FIFO and LIFO, 851–53
Serial shift registers, 575–83
SET, 411–12
Settling time, 778
Setup time, 676–77
Seven-segment decoders, 281–91
Seven-segment display, 281, 282
74LS driving 74HC, 691
7402 NOR gate, 712–13
74LS driving 74HCT, 691–92
74LS driving low-voltage CMOS, 692
Shared logic expanders, 471–72
Shift registers, 575–611
    bidirectional, 575, 583–85
    counters, 601–11
        Johnson counters, 601, 604–11
        ring counters, 601–4
    defined, 575
    parallel-load, 575, 585–89
    programming in VHDL, 589–601
        behavioral design, 589, 592–94
        dataflow design, 589, 592
        of generic width, 594–97
        Library of Parameterized Modules (LPM) and, 597–601
        structural design, 589–91
    serial shifting, 575–83
Sigma-delta ADC, 793–800
Sign bit overflow, 357–60
Sign bits, 351
Sign extension, 357
Signal attribute, 330, 333–34
Signal concurrency, 234, 237
Signals, in VHDL, 254–59
Signed addition, 353
Signed binary arithmetic, 353–60
    negative sum or difference, 355
    range of signed numbers, 355, 357
    sign bit overflow, 357–60
        in negative sums, 358
    sign extension, 357
    signed addition, 353

subtraction, 354–55
    1's complement methods, 354
    2's complement method, 354
    true-magnitude method, 354–55
Signed binary numbers, 348, 351–53
Signed subtraction, 354–55
SIMM. *See* Single-in-line memory module
Simulation, 179
    criteria for *n*-line-to-*m* line decoders, 275–79
    in Quartus II, 196–204
    as a troubleshooting tool, 204–5
    in VHDL, 239–43
Simultaneous converter. *See* Flash A/D converter
Single-bit signals, 257–58
    combining with multi-bit, 258–59
Single-channel data selection, 302–3
Single-gate decoders, 268–70
Single-in-line memory module (SIMM), 853–54
Single-pulse generator, 641–45
Single-variable theorems, 89–93
    See also Boolean algebra theorems
Sinking (current), 679, 681
Small outline IC (SOIC), 56, 58
Small scale integration (SSI), 55–56
    practical implementation in, 128–31
Software, defined, 844
Software instructions, 845
SOP. *See* Sum-of-products
Sourcing (current), 679, 681
Speed grade, 393–94
Speed-power product, 726–27, 729
SRAM. *See* Static random-access memory
SRG*n*, 575
SSI. *See* Small scale integration
Stack (LIFO memory), 851–52
Staic random-access memory (SRAM), cell arrays, 840–42
Standard circuits, 31
    See also Logic functions and gates
State diagrams, 503–4, 513, 519
State machines, 514, 626–67
    with control inputs, 636–45
        classical design techniques, 637–39
        VHDL design, 640–45
    Mealy machine, 628–29
    Moore machine, 628
    with no control inputs, 629–36
        classical design techniques, 630–31
        VHDL design, 631, 633–36
    state variables, 628
    switch debouncer for a normally open pushbutton switch, 645–52
        based on a 4–bit shift register, 646–50
        behaviorally designed, 650–52
        VHDL code for, 651–52
    traffic light controller, 659–60
    unused states in, 653–59
State variables, 628
Static random-access memory (SRAM), 166, 838–40
Status lines, 508
Steering gates, 429, 431
Stimulus waveforms, 275
Storage time, 708
Strings, 244, 246
Substrate, 715

Subtraction
    hexadecimal, 361–62
    signed binary arithmetic, 354–55
    unsigned binary arithmetic, 350–51
    See also Binary adders and subtractors
Subtrahend, 350
Successive approximation A/D converter, 782–88
Suite of software tools, 179
Sum, 349
Sum bit (single-bit addition), 349
Sum correction, 390–91
Sum of powers of 2, 10–11
Sum term, 80, 83
Sum of weighted hexadecimal digits, 16
Sum-of-products (SOP), 80–86
    Karnaugh map method, 118–20
    programmable, 169–70
Sum-of-products (SOP) simplification, 100–104
Surface-mount technology (SMT), 56, 58
Switch debouncer for a normally open pushbutton switch, 645–52
Switching noise, 708–9
Symbol files, 216
Synchronous counters, 508–14
    block diagram, 508
    control options for, 534–51
    determining the modulus of, 511
    flip-flop excitation tables, 514–15
    mod-12, 515–22
    See also Counters
Synchronous DRAM (SDRAM), 853, 855
Synchronous inputs, 457
Synchronous vs. asynchronous circuits, 455–57
Syntax, 234–35
Synthesis, 75

Target device, 179
TDM. *See* Time division multiplexing
Terminal count, 534–35
    and ripple carry out or ripple clock out (RCO), 548–50
Test Clock (JTAG), 209, 211
Test Data In (JTAG), 209–10
Test Data Out (JTAG), 209–10
Test Mode Select (JTAG), 209–10
Text files, 171, 173
Thevenin equivalent, 766
Thin shrink small outline package (TSSOP), 56, 58
Threshold voltage, 716
Through-hole, 56–57
Time division multiplexing (TDM), 310–13
    demultiplexing a TDM signal, 314–17
Time HIGH (periodic waveform), 18–19
Time LOW (periodic waveform), 18–19
Time slot, 310–11
Time-dependent multiplexer applications, 305–10
Time-varying digital waveforms, 3
Time-varying signals, multiplexing of, 299–300
Timing diagram, 275, 505–8, 513
Toggle, 452
Toggle flip-flops, 460–62
Top boot block, 850–51
Top level, of a hierarchy, 216–17
Totem pole outputs, 705–10
Traffic light controller, 659–60

Trailing edge, 20–23
Transfer characteristic, 752
Transistor-transistor logic (TTL), 55–56
  7402 NOR gate, 712–13
  7408 AND gate and 7432 OR gate, 713–14
  bipolar transistors as logic devices, 693–95
  compatibility, 688–89
  interfacing with CMOS gates, 690–92
  logic families, 726–29
  open-collector applications, 701–5
  open-collector inverter and NAND gate,
    696–701
  power dissipation in, 685–86
  totem pole outputs, 705–10
  tristate gates, 710–12
Transparent latch (gated D latch), 429, 431–32
Tristate buffers, 52–55
Tristate busses in Altera CPLDs, 885–96
  See also Microprocessors
Tristate bussing, 867–70
Tristate gates, 710–12
Tristate outputs, 710, 867
True form, 49
True-magnitude form, 351–52, 354–55
Truth tables
  defined, 33
  enable and inhibit properties, 50–52
  from Boolean expressions, 33–36
  from derived logic functions, 38–39, 41–42
  from logic diagrams or Boolean expressions,
    78–80
  loading a Karnaugh map from, 108–10
  See also NOT function
TTL. See Transistor-transistor logic; Logic gate
    circuitry
Two-pulse generator, 654–58
Two-variable map, 105–6
2's complement, 351, 353, 354
2's complement coding
  ADC, 756–57
  DAC, 759–61
2's complement subtractor, 380–81
Types (in VHDL), 234, 236, 244, 246

Unconditional transition, 636–37
Unipolar analog-to-digital converter (ADC),
    752–54
Unipolar digital-to-analog converter (DAC),
    757–58
Universal shift register, 575, 585
Unsigned binary arithmetic, 349–51
Unsigned binary numbers, 348
Unused states in state machines, 653–59
UP counter, 503–4
User libaries, 216

Valid names, 244–45
V_{CC}, 45–46
Vector Waveform File (VWF), 196
Very High Speed Integrated Circuit. See VHSIC
    Hardware Desciption Language
Very large scale integration (VLSI), 56–57
VHDL. See VHSIC Hardware Description
    Language
VHSIC Hardware Description Language
    (VHDL), 232–65
  architecture body, 235–37
  array, 246
  binary decoder, 279–81
  BIT and BIT_VECTOR, 246–48
  code for address bus components, 891–96
  comments, 245
  counters and, 523–27, 559–62
  entity declaration, 235
  implemention of multiplexers, 300–302
  INTEGER, NATURAL, and POSITIVE,
    253–54
  magnitude comparators, 322–25
  modes, 245–46
  multibit latches in, 435–38
  parallel adders, 373–86
  parity circuits in VHDL using the
    GENERATE statement, 330–34

  priority encoder, 294–96
  programming shift registers in, 589–601
  in Quartus II, 237–43
    design entry, 238–39
    pin assignments, 243
    simulation, 239–43
  range, 246
  selected signal assignment statement, 248–50
  signal concurrency, 237
  signals, 254–59
    combining single- and multiple-bit, 258–59
    multi-bit, 255–57
    single-bit, 257–58
  state machine design, 631, 633–36
  state machines with control inputs, 640–45
  state machines with no control inputs, 631,
    633–36
  switch debouncer, 648–52
  syntax, 244–54
  type, 246
  valid names, 244–45
Volatile memory, 838
VWF. See Vector Waveform File

Waveform generation, 196–204, 306–10
Waveforms, 18–23
  forms of, 3
  See also Digital waveforms
Weighted resistor D/A converter, 763–65
"WHEN others," 280
Wire-wrap, 56, 58
Wired-AND, 701–3
Word, 840, 842
Word length, 840, 842
Word-organized, 840, 842
Write, 830, 833

x AND/OR/XOR 0/1, 90–91
x AND/OR/XOR $x/\bar{x}$, 91–93